Aerial Vehicles

Aerial Vehicles

Edited by
Thanh Mung Lam

Aerial Vehicles
Edited by Thanh Mung Lam

CBS, Edition 2016

Published by InTech
Janeza Trdine 9, 51000 Rijeka, Croatia

Technical Editor Teodora Smiljanic

Cover Designer Dino Smrekar

Additional hard copies can be obtained from orders@intechopen.com

Aerial Vehicles, Edited by Thanh Mung Lam
p. cm.
ISBN 978-953-7619-41-1

Preface

Unmanned and micro aerial vehicles (UAV and MAV) have the potential to enable low-cost and safe operation. Due to their small and lightweight platform the aerial vehicles can be used for surveillance, search and rescue, and scientific research missions in unknown, dangerous environments and operations where the use of manned air vehicles is not suitable or too expensive. However, in order to actually enable safe operation and low operation and maintenance cost, many challenges exist in various fields, such as robust autonomous control, vehicle dynamics modeling, efficient energy use, robust communication, intuitive guidance and navigation, and the ability to carry payload.

With the current availability of faster computers, sophisticated mathematical tools, lighter materials, wireless communication technology, efficient energy sources, and high resolution cameras, the abilities offered to face these challenges have captured the interest of the scientific and operational community. Thanks to the efforts of many, aerial vehicles have become more and more advanced, and today's aerial vehicles are able to meet requirements that would not have been feasible just less than one decade ago. Besides the already quite extensive military use, currently also civil users such as police, fire fighters, and life guards are starting to recognize the many possibilities that low-cost, safe, and user-friendly aerial vehicles may offer to their operations.

This book contains 35 chapters written by authors who are experts in developing techniques for making aerial vehicles more intelligent, more reliable, more flexible in use, and safer in operation. I hope that when you read this book, you will be inspired to further improve the design and application of aerial vehicles. The advanced techniques and research described here may also be applicable to other high-tech areas such as robotics, avionics, vetronics, and space.

I would like to thank the authors for their excellent research and contribution to this book.

Editor

Thanh Mung Lam
Delft University of Technology
Faculty of Aerospace Engineering
Control and Simulation Division
The Netherlands
t.m.lam@tudelft.nl

Contents

1

Design and Development of a Fly-by-Wireless UAV Platform

Paulo Carvalhal, Cristina Santos, Manuel Ferreira, Luís Silva and José Afonso
University of Minho
Portugal

1. Introduction

The development of unmanned aerial vehicles (UAVs) has become an active area of research in recent years, and very interesting devices have been developed. UAVs are important instruments for numerous applications, such as forest surveillance and fire detection, coastal and economic exclusive zone surveillance, detection of watershed pollution and military missions.

The work described in this chapter is part of a larger project, named AIVA, which involves the design and development of an aerial platform, as well as the instrumentation, communications, flight control and artificial vision systems, in order to provide autonomous takeoff, flight mission and landing maneuvers. The focus of the chapter is on one of the main innovative aspects of the project: the onboard wireless distributed data acquisition and control system. Traditionally, UAVs present an architecture consisting of one centralized and complex unit, with one or more CPUs, to which the instrumentation devices are connected by wires. At the same time, they have bulky mechanical connections. In the approach presented here, dubbed "fly-by-wireless", the traditional monolithic processing unit is replaced by several less complex units (wireless nodes), spread out over the aircraft. In that way, the nodes are placed near the sensors and controlled surfaces, creating a network of nodes with the capacity of data acquisition, processing and actuation.

This proposed fly-by-wireless platform provides several advantages over conventional systems, such as higher flexibility and modularity, as well as easier installation procedures, due to the elimination of connecting cables. However, it also introduces several challenges. The wireless network that supports the onboard distributed data acquisition and control system needs to satisfy demanding requirements in terms of quality of service (QoS), such as sustainable throughput, bounded delay and reliable packet delivery. At the same time, it is necessary to guarantee that the power consumption of the battery powered wireless nodes is small, in order to increase the autonomy of the system. Currently there are many different wireless network technologies available in the market. Section 2 presents an overview of the most relevant technologies and discusses their suitability to meet the above requirements. Based on this analysis, we chose the Bluetooth wireless network technology as the basis for the design and development of a prototype of the fly-by-wireless system. The system was implemented using commercial off-the-shelf components, in order to provide a good trade-off between development costs, reliability and performance. Some other objectives were also pursued in the development of the system, namely the design of a framework where communication between nodes is effective and independent of the technology adopted, the development of a design approach to model the embedded system and the development of an application oriented operating system with a modular structure.

The following sections are organized as follows: section 2 presents an overview of available wireless network technologies, taking into account the requirements of the application; section 3 presents the global electronics architecture of the UAV platform, while section 4 describes the developed onboard wireless system; section 5 presents experimental performance results obtained with this system, and section 6 presents the conclusions and addresses the future work.

2. Wireless Network Technologies

Most wireless networks technologies available nowadays can be subdivided in a few categories: satellite networks, mobile cellular networks, broadband wireless access, wireless local networks (WLAN) and wireless personal networks (WPAN). The former three differ substantially from the latter two. One difference is that the network infrastructure does not belong to the user, but to the network operator, which charges the user for the services provided. Other difference is that they provide coverage over a large area. On the other hand, WLAN and WPAN are short range technologies in which all the communications equipment usually belongs to the user. These characteristics are more adequate for the intended application, so the remainder of this section will focus on wireless network technologies belonging to these two categories.

The most widespread type of WLAN nowadays is the IEEE 802.11 (IEEE, 2007), also known as WiFi. These networks are available on multiple physical options and operating frequency bands. However, all these versions use the same MAC (Medium Access Control) protocol; a contention based CSMA/CA (Carrier Sense Multiple Access/Collision Avoidance) mechanism known as DCF (Distributed Control Function). Given its statistical nature, this protocol is not adequate to provide the QoS guarantees required by the onboard wireless data acquisition and control system due to the probability of collisions.

In order to support real-time traffic, the 802.11 standard defines an alternative MAC protocol know as PCF (Point Coordination Function), based on a polling scheme, which is capable of providing QoS guarantees. However, unlike the DCF protocol, the implementation of PCF is not mandatory, and the availability of products that support it is scarce. More recently, a newer standard, the IEEE 802.11e (IEEE, 2007), designed to improve the efficiency and QoS support of 802.11 networks was released, but its availability on the market is also low.

Concurrently to the development of the 802.11, the European Telecommunications Standards Institute (ETSI) has developed another WLAN standard: HIPERLAN/2 (ETSI, 2002). HIPERLAN/2 networks are designed to operate at the 5 GHz band using OFDM (Orthogonal Frequency Division Modulation). Its physical layer is similar to the one used by the IEEE 802.11a due to agreements made by the two standard bodies. On the other hand, the MAC protocols used by these networks are radically different. HIPERLAN/2 uses a demand based dynamic TDMA (Time Division Multiple Access) protocol, which is able to provide extensive support of QoS to multiple types of traffic, including those generated by data acquisition and control systems (Afonso & Neves, 2005). However, the 802.11 standard won the battle for the wireless LAN market and as such no available HIPERLAN/2 products are known at the moment.

Due to its design characteristics, 802.11 and HIPERLAN/2 modules present relatively high power consumption. Although these networks can be suitable to interconnect devices like computers, there is an enormous potential market to provide wireless communication capabilities to smaller and cheaper devices running on batteries without the need of frequent recharging. Such devices include computer peripherals, biomedical monitoring appliances, surveillance units and many other sensing and actuation devices. To provide communication capabilities to such devices, various low cost short range networks, known collectively by the term wireless personal area network (WPAN), are being developed.

At the IEEE, the task of standardization of WPAN networks is under the scope of the IEEE 802.15 group. One of these standards, the IEEE 802.15.4 (IEEE, 2006) defines the physical and MAC layer of ZigBee (ZigBee, 2006), which aims to provide low power and low bit rate WPANs with the main purpose of enabling wireless sensor network applications. At the physical layer, the IEEE 802.15.4 relies on direct sequence spread spectrum (DSSS) to enhance the robustness against interference, and provides gross data rates of 20/40 kbps, at the 868/915 band, and 250 kbps, at the 2.4 GHz band. As in 802.11 networks, the basic ZigBee MAC protocol is a contention based CSMA/CA mechanism. A complementary mechanism defined in the 802.15.4 standard, the guaranteed time slot (GTS), enables the provision of some QoS guarantees to real-time traffic.

Bluetooth (Bluetooth, 2003) is another WPAN technology. It operates in the 2.4 GHz band using frequency hopping spread spectrum (FSSS) and provides a gross data rate of 1 Mbps. Bluetooth operates using a star topology, called piconet, formed by one master and up to seven active slaves. Transmissions can achieve a range of 10 or 100 m, depending of the class of the device. At the MAC layer, the Bluetooth devices uses a polling based protocol that provides support for both real-time and asynchronous traffic.

Bluetooth provides better overall characteristics than the other networks discussed here for the desired application. It drains much less power than 802.11 and HIPERLAN/2, uses a MAC protocol that provides support for real-time traffic, and provides a higher gross data rate than ZigBee. Bluetooth spread spectrum covers a bandwidth of 79 MHz while ZigBee operates in a band of less than 5 MHz, what makes the former more robust against interference. Moreover, Bluetooth provides an adaptive frequency hopping mechanism that avoids frequency bands affected by interference. Given these characteristics and the availability of the technology at the time of development, Bluetooth was chosen as the supporting wireless network technology for the development of the prototype of the system described in the following sections.

3. Global Electronics Architecture

The global view of architectural model of the onboard computing and communication system of the AIVA fly-by-wireless UAV platform is presented in Figure 1. It is a multitasking/multiprocessor based system connected by an asynchronous local bus that allows for speed adaptation of different tasks/processors. The system architecture supports one processing unit for a Bluetooth piconet master node, one flight controller unit, one data logger and earth link, and one embedded vision system (EVS). In each of these nodes many critical processes are permanently running.

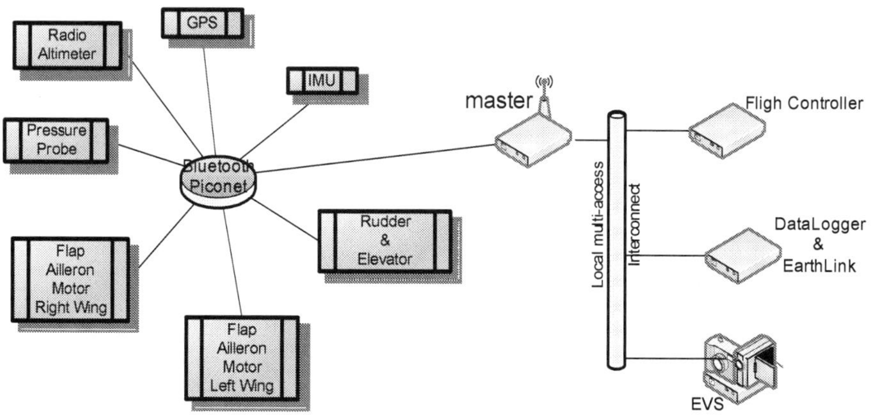

Figure 1. Global electronics architecture of the AIVA UAV platform

This architecture allows an easy way to introduce or remove processing units from the platform. For instance new sensors or new vision units can be included. In the first case a new module must be connected to the Bluetooth piconet, and in the second case the new module is connected to the local multi-access bus.

4. Onboard Wireless System

The AIVA UAV platform implements an onboard wireless distributed data acquisition and control system based on Bluetooth (BT) wireless network technology, represented by the Bluetooth piconet of Figure 1. The general architecture of a wireless node is presented on Figure 2. Each node is composed by a commercial off-the-shelf Bluetooth module that contains the radio electronics, a microcontroller that runs the code that controls the behavior of the node, and a local bus that provides interfacing between the node components, as well as specific sensors and/or actuators according to the purpose of the node.

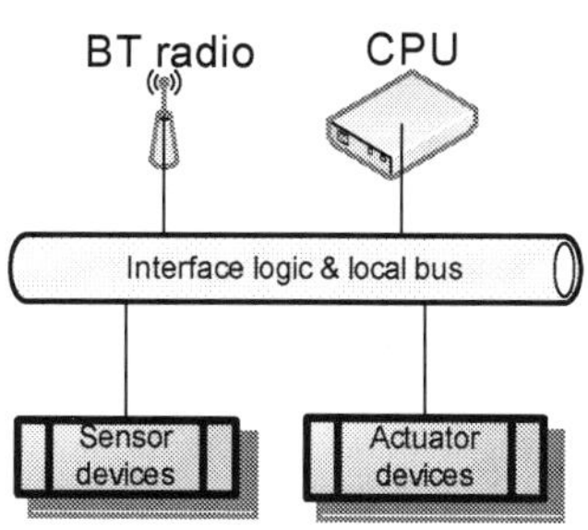

Figure 2. Architecture of a Bluetooth wireless node

4.1 Physical Architecture

The physical part of the platform is built around a low power Texas Instruments MSP430 microcontroller, a Von-Neumann 16 bit RISC architecture with mixed program, data and

I/O in a 64Kbytes address space. Besides its low power profile, which uses about 280 μA when operating at 1 MHz @ 2.2 Vdc, MSP430 offers some interesting features, like single cycle register operations, direct memory-to-memory transfers and a CPU independent hardware multiplication unit. From the flexibility perspective, a flexible I/O structure capable of independently dealing with different I/O bits, in terms of data direction, interrupt programming, and edge triggering selection; two USARTs supporting SPI or UART protocols; an onboard 12 bit SAR ADC with 200 kHz rate; and PWM capable timers, are all relevant features.

The Bluetooth modules chosen for the implementation of the wireless nodes are OEM serial adapter devices manufactured by connectBlue. The master node uses an OEMSPA33i module and the slave nodes use OEMSPA13i modules (connectBlue, 2003). These modules include integrated antennas; nevertheless, we plan to replace them with modules with external antennas in future versions of the platform, to be able to shield the modules in order to increase the reliability of the system against electromagnetic interference.

While the module used on the master (OEMSPA33i) allows up to seven simultaneous connections, the module used on the slaves (OEMSPA13i) has a limitation of only three simultaneous connections. However, this limitation does not represent a constraint to the system because the slaves only need to establish one connection (to the master).

The connectBlue modules implement a virtual machine (VM) that enables the provision of a serial interface abstraction to the microcontroller, so Bluetooth stack details can be ignored and focus can be directed to the application. The manufacturer's virtual machine implements a wireless multidrop access scheme where the master receives all frames sent by the slaves and all slaves can listen to the frames sent by the master, in a point-to-multipoint topology.

The AIVA onboard wireless system is composed by one Bluetooth piconet containing seven nodes: one master (MM - Master Module) and six slaves (SAM - Sensing & Actuation Modules). The nodes are spread over the aircraft structure, as shown in Figure 3. The master node (MM) is placed at the fuselage body, and acts as the network and flight controller, onboard data logger, and communications controller for the link with the ground station. On each wing, there is a SAM node for an electrical propulsion motor and for control surfaces (ailerons and flaps). These wing nodes are responsible for motor speed control and operating temperature monitoring, as well as control surfaces actuation and position feedback.

In the fuselage body, there are other two SAM nodes, one for a GPS module and other for an inertial measurement unit (IMU), which provide information assessment for navigational purposes. At the tail, there is another SAM node for elevator and rudder control, and position feedback. Finally, at the nose there is a SAM node connected to a nose probe consisting of a proprietary design based on six independent pressure sensors that give valuable digital information for flight control. This node also contains an ultrasonic probe that provides information for support of the automatic take-off and landing system. Figure 4 displays the physical layout of the nose node. The Bluetooth module is in the lower corner, the microcontroller is on the left hand side and the sensor hardware on the right hand side of the board.

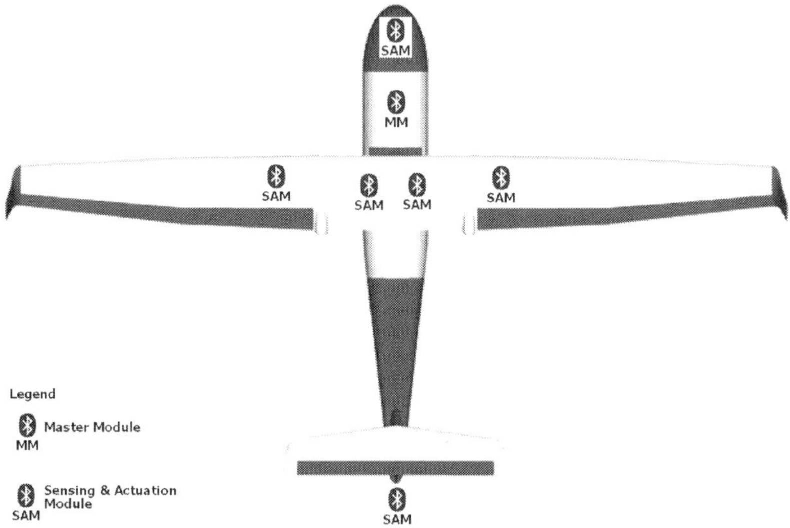

Figure 3. Node distribution on the aircraft structure

Figure 4. Physical layout of the nose node

4.2 Logical Architecture

The logical architecture of the developed system is a two layered state machine implementation, composed by a transport layer and an application layer. The transport

layer provides a packet delivery service, under control of master node, capable of transparent delivery of data packets across the network.
The transport layer is application independent, and interfaces with the top level application layer by means of a data space for buffering and a set of signaling control bits that allow total synchronization between the two layers. The hierarchy and signaling between the two layers is represented in Figure 5.

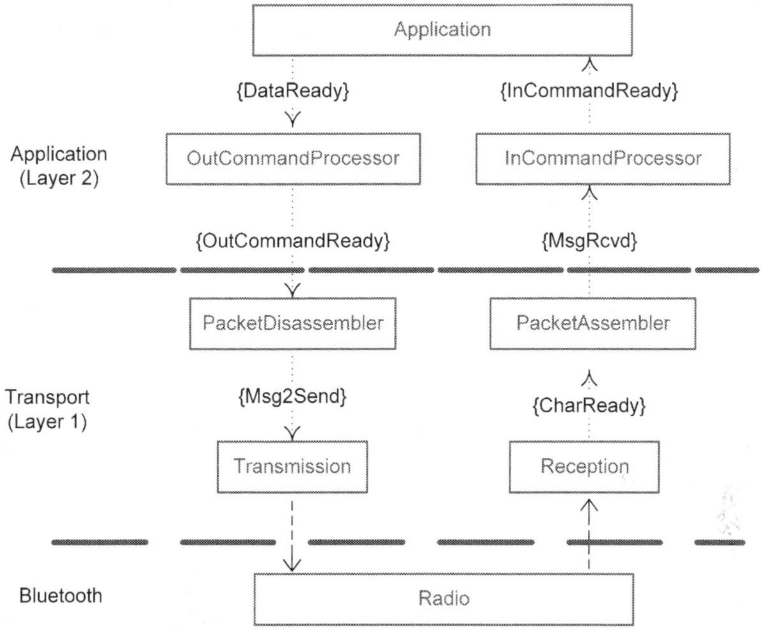

Figure 5. Hierarchy and signaling at the logical level of the platform

The asynchronous reception process delivers characters to upper processes. Analyzing the hierarchy from the lower level to the upper level, CharReady condition goes TRUE every time a new character arrives to the interface. The next process in the hierarchy is PacketAssembler, a state machine that performs packet re-assembly, reconstructing the original packet from a group of segments, and delivers packets for the next process in the hierarchy. When MsgRcvd (message received) goes TRUE, a new message is ready for processing. Thus, for incoming data, the model at layer 1 receives characters and delivers ready-to-process messages to the application layer. When the application layer understands that the message is ready to process, a command processor for incoming messages is activated in order to decode the embedded command and semantics contained in the message, to eventually execute some action, and to pass relevant information for the final application.
For outgoing data, the resident application eventually makes available some data to transmit to the master, signaling this event with a DataReady signal. This causes the output command processor to execute its cycle, preparing one message to be sent. When the message is ready, OutCommandReady goes TRUE, signaling to the lower layer that there is

a message to send to the network. At this phase, frame segmentation starts (if needed) by means of a state machine for packet disassembly. This state machine breaks the original message in smaller segments prepared to be serialized. Each time a segment is ready to be sent, Msg2Send goes TRUE and serialization is triggered. So, for outgoing data, the transport layer receives messages from application layer, and sends segments to the radio module in order to be sent over the wireless medium.

Layer 2 is application dependent, and has no knowledge of the lower layer internal mechanisms, just the services and interfaces available from it. That means that its logical architecture can be used in other applications. For the fly-by-wireless application, its main goal is to replicate a system table among all network nodes, at the maximum possible frequency. This system table maintains all critical system values, describing the several sensors and actuators, status parameters, and loop feedback information. Each network node is mapped to a table's section, where all related variables from sensing, actuators and metering are located. This layer is responsible for cyclic refreshing the respective table contents, based on local status, and also for cyclic actuation according to data sent from the master node (flight controller orders). This way, the whole system is viewed as a resident two-dimensional array located at master, with different partial copies distributed and synchronized among the slave nodes.

4.3 Other Design Issues

All the Bluetooth modules in the developed platform are configured in non discoverable mode, which contributes to the security of the system. The node discovery process of Bluetooth is a slow process, in the order of seconds, however it is not a problem since the master stores the addresses of all slave nodes that should participate in the piconet, so this process is avoided. The piconet formation is performed on the ground before the takeoff procedure, so the associated delay does not constitute a problem as well.

The use of Bluetooth technology limits the piconet operation to a maximum of seven active slaves; however, this limitation is not of major concern on the developed system, since only six slaves are used, and could only impose some restrains if node number should be raised.

The number of slaves in the network could be increased by interconnecting a number of piconets to form a scatternet. That way, a device participating in two piconets could relay traffic between both piconets. However, this architecture would probably have a negative impact in the performance of the network, making it more difficult to provide QoS guarantees to the application. Moreover, currently there are very few actual implementations of scatternets available.

Given that free space propagation loss is proportional to the square of the distance, it is not expected that the onboard wireless network will either suffer or induce interference on other networks operating in the same frequency band, such as the widely deployed WiFi networks, since the former operates in the sky most of time, while the later are normally based on the ground.

5. Experimental Results

The performance of the developed wireless system was evaluated in laboratory. The experimental setup used to achieve the results presented in this section is composed by 6 slaves sending data periodically to the master (uplink direction) at the same predefined sampling rate. Each sampling packet has a length of 15 octets, which is the maximum packet length supported by the transport layer due to a limitation imposed by the virtual machine used by the Bluetooth module.

Figure 6 presents the aggregated uplink throughput that reaches the master node as a function of the sampling rate used by the 6 slaves. Since Bluetooth uses a contention-free MAC protocol, the uplink transmissions are not affected by collisions, so the network throughput increases linearly with the offered load until the point it reaches saturation, which in this scenario corresponds to the situation where the slaves transmit data at sampling rates higher than 200 Hz. As this figure shows, the maximum throughput available to the application is about 160 kbps, which is significantly lower than the gross data rate provided by Bluetooth (1 Mbps). This difference can be explained by the overhead introduced by the Bluetooth protocol and the virtual machine, including the gap between the packets, the FEC (Forward Error Correction) and ARQ (Automatic Repeat reQuest) mechanisms, the packet headers, as well as the overhead introduced by control packets such as the POLL packet, that is sent by the master to grant permission to slaves to transmit.

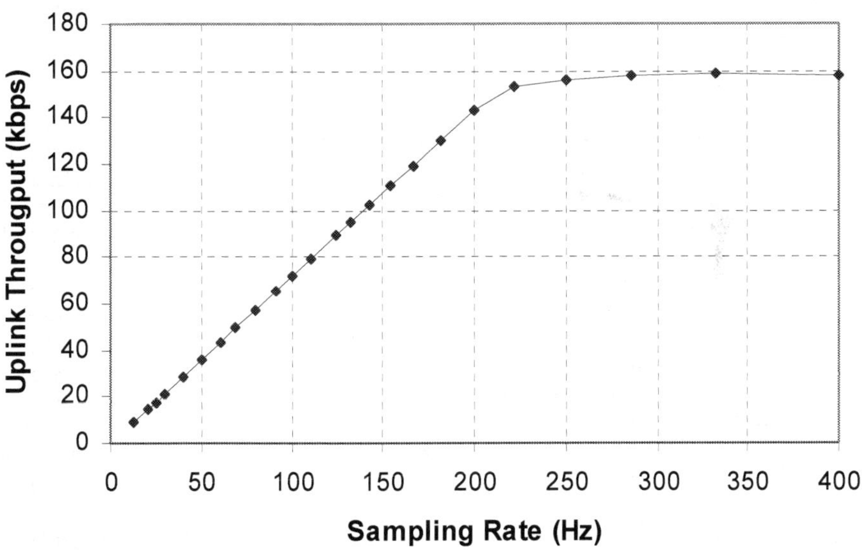

Figure 6. Uplink throughput as a function of the sampling rate

Figure 7 presents the packet loss ratio (PLR) averaged over the 6 slaves as a function of the sampling rate. As the figure shows, the PLR is limited to less than 0.5 % in the region where the network is not congested, but increases rapidly after the saturation point. The flight control application should be able to tolerate such small losses; otherwise a change in the supporting wireless technology should be made in the attempt to obtain higher link reliability.

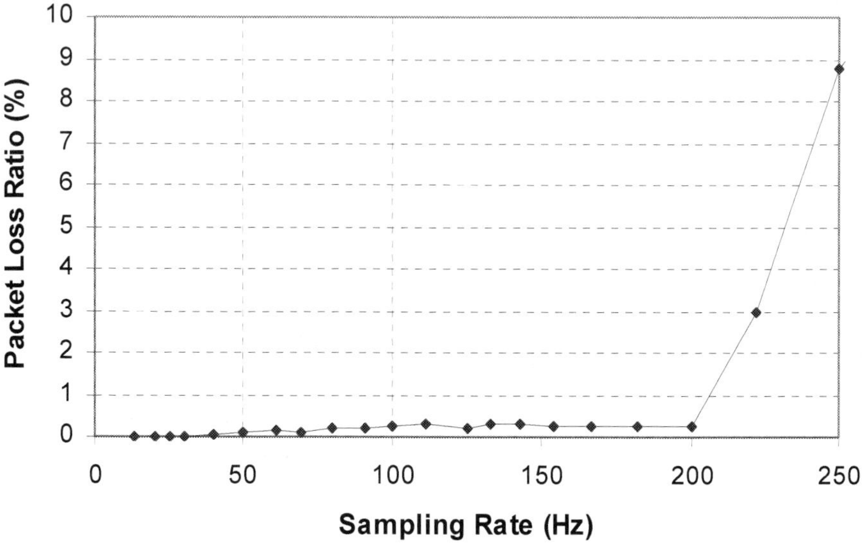

Figure 7. Packet loss ratio as a function of the sampling rate

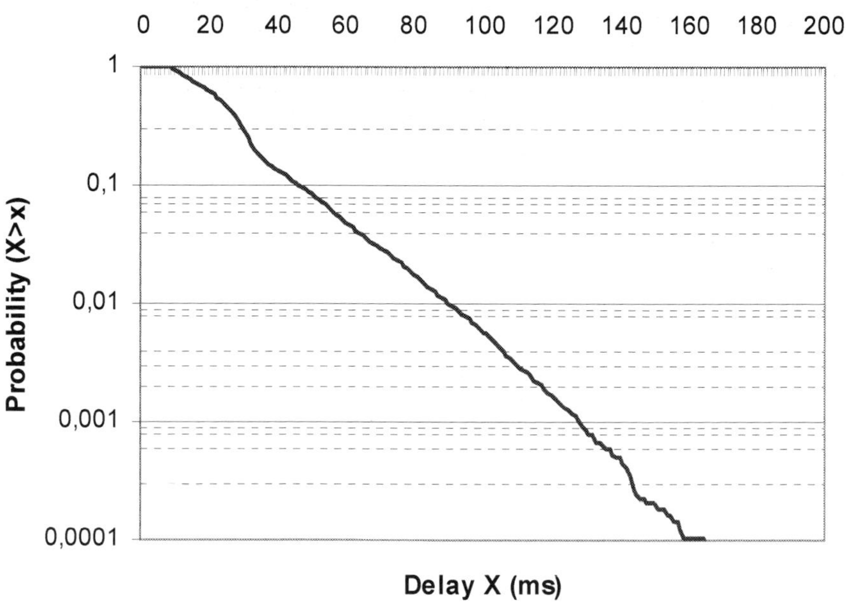

Figure 8. Complementary cumulative distribution of the delay

Concerning to the delay experienced by the packets as they travel from the slaves to the master, results showed that the delay is not adversely affected by the rise in the offered load, as long as the network operates below the saturation point. For sampling rates up to 200 Hz, the registered average delay was 27 ms and the standard deviation was 16 ms.
Figure 8 presents the complementary cumulative distribution (P{X>x}) for the random variable X, which represents the delay for all the samples collected using sampling rates in the range from 0 to 200 Hz. With this chart, it is possible to see the probability that the delay exceeds a given delay bound, which is an important metric for real-time applications such as the one considered in this chapter. The chart shows, for instance, that less than 1 % of the sample packets suffer a delay higher than 90 ms, while less than 0.1 % of the packets suffer a delay higher than 120 ms.
Experimental tests were also made with a varying number of slaves in the piconet (from 1 to 6), both in the uplink and downlink direction. The average delay measured in the downlink direction (from the master to the slaves) was slightly higher than the one registered in the uplink direction, but below 40 ms, for the measurements made with up to 4 slaves. However, the average master-to-slave delay with 5 slaves in the network ascended to 600 ms, while with 6 slaves the performance was even worse, with the average delay reaching 1000 ms.

6. Conclusion

This chapter presented the design and development of a fly-by-wireless UAV platform built on top of Bluetooth wireless technology. The developed onboard wireless system is composed by one master node, connected to the flight controller and six slave nodes spread along the aircraft structure and connected to several sensors and actuators.
In order to assess the suitability of the developed system, several performance evaluation tests were carried out. The experimental results showed that, for the slave-to-master direction, the system prototype is able to support a sampling rate of up to 200 Hz for each of the 6 slaves simultaneously without significant performance degradation in terms of throughput, loss or delay. On the other hand, although the master-to-slave delay with 1 to 4 slaves in the network is low, its value increases significantly with 5 and 6 slaves, which is unacceptable given the real-time requirements of the desired application. This problem is caused by implementation issues related to the proprietary embedded virtual machine provided by the manufacturer of the Bluetooth module that is used in the master node of the prototype.
The approach of relying on the virtual machine provided by the manufacturer, which hides the Bluetooth protocol stack functionality, allowed the development focus to be directed to the application, reducing the development costs. The disadvantage, however, is the lack of control of the behavior of the system at the Bluetooth stack level, which impedes the optimization of the performance of the system at this level and the correction of problems such as the verified with the master-to-slave delay. The solution to the detected problem can pass either by the replacement of the Bluetooth module by a newer version (already available) from the same manufacturer or by the direct interaction with the Bluetooth stack, with the bypass of the virtual machine.
Despite the limitations of the current prototype, the overall results provided by the experimental tests are satisfactory. Nevertheless, further tests are needed in order to

evaluate the behavior of the system under more harsh interference conditions, as well as in a real scenario onboard the aircraft.

7. References

Afonso, J. A. & Neves, J. E. (2005), Fast Retransmission of Real-Time Traffic in HIPERLAN/2 Systems, *Proceedings of Advanced Industrial Conference on Telecommunications (AICT2005)*, pp. 34-38, ISBN 0-7695-2388-9, Lisbon, Portugal, July 2005, IEEE Computer Society.

Bluetooth SIG (2003), Specification of the Bluetooth system. Available at http://www.bluetooth.org/.

connectBlue (2003), Serial Port Adapter - 2nd Generation, *User Manual*. Available at http://www.connectblue.com/.

ETSI TR 101 683 V1.1.1 (2000), Broadband Radio Access Networks (BRAN)—HIPERLAN Type 2—Data Link Control (DLC) Layer—Part 1: Basic Data Transport Functions.

IEEE Std 802.11 (2007), IEEE Standard for Information Technology—Telecommunications and information exchange between systems—Local and metropolitan area networks—Specific requirements—Part11: Wireless LAN Medium Access Control (MAC) and Physical Layer (PHY) Specifications.

IEEE Std 802.15.4 (2006), IEEE Standard for Information Technology—Telecommunications and information exchange between systems—Local and metropolitan area networks—Specific requirements—Part 15.4: Wireless Medium Access Control (MAC) and Physical Layer (PHY) Specifications for Low-Rate Wireless Personal Area Networks (WPANs).

ZigBee Standards Organization (2006), ZigBee Specification. Available at http://www.zigbee.org/.

2

Combining Occupancy Grids with a Polygonal Obstacle World Model for Autonomous Flights

Franz Andert and Lukas Goormann
Institute of Flight Systems, Unmanned Aircraft, German Aerospace Center (DLR)
Germany

1. Introduction

1.1 Overview

This chapter presents a mapping process that can be applied to autonomous systems for obstacle avoidance and trajectory planning. It is an improvement over commonly applied obstacle mapping techniques, such as occupancy grids. Problems encountered in large outdoor scenarios are tackled and a compressed map that can be sent on low-bandwidth networks is produced. The approach is real-time capable and works in full 3-D environments. The efficiency of the proposed approach is demonstrated under real operational conditions on an unmanned aerial vehicle using stereo vision for distance measurement.

1.2 The Problem of Mapping and Obstacle Representation

To be autonomous, vehicles must know the environment in which they are to move. Like humans or animals, they have to sense, understand and remember at least those parts of the environment that are in the vicinity. Only then can the vehicles operate in their environment - without manual remote control. Successful results of autonomous flights in urban areas with unmanned aircraft have been presented in the past (Hrabar et al., 2005; Zufferey & Floreano, 2005; Griffiths et al., 2007; Scherer et al., 2007). Beside that, the majority of obstacle detection, mapping, and avoidance research are carried out with ground vehicles and many approaches used in flight applications are based on practices derived from that wealth of knowledge.

A main requirement for autonomous vehicles is to detect obstacles and to generate environmental maps from sensor data and a large number of approaches have been developed over the years (Thrun, 2002). One approach to represent the environment is the use of grid-based maps (Moravec & Elfes, 1985; Konolige, 1997). They allow an easy fusion of data from different sensors; including noise reduction and simultaneous pose estimation, but they have large memory requirements. Further, they do not separate single objects. A second approach, called feature-based maps, focuses on individual objects. An early work (Chatila & Laumond, 1985) uses lines to represent the world in 2-D. Later approaches use planar (e.g. Hähnel et al., 2003) or rectangular (Martin & Thrun, 2002) surfaces for 3-D modeling - but mostly to rebuild the world with details and possible texture mapping. A suitable model for autonomous behavior is the velocity obstacle paradigm (Fiorini & Shiller, 1998) that can be added with the introduced specifications on how to measure the obstacles.

These map types, and others not discussed here, have their respective advantages and disadvantages. As a result, there is a need to have different map types for different robot tasks (Kuipers, 2000). In many cases, it is advantageous to use grid-based maps for sensor fusion and feature-based polygonal metric maps for local planning, e.g. in order to avoid obstacles (Fulgenzi et al., 2007). Additionally, non-metric topological maps are most suitable for global search tasks like route planning. In a complex scenario, a robot must deal with all of these different maps and keep them updated. These tasks need the usual information exchange.
To generate maps from sensor data that are applicable to autonomous applications, it is a straightforward procedure to generate a grid map from sensor data and extract out the features. Outdoor scenarios, however, can be too large to store the whole scene in a data array with reasonably accurate resolution. Additionally, the area boundaries may be unknown before mapping.
The approach presented here combines grid maps and polygonal obstacle representations and tackles the problem of large environments by using small grid maps that cover only essential parts of the environment for sensor fusion. Characteristic features are recognized, their shapes are calculated, and inserted to a global map that takes less memory and is easily expandable. This map is not restricted to the sensor environment and is used for path planning and other applications.

2. Flight Testbed

2.1 ARTIS – A Flying Robot

The presented mapping and world modeling approach is developed within the ARTIS (Autonomous Rotorcraft Testbed for Intelligent Systems) research project that deals with mid-sized unmanned helicopters (Dittrich et al., 2003). One of the helicopters is shown in figure 1. It has a main rotor diameter of 3 meters and a total weight of up to 25 kg. Flights of more than 30 minutes are possible.

Figure 1. The unmanned helicopter ARTIS

The 5 kW turbine engine has enough power to carry more than six kilograms of experimental payload in addition to the avionics system and power supply. The actual configuration is a dedicated image processing computer and a stereo camera system weighing 2 kg so that additional sensors like multiple cameras or laser scanners can be used in future applications.

2.2 Sense and Avoid Setup

For applications that need environmental sensing capabilities, a vision system separated from the flight controller is installed at the helicopter. The actual configuration uses only cameras because they are lightweight, passive, and have low power consumption. A dedicated computer is used to process image information. This results in improved speed and no influence on the real-time behavior of the flight control computer. For interaction between image-based results and flight control, data exchange is provided via a local network. A mission planning and automatic control system is installed on the flight control computer (Dittrich et al., 2008). It calculates trajectories around obstacles, considers changes due to actual image-based map updates, and instructs the helicopter to fly these paths. The autopilot ensures stabilized hover so that the vehicle can completely fly autonomously.

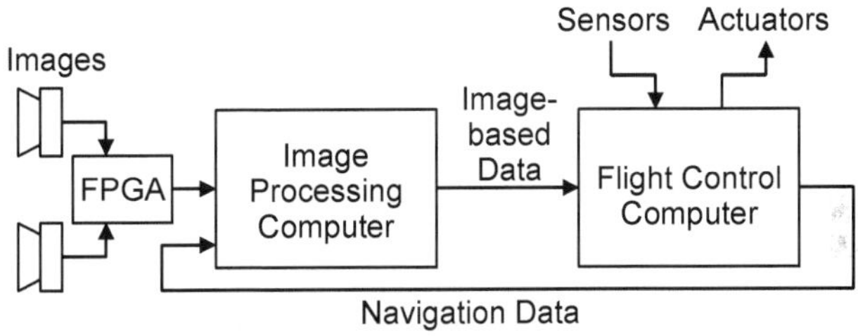

Figure 2. Overview of the onboard hardware for vision applications

Figure 2 illustrates the connection between vision hardware and flight controller. Since obstacle mapping and other image-based algorithms require flight information, a navigation solution provides the global position and attitude of the helicopter.

A stereo camera (Videre Design STOC, fig. 3) with a baseline of 30 cm and a field of view of approximately 51° x 40° is used. It creates images with 640 x 480 pixels and has an inbuilt FPGA processor that calculates a depth image out of the two input images in real-time with 30 Hz. This image is a result of a complex processing step where regions of the two camera images are matched (e.g. Scharstein & Szeliski, 2002), and it acts as a depth sensor in the mapping process. Basic pre-processing to enhance the depth image quality is already done by the camera. In the depth images shown in the figures of this article, near distances are represented by light colors and farther distances by darker colors. White space indicates that depth values are missing or have been filtered due to bad image quality, e.g. low texturing.

Figure 3. Stereo camera mounted at the helicopter (left), left onboard camera image (center), depth image (right)

The helicopter's position and attitude is provided in six degrees of freedom by the flight control computer using a differential GPS sensor, a magnetometer and an inertial measurement unit. The raw data of these sensors are integrated by an Extended Kalman filter (Koch et al., 2006) to provide an accurate solution in all six degrees of freedom. Filtered navigation data is sent with a rate of 100 Hz to the vision computer. All computer clocks are synchronized so that a fitting recording pose of an image with a given timestamp can be obtained.

3. Model Overview

3.1 Grid Maps

The basic method to interpret data from depth image sequences follows classical approaches with occupancy grids (Moravec & Elfes, 1985; Raschke & Borenstein, 1990). These grids have turned out to be very useful for obstacle mapping since they allow easy sensor fusion, reduce sensor noise, and are also applicable to multiple vehicles. Here, a world-centric 3-D grid represents the map. Each cell consists of a value describing the presence of obstacles: the log odd of the occupancy probability. Higher values refer to a higher probability of the cell being occupied. The map is created incrementally by starting with an empty grid and writing the actual sensor information with each new depth image.

3.2 Feature Maps

As already denoted, occupancy grid maps are advantageous for sensor fusion and will be used to interpret data from depth image sequences. In addition to that, feature maps are built out of these occupancy grids to store global obstacle information in a compressed way and to be an input for applications that use the map.

For this reason, it is a primary requirement to determine the required level of detailing to represent objects in a feature map. For obstacle avoidance applications, the important criteria are:

1. Small details are not needed,
2. an identification of planes for texture projection is not needed, and
3. real-time capabilities are more important than centimeter accuracy.

These criteria imply that it is sufficient to mark a box around an object and simply avoid this area.

The simplest way of modeling potentially danger areas is the usage of cuboidal bounding boxes that are aligned with the coordinate axes. Unfortunately, they are too rough for object modeling, e.g. a set of objects like houses with a rectangular base shape and an arbitrary angle to the coordinate axes. Aligned bounding boxes will be larger than the houses and narrow paths between them will not be found because they exist inside the area marked by the occupied box.

In a real scenario like a city, objects can have any ground shape from the top view, but it is likely that they have vertical walls. This assumption is made in a lot of mapping approaches (e.g. Iocchi et al., 2000). For this reason, prism shapes are used here. They do not require an axis alignment like the bounding boxes. Additionally, they can deal with any ground shape without dramatically increasing the complexity. If walls are not vertical like roofs, the prism's volume will be much larger than the real object. To improve the modeling of objects, they can be subdivided vertically and represented by multiple prisms. In other words, the world is split into layers in different heights, and each layer has its own polygonal 2-D obstacle map. This is sufficient for a lot of applications like flight trajectory planning in urban scenarios (Dittrich et al., 2007).

Nevertheless, the prism model can only handle complex shapes in 2-D as seen from the top view. A gabled roof or a tunnel with an up- or downhill road must be modeled by multiple obstacle prisms. These scenarios are more complex than a simplified urban canyon with a horizontal ground plane and obstacles with vertical walls. Avoiding roofs that are represented by bounding prisms will be only a small restriction and a tunnel, however, will be an unusual case, at least with respect for flying robots.

3.3 The Modeling Process

As already mentioned, advantages of both grid and feature maps are combined, see figure 4. The mapping process works as follows:

1. Create an occupancy grid around the vehicle's position if not existing in the map.
2. If a new grid is allocated or the grid is extended, check whether previously stored features can be inserted.
3. Insert the actual sensor data information to the grid.
4. Find clusters of occupied grid cells and mark them as single obstacle features.
5. Find out which obstacle features are new and which are updates of objects that have already been identified in the previous loop cycle. Preexisting objects may also be removed.
6. Calculate the shape of each new or updated feature.
7. To insert the next sensor data, go back to step 1.

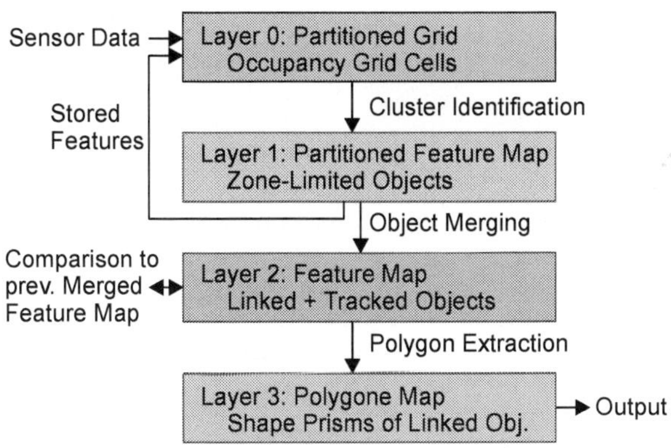

Figure 4. The general mapping process with different map layers

This approach uses different map types as illustrated. The occupancy grid map for sensor inputs (layer 0) is partitioned into zones and each zone is searched for obstacles separately (layer 1). This method is described in the following section. Next, a map with separate obstacles, but without zone separation is generated (layer 2), and finally, the prism shapes are extracted out of them (layer 3). Grid resolution and zone sizes are user-defined but may not change over time when processing image series.

4. Occupancy Grid Mapping

4.1 Sensor Interpretation and Local Mapping

To create an occupancy grid incrementally out of depth image data sequences, a local grid map is built. This local map is world-centric like the global output maps but includes only the occupancy data of a single depth image taken from its corresponding camera position and attitude.

The local map is the interpretation of a single depth image as follows: Using projective geometry and the pinhole camera model, each pixel refers to an object coordinate in the world. By assuming no transparent objects in front of the object, there is free space along a ray between the camera center and the object coordinate. Behind the opaque obstacle, no information is available. If the distance encoded by a pixel exceeds a threshold, this image point leads to free space. A line is drawn for each pixel, illustrated by figure 5 that shows the interpretation of an ideal sensor (left) or considers measurement noise, respectively (right). The uncertainty of stereo-based depth measurement increases quadratically with higher distance values. With this information, the viewable area is spanned and the obstacles are drawn.

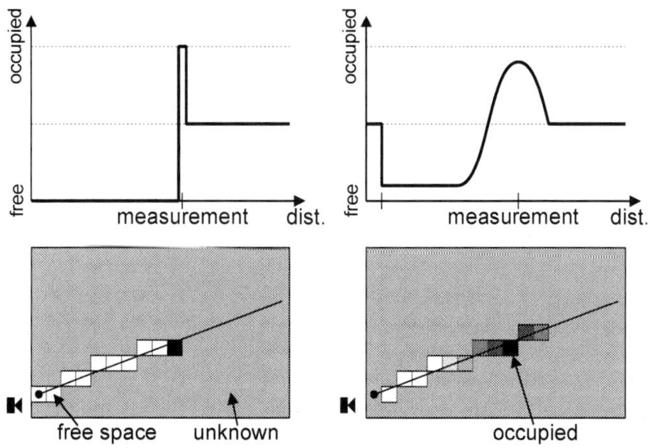

Figure 5. Interpretation of an ideal depth image pixel (left) and a more realistic model that considers depth uncertainty (right)

The result of processing all rays of a single image is illustrated in figure 6. Obstacles and free areas of the image are also visible in the corresponding map. The thickness of an obstacle cannot be derived from a single image so it depends on the depth uncertainty first. Information about the true size and the area behind obstacles are included when the vehicle moves there.

The size of the local grid is determined by considering that the data fusion of the empty local map with one image will only affect grid cells inside the sensor range, i.e. a small environment of the actual position. There is no need to update cell values outside this area. Hence, it is satisfactory to have only an occupancy grid representation of the map inside the sensor range. Further, the local map definition is independent to the outer camera orientation angles to get a fast implementation. With that, the camera is set to the center of the local map, with maximal half a grid cell deviation to keep the grid cells aligned to a global rasterization. To give an example, a sensor range of 30m implies zones with a size of each $60 \times 60 \times 60$m to cover all attitude angles without truncating far distances. In applications where looking and moving straight upwards or downwards is restricted, smaller zones with a height of e.g. 15m are satisfactory. If the grid resolution is set to 0.5m, grid arrays with 120 x 120 x 30 cells are created to represent a local map.

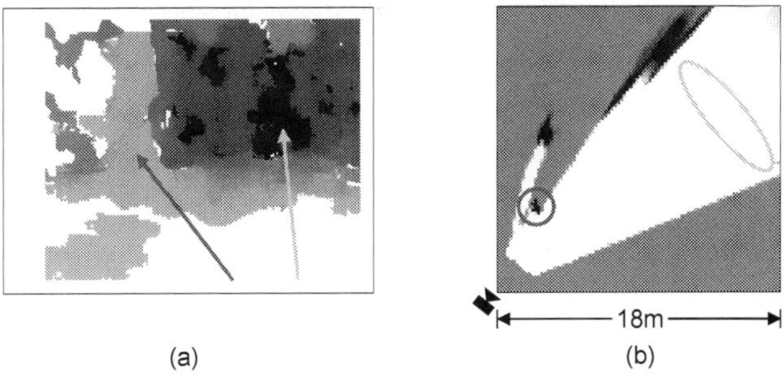

Figure 6. Depth image example (a) and a 2-D view of the local map that was built out of it (b)

4.2 Creating Global Maps

The data fusion between succeeding image frames is done by integrating the local grid with a global map by adding the values of cells at the same global position. Similar to other approaches, free or occupied cells become more significant if measured several times. Noise is filtered when occupancy information from one image is disproved by free space information from another image. With that, especially static objects can be easily determined. In practice, the map values are truncated to a specified range so that integers can be used for the cell array. The range must be large enough to ensure the robustness to failures.

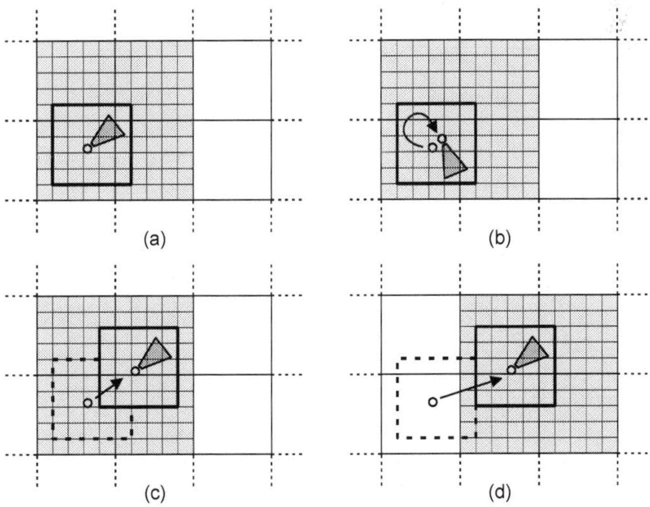

Figure 7. 2-D view of the global map that is divided into zones. For zones around the actual camera position, an occupancy grid representation exists

Since grids are usually stored as continuous data blocks in the computer memory, the boundaries of the map must be known a priori. If the vehicle moves outside, extensive reallocation or shifting methods must be applied because the map boundaries change.

To avoid shifting a large number of map cells when moving, the implementation divides the global map into cuboidal zones. Their position is fixed. The size of each zone is equal to the local map so each zone overlaps a maximum of eight zones in the global 3-D map. The environmental cuboid moves through the global map with the movement of the camera, i.e. with the helicopter. Only those global zones overlapping with the local grid are represented by an occupancy grid array.

Figure 7 illustrates how the zone partitioning works. The actual camera and local map position is shown in figure 7a, the other graphics show possible effects on the map, caused by the next measurement. Often, rotations (7b) or movements (7c) will not have an effect on the zone boundaries. But if the movement is larger so that boundaries are crossed (7d), new memory is allocated for these zones. Grid information is discarded for zones that fall outside the immediate vicinity.

5. Extracting Objects from the Grid Map

5.1 Determining Separate Objects and Bounding Boxes

Object features are detected by segmenting the global map into occupied and free areas applying a threshold. A single object is a cluster of connected occupied cells of the 3-D array. These objects are recognized with a flood fill algorithm. By saving the minimal and maximal values of the coordinates of cells belonging to the object, the bounding box is calculated and put into the global feature map as it is illustrated in figure 8. Unlike the cell array, these boxes need much less memory and it is easy to make them available to further applications, independent of the existence of a grid. For each object, the binary cell shape is stored in addition to the bounding box so that this shape can be re-inserted. Compression with an oct-tree structure is applicable here. The polygonal shape is calculated later.

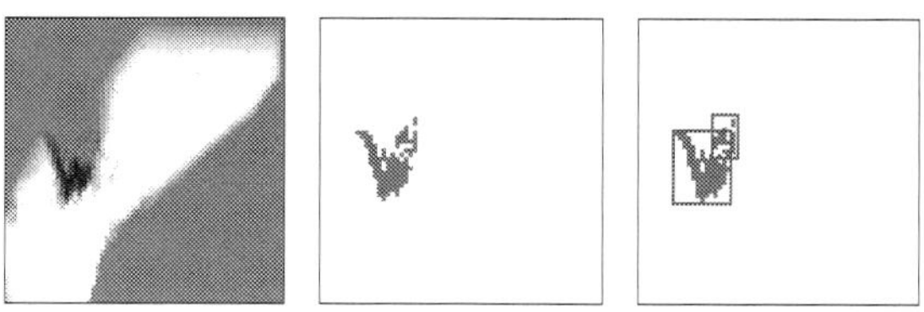

(a) (b) (c)

Figure 8. Extracting features from an occupancy grid (a) with thresholding (b) and bounding box calculation (c). Bounding boxes can overlap

5.2 Integration Between Features and Grid

Unlike the occupancy grid, the feature map is not limited to an environment around the actual vehicle position. Objects are stored independently from the presence of a grid. If some grid data is removed, the corresponding features will remain (fig. 9). Vice versa, objects can be inserted into the grid (fig. 10). It is possible to include a-priori knowledge about obstacles here.

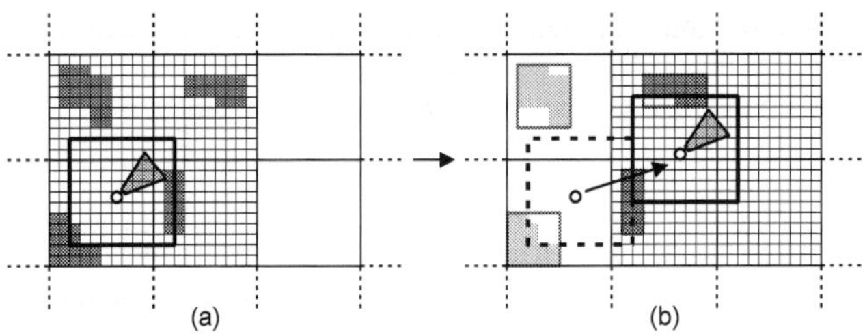

Figure 9. Vehicle movement from (a) to (b). Features are stored when the grid data is discarded due to helicopter movement

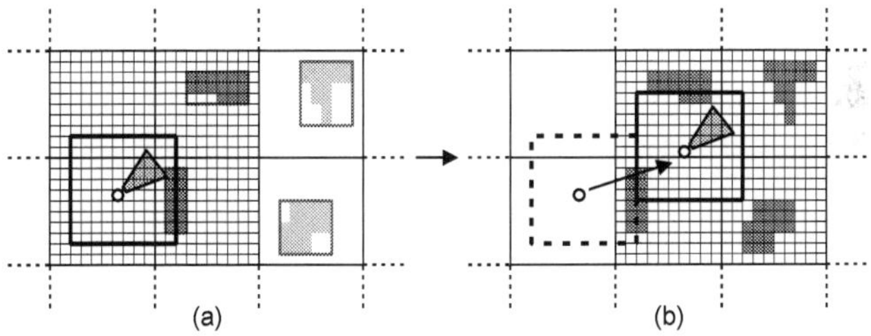

Figure 10. Vehicle movement from (a) to (b). Inserting features to the grid if there are objects inside a new zone

5.3 Merging and Tracking Obstacles

Single objects found in the temporary grid zones are limited to the zone boundaries since each zone is processed separately. To build an output map for the application that has no zone partitioning, the features of different zones are merged if a connection exists as illustrated in figure 11. First, it is checked whether the bounding box of an object is located at a zone boundary. If another object box is located at the same boundary from a different zone, the cell shapes of both objects are tested for connection. Connected objects are linked together, and the linkage is not limited to only two objects.

Figure 11. Merging tangent objects in the partitioned map (a) to linked objects (b) that are not partitioned into zones. Simplified 2-D view

After an update step where sensor data is inserted, the zones are checked for obstacles again without reference to the previous state. Since it is useful to know which specific object has been updated with the actual sensor information, the linked objects are tracked. This is done as follows:

1. Calculate the center of mass of each linked object.
2. For each linked object, try to find a matching object in the map with linked objects of the previous state. This is the object with the nearest center of mass determined by the Euclidean distance (Prassler et al., 2000). To avoid the matching of objects that are too far away, distances above a threshold will not result in a match.
3. If no match exists for an object of the actual state, give it a new unique ID. Otherwise, copy the ID from the matched object.

With object tracking, each linked obstacle in the world gets a unique ID that remains constant over time, if the tracking is successful. The merging and tracking step can be skipped for linked objects whose coordinates are completely outside the zones where a grid exists since there will be no update.

Due to sensor updates it is possible that a linked object becomes split. Usually, one of the new objects gets the old ID, the others receive the new IDs. If multiple linked objects are merged together over time, one ID is kept and the others removed.

5.4 Horizontal Shape Slicing

The approach presented in this paper models the shape of each linked object with a prism (see section 5.5) with horizontal bases. Since a bounding prism of complex shapes may be too rough, the cell-based shape $s(x, y, z)$ is partitioned into horizontal slices with the height of one cell. Similar slices are merged and a polygonal shape prism is calculated for each multi-slice.

A single slice in height z is denoted as $s_z(x, y)$ with $s_z(x, y) = s(x, y, z)$. Without loss of generality, z is valid from 1 to n. Similar consecutive slices are put together to multi-slices S using the following algorithm:

1. Set $i = 1$, set $z = 1$.
2. Create a multi-slice S_i: $S_i = \{s_z\}$.
3. If $z = n$, break.
4. If the slices s_z and s_{z+1} are similar given by the equation

$$\sum_x \sum_y | s_z(x,y) - s_{z+1}(x,y) | < t \qquad (1)$$

with a similarity threshold t,
set $S_i = S_i \cup \{s_{z+1}\}$,
set $z = z + 1$ and go to step 3.
Otherwise,
set $i = i + 1$;
set $z = z + 1$ and go to step 2.

The result is a set of multi-slices $\{S_1, \ldots, S_m\}$. The parameter t of equation 1 controls the degree of similarity consecutive single slices must have in order to be unified. The extreme case $t = 0$ puts every single slice into a separate multi-slice and $t = \infty$ merges all single slices together. As illustrated in figure 12, grid-based object shapes are splitted into parts that have approximatively vertical walls. They can be approximated by their 2-D shape from top view and easily modeled by prisms without forfeiting large irregularities in the shape.

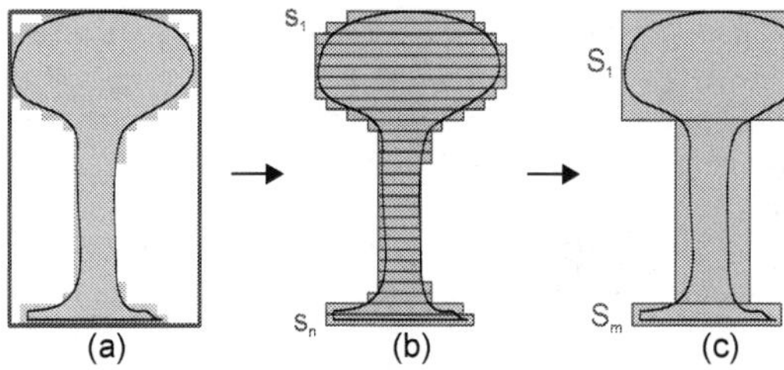

Figure 12. Grid shape and bounding box of linked feature (a), single slices (b) and merged multi-slices (c). Side view of a tree as an example

In the next step, the 2-D shape is calculated for each multi-slice S_i ($1 \le i \le m$). Occupied cells of a multi-slice are calculated through the logical disjunction of all single slices s_z inside S_i. It is

$$S_i(x,y) = \begin{cases} 0, \text{ if } \sum_{s_z \in S_i} s_z(x,y) = 0; \\ 1, \text{ otherwise.} \end{cases} \qquad (2)$$

5.5 Extraction and Approximation of the Polygonal Shape

Now, algorithms developed for binary images can be applied to a multi-slice S_i since its shape is represented by a binary 2-D array. Figure 13 illustrates the process. The polygonal extraction is done in two steps. First, the contour is calculated with a tracing algorithm (Ren et al., 2002). The main idea is to start at one edge pixel and search incrementally for the next edge pixel until the whole contour is covered. Its output is a list of pixels sorted counterclockwise for outer contours and clockwise for inner contours. Multiple lists are possible. Each contour pixel can be interpreted as a polygon vertex by using the corresponding grid cell center coordinate.

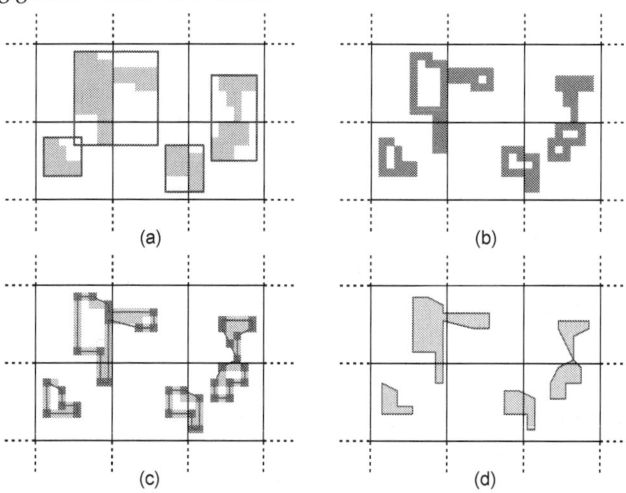

Figure 13. Shape Extraction, top view. Cell-based shapes (a), contours (b), approximated polygons with its vertices (c) and final ground shapes (d)

The second step is for data compression and to accelerate later applications. A polygon with a lower number of vertices is calculated here with the approximation algorithm presented by Ramer (1972). It is parameterized by a maximal distance value that specifies the accuracy of the algorithm. Every point of the original shape has this maximal distance to the lines defining the approximated shape.

This 2-D polygon acts as the ground shape of the right prism that is calculated for each S_i. The prism's height and vertical position is determined by the span of single slices $s_z \in S_i$. As seen in figure 13, the prism shapes can be smaller than the grid-based shapes since the center coordinates of the grid cells are used for the shape vertices and an approximation is performed. In obstacle avoidance applications, a safety distance to the objects is added; it must be larger than half a cell size plus the approximation accuracy to ensure that the grid cells of each object are completely covered by the polygon hull.

6. Estimating Ground Planes

6.1 The Height Histogram

In addition to the prism-based shape detection of obstacles, the floor plane is extracted out of the sensor data. This simplifies the resulting world model. The floor will not be a shape prism that has to be tracked. Furthermore, obstacles and the floor are not merged which helps to identify objects that hit the ground.

The floor plane detection is similar to classical Hough Transform based approaches (e.g. Okada et al., 2001) with the limitation that only the horizontal planes are searched.

The actual sensor data leads to a cloud of points (x, y, z) in global coordinates. The vehicle's position must be known and a calibrated pitch and roll angle measurement is assumed to ensure that a horizontal ground will lead to a horizontal plane in map coordinates. A sampled histogram $n(z)$ of all z-values of these points is built. The sampling should be more precise than the occupancy grid. Points are inserted with blurring considering their depth-dependent uncertainty.

Since all measurements of a horizontal plane have approximately the same z-value, planes lead to significant peaks in the histogram and can be detected there. If more than one plane is found, the floor is the one with the lowest height, i.e. with the largest z-value. There will be no ground plane if no peak with a high confidence is found.

It can be useful to increase the floor plane height, e.g. to force a minimal distance to the ground in urban flight scenarios. First, the peak maximum $n_{max} = n(z_{max})$ is not taken. Rather, z is decreased while $n(z) > t \cdot n_{max}$, starting at $z = z_{max}$. The result is denoted as z_{ground}. In the experiments, t is set to 0.5. Second, an offset $z_{offset} < 0$ can be added to this height.

Figure 14. Depth image (left), depth image with marked floor pixels (center) and original camera image with floor pixels for comparison (right)

Figure 14 shows an example of the ground estimation. Pixels that lead to a larger z-value are marked in the depth image and for comparison in the original camera image. As seen in the right subfigure, the floor is marked contrary to the house at the right side. It is not necessary to insert the marked measurements into the obstacle map.

6.2 A Floor for each Zone

To combine the data of multiple images, the ground plane estimation is done separately for map zones. The map partitioning as described in section 4.2 is used but without the vertical separation. This leads to rectangular 2-D zones from the top view and each zone can have a floor plane height. A height histogram is calculated for each zone by accumulating these sensor data elements that lead to object points in that zone. The histogram is built incrementally over time so that multiple sensor updates can be stored in one zone if the vehicle stays there. The result is a kind of an elevation map with an estimated height value for each zone.

7. Tests and Results

A first experiment tests the mapping algorithm in a simulation environment where two views are generated and captured with cameras (fig. 15). The grid resolution is 0.25m and each map zone has a size (x, y, height) of 128 x 128 x 32 cells. Figures 15 and 16 show the result of the simulation experiment where two obstacle arches of 6 x 6m size are placed with a distance of 50m.

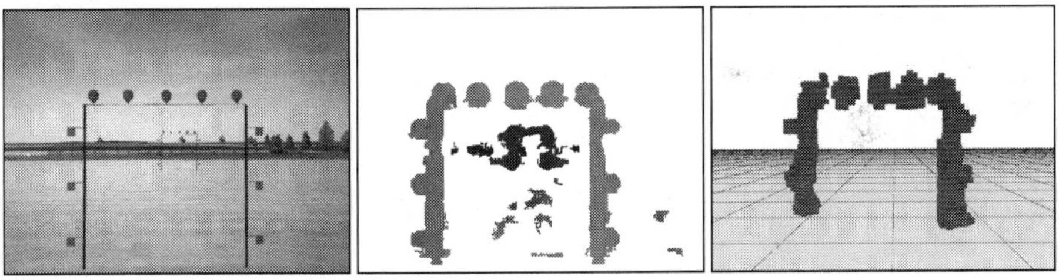

Figure 15. Original image from left camera (left), depth image (center) and extracted occupancy grid (right)

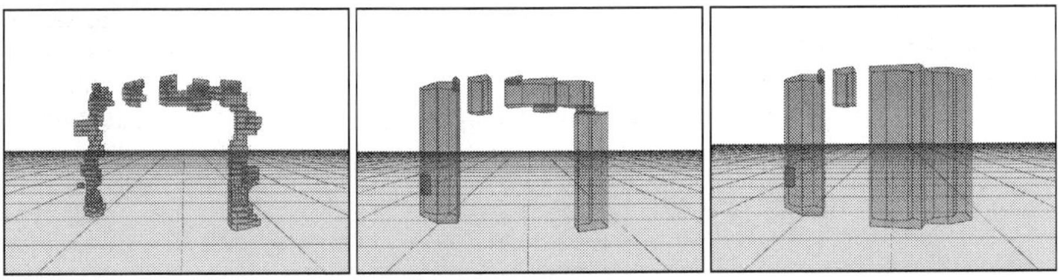

Figure 16. Prism shapes calculated from occupancy grid of figure 15 with different similarity thresholds for shape slice merging: zero (left), 100 (center) and infinite (right)

As shown in the image, the first arch is mapped while the second arch in the background is filtered due to its far distance from the viewpoint. The subfigures include the occupancy grid map (fig. 15, right) and the resulting shape prisms (fig. 16). As shown in the figures, the final map representation is only a small set of simple shapes. The similarity threshold for horizontal shapes should not be too large to allow multiple prisms for features generated from grid cell clusters in order to represent non-vertical walls. For a better illustration, floor planes marked with lines every 8 meters are shown in the images. The computation speed on the 3 GHz test computer was generally in the interval from 15 to 20 frames per second.

In a second test series, helicopter flights are performed outdoors. The grid array size is the same as in the simulation and its resolution is 0.5m, so that the map size of each zone is doubled. The helicopter is manually directed through obstacle posts and in an urban environment near house walls.

Figure 17. Obstacle posts on a flight field (left), example of a depth image (center), and the resulting map with obstacles and flight path (right)

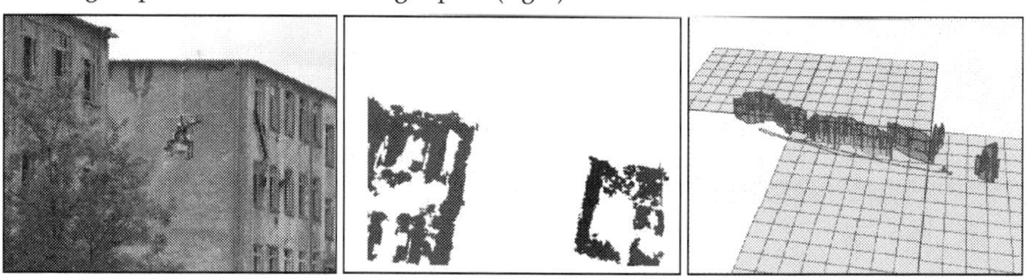

Figure 18. An urban environment (left), example of a depth image (center), and the resulting map (right) with obstacles, detected ground planes, and flight path

As seen in figure 17, obstacle posts are detected and each post can be represented by just one polygon. The ground plane was not detected here; the image shows only a plane in the height the helicopter took off. The prisms are larger than the real obstacles because they include the wires that hold the posts. The flight trajectory is marked red. Figure 18 shows the result of mapping houses and as seen in the resulting map, the walls and the gap between the houses have been recognized correctly. Ground plane detection was enabled and successful.

8. Conclusions

This work describes a method to extract obstacles from sensor data to be used for autonomous applications. The focus is on the creation of a compact representation of the bounding shapes required for obstacle avoidance, as opposed to detail object representation. Sensor data fusion is performed with an occupancy grid map, and this map is the basis for the shape extraction. A shape defined through one or multiple right prisms is calculated for each cluster of occupied grid cells. The shapes are calculated with each new sensor data in real-time. In addition to that, ground planes are identified. The output of this algorithm is very compact map representations of obstacles, and it is possible to send actual map updates through low-bandwidth networks which can serve as an input to other external applications.

To prove the approach, tests were performed in a simulation environment in a laboratory and under real conditions where a helicopter flies through obstacle posts and in an urban environment. The results show that it is possible to map obstacles with GPS/INS-based positioning and stereo vision, and that feature extraction functions such that the resulting map is suitable for obstacle avoidance.

9. References

Chatila, R. & Laumond, J.-P. (1985). Position referencing and consistent world modeling for mobile robots. In *IEEE International Conference on Robotics and Automation*, pp. 138-145.

Dittrich, J.; Bernatz, A. & Thielecke, F. (2003). Intelligent systems research using a small autonomous rotorcraft testbed. In *2nd AIAA Unmanned Unlimited Conference, Workshop and Exhibit*, paper 6561, San Diego.

Dittrich, J.; Adolf, F.; Langer, A. & Thielecke, F. (2007). Mission planning for small VTOL UAV systems in unknown environments. In *AHS International Specialists' Meeting on Unmanned Rotorcraft*, Chandler.

Dittrich, J.; Andert, F. & Adolf, F. (2008). An obstacle avoidance concept for small unmanned rotorcraft in urban environments using stereo vision. In *64th Annual Forum of the American Helicopter Society*, Montréal.

Fiorini, P & Shiller, Z. (1998). Robot motion planning in dynamic environments using velocity obstacles. *International Journal of Robotics Research*, Vol. 17, No. 7, pp. 760-772. ISSN 0278-3649.

Fulgenzi, C.; Spalanzani, A.; Laugier, C. (2007). Dynamic obstacle avoidance in uncertain environment combining PVOs and occupancy grid. In *IEEE International Conference on Robotics and Automation*, pp. 1610-1616, Roma.

Griffiths, J.; Saunders, A.; Barber, B.; McLain, T.; Beard, R. (2007). Obstacle and terrain avoidance for miniature aerial vehicles. Valavanis, K. P. (ed.), *Advances in Unmanned Aerial Vehicles*, pp. 213-244. ISBN 978-1-4020-6113-4.

Hähnel, D.; Burgard, W. & Thrun, S. (2003). Learning compact 3D models of indoor and outdoor environments with a mobile robot. *Robotics and Autonomous Systems*, Vol. 44, No. 1, pp. 15-27. ISSN 0921-8890.

Hrabar, S.; Corke, P.; Sukhatme, G.; Usher, K. & Roberts, J. (2005). Combined optic flow and stereo-based navigation of urban canyons for a UAV. In *IEEE/RSJ International Conference on Intelligent Robots and Systems*, pp. 302-309, Edmonton.

Iocchi, L.; Konolige, K. & Bajracharya, M. (2000). Visually realistic mapping of a planar environment with stereo. In *Seventh International Symposium on Experimental Robotics*, Waikiki.

Koch, A.; Wittich, H. & Thielecke, F. (2006). A vision-based navigation algorithm for a VTOL UAV. In *AIAA Guidance, Navigation and Control Conference and Exhibit*, paper 6546, Keystone.

Konolige, K. (1997). Improved occupancy grids for map building. *Autonomous Robots*, Vol. 4, pp. 351-367. ISSN 0929-5593.

Kuipers, B. J. (2000). The spatial semantic hierarchy. *Artificial Intelligence*, Vol. 119, pp. 191-233. ISSN 0004-3702.

Martin, C. & Thrun, S. (2002). Real-time acquisition of compact volumetric 3D maps with mobile robots. In *IEEE International Conference on Robotics and Automation*, pp. 311-316, Washington D.C.

Moravec, H. P. & Elfes, A. (1985). High resolution maps from wide angle sonar. In *IEEE International Conference on Robotics and Automation*, pp. 116-121.

Okada, K.; Kagami, S.; Inaba, M. & Inoue, H. (2001). Plane segment finder: Algorithm, implementation and applications. In *IEEE International Conference on Robotics and Automation*, pp. 2120-2125, Seoul.

Prassler, E.; Scholz, J. & Elfes, A. (2000). Tracking multiple moving objects for real-time robot navigation. *Autonomous Robots*, Vol. 8, No. 2, pp. 105-116. ISSN 0929-5593.

Ramer, U. (1972). An iterative procedure for the polygonal approximation of plane curves. *Computer Graphics and Image Processing*, Vol. 1, No. 3, pp. 244-256. ISSN 0146-664X.

Raschke, U. & Borenstein, J. (1990). A comparison of grid-type map-building techniques by index of performance. In *IEEE International Conference on Robotics and Automation*, pp. 1828-1832, Cincinnati.

Ren, M.; Yang, J. & Sun, H. (2002). Tracing boundary contours in a binary image. *Image and Vision Computing*, Vol. 20, pp. 125-131. ISSN 0262-8856.

Scharstein, D. & Szeliski, R. (2002). A taxonomy and evaluation of dense two-frame stereo correspondence algorithms. *International Journal of Computer Vision*, Vol. 47, pp. 7-42. ISSN 0920-5691.

Scherer, S.; Singh, S.; Chamberlain, L. & Saripalli, S. (2007). Flying fast and low among obstacles. In *IEEE International Conference on Robotics and Automation*, pp. 2023-2029, Roma.

Thrun, S. (2002). *Robotic mapping: A survey*. Tech. Rep., Carnegie Mellon University. No. CMU-CS-02-111.

Zufferey, J.-C. & Floreano, D. (2005). Toward 30-gram autonomous indoor aircraft: Vision-based obstacle avoidance and altitude control. In *IEEE International Conference on Robotics and Automation*, pp. 2594-2599, Barcelona.

3

Field Programmable Gate Array (FPGA) for Bio-inspired visuo-motor control systems applied to Micro-Air Vehicles

Fabrice Aubépart, Julien Serres, Antoine Dilly, Franck Ruffier and Nicolas Franceschini

Institute of Movement Science., University of the Mediterranean and CNRS France

1. Introduction

Nature provides us with many examples of ingenious sensors and systems at the service of animal behavior, which can be transferred into innovative systems for the control of Micro-Air Vehicles (MAVs). Winged insects demonstrate daunting behaviors in spite of the poor resolution of their visual system. For more than 100 million years, they have been navigating in unfamiliar 3D environments by relying on *optic flow* (OF) cues. To sense and react to the flowing image of the environment, insects are equipped with smart sensors called Elementary Motion Detectors (EMDs) that can act as optic flow sensors (figure 1). The principle of our bio-inspired optic flow sensors is based on findings obtained at our laboratory on the common housefly's EMDs, by performing electrophysiological recordings on single neuron while applying optical microstimuli to two single photoreceptors cells within a single ommatidium (Franceschini, 1985, Franceschini et al. 1989).

The OF-field gives the *angular velocity* (in rad/s) at which any contrasting object in the environment is moving past the eye (Koenderink, 1986). One lesson we have learned from insects is that they are able to navigate swiftly through the most unpredictable environments without using any velocimeters or rangefinders. Insects rely on optic flow to avoid collisions (Collett, 1980; Wagner, 1982; Tammero and Dickinson, 2002), to follow a corridor (Kirchner and Srinivasan, 1989; Baird et al. 2005; Ruffier et al., 2007; Serres et al., 2007; Serres et al. 2008), to follow the terrain (William, 1965; Srygley and Oliveira, 2001), to fly against wind (Kennedy 1939, Kennedy 1951), and to cruise and land (Srinivasan et al., 1996), for example.

Interestingly, insects seem to maintain a constant optic flow with respect to their surrounding environment while cruising and landing (Kennedy, 1951; David, 1978, Srinivasan et al. 1996, 2000). Several MAV autopilots were built in recent years which show how insects could achieve this feat by using a feedback control system called the optic flow regulator (Ruffier and Franceschini, 2003, 2005, Serres et al. 2008). Future MAVs' visual guidance systems may have to incorporate optic flow sensors covering various parts of the visual field, like insects do (Figure 2).

The biorobotic approach developed at our laboratory over the past 20 years enabled us to construct several terrestrial and aerial robots based on OF sensing (Pichon et al., 1989, Franceschini et al., 1992, 1997; Mura and Franceschini, 1996; Netter and Franceschini., 2002,

Ruffier et al., 2004). The robot Fly ('le robot-mouche') started off as a small, completely autonomous (terrestrial) robot equipped with 114 optic flow sensors. This robot was able to steer its way to a target through an unknown field of obstacles at a relatively high speed (50cm/s) (Franceschini et al. 1992, 1997). Over the last 10 years, we developed small (mass < 1kg) optic flow based aerial demonstrators with limited degrees of freedom (Viollet and Franceschini, 1999, 2001 ; Netter and Franceschini, 2002; Ruffier and Franceschini, 2003, 2005; Kerhuel et al., 2007; Serres et al., 2008,) called FANIA, OSCAR, OCTAVE, and LORA, respectively.

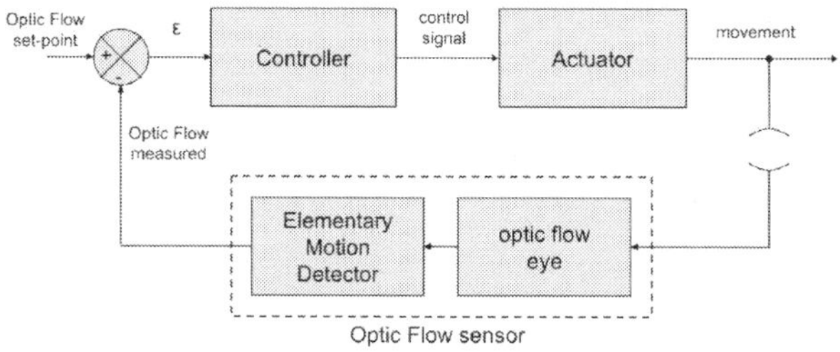

Figure 1. Principle of an optic flow regulator. The feedback control loop aims at maintaining the OF measured constant and equal to an optic flow set point

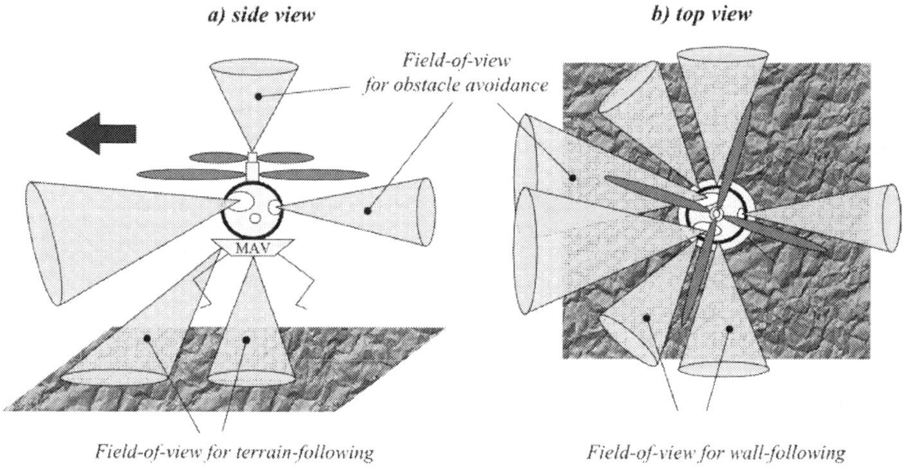

Figure 2. Several optic flow sensors covering various, strategic fields of view for an insect-like visual guidance of Micro-Air Vehicles (MAV)

In addition to visuo-motor control loops, inner control loops are necessary (i) to lock the micro-aircraft in certain desired attitudes, and (ii) to improve the control performances and robustness.

(i) Attitude control systems laid out in parallel with the *visuo-motor control loops* may use inertial and/or magnetic angular sensors (Ruffier et al., 2005; Viollet et al., 2001; Serres et al., 2008)

(ii) Control speed and accuracy can be improved with advanced actuators and more classical control loops based on proprioceptive sensors.

MAVs suffer, however, from stringent constraints on avionic payload, which requires highly miniaturized sensors, actuators and control systems (Viollet et al., 2008, Ahmad and Taib, 2003, Van Nieuwstadt and Morris, 1995).

Some of the aerial robots we developed thus far (FANIA and LORA) embedded only the optic flow sensors, the electromechanical actuators, and some internal stabilizing loops. The *visuo-motor control* system therefore operated off-board, yet real time experimentation was made possible by the joint use of *Simulink* (The Mathworks) and *Dspace* (Dspace) softwares. This permanent exchange between simulation (from the high level models) and experimental tests (from the robot) is appealing because it permits quick implementation of any new visuo-motor control systems onto a physical demonstrator, in addition to easy monitoring, and validation and tuning procedures. In this approach, the robot's behavior may be limited by the wire umbilical, however, unless a wireless link is established between the robot and the real-time board.

In our quest to achieve robot's complete - computational and energetic – autonomy, we now designed a complete digital system that integrates both optic flow sensors and robot's control systems within the same target: a 0.5 gram Field Programmable Gate Array (FPGA). An FPGA offers significantly more computational capabilities-per-gram than an embedded microprocessor or Digital-Signal Processor (DSP) because it supports custom, application-specific logic functions that may accelerate processing.

In the next sections, we describe our autopilot project based on an FPGA that meets three constraints: (i) the complete autonomy requirements, (ii) the computational requirement for real-time processing, and (iii) the requirement for light weight imposed by the very limited avionic payload (Chalimbaud and Berry, 2007).

Our project consists in the FPGA implementation of a *visuo-motor control system,* called LORA (LORA stands for *Lateral Optic flow Regulation Autopilot*), that was developed for a miniature hovercraft (section 3). The FPGA integration work was performed from a top-down design methodology using Intellectual Property (IP) cores and VHDL descriptions imported in the *Simulink* high-level graphical environment (The Mathworks) and the *System Generator* software interface (Xilinx) (section 2). The high-level *Simulink* blocks of the designed autopilot are then substituted for digital Xilinx blocks. Digital specifications of several functions, such as sampling time, fixed-point binary formats (section 4) and architectures (section 5), were defined from behavioral studies. Only models built from Xilinx blocks were translated into hardware logic devices using *System Generator*. Finally, the overall behavior of the integrated systems was analyzed using a hardware/software simulation based on both the *Simulink* environment (on a PC) and the FPGA (via a JTAG protocol). This co-simulation allowed us to analyze the hovercraft's behavior in various visual environments (section 6).

2. Design methodology for FPGA

2.1 FPGA general description

A field programmable gate array (FPGA) is a general-purpose integrated circuit that is programmed by the designer rather than the device manufacturer. Unlike an application-specific integrated circuit (ASIC), which can perform a similar function as in an electronic system, a FPGA can be reprogrammed, even after it has been deployed into a system. A FPGA is programmed by downloading a configuration program called a *bitstream* into static

on-chip random-access memory (RAM). Much like the object code for a microprocessor, this *bitstream* is the product of compilation tools that translate the high-level abstractions produced by a designer in an equivalent logic gate level.
A platform FPGA is developed from low-density to high-density designs that are based on IP cores and customized modules. These devices are user-programmable gate arrays with various configurable elements. The programmable device is comprised of I/O blocks and internal configurable logic blocks. Programmable I/O blocks provide the interface between package pins and the inside *configurable logic*.
The internal configurable logic includes elements organized in a regular array: *configurable logic blocks* (CLB) provide functional elements for combinatorial and synchronous logic, including basic storage elements; memory modules provide large storage elements of single or dual-port; Multiplier blocks integrating adder and accumulator (MAC); digital clock managers provide self-calibrating, fully digital solutions for clock distribution delay compensation, clock multiplication and division, coarse- and fine-grained clock phase shifting.
Several FPGAs also support the embedded system functionality such as high-speed serial transceivers, one or some hard embedded processors, Ethernet media-access control cores or others integrated high-functionality blocks.
A general routing matrix provides an array of routing switches between each component. Each programmable element is tied to a switch matrix, allowing multiple connections to the general routing matrix.
All programmable elements, including the routing resources, are controlled by values stored in memory cells. These values are loaded in the memory cells during configuration and can be reloaded to change the functions of the programmable elements.
FPGAs especially find applications in any area or algorithm that can use the massive parallelism offered by their architecture. They begin to take over larger and larger functions to the state where some are now marketed as full systems on chips (SOC). In this sense, they can satisfy the computational needs of real-time processing onboard autonomous MAVs.

2.2 FPGA design process

There are many design entry tools used for FPGA design. The easiest and most intuitive is the schematic entry. In this case, the required functionality is drawn using a set of library components. These one include the components primitives available on the FPGA as well as some higher level functions. Intellectual Property cores (IP) can be configured and placed in the schematic as a black box.
A trend in the digital hardware design world is the migration from graphical design entries to Hardware Description Languages (HDLs). They allow specifying the function of a circuit using a specific language such as VHDL or Verilog. These languages are specially made to describe the inherently parallel nature of digital circuits (behavioural and data flow models) and to wire different components together (structural models). In addition, HDLs are well adapted for designs with synchronous Finite State Machine (FSM) (Golson, 1994, Chambers, 1997). State machines impose a strict order and timing for operations making them similar to programs for CPUs.
However, HDL descriptions are difficult to design in the case of complex systems using digital signal processing functions or control applications. Indeed, the high level mathematical modelling tools are often used to validate a system model and to make a floating or fixed point model. Developments in the simulation capabilities of high-level

mathematical modeling tools have opened new design flow possibilities. These tools are includes in methodologies that help to hand complex design efficiently, minimize design time, eliminate many sources of errors, reduce the manpower required to complete the design and to produce optimal solutions. Nowadays, we can integrate complex system using IP cores and HDL descriptions imported in a high-level graphical environment. A software interface converts a fixed point model into hardware using traditional low level design implementation tools. Our design approach is based on this latter method.

In our project the high-level environment is *Matlab/Simulink* (The Mathworks), and the used interface tool is *System Generator for DSP* (Xilinx). *Matlab* is interactive software for doing numerical computations that simplifies the implementation of linear algebra routines. Powerful operations can be performed by utilizing the provided *Matlab* commands. *Simulink* is an additional toolbox of the *Matlab* software that provides a graphical environment for modeling, simulating and analyzing dynamic systems. *System Generator for DSP* toolbox was specifically designed to allow the fast development of complex system requiring powerful digital signal operations. It is presented in the form of a 'toolbox' added to the *Simulink* graphic environment, which allows the interfacing of certain functions fulfilled with the others 'toolbox' dedicated to *Simulink* (Turney, 1999, Ownby, 2003).

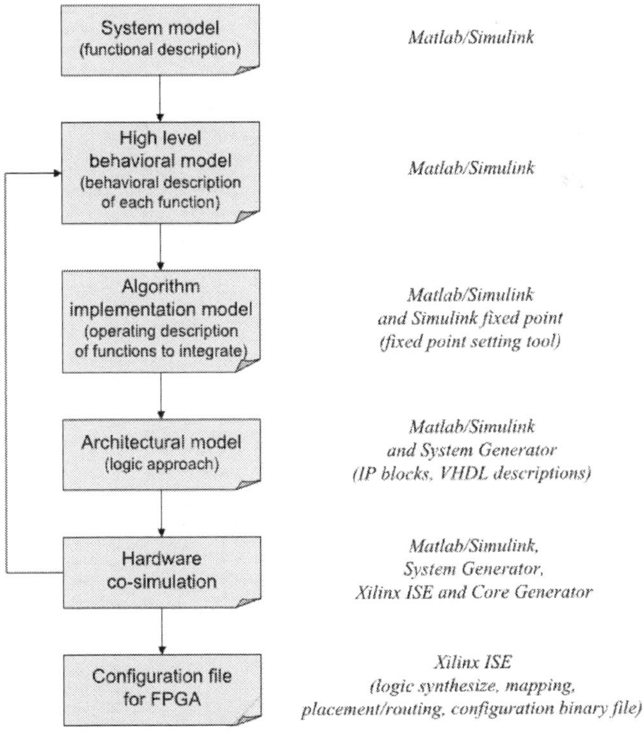

Figure 3. Top-down design methodology with a view to designing, simulating and implementing 'visuo-motor control loops' in FPGA (from Aubépart and Franceschini, 2007)

The design flow is presented in Figure 3. This methodology requires design various stages as well as the linking framework. This top-down methodology simplifies the integration problems by using detailed descriptions (Browny et al., 1997; Aubépart & Franceschini, 2005, 2007, Murthy et al., 2008).

Firstly, we start to create a *'system model'* as well as one or more simulated environment configurations for validating its principle. In this state, a functional approach involves dividing the system into elementary function blocks. The elements used in constructing the *'system model'* are all capable of operating on real (double precision, floating point) or integer (quantized, fixed point binary) data types. When the model is first entered, simulation is typically performed using floating data types to verify that its theoretical performance is as desired.

Secondly, we define the *'high-level behavioral model'* which describes the computation sequences and timing data (sampling frequency, delays, etc.). For example, mathematical operators that occur within an algebraic feedback loop may have an associated delay to avoid the system instability. The fixed time step, insert delays and/or rate changes need to be also considered to ensure a stability of a system based on a feedback loop.

Thirdly, an *'algorithm implementation model'* must be defined for each function to integrate. This model identifies the data type and the procedures used in each function. It takes account some factors, such as the binary format, in order to optimize the digital calculations. The internal data types are then converted to the bit true representations that will be used in the hardware implementation, and the model is re-simulated to verify its performance with quantized coefficient values and limited data bit widths, which can lead to overflow, saturation and scaling problems.

Fourthly, the hardware constraints are study in the *'architectural model'* that defines one or several implementation architectures for each function to integrate. They are replaced by IP blocks available in the *System Generator* toolbox. We can also define black boxes, such as VHDL descriptions, which can be incorporated into the model and the elaboration process. The importation of VHDL descriptions will be limited to complex synchronous descriptions, such as Finite State Machines.

Fifthly, the final verification will be completed by implementing the hardware co-simulation of the *System Generator 'architectural models'*. The co-simulation process uses *Xilinx ISE* and *core generator* to synthesize and generate and FPGA programming bit file. A new *Simulink* library was created containing the hardware co-simulation blocks. These blocks were copied into the *Simulink* project file for replacing all the *System Generator 'architectural models'*. The port names, types and rates are matching the original design. The hardware implementation is then executed by connecting the FPGA board to the computer and using a standard JTAG connection. When the simulation is run, stimuli are send to FPGA and output signals are receive by *Matlab/Simulink* environment, closing the loop.

Finally, the designer can invoke the *netlister* and test-bench generator available from *System Generator*. The *netlister* extracts a hierarchical VHDL representation of the model's structure annotated with all element parameters and signal data types that will integrate into FPGA from traditional low level tools include in *Xilinx ISE*: logic synthesize, mapping, placement and routing, generate programming file.

3. LORA III autopilot

In this part, we present the 'system model' and the implementation of a full control system using *System Generator* toolbox. This study integrate a complete *visuo-motor control system*, called LORA III (LORA stands for Lateral Optic flow Regulation Autopilot, Mark 3), which was designed for a particular kind of aerial vehicle: a fully actuated miniature hovercraft that is able to "fly" at a few millimetres above the ground and to control both its speed and its distance from the walls - without any measurements of speed and distance (Serres et al., 2008).

3.1 Robot position in travel environment

The hovercraft is simulated to travel through an unknown textured corridor at a ground speed vector $\vec{V}$ over a flat surface, Figure 4. The walls are lined with vertical stripes with random spatial frequency and contrast that mimic a richly textured environment (Iida, 2001, Ruffier & Franceschini, 2005). The corridors are of two types: right or tapering.

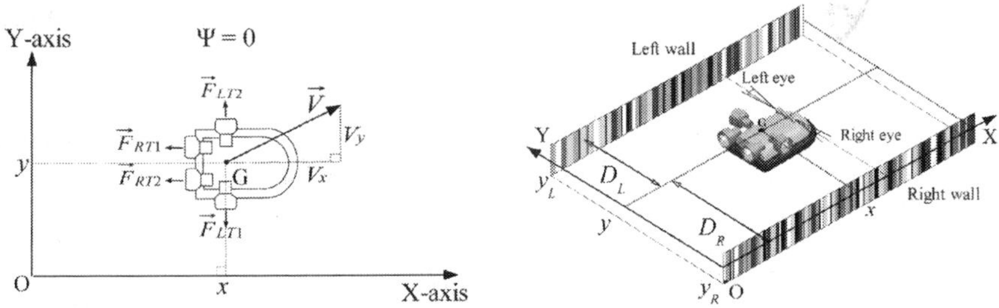

Figure 4. Miniature hovercraft moving through an unknown textured corridor: the groundspeed vector $\vec{V}$ is projected onto the corridor-fixed coordinate frame. Four thrusters (two rear thrusters and two lateral thrusters) allow the hovercraft to be fully actuated in the plane. (Adapted from Serres et al., 2008)

The robot's position (x, y) is computed in including side forward speed V_x and the side speed V_y. The hovercraft is fully actuated because in addition to the pair of rear thrusters providing forward motion (surge axis) and heading control (yaw axis), the vehicle is equipped with a pair of lateral thrusters generating independent side-slip motion (sway). The angle ψ defines the hovercraft's yaw angle with respect to the corridor axis, which is kept at a reference value equal to zero ($\psi = 0$). In this study, the hovercraft's heading ψ is assumed to be stabilized along the X-axis of the corridor.

The hovercraft's motion is defined by dynamic equations involving the forward thrust ($F_{Fwd} = F_{RT1} + F_{RT2}$) produced by the rear thrusters (left: RT1, right: RT2) and the lateral thrust ($F_{Side} = F_{LT2} - F_{LT1}$) produced by the lateral thrusters (left: LT1, right: LT2). In the simulations, the maximum forward speed is 2m/s and the maximum side speed is 0.5m/s. At such low speeds, the drag-versus-speed function can be linearized. The following equations referred to the center of gravity G define the dynamics of the simulated hovercraft (Figure 5):

$$\begin{aligned} m.\frac{dV_x}{dt} + \xi_x.V_x &= F_{RT1} + F_{RT2} = K_T.(U_{RT1} + U_{RT2}) \\ m.\frac{dV_y}{dt} + \xi_y.V_y &= F_{LT2} - F_{LT1} = K_T.(U_{LT2} - U_{LT1}) \end{aligned} \quad (1)$$

Where m = 0,73 kg is the total mass of the hovercraft, and ξ_x and ξ_y are translational viscous friction coefficients along the X-axis and Y-axis, respectively. K_T (0.10 N/V) is a simple gain that relates the thrust to the applied voltage: U_{RT1} and U_{RT2} are the forward control signals received by the rear thrusters, U_{LT2} and U_{LT1} are the side control signals received by the lateral thrusters.

3.2 Dual optic-flow regulator

The system addresses both issues of automatic speed control and side wall avoidance of the miniature hovercraft simultaneously. LORA is a dual optic flow regulator that consists of two interdependent control loops: a forward control loop and a side control loop. Figure 5 shows the block diagram involving multiple processing stages. This scheme is composed of two parts. In first, all functions that we would integrate into FPGA (blue, green and red functions). Secondly, hovercraft dynamics, lens/photoreceptors system and visual environments simulate the trajectories resulting from the LORA dual regulator scheme (Cyan and cyan hatched functions). These *'high level behavioural models'* will be used when digital, timing and architecture specifications will be defined for the functions to integrate.

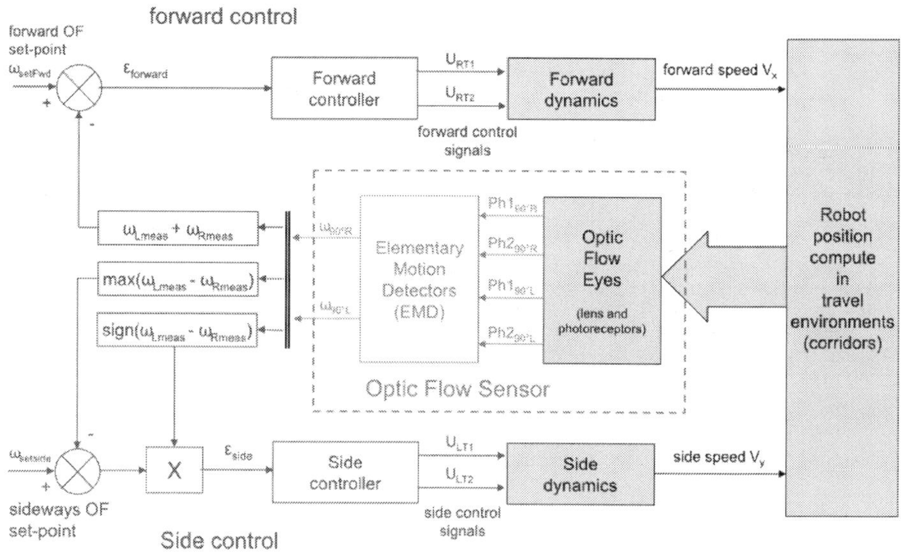

Figure 5. LORA III functional block diagram. LORA III autopilot is based on two interdependent visual feedback loops working in parallel with their own optic flow set-point (the *forward control system is the upper one, and the side control* system is the bottom one). *Optic flow sensors measure the right and left* optic flows in accordance with the hovercraft's speed and position in the environment (adapted from Serres et al., 2008a)

In Serres et al. (2008), the full control system is described in detail. We summarize here the principal aspects of them. That the hovercraft is fully actuated means that each groundspeed component V_x and V_y can be controlled independently. LORA III regulates (i.e., maintains constant) the lateral OF by side and forward controls, according to the following principles:

(i) The first *lateral OF regulator* adjusts the air vehicle's lateral thrust (which determines the lateral speed V_y, i.e., the sway speed) so as to keep the lateral optic flow equal to the *sideways OF set-point*. The outcome is that the distance to the wall becomes proportional to the vehicle's forward speed V_x which is defined in (ii): the faster the air vehicle travels, the further away from the walls it will be. The clearance from the walls will depend directly on the *sideways OF set-point*.
(ii) The second *lateral OF regulator* adjusts the air vehicle's forward thrust (which determines the forward speed V_x, i.e., the surge speed) so as to maintain the sum of the two (right and left) optic flows equal to the *forward OF set-point*. The outcome is that the air-vehicle travels all the faster as the environment is less cluttered. The forward speed attained by the vehicle will depend directly on the *forward OF set-point and will become proportional to the local width of the corridor*.
The first lateral optic flow regulator is based on a feedback signal that takes into account the left or right optic flow measured. The feedback is simply the larger of the two optic flows measured (left or right): $\max(\omega_{Lmeas}, \omega_{Rmeas})$: it corresponds to the nearest wall: $\min(D_L, D_R)$. This optic flow regulator was designed to keep the lateral optic flow constantly equal to the side optic flow set-point $\omega_{SetSide}$. The hovercraft then reacts to any deviation in the lateral optic flow (left or right) from $\omega_{SetSide}$ by adjusting its lateral thrust, which determines the hovercraft's side speed V_y: this eventually leads to a change in the distance to the left (D_L) or right (D_R) wall. A sign function automatically selects the wall to be followed, and a maximum criterion is used to select the higher optic flow value measured between ω_{Rmeas} and ω_{Lmeas}. This value is then compared with the sideways optic flow set-point $\omega_{SetSide}$. The error signal ε_{side} feeding the side controller is therefore calculated as follows:

$$\varepsilon_{side} = \mathrm{sign}(\omega_{Lmeas} - \omega_{Rmeas}) \times (\omega_{SetSide} - \max(\omega_{Lmeas}, \omega_{Rmeas})) \quad (2)$$

The identified transfer function of the side dynamics $G_y(s)$ relating the hovercraft's ordinate y to the control signal approximates a first-order low-pass filter (with a time constant of 0.5s) in series with an integrator:

$$G_y(s) = \frac{Y(s)}{(U_{LT1} - U_{LT2})(s)} = \frac{1}{s} \times \frac{\frac{K_T}{\xi_y}}{1 + \frac{m}{\xi_y} s} = \frac{1}{s} \times \frac{0.1}{1 + 0.5s} \quad (3)$$

A lead controller $C_y(s)$ was introduced into this feedback loop to increase the damping, thus improving the stability and enhancing the response dynamics. The lead controller $C_y(s)$ (Eq.3) is tuned to reach a phase margin of 45° and a crossover frequency of 4rad/s (0.64Hz):

$$C_y(s) = \frac{(U_{LT1} - U_{LT2})(s)}{\varepsilon_{side}(s)} = 10 \times \frac{1 + 1.5s}{1 + 0.5s} \quad (4)$$

The second optic flow regulator is the forward control system. It is intended to keep the sum of the two lateral optic flows measured ($\omega_{Rmeas} + \omega_{Lmeas}$) constant and equal to a *forward optic flow set-point* ω_{SetFwd} by adjusting the forward thrust, which will determine the hovercraft's forward speed V_x. At a given corridor width, any increase in the sum of the two lateral optic flows is assumed here to result from the hovercraft's acceleration. This control scheme thus automatically ensures a 'safe forward speed' that is commensurate with the local corridor width. The sum of the two optic flows measured is compared with a *forward OF set-point* ω_{SetFwd}, and the error signal ε_{Fwd} (the input to the forward controller) is calculated as follows:

$$\varepsilon_{Fwd} = \omega_{SetFwd} - (\omega_{Rmeas} + \omega_{Lmeas}) \quad (5)$$

The model $G_{Vx}(s)$ for the dynamics of our hovercraft is described by a first order low-pass filter with a time constant of 0.5s:

$$G_{Vx}(s) = \frac{V_x(s)}{(U_{RT1} + U_{RT2})(s)} = \frac{\frac{K_T}{\xi_X}}{1 + \frac{m}{\xi_X} s} = \frac{0.1}{1 + 0.5s} \quad (6)$$

A proportional-integral (PI) controller $C_{Vx}(s)$ (Eq. 7) is tuned to cancel the dominant (aeromechanical) pole of the hovercraft and to reduce the forward time constant computed in the closed-loop by a factor of 1.57. The integral action is introduced to cancel the steady state error:

$$C_{Vx}(s) = \frac{(U_{RT1} + U_{RT2})(s)}{\varepsilon_{Fwd}(s)} = 10 \times \frac{1 + 0.5s}{s} \quad (7)$$

3.3 Bio-inspired visual system

The visual system of the hovercraft consists of two optic flow sensors. Each optic flow sensor is designed by a lens/photoreceptor assembly (the eye) including at least two photoreceptors (i.e. two pixels) driving an Elementary Motion Detector (EMD) circuit.

Each eye consists of two photoreceptors mounted slightly defocused behind a lens, which creates a bell-shaped Angular Sensitivity Function (ASF) for each of them (Hardie, 1985). The ASF, which is often modelled in the form of a truncated Gaussian curve (Netter & Franceschini, 2002) and characterized by the 'acceptance angle' $\Delta\rho=4°$ (i.e. the angular width at half height). The ASF plays an important role in the visual processing chain, because it serves as an effective low pass anti-aliasing spatial filter. The visual axes of each pixel are separated by an inter-receptor angle $\Delta\varphi = 4°$.

The EMD principle was originally based on the results of experiments in which a combined electrophysiological and micro-optical approach was used. The activity of a large field motion detecting neuron in the housefly's eye was recorded with a microelectrode while applying optical microstimuli to a single pair of photoreceptor cells located behind a single facet (Franceschini, 1985, Franceschini et al., 1989). Based on the results of these experiments, a principle was drawn up for designing an artificial EMD capable of measuring the angular speed ω of a contrasting object (Franceschini et al., 1986, Blanes, 1986).

Thus, in each of EMDs, the lens/photoreceptor combination transforms the motion of a contrasting object into two successive photoreceptor signals separated by a delay Δt:

$$\Delta t = \frac{\Delta\varphi}{\omega} \quad (8)$$

Where $\Delta\varphi$ is the inter-receptor angle and ω is the relative angular speed (the optic flow). An electronic device based on some linear and nonlinear functions estimates the angular speed ω_{EMD}:

$$\omega_{EMD} = e^{-\left(\frac{\Delta t}{\tau}\right)} \Leftrightarrow K.\omega = \frac{K.\Delta\varphi}{\Delta t} = \frac{K'}{\Delta t} \tag{9}$$

Our original EMD functional scheme (Franceschini et al., 1986, Blanes 1986, Aubépart & Franceschini, 2007) consists of five processing steps giving ω_{EMD} (Figure 6):

1. A first-order high-pass temporal filter (f_c = 20Hz) produces a transient response whenever a contrasting border crosses the photoreceptors' visual field. This filter enhances the contrast information while eliminating the DC components of the photoreceptor signals.
2. A higher order low-pass temporal filter (f_c = 30Hz) attenuates any high frequency noise, as well as any interferences brought about by the artificial indoor lighting (100Hz) used.
3. A thresholding device/step normalizes the signals in each channel.
4. A time delay circuit is triggered by one channel and stopped by the neighboring channel. This function measures the time Δt elapsing between similar transitions occurring in two adjacent photoreceptors.
5. A converter translates the delay Δt measured into a monotonic function that will approximate the angular speed ω_{EMD}. A simple inverse exponential function makes for a relatively large dynamic range (eq. 9).

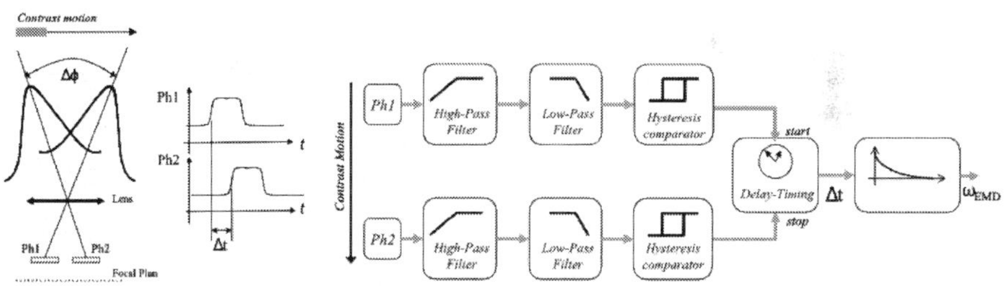

Figure 6. Principle scheme of Elementary Motion Detector (EMD). (Aubépart & Franceschini, 2007)

4. Implementation of visual feedback loops

4.1 Sampling time consideration

In aerial robotic applications, the sampling time must comply with the requirements imposed by the digital control system so that the MAV can be controlled throughout its safe flight envelope. The maximum sampling time T_{SMAX} will depend on the minimum delay Δt_{min} encountered by the robot's EMDs during the fastest manoeuvres in the most critical applications. Indeed, in an extreme case the smallest delay Δt_{min} which could be determined by the EMD will be that corresponding to one only counter clock period, whose clock signal corresponds to the sampling time.

One example of a fast maneuver is automatic wall-following, which is performed by measuring the side optic flow in the side direction. When the eye-bearing hovercraft is in pure translation at speed V_x and follows a textured wall at distance D, the image of the wall slips at an angular speed ω that depends on both V_x and D:

$$\omega = \frac{V_X}{D} = \frac{\Delta\varphi}{\Delta t} \tag{10}$$

If we take an extreme case where the hovercraft is allowed to follow the wall at the minimum distance D = 0.05m at the maximum speed V_x = 2m/s, equations 8 and 10 show that the 90° oriented EMD onboard the hovercraft, with its inter-receptor angle $\Delta\varphi = 4°$, will be subject to a minimum delay $\Delta t_{min} \approx 1.8$ms. Accordingly, the sampling frequency f_{Smin} will have to be set at values of at least 500Hz. When considering a MAV flying over an unknown terrain, we selected a minimum sampling frequency of 1kHz (Aubépart & Franceschini, 2007).

The maximum sampling frequency f_{SMAX} is less constraining to choose. It must be in keeping with the timing specifications to which the lens/photoreceptors devices are subject, especially in the case photoreceptors using the current-integrator mode (Kramer et al., 1997). On the other hand, the maximum sampling frequency f_{SMAX}, is limited by the lower end of the illuminance range over which the sensor is intended to operate, because at low illuminance levels, the integration of the photoreceptor signal takes a relatively long time and the sampling procedure will have to wait for this integration process to be completed (Aubépart & Franceschini, 2007). Taking the range [100Lux–2000Lux] to be a reasonable working illuminance range for the hovercraft, this gives f_S = 2.5kHz, which we call the 'nominal sampling frequency' (Aubépart & Franceschini, 2005). At twice this sampling frequency (5kHz), the hovercraft would still operate efficiently in the [200Lux - 2000Lux] range, but it would then be difficult for it to detect low contrasts under artificial indoor lighting conditions. In addition, we avoided CMOS cameras equipped with digital outputs, which have not high frame rates, which still need to scan every pixel internally (Yamada et al., 2003, Zufferey et al., 2003).

As regards the forward and side controllers or others functions ('sign', 'add' and 'max' functions) present in the visual feedback loops, the sampling time should be adapted to the maximum frequency of the processed signals. Generally, the highest frequency corresponds to the smallest time-constant of control loops. In our loops, it is equal to 0.5-second that corresponds to a cut-off pulsation of 2rad/s. In discrete time we may consider that a sampling frequency 100 times higher than the system cut-off frequency would give results similar to continuous time. An equivalent sampling frequency of about 32 Hz could be sufficient. Since this value is in the lower part of the minimal value necessary to EMDs, we chose 2.5kHz as the nominal sampling.

4.2 Digital specifications

In feedback control loops, the digital specifications were mainly defined for EMDs design (i) and controllers design (ii). The secondary functions, such as 'sign', 'add' and 'max' functions, are easily adapted to the digital choices (binary format, implementation, etc.) without modifying the control loops operation.

In EMDs (i), the digital specifications were found during the filter design. Due to the low values of the high-pass and low-pass filter corner frequencies (f_{CHP} = 20Hz, f_{CLP} = 30Hz) in comparison with the sampling frequency (f_S = 2,5kHz), it was not possible to obtain a digital band-pass filter meeting the Bode specifications. The high-pass filter section and low-pass filter section were therefore designed separately and cascaded.
Infinite Impulse Response (IIR) filters were synthesized (see eq. 5 below) because they require far fewer coefficients than Finite Impulse Response (FIR) filters, given the low cut-off frequencies and short sampling times involved:

$$y(n)=\sum_{i=1}^{n}b(i)\cdot x(i)-\sum_{i=1}^{n-1}a(i)\cdot y(i) \tag{11}$$

A transposed Direct-Form II structure was used because this structure reduces the number of delay-cells and decreases the quantization errors. Ripples on the low-pass filter temporal response were prevented by using a 4th-order Butterworth Filter, the phase of which was linearized over the frequency range of interest. The filters require 17 coefficients in all (4 coefficients for the 1st-order high-pass section, 12 for the 4th-order low-pass section, and 1 for the adjustment between the two filters). Three Direct-Form II filters suffice in fact to perform all the filtering, including that carried out by the two cascaded 2nd order low-pass filters.
A specific binary format was developed and used to prevent offset and stabilization problems. A two-complement fixed-point binary format, denoted $[s,m_I,m_D]$, was defined. The bit number of integer parts, m_I, and the decimal part, m_D, were defined so as to ensure maximum accuracy and to eliminate overflow from the filter calculations. Based on the results of a study carried out with *Filter Design and Analysis* and *Fixed-point Blockset* of the Mathworks tools, 6 bits were selected for the integer part m_I and 29 bits for the decimal part m_D. The large m_D bit number is due to the low value required to make the coefficients in the low-pass filter section comply with a Bode template characterized by a low cut-off frequency at high sampling frequencies.
Other digital specifications were defined in EMDs as regards the bit number of the counter output giving the delay time Δt, and the inverse exponential function giving the angular speed ω_{EMD}.
The delay time Δt is measured in terms of a count number at a given clock period. The minimum delay to be measured determines the minimum clock period (400µs for f_S = 2.5kHz). The maximum delay to be measured is taken to be ~100ms, which is compatible with the wide range of angular speed values encountered by the hovercraft eyes (eq.4): ω ~40°/s to ~10000°/s, for $\Delta\varphi$ = 4°. Using an 8-bit counter at f_S = 2.5kHz gives an delay of 102.4ms.
The measured angular speed ω is a hyperbolic function of Δt (eq. 8), but we used a function that decreases more slowly: an inverse exponential function with a time constant τ = 30ms. A Look-Up Table (LUT) was used to convert the delay Δt into an output that decreases monotonically (exponentially) with the delay and therefore approximately reflects the angular speed ω_{EMD} (eq. 9). The Look-Up Table features an 8-bit input resolution (at f_S = 2.5kHz) and a 12-bit output resolution for memorizing the results of the conversion.
The digital correctors design (ii) is easy in using 'c2d' Matlab function (discretize continuous-time system). The correctors' continuous-time models have been converted to

discrete-time models with sampling time T_S. The discretization method used was the bilinear Tustin approximation. Table 1 presents the discretized polynomial transfer functions.

Correctors	*Continous-time*	*Discrete-time*
side	$C_y(s) = 10 \times \frac{1+1.5s}{1+0.5s}$	$C_y(z) = \frac{29.99z - 29.98}{z - 0.9992}$
forward	$C_{Vx}(s) = 10 \times \frac{1+0.5s}{s}$	$C_{Vx}(z) = \frac{5.002z + 4.998}{z - 1}$

Table 1. Correctors' discretization using Tustin bilinear approximation with f_S = 2.5kHz

As during EMD design, a specific binary format was defined to avoid offset and stability problems. Once again, we used the '*Filter Design and Analysis*' tool as well as the '*Fixed-point Blockset*' tool. A first approximation gave two signed fixed-point binary formats: 9 bit and 8 bit for the integer part m_I of the forward and side correctors, respectively, 10 bit for their decimal parts m_D. However, simulations of the full feedback control system showed that accuracy was too low. The decimal parts m_D were then increased to 13 bit for best accuracy in computation.

5. Architecture

5.1 Optic flow sensor

Figure 7 shows the "EMD *architecture*". This architecture has several important features, such as the optimization of the digital filters, the simplicity of design owing to the use of Intellectual Property (IP) cores, and the added flexibility in the circuit design. Special care was taken to restrict the space taken by the digital filter implementation. Instead of implementing eight IIR digital filters in parallel - which would each require a high number of mathematical operators (adders, subtracters and multipliers) - a single structure called the *"Filter Compute Unit"* was developed, with which high speed sequential processing can be performed, as shown in figure 8. This unit consists of only one multiplier, one adder, one Read Only Memory (ROM), one Random Access Memory (RAM), two multiplexers, three registers and two binary transformation functions. The ROM contains the 17 filter coefficients obtained at a sampling frequency f_S = 2.5kHz. The RAM is used to store the intermediate values computed. The multiplexers minimize the number of operators.

In this unit, each photoreceptor signal is processed during the sampling time T_S. A Finite State Machine (FSM) was written in VHDL language and imported into the *Simulink* environment. The FSM specifies the filtering sequence for each photoreceptor channel. Even though this solution somehow complicates the design (due to the mixing of VHDL description and System Generator IP blocks), it has the advantage to minimize the number of logical gates necessary for integration in the FPGA.

Once the filtering process is completed, the delay Δt between the excitations of two neighboring photoreceptors starts being measured. A comparator determines the instant at which the band-pass filtered signal from each photoreceptor channel reaches the threshold value. The resulting logical signal is used to trigger the measurement of the delay Δt corresponding to each of the two eyes (an eye consists of a lens and only two photoreceptors). Specifically, the logical signal delivered by a photoreceptor starts a counter that will be stopped by the logical signal delivered by the neighboring photoreceptor.

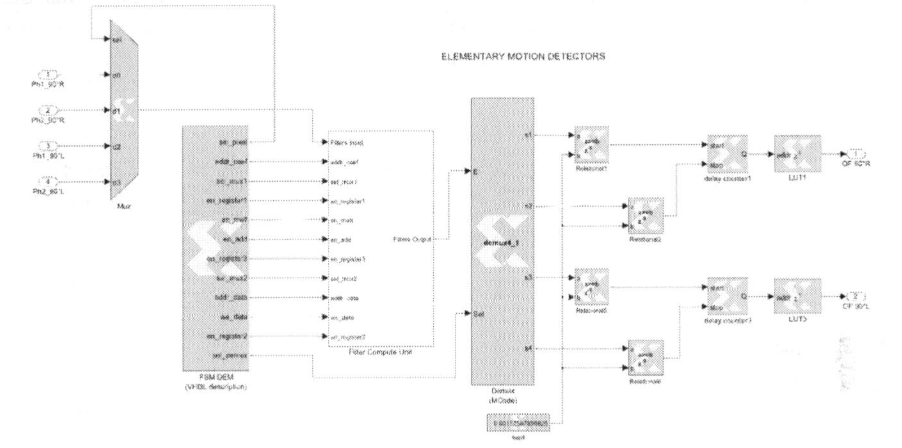

Figure 7. Elementary Motion Detector Architecture

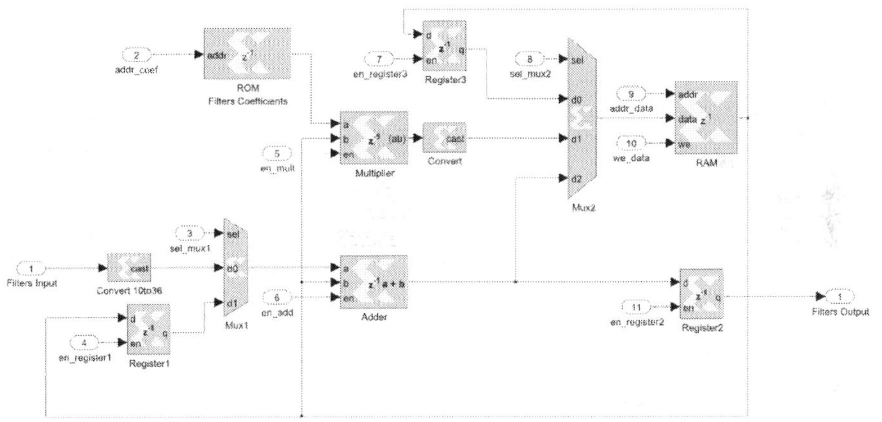

Figure 8. "Filter Compute Unit" architecture. The Filters' input is at port Nb 1. Others inputs (from 2 to 11) are control signals from a Finite State Machine that specifies the sequence of the filtering process for each photoreceptor channel

The final piece of the architecture is the inverse exponential function with a time constant $\tau = 30$ms, which allows for a wide range of delays ranging up to 102.4ms. This component was implemented in the form of Look-Up Tables (ROM) to facilitate the conversion of each delay data into an estimated angular speed ω_{EMD} (i.e., the optic flow).

5.2 Forward and Side correctors

The forward and side controllers (see Figure 5) have discretized polynomial transfer functions that can be designed as first-order IIR filters. For their realization, we chose the transposed-Direct Form-II architecture. *System generator* IP blocks allow for graphical integration of IIR filters because they use only multipliers, registers and adders. Using the Xilinx Multiplier Block called "Mult" (instead of the "CMult Xilinx Constant Multiplier") allowed this operation to be integrated via the multipliers embedded in the FPGA.

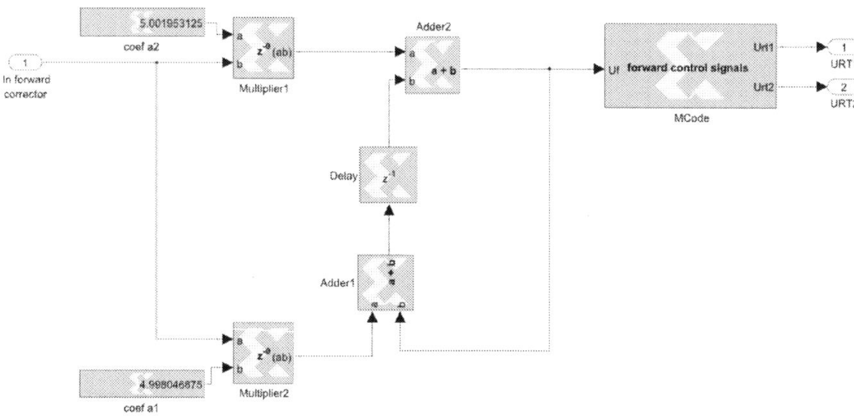

Figure 9. Transposed-Direct-Form-II implementation for the *forward controller* (input to the left, outputs to the right)

Figure 9 shows the forward controller, in which only two multipliers are implemented because the denominator coefficients are equal to one (see Table 1). A Xilinx "MCode" block allows the two forward control signals (U_{RT1} + U_{RT2}) to be determined, for driving the two rear thrusters RT1 and RT2. The Xilinx "MCode" block is a container that executes a user-supplied MATLAB function within *Simulink*. A parameter of this block specifies the M-code function name. The block executes the M-code and calculates the block outputs during the *Simulink* simulation. When hardware is generated, the same code is translated into equivalent behavioral VHDL in a straightforward manner.

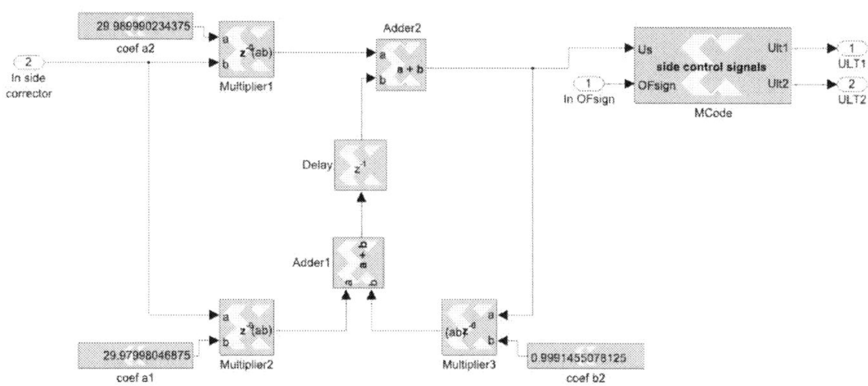

Figure 10. Transposed-Direct-Form-II implementation for the *side controller* (input to the left, outputs to the right)

The *side controller* was designed in the same way as the *forward controller*, figure 10. An additional multiplier was needed to perform calculation with the denominator coefficient b2. The two side control signals (U_{LT1} - U_{LT2}) controlling the two lateral thrusters LT1 and LT2 were again obtained by a Xilinx "MCode" block in which the sign of the difference between the right and left lateral optic flows measured was used.

5.3 Complete visuo-motor control system

The complete control system based on *System Generator* is presented in Figure 11. The secondary functions such as the determination of the larger value (OF max) and sign between right and left optic flows measured are easily realized with Xilinx "MCode" blocks. An adder and two subtracters are put in to close the loops. Xilinx "Gateway Out" blocks and "Gateway In" blocks limit the I/O from the Xilinx portion of the user's *Simulink* design. "Gateway Out" blocks convert the *System Generator* fixed point data type into *Simulink* double format while "Gateway In" blocks convert *Simulink* integer, double and fixed point data types into the *System Generator* fixed point type. Each block will define a top-level I/O port in the HDL design generated by *System Generator*.
Forward dynamics, side dynamics and "robot position compute" are defined by high level *Simulink* models linked to Matlab files that set several constants before simulation. "Lens/photoreceptors model" is designed with S-functions. An S-function is a computer language description of a *Simulink* block and is used to add our own blocks to *Simulink* models. Generally, such functions make it possible to accelerate simulations.

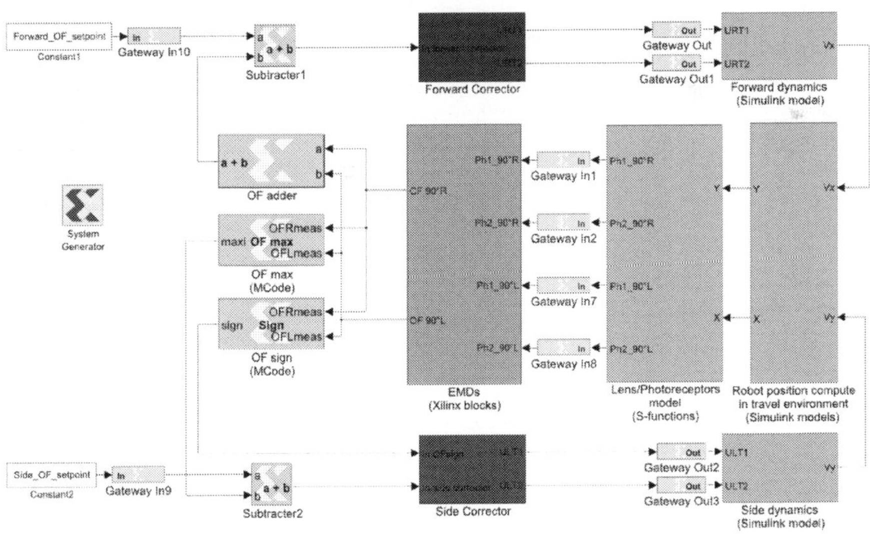

Figure 11. Complete control system based on *System Generator*. Xilinx "Gateway Out" blocks and "Gateway In" blocks limit the I/O from the Xilinx portion of the user's *Simulink* design

The *System Generator* block provides for control of system and simulation parameters, and is used to execute the code generator. Every *Simulink* model containing any element from the Xilinx Blockset must contain at least one *System Generator* block. Once a *System Generator* block is added to a model, it is possible to specify how code generation and simulation should be handled.

6. Hardware co-simulation

The final tests were achieved by making a hardware co-simulation of the system. The various functions (the EMDs and the two complete visuo-motor control loops) were implemented on a small Virtex-4 FPGA (type XC4FX12, size 17mm X 17mm, mass 0.5gram) whose digital processing capacities are largely sufficient. We designed a specific electronic board, figure 12, embedding the FPGA to validate the various digital properties of the whole system in co-simulation. This board was realized with an aim to install it on-board the miniature hovercraft and therefore also included the required interface components (ADC, DAC, etc.). The complete board weighs only 17.3grams and measures 90mm x 50mm and it is well suited to an embedded technological solution.

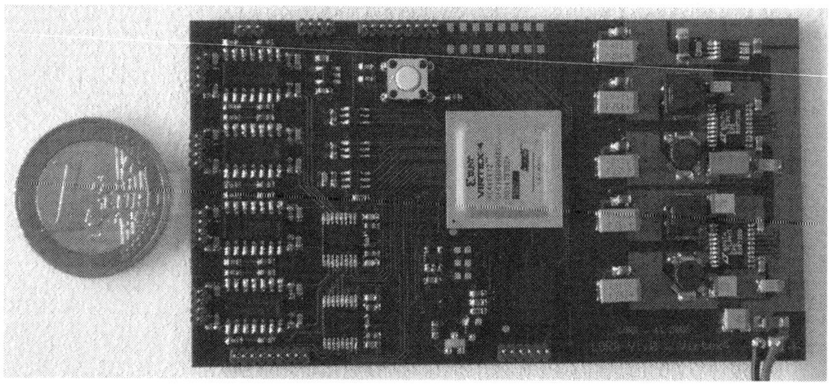

Figure 12. Specific electronic board based on a small FPGA Virtex-4: (i) right hand side: components for the power supply; (ii) middle part: FPGA XC4FX12 (top) and the Flash-memory to configure it (bottom); (iii) left hand side: electronic front-end related to the photosensors and several Analog-Digital and Digital-Analog Converters

The FPGA power consumption was evaluated using Xilinx "XPower" tool. The consumption is estimated at 149mW for a 2.5 kHz sampling frequency of the visuo-motor control loops and for the FPGA running at a clock frequency of 4MHz. The total power consumption of the electronic board, figure 12, is estimated at ~500mW

6.2 Hardware co-simulation results

A *Simulink* library was created from *System Generator* and copied into the *Simulink* project file replacing all the Xilinx System Generator blocks (i.e. between "Gateway In" and "Gateway Out" blocks, figure 11). The simulated robot is equipped with two lateral eyes oriented at ±90° to the walls (the inter-receptor angle is $\Delta\varphi=4°$ and the acceptance angle is $\Delta\rho=4°$). At 2.5kHz sampling frequency, the computing temporal step is $\delta t = 400\mu s$ and the simulation spatial accuracy is $\delta\theta = 0.005°$. The two OF set-points were chosen according to the results of behavioural studies on honeybees that were video-filmed when flying through a straight corridor (Serres et al., 2008b). The forward OF set-point was set to $\omega_{setFwd} = 314°/s$ and the side OF set-point was set to $\omega_{setSide} = 238°/s$.

6.2.1 Automatic speed control and lateral positioning in a straight corridor

The simulated visual environment is a 3-meter long, 1-meter wide straight corridor with textured walls. The right and left walls are lined with a random pattern of various grey vertical stripes covering a 1-decade contrast range (from 4% to 38%) and a 1.1-decade angular frequency range (from 0.068 c/° to 0.87 c/° reading from the corridor midline).
In figure 13, the hovercraft can be seen to follow either the right (red or black trajectory) or the left wall (blue trajectory), depending on the sign of the error signal ε_{side}, (Eq. 1). The robot can be seen to generate a steady state clearance of 0.24m from either wall (figure 13a), while reaching a steady state forward speed of $V_x = 1m/s$ (figure 13b). Thus, the hovercraft adopts a wall-following behavior in much the same way as honeybees do in a similar situation (Serres et al., 2008a).

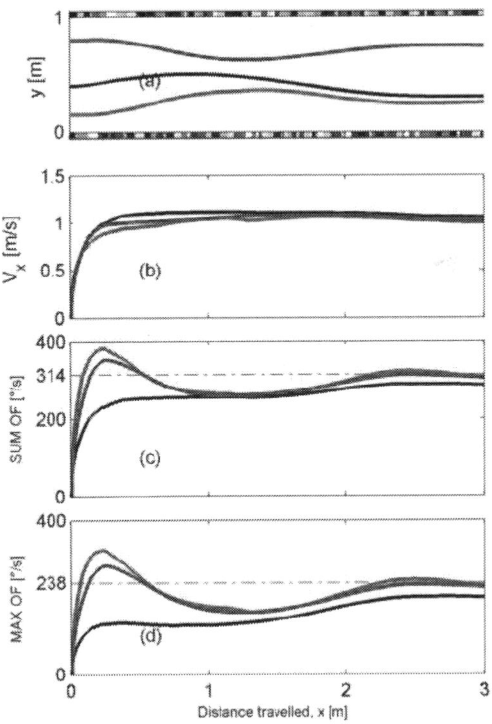

Figure 13. Hovercraft automatic wall-following behavior: (a) Simulated trajectories starting from three different initial positions y_0 (red: $y_0 = 0.15m$; black: $y_0 = 0.40m$; blue: $y_0 = 0.80m$). (b) Forward speed profiles. In the steady state, the forward speed can be seen to have reached $V_x = 1m/s$ in the three cases. (c) Sum of the two lateral optic flows ($\omega_R + \omega_L$). (d) Larger value of the two lateral optic flows, right and left

6.2.2 Automatic response in a tapered corridor

For the dual OF regulator, a tapered corridor acts like a *non-constant OF disturbance*. The forward control system adjusts the forward speed proportionally to the local corridor width (the width varies from 1.24m to 0.50m). The lateral control system controls the distance to the right wall in proportion to the forward speed at all times. This simulation experiment shows that the dual OF regulator is able to cope with the major disturbance caused by a tapered corridor, by making the robot decelerate or accelerate appropriately, figure 14.
The hardware co-simulation results presented here closely match the *Simulink* results presented by Serres et al. (2008a).

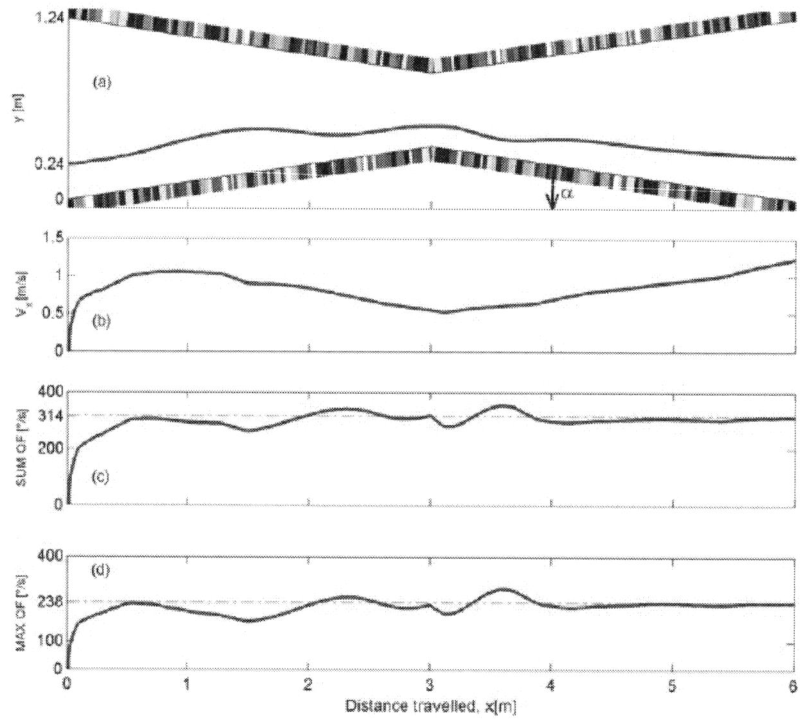

Figure 14. Automatic response in a *tapered* corridor: (a) Simulated trajectory of the hovercraft moving to the right in a tapered corridor starting at the initial position $y_0 = 0.24$m. This trajectory shows that the hovercraft automatically slows down when the local corridor width decreases and accelerates again when it widens again after the constriction.
(b) Corresponding forward speed profile V_x. The forward speed turns out to be a quasi linear function of the distance travelled x, and hence of the local corridor width. (c) Sum of the two optic flow ($\omega_R + \omega_L$). The speed control system succeeds to keep the sum of the optic flows measured virtually constant. (d) Larger value ("max") between the two optic flows measured. The side control system succeeds to keep the larger value of the two optic flows measured virtually constant

6.3 Hardware integration

In order to make the interface between the peripheral components and the FPGA-integrated system, several pilots have being developed and tested: driver ADC108S102, driver ADC101S101, driver DAC1101S101, PWM generator. These drivers link the components or the corresponding boards to the integrated *visuo-motor control loops* (generated by *System Generator*). They are integrated in the Virtex-4 FPGA and designed in VHDL because this language is more adapted to the realization of both the PWM generator and the communications protocols, such as the SPI protocol (*Standard Peripherical Interface*) that is used by ADC and DAC. For example, it is easier to manage the timing aspects of these low level hardware functions by a VHDL Finite State Machine synchronized by the FPGA clock.

The complete description can be performed in two different ways: (i) In *Simulink*, the *visuo-motor control system* is associated to import some drivers' descriptions (using Xilinx black box in the *System Generator* library); (ii) in Xilinx ISE environment, the VHDL generated description of the *visuo-motor control system* is connected to drivers descriptions (using ISE schematic description). The first solution (i) is difficult to apply because it requires the addition of peripheral components or functions high level models. This solution is not likely to improve the digital integration of control loops (validated in co-simulation) or drivers (validated in low level simulations and/or tested with peripheral components). The second solution does not allow any simulations of the closed loop system, but facilitates the integration stages (synthesis, mapping, placement and routing) and the generation of FPGA configuration binary file.

Table 2 shows the working characteristics of the Virtex-4 FPGA obtained after the integration of the LORA III autopilot and drivers.

DSP48	15 out of 32	42%
Slices	2569 out of 5472	47%
Block RAM 18kb	9 out of 36	25%

Table 2. Working characteristics of the XC4FX12 Virtex-4 FPGA

The DSP48 is a DSP-oriented component. The DSP48 is basically a multiplier followed by an adder with several optional registers on the ports and between the multiplier and adder. The multiplier takes two 18-bit signed signals and multiplies them, giving a 36-bit result. This is then sign extended to 48 bits and can be fed into the adder or routed directly to the outputs of the DSP48. The adder, which can be configured either as an adder or a subtractor, can accept the sign-extended output of the multiplier and 48-bit input to the DSP48. In addition, the adder can also accept itself as an input, to form an accumulator.

A slice is a basic element of Configurable Logic Block (CLB). A CLB element contains four interconnected slices. These slices are grouped in pairs. The elements common to both slice pairs are two logic-function generators (or look-up tables), two storage elements, wide-function multiplexers, carry logic, and arithmetic gates. The slices are used to provide logic, arithmetic, and ROM functions. Virtex-4 device features a large number of 18 Kb block RAM memories. True Dual-Port RAM offers fast blocks of memory in the device. Block RAMs are placed in columns, and the 18 Kb blocks are cascadable to enable a deeper and wider memory implementation, with a minimal timing penalty.

7. Conclusion and future work

We have developed a digital integration of an autopilot called LORA III (Serres et al. 2008a) onto a Virtex-4 FPGA. The autopilot consists of two interdependent visuo-motor control loops that are meant to control the visual guidance of a miniature hovercraft in a corridor without any measurements of speed and distance from the walls.

A top-down methodology was used to design and simulate the overall visuo-motor control system. The latter was studied using both a high-level graphical environment: *Simulink from Mathworks,* and *System Generator for DSP* toolbox, from Xilinx. The remarkable analysis capability is due to the fact that *System Generator* allows the designed *sensory-motor control loop* to be implemented from within the *Simulink* environment. The design flow has simplified the integration problems in using several levels of abstraction that were validated at each stage of development. According to this methodology, digital specifications and architectures of each control loop were optimized for the LORA III autopilot. Moreover, final tests were performed by exploiting the hardware co-simulation. We were therefore able to test final descriptions in FPGA from *Matlab/Simulink* environment with a JTAG connection. In this way, integrated architectures were validated by considering the hovercraft "flying" in straight or tapered corridors.

Integrating the *visuo-motor control loops* also required designing a specific electronic board based on a Virtex-4 FPGA (Figure 12). Linking the control system to the external components (ADC, DAC, motors control) also required designing several drivers that were embedded into the same 0.5 gram FPGA.

Future work will consist in installing the FPGA based sensory-motor control board into the miniature hovercraft for which it was built. Tests similar to those made in co-simulation will then be carried out. Additional improvements are also planned to increase the robustness of LORA III control system and make the robot negotiate more challenging corridors. The passive OF sensors and the simple processing system described here are particularly suitable for use with Micro-Air Vehicles (MAVs), in which highly stringent constraints are imposed in terms of the permissible avionic payload and onboard energy resources. FPGA implementation has recently been envisioned not only for the visuo-motor control of micro-air vehicles but also for the automatic visual guidance and retrorocket control of future planetary landers.

8. References

Ahmad, Z. & Taib, M.N. (2003). A study on the dc motor speed control by using back-emf voltage, *Proceeding of IEEE Asian conference on Sensors (AsiaSense)*, pp. 359-364

Aubépart, F. & Franceschini, N. (2005). Optic flow sensors for robots: Elementary Motion Detectors based on FPGA, *Proceedings of IEEE International Workshop on Signal Processing Systems*, pp.182-187, Athens, Greece, 2-4 November, 2005

Aubépart, F. & Franceschini, N. (2007). Bio-inspired optic flow sensors based on FPGA: Application to Micro-Air-Vehicles. *Journal of Microprocessors and Microsystems*, Vol. 31, No.6, (2007) (408-419) , ISSN°0141-9331

Baird, E.; Srinivasan, M. V.; Zhang, S. & Cowling, A. (2005). Visual control of flight speed in honeybees. *The Journal of experimental Biology,* Vol.208, No.20, (3895-3905), ISSN°0022-0949

Blanes, C. (1986). Appareil visuel élémentaire pour la navigation à vue d'un robot mobile autonome, DEA thesis (in french), Aix-Marseille University

Browy, C.; Gullikson, G & Indovina, M. (1997). A Top-Down approach to IC design, www.indovina.us/~mai/a_top_down_approach_to_ic_design.pdf

Chambers, P. (1997). The ten commandments of excellent design: VHDL code examples, *Electronic design,* Vol. 45, No.8, (123-126)

Chalimbaud, P. & Berry, P. (2007). Embedded active vision system based on an FPGA architecture. *EURASIP Journal on embedded systems,* Vol.2007, No.1 (January 2007) (12p.), ISSN°1687-3955

Collett, T.S. (1980). Some operating rules for the optomotor system of a hoverfly during voluntary flight. *Journal of Comparative Physiology A,* Vol.138, No.3, (September 1980) (271-282), ISSN°0340-7594.

David, C. (1978). The relationship between body angle and flight speed in free-flying *Drosophila. Physiological Entomology,* Vol.3, No.3, (191-195), ISSN°0307-6962

Franceschini, N. (1985). Early processing of color and motion in a mosaic visual system, *Neurosciences. Res. Suppl.* 2, (17-49)

Franceschini, N.; Blanes, C & Oufar, L. (1986). Passive non-contact optical velocity sensor, *Dossier technique N° 51549* (in French), Agence Nationale pour la Valorisation de la Recherche/Direction de la Valorisation, Paris

Franceschini, N.; Riehle, A. & Le Nestour, A. (1989). Directionally Selective Motion Detection by insect neurons, In: *Facets of Vision,* Stavenga D.G. & Hardie R.C. (Eds.), (360-390), Springer, ISBN°0387503064, Berlin

Franceschini, N.; Pichon, J-M. & Blanes, C. (1992). From insect vision to robot vision, *Philosophical transactions of the Royal Society of London. Series B, Biological sciences* Vol. 337, No.1281, (283-294), ISSN°0080-4622

Franceschini, N.; Pichon, J-M. & Blanes, C. (1997). Bionics of visuomotor control, *In Evolutionary robotics: from intelligent robots to artificial life, AAAI books,* : T. Gomi (Ed.), (49-67), Ottawa

Franceschini, N. (1998). Combined optical, neuroanatomical, electrophysiological and behavioural studies on signal processing in the fly compound eye, In: *Biocybernetics of vision: Integrative Mechanisms and Cognitive Processes,* C. Taddei-Ferretti, (Ed.), (341-361), World Scientific, London

Golson, S. (1994). State Machine design techniques for Verilog and VHDL, *Synopsys Journal of High-Level Design,* (September 1994), (1-48)

Hardie, R.C. (1985). Functional organisation of fly retina, In: *Progress in Sensory Physiology,* D. Ottoson (Ed.), Vol.5 (1-79), Springer, Berlin

Iida, F. (2001). Goal-Directed Navigation of an Autonomous Flying Robot Using Biologically Inspired Cheap Vision, Proceedings of the 32nd International Symposium on Robotics (ISR), pp.1404-1409, 19-21 April 2001

Kennedy, J.S. (1939). Visual responses of flying mosquitoes. *Proceedings of Zoological Society of London,* Vol.109, 221-242.

Kennedy, J.S. (1951). The migration of the desert locust (Schistocerca gregaria Forsk.) I. The behaviour of swarms. *Philosophical Transactions of the Royal Society of London. Series B, Biological Sciences,*. Vol.235, No.625 (May 31, 1951), (163-290), ISSN°02610523.

Kerhuel, L.; Viollet, S. & Franceschini, N. (2007). A sighted aerial robot with fast gaze and heading stabilization, *Proceedings of the IEEE International Conference on Intelligent Robots and Systems (IROS)*, pp.2634-2641, ISBN°978-1-4244-0912-9, San Diego, USA, October 29-November 2 2007

Kirchner, W.H. & Srinivasan, M.V. (1989). Freely flying honeybees use image motion to estimate object distance. *Naturwissenschaften*, Vol.76, No.6, (281-282), ISSN 0028-1042.

Koenderink, J.J. (1986) Optic flow, *Vision Research*, Vol.26, No.1, (161-179), ISSN°0042-6989

Kramer, J.; Sarpeshkar, R. & Kock, C. (1997). Pulse-Based Analog VLSI Velocity Sensors, *IEEE Transactions on Circuits and Systems* II, Vol.44, No.2 (February 1997) (86-101), ISSN° 1057-7130

Liu, S.C. & Usseglio-Viretta, A. (2001). Fly-like visuomotor responses of a robot using a VLSI motion-sensitive chips, *Biological Cybernetics*, Vol.85, No.6 (december 2001) (449-457), ISSN°0340-1200

Mura, F. & Franceschini, N. (1996). Obstacle avoidance in a terrestrial mobile robot provided with a scanning retina, In: *Intelligent Vehicles Vol.II*, M. Aoki and I. Masaki (Eds.), (47-52), MIT Press, Cambridge

Murthy, S. N.; Alvis, W.; Shirodkar, R.; Valavanis, K. & Moreno, W. (2008). Methodology for implementation of Unmanned vehicle control on FPGA using System generator, *Proceedings of the 7th international Caribbean conference on devices, circuits and systems (ICCDCS)*, pp.1-6, ISBN°978-1-4244-1956-2, Mexico, April 28-30, 2008

Netter, T. & Franceschini, N. (1999). Neuromorphic optical flow sensing for nap-of-the-earth flight, *Proceeding of SPIE Conference on Mobile Robots XIV*, Vol.3838, pp.208-216, Boston, USA, 20 September 1999

Netter, T. & Franceschini, N. (2002). A Robotic aircraft that follows terrain using a neuromorphic eye, *Proceeding IEEE International Conference on Robots and Systems (IROS)*, pp.129-134, ISBN 0-7803-7398-7, Lausanne, Swiss, 30 September - 4 October 2002.

Ownby, M. & Mahmoud W. H. (2003). A design methodology for implementing DSP with Xilinx System Generator for Matlab, *Proceedings of the 35th Southeastern Symposium on System Theory*, pp. 404-408, 16-18 March 2003

Pichon, J-M.; Blanes, C. & Franceschini, N. (1989). Visual guidance of a mobile robot equipped with a network of self-motion sensors, *Proceeding of SPIE Mobile RobotsIV.*, Vol.1195, (44-53), ISBN°9780819402349

Ruffier, F. & Franceschini, N. (2003). OCTAVE, a bioinspired visuo-motor control system for the guidance of Micro-Air-Vehicles, *Proceeding of SPIE Conference on Bioengineered and Bioinspired Systems*, Vol.5119, pp.1-12, Bellingham, USA

Ruffier, F.; Viollet, S. & Franceschini, N. (2004). Visual control of two aerial micro-robots by insect-based autopilots, *Advanced Robotics*, Vol.18, No.8, (771-786)

Ruffier, F. & Franceschini, N. (2005) Optic flow regulation: the key to aircraft automatic guidance, *journal of Robotics and Autonomous Systems*, Vol.50, No.4, (177-194), ISSN° 0921-8890

Ruffier, F., Serres J., Masson, G.P. and Franceschini N. (2007). A bee in the corridor: regulating the optic flow on one side. *Proceedings of the 7th Meeting of the German Neuroscience Society - 31st Göttingen Neurobiology Conference*, Göttingen, Germany, Abstract T14-7B.

Serres, J., Ruffier, F. & Franceschini, N. (2006). Two optic flow regulators for speed control and obstacle avoidance, *Proceedings of the first IEEE International Conference on Biomedical and Biomechatronics (Biorob)*, pp.750-757, Pisa, Italy.

Serres, J., Ruffier, F., Masson, G.P. & Franceschini N. (2007). A bee in the corridor: centring or wall-following? In *proceedings of the 7th Meeting of the German Neuroscience Society - 31st Göttingen Neurobiology Conference*, Göttingen, Germany, Abstract T14-8B.

Serres, J., Dray, D., Ruffier, & Franceschini N. (2008a). A vision-based autopilot for a miniature air vehicule: joint speed control and lateral obstacle avoidance. *Autonomous Robots*, Vol.25, No.1-2 (August 2008) (103-122). ISSN° 0929-5593

Serres, J., Masson, G.M., Ruffier, F. & Franceschini, N. (2008b). A bee in the corridor: centering and wall-following. Naturwissenchaften. (in press)

Srinivasan, M.V., Zhang, S.W., Lehrer, M. and Collett T.S. (1996). Honeybee navigation *en route* to the goal: visual flight control and odometry. *Journal of Experimental Biology*, Vol.199, (237-244)., ISSN°0022-0949.

Srinivasan, M.V., Zhang, S.W., Chahl, J.S., Barth, E., Venkatesh, S. (2000) How honeybees make grazing landings on flat surface. Biological Cybernetics, Vol.83, No.3, (171-183), ISSN°0340-1200

Srygley, R.B., and Oliveira, E.G. (2001). Orientation mechanisms and migration strategies within the flight boundary layer. In Insect Movements: Mechanisms and Consequences, T.P. Woiwod, D.R. Reynolds, and C.D. Thomas, (Eds). (183–206), Wallingford, Oxon, UK: CABI Publishing, CAB International), ISBN 0851994563.

Tammero, L.F. and Dickinson, M.H. (2002). The influence of visual landscape on the free flight behavior of the fruit fly *drosophila melanogaster*. *Journal of Experimental Biology*, Vol.205, No3 (327-343), ISSN°0022-0949

Turney, R. D.; Dick, C.; Parlour, D. P. & Hwang, J. (1999). Modeling and implementation of DSP FPGA Solutions, *Proceedings of International Conference on Signal Processing Applications and Technology (ICSPAT)*, November 1-4, 1999, Orlando, USA (http://www.xilinx.com/products/logicore/dsp/matlab_final.pdf)

Van Nieuwstadt, M. & Morris, J.C. (1995). Control of rotor speed for a model helicopter: a design cycle, *Proceedings of American Control Conference*, Vol.1, pp. 668-672

Viollet, S. & Franceschini, N. (2001). Aerial Minirobot that stabilizes and tracks with a bio-inspired visual scanning sensor, In: *Biorobotics: Methods and applications*, B. Webb and T. Consi (Eds.), (67-83), MIT Press, ISBN°026273141X, Cambridge

Viollet, S.; Kerhuel, L. & Francechini, N. (2008). A 1-gram dual sensorless spedd governor for Micro-Air-Vehicles, *Proceedings of 16th Mediterranean Conference on Control and Automation*, pp.1270-1275, ISBN°978-1-4244-2504-4, Ajaccio, France, June 25-27, 2008

Wagner, H. (1982). Flow-field variables trigger landing in flies. *Nature*, Vol.297, (147-148).

William C. (1965). *Insect migration*, Second edition ed. Collins editor, London.

Yamada, H.; Tominaga, T. & Ichikawa, M. (2003). An autonomous flying object navigated by real-time optical flow and visual target detection, *Proceedings of the IEEE International Conference on Field Programmable Technology*, pp.222-227, ISBN°0-7803-8320-6, Tokyo, 15-17 December 2003

Zufferey, J.C.; Beyeler, A. & Floreano, D. (2003). Vision-based Navigation from Wheels to Wings, *Proceedings of IEEE International Conference on intelligent Robots and Systems*, pp.2968-2973, ISBN°0-7803-7860-1, Las-Vegas, USA, 27-31 October 2003

4

Advanced UAV Trajectory Generation: Planning and Guidance

Antonio Barrientos, Pedro Gutiérrez and Julián Colorado
Universidad Politécnica de Madrid – (Robotics and Cybernetics Group)
Spain

1. Introduction

As technology and legislation move forward (JAA & Eurocontrol, 2004) remotely controlled, semi-autonomous or autonomous Unmanned Aerial Systems (UAS) will play a significant role in providing services and enhancing safety and security of the military and civilian community at large (e.g. surveillance and monitoring) (Coifman et al., 2004). The potential market for UAVs is, however, much bigger than just surveillance. UAVs are ideal for risk assessment and neutralization in dangerous areas such as war zones and regions stricken by disaster, including volcanic eruptions, wildfires, floods, and even terrorist acts. As they become more autonomous, UAVs will take on additional roles, such as air-to-air combat and even planetary science exploration (Held et al., 2005).

As the operational capabilities of UAVs are developed there is a perceived need for a significant increase in their level of autonomy, performance, reliability and integration with a controlled airspace full of manned vehicles (military and civilian). As a consequence researchers working with advanced UAVs have moved their focus from system modeling and low-level control to mission planning, supervision and collision avoidance, going from *vehicle* constraints to *mission* constraints (Barrientos et al., 2006). This mission-based approach is most useful for commercial applications where the vehicle must accomplish tasks with a high level of performance and maneuverability. These tasks require flexible and powerful trajectory-generation and guidance capabilities, features lacking in many of the current commercial UAS. For this reason, the purpose of this work is to extend the capabilities of commercially available autopilots for UAVs. Civil systems typically use basic trajectory-generation algorithms, capable only of linear waypoint navigation (Rysdyk, 2003), with a minimum or non-existent control over the trajectory. These systems are highly constrained when maneuverability is a mission requirement. On the other hand, military researchers have developed algorithms for high-performance 3D path planning and obstacle avoidance (Price, 2006), but these are highly proprietary technologies that operate with different mission constraints (target acquisition, threat avoidance and situational awareness) so they cannot be used in civil scenarios.

This chapter presents a robust Trajectory Generation and Guidance Module (TG^2M), a software tool capable of generating complex six-degrees-of-freedom trajectories in 3D space with velocity, acceleration, orientation and time constraints. The TG^2M is an extension module to the Aerial Vehicle Control Language (AVCL), a software architecture and interpreted language specification that address the issues of mission definition, testing, validation and supervision for UAVs (Barrientos et al., 2006). The AVCL platform incorporates a 3D visual simulator environment that uses a Geographic Information System (GIS) as the world's data model. The GIS backend means all objects are geo-referenced and that several official and commercial databases may be used for mission planning (roads, airports, power lines, crop fields, etc.). The language specification contains a wide set of instructions that allow the off/on-line supervision and the creation of vehicle-independent missions.

The TG²M is designed to model 3D cartesian paths with parametric constraints. The system uses two types of mathematical descriptors for trajectories: analytical functions and polynomial interpolation. The two main contributions of the module are its geometrical representation of the trajectory and its parametric definition. Simple maneuvers like lines and circumference arcs are created with analytical functions that constrain the geometry of the desired path; then the parametric constraints are *applied*. These constraints are typically kinematic: constant velocity and acceleration, trapezoidal curves to accelerate or stop, etc. More complex maneuvers are described with polynomial interpolation and fitted to the critical control path points, meeting desired position, time and velocity constraints. These polynomial functions are based on third and fourth order splines with fixed boundary conditions (for example initial and final velocities), which join all control points with a continuous and smooth path (Jaramillo-Botero et al., 2004).

The section 2 of this chapter presents some of the current techniques extracted from the military aerial survey applications, showing *complex* mission-planning tools capable of addressing the mission-specific constraints required. Within those approaches, this section also introduces a brief description of the AVCL architecture, describing its components, modules, and the main features provisioned, addressing a novel way of human-mission planning definition and testing. The section 3 introduces the AVCL built-in TG²M framework, describing the different techniques used for the trajectory planning and guidance of UAVs, as well as the mathematical treatment of these methods (analytical functions and polynomial interpolation). On the other hand, simulation-based results (see section 4) using a mini-helicopter simulator embedded into the AVCL environment will show the capabilities of the TG²M while flying aggressive and simple maneuvers. Last but not least, "final observations" in section 5 includes comments about the TG²M framework and the upcoming additions to the TG²M under current development.

2. Motion-planning Methodologies

A planning algorithm should provide feasible and flyable optimal trajectories that connect starting with target points, which should be compared and valued using specific criteria. These criteria are generally connected to the following major concerns, which arise during a plan generation procedure: feasibility and optimality. The first concern asks for the production of a plan to safely "move" the UAV to its target state, without taking into account the quality of the produced plan. The second concern asks for the production of optimal, yet feasible, paths, with optimality defined in various ways according to the problem under consideration (LaValle, 2006). Even in simple problems searching for optimality is not a trivial task and in most cases results in excessive computation time, not always available in real-world applications. Therefore, in most cases we search for suboptimal or just feasible solutions. The simplest way to model an UAV path is by using straight-line segments that connect a number of waypoints, either in 2D or 3D space (Moitra et al., 2003, Zheng et al., 2005). This approach takes into account the fact that in typical UAV missions the shortest paths tend to resemble straight lines that connect waypoints with starting and target points and the vertices of obstacle polygons. Although waypoints can be efficiently used for navigating a flying vehicle, straight-line segments connecting the corresponding waypoints cannot efficiently represent the real path that will be followed by the vehicle due to the own kinematics of the traced path. As a result, these simplified paths cannot be used for an accurate simulation of the movement of the UAV in an optimization procedure, unless a large number of waypoints is adopted. In that case the number of design variables in the optimization procedure explodes, along with the computation time. This is why this section presents some background based on the state-of-the-art mission/path planning for UAVs. The purpose (apart from the state-of-the-art survey) is to compare to the AVCL architecture in order to observe how its TG^2M embedded framework (see section 3 for details) is a complement of some lacking approaches found in this specialized literature.

2.1 Background

Researchers at top research centers and universities (e.g. JPL at Caltech, Carnegie Mellon, NASA, ESA, among others) are dealing with the development of new strategies that allow high-level mission UAV planning within the desired flight performance. As examples of these approaches we present the NASA Ames Research Center, the Georgia Institute of Technology, and the University of Illinois.

For the past decade NASA has focused on developing an efficient and robust solution to Obstacle Field Navigation (OFN), allowing a fast planning of smooth and flyable paths around both known and unknown obstacles (Herwitz, 2007). Evolutionary algorithms have been used as a viable candidate to solve path-planning problems effectively, providing feasible solutions within a short time. Given a number of UAVs that are launched from different and known initial locations, the algorithm must create 2-D trajectories with a smooth velocity distribution along each trajectory and with the goal of reaching a predetermined target location, while ensuring collision avoidance and satisfying specific route and coordination constraints and objectives. B-Splines curves are used in order to model both the 2-D trajectories and the velocity distribution along each flight path. A flying laboratory for autonomous systems, based on the Yamaha RMAX helicopter is being developed, which it incorporates the OFN planner and one all-digital camera system with a state-of-the-art tracking and passive ranging capabilities. Machine stereo-vision is being used to determine safe landing areas and monocular vision is used to track the landing location without access to GPS. A Ground Control Station (GCS) is being integrated into the simulation environment to investigate the issues related to high altitude and long endurance flights.

As an alternative approach other researchers do not focus on generating complex trajectories profiles; their aim is to develop robust mission management, capable of successfully integrating the available onboard hardware (e.g. cameras, sensors, etc). The Georgia Tech's UAV Lab developed the GTMax architecture (Alison et al., 2003) based on a Yamaha R-Max mini-helicopter, is an example of this alternative approach. The GTMax system is capable of fully autonomous flight with decentralized software modules for trajectory generation, offline simulation, supervision, guidance and control. The trajectory module generates lineal waypoints between targets with initial and final velocity parameterization. Commands to the helicopter take the form of different types of waypoints. All trajectories generated are assumed to be physically feasible by the helicopter, the kinematic model used for the trajectory generation uses specifiable limits on the maximum speed and acceleration the aircraft may have during a maneuver. From a high-level mission management perspective the GTMax architecture allows the UAV to autonomously locate targets with an identifying symbol within a designated search area. Once the correct target is identified the mission coordination and flight path generation is done at a centralized location on the ground, and the commands are transmitted to the UAV via a wireless datalink. The GTMax GCS interfaces with the primary flight computer and displays vehicle status, the object tracker information and the flight plans generated by the mission planner. The Vision Monitoring Station receives streaming video from the camera and the output of the image processor. This allows the operator to monitor the efficiency of the image processing as well as visually document the results of the search in the final phase of the mission.

For other projects the main goal is the unification of high-level mission commands and development of generic languages that support online UAV mission planning without constraining the system to a single vehicle. The University of Illinois at Urbana-Champaign (Frazzoli, 2002) presents a new framework for UAV motion planning and coordination. This framework is based on the definition of a modelling language for UAV trajectories, based on the interconnection of a finite number of suitably defined motion primitives, or maneuvers. Once the language is defined, algorithms for the efficient computation of feasible and optimal motion plans are established, allowing the UAV to fulfill the desired path.

If we analyze the UAS described above, almost all of them share one important limitation: their software architecture is tightly coupled to one vehicle and the capabilities of its low-level controller. Civil applications require open and extendable software architectures capable of *talking* to vehicles from different suppliers. The AVCL addresses those limitations, allowing to model different vehicles into a single common language (vehicle-independent missions). In the same fashion the described vehicles show that complex and simple maneuvers could be a suitable solution depending on the kind of mission to fulfill. For this reason the AVCL is extended with the TG^{2}M framework, capable of generating simple and complex 3D paths with the necessary vehicle constrains. The next sub-section introduces the AVCL architecture and some of the features provisioned.

2.2 The Aerial Vehicle Control Language - AVCL

The AVCL is not just a language capable of describing the missions and capabilities of an heterogeneous group of vehicles, it is part of a bigger framework that includes its interpreter, a definition of a base-vehicle, and a Mission Planner that uses GIS as the data-model for the world (Barrientos et al., 2006). The Mission Planner (MP) is not tied to a particular set of vehicles, sensors or commands. At any given time new functionality can be loaded and displayed to the human operator as new options and commands. This means that the MP tool is to be extended through Vehicle and Command Libraries without recompiling, and those new capabilities and better vehicles can be added easily. The Mission Planner is a great tool for simulation and direct comparison of various trajectory trackers, UAV models and controllers, because it can display *N* missions at the same time.

When considered just as a language the AVCL concept is the abstraction layer that allows the human supervisor to create missions that are vehicle and payload independent, promoting code reuse. At the same time the AVCL statements and commands hide device specific software/hardware, and serve as mission definition and storage. As an example of the code used to define operations within a mission:

```
uav.Sensors(0) = parser.loadObject ('camera.lib')
uav.Sensors(0).LookAt (p1)
uav.Sensors(1) = parser.loadObject ('laser.lib')
uav.Sensors(1).TurnOn()
uav.doLine (way_points = {p1, p2, p3, p4}, vel = 0.9 m_s)
```

Compared to previous vehicle *programming* languages the AVCL and its interpreter provide several advantages: intuitive handling of different systems of units; its use of the object-oriented-programming paradigm; facilities for inter-vehicle communications; run-time definition of relations between vehicles, sensors and other equipment; it may be extended easily through C, C++ or C# code; the interpreter is a light-weight application written in C++, therefore it may be deployed in many SW/HW architectures. Before the development of the TG²M module the AVCL framework relied on a simpler guidance module to connect waypoints with straight-line segments, and while the language could describe complex maneuvers and mission constraints the framework lacked the capacity to *fly* a vehicle through complex paths.

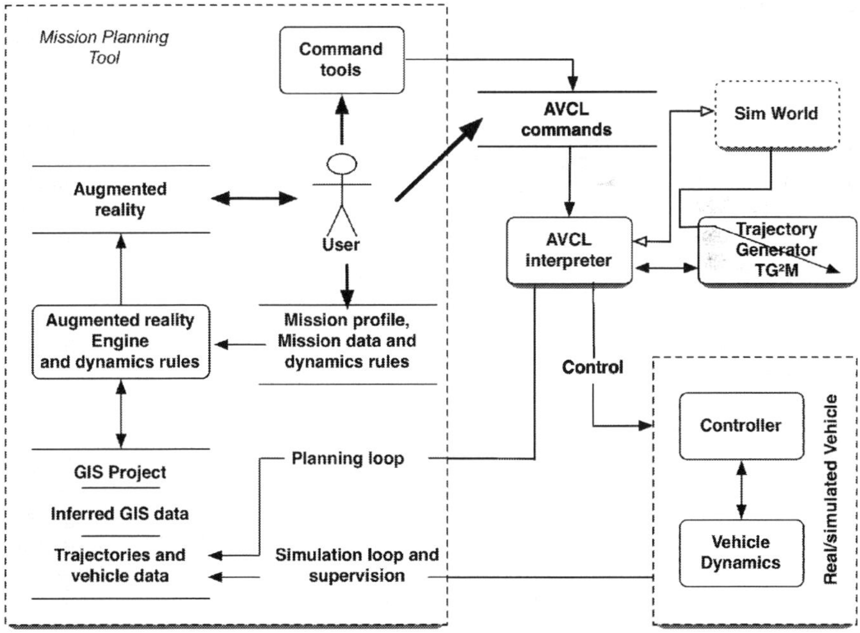

Figure 1. The AVCL simplified diagram for mission planning

3. Trajectory Generation and Guidance Module - TG²M

The TG²M is designed to model 3D cartesian paths with parametric constraints. The system uses two types of mathematical descriptors for trajectories: polynomial interpolation and analytical functions. Complex maneuvers are described with polynomial interpolation based on third and fourth order splines with fixed boundary conditions (e.g. initial and final velocities user definition), which join all control points with a continuous and smooth path. Likewise, simple maneuvers like lines and circumference are created with analytical functions that constrain the geometry of the desired path. These constraints are typically kinematic: constant velocity and acceleration, trapezoidal curves to accelerate or stop, etc.

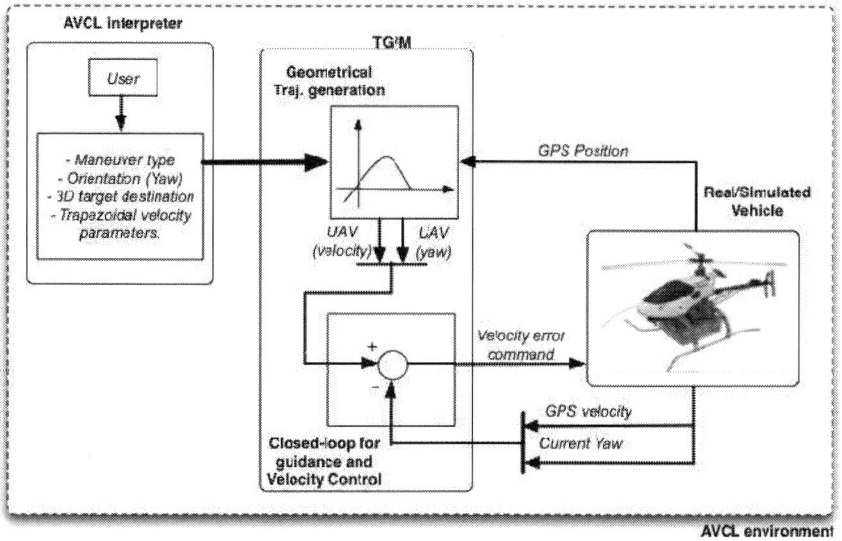

Figure 2. The TG²M framework

As shown in Fig. 2, two main modules compose the TG²M framework: the geometrical trajectory generation and the online closed-loop guidance. All the parameters and fundamental data (e.g. the desired maximum speed, etc) are provided by the user via the AVCL interpreter.

3.1 Polynomial interpolation using splines

The fundamental idea behind the spline interpolation is based on the definition of smooth curves through a number of points. These curves are represented by a polynomial function (normally of third-grade) defined by some coefficients that determine the spline used to interpolate the numerical data points. These coefficients *bend* the line so that it passes through each of the data points without any erratic behavior or breaks in continuity. The essential idea is to define a third-degree polynomial function $S(t)$ of the form:

$$S(t) = a_i t^3 + b_i t^2 + c_i t + d_i \tag{1}$$

This third-degree polynomial needs to conform to the following conditions in order to interpolate the desired knot-points as depicted in Fig. 3:

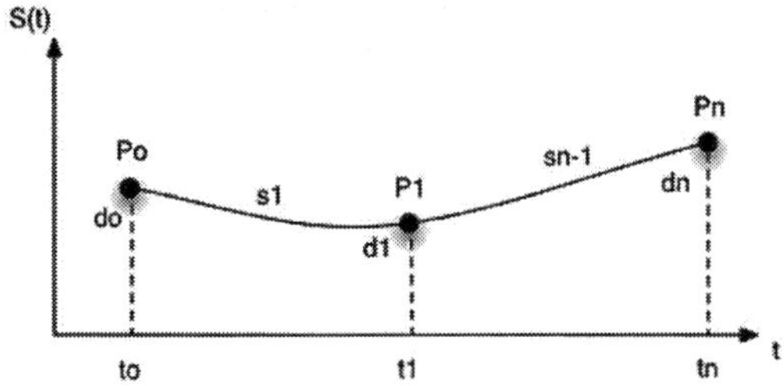

Figure 3. Knot-points to interpolate using 3D splines (with free boundary conditions)

- $S(t)$, $S'(t)$ and $S''(t)$ will be continuous on the interval $[t_0, t_n]$.
- Since the curve $S(t)$ must be continuous across its entire interval, it can be concluded that each sub-function must joint at the data points, so: $s_i(t_i) = s_{i-1}(t_i)$.

Taking into account the set of conditions previously described, for each $i = 1,2.....n-1$, $t_i \in [t_i, t_{i+1}]$ $S(t_i) = s_i(t_i)$, and letting $h_i = \sum_{i=1}^{n-1} t_{i+1} - t_i$, Eq. (1) is re-writing as:

$$s_i(t_i) = a_i h_i^3 + b_i h_i^2 + c_i h_i + d_i \tag{2}$$

Also, to make the curve smooth across the interval, the first derivative of the $S(t)$ function must be equal to the derivative of the position reference points; this yields:

$$\begin{aligned} s_i'(t_i) &= 3a_i h_i^2 + 2b_i h_i + c_i \\ s_i'(t_i) &= s_{i-1}'(t_i) \\ s_i'(t_i) &= c_i = 3a_{i-1} h_i^2 + 2b_{i-1} h_i + c_{i-1} \end{aligned} \tag{3}$$

Applying the same approach for the second derivative:

$$\begin{aligned} s_i''(t_i) &= 6a_i h_i + 2b_i \\ s_i''(t_i) &= s_{i+1}''(t_i) \\ s_i''(t_i) &= 2b_{i+1} = 6a_i h_i + 2b_i \end{aligned} \tag{4}$$

For the solution of the polynomial coefficients in Eq. (1), the system in Eq. (5) must be solved as:

$$A \cdot Y = f \tag{5}$$

where the matrix $A \in \Re^{nxn}$ (n is the number of knot-points to interpolate) corresponds to:

$$A = \begin{bmatrix} 1 & 0 & 0 & \cdots & \cdots & \cdots & 0 \\ h_0 & 2(h_0+h_1) & h_1 & \ddots & & & \vdots \\ 0 & h_0 & 2(h_1+h_2) & h_2 & \ddots & \ddots & \vdots \\ \vdots & \ddots & \ddots & \ddots & \ddots & \ddots & \vdots \\ \vdots & & \ddots & \ddots & \ddots & \ddots & \vdots \\ \vdots & & & \ddots & h_{n-2} & 2(h_{n-2}+h_{n-1}) & h_{n-1} \\ 0 & \cdots & \cdots & \cdots & 0 & 0 & 1 \end{bmatrix} \tag{6}$$

The term $h \in \Re^{n-1}$ in Eq. (6) is the time vector defined for each point $(P_0, P_1, \ldots P_n)$. The b_i coefficients in Eq. (1) are stacked in $Y \in \Re^{nx3}$, which yields the term $f \in \Re^{nx3}$ as:

$$f = \begin{bmatrix} 0 \\ \frac{3}{h_1}(d_2 - d_1) - \frac{3}{h_0}(d_1 - d_0) \\ \vdots \\ \frac{3}{h_{n-1}}(d_n - d_{n-1}) - \frac{3}{h_{n-2}}(d_{n-1} - d_{n-2}) \\ 0 \end{bmatrix} \tag{7}$$

From Eq. (2) and (3), the a_i and c_i coefficients are respectively obtained as:

$$\begin{aligned} a_i &= \frac{b_{i+1} - b_i}{3h_i} \\ c_i &= \frac{1}{h_i}(d_{i+1} - d_i) - \frac{h_i}{3}(2b_i - b_{i+1}) \\ d_i &= S(t_i) \end{aligned} \tag{8}$$

Example 1. *Simple UAV trajectory planning using 3D splines with fixed boundary condition.*

Let's define three knot-points as: $P_0 = [0,0,0]$ $P_1 = [5,10,10]$ $P_2 = [0,10,20]$ *with the following time condition for each point:* $t = [0,10,20]$ *(note that the time vector components are given in seconds). Calculate a 3D spline considering zero initial and final velocities.*

The natural 3D splines with free boundary conditions may generate smooth paths but without control over the velocities over the knot-points. Because of this, Eq. (6) and (7) must be complemented in order to generate a 3D spline with the fixed boundary conditions, in this case, with zero initial and final velocities. Re-writing those equations yield:

$$A = \begin{bmatrix} 2h_0 & h_0 & 0 & \cdots & \cdots & \cdots & 0 \\ h_0 & 2(h_0+h_1) & h_1 & \ddots & & & \vdots \\ 0 & h_0 & 2(h_1+h_2) & h_2 & \ddots & \ddots & \vdots \\ \vdots & \ddots & \ddots & \ddots & \ddots & \ddots & \vdots \\ \vdots & & \ddots & \ddots & \ddots & \ddots & \vdots \\ \vdots & & & \ddots & h_{n-2} & 2(h_{n-2}+h_{n-1}) & h_{n-1} \\ 0 & \cdots & \cdots & \cdots & 0 & h_{n-1} & 2h_{n-1} \end{bmatrix} \tag{9}$$

The same system $A \cdot Y = f$ must be solved with the new $A \in \Re^{nxn}$ from Eq. (9) and $f \in \Re^{nx3}$ defined in Eq. (10). Note that the first derivative of the $S(t)$ function has been added in the first and last position of the f vector, allowing the control of the initial and final velocities of the curve.

$$f = \begin{bmatrix} \frac{3}{h_0}(d_1 - d_0) - 3S'(t_0) \\ \frac{3}{h_1}(d_2 - d_1) - \frac{3}{h_0}(d_1 - d_0) \\ \vdots \\ \frac{3}{h_{n-1}}(d_n - d_{n-1}) - \frac{3}{h_{n-2}}(d_{n-1} - d_{n-2}) \\ 3S'(t_n) - \frac{3}{h_{n-1}}(d_n - d_{n-1}) \end{bmatrix} \tag{10}$$

The first procedure is to obtain the time vector as follows: $h = \sum_{i=1}^{n-1} t_{i+1} - t_i = [10 \quad 10]$, then the system: $Y = A^{-1} f$ must be solved to obtain the polynomial coefficients as:

$$A = \begin{bmatrix} 20 & 10 & 0 \\ 10 & 40 & 10 \\ 0 & 10 & 20 \end{bmatrix}, \quad Y_x = \begin{bmatrix} b_{0x} = 0.15 \\ b_{1x} = -0.15 \\ b_{2x} = 0.15 \end{bmatrix}, \quad Y_y = \begin{bmatrix} b_{0y} = 0.15 \\ b_{1y} = 0 \\ b_{2y} = -0.15 \end{bmatrix}, \quad Y_z = \begin{bmatrix} b_{0z} = 0.1125 \\ b_{1z} = 0.0750 \\ b_{2z} = -0.2625 \end{bmatrix}$$

$$f_x = \begin{bmatrix} 1.5 \\ -3 \\ 1.5 \end{bmatrix} \quad f_y = \begin{bmatrix} 3 \\ 0 \\ -3 \end{bmatrix} \quad f_z = \begin{bmatrix} 3 \\ 1.5 \\ -4.5 \end{bmatrix}$$

Finally, the two polynomials for each (x, y, z) component are evaluated from the time [0, 10] to [10, 20]. Figure 4 shows the results.

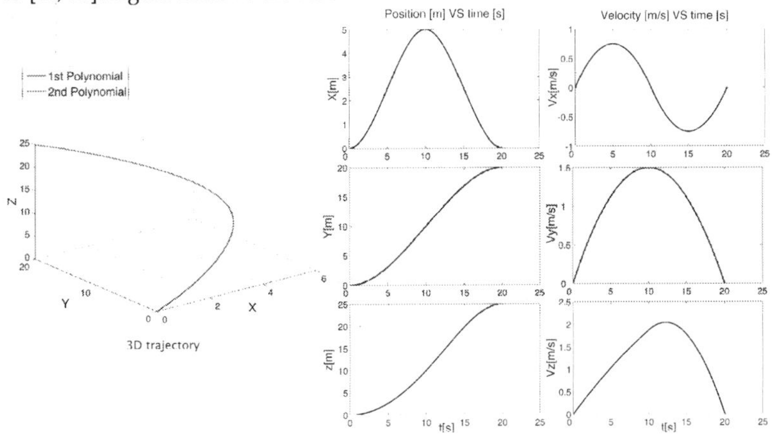

Figure 4. 3D spline with fixed boundary condition

The end of Example 1: Simple UAV trajectory planning using 3D splines

In the previous example, the 3D splines with fixed boundary conditions allowed to define a smooth curve across the knot-points. Nevertheless, two basic problems must be taken into account: the 3D spline only allows the user to establish the initial and the final velocities of the whole trajectory, limiting the user to have total control over the other points. The second problem is the smoothness of the acceleration curves (linear). To solve these problems 4D splines must be used. These address the user total control over the velocity profile across the whole trajectory and its results in additional *smoothness* for position, velocity and even acceleration curves. This could be more effective for UAVs tasks where the mission requires a strong control of the vehicle acceleration and velocities at each defined knot-point.

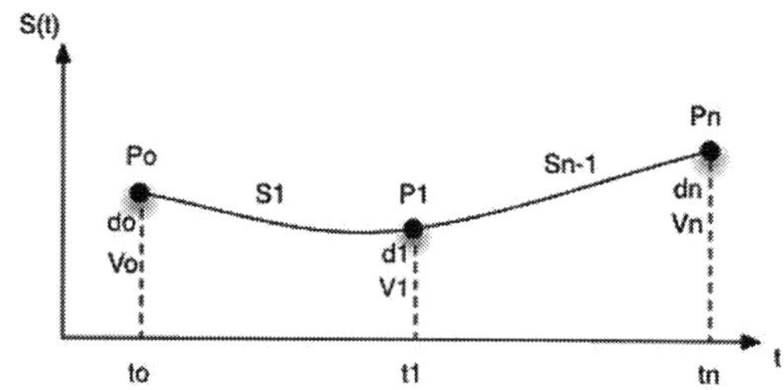

Figure 5. Knot-points to interpolate with 4D splines

The fourth-degree polynomial is defined by:

$$S(t) = e_i t^4 + a_i t^3 + b_i t^2 + c_i t + d_i \tag{11}$$

The same 3D-spline conditions previously described also apply to the 4D-spline. To obtain a generalized solution of the system to solve ($A \cdot Y = f$), we start from the three-point case as depicted in Fig. 5. The polynomials for each trajectory segment as a function of time t are:

$$\begin{aligned} s_n(t_0) &= e_0 t_0^4 + a_0 t_0^3 + b_0 t_0^2 + c_0 t_0 + d_0 = f(t_0) \\ s_n(t_1) &= e_0 t_1^4 + a_0 t_1^3 + b_0 t_1^2 + c_0 t_1 + d_0 = f(t_1) \\ s_{n-1}(t_1) &= e_1 t_1^4 + a_1 t_1^3 + b_1 t_1^2 + c_1 t_1 + d_1 = f(t_1) \\ s_{n-1}(t_n) &= e_1 t_n^4 + a_1 t_n^3 + b_1 t_n^2 + c_1 t_n + d_1 = f(t_n) \end{aligned} \tag{12}$$

Taking the first and second derivatives (velocities and accelerations), we obtain:

$$\begin{aligned} s_1'(t_0) &= 4e_0 t_0^3 + 3a_0 t_0^2 + 2b_0 t_0 + c_0 = V_0 \\ s_1'(t_1) &= 4e_0 t_1^3 + 3a_0 t_1^2 + 2b_0 t_1 + c_0 = V_1 \\ s_{n-1}'(t_1) &= 4e_1 t_1^3 + 3a_1 t_1^2 + 2b_1 t_1 + c_1 = V_1 \\ s_{n-1}'(t_n) &= 4e_1 t_n^3 + 3a_1 t_n^2 + 2b_1 t_n + c_1 = V_n \end{aligned} \tag{13}$$

The second derivatives of Eq. (12) yield a set of accelerations. Equaling the acceleration functions for the intermediate points (t_1 for each case) and setting to zero the initial and final acceleration of the path segment yields:

$$s_0''(t_0) = 12e_0t_0^2 + 6a_0t_0 + 2b_0 = s_{n-1}'(t_1) = 12e_1t_1^2 + 6a_1t_1 + 2b_1$$
$$s_0''(t_1) = 12e_0t_1^3 + 6a_0t_1 + 2b_0 = s_{n-1}''(t_n) = 12e_1t_n^2 + 6a_1t_n + 2b_1 \tag{14}$$

Equations (12), (13) and (14) conform the complete system in order to obtain the ten polynomial coefficients. Solving $A \cdot Y = f$, the matrix $A \in \Re^{5(n-1)\,x\,5(n-1)}$ corresponds to:

$$A = \begin{bmatrix} 1 & t_0 & t_0^2 & t_0^3 & t_0^4 & 0 & 0 & 0 & 0 & 0 \\ 1 & t_1 & t_1^2 & t_1^3 & t_1^4 & 0 & 0 & 0 & 0 & 0 \\ 0 & 0 & 0 & 0 & 0 & 1 & t_1 & t_1^2 & t_1^3 & t_1^4 \\ 0 & 0 & 0 & 0 & 0 & 1 & t_2 & t_2^2 & t_2^3 & t_2^4 \\ 0 & 1 & 2t_0 & 3t_0^2 & 4t_0^3 & 0 & 0 & 0 & 0 & 0 \\ 0 & 1 & 2t_1 & 3t_1^2 & 4t_1^3 & 0 & 0 & 0 & 0 & 0 \\ 0 & 0 & 0 & 0 & 0 & 0 & 1 & 2t_1 & 3t_1^2 & 4t_1^3 \\ 0 & 0 & 0 & 0 & 0 & 0 & 1 & 2t_2 & 3t_2^2 & 4t_2^3 \\ 0 & 0 & 2 & 6t_1 & 12t_1^2 & 0 & 0 & -2 & -6t_1 & -12t_1^2 \\ 0 & 0 & 2 & 6t_0 & 12t_0^2 & 0 & 0 & -2 & -6t_2 & -12t_2^2 \end{bmatrix} \tag{15}$$

The polynomial coefficients are stacked into the $Y \in \Re^{5(n-1)}$ vector, and with the $f \in \Re^{5(n-1)}$ term, the total system is defined from Eq. (14) and (15):

$$Y = \begin{bmatrix} d_0 & c_0 & b_0 & a_0 & e_0 & d_1 & c_1 & b_1 & a_1 & e_1 \end{bmatrix}^T$$
$$f = \begin{bmatrix} f(t_0) & f(t_1) & f(t_1) & f(t_n) & V_0 & V_1 & V_1 & V_n & 0 & 0 \end{bmatrix} \tag{16}$$

Example 2. *UAV trajectory planning using 4D splines with total velocity control*

Lets define three knot-points as: $P_0 = [0,0,0]$, $P_1 = [5,7,5]$, $P_2 = [4,2,10]$ *in the time* $t = [0,4,8]$ *(note that the time vector components are given in seconds). Calculate the 4D spline considering the following velocity profile:* $V_0 = [0,0,0]$, $V_1 = [0.5,0.8,1]$, $V_2 = [1,1.5,1]$.

Solving the system: $Y = A^{-1}f$, we obtain the following numerical data:

$$A = \begin{bmatrix} 1 & 0 & 0 & 0 & 0 & 0 & 0 & 0 & 0 & 0 \\ 1 & 4 & 16 & 64 & 256 & 0 & 0 & 0 & 0 & 0 \\ 0 & 0 & 0 & 0 & 0 & 1 & 4 & 16 & 64 & 256 \\ 0 & 0 & 0 & 0 & 0 & 1 & 8 & 64 & 512 & 4096 \\ 0 & 1 & 0 & 0 & 0 & 0 & 0 & 0 & 0 & 0 \\ 0 & 1 & 8 & 48 & 256 & 0 & 0 & 0 & 0 & 0 \\ 0 & 0 & 0 & 0 & 0 & 0 & 1 & 8 & 48 & 256 \\ 0 & 0 & 0 & 0 & 0 & 0 & 1 & 16 & 192 & 2048 \\ 0 & 0 & 2 & 24 & 192 & 0 & 0 & -2 & -24 & -192 \\ 0 & 0 & 2 & 0 & 0 & 0 & 0 & -2 & 0 & 0 \end{bmatrix}, \quad Y_{x,y,z} = \begin{bmatrix} d_{0x} & d_{0y} & d_{0z} \\ c_{0x} & c_{0y} & c_{0z} \\ b_{0x} & b_{0y} & b_{0z} \\ a_{0x} & a_{0y} & a_{0z} \\ e_{0x} & e_{0y} & e_{0z} \\ d_{1x} & d_{1y} & d_{1z} \\ c_{1x} & c_{1y} & c_{1z} \\ b_{1x} & b_{1y} & b_{1z} \\ a_{1x} & a_{1y} & a_{1z} \\ e_{1x} & e_{1y} & e_{1z} \end{bmatrix} \quad f_{x,y,z} = \begin{bmatrix} 0 & 0 & 0 \\ 5 & 7 & 5 \\ 5 & 7 & 5 \\ 4 & 2 & 10 \\ 0 & 0 & 0 \\ 0 & 0 & 0 \\ 0 & 0 & 0 \\ 0 & 0 & 0 \\ 0 & 0 & 0 \\ 0 & 0 & 0 \end{bmatrix} \quad f_{x,y,z} = \begin{bmatrix} 0 & 0 & 0 \\ 5 & 7 & 5 \\ 5 & 7 & 5 \\ 4 & 2 & 10 \\ 0 & 0 & 0 \\ 0 & 0 & 0 \\ 0 & 0 & 0 \\ 0 & 0 & 0 \\ 0 & 0 & 0 \\ 0 & 0 & 0 \end{bmatrix}$$

Finally, the two polynomials for each (x, y, z) component are evaluated from the time [0, 4] to [4, 8] seconds. Figure 6 shows the results:

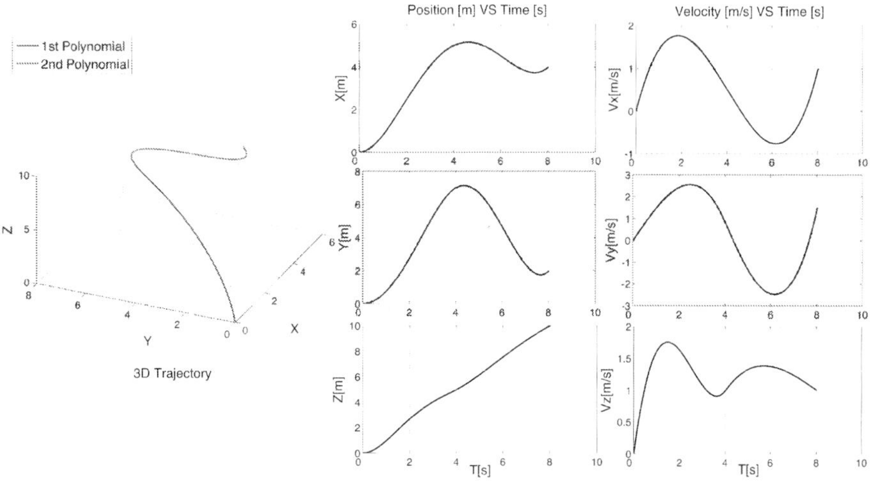

Figure 6. 4D spline with total control of the knot-points velocities

The end of Example 2: UAV trajectory planning using 4D splines with total velocity control

3.2 Simple Maneuvers with Analytical Functions

For simple maneuvers the TG²M framework also supports the definition of straight-lines and circumferences via analytical functions that constrain the geometry of the desired path. These constraints are typically kinematic: constant velocity and acceleration, trapezoidal curves to accelerate or stop, etc. This kind of parameterization is useful when the mission requires the UAV stops at the desired end-point of the trajectory effectively, due to the user-control of the acceleration slope tilt level. For both (lines and circumferences) the desired set of velocities must fulfill the following trapezoidal velocity profile:

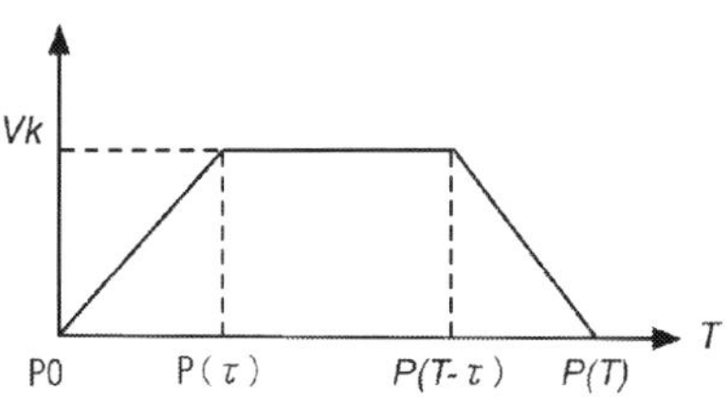

Figure 7. Trapezoidal velocity profile used for simple UAV maneuvers

From Fig. 7, three fundamental segments compose the total function to define the straight-line. The two intermediate points of the trapezoidal curve: $P(\tau)$ $P(T-\tau)$ are calculated as [1]:

[1] Sub-indices x,y,z refer to each coordinate of motion.

$$
\begin{aligned}
P(\tau)_{x,y,z} &= \frac{1}{2\left\| P(T)_{x,y,z} - P_{0x,y,z} \right\|} \frac{V_k}{t} t^2 \left(P(T)_{x,y,z} - P_{0x,y,z} \right) + P_{0x,y,z} \\
P(T-\tau)_{x,y,z} &= \frac{V_k (T - 2t)}{\left\| P(T)_{x,y,z} - P_{0x,y,z} \right\|} \left(P(T)_{x,y,z} - P_{0x,y,z} \right) + P(\tau)_{x,y,z}
\end{aligned} \tag{17}
$$

Once that the intermediate points have been defined, the three segments that compose the total function are defined by Eq. (18). In the first section (from P_0 to $P(\tau)$) the initial velocity is set $V_i = 0$ and progresses toward a final velocity V_k. The second segment (from $P(\tau)$ to $P(T-\tau)$) is traced at constant maximum velocity V_k. Finally the last segment (from $P(T-\tau)$ to $P(T)$) drives the vehicle from V_k to zero velocity.

$$
f(t) = \begin{cases}
\dfrac{V_k}{2\tau\left(P(\tau) - P_0\right)} t^2 \left(P(\tau) - P_0\right) + P_0 & 0 \le t \le \tau \\
\dfrac{V_k}{\left(P(T-\tau) - P(\tau)\right)} (t-\tau)\left(P(T-\tau) - P(\tau)\right) + P(\tau) & \tau < t \le T - \tau \\
\dfrac{(t - T + \tau)}{\left(P(T) - P(T-\tau)\right)} \left(-\dfrac{V_k}{2\tau}(t - T + \tau) + V_k\right)\left(P(T) - P(T-\tau)\right) + P(T-\tau) & T - \tau < t \le T
\end{cases} \tag{18}
$$

The same approach is applied to generate circumferences with the trapezoidal profile used for the straight-lines. Three knot-points define the circumference path and the objective is to find the trace angle across the trajectory.

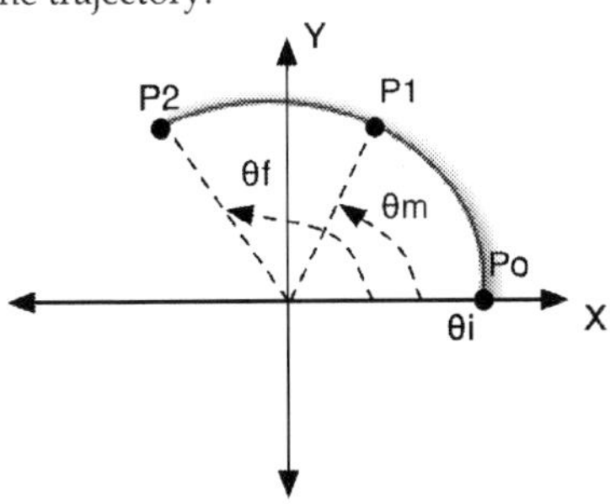

Figure 8. X-Y plane for circumference maneuver

The first step is to determine the center of the circle that joins the three knot-points. The equation of the trajectory plane is defined by the cross product between the vectors formed by points P_0, P_1 and P_2, as shown in Eq. (19).

$$
\begin{aligned}
\vec{P_0P_1} \times \vec{P_0P_2} &= N(A,B,C) \\
Ax + By + Cz &= cte
\end{aligned} \tag{19}
$$

The center of the circle $c(x_c, y_c, z_c)$ is always equidistant to any point over the circle[2], then:

$$
\begin{aligned}
&(x_{0,1,2} - x_c)^2 + (y_{0,1,2} - y_c)^2 + (z_{0,1,2} - z_c)^2 = K^2 \\
&Ax_c + By_c + Cz_c = cte
\end{aligned}
\tag{20}
$$

To obtain the relation between the angle motion (θ), the angular velocity (ω) and the tangential velocity (v_t), constant velocity is assumed across the path, yielding:

$$
\begin{aligned}
&\omega = \frac{d\theta}{dt} = \frac{v_t}{r} \\
&d\theta = \frac{v_t}{r} dt \\
&\int d\theta = \int \frac{v_t}{r} dt \\
&\theta(t) = \frac{v_t}{r} t + \theta_i \quad \text{where} \quad t \in [0, T]
\end{aligned}
\tag{21}
$$

Likewise, the relation between the angle motion (θ) and the angular acceleration of the curve is given by:

$$
\begin{aligned}
&at + v_0 = r\frac{d\theta}{dt} \\
&d\theta = \frac{at + v_0}{r} dt \\
&\int d\theta = \int \frac{at + v_0}{r} dt \\
&\theta(t) = \frac{a}{2r} t^2 + \frac{v_0}{r} t + \theta_i \quad \text{where} \quad t \in [0, T]
\end{aligned}
\tag{22}
$$

If the known parameter is the total motion time (T) of the trajectory, we set the acceleration term to the left-side of the equation, yielding:

$$
\begin{aligned}
&\theta(t) = \theta_f = \frac{a}{2r} t^2 + \frac{v_0}{r} t + \theta_i \\
&a = \frac{2[(\theta_f - \theta_i) - v_0 T]}{T^2}
\end{aligned}
\tag{23}
$$

Otherwise, if the constrained parameters are the initial and final velocity, the acceleration function is given by:

$$
\begin{aligned}
&a = \frac{v_{ft} - v_{0t}}{T} \\
&T = \frac{2(\theta_f - \theta_i)}{v_{0t} + v_{ft}}
\end{aligned}
\tag{24}
$$

[2] Sub-indices *0,1,2* refers to each vector component for Po,P1,P2 (see Fig. 8).

Example 3. UAV straight-line trajectory with trapezoidal velocity profile

Lets defined a straight-line trajectory starting from $P_0 = [0,0,20]$ *to* $P(T) = [10,0,20]$ *with 10 meters of displacement in the X coordinate at a constant altitude of 20 meters. Define the 3D cartesian trajectory with a maximum velocity of 1m/s taking into account a trapezoidal velocity profile with 30% of acceleration time.*

The distance vector to trace is: $dist = P(T) - P_0 = [10,0,0]$.

The total velocity is obtained from the norm of the vector *dist:* $V_T = \|dist\| = 10$.

The total motion time is calculated as: $T = \frac{V_T}{V_k(1-pt)} = 7$, where V_k is the maximum velocity at 1m/s and pt is the percentage of acceleration (30%=0.3).

The acceleration motion is also obtained as: $a = \frac{V_k}{T * pt} = 0.2381$.

The number of the total points to generate is: $n = 28$ (the user defines this parameter depending on the UAV controller data rate). In the same way, the number of points in the first and last segment of the trapezoidal profile is calculated as: $k = 2(pt \cdot n) = 16$, and for the constant velocity segment: $k_c = n - 2k = 12$. Once that the preliminary data have been calculated, Eq. (16) is used to obtain the intermediate points (see Fig. 7):

$$Px(\tau) = \frac{1}{2V_t} at^2 (dist)_x + P_{0x} = 2.10 \qquad Px(T-\tau) = \frac{V_k(T-2t)}{V_T}(dist)_x + P(\tau)_x = 7.7$$

$$Py(\tau) = \frac{1}{2V_t} at^2 (dist)_y + P_{0y} = 0 \qquad Py(T-\tau) = \frac{V_k(T-2t)}{V_T}(dist)_y + P(\tau)_y = 0$$

$$Pz(\tau) = \frac{1}{2V_t} at^2 (dist)_z + P_{0z} = 20 \qquad Pz(T-\tau) = \frac{V_k(T-2t)}{V_T}(dist)_z + P(\tau)_z = 20$$

The first segment of the path (from P_0 to $P(\tau)$) is calculated using Eq. (17):

$$fx(t) = \sum_{i=1}^{k/2} \frac{V_k}{2t\|P(t) - P_0\|_x} t_i^2 \left(P(t)_x - P_{0x}\right) + P_{0x}$$

$$fy(t) = \sum_{i=1}^{k/2} \frac{V_k}{2t\|P(t) - P_0\|_y} t_i^2 \left(P(t)_y - P_{0y}\right) + P_{0y} \quad 0 \le t \le \tau$$

$$fz(t) = \sum_{i=1}^{k/2} \frac{V_k}{2t\|P(t) - P_0\|_y} t_i^2 \left(P(t)_z - P_{0z}\right) + P_{0z}$$

The time vector t is obtained as: $t = \left[0 : \frac{1}{V_k} \frac{2\|P(\tau) - P_0\|}{k} : \frac{1}{V_k} 2\|P(\tau) - P_0\|\right]$, which is:

$t = [0 \quad 0.5250 \quad 1.0500 \quad 1.5750 \quad 2.1000 \quad 2.6250 \quad 3.1500 \quad 3.6750 \quad 4.2000]$.

For the second segment (at constant velocity V_k), the time vector is calculated as: $t=\left[0:\dfrac{1}{n-2k}:1\right]=\left[0\ 0.0833\ 0.1667\ 0.25\ 0.3333\ 0.4167\ 0.5\ 0.5833\ 0.6667\ 0.75\ 0.83\ 0.9167\right]$

$$\begin{aligned}
fx(t) &= \sum_{i-1}^{kc} t_i\left(P(T-\tau)_x - P(\tau)_x\right) + P(\tau)_x \\
fy(t) &= \sum_{i-1}^{kc} t_i\left(P(T-\tau)_y - P(\tau)_y\right) + P(\tau)_y \qquad \tau < t \le T-\tau \\
fz(t) &= \sum_{i-1}^{kc} t_i\left(P(T-\tau)_z - P(\tau)_z\right) + P(\tau)_z
\end{aligned}$$

Finally the last segment drives the vehicle from V_k to zero end velocity. The time vector is:

$$t=\left[0:\frac{1}{V_k}\frac{2\left\|P(T-\tau)-P(\tau)\right\|}{k}:\frac{1}{V_k}2\left\|P(T-\tau)-P(\tau)_0\right\|\right]=\begin{bmatrix}0\ 0.287\ 0.575\ 0.862\ 1.15\ 1.43\\ 1.72\ 2.01\ 2.3\end{bmatrix},$$

yielding the following function for each coordinate of motion:

$$\begin{aligned}
f(t)_x &= \sum_{i=1}^{k/2}\frac{-V_k t_i^2 + V_k t_i}{2\left\|P(T)-P(T-t)\right\|_x}\left(P(T)_x - P(T-t)_x\right) + P(T-t)_x \\
f(t)_y &= \sum_{i=1}^{k/2}\frac{-V_k t_i^2 + V_k t_i}{2\left\|P(T)-P(T-t)\right\|_y}\left(P(T)_y - P(T-t)_y\right) + P(T-t)_y \qquad T-t < t \le T \\
f(t)_z &= \sum_{i=1}^{k/2}\frac{-V_k t_i^2 + V_k t_i}{2\left\|P(T)-P(T-t)\right\|_z}\left(P(T)_z - P(T-t)_z\right) + P(T-t)_z
\end{aligned}$$

The results are presented in Fig. 9:

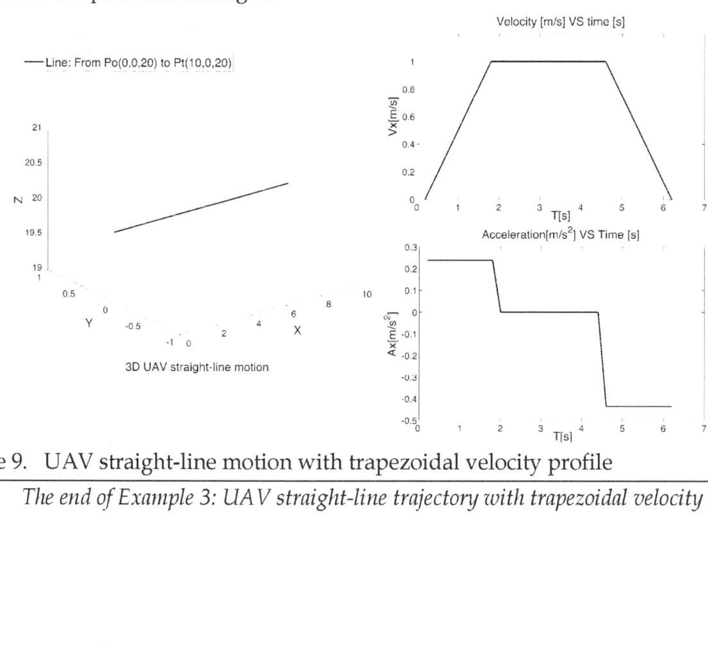

Figure 9. UAV straight-line motion with trapezoidal velocity profile

The end of Example 3: UAV straight-line trajectory with trapezoidal velocity profile

3.3 UAV Guidance

The geometrical representation of a UAV trajectory has been presented in the previous sub-section. Complex trajectories may be described using third-fourth order degree splines, or simple maneuvers may be generated using common lines and circumferences functions with some parametric features defined by the end-user. But in order to complete the generation of trajectories for a single UAV a guidance module is required.

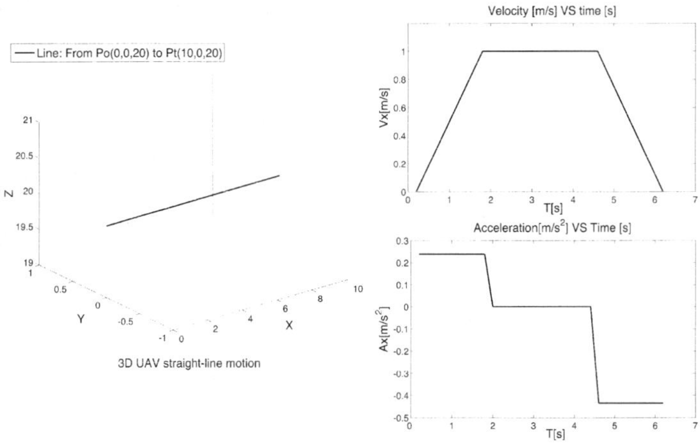

Figure 10. TG²M guidance scheme

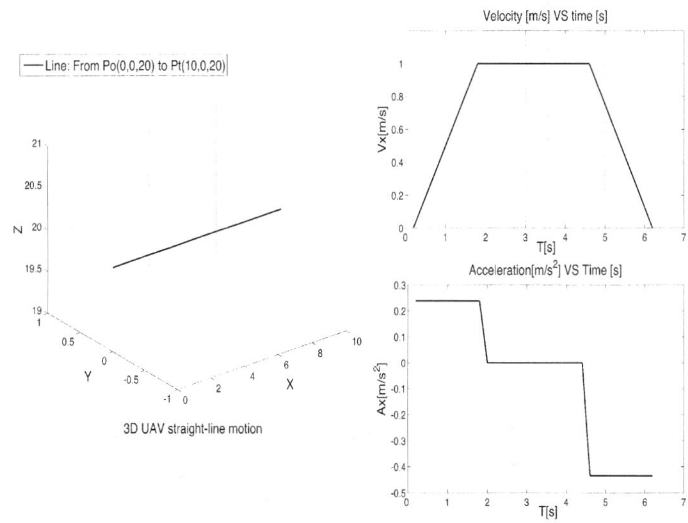

Figure 11. Velocity error command

The AVCL simulation module contains a dynamic model of a mini-helicopter (Valero, 2005) and its associated controllers: attitude, velocity and position. The velocity controller is based on a Proportional-Integrative "PI" design; it receives the *velocity error command* generated by the guidance module shown in Fig. 10. These velocity error references are named *Vxref* and *Vyref*, with respect to either the world or vehicle's frame, and are calculated with the UAV's error position. In Fig. 11 the theoretical (or ideal) UAV velocity vector is represented by the *Vt* term. Due to wind and other perturbations during flight, a velocity error vector *Ve* must be considered in order to set the final velocity references to send to the vehicle's controller. This vector is derived from the UAV position error between the current and desired positions. Finally, the velocity references *Vxref* and *Vyref* are the components of the vector $Vt + Ve$.

The guidance module must take into account that the helicopter's orientation (yaw) is *independent* from the direction of travel. The high-level modules must generate a set of orientations across the vehicle's trajectory. The built-in AVCL control module (see Fig. 10) is capable of receiving *velocity* and *yaw* angle orientation commands from the TG²M module and generating the needed commands for the attitude controller that governs the vehicle's roll and pitch. The TG²M's guidance module focuses on the generation of *yaw* references, and use a simple Proportional "P" controller for smooth *yaw* angle transition during the flight. Two methods are use to generate the *yaw* angle references: for simple maneuvers the *yaw* angle is calculated using simple trigonometric relations due to the path displacement. For complex trajectories using splines we introduce a feasible method to calculate a smooth *yaw* angle evolution. This method also shows how to calculated roll and pitch references due to the vehicle trajectory and its velocity. For *roll, pitch* and *yaw* angles calculation, the following frame of reference is used:

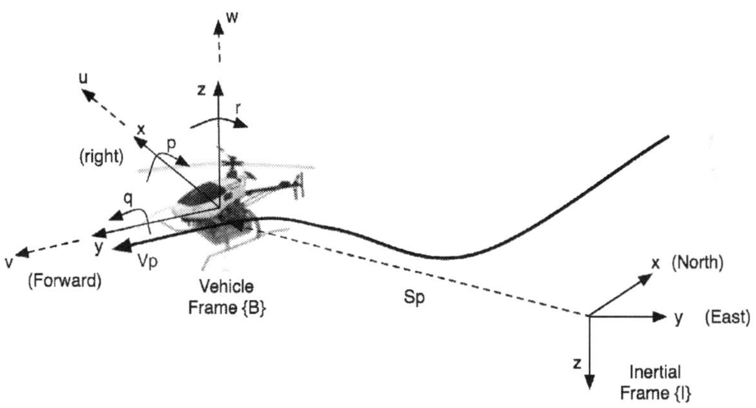

Figure 12. Frames of reference for UAV guidance and control

From Fig. 12, the following vectors can describe the motion of the UAV vehicle in 6 DOF:

$$\eta = [\eta_1, \eta_2]^T = [x, y, z, \phi, \theta, \psi]^T$$
$$v = [v_1, v_2]^T = [u, v, w, p, q, r]^T \quad (25)$$

In Eq. (25), η_1 denotes the position of the center of mass CM of the vehicle and η_2 its orientation described by the *Euler angles* with respect to the inertial frame {I}. The vector v_1 refers to the linear velocity and v_2 to the angular velocity of vehicle frame {B} with respect to inertial frame {I}. In order to express the velocity quantities between both frames of references (from {B} to {I} and vice versa), the following transformation matrix is used[3]:

$$\eta_1' = R_B^I v_1$$
$$\eta_1' = \begin{bmatrix} c\psi c\theta & -s\psi c\phi + c\psi s\theta s\phi & s\psi s\phi + c\psi c\theta s\phi \\ s\psi c\theta & c\psi c\phi + s\phi s\theta s\psi & -c\psi s\phi + s\theta s\psi c\phi \\ -s\theta & c\theta s\phi & c\theta c\phi \end{bmatrix} v_1 \tag{26}$$

The body-fixed angular velocities and the rate of the Euler angles are related through:

$$\eta_2' = \begin{bmatrix} 1 & s\phi t\theta & c\phi t\theta \\ 0 & c\phi & -s\phi \\ 0 & s\phi / c\theta & c\phi / c\theta \end{bmatrix} v_2 \tag{27}$$

The position and the magnitude of the velocity vector at a point *P* on the trajectory are given by [4]:

$$S_p = [x, y, z]^T$$
$$V_p = \|S_p'\| = \sqrt{x'^2 + y'^2 + z'^2} \tag{28}$$

The method to define the Euler angles is based on the Frenet-Serret theory (Angeles, 1997). To every point of the curve we can associate an orthonormal triad of vectors (a set of unit vectors that are mutually orthogonal) namely the tangent e_t, the normal e_n and the bio-normal e_b (see Fig. 13). The Frenet-Serret theory says that by properly arranging these vectors in a matrix $\in \Re^{3x3}$, we obtain a description of the curve orientation due to the position, velocity and acceleration of the UAV while tracing out the path. The unit vectors are then defined as:

$$e_t = \frac{S_p'}{V_p}, \quad e_b = \frac{(S_p' \times S_p'')}{\|S_p' \times S_p''\|}, \quad e_n = e_b \times e_t \tag{29}$$

In the definition of a frame associated with the curve the original definition of the Frenet frame for counterclockwise rotating curves is used; in the case of a clockwise rotating curve, the z□axis of the Frenet frame points in the opposite direction upwards than the inertial {I} frame. So in order to define small relative rotation angles for the orientation of a vehicle rotating clockwise and having its z_b axis pointing downwards, we define a reference frame

[3] where $c = \cos()$; $s = \sin()$; $t = \tan()$

[4] The terms (x,y,z) are computed due to the spline methodology previously exposed.

associated with the curve as previously, but rotated with respect to the Frenet by an angle of 180 degrees about the x–axis of the Frenet frame (see Fig. 13).

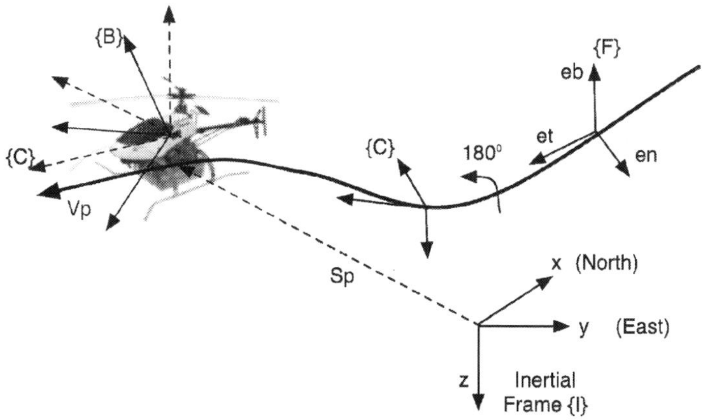

Figure 13. The inertial, the Frenet, the vehicle and the curve Frames

Collectively we denote the Frenet and the rotated frame as the "curve" frame {C}. According to the notation of rotational transformations used in robotics literature, we can express the coordinates of a vector given in the curve frame {C} to the {I} frame with the matrix:

$$R_C^I = \begin{bmatrix} e_t & e_n & e_b \end{bmatrix}$$
$$R_I^C = R_C^{I\,T} \tag{30}$$

For a counterclockwise rotation: $R_C^I = R_x\left(180^\circ\right)\begin{bmatrix} e_t & e_n & e_b \end{bmatrix}$. Likewise, the rotation of the {B} frame from the {C} frame to the reference {R} frame can be expressed using customary aeronautical notation by considering the sideslip angle β and angle of attack α, (Siouris, 2004):

$$\beta = \sin^{-1}\left(\frac{v_R}{V_p}\right)$$
$$\alpha = \tan^{-1}\left(\frac{w_R}{u_R}\right) \tag{31}$$

The vector v_R refers to the y-axis velocity component in the reference frame and $w_{R,}u_R$ to the z and x - axis respectively. The overall rotation is composed by a rotation about body z_B axis through the angle β, followed by a rotation about the body y_B through the angle α, which is expressed as:

$$R_C^R = R_y^T(\alpha)R_z^T(-\beta) \tag{32}$$

Finally, the *roll, pitch* and *yaw* angles can be deduced as follows:

$$
\begin{aligned}
R_I^R &= R_C^R R_I^C \\
\phi &= \operatorname{atan2}(r_{23}, r_{33}) \\
\theta &= \operatorname{atan2}\left(-r_{13}, \sqrt{r_{23}^2 + r_{33}^2}\right) \\
\psi &= \operatorname{atan2}(r_{12}, r_{11})
\end{aligned}
\tag{33}
$$

Where $r_{i,j}$ represent the components of the rotation matrix $R_I^R \in \Re^{3x3}$. Computing the previous methodology, we use 3D splines with fixed boundary conditions in order to generate a complicated path as shown in Fig. 11. Seven knot-points have been used (distance are in meters): P=[0 0 0; 5 1 2; 10 5 5; 15 10 10; 10 15 15; 5 10 20; 0 8 20; 5 0 20]. Eq. (33) has been used to obtain the UAV orientation with respect to the Inertial Frame as:

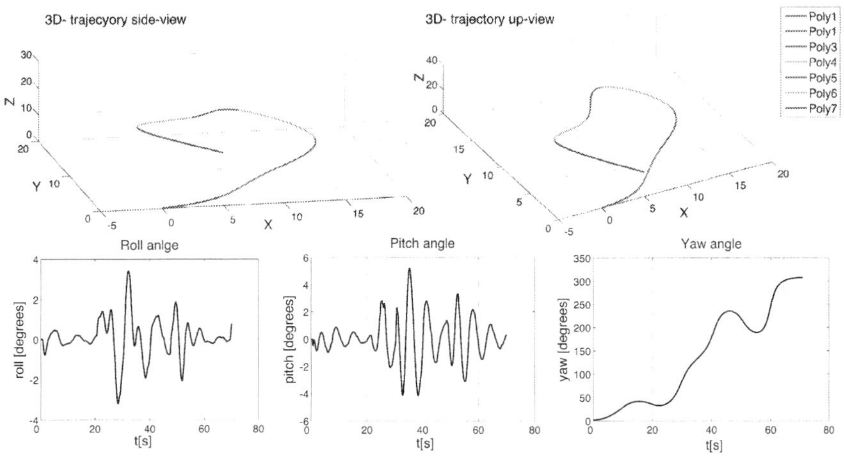

Figure 14. Roll, Pitch and Yaw angle references for UAV guidance

For complex maneuvers the Frenet theory allowed to define smooth *yaw* references as well as *roll* and *pitch* angles if it is required. Nevertheless, the computational cost of calculating those equations could decrease the system performance if the number of knot-points of the path is large. For this reason, normal maneuvers such as straight-lines use simple trigonometric theory to obtain the UAV orientation. On the other hand, the end-user is able to define the kind of orientation of the vehicle, this means that the vehicle is not constrained to be oriented just by the trajectory direction, hence, it will be able to trace out the trajectory oriented towards to any fixed point defined.

The *yaw* angle defined by the term ψ is given by:

$$
\psi = \tan^{-1}\left(\frac{y_{diff}}{x_{diff}}\right) \tag{34}
$$

Where x_{diff}, y_{diff} correspond to the difference between the target fixed point and the current position of the UAV. In addition, depending of the motion quadrant, the term ψ in Eq. (34) must be fixed, this means that for the {x-,y+} and the {x-,y-} quadrant, the *yaw* angle is $\psi = \psi + \pi$, otherwise, for the {x+,y-} quadrant, $\psi = \psi + 2\pi$.
Figure 15 shows a circumference-arc generated using Eq. (24) as well as the *yaw* evolution of the UAV oriented towards the center of the arc located at [0,0] coordinate in the *x-y* plane using the previosly theory described in Eq. (34).
This section has successfully introduced the mathematical treatment and methods for the generation of complex trajectories and simple maneuvers using the available theory reported in specialized literature (LaValle S.M, 2006), (Jaramillo-Botero et al., 2004). Geometrical trajectory generation and some techniques for its parameterization based on polynomial splines and function with trapezoidal velocity profile are an interest solution for this problem, actually, some of these methodologies are used for complex UAV trajectory definition nowadays. The novel solution presented in this book is the integration of these methods into a powerful environment that allows high-level user control of complex missions. The AVCL definitively brings those features and the next section will introduce some tests using the AVCL interpreter and the simulation environment.

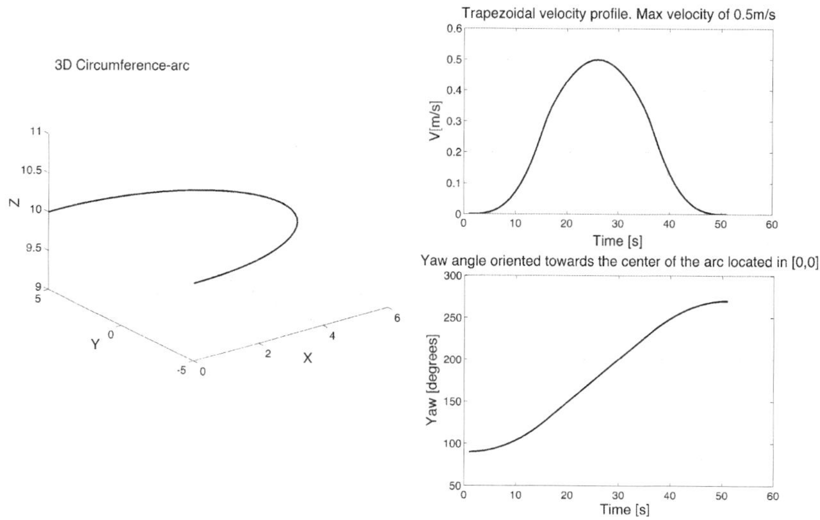

Figure 15. Circumference 3D motion, trapezoidal velocity profile and *yaw* angle evolution

4. TG^{2}M Simulation Results

As shown in Fig. 1 the Mission Planner (MP) has two similar loops for mission planning and simulation/supervision. The difference is that in the Planning Loop the interpreter sends the projected waypoints back to the MP's Enhanced Reality, while in the Simulation Loop the interpreter commands the simulated vehicle, which in turn sends the simulated positions to the MP. Our research group has developed a Simulink-based model of a UAV helicopter named *Vampira*, which includes a position controller and is capable of real-time simulation. This simulator has been used with the Mission Planning and Simulation tool to test the TG²M. For Mission Supervision the AVCL commands would be sent to the real vehicle, and its position plotted alongside the projected and/or simulated paths. The *Vampira* helicopter was built within the framework of the project: "Guidance and Control of an Unmanned Aerial Vehicle" DPI 2003 01767 (VAMPIRA), granted by the Spain Ministry of Science and Technology, and it will be used for the real-world tests of the built-in TG²M framework. Figure 16 shows the *Vampira* prototype, which includes: a GPS, Wi-Fi link, IMU, and a PC104 computer for the low-level control (main rotor and tail servos). The Vampira's dynamics model has been obtained, identified and validated in previous works (Valero, 2005), (del Cerro et al., 2004). This work takes advantage from the AVCL simulation capabilities in order to validate the TG²M framework theory for trajectory planning.

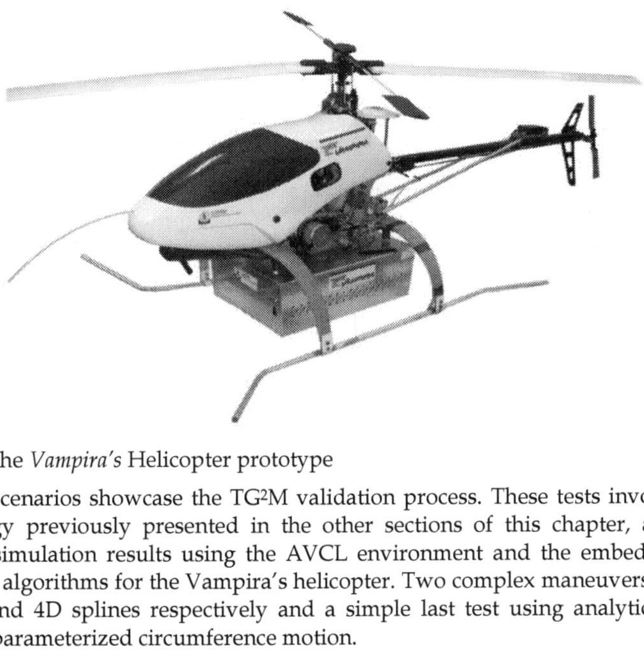

Figure 16. The *Vampira's* Helicopter prototype

Three test scenarios showcase the TG²M validation process. These tests involve the whole methodology previously presented in the other sections of this chapter, as well as the numerical simulation results using the AVCL environment and the embedded dynamics and control algorithms for the Vampira's helicopter. Two complex maneuvers are presented using 3D and 4D splines respectively and a simple last test using analytical function to generate a parameterized circumference motion.

1). *Semi-spiral using 3D splines for the velocity profile generation and the Frenet theory for UAV orientation:* In this first test, we used a 3D spline to joint three knot control points: (P_0(0, 0, 0), P_1(3, 5, 10), P_2(6, -7, 20)) at the desire time (given in seconds) for each point: (t(0, 10, 20)) and the desire initial and final speed (given in m/s): (V_0(0, 0, 0), V_f(0, -0.2, 0.4)):

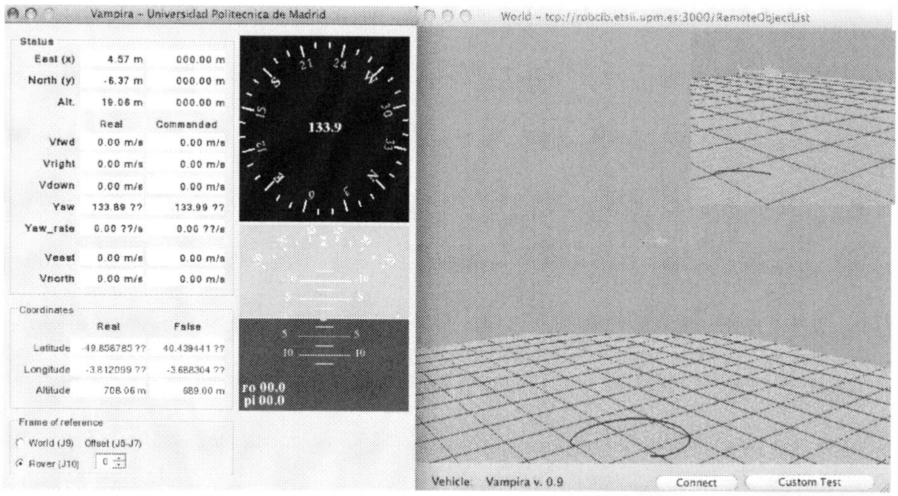

Figure 17. The AVCL simulation environment: Vampira's helicopter executing a semi-spiral motion using 3D splines

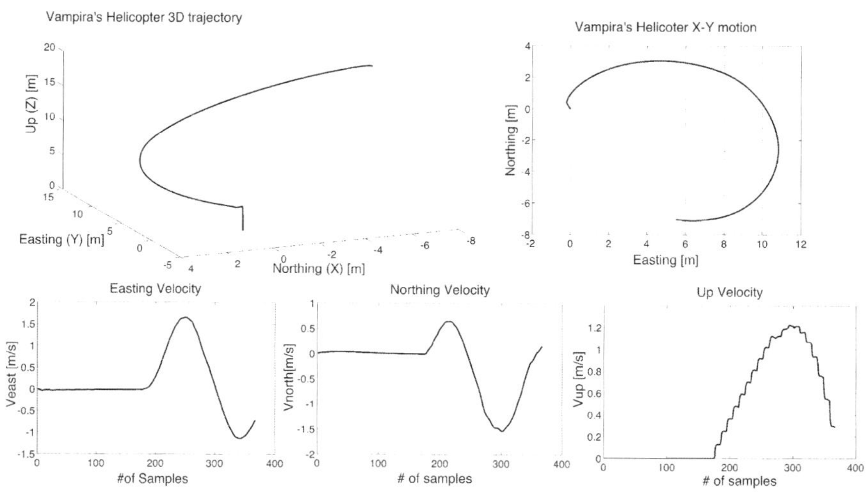

Figure 18. Test1: Cartesian UAV position and velocities with respect to the Inertial Frame

The UAV started from initial point located at (0, 0, 0) coordinate and finished its trajectory at (6, -7, 20). Visual simulation depicted in Fig 17, showed smooth motion across the trajectory due to the 3D spline approach. Nonetheless, 3D splines just allow the user to define the initial and final velocities of the motion, lacking of velocity control for the rest of the knot-points.

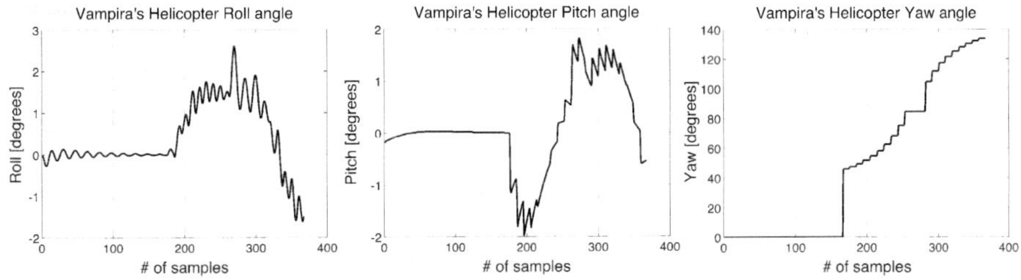

Figure 19. Test1: UAV orientation (Euler angles evolution)

To solve this problem, the following test introduces a more complex trajectory generation using 4D splines, addressing total user control of the UAV velocity profile.

2). *Complex trajectory using 4D splines for the velocity profile generation and the Frenet theory for UAV orientation.* This trajectory includes different kind of maneuvers joined into a single polynomial function (take-off, circumference-type motion and slow down in spiral-type motion). This test includes UAV long-endurance to high altitude (150 meters above ground) and a maximum easting displacement about of 60 meters. The following knot-control points (given in meters) have been defined: (P_0(0, 0, 0), P_1(0, 0, 20), P_2(0, 0, 40), P_3(0, 0, 60), P_4(10, 2, 80) , P_5(20, 4, 110) , P_6(25, -7, 130) , P_7(30, -10, 150), P_8(35, -5, 140) , P_9(30, 16, 125) , P_{10}(20, 5, 130) , P_{11}(33, -10, 145) , P_{12}(40, -5, 135) , P_{13}(55, -6, 125)):

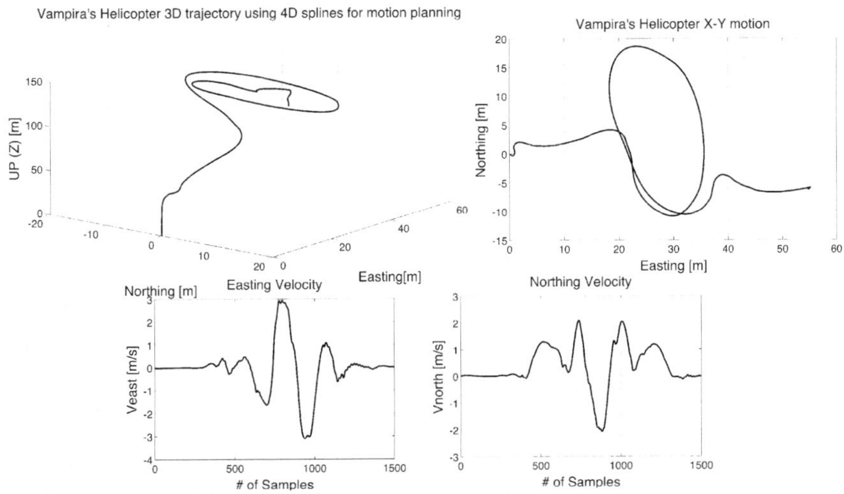

Figure 20. Test2: smooth 4D spline for complex maneuve

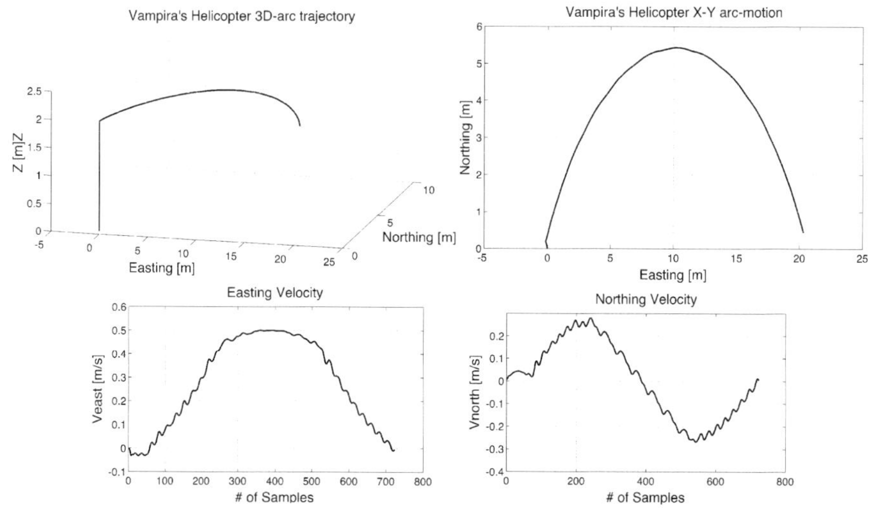

Figure 21. Test3: Arc-type motion using analytical functions

The advantage about using 4D splines relies in the possibility of defining feasible paths that matches with the knot-control points defined (with less match error percentage than the 3D polynomial splines). In addition, the user is able to define the set of velocities for each of the knot-points during the motion. The set of velocities (given in m/s) are: (V_0(0, 0, 0), V_1(0, 0, 0.8), V_2(0, 0, 1), V_3(0, 0, 1.2), V_4(0.5, 1, 1.4) , V_5(0, 0.5, 1.7) , V_6(2, 1.5, 2.5), V_7(3, 2.2, 3), V_8(2, 1.2, 2) , V_9(1, 0.5, 1.5) , V_{10}(-0.5, -1, 0.8) , V_{11}(-3, -2, 0.4) , V_{12}(-1, -0.5, 0.8), V_{13}(0, 0, 0)).

3). *Simple arc-type maneuver with trapezoidal velocity profile parameterization:* for the analytical AVCL feature of trajectory planning, the TG^{2}M module supports straight-lines and circumferences motions. An arc defined by: P_0(0, 0, 2), P_1(10, 5.5, 2), P_2(20, 0, 2)) with a maximum velocity of 0.5m/s is tested using the AVCL interpreter that allows the user to define the trapezoidal velocity profile configuration. For this case, the acceleration slopes of the curves (see Fig. 21) have been set to the 30% of the total motion.

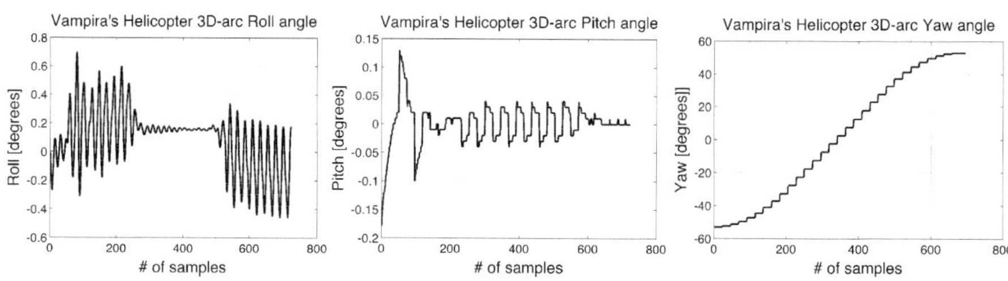

Figure 22. Test3: UAV orientation (Euler angles evolution)

5. Final Observations

For modeling continuous cartesian trajectories in the AVCL, several analytical functions and polynomial interpolation methods are available; all of which can be used in any combination. The TG^2M module handles the definition of trajectories using knot control points as well as the incorporation of path constraints. It also supports the definition of complex tasks that require the construction of trajectories made up of different primitive paths in any combination of analytical and interpolated functions. The user-designed spatial trajectories can be visualized in three dimensions on the display window or plotted versus time using the embedded plotting library.

Simulation results have shown that the TG^2M module works perfectly for the definition and testing of wide kind of smooth trajectories, allowing the user a high-level control of the mission due to the AVCL interpreter. The three different scenarios used for testing, allowed verifying that the mathematical framework used for the trajectory generation and guidance was really working during simulation flight. Percentage errors during maneuver execution were minimal, maintaining the UAV at the desired velocity limits and within the established path. We also incorporated velocity error fixing during flight. For high altitude tests, the velocity of the wind plays a mandatory role as a main disturbance external force. The TG^2M module includes wind perturbation compensation. The Guidance module fixes the velocity commands in real-time flight maneuver, decreasing the error position tracking. For the three scenarios tests, the AVCL simulation environment includes normal wind conditions during simulation flight, introducing small perturbations into the UAV equations of motion. As shown in the obtained results, those perturbations were compensated, allowing the UAV to follow the desired trajectory within the less error as possible.

The Frenet-Serret formulas included for the UAV orientation also presented a good approach in order to obtain smooth UAV rotation rate during flight. The use of simple trigonometric theory to obtain and define the UAV orientation profile (*Yaw angle*) is not convenient for complex maneuvers. Splines sometimes require a lot of know-points for feasible trajectory guidance, hence, using these polynomial equations, the Frenet approach allowed smooth angle changes between knot-points, which it had not been obtained with the simple trigonometric angle calculation.

6. References

JAA & Eurocontrol,.A concept for the European Regulations for Civil Unmanned Aerial Vehicles. *UAV Task-Force Final Report.* 2004

Coifman, B., McCord, M., Mishalani, M., Redmill, K., Surface Transportation Surveillance from Unmanned Aerial Vehicles. *Proc. of the 83rd Annual Meeting of the Transportation Research Board,* 2004.

Held, Jason M; Brown, Shaun and Sukkarieh, Salah. Systems for Planetary Exploration. *I5th NSSA Australian Space Science Conference,* pp. 212-222, ISBN: 0864593740. RMIT University, Melbourne, Australia, from 14 to 16 September 2005.

Gutiérrez, P., Barrientos, A., del Cerro, J., San Martin., R. Mission Planning and Simulation of Unmanned Aerial Vehicles with a GIS-based Framework; *AIAA Guidance,* Navigation and Control Conference and Exhibit. Denver, EEUU, 2006.

Rysdyk, R. UAV path following for constant line-of-sight. In 2th AIAA Unmanned Unlimited. *Conf. and Workshop and Exhibit*, San Diego, CA, 2003.

Price, I.C, Evolving self organizing behavior for homogeneous and heterogeneous swarms of UAVs and UCAVS. *PhD Thesis*, Graduate School of Engineering and Management, Air Force Institute of Technology (AU), Wright-Patterson AFB, OH, March 2006.

Herwitz, S. Developing Military COEs UAV applications. *UAV Applications Center - NASA Ames Research Center*, Moffett Field, CA. 2007.

Alison A. P., Bryan G., Suresh K. K., Adrian A. K., Henrik B. C. and Eric N. J. Ongoing Development of an Autonomous Aerial Reconnaissance System at Georgia Tech . *UAV Laboratory*, School of Aerospace Engineering. Georgia Institute of Technology, 2003

Jaramillo-Botero A., Correa J.F., and Osorio I.J., Trajectory planning in ROBOMOSP, *Robotics and Automation Group* GAR, Univ. Javeriana, Cali, Colombia, Tech. Rep. GAR-TR-10-2004, 2004.

LaValle, S.M.: *Planning Algorithms*. Cambridge University Press (2006)

Moitra, A., Mattheyses, R.M., Hoebel, L.J., Szczerba, R.J., Yamrom, B.: Multivehicle 5reconnaissance route and sensor planning. *IEEE Transactions on Aerospace and Electronic Systems*, 37 (2003).

Zheng, C., Li, L., Xu, F., Sun, F., Ding, M.: Evolutionary Route Planner for Unmanned Air Vehicles. *IEEE Transactions on Robotics* 21 (2005) 609–620

Frazzoli, E. Maneuver-based motion planning and coordination for single and multiple UAVs. *AIAA's 1st technical conference and workshop on unmanned aerospace vehicles*. University of Illinois at Urbana-Champaign, il 61801. S. Portsmouth, Virginia. May 2002.

Angeles J., Fundamentals of Robotic Mechanical Systems. *Theory, Methods, and Algorithms*. Springer, New York, 1997.

Siouris, G. M. Missile Guidance and Control Systems. Springer, New York, 2004.

Valero, Julio. Modelo dinámico y sistema de supervisi6n de vuelo de un helic6ptero aut6nomo. *Proyecto de fin de carrera*. ETSIIM-UPM. 2005.

5

Modelling and Control Prototyping of Unmanned Helicopters

Jaime del-Cerro, Antonio Barrientos and Alexander Martínez
Universidad Politécnica de Madrid – Robotics and Cybernetics Group
Spain

1. Introduction

The idea of using UAV's (Unmanned Aerial Vehicles) in civilian applications has created new opportunities for companies dedicated to inspections, surveillance or aerial photography amongst others. Nevertheless, the main drawback for using this kind of vehicles in civilian applications is the enormous cost, lack of safety and emerging legislation. The reduction in the cost of sensors such as Global Positioning System receivers (GPS) or non-strategic Inertial Measurement Units (IMU), the low cost of computer systems and the existence of inexpensive radio controlled helicopters have contributed to creating a market of small aerial vehicles within an acceptable range for a wide range of applications. On the other hand, the lack of safety is mainly caused by two main points: Mechanical and control robustness.

The first one is due to the platform being used in building the UAV in order to reduce the cost of the system, which is usually a radio controlled helicopter that requires meticulous maintenance by experts.

The second is due to the complexity of the helicopter dynamics since it is affected by variations in flying altitude, weather conditions and changes in vehicle's configuration (for example: weight, payload or fuel quantity). These scenarios disrupt the modeling process and, consequently, affect the systematic development of control systems, resulting to tedious and critical heuristic adjustment procedures.

Researchers around the World propose several modeling techniques or strategies for dynamic modeling of helicopters. Some works on helicopter modeling such as (Godbole et al or Mahony et al,2000), (Gavrilets et al, 2001), (Metler et al and La Civita et al, 2002), and (Castillo et al, 2007) show broad approaches that have been done in this field of engineering. The lack of an identification procedure in some cases and the reduced field of application in others, make sometimes difficult to use them.

In this chapter, not only a modeling is described, but also the identification procedure that has been successfully tested. The proposed model has been defined by using a hybrid (analytical and heuristic) algorithm based on the knowledge of flight dynamic and by resolving some critical aspects by means of heuristic equations that allow real time simulations to be performed without convergence problems. Evolutionary algorithms have been used for identification of parameters. The proposed model has been validated in the different phases of the aircraft flight: hovering, lateral, longitudinal or vertical using a

Benzin Trainer by Vario which relies on a 1.5 m of main rotor diameter and a payload of about five kilograms.

Figure 1. Benzin trainer by Vario with GNC (Guidance Navigation & Control) system onboard

A full structure of control has been developed and tested by using a proposed fast design method based on Matlab-Simulink for simulation and adjustment in order to demonstrate the usefulness of the developed tool. The proposed architecture describes low level control (servo actuators level), attitude control, position, velocity and maneuvers. Thus there are commands such as straight flying by maintaining a constant speed or maneuvers such as flying in circles i.e.

The results have confirmed that hybrid model with evolutionary algorithms for identification provides outstanding results upon modeling a helicopter. Real time simulation allows using fast prototyping method by obtaining direct code for onboard controllers and consequently, reducing the risk of bugs in its programming.

2. Helicopters flight principles

The required thrust for flying helicopters (elevation and advance) is only provided by the main rotor. In order to obtain a comprehensive knowledge about the forces involved in the main rotor, an exhaustive explanation would be required. The forces involved in the main rotor generate twisting and bending in the blades, as well as elevation forces and different kinds of aerodynamic resistance.

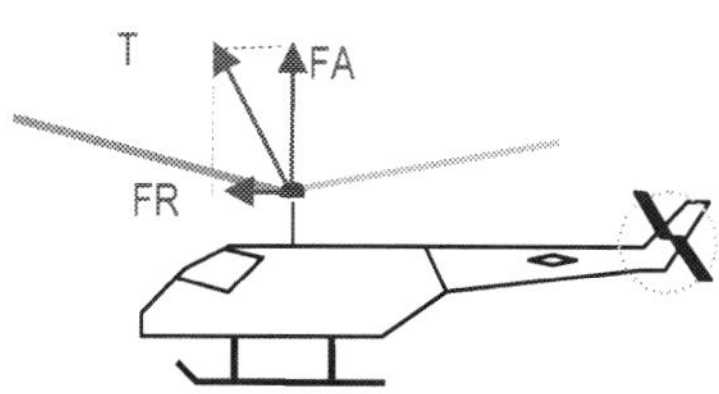

Figure 2. Basic Helicopter Thrust analysis

In a first approach, only a sustentation and a resistance force isolated from any other influence could be taken into account. The thrust (T) generated by the air pressure against the blade has an inclination in relation to the rotation plane. This force can be divided in vertical force (FA) and resistance (FR) that is applied in the horizontal plane against rotation direction (Figure 2).

Helicopters rely on mechanisms to modify the attack angle of the blade in the main rotor. It allows controlling the movement of the fuselage through the inclination of the rotation plane. The fact is that the attack angle of the blade is not constant neither in time nor space. It continuously changes while the blade is rotating as azimuth angle indicates. It can be assumed that the attack angle is the addition of two components: The first is an average attack angle during one complete rotation of the blade, called collective angle. The second component depends on the azimuth angle. When the azimuth angle is 0° or 180°, the blade has the roll cyclic angle. When 90 ° or 270 °, the blade has the pitch cyclic angle (Figure 3).

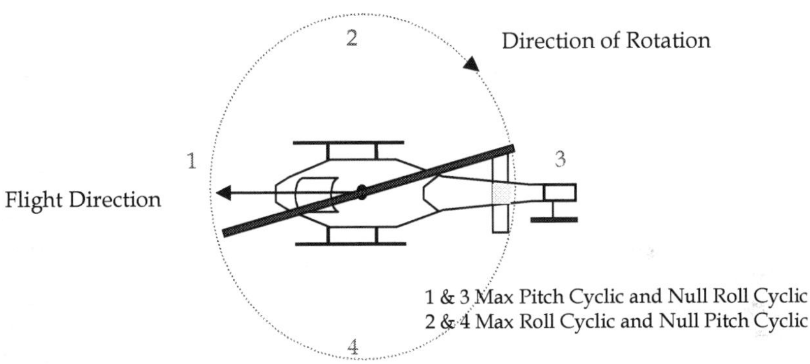

Figure 3. Attack angle during a blade revolution

Using these three signals (collective, roll and pitch cyclic), a pilot is able to control the main rotor. In addition to these signals, the pilot also controls the attack angle of tail rotor blades and the engine throttle.

Typically, radio-controlled helicopters rely upon a commercial control system to maintain the speed of the main rotor constant. The vertical control of the helicopter is done by changing the collective attack angle in the main rotor.

On other hand, the mission of the tail rotor is to compensate the torque that main rotor creates on the helicopter fuselage. The compensation torque can be adjusted by changing the attack angle of its blades. The tail rotor typically requires values between 5 to 30 per cent of the total power of the helicopter.

A lot of physical principles such as ground effect, downwash, flapping or atmospheric effects have to be considered in an in-depth study of helicopter flight dynamics. Taking into account the application scope for the proposed model, which is a small helicopter with small capabilities, no change of air density is considered. Moreover, the blades of the proposed model are also considered as solid bodies.

3. Model Description

Mathematical models of helicopter dynamics can be either based on analytic or empirical equations. The former applies to the physical laws of aerodynamics, and the latter tries to define the observed behavior with simpler mathematical expressions.
The proposed model is a hybrid analytic-empirical model that harnesses the advantages of both: high-velocity and simplicity of the empirical method, as well as fidelity and the physical meaning of the analytic equations.

3.1 Inputs and outputs

The proposed model tries to replicate the behavior of a small radio controlled helicopter, therefore the inputs and outputs have been selected as the controls that the pilot relies when using a commercial emitter. Cyclic (Roll and Pitch) and collective controls have been considered as inputs.

Denomination	I/O	Symbol	Units
Collective	Input	θcol	Degrees
Roll Cyclic	Input	θRoll	Degrees
Pitch Cyclic	Input	θPitch	Degrees
Rotational speed over a main rotor shaft.	Input	ω_z^h	Degrees/s
Acceleration (Helicopter reference frame)	Output	$\overline{a}$	m/s2
Velocity (Helicopter reference frame)	Output	$\overline{v}$ (ua,va,wa)	m/s

Table 1. Model inputs and outputs

Radio controlled helicopters usually rely on electronic systems based on gyroscopes to tail stabilization. Based on this fact, the yaw angle is controlled by giving rotational speed commands. Thus, the yaw rate has been considered as one of the inputs to the model.
It is also common that helicopters rely on a main rotor speed hardware controller. In such manner, the rotor maintains the speed and consequently, the vertical control is then performed by the changing of the collective angle. The assumption in considering constant speed reduces the complexity of the model and maintains its quality. Furthermore, the use of this hardware controller decreases the number of inputs since no throttle command is required.
Accelerations and rotational speeds are the outputs of the model. Table 1 summarizes the inputs and outputs of the model.

3.2 Model block diagram

A block diagram of the proposed model is described in Figure 5. A brief definition of every part is also shown in the following sections.

3.2.1 Main rotor

It is modeled with analytic equations, derived form research on full-scale helicopters [Heffley 2000], which can be used to model small helicopters without fly-bars. The iterative algorithm computes thrust and induced wind speed while taking into account geometrical parameters, speed of the blades, commanded collective, roll and pitch cyclic.

One of the most important concepts to be known is the relationship among the Torque, Power and Speed. Such response arises from the induced velocity of the air passing through the disk of the rotor.

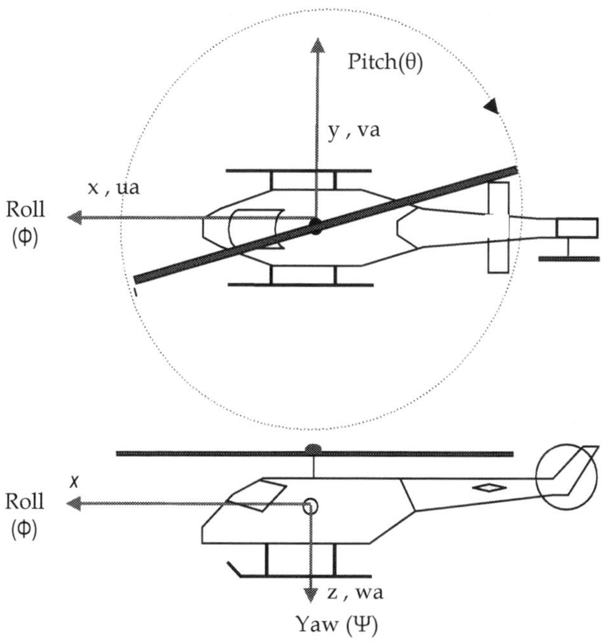

Figure 4. Outputs of the model

The airflow goes downward, and due to the action-reaction principle, it generates a vertical force that holds the helicopter in the air. The engine provides the torque to make the blades rotate and create the airflow.

Although there are many factors that make difficult to exactly determination of the relative speed between the helicopter and the airflow through the disk that the rotor creates when rotating, it is possible to work with a first order approach. In this way, it is possible to model using the momentum classic theory and estimating the force and induced velocity using an feedback aerodynamic block. Nevertheless, calculus turns to be difficult because the feedback is highly non-linear.

Another aspect regarding the induced speed is its influence on the surrounding surfaces, thus it can be affected by the ailerons and others fuselage parts and it changes depending on the speed and direction of the flight. In this model, torque and induced speed have been modeled assuming a uniform distribution of the air passing through the rotor disk.

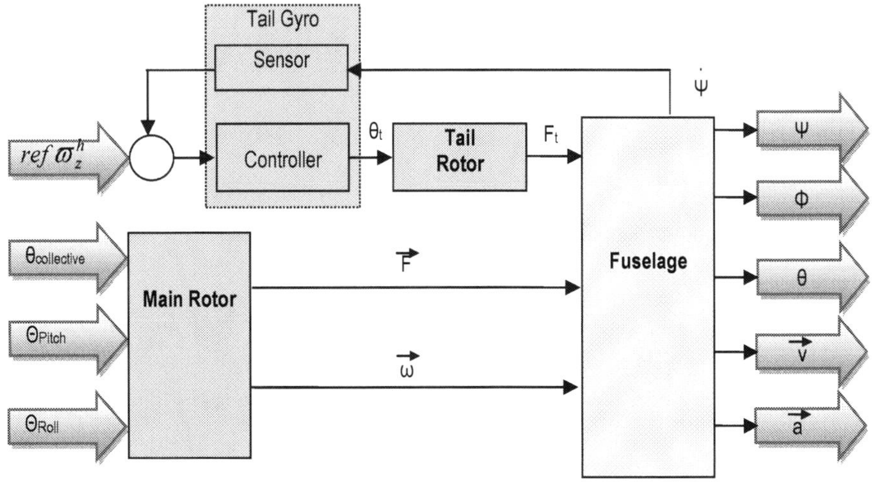

Figure 5. Model Block Diagram

The computation of the torque and induced speed is based on the classic momentum theory but using a recursive scheme that allows to reach a fast convergence. The equations used to model the main rotor have two groups of inputs, as Figure 6 shown.

The collective step (θ_{Col}) and the blade torsion (θ_{Twist}), compose the first branch. The rotor axis inclination (Is) and the cyclic roll and pitch the second one.

The attack angles a1 and b1 are derived from the cyclic roll and pitch references. The wind speed relative to the helicopter is present through its three components: ua, va, and wa.

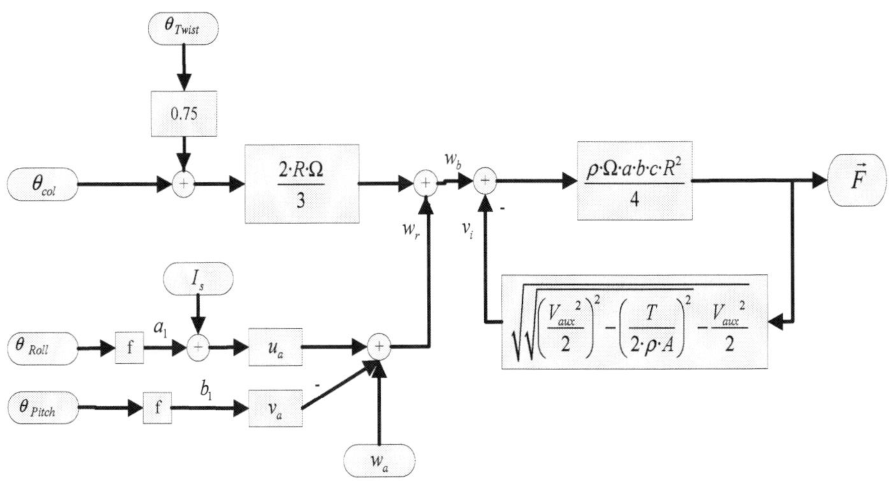

Figure 6. Main rotor block diagram

The equations corresponding to this part are:

$$w_r = w_a + (a_1 + I_s)u_a - b_1 \cdot v_a \tag{1}$$

$$w_b = w_r + \frac{2 \cdot \Omega \cdot R}{3}(\theta_{col} + \frac{3}{4}\theta_{Twist}) \tag{2}$$

The output of the block is the rotor's thrust (F). R and Ω represent the radius of the rotor and its angular speed respectively; ρ is the air density, and a, b and c are geometrical blade factors. The relationship between thrust and angular rates has been derived from observation; therefore the model is also empirical.

$$V_{aux}^2 = u_a^2 + v_a^2 + w_r(w_r - 2v_i) \tag{3}$$

$$T = (w_b - v_i)\frac{\rho \cdot \Omega \cdot R \cdot a \cdot b \cdot c \cdot R}{4} \tag{4}$$

$$v_i^2 = \sqrt{(\frac{V_{aux}^2}{2})^2 - (\frac{T}{2 \cdot \rho \cdot \Pi \cdot R^2})^2} - \frac{V_{aux}^2}{2} \tag{5}$$

When the helicopter flies close to the ground (distances less than 1.25 times the diameter of the rotor) the ground effect turns to be very important. This effect has been modeled using the parameter η defined in (6) where h is the distance from the helicopter to the ground.

$$\eta = \frac{h}{2R} \tag{6}$$

In these cases, the thrust is modified using (7) where Th' is the resulting thrust after correcting T_h. Values of To, T1 and T2 have been calculated for no creating a discontinuity when h is 1.25 times the diameter of the rotor.

$$T_h' = T_h\left(T_0 + T_1\eta + T_2\eta^2\right) \tag{7}$$

By other hand, the commanded Roll and Pitch cyclic have been considered as reference for rotations in x and y axes considering the helicopter frame. In addition to this, a coupling effect has been considered for simulating real coupling in the helicopter.

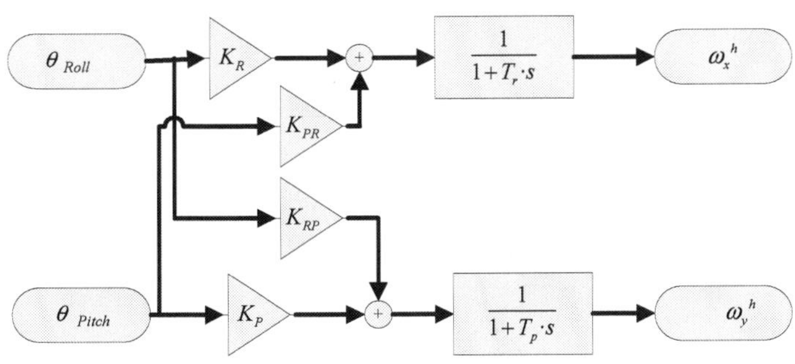

Figure 7. Roll and Pitch dynamic

As it was mentioned in the last section, RC helicopters usually rely upon commercial speed controllers in the main rotor. These devices have been modeled using an ideal dynamic response. On the other hand, engine has been modeled as a first order system. Therefore, variations in the speed of the rotor have been considered only due to changes in the collective angle as
Figure 8 shows.

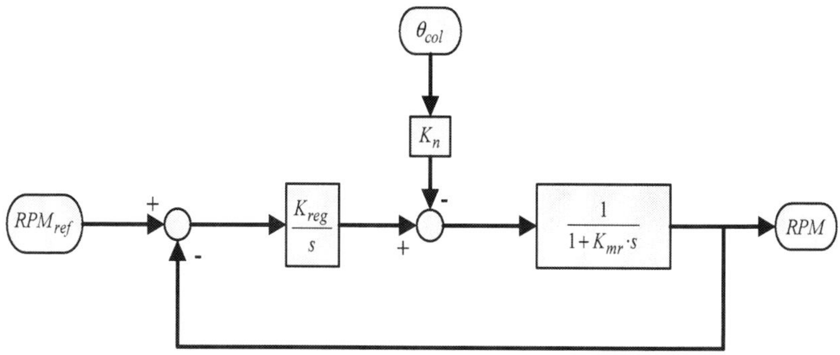

Figure 8. Engine model

3.2.2 Tail rotor

The algorithm for estimating the thrust provided by tail rotor is similar to the main one but only the pitch angle of the blades has been considerated as input. This signal is provided by the hardware controller that is in charge of stabilization of the tail. A PI classical controller has been used to model the controller and the sensor has been considered as no dynamic as Figure 9 shows.

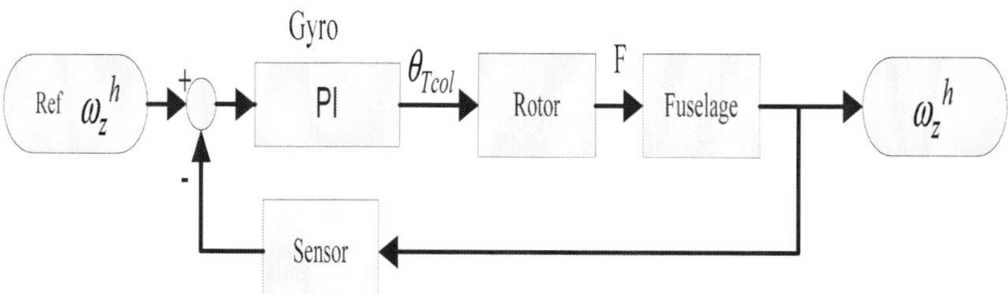

Figure 9. Tail rotor model

3.2.3 Fuselage

In a first step, all the forces (main and tail rotor, gravity, friction and aerodynamic resistances) have to be taking into account for computing the movement of the fuselage of the helicopter. After that, accelerations and velocities can be estimated.
Forces due to the aerodynamic frictions are estimated using the relative velocity of the helicopter with respect to the wind applying (8), where A_r is the equivalent area and v_w is the wind velocity.

$$F = \frac{1}{2} A_r \rho v_w{}^2 \tag{8}$$

The resulting equations are summarized in (9).

$$\begin{aligned} F_x^h &= G_x^h + Fus_x^h \\ F_y^h &= G_y^h + Fus_y^h \\ F_z^h &= G_z^h + Fus_z^h + T^h \\ M_z^h &= M_d^h + M_t^h + M_{vt}^h \end{aligned} \tag{9}$$

Where the prefix Fus means aerodynamic forces on fuselage and T the main rotor thrust. The prefix M means Torque where suffix d denotes main rotor, the t denotes tail rotor and the v is the effect of the air in the tail of the helicopter. The prefix G means the gravity components.

Once forces and torques have been calculated, accelerations can be obtained and therefore velocities. Once the velocity referred to helicopter frame (v^h) is calculated, a transformation is required in order to obtain an absolute value by using (10).

$$\begin{bmatrix} v_x \\ v_y \\ v_z \end{bmatrix} = \begin{bmatrix} \cos\theta\cos\psi & -\cos\phi\sin\psi+\sin\theta\sin\phi\cos\psi & \sin\phi\sin\psi+\cos\phi\sin\theta\cos\psi \\ \cos\theta\sin\psi & \cos\phi\cos\psi+\sin\theta\sin\phi\sin\psi & -\sin\phi\cos\psi+\cos\phi\sin\theta\sin\psi \\ -\sin\theta & \sin\phi\cos\theta & \cos\phi\cos\theta \end{bmatrix} \cdot \begin{bmatrix} v_x{}^h \\ v_y{}^h \\ v_z{}^h \end{bmatrix} \tag{10}$$

Flybars are important elements to take into account in the dynamic model. In this work, they have been modeled by using empiric simple models, and the stabilization effect that they produced on the helicopter has been simulated using (11). A more realistic model of flybars can be obtained in [Mettler-2003].

$$\begin{aligned} \dot{\phi}' &= \dot{\phi} - k_1\phi \\ \dot{\theta}' &= \dot{\theta} - k_2\theta \end{aligned} \tag{11}$$

3.3 Identification of the parameters

Once the mathematic equations have been obtained, a procedure to give values to the parameters that appear into the model is required. Some of the parameters can be easily obtained by simple measurements or weighs but some of them turn to be very difficult to obtain or estimate.

Table 2 describes the parameter list to be identified.

Some methods have been studied for performing the parameters identification, such as multi-variable systems procedures (VARMAX), but they are difficult to apply because there is an iterative process into the model. Due to this, the selected method was evolutionary algorithms, after trying unsuccessfully stochastic methods.

Genetic algorithms may be considered as the search for a sub-optimal solution of a specific cost-based problem. The parameters are codified as chromosomes that are ranked with a fitness function, and the best fits are reproduced through genetic operators: crossover and mutation. This process continues until the fitness of the best chromosome reaches a preset threshold.

#	Name	Meaning
1	TwstMr	Main rotor blade twist.
2	TwstTr	Tail rotor blade twist.
3	KCol	Collective step gain.
4	IZ	Moment of inertia around the z axis.
5	HTr	Vertical distance from tail rotor to centre of mass of the helicopter.
6	WLVt	Vertical position of the aerodynamic centre of the tail.
7	XuuFus	Frontal effective area of the helicopter.
8	YvvFus	Lateral effective area of the helicopter.
9	ZwwFus	Effective area of the helicopter
10	YuuVt	Frontal area of the tail.
11	YuvVt	Lateral area of the tail.
12	Corr1	Correction parameter for Roll.
13	Corr2	Correction parameter for Pitch.
14	Tproll	Time constant for Roll response.
15	Tppitch	Time constant for Pitch response.
16	Kroll	Gain for Roll input.
17	Kpitch	Gain for Pitch input.
18	Kyaw	Gain for Yaw input.
19	DTr	Horizontal distance from centre of the tail rotor to mass centre of the helicopter.
20	DVt	Horizontal distance from the aerodynamic centre of the tail and mass centre of the helicopter.
21	YMaxVt	Saturation parameter (no physical meaning).
22	KGyro	Parameter of commercial gyro controller. (gain)
23	KdGyro	Parameter of commercial gyro controller. (derivative)
24	Krp	Cross gain for Roll and Pitch coupling.
25	Kpr	Cross gain for Pitch and Roll coupling.
26	OffsetRoll	Offset of Roll input (trimmer in radio transmitter).
27	OffsetPitch	Offset of Pitch input (trimmer in radio transmitter).
28	OffsetCol	Offset of Collective input (trimmer in radio transmitter).

Table 2. Parameter to identify list

The main steps of the process for identification using GA's have been:

3.3.1 Parameters codification

The parameters are codified with real numbers, as it is more intuitive format than large binary chains. Different crossover operations have been studied:

- The new chromosome is a random combination of two chains.

Asc1	0	1	2	3	4	5	6	7	8	9
Asc2	10	11	12	13	14	15	16	17	18	19
Desc	0	11	12	3	14	5	6	17	18	9

- The new chromosome is a random combination of random genes with values in the range defined by the ascendant genes.

Asc1	0	1	2	3	4	5	6	7	8	9
Asc2	10	11	12	13	14	15	16	17	18	19
Desc	6.5	7.6	2	13	4	15	9.2	8.5	8	19

The first operator transmits the positive characteristics of its ascendants while the second one generates diversity in the population as new values appear. In addition to the crossover operator there is a mutation algorithm. The probability of mutation for the chromosomes is 0.01, and the mutation is defined as the multiplication of random genes by a factor between 0.5 and 1.5.

When the genetic algorithms falls into a local minimum, (it is detected because there is no a substantial improvement of the fitness in the best chromosome during a large number of iterations), the probability of mutation have to be increased to 0.1. This improves mutated populations with increased probability of escaping from the local minimum.

3.3.2 Initial population

The initial population is created randomly with an initial set of parameters of a stable model multiplied by random numbers selected by a Monte-Carlo algorithm with a normal distribution with zero mean and standard deviation of 1.

The genetic algorithm has been tested with different population sizes, between thirty and sixty elements. Test results showed that the bigger population did not lead to significantly better results, but increased the computation time. Using 20 seconds of flight data and a population of 30 chromosomes, it took one hour to process 250 iterations of the genetic algorithm on a Pentium IV processor. Empirical data suggests that 100 iterations are enough to find a sub-optimal set of parameters.

3.3.3 Fitness function

The fitness function takes into consideration the following state variables: roll, pitch, yaw and speed. Each group of parameters (chromosome) represents a model, which can be used to compute simulated flight data from the recorded input commands. The difference between simulated and real data is determined with a weighted least-squares method and used as the fitness function of the genetic algorithm.

In order to reduce the effect of the error propagation to the velocity due to the estimated parameters that have influence in attitude, the global process has been decomposed in two steps: Attitude and velocity parameters identification.

The first only identifies the dynamic response of the helicopter attitude, and the genetic algorithm modifies only the parameters related to the attitude. Once the attitude-related parameters have been set, the second process is executed, and only the velocity-related parameters are changed. This algorithm uses the real attitude data instead of the model data so as not to accumulate simulation errors. Using two separate processes for attitude and velocity yields significantly better results than using one process to identify all parameters at the same time.

The parameters related to the attitude process are: TwstTr, IZ, HTr, YuuVt, YuvVt, Corr1, Corr2, Tproll, Tppitch, Kroll, Kpitch, Kyaw, DTr, DVt, YMaxVt, KGyro, KdGyro, Krp, Kpr,

OffsetRoll, OffsetPitch and OffsetYaw. The parameters related to the vehicle's speed are TwstMr, KCol, XuuFus, YvvFus, ZwwFus and OffsetCol.

The fitness functions are based on a weighted mean square error equation, calculated by comparing the real and simulated responses for the different variables (position, velocity, Euler angles and angular rates) applying a weighting factor.

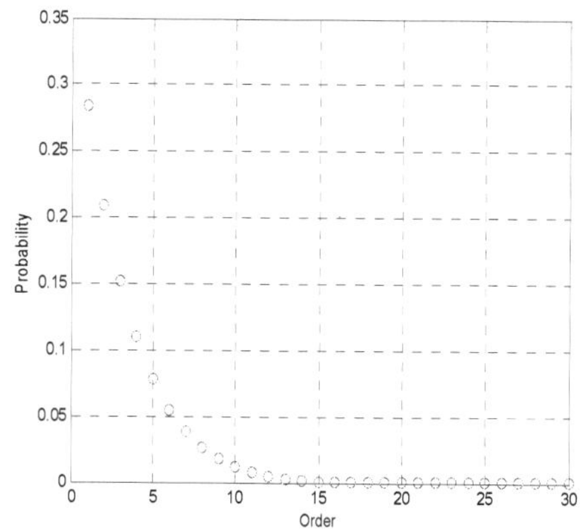

Figure 10. Probability function for selection of elements

The general process is described bellow:

- To create an initial population of 30 elements.
- To perform simulations for every chromosome;
- Computation of the fitness function.
- Classification the population using the fitness function as the index. The ten best elements are preserved. A Monte-Carlo algorithm with the density function shown in

Figure 10. Probability function for selection of elements is used to determine which pairs will be combined to generate 20 more chromosomes. The 10 'better' elements are more likely (97%) to be combined with the crossover operators.

- To repeat from step 2 for a preset number of iterations, or until a preset value for the fitness function of the best element is reached.

3.3.4 Data acquisition

The data acquisition procedure is shown in Figure 11. Helicopter is manually piloted using the conventional RC emitter. The pilot is in charge to make the helicopter performs different maneuvers trying to excite all the parameters of the model. For identification data is essential to have translational flights (longitudinal and lateral) and vertical displacements.

All the commands provided by the pilot are gathered by a computer trough to a USB port by using a hardware signal converter while onboard computer is performing data fusion from sensors and sending the attitude and velocity estimation to the ground computer using a WIFI link.

In this manner inputs and outputs of the model are stored in files to perform the parameters identification.

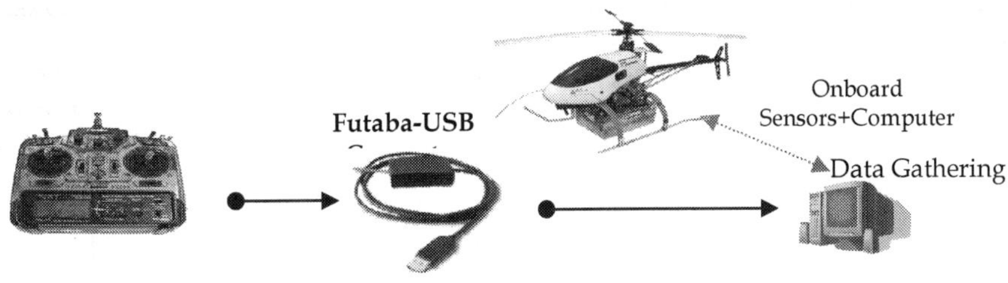

Figure 11. Data acquisition architecture

4. Identification Results

The identification algorithm was executed several times with different initial populations and, as expected, the sub-optimal parameters differ. Some of these differences can be explained by the effect of local minimum. Although the mutation algorithm reduces the probability of staying inside a local minimum, sometimes the mutation factor is not big enough to escape from it.
The evolution of the error index obtained with a least-squares method in different cases is shown in Figure 12. The left graph shows the quick convergence of the algorithm in 50 iterations. On the other hand the right graph shows an algorithm that fell in a local minimum and had an overall lower convergence speed. Around the 50th step a mutation was able to escape the local minimum, and the same behavior is observed in the 160th step.

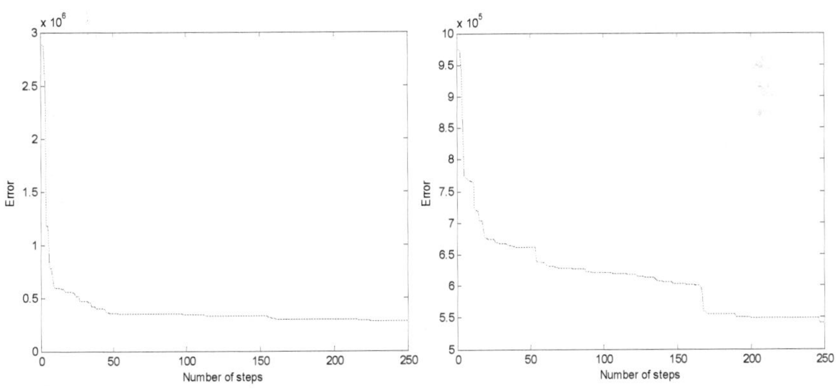

Figure 12. Error evolution for two different cases

The result may be used as the initial population for a second execution of both processes. In fact, this has been done three times to obtain the best solution.
The analysis of cases where mutation was not able to make the algorithm escaped from local minimum, led to the change of the mutation probability from 0.1 to 1 when detected.
On the other hand, not all the parameters were identified at the same time, actually two iterative processes were used to identify all parameters. The first process used 100 steps to identify the parameters related to the modeling of the helicopter's attitude, beginning with a random variation of a valid solution. The second process preserved the best element of the

previous process' population, and randomly mutated the rest. After 100 steps the parameters related to the helicopter's speed were identified.
The simulated attitude data plotted as a blue line against real flight data in red, is shown for roll, pitch, yaw angles in Figure 13. They give you an idea about how simulations follow real tendencies even for small attitude variations.

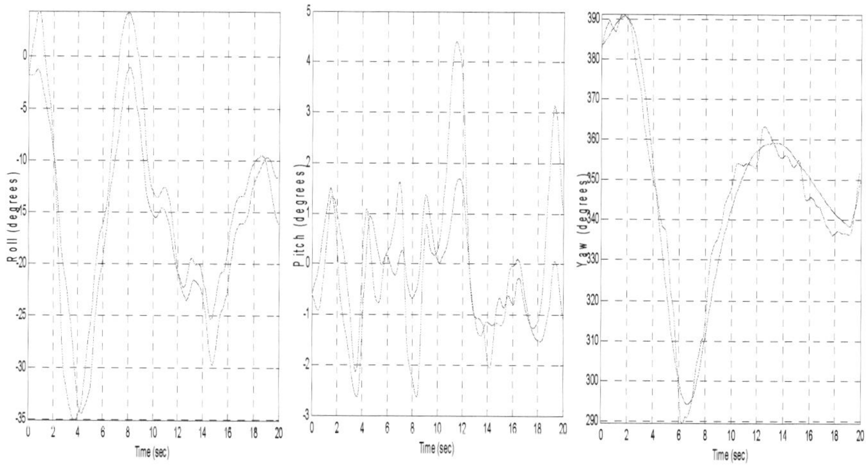

Figure 13. Roll, Pitch and Yaw real vs simulated

On the other hand, the results obtained for the velocity analysis are shown in Figure 14.

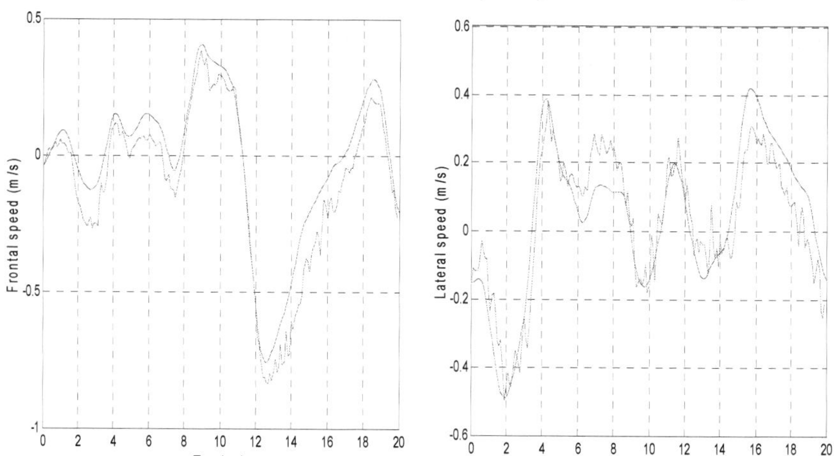

Figure 14. Velocity Simulation Analysis

The quality of the simulation results is the proof of a successful identification process both for attitude and speed.

The simulated roll and yaw fit accurately the registered helicopter response to input commands. The simulated pitch presents some problems due to the fact that the flight data did not have a big dynamic range for this signal. Nevertheless the error is always below three degrees.

For the simulation of the vehicle's velocity good performance was obtained for both frontal and lateral movement. The unsuccessful modeling of vertical velocity can be linked to the sensors available onboard the helicopter. Vertical velocity is registered only by the differential GPS, and statistical studies of the noise for this sensor show a standard deviation of 0.18 meters per second. In other words, the registered signal cannot be distinguished from the noise because the registered flight did not have a maneuver with significant vertical speed; therefore modeling this variable is not possible.

It is important to analyze the values of the obtained parameters since the parameters are identified using a genetic algorithm. The dispersion of the value of the parameters for five different optimization processes was analyzed. Most of the parameters converge to a defined value, which is coherent with the parameter's physical meaning, but sometimes, some parameters were not converged to specific values, usually when no complete set of flights was used. Thus, if no vertical flights were performed, the parameters regarding vertical movements turned to be with a great dispersion in the obtained values.

This conclusion can be extended to different optimization processes for different groups of flight data. In other words, several flights should be recorded, each with an emphasis on the behaviors associated to a group of parameters. With enough flight data it is trivial to identify all the parameters of the helicopter model.

5. Control Prototyping

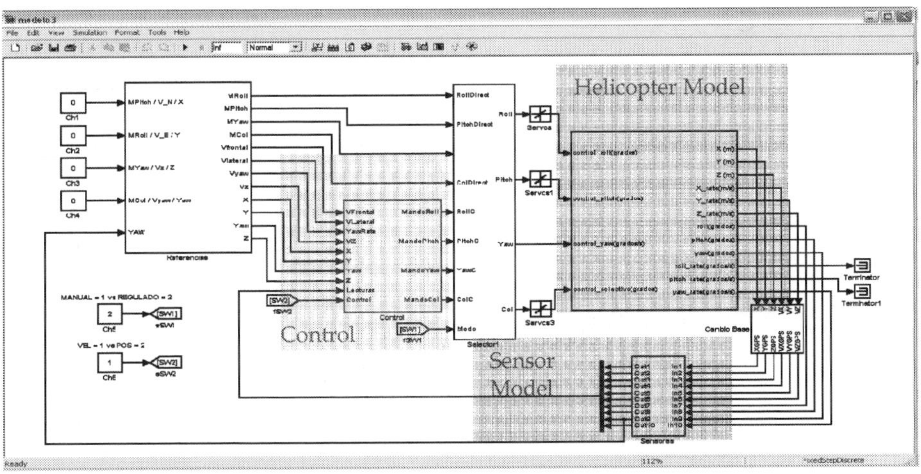

Figure 15. Control Developing template

In order to develop and test control algorithms in a feasible way, the proposed model has been encapsulated into a Matlab-Simulink S-Function, as a part of a prototyping template as Figure 15 shows. Others modules have been used for performing realistic simulations, thus

a sensor model of GNC systems has also been created. Auxiliary modules for automatic-manual switching have been required for testing purposes.
The proposed control architecture is based on a hierarchic scheme with five control levels as Figure 16 shows. The lower level is composed by hardware controllers: speed of the rotor and yaw rate. The upper level is the attitude control. This is the most critical level for reaching the stability of the helicopter during the flight. Several techniques have been tested on it (classical PI or fuzzy controllers). The next one is the velocity and altitude level. The fourth is the maneuver level, which is responsible for performing a set of pre- established simple maneuvers, and the highest level is the mission control.
Following sections will briefly describe these control levels from the highest to the lowest one.

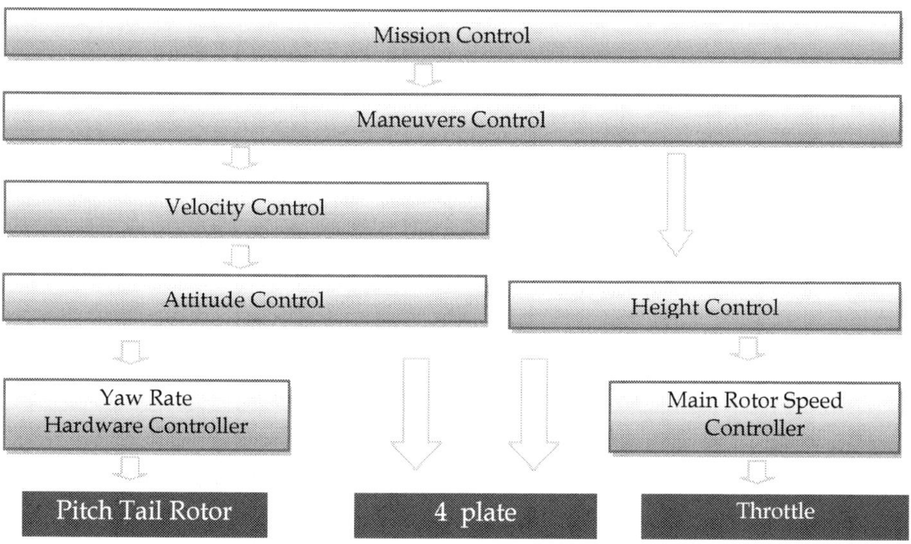

Figure 16. Control Architecture

5.1 Mission control

In this level, the operator designs the mission of the helicopter using GIS (Geographic Information System) software. These tools allow visualizating the 3D map of the area by using Digital Terrain Model files and describing a mission using a specific language. (Gutiérrez et al -06). The output of this level is a list of maneouvers (parametric commands) to be performed by the helicopter.

5.2 Maneuver Control

This control level is in charge of performing parametric manoeuvres such as flight maintaining a specific velocity during a period of time, hovering with a fix of changing yaw, forward, backward or sideward flights and circles among others. The output of this level are velocity commands.

Internally, this level performs a prediction of the position of the helicopter, applying acceleration profiles (Figure 17).
Velocity and heading references are computed by using this theoretical position and the desired velocity in addition to the real position and velocity obtained from sensors.

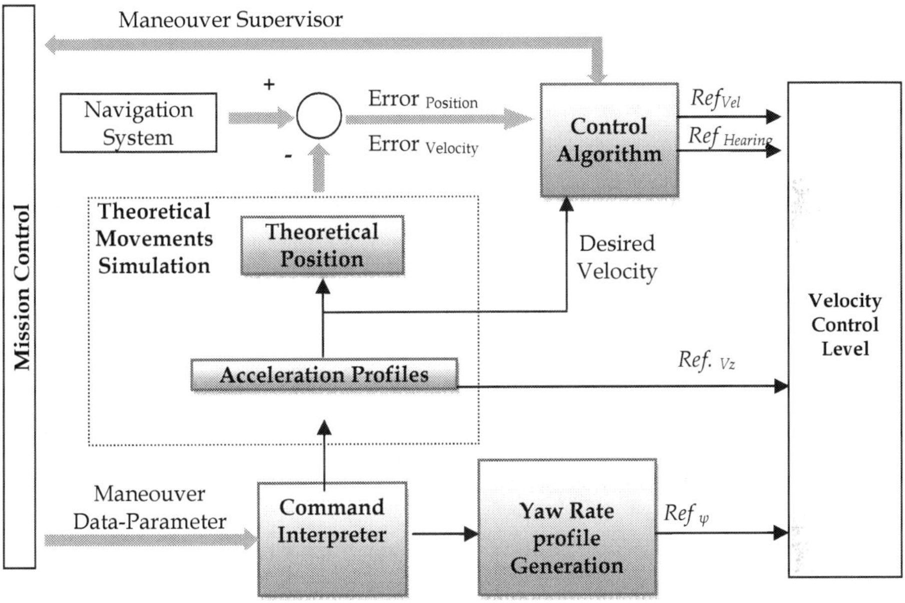

Figure 17. Maneuvers Control Scheme

A vertical control generates references for altitude control and Yaw control.
In manoeuvres that no high precision of the position is required, i.e. forwared flights, the position error is not taken into account, because the real objective of the control is maintain the speed. This system allow manage a weighting criteria for defining if the main objetive is the trajectory or the speed profile. The manoeuvres that helicopter is able to perform are an upgradeable list.

5.3 Velocity Control

The velocity control is in charge of making the helicopter maintains the velocity that maneouvers control level computes.
The velocity can be referred to a ENU (East-North-Up) frame or defined as a module and a direction. Therefore, a coordinate transformation is required to obtain lateral and frontal velocities (vl, vf) that are used to provide Roll and Pitch references. This transformation is very sensitive to the yaw angle estimation.

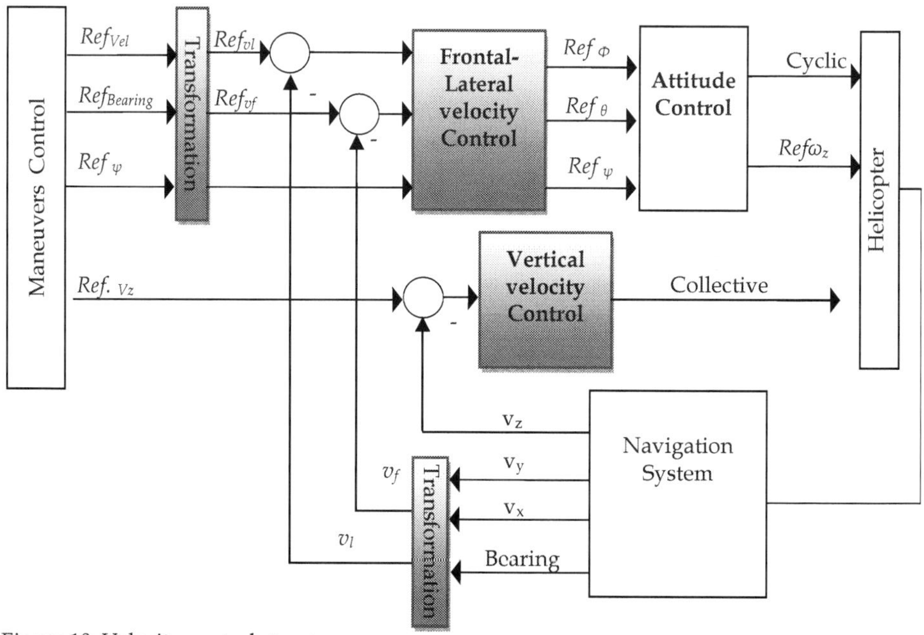

Figure 18. Velocity control structure

Control maneuver has also to provide with the yaw reference due to the capability of helicopters for flying with derive (different bearing and heading).
Vertical velocity is isolated from the horizontal because it is controlled by using the collective command.

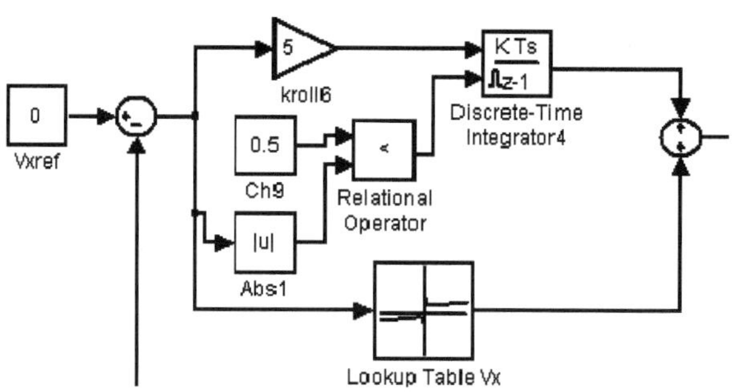

Figure 19. Velocity control example

Concerning to the control algorithms used to test the control architecture, two main problems have been detected and solved:
The first one is based on the fact that the speed is very sensitive to the wind and payload. Due to this, a gain-scheduller scheme with a strong integration efect has been requiered in PI algorithm. (Figure 19)

The second one arises from the solution of the first problem. Thus, when operator performs a manual-automatic swiching by using channel nine of the emitter, integral action needs to be reset for reducing the effect of the error integration when manual mode is active.

5.4 Attitude Control

The attitude control is in charge of providing with commands to the servos (or hardware controllers). Several control techiques were tested in this level, but probably fuzzy controllers turned out to have the best performance. Figure 20 shows the proposed architecture and Figure 21 the control surfaces used by fuzzy controller.
As it can be observed, the Yaw have been isolated from the multivariable Roll-Pitch controler with a reasonable quality results.

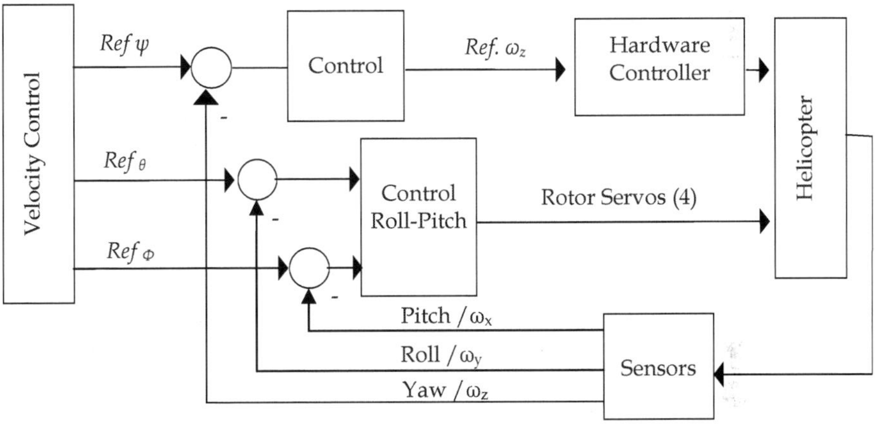

Figure 20. Attitude control structure

Once the control structure has been designed, the first step is to perform simulations in order to analyze if the control fulfills all the established requirements.
Considering the special characteristics of the system to control, they main requirements to be taken into account are:

- No (small) oscillations for attitude control. In order to transmit a confidence in the control system to the operator during the tests.
- No permanent errors in velocity control. This is an important concept to consider for avoiding delays in long distance missions.
- Adjustable fitness functions for maneuvers control. This allows characterizing the mission where position or velocity is the most important reference to keep.

The operator relies upon all the tools that Matlab-Simulink provides and an additional 3D visualization of the helicopter as Figure 22 shows for test performing.

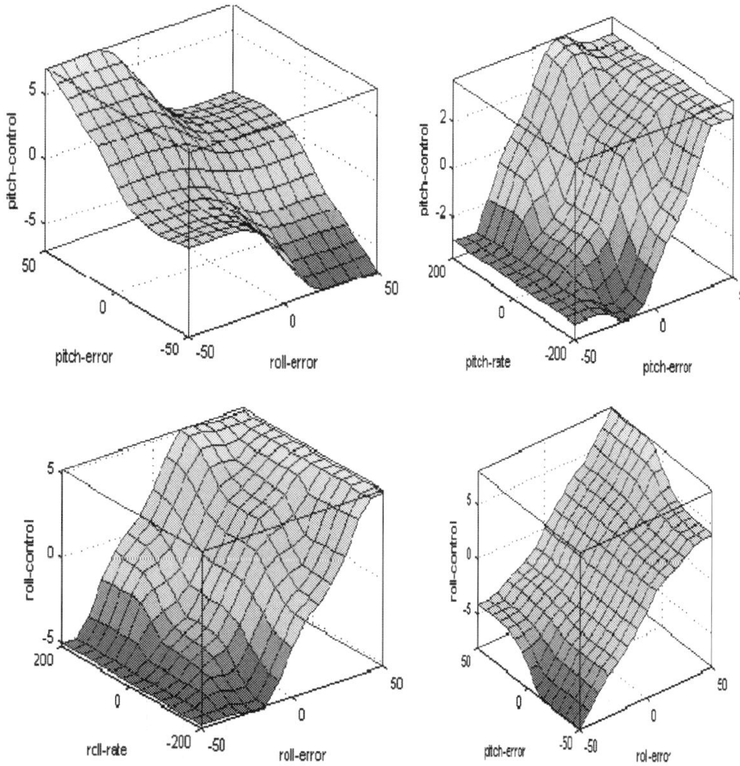

Figure 21. Control surfaces for aAttitude level

The controller block has to be isolated for automatic C codification by using the embedded systems coder of Matlab-Simulink when simulation results are satisfactory.Then, the obtained code can be loaded into onboard computer for real testing (Del-Cerro 2006).

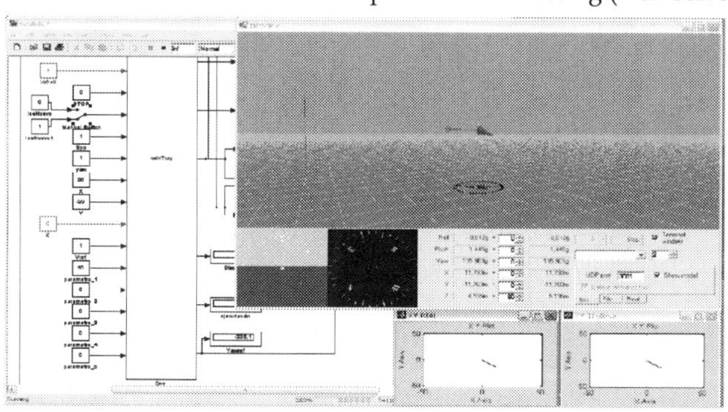

Figure 22. Simulation framework

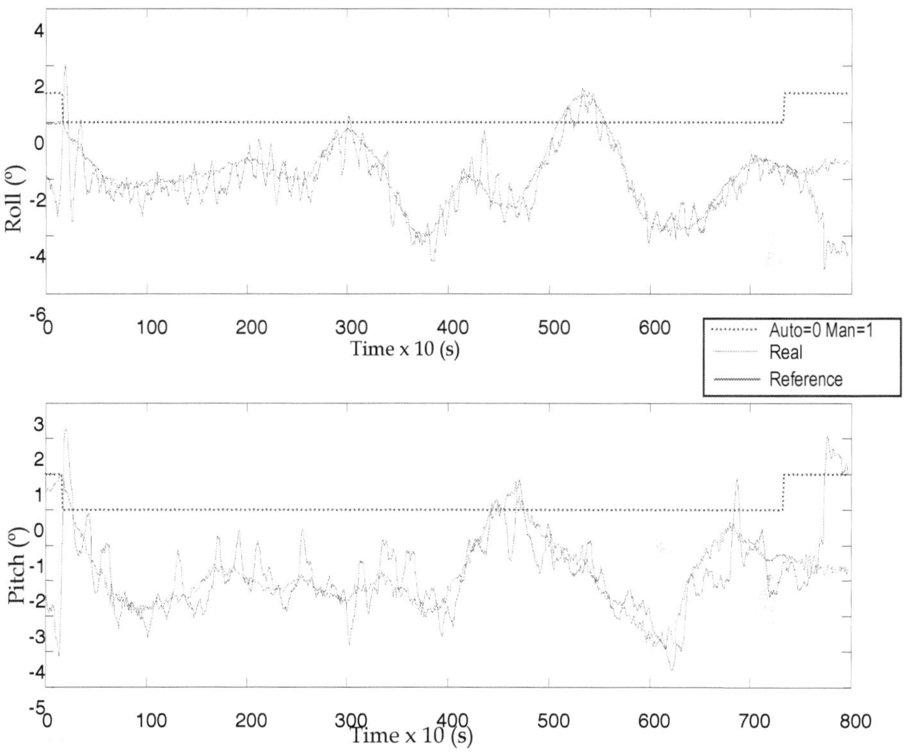

Figure 23. Roll and pitch Control results during real flight

6. Control Results

Figure 23 shows the results obtained for attitude control during an eighty-seconds real flight with the Vario helicopter.

Lines in red show the real attitude of the helicopter while green ones are the references provided by the velocity control level.

Dotted blue line indicates when the helicopter is in manual (value of 1) or automatic (value of 0). It can be observed that, when helicopter is in manual mode, the red line does not follow the references of the automatic control, because the pilot is controlling the helicopter.

Figure 24 shows the obtained results for velocity control. The same criteria that in last graphic has been used, thus red line indicates real helicopter response and green means references from maneuvers control level. Dotted blue line is used to show when the helicopter flies autonomously or remote piloted. Small delays can be observed.

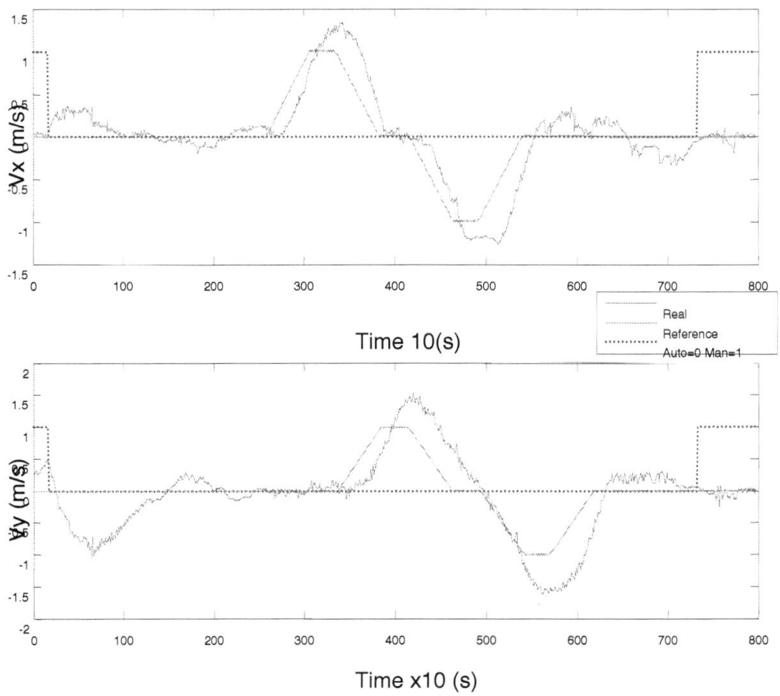

Figure 24. Velocity control results

Figure 25 shows two examples of the maneuvers control performance.

The right side graph shows the position when helicopter is moving in a straight line. It can be observed that maximum error in transversal coordinates is less than one meter. This result can be considered as very good due to is not a scalable factor. The precision of the position control reduces the error when trajectory is close to the end by weighting the error position against the velocity response.

On the other hand, the left side shows a square trajectory. The speed reference performs smooth movementes and therefore the position control is less strong that in the first case. Even in this case, position errors are smaller than two meters in the worst case.

In general, a high precision flight control has been developed, because sub-metric precision has been reached during the tests with winds slower than twenty kilometers per hour.

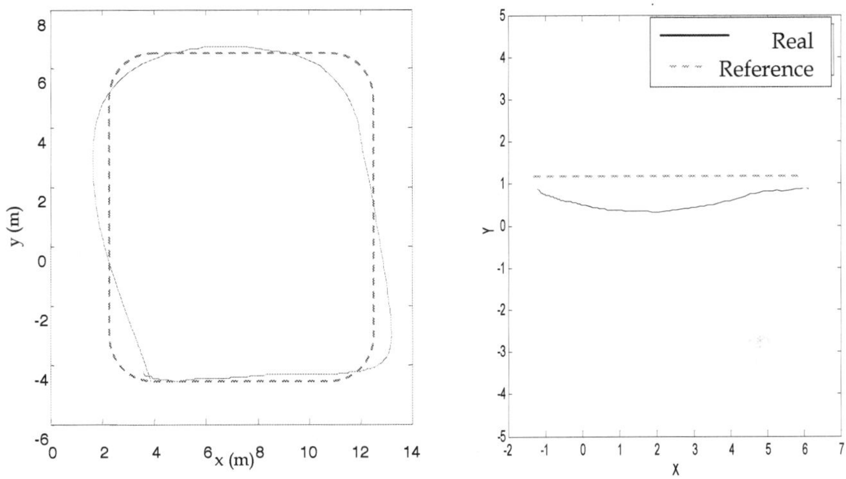

Figure 25. Maneuvers control results

7. Final Observations and Future Work

The first part of this chapter describes a procedure for modeling and identification a small helicopter. Model has been derived from literature, but changes introduced in the original model have contributed to improve the stability of the model with no reduction of the precision.

The proposed identification methodology based on evolutionary algorithms is a generic procedure, having a wide range of applications.

The identified model allows performing realistic simulations, and the results have been validated.

The model has been used for designing real controllers on a real helicopter by using a fast prototyping method detailed in section 4.

The model has confirmed a robust behavior when changes in the flight conditions happen. It does work for hover and both frontal and lateral non-aggressive flight.

The accuracy and convenience of a parametric model depends largely on the quality of its parameters, and the identification process often requires good or deep knowledge of the model and the modeled phenomena. The proposed identification algorithm does away with the complexity of model tuning: it only requires good-quality flight data within the planned simulation envelope. The genetic algorithm has been capable of finding adequate values for the model's 28 parameters, values that are coherent with their physical meaning, and that yields an accurate model.

It is not the objective of this chapter to study what the best control technique is. The aim is to demonstrate that the proposed model is good enough to perform simulations valid for control designing and implementation using an automatic coding tool.

Future work will focus on the following three objectives:

- To develop a model of the engine in order to improve the model behavior for aggressive flight maneuvers.
- To compare this model with others in the literature in a wide range of test cases.
- To validate the proposed model with different helicopters.

8. References

Castillo P, Lozano R, Dzul A. (2005) Modelling and Control of Mini-Flying Machines. Springer Verlag London Limited 2005. ISBN:1852339578.

Del-Cerro J, Barrientos A, Artieda J, Lillo E, Gutiérrez P, San- Martín R, (2006) Embedded Control System Architecture applied to an Unmanned Aerial Vehicle International Conference on Mechatronics. Budapest-Hungary

Gavrilets, V., E. Frazzoli, B. Mettler. (2001) Aggresive Maneuvering of Small Autonomous Helicopters: A. Human Centerred Approach. *The International Journal of Robotics Research.* Vol 20 No 10 , October 2001. pp 795-807

Godbole et al (2000) Active Multi-Model Control for Dynamic maneuver Optimization of Unmanned Air Vehicles. *Proceedings of the 2000 IEEE International conference on Robotics and Automation.* San Francisco, CA. April 2000.

Gutiérrez, P., Barrientos, A., del Cerro, J., San Martin., R. (2006) Mission Planning and Simulation of Unmanned Aerial Vehicles with a GIS-based Framework; *AIAA Guidance, Navigation and Control Conference and Exhibit.* Denver, EEUU, 2006.

Heffley R (1988). Minimum complexity helicopter simulation math model. *Nasa center for AerosSpace Information* . 88N29819.

La Civita M, Messner W, Kanade T. Modeling of Small-Scale Helicopters with Integrated First-Principles and System-Identification Techniques. *American Helicopter Society 587h Annual Forum,* Montreal, Canada, June 11-13,2002.

Mahony-R Lozano-R, Exact Path Tracking (2000) Control for an Autonomous helicopter in Hover Manoeuvres. *Proceedings of the 2000 IEEE International Conference on Robotics & Automation.* San Francisco, CA. APRIL 2000. P1245-1250

Mettler B.F., Tischler M.B. and Kanade T.(2002) System Identification Modeling of a Model-Scale Helicopter. *T. Journal of the American Helicopter Society,* 2002, 47/1: p. 50-63. (2002).

Mettler B, Kanade T, Tischler M. (2003) System Identification Modeling of a Model-Scale Helicopter. *Internal Report* CMU-RI-TR-00-03.

6

Stabilization of Scale Model Vertical Take-Off and Landing Vehicles without Velocity Measurements

Bertrand Sylvain[1], Hamel Tarek[2] and Piet-Lahanier Hélène[1]
[1] *ONERA-DPRS*, [2] *I3S-UNSA-CNRS*
France

1. Introduction

Miniature rotorcraft-based Unmanned Aerial Vehicles (UAVs) have received a growing interest in both industrial and academic research. Thanks to their hover and vertical take-off and landing (VTOL) capabilities, they are indeed particularly well suited for many civil missions such as video supervision of road traffic, surveillance of urban districts, victims localization after natural disasters, fire detection or building inspection for maintenance.

Design of guidance navigation and control algorithms for the autonomous flight of small rotorcraft-based UAVs is a challenging research area because of their nonlinear dynamics and their high sensitivity to aerodynamic perturbations. Various control strategies such as backstepping (Bouabdallah & Siegwart, 2005), (Frazzoli et al., 2000), (Mahony & Hamel, 2004), nonlinear model predictive control (Kim et al., 2002), (Bertrand et al., 2007a) or sliding modes (Bouabdallah & Siegwart, 2005) have been successfully applied to stabilization or trajectory tracking of UAV models. Nevertheless most of them require full state knowledge for feedback control design.

For robotic systems it may be useful, for cost or payload reasons, to limit the number of embedded sensors. For a miniature UAV, the nature of the mission itself may also directly impact the choice of the sensors that will be used, and therefore the type of measurements that will be available for the vehicle control.

In constrained environments, for example, the use of a vision based sensor may be preferred to a GPS to estimate the relative position of the vehicle with respect to its environment. In that case, linear velocity measurements may not be available. Another example is the case of a test bench design, where a "ready-to-use" radio controlled vehicle is used along with external sensors that do not require structural modifications of the vehicle. Such external sensors are for example motion capture systems (Kondak et al., 2007), (Kundak & Mettler, 2007), (Valenti et al., 2006), or magnetic field based sensors (Castillo et al., 2004). With such equipments, only the position and the attitude angles of the vehicle can be directly measured.

Nevertheless, knowledge of the vehicle state components (positions, linear velocities, attitude angles and angular velocities) is required for control.

A practical approach may consist in computing the velocities from the position measurements by finite differentiations. This method is used in (Kondak et al., 2007) to compute the linear velocity of rotorcraft-based miniature UAVs, and in (Castillo et al., 2004) to compute both linear and angular velocities to control a four-rotors vehicle. However, no theoretical stability guarantee is provided.

One way to theoretically deal with partial state measurement is to define an observer. In (Do et al., 2003) the problem of trajectory tracking for a planar Vertical Take-Off and Landing (VTOL) aircraft with only position and attitude angle measurements is solved by designing a full-order observer. Changes of coordinates are then used to put the system in a triangular form so that a backstepping technique can be used to develop a velocity-independent stabilizing controller.

However, the use of an observer may introduce additional computational burden in the control loop. It is also necessary to prove firstly its own convergence. In addition, compatibility between the frequency of the observer and the frequency of the controller must be checked to ensure the closed loop stability of the complete observer-based controlled system.

Another approach that can be used to avoid computational burden or complexity due to the introduction of an observer is partial state feedback: the controller is designed directly from the available measurements.

Early work on partial state feedback has been done in the context of rigid-link robot manipulators when no velocity measurement is available. In (Burg et al., 1996) and (Burg et al., 1997) the velocity measurement is replaced by a velocity-related signal generated by a linear filter based only on link position measurement. An extension of this work, using a nonlinear filter, can be found in (Dixon et al., 2000). The same method has been applied to solve the problem of attitude tracking of rigid bodies. A velocity-related signal generated by a linear filter is indeed employed in (Wong et al., 2000), where a kinematic representation using modified Rodrigues parameters has been chosen. In (Costic et al., 2000), a unit-quaternion-based representation is adopted and a nonlinear filter generates a signal replacing the angular velocity measurement in the feedback controller.

First-order dynamic attitude feedback controllers have been proposed in (Arkella, 2001) and (Astolfi & Lovera, 2002) to respectively solve the attitude tracking problem for rigid bodies and spacecrafts with magnetic actuators. The kinematic representations that are used in these two works are respectively based on modified Rodrigues parameters and quaternions. A unit-quaternion representation is also used in (Tayebi, 2007) where a feedback controller depending on an estimation error quaternion is designed to solve the problem of a rigid spacecraft attitude tracking.

Attitude control of rigid bodies without angular velocity measurement is also addressed in (Lizarralde & Wen, 1996) and (Tsiotras, 1998) where a passivity-like property of the system is used to design feedback controllers for kinematic representations respectively based on unit-quaternions and Rodrigues parameters. Both of them use a filtering technique to avoid the use of velocity measurement.

In this chapter, we deal with the problem of position and attitude stabilization of a six degrees of freedom VTOL UAV model when no measurement of the linear velocity nor of the angular velocity is available. Contrary to the previous works, the kinematic representation we use exploits the $SO(3)$ group and its manifold. The method we present is based on the introduction of virtual states in the system dynamics; no observer design is required.

The rest of the chapter is organized as follows. Section 2 introduces the notations and the mathematical identities that will be used in the chapter. Section 3 presents the VTOL UAV model dynamics and the cascaded structure of the controller. The design of the position controller is detailed in Section 4 whereas the attitude controller is presented in Section 5. In Section 6, the closed loop stability of the system is analyzed, and simulations results are provided in Section 7. Concluding remarks are finally given in the last part of this chapter.

2. Notations and Mathematical Background

Let $SO(3)$ denote the special orthogonal group of $\mathbb{R}^{3\times 3}$ and $\mathfrak{so}(3)$ the group of antisymmetric matrices of $\mathbb{R}^{3\times 3}$. We define by $(.)_\times$ the operator from $\mathbb{R}^3 \rightarrow \mathfrak{so}(3)$ such that

$$\forall b \in \mathbb{R}^3,\ b_\times = \begin{bmatrix} 0 & -b_3 & b_2 \\ b_3 & 0 & -b_1 \\ -b_2 & b_1 & 0 \end{bmatrix} \tag{1}$$

where b_i denotes the ith component of the vector b.

Let $V(.)$ be the inverse operator of $(.)_\times$, defined from $\mathfrak{so}(3) \rightarrow \mathbb{R}^3$, such that

$$\forall b \in \mathbb{R}^3, V(b_\times) = b \qquad \forall B \in \mathfrak{so}(3), V(B)_\times = B \tag{2}$$

For a given vector $b \in \mathbb{R}^3$ and a given matrix $M \in \mathbb{R}^{3\times 3}$, let us consider the following notations and identities:

$$P_a(M) = \frac{M - M^T}{2} \qquad P_s(M) = \frac{M + M^T}{2} \tag{3}$$

$$\mathrm{tr}(P_a(M) P_s(M)) = 0 \tag{4}$$

$$\frac{1}{2}\,\mathrm{tr}(b_\times M) = -b^T V(P_a(M)) \tag{5}$$

The following identity will also be used:

$$\forall (A_a, B_a) \in \mathfrak{so}(3)^2,\ \frac{1}{2}\,\mathrm{tr}(A_a^T B_a) = V(A_a)^T V(B_a) \tag{6}$$

Denote by (γ_R, n_R) the angular-axis coordinates of a given matrix $R \in SO(3)$, and by I_d the identity matrix of $\mathbb{R}^{3\times 3}$. One has:

$$\forall R \in SO(3),\ \mathrm{tr}(I_d - R) = 2(1 - \cos(\gamma_R)) \tag{7}$$

3. UAV Model and Control Strategy

3.1 VTOL UAV Model

The VTOL UAV is represented by a rigid body of mass m and of tensor of inertia $I= diag(I_1,I_2,I_3)$ with I_1, I_2 and I_3 strictly positive. To describe the motion of the UAV, two reference frames are introduced: an inertial reference frame $(\mathcal{I})$ associated with the vector basis (e_1, e_2, e_3) and a body frame $(\mathcal{B})$ attached to the UAV and associated with the vector basis (e_1^b, e_2^b, e_3^b) (see Figure 1).

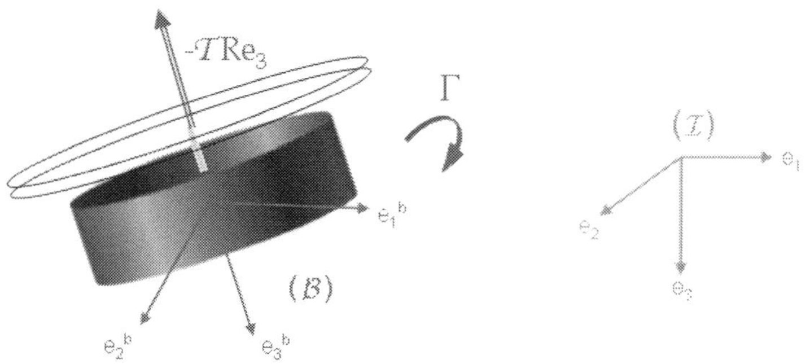

Figure 1. Reference frames

The position and the linear velocity of the UAV in $(\mathcal{I})$ are respectively denoted $\chi = \left[\chi_x\, \chi_y\, \chi_z\right]^T$ and $v=\left[v_x\, v_y\, v_z\right]^T$. The orientation of the UAV is given by the orientation matrix $R\in SO(3)$ from $(\mathcal{I})$ to $(\mathcal{B})$, usually parameterized by Euler's pseudo angles ψ, θ, ϕ (yaw, pitch, roll):

$$R=\begin{bmatrix} c_\theta\, c_\psi & s_\phi\, s_\theta\, c_\psi - c_\phi\, s_\psi & c_\phi\, s_\theta\, c_\psi + s_\phi\, s_\psi \\ c_\theta\, s_\psi & s_\phi\, s_\theta\, s_\psi + c_\phi\, c_\psi & c_\phi\, s_\theta\, s_\psi - s_\phi\, c_\psi \\ -s_\theta & s_\phi\, c_\theta & c_\phi\, c_\theta \end{bmatrix} \tag{8}$$

with the trigonometric shorthand notations $c_\alpha=\cos(\alpha)$ and $s_\alpha=\sin(\alpha)$, $\forall\alpha\in \mathbb{R}$.

Let $\Omega=\left[\omega_p\, \omega_q\, \omega_r\right]^T$ be the angular velocity of the UAV defined in $(\mathcal{B})$.

The dynamics of a rigid body can be described as:

$$\begin{cases} \dot{\chi}=v \\ m\dot{v}=F \\ \dot{R}=R\Omega_\times \\ I\dot{\Omega}=-\Omega\times I\Omega+\Gamma \end{cases} \tag{9}$$

where the inputs are a translational force $F \in \mathbb{R}^3$ and a control torque $\Gamma \in \mathbb{R}^3$.

For the VTOL UAV, the translational force F combines thrust, lift, drag and gravity components. In quasi-stationary flight we can reasonably assume that the aerodynamic forces are always in direction e_3^b $(=Re_3)$, since the lift force predominates the other components (Hamel & Mahony, 2004).
By separating the gravity component mge_3 from the combined aerodynamic forces, the dynamics of the VTOL UAV are rewritten as:

$$\begin{cases} \dot{\chi}=v \\ m\dot{v}=-T\,Re_3+mge_3 \\ \dot{R}=R\Omega_\times \\ I\dot{\Omega}=-\Omega\times I\Omega+\Gamma \end{cases} \tag{10}$$

where the inputs are the scalar $T\in\mathbb{R}$ representing the magnitude of the external forces applied in direction e_3^b, and the control torque $\Gamma=[\Gamma_1\ \Gamma_2\ \Gamma_3]^T$ defined in $(\mathcal{B})$.

3.2 Control Strategy

In this chapter, we consider the problem of the vehicle stabilization around a desired position χ^d assumed to be constant $\left(\dot{\chi}^d=0\right)$.

For control design, let us define the position error $\xi=\left(\chi-\chi^d\right)$. The system (10) becomes:

$$\begin{cases} \dot{\xi}=v \\ m\dot{v}=-T\,Re_3+mge_3 \\ \dot{R}=R\Omega_\times \\ I\dot{\Omega}=-\Omega\times I\Omega+\Gamma \end{cases} \tag{11}$$

Designing a controller for the model (11) can be realized by a classical backstepping approach applied to the whole dynamical system. In that case, the input vector $-(T/m)Re_3$ must be dynamically extended (Frazzoli et al., 2000), (Mahony et al., 1999). To avoid such a dynamical extension, the singular perturbation theory can be used to split the system dynamics into two reduced order subsystems (Khalil, 2002), (Calise, 1976). This approach leads to a time-scale separation between the translational dynamics (slow time-scale) and the orientation dynamics (fast time-scale). Reduced order controllers can therefore be designed to stabilize the system dynamics (Njaka et al., 1994).
We introduce the scaling parameter $\varepsilon\in(0,1]$ such that:

$$\begin{cases} \dot{\xi}=v \\ \dot{v}=-\frac{T}{m}Re_3+ge_3 \end{cases} \tag{12}$$

$$\begin{cases} \varepsilon\dot{R}=R\Omega_\times \\ \varepsilon I\dot{\Omega}=-\Omega\times I\Omega+\Gamma \end{cases} \tag{13}$$

For $\varepsilon=1$, we obtain the full order system. Setting $\varepsilon=0$ leads to the slow time-scale reduced-order system, where the orientation dynamics satisfy a quasi steady state condition $\Omega=0$.
For the translational dynamics (12), we will define the full vectorial term TRe_3 as the position control input. We will assign its desired value

$$\left(TRe_3\right)^d = f(\xi, v) \tag{14}$$

Assuming that the actuator dynamics can be neglected with respect to the rigid body dynamics of the UAV, the value T^d is considered to be instantaneously reached by T. Therefore, we have $(TRe_3)^d = TR^d e_3$, where R^d is the desired orientation of the vehicle. The vector $TR^d e_3$ will then be split into its magnitude

$$T=\|f(\xi,v)\| \tag{15}$$

representing the first control input, and its direction

$$R^d e_3=\frac{1}{T}f(\xi,v) \tag{16}$$

representing the desired orientation.
Remark 1: *The desired orientation* R^d *can then be deduced from the given direction* $R^d e_3=\frac{1}{T}f(\xi,v)$, *solving for* (ψ, θ, ϕ) *for a given specified yaw value* ψ^d *and using (8) (Hamel, 2002).*
Since $\varepsilon \ll 1$ for the considered system, the design of the position controller can be done in the slow time-scale, i.e. for $R \equiv R^d$.
For the orientation dynamics (13), we will assign the control torque Γ such that the orientation R of the UAV converges asymptotically to the desired orientation R^d, and such that the angular velocity Ω converges to Ω^d defined by:

$$\dot{R}^d = R^d \Omega_\times^d \tag{17}$$

Therefore, the assigned control law will be of the form

$$\Gamma = h(R, R^d, \Omega, \Omega^d) \tag{18}$$

The design of the attitude controller can be done in the fast time-scale, assuming $\Omega^d=0$. Indeed, defining $\Lambda=R-R^d$ and using the singular perturbation theory, we get

$$\varepsilon\dot{\Lambda}=R\Omega_\times-\varepsilon R^d\Omega_\times^d \tag{19}$$

Since $\varepsilon \ll 1$ for the considered system (the translational dynamics are characterized by a slow time-scale with respect to the orientation dynamics), the term $\varepsilon R^d \Omega^d_\times$ can therefore be ignored.
The structure of the controller we will develop is summarized in Figure 2. It is defined as a cascaded combination of the position controller and the attitude controller.

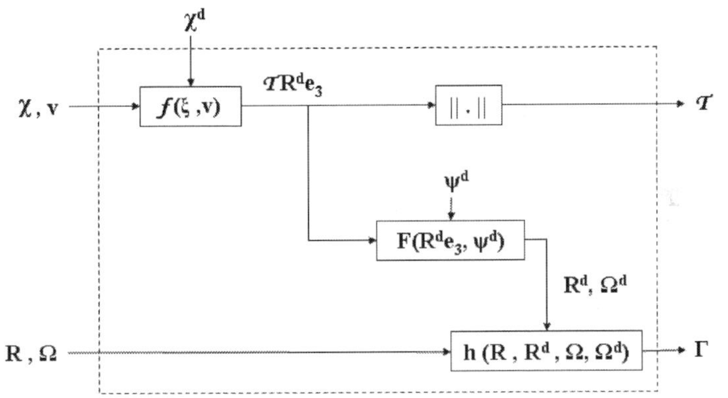

Figure 2. Cascaded structure of the controller

Remark 2: *Note that in the considered case where no velocity measurements are available, the functions f and h defining the control laws will not depend on v nor on Ω.*

4. Position Controller

Consider the translational dynamics. According to the above discussion of Section 3.2, we assume, for control design, that $\mathcal{T}Re_3 \equiv \mathcal{T}R^d e_3$ is the control input of the translational dynamics.
Let $q, w \in \mathbb{R}^3$ be two virtual states and let $\delta \in \mathbb{R}^3$ be a virtual control such that:

$$\begin{cases} \dot{\xi} = v \\ \dot{v} = -\frac{\mathcal{T}}{m} R^d e_3 + g e_3 \\ \dot{q} = -w \\ \dot{w} = \delta \end{cases} \tag{20}$$

Lemma 1: *Consider the system dynamics (20). Let us define the control vector*

$$\mathcal{T}R^d e_3 = \frac{m}{k_v}\{k_x\, \xi + k_1(\xi - q) + k_2(\xi - q + w)\} + m g e_3 \tag{21}$$

and the virtual control

$$\delta = -\frac{1}{k_2}\left\{k_2 w + k_1(\xi - q) + k_2(\xi - q + w)\right\} \tag{22}$$

where k_x, k_v, k_1 *and* k_2 *are strictly positive gains.*
Consider the Lyapunov function candidate

$$\mathcal{S} = \frac{1}{2}k_x\|\xi\|^2 + \frac{1}{2}k_v\|v\|^2 + \frac{1}{2}k_1\|\xi - q\|^2 + \frac{1}{2}k_2\|\xi - q + w\|^2 \tag{23}$$

Define $k_{\min} = \min(k_x, k_v, k_1, k_2)$ *and* $k_{max} = \max(k_x, k_1, k_2)$.
Then, for any initial conditions $\xi(0)$, $v(0)$, *and* $q(0) = \xi(0)$ *and* $w(0) = 0$ *verifying*

$$\mathcal{S}(0) < \frac{1}{18}\frac{g^2 k_v^2 k_{min}}{k_{max}^2} \tag{24}$$

the control vector (21) along with the virtual control (22) exponentially stabilizes the translational dynamics (20), and the input $\mathcal{T}$ *is strictly positive and bounded.*
Proof:
The time derivative of the Lyapunov function candidate $\mathcal{S}$ is

$$\dot{\mathcal{S}} = k_x \xi^T v + k_v v^T\left(-\frac{\mathcal{T}}{m}R^d e_3 + g e_3\right) + k_1(\xi - q)^T(v + w) + k_2(\xi - q + w)^T(v + w + \delta) \tag{25}$$

which can be expressed as:

$$\dot{\mathcal{S}} = v^T\left\{k_x\xi - k_v\frac{\mathcal{T}}{m}R^d e_3 + k_v g e_3 + k_1(\xi - q) + k_2(\xi - q + w)\right\} + k_1 w^T(\xi - q) + k_2(\xi - q + w)^T(w + \delta) \tag{26}$$

Taking the control vector $\mathcal{T}R^d e_3$ as defined in (21), we get:

$$\dot{\mathcal{S}} = k_1 w^T(\xi - q) + k_2(\xi - q + w)^T(w + \delta) \tag{27}$$

We introduce $(\xi - q)$ in the first term to obtain:

$$\dot{\mathcal{S}} = k_1(w + (\xi - q) - (\xi - q))^T(\xi - q) + k_2(\xi - q + w)^T(w + \delta) \tag{28}$$

which leads to:

$$\dot{\mathcal{S}} = -k_1\|\xi - q\|^2 + (\xi - q + w)^T(k_1(\xi - q) + k_2(w + \delta)) \tag{29}$$

Choosing the virtual control δ according to (22) makes $\dot{\mathcal{S}}$ become:

$$\dot{\mathcal{S}} = -k_1\|\xi - q\|^2 - k_2\|\xi - q + w\|^2 \tag{30}$$

The application of La Salle's principle leads to $\xi \to q$ and $w \to -(\xi - q)$, i.e. $w \to 0$. By continuity, we get $\dot{\xi} \to \dot{q}$, that is $v \to -w$. Since $w \to 0$, it yields $v \to 0$ and $\xi \to 0$, and by continuity $\dot{v} \to 0$. Combining the second equation of system (20) with (21) and the fact that $\dot{v} \to 0$ leads to:

$$\left(k_x\xi + k_1(\xi - q) + k_2(\xi - q + w)\right) \to 0 \quad (31)$$

Using $(\xi - q) \to 0$ and $w \to 0$, we finally get $\xi \to 0$ and $q \to 0$. Therefore, the closed-loop system (20) is asymptotically stable, and since it is linear, we can conclude that it is exponentially stable.

It remains to show that the input $\mathcal{T}$ is strictly positive and bounded.

From (21), we have:

$$\mathcal{T} = \left\| \frac{m}{k_v}\{k_x\,\xi + k_1(\xi - q) + k_2(\xi - q + w)\} + mge_3 \right\| \quad (32)$$

By triangular inequality, we get:

$$\mathcal{T} \geq mg - \frac{m}{k_v} k_{max} \{\|\xi\| + \|\xi - q\| + \|\xi - q + w\|\} \quad (33)$$

$$\mathcal{T} \geq mg - 3\frac{m}{k_v} k_{max} \left\{\|\xi\|^2 + \|v\|^2 + \|\xi - q\|^2 + \|\xi - q + w\|^2\right\}^{\frac{1}{2}} \quad (34)$$

Using the definition of $\mathcal{S}$ and k_{min}, we have:

$$\frac{1}{2}k_{min}\left\{\|\xi\|^2 + \|v\|^2 + \|\xi - q\|^2 + \|\xi - q + w\|^2\right\} \leq \mathcal{S} \quad (35)$$

Since $\mathcal{S}$ is decreasing, we have:

$$\mathcal{S} \leq \mathcal{S}(0) \quad (36)$$

where, taking $q(0) = \xi(0)$ and $w(0) = 0$, the initial value $\mathcal{S}(0)$ is defined by:

$$\mathcal{S}(0) = \frac{1}{2}k_x\|\xi(0)\|^2 + \frac{1}{2}k_v\|v(0)\|^2 \quad (37)$$

Using (34) along with (35) and (36), we get:

$$\mathcal{T} \geq mg - 3\frac{m}{k_v} k_{max}\sqrt{\frac{2\mathcal{S}(0)}{k_{min}}} \quad (38)$$

Using condition (24), we obtain $\mathcal{T} > 0$.

Let us finally show that $\mathcal{T}$ is bounded.

From equation (32) and by triangular inequality, we also get:

$$\mathcal{T} \leq mg + \frac{m}{k_v} k_{max}\{\|\xi\| + \|\xi - q\| + \|\xi - q + w\|\} \quad (39)$$

$$\mathcal{T} \le mg + 3\frac{m}{k_v}k_{max}\left\{\|\xi\|^2+\|v\|^2+\|\xi-q\|^2+\|\xi-q+w\|^2\right\}^{\frac{1}{2}} \tag{40}$$

Using (35) and (36) leads to:

$$\mathcal{T} \le mg+3\frac{m}{k_v}k_{max}\sqrt{\frac{2\mathcal{S}(0)}{k_{min}}} \tag{41}$$

and, therefore, the input $\mathcal{T}$ is bounded. ■

Lemma 1 ensures that the control vector (21) along with the virtual control (22) exponentially stabilizes the translational dynamics (20) without using any measurement of the linear velocity.

Remark 3: *From our previous discussion, Section 3.2, equation (32) can be directly used to provide the control input* $\mathcal{T}$ *. Furthermore, since* $\mathcal{T}$ *is strictly positive, the direction given by*

$$R^d e_3 = \frac{1}{\mathcal{T}}\left(mge_3+\frac{m}{k_v}\{k_x\,\xi+k_1(\xi-q)+k_2(\xi-q+w)\}\right) \tag{42}$$

is well defined and can be used to compute the desired orientation R^d .

Due to the position controller we developed, the closed-loop translational dynamics are exponentially stable for $R=R^d$. However, since the orientation R will not converge instantaneously to the desired value R^d , an orientation error term is introduced in the translational dynamics:

$$m\dot{v}=-\mathcal{T}R^d e_3+mge_3-\mathcal{T}(R-R^d)e_3 \tag{43}$$

Therefore an attitude controller must be designed to allow, at least, asymptotic convergence of R to R^d .

5. Attitude Controller

Consider the orientation dynamics (13) and assume that measurements on the angular velocity Ω are not available. Let us introduce $\tau=t/\varepsilon$. In the fast time-scale, the time derivative of a given function g will be denoted by

$$\overset{\circ}{g} = \frac{d}{d\tau}g = \varepsilon\frac{d}{dt}g \tag{44}$$

Similarly to the translational dynamics, we introduce two virtual states $Q\in SO(3)$, $W\in\mathbb{R}^3$ and a virtual control $\Delta\in\mathbb{R}^3$ for the orientation dynamics, such that:

$$\begin{cases} \overset{\circ}{R}=R\Omega_{\times} \\ I\overset{\circ}{\Omega}=-\Omega\times I\Omega+\Gamma \\ \overset{\circ}{Q}=-QW_{\times} \\ \overset{\circ}{W}=\Delta \end{cases} \tag{45}$$

For a given desired orientation R^d we define

$$\tilde{R}=(R^d)^T R \tag{46}$$

$$\tilde{Q}=Q^T\tilde{R} \tag{47}$$

According to the previous discussion of Section 3.2, we assume $\Omega^d=0$ for control design, *i. e.* $\overset{\circ}{R^d}=0$.
Using (46) and (47), we rewrite (45) as:

$$\begin{cases} \overset{\circ}{\tilde{R}}=\tilde{R}\Omega_{\times} \\ I\overset{\circ}{\Omega}=-\Omega\times I\Omega+\Gamma \\ \overset{\circ}{\tilde{Q}}=W_{\times}\tilde{Q}+\tilde{Q}\Omega_{\times} \\ \overset{\circ}{W}=\Delta \end{cases} \tag{48}$$

Lemma 2: *Consider the orientation dynamics (48).*
Define the control torque

$$\Gamma=\frac{1}{k_\omega}\left\{-k_r V(P_a(\tilde{R}))-k_3 V(P_a(\tilde{Q}))+k_4 V(P_a(M))+k_4 V(P_a(N))\right\} \tag{49}$$

and the virtual control

$$\Delta=-\frac{1}{k_4}V\left(\frac{1}{2}k_3 P_a(\tilde{Q})+\frac{1}{2}k_4(W_{\times}\tilde{Q}+\tilde{Q}^T W_{\times})+\frac{1}{2}k_5(W_{\times}+P_a(\tilde{Q}))\right) \tag{50}$$

where

$$M=(W_{\times}+P_a(\tilde{Q}))^T\tilde{Q} \tag{51}$$

$$N=\tilde{Q}^T(W_{\times}+P_a(\tilde{Q}))^T \tag{52}$$

and k_r, k_ω, k_3, k_4 and k_5 are strictly positive gains with

$$k_r < k_3 \tag{53}$$

Consider the control Lyapunov function candidate

$$\mathcal{L}=\frac{1}{2}k_r\ \mathrm{tr}(I_d-\tilde{R})+\frac{1}{2}k_\omega \Omega^T I\Omega+\frac{1}{2}k_3\ \mathrm{tr}(I_d-\tilde{Q})+\frac{1}{2}k_4\ \mathrm{tr}\left\{(W_\times+P_a(\tilde{Q}))^T (W_\times+P_a(\tilde{Q}))\right\} \tag{54}$$

Then, for any initial condition $\tilde{R}(0)$ *,* $\Omega(0)$ *, with* $Q(0)=\tilde{R}(0)$ *and* $W(0)=0$ *, such that*

$$\mathcal{L}(0)<2k_r \tag{55}$$

the control torque (49) along with the virtual control (50) asymptotically stabilizes the orientation dynamics (48).

Proof :

Consider the Lyapunov function candidate $\mathcal{L}$ defined by (54). It's time derivative along the trajectories of (48) is given by:

$$\overset{\circ}{\mathcal{L}}=-\frac{1}{2}k_r\ \mathrm{tr}(\overset{\circ}{\tilde{R}})+k_\omega\Omega^T\{-\Omega\times I\Omega+\Gamma\}-\frac{1}{2}k_3\ \mathrm{tr}(\overset{\circ}{\tilde{Q}})+k_4\ \mathrm{tr}\left\{(W_\times+P_a(\tilde{Q}))^T (\Delta_\times+\overset{\circ}{\overbrace{P_a(\tilde{Q})}})\right\} \tag{56}$$

where

$$\overset{\circ}{\overbrace{P_a(\tilde{Q})}}=\frac{1}{2}\left\{W_\times\tilde{Q}+\tilde{Q}\Omega_\times+\tilde{Q}^T W_\times+\Omega_\times\tilde{Q}^T\right\} \tag{57}$$

Using (57) and the fact that $\Omega^T(\Omega\times I\Omega)=0$, it yields:

$$\begin{aligned}\overset{\circ}{\mathcal{L}}= &-\frac{1}{2}k_r\ \mathrm{tr}(\tilde{R}\Omega_\times)+k_\omega\Omega^T\Gamma-\frac{1}{2}k_3\ \mathrm{tr}(W_\times\tilde{Q}+\tilde{Q}\Omega_\times)\\ &+k_4\ \mathrm{tr}\left\{(W_\times+P_a(\tilde{Q}))^T\left(\Delta_\times+\frac{1}{2}\left\{W_\times\tilde{Q}+\tilde{Q}\Omega_\times+\tilde{Q}^T W_\times+\Omega_\times\tilde{Q}^T\right\}\right)\right\}\end{aligned} \tag{58}$$

Using identity (4) we get:

$$\begin{aligned}\overset{\circ}{\mathcal{L}}= &-\frac{1}{2}k_r\ \mathrm{tr}(\Omega_\times P_a(\tilde{R}))+k_\omega\Omega^T\Gamma-\frac{1}{2}k_3\ \mathrm{tr}(W_\times P_a(\tilde{Q}))-\frac{1}{2}k_3\ \mathrm{tr}(\Omega_\times P_a(\tilde{Q}))\\ &+k_4\ \mathrm{tr}\left\{(W_\times+P_a(\tilde{Q}))^T\left(\Delta_\times+\frac{1}{2}\left\{W_\times\tilde{Q}+\tilde{Q}\Omega_\times+\tilde{Q}^T W_\times+\Omega_\times\tilde{Q}^T\right\}\right)\right\}\end{aligned} \tag{59}$$

Also using (6) we obtain:

$$\begin{aligned}\overset{\circ}{\mathcal{L}}= &\ \Omega^T\left\{k_r V(P_a(\tilde{R}))+k_\omega\Gamma+k_3 V(P_a(\tilde{Q}))\right\}-\frac{1}{2}k_3\ \mathrm{tr}(W_\times P_a(\tilde{Q}))+\frac{1}{2}k_4\ \mathrm{tr}\left\{(W_\times+P_a(\tilde{Q}))^T\tilde{Q}\Omega_\times\right\}\\ &+\frac{1}{2}k_4\ \mathrm{tr}\left\{(W_\times+P_a(\tilde{Q}))^T\Omega_\times\tilde{Q}^T\right\}+k_4\ \mathrm{tr}\left\{(W_\times+P_a(\tilde{Q}))^T(\Delta_\times+\frac{1}{2}(W_\times\tilde{Q}+\tilde{Q}^T W_\times))\right\}\end{aligned} \tag{60}$$

Using (51) we have:

$$\frac{1}{2}k_4\,\mathrm{tr}\left\{(W_\times+P_a(\tilde{Q}))^T\tilde{Q}\Omega_\times\right\}=\frac{1}{2}k_4\,\mathrm{tr}\{M\Omega_\times\}=-k_4\Omega^T V(P_a(M)) \tag{61}$$

In the same way, we use (52) to get:

$$\frac{1}{2}k_4\,\mathrm{tr}\left\{(W_\times+P_a(\tilde{Q}))^T\Omega_\times\tilde{Q}^T\right\}=-k_4\Omega^T V(P_a(N)) \tag{62}$$

Therefore, the time derivative of $\mathcal{L}$ can be simplified:

$$\begin{aligned}\overset{\circ}{\mathcal{L}}=\Omega^T\{k_r V(P_a(\tilde{R}))+k_\omega\Gamma+k_3V(P_a(\tilde{Q}))-k_4V(P_a(M))-k_4V(P_a(N))\}-\frac{1}{2}k_3\,\mathrm{tr}(W_\times P_a(\tilde{Q}))\\+k_4\,\mathrm{tr}\left\{(W_\times+P_a(\tilde{Q}))^T(\Delta_\times+\frac{1}{2}(W_\times\tilde{Q}+\tilde{Q}^TW_\times))\right\}\end{aligned} \tag{63}$$

Choosing Γ according to (49) leads to:

$$\overset{\circ}{\mathcal{L}}=-\frac{1}{2}k_3\,\mathrm{tr}(W_\times P_a(\tilde{Q}))+k_4\,\mathrm{tr}\left\{(W_\times+P_a(\tilde{Q}))^T(\Delta_\times+\frac{1}{2}(W_\times\tilde{Q}+\tilde{Q}^TW_\times))\right\} \tag{64}$$

As $W_\times^T=-W_\times$, and introducing $P_a(\tilde{Q})$ in the first term:

$$\overset{\circ}{\mathcal{L}}=\frac{1}{2}k_3\,\mathrm{tr}\left\{\left(W_\times-P_a(\tilde{Q})+P_a(\tilde{Q})\right)^TP_a(\tilde{Q})\right\}+k_4\,\mathrm{tr}\left\{(W_\times+P_a(\tilde{Q}))^T(\Delta_\times+\frac{1}{2}(W_\times\tilde{Q}+\tilde{Q}^TW_\times))\right\} \tag{65}$$

$$\overset{\circ}{\mathcal{L}}=-\frac{1}{2}k_3\,\mathrm{tr}(P_a(\tilde{Q})^TP_a(\tilde{Q}))+\mathrm{tr}\left\{(W_\times+P_a(\tilde{Q}))^T\left(\frac{1}{2}k_3P_a(\tilde{Q})+k_4\Delta_\times+\frac{1}{2}k_4(W_\times\tilde{Q}+\tilde{Q}^TW_\times)\right)\right\} \tag{66}$$

Taking Δ as defined in (50), one has:

$$\overset{\circ}{\mathcal{L}}=-\frac{1}{2}k_3\,\mathrm{tr}\left\{P_a(\tilde{Q})^TP_a(\tilde{Q})\right\}-\frac{1}{2}k_5\,\mathrm{tr}\left\{(W_\times+P_a(\tilde{Q}))^T(W_\times+P_a(\tilde{Q}))\right\} \tag{67}$$

Using again identity (6), we finally have:

$$\overset{\circ}{\mathcal{L}}=-k_3\left\|V(P_a(\tilde{Q}))\right\|^2-k_5\left\|V(W_\times+P_a(\tilde{Q}))\right\|^2 \tag{68}$$

ensuring that $\mathcal{L}$ is strictly decreasing until $P_a(\tilde{Q})\to 0$ and $W_\times\to -P_a(\tilde{Q})$, i.e. $W_\times\to 0$. Denote by $(\gamma_{\tilde{Q}},n_{\tilde{Q}})$ the angle-axis coordinates of $\tilde{Q}$. Using (7), one has:

$$k_3(1-\cos(\gamma_{\tilde{Q}}))=\frac{1}{2}k_3\,\mathrm{tr}(I_d-\tilde{Q})\le\mathcal{L} \tag{69}$$

Since $\mathcal{L}$ is decreasing, we have $\mathcal{L}\le\mathcal{L}(0)$. Using (55) it yields:

$$k_3(1-\cos(\gamma_{\tilde{Q}}))\le\mathcal{L}\le\mathcal{L}(0)<2k_r \tag{70}$$

Using (53), we get:

$$1-\cos(\gamma_{\tilde{Q}})<2\frac{k_r}{k_3}<2 \tag{71}$$

From $P_a(\tilde{Q})\to 0$, we have $\gamma_{\tilde{Q}}=0$ or $\gamma_{\tilde{Q}}=\pm\pi$. The second possibility is excluded by (71). Therefore we have $\tilde{Q}\to I_d$. By (47), it yields $\tilde{R}\to Q$. By continuity and using La Salle's principle, we get $\overset{\circ}{\tilde{R}}\to\overset{\circ}{Q}$. Using the first equation of (48) and the third equation of (45), one has $\tilde{R}\Omega_\times\to -QW_\times$. Since $\tilde{R}$ is orthogonal, we get $\Omega_\times\to -\tilde{R}^T QW_\times$. Using $W_\times\to 0$ it yields $\Omega_\times\to 0$ and then $\Omega\to 0$. Therefore, using the first equation of (48), we get $\overset{\circ}{\tilde{R}}\to 0$. By continuity $\overset{\circ}{\Omega}\to 0$ and then, by the second equation of system (48), $\Gamma\to 0$. Knowing that $P_a(\tilde{Q})$ and $W_\times$ converge to zero, one can ensure that, respectively from (51) and (52), M and N converge to zero. Combining the above discussion with the fact that $\Gamma\to 0$, equation (49) ensures that $P_a(\tilde{R})\to 0$.

Similarly to the previous analysis on $\tilde{Q}$, let us denote $(\gamma_{\tilde{R}}, n_{\tilde{R}})$ the angle-axis coordinates of $\tilde{R}$. One has:

$$k_r(1-\cos(\gamma_{\tilde{R}}))=\frac{1}{2}k_r\ \mathrm{tr}(I_d-\tilde{R})\leq\mathcal{L}\leq\mathcal{L}(0)<2k_r \tag{72}$$

It yields

$$1-\cos(\gamma_{\tilde{R}})<2 \tag{73}$$

From $P_a(\tilde{R})\to 0$, we have $\gamma_{\tilde{R}}=0$ or $\gamma_{\tilde{R}}=\pm\pi$. The second possibility is excluded by (73). Therefore, we finally have $\tilde{R}\to I_d$ and $\tilde{R}\to R^d$. ∎

Lemma 2 ensures that the control (49) along with the virtual control (50) asymptotically stabilizes the orientation dynamics (48) without using any measurement of the angular velocity.

Remark 4: *The time-scale parameter $\varepsilon>0$ is chosen such that the deviation $\tilde{R}$ converges to I_d faster than the translational dynamics ($\varepsilon\ll 1$).*

Remark 5: *The condition (55) is not very conservative. Choosing $W(0)=0$ and $Q(0)=\tilde{R}(0)$, the condition (55) can be simplified:*

$$\frac{k_r}{k_\omega}>\frac{E_\omega(0)}{(2-E_r(0))} \tag{74}$$

with $E_r(0)=\frac{1}{2}\,\mathrm{tr}(I_d-\tilde{R}(0))$ *and* $E_\omega(0)=\frac{1}{2}\Omega(0)^T I\Omega(0)$.

6. Stability Analysis

We consider the full dynamics of the system along with virtual states and with the orientation error term in the translational dynamics:

$$\begin{cases} \dot{\xi}=v \\ \dot{v}=-\frac{T}{m}R^d e_3+ge_3-\frac{T}{m}(R-R^d)e_3 \\ \dot{q}=-w \\ \dot{w}=\delta \\ \overset{\circ}{\tilde{R}}=\tilde{R}\Omega_\times \\ I\overset{\circ}{\Omega}=-\Omega\times I\Omega+\Gamma \\ \overset{\circ}{\tilde{Q}}=W_\times\tilde{Q}+\tilde{Q}\Omega_\times \\ \overset{\circ}{W}=\Delta \end{cases} \tag{75}$$

with $\tilde{R}$ and $\tilde{Q}$ respectively defined by (46) and (47).

Proposition 1: *Consider the system dynamics (75). Under the conditions (24), (53) and (55), the control laws (21) and (49), along with the virtual controls (22) and (50), asymptotically stabilize the system (75).*

Sketch of the proof:

By Lemma 2, under the conditions (53) and (55), the closed loop orientation dynamics are asymptotically stable when (49) and (50) are respectively used as control and virtual control. By Lemma 1, under the condition (24), the input T is bounded. Therefore, the orientation error term $-\frac{T}{m}(R-R^d)e_3$ asymptotically converges to zero.

Since, from Lemma 1, the control of the translational dynamics is exponentially stabilizing for $R\equiv R^d$, we can use (Khalil, 2002) to conclude that the control of the translational dynamics is asymptotically stabilizing in presence of the orientation error term.

Therefore, the system (75) is asymptotically stable when the control laws (21) and (49) are used along with the virtual control laws (22) and (50). ■

By introducing virtual states, we have been able to design stabilizing controllers for the position and attitude of the VTOL UAV model using no measurement of the linear velocity v nor of the angular velocity Ω.

Remark 6: *In the case where only the linear velocity* v *of the vehicle is not measured, a detailed proof using the singular perturbation theory can be found in (Bertrand et al., 2008).*

Remark 7: *Control laws for trajectory tracking, in the case where the linear and angular velocities are not measured, have been proposed in (Bertrand et al., 2007b).*

7. Simulation Results

The VTOL UAV is described by the following parameters: $m = 2.5\,\text{kg}$, $I_1 = I_2 = 0.13\,\text{kg.m}^2$ and $I_3 = 0.16\,\text{kg.m}^2$. The gravitational acceleration is $g = 9.81\,\text{m.s}^{-2}$.
Simulation results are provided for stabilization at hover around the desired position $\chi^d = [3\ -1\ 1]^T$ (m), starting from the initial condition $\chi_0 = \begin{bmatrix}5 & -3 & 4\end{bmatrix}^T$ (m), $\begin{bmatrix}\psi_0 & \theta_0 & \phi_0\end{bmatrix} = \begin{bmatrix}0 & -8 & -10\end{bmatrix}^T$ (deg), $v_0 = 0$ and $\Omega_0 = 0$. The desired yaw ψ^d was chosen to be equal to zero.
The values of the gains are: $k_x = 0.2$, $k_v = 3.0$, $k_1 = 0.8$, $k_2 = 0.8$, $k_r = 0.74$, $k_\omega = 3.3$, $k_3 = 12$, $k_4 = 0.25$, $k_5 = 6.1$.

Figure 3 presents the coordinates of the position error $\xi = [x\ y\ z]^T$ and attitude angles. Stabilization of the UAV model is achieved from the given initial condition with satisfying behaviour performances. The input T and the components of the control torque Γ are plotted in Figure 4.

The evolution of the angular deviation terms $\tilde{\phi} = \phi - \phi^d$, $\tilde{\theta} = \theta - \theta^d$ and $\tilde{\psi} = \psi - \psi^d$ are presented in Figure 5. It can be verified that these terms converge faster than the closed loop translation, hence validating the time scale separation approach used for the design of the controllers.

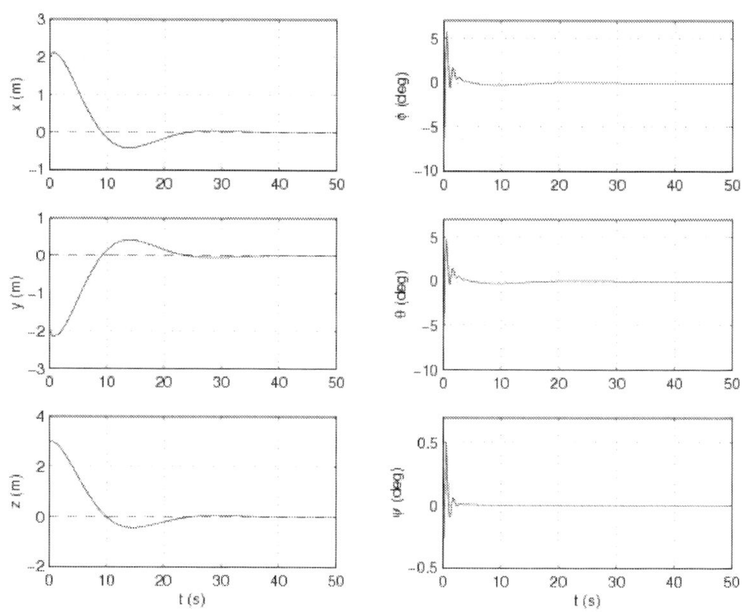

Figure 3. Position error and attitude angles

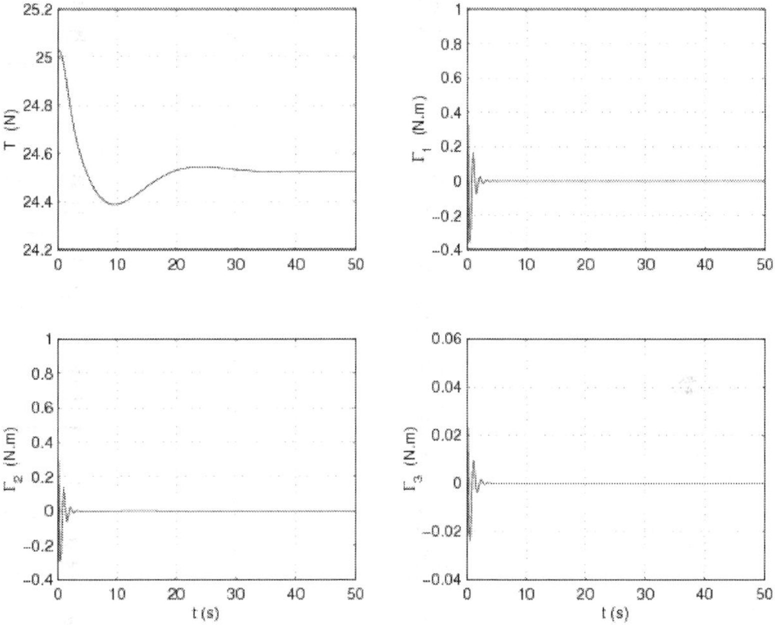

Figure 4. Control inputs

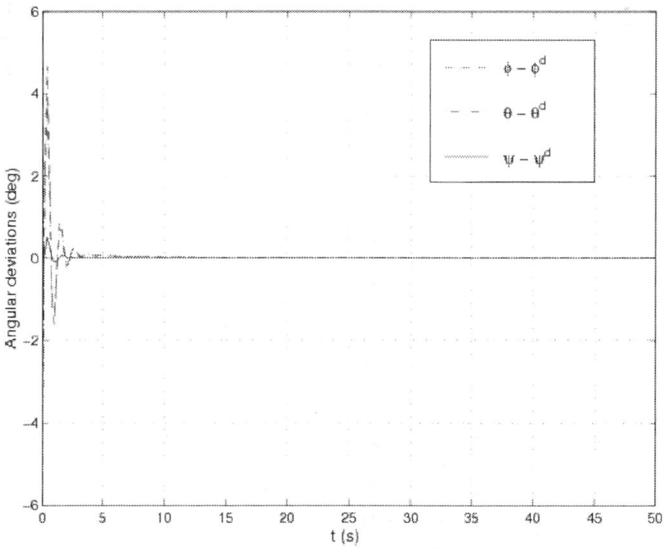

Figure 5. Angular deviation terms

8. Conclusion

In this chapter, we have presented a method to design guidance and control laws for the stabilization of a scale-model VTOL UAV when no measurements of the linear velocity nor of the angular velocity are available.

Motivated by the cascade structure of the model and by the singular perturbation approach, the controller is designed in two steps by considering a time-scale separation between translational and orientation dynamics. The position controller computes the magnitude of external forces, considered as a control input for the translational dynamics, and the desired orientation of the UAV. The attitude controller delivers the control torque ensuring asymptotic convergence of the actual orientation to the desired one.

By the proposed approach, these two feedback controllers have been designed by introducing virtual states in the system dynamics, and without using any observer. It is also worth noticing that this work is based on a kinematic representation exploiting the $SO(3)$ group and its manifold.

Elements for stability analysis have been given and simulation results have been provided to illustrate the good performances of the proposed approach.

9. References

Arkela, M. R. (2001). Rigid Body Attitude Tracking without Angular Velocity Feedback. *Systems & Control Letters*, Vol. 42, pp. 321-326, 2001

Astolfi, A. & Lovera, M. (2002). Global Spacecraft Attitude Control using Magnetic Actuators, *Proceedings of the American Control Conference*, Vol. 2, pp. 1331-1335, Anchorage, USA, 2002

Bertrand, S.; Piet-Lahanier, H. & Hamel, T. (2007a). Contractive Model Predictive Control of an Unmanned Aerial Vehicle, *17th IFAC Symposium on Automatic Control in Aerospace*, Toulouse, France, 2007

Bertrand, S.; Hamel, T. & Piet-Lahanier, H. (2007b). Trajectory Tracking of an Unmanned Aerial Vehicle Model using Partial State Feedback, *Proceedings of the European Control Conference 2007*, pp. 307-314, Kos, Greece, 2007

Bertrand, S.; Hamel, T.; Piet-Lahanier, H. (2008). Stability Analysis of an UAV Controller Using Singular Perturbation Theory, *Proceedings of the 17th IFAC World Congress*, pp. 5706-5711, Seoul, Korea, 2008

Bouabdallah, S. & Siegwart, R. (2005). Backstepping and Sliding-Mode Techniques Applied to an Indoor Micro Quadrotor, *Proceedings of the 2005 IEEE International Conference on Robotics and Automation*, pp 2259-2264, Barcelona, Spain, 2005

Burg, T.; Dawson, D.; Hu, J. & de Queiroz, M. (1996). An Adaptive Partial State-Feedback Controller for RLED Robot Manipulators. *IEEE Transactions on Automatic Control*, Vol. 41, No. 7, pp. 1024-1030, 1996

Burg, T.; Dawson, D. & Vedagarbha, P. (1997). A Redesigned DCAL Controller without Velocity Measurements: Theory and Demonstration. *Robotica*, Vol. 15, pp. 337-346, 1997

Calise, A. J. (1976). Singular Perturbation Methods for Variational Problems in Aircraft Flight. *IEEE Transactions on Automatic Control*, Vol. 41, No. 3, pp. 345-353, 1976

Castillo, P.; Dzul, A. & Lozano, R. (2004). Real-Time Stabilization and Tracking of a Four-Rotor Mini Rotorcraft. *IEEE Transactions on Control Systems Technology*, Vol. 12, No. 4, pp. 510-516, 2004

Costic, B. T.; Dawson, D. M.; de Queiroz, M. S. & Kapila, V. (2000). A Quaternion-Based Adaptive Attitude Tracking Controller Without Velocity Measurements, *Proceedings of the 39th IEEE Conference on Decision and Control*, Vol. 3, pp. 2424-2429, Sydney, Australia, 2000

Dixon, W. E.; Zergeroglu, E.; Dawson, D. M. & Hannan, M. W. (2000). Global Adaptive Partial State Feedback Tracking Control of Rigid-Link Flexible-Joints Robots. *Robotica*, Vol. 18, pp. 325-336, 2000

Do, K. D.; Jiang, Z. P. & Pan, J. (2003). On Global Tracking Control of a VTOL Aircraft without Velocity Measurements. *IEEE Transactions on Automatic Control*, Vol. 48, No. 12, pp. 2212-2217, 2003

Frazzoli, E.; Dahleh, M.A. & Feron, E. (2000). Trajectory Tracking Control Design for Autonomous Helicopters using a Backstepping Algorithm, *Proceedings of the American Control Conference*, Vol. 6, pp. 4102-4107, Chicago, USA, 2000

Hamel, T. & Mahony, R. (2004). Pure 2D Visual Control for a Class of Under-Actuated Dynamic Systems, *Proceedings of the 2004 IEEE International Conference on Robotics and Automation*, Vol. 3, pp. 2229-2235, New Orleans, USA, 2004

Hamel, T.; Mahony, R; Lozano, R. & Ostrowski, J. (2002). Dynamic Modelling and Configuration Stabilization for an X4-Flyer, *Proceedings of the 15th IFAC World Congress*, Vol. 15, Barcelona, Spain, 2002

Khalil, H. K. (2002). *Nonlinear Systems*, Prentice Hall, ISBN: 0130673897, New Jersey, USA

Kim, H. J.; Shim, D. H. & Sastry, S. (2002). Nonlinear Model Predictive Tracking Control for Rotorcraft-Based Unmanned Aerial Vehicles, *Proceedings of the American Control Conference*, Vol. 5, pp. 3576-3581, Anchorage, USA, 2002

Kondak, K.; Bernard, M.; Meyer, N. & Hommel, G. (2007). Autonomously Flying VTOL-Robots: Modeling and Control, *Proceedings of the 2007 IEEE International Conference on Robotics and Automation*, pp. 736-741, Roma, Italy, 2007

Kundak, N. & Mettler, B. (2007). Experimental Framework for Evaluating Autonomous Guidance and Control Algorithms for Agile Aerial Vehicles, *Proceedings of the European Control Conference 2007*, pp. 293-300, Kos, Greece, 2007

Lizarralde, F. & Wen, J. T. (1996). Attitude Control without Angular Velocity Measurement: A Passivity Approach. *IEEE Transactions on Automatic Control*, Vol. 41, No. 3, pp. 468-472, 1996

Mahony, R. & Hamel, T. (2004). Robust Trajectory Tracking for a Scale Model Autonomous Helicopter. *International Journal of Robust and Nonlinear Control*, Vol. 14, pp. 1035-1059, 2004

Mahony, R.; Hamel, T. & Dzul, A. (1999). Hover Control via Approximate Lyapunov Control for a Model Helicopter, *Proceedings of the 38th IEEE Conference on Decision and Control*, Vol. 4, pp. 3490-3495, Phoenix, USA, 1999

Njaka, C. E.; Menon, P. K. & Cheng, V. H. L. (1994). Towards an Advanced Nonlinear Rotorcraft Flight Control System Design, *13th AIAA/IEEE Digital Avionics Systems Conference*, Phoenix, USA, 1994

Tayebi, A. (2007). A Velocity-free Attitude Tracking Controller for Rigid Spacecraft, *Proceedings of the 46th IEEE Conference on Decision and Control*, pp. 6430-6434, New Orleans, USA, 2007

Tsiotras, P. (1998). Further Passivity Results for the Attitude Control Problem. *IEEE Transactions on Automatic Control*, Vol. 43, No. 11, pp. 1597-1600, 1998

Valenti, M.; Bethke, B.; Fiore, G.; How, J. P. & Feron, E. (2006). Indoor Multi-Vehicle Flight Testbed for Fault Detection, Isolation, and Recovery, *AIAA Guidance, Navigation, and Control Conference and Exhibit*, Keystone, USA, 2006

Wong, H.; de Queiroz, M. S. & Kapila, V. (2000). Adaptive Tracking Control Using Synthesized Velocity from Attitude Measurements, *Proceedings of the American Control Conference*, Vol. 3, pp. 1572-1576, Chicago, USA, 2000

7

Flight Control System Design Optimisation via Genetic Programming

Anna Bourmistrova and Sergey Khantsis
Royal Melbourne Institute of Technology
Australia

1. Introduction

This chapter presents a methodology which is developed to design a controller that satisfies the objectives of shipboard recovery of a fixed-wing UAV. The methodology itself is comprehensive and should be readily applicable for different types of UAVs and various task objectives. With appropriate modification of control law representation, the methodology can be applied to a broad range of control problems. Development of the recovery controller for the UAV Ariel is a design example to support the methodology.

This chapter focuses on adaptation of Evolutionary Algorithms for aircraft control problems. It starts from analysis of typical control laws and control design techniques. Then, the structure of the UAV controller and the representation of the control laws suitable for evolutionary design are developed. This is followed by the development of the general evolutionary design algorithm, which is then applied to the UAV recovery problem. Finally the presented results demonstrate robust properties of the developed control system.

2. Aircraft flight control

2.1 Overview of types of feedback control

Not unlike the generic control approach, aircraft flight control is built around a feedback concept. Its basic scheme is shown in Fig. 1. The controller is fed by the difference between the commanded reference signal r and the system output y. It generates the system control inputs u according to one or another algorithm.

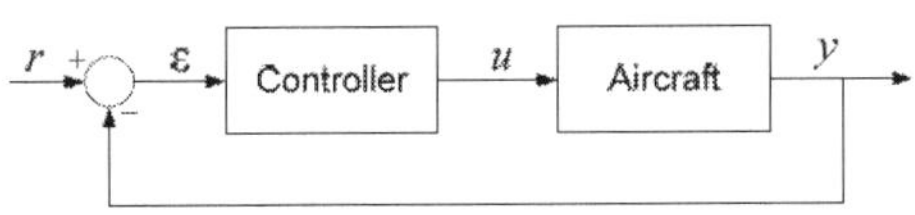

Figure 1. Feedback system (one degree of freedom)

One of the main tasks of flight control as an engineering discipline is design of the controllers which enable a given aircraft to complete a defined mission in the most optimal manner, where optimality is based on mission objective. A number of techniques of producing a required controller have been developed over the last hundred years since the first altitude hold autopilot was introduced by Sperry Gyroscope Company in 1912. Most of the techniques make certain assumptions about the controlled system (i.e. the aircraft), most notably linearity of its dynamics and rigidity of the airframe, to simplify synthesis of the controller. A brief overview of types of feedback control is presented below.

2.1.1 On-Off control

The simplest feedback control is the *on-off control*, also referred to among engineers as *bang-bang control*. This control law can be expressed as follows:

$$u = u_0 + \begin{cases} k, z < a \\ 0, z > a \end{cases} \tag{1}$$

where u_0 and k are arbitrary offset and gain suitable for the given system, and a is the threshold (set point). It can be extended to a multiple choice situation.

This law is not particularly suitable for direct flight control in normal situations. Indeed, with only two (or several) discrete control input levels, the system output will tend to oscillate about the set point, no matter how well damped the system is, because the control signal will not switch until the set point is already passed. Moreover, if the dynamics of the system is fast or a significant amount of noise in the system output is present, the controller will be switching also fast ('hunting'), possibly causing extensive wear to the control actuators.

To prevent such behaviour, a special 'dead zone' (or 'deadband') is established around the set point where no switching occurs. Obviously, this reduces the accuracy of control. However, the on-off control comes into play when event handling is required. A classic example of application of such control for aircraft is stall recovery. When angle of attack exceeds a specified value, a nose-down elevator command is issued. A similar logic can be implemented for overload protection or ground collision avoidance.

Another important area in the aerospace field where on-off rules can be successfully applied is internal switching between the controllers (or controller parameters). This approach is known as Gain scheduling (Rugh, 2000). The technique takes advantage of a set of relatively simple controllers optimised for different points of the flight envelope (or other conditions). However, it was found that rapid switching may cause stability and robustness problems (Shamma & Athans, 1992). One of the popular simple solutions is to 'blend' (interpolate) the output of two or more controllers, which effectively turns simple on-off switching into a complicated control method.

2.1.2 PID control

The proportional-integral-derivative (PID) control is probably the most widely used type of control, thanks to the simplicity of its formulation and in most cases, predictable characteristics. In a closed-loop system like that in Fig. 1, its control law is

$$u = K_P\varepsilon + K_I\int \varepsilon dt + K_D\dot{\varepsilon} \tag{2}$$

where the parameters K_P, K_I and K_D are coefficients for proportional, integral and derivative components of the input error signal respectively. By adjusting these three parameters, the desired closed-loop dynamics can be obtained.

Probably the most problematic issue with the PID control is due to the differential term. While being important for good response time and high-speed dynamics, the differential component keenly suffers from both instrumental noise and calculation errors. Indeed, even a small amount of noise can greatly affect the slope of the input signal. At the same time, numerical calculation of the derivative (for a digital controller in particular) must be done with a fairly small time step to obtain correct slope at a given point. Although lowpass filtering applied to the input signal smoothens the signal, it severely compromises the usefulness of the derivative term because the low-pass filter and derivative control effectively cancel each other out.

In contrast, the integral term averages its input, which tends to eliminate noise. However, a common problem associated with the integral control owes exactly to its 'memory'. When a large input value persists over a significant amount of time, the integral term becomes also large and remains large even after the input diminishes. This causes a significant overshoot to the opposite values and the process continues.

In general, integral control has a negative impact on stability and care must be taken when adjusting the integral coefficient. Limiting the integrator state is a common aid for this problem. A more elegant approach involves 'washing out' the integrator state by incorporating a simple gain feedback, effectively implementing a low-pass filter to the input signal. PID control found wide application in the aerospace field, especially where near-linear behaviour takes place, for example, in various hold and tracking autopilots such as attitude hold and flight path following. The techniques of selecting the optimal PID coefficients are well established and widely accepted. Typically, they use the frequency domain as a design region and utilise phase margins and gain margins to illustrate robustness. However, design of a PID controller for nonlinear or multi-input multi-output (MIMO) systems where significant coupling between the different system inputs and outputs exists is complicated. As the aircraft control objectives evolved, new design and control techniques were developing. Although many of them essentially represent an elaborated version of PID control, they are outlined in the following sections.

2.1.3 Linear optimal control

The concept of optimality in mathematics refers to minimisation of a certain problem dependent cost functional:

$$J = \int_0^T g(t, x, u)dt + h(x(T)) \tag{3}$$

where T is the final time, u is the control inputs and x is the state of the controlled system. The last term represents the final cost that depends on the state in which the system ends up. Optimal control is, therefore, finding the control law $u(t)$ that minimises J. In general, for an arbitrary system and cost functional, only numerical search can find the optimal (or near optimal) solution.

There is a number of types of linear optimal control, such as the *Linear Quadratic Regulator (LQR)*, the *Linear Quadratic Gaussian with Loop Transfer Recovery (LQG/LTR)* and the extended Kalman filter (EKF). The theory and application of these control techniques and Kalman filtering is detailed in many common control and signal processing textbooks, such as for example (Brown & Hwang, 1992; Bryson & Ho, 1975; Kalman, 1960; Maciejowski, 1989).

The Kalman filter relies on the model of the system (in its predicting part), and the guaranteed performance of the controller can easily be lost when unmodelled dynamics, disturbances and measurement noise are introduced. Indeed, robustness is a challenging issue of modern control designs. Also, the well known in classic linear design phase and gain margin concepts cannot be easily applied to the multivariable systems that modern control is so suited for. These problems led to the introduction of robust modern control theory.

2.1.4 Robust modern control

Unlike traditional optimal control, robust optimal control minimises the influence of various types of uncertainties in addition to (or even instead of) performance and control energy optimisation. Generally, this implies design of a possibly low gain controller with reduced sensitivity to input changes. As a consequence, robust controllers often tend to be conservative and slow. On the other hand, they may be thought as the stabilising controllers for the whole set of plants (which include a range of uncertainties) and not only for the modelled system, which is a more demanding task.

The majority of modern robust control techniques have origins in the classical frequency domain methods. The key modification of the classic methods is shifting from eigenvalues to singular values (of the transfer function that describes the system), the singular value Bode plot being the major indicator of multivariable feedback system performance (Doyle & Stein, 1981).

Probably the most popular modern robust control design techniques (particularly in the aerospace field) are H_2 and H_∞ control, also known as the frequency-weighted LQG synthesis and the small gain problem respectively. These techniques and the underlying theories are thoroughly described in several works, notably (Doyle at al., 1989; Kwakernaak, 1993; Zhou & Doyle, 1998).

The H_∞ control design found extensive use for aircraft flight control. One of the first such applications was the development of controllers for the longitudinal control of a Harrier jump jet (Hyde, 1991). This work has been subsequently extended in (Postlethwaite & Bates, 1999) to fully integrated longitudinal, lateral and propulsive control. Other works include (Kaminer et al., 1990), where a lateral autopilot for a large civil aircraft is designed, and (Khammash & Zou, 1999), where a robust longitudinal controller subject to aircraft weight and c.g. uncertainty is demonstrated.

A mixed H_2 / H_∞ approach is applied in (Shue & Agarwal, 1999) to design an autoland controller for a large commercial aircraft. The method employed here utilises the H_2 controller for slow trajectory tracking and the H_∞ controller for fast dynamic robustness and disturbance rejection.

Several H_∞ controllers have been tried to accomplish the UAV shipboard launch task (Crump, 2002). It has been found that these controllers perform quite well in nominal situations. However, in the presence of large disturbances which place the aircraft well beyond its linear design operating point, the controllers performed poorly (sometimes extremely). At the same time, inability to include even simple static nonlinearities such as time delays and saturations made it difficult to synthesise a practical controller for this task within linear approach. Another deficiency found is common to all frequency domain techniques: the frequency domain performance specifications cannot be rigidly translated into time and spatial domain specifications. Meanwhile, time and especially spatial (trajectory) constrains are crucial for both launch and recovery tasks.

The time domain performance can be accounted for directly in the time domain l_1 design (Blanchini & Sznaier, 1994; Dahleh & Pearson, 1987).

It is often extended to include the frequency domain objectives, resulting in a mixed norm approach. However, l_1 design is plagued by the excessive order of the generated controllers. This is usually solved by reducing the problem to suboptimal control, imposing several restrictions on the system and performance specifications (Sznaier & Holmes, 1996). The application of l_1 approach to flight control is discussed in (Skogestad & Postlewaite, 1997), with the conclusion that controllers with excessive order will generally be produced when using practical constraints.

There have been attempts to solve the H_∞ optimal control problems for nonlinear systems. However, these methods usually rely on very limiting assumptions about the model, uncertainties and disturbance structure. The mathematical development of nonlinear H_∞ control can be found in (Dalsamo & Egeland, 1995). An example of nonlinear control of an agile missile is given in (Wise, 1996).

Despite a very complicated solution algorithm, this work is limited by a linear assumption on the vehicle aerodynamics, reducing the benefits gained from the use of nonlinear control.

2.1.5 Nonlinear control

Linear control design techniques have been used for flight control problems for many years. One of the reasons why aircraft can be controlled quite well by linear controllers is that they behave almost linearly through most of their flight envelope. However, when the aircraft is required to pass through a highly nonlinear dynamic region or when other complicated control objectives are set, it has been found by several researchers that it is difficult to obtain practical controllers based on linear design techniques. The UAV shipboard launch and recovery tasks are substantially nonlinear problems. The sources of nonlinearities are the aerodynamic forces generated at low airspeeds and high angles of attack (especially when wind disturbances are present); trajectory constraints imposed due to proximity of ground (water) and ship installations; kinematic nonlinearities when active manoeuvring is required; actuator saturations and some more.

In contrast to the linear systems, the characteristics of the nonlinear systems are not simply classified and there are no general methods comparable in power to those of linear analysis. Nonlinear techniques are quite often designed for individual cases, regularly with no mathematical justification and no clear idea of re-applicability of the methods.

Some of the more popular nonlinear control techniques are covered in textbooks (Atherton, 1975; Graham & McRuler, 1971).

The most basic nonlinear control laws and the On-off control and Gain scheduling, noting that these controllers often lack robustness when the controllers are scheduled rapidly.

Another modern control technique remotely related to the on-off control is *variable structure* (also known as *sliding mode*) control (DeCarlo et al., 1988; Utkin, 1978). In this approach, a hypersurface (in state space) called *sliding surface* or *switching surface* is selected so that the system trajectory exhibits desirable behaviour when confined to this hypersurface. Depending on whether the current state is above or below the sliding surface, a different control gain is applied. Unlike gain scheduling, the method involves high speed switching to keep the system on the sliding surface.

A completely different approach is to enable applicability of the well known linear control methods to control nonlinear systems. This can be achieved using *nonlinear dynamic inversion*. This process, also known as *feedback linearisation*, involves online approximate linearisation of a nonlinear plant via feedback. Dynamic inversion gained particular attention in aviation industry in the late 1980s and 90s, aiming to control high performance fighters during high angle of attack manoeuvres (known as *supermanoeuvres*). One of the early applications is NASA High Angle of Attack Vehicle (HARV) (Bugajski & Enns, 1992). In this work, quite good simulation performance results are obtained; however, with the inversion based on the same simulation model, any possible discrepancies are transferred into the controller, leading to questionable results in physical implementation with respect to incorrectly modelled dynamics.

2.1.6 Intelligent control

Intelligent control is a general and somewhat bold term that describes a diverse collection of relatively novel and non-traditional control techniques based on the so called *soft computing* approach. These include neural networks, fuzzy logic, adaptive control, genetic algorithms and several others. Often they are combined with each other as well as with more traditional methods; for example, fuzzy logic controller parameters being optimised using genetic algorithms or a neural network driving a traditional linear controller.

Neural network (NN), very basically, is a network of simple nonlinear processing elements (neurons) which can exhibit complex global behaviour determined by element parameters and the connections between the processing elements. The use of artificial neural networks for control problems receives an increased attention over the last two decades. It has been shown that a certain class of NN can approximate any continuous nonlinear function with any desired accuracy (Spooner, 2002). This property allows to employ NN for *system identification* purposes, which can be performed both offline and online. This approach is used in (Coley, 1998; Kim & Calise, 1997) for flight control of a tilt-rotor aircraft and the F-18 fighter.

Another area of intensive application of NN is fault tolerant control, made possible due to online adaptation capability of NN. In the recent work (Pashilkar et al., 2006), an 'add-on' neural controller is developed to increase auto-landing capabilities of a damaged aircraft with one of two stuck control surfaces. A significant increase in successful landings rate is shown, especially when the control surfaces are stuck at large deflections.

Fuzzy logic control also gained some popularity among flight control engineers. This type of control relies on approximate reasoning based on a set of rules where intermediate positions between 'false' and 'truth' are possible. Fuzzy logic control may be especially useful when a decision must be made between several controversial conditions. For example, in (Fernandez-Montesinos, 1999) fuzzy logic is used for windshear recovery in order to decide whether energy should be given to altitude or airspeed, based upon the current situation.

Adaptive control term covers a set of various control techniques that are capable of online adaptation. A good survey of adaptive control methods is given in (Astrom & Wittenmark, 1995). The applications of adaptive control is generally biased towards control for large time scales so that the controller has sufficient time to learn how to behave. This makes the relatively short-time recovery process unsuitable for online adaptation.

Evolutionary and *genetic algorithms* (EAs, GAs) are global optimisation techniques applicable to a broad area of engineering problems. They can be used to optimise the parameters of various control systems, from simple PID controllers (Zein-Sabatto & Zheng, 1997) to fuzzy logic and neural network driven controllers (Bourmistrova, 2001; Kaise & Fujimoto, 1999). Another common design approach is evolutionary optimisation of trajectories, accompanied by a suitable tracking controller (e.g. (Wang & Zalzala, 1996)). An elaborated study of applications of EAs to control and system identification problems can be found in (Uzrem, 2003).

Unlike the majority of other techniques, Genetic Algorithms (in the form of *Genetic Programming*) are able to evolve not only the parameters, but also the *structure* of the controller.

In general, EAs require substantial computational power and thus are more suitable for offline optimisation. However, online evolutionary-based controllers have also been successfully designed and used. The *model predictive control* is typically employed for this purpose, where the controller constantly evolves (or refines) control laws using an integrated simulation model of the controlled system. A comprehensive description of this approach is given in (Onnen et al., 1997).

2.2 Flight control for the UAV recovery task

Aircraft control at recovery stage of flight can be conventionally separated into two closely related, but distinctive tasks: guidance and flight control. Guidance is the high-level ('outer loop') control intended to accomplish a defined mission. This may be path following, target tracking or various navigation tasks. Flight control is aimed at providing the most suitable conditions for guidance by maintaining a range of flight parameters at their optimal levels and delivering the best possible handling characteristics.

2.2.1 Traditional landing

In a typical landing procedure, the aircraft follows a defined glide path. The current position of the aircraft with respect to the glidepath is measured in a variety of ways, ranging from pilot's eyesight to automatic radio equipment such as the Instrument Landing System (ILS). Basically, the objective of the pilot (or autopilot) is to keep the aircraft on the glidepath, removing any positioning error caused by disturbances and aircraft's dynamics. This stage of landing is known as *approach*. Approach may be divided into *initial approach*, in which the aircraft makes contact with ('fixes') the approach navigation system (or just makes visual contact with the runway) and aligns with the runway; and *final approach*, when the aircraft descends along a (usually) straight line. In conventional landing, final approach is followed by a flare manoeuvre or nose-up input to soften the touchdown; however, flare is typically not performed for shipboard landing due to random ship motion and severe constraints on the landing space.

The guidance task during final approach involves trajectory tracking with both horizontal and vertical errors (and their rates) readily available. It is important to note that the errors being physically measured are *angular* deviations of the aircraft position (Fig.2) as seen from the designated touchdown point (or more precisely, from where the radars or antennas or other guidance systems are located). They can be converted to linear errors Δh if the *distance* L to the aircraft is known.

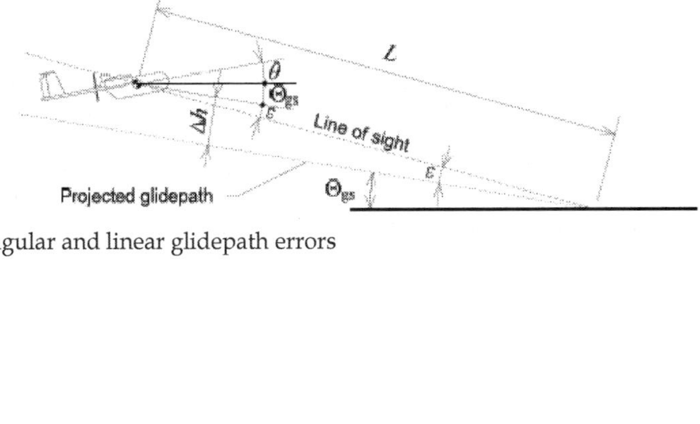

Figure 2. Angular and linear glidepath errors

However, in many applications precise distance measurement is unavailable. Nevertheless, successful landing can be carried out even without continuous distance information. This comes from the fact that exactly the angular errors are relevant to precise landing. Indeed, the goal is to bring the aircraft to a specified point on the runway (or on the deck) in a certain state. The choice of the landing trajectory only serves this purpose, considering also possible limitations and secondary objectives such as avoiding terrain obstacles, minimising noise level and fuel consumption and so on. The angular errors provide an adequate measurement of the current aircraft position with respect to the glidepath.
However, if the landing guidance system takes no account of distance and is built around the angular error only, it may cause stability problems at close distances, because the increasing sensitivity of the angular errors to any linear displacement effectively amplifies the system gain.

2.2.2 UAV control for shipboard recovery

As it was seen from the discussion in the previous section, landing of an aircraft is a well established procedure which involves following a predefined flight path. More often than not, this is a rectilinear trajectory on which the aircraft can be stabilised, and the control interventions are needed only to compensate disturbances and other sources of errors.
The position errors with respect to the ideal glidepath can be measured relatively easily. Shipboard landing on air carriers is principally similar; the main differences are much tighter error tolerances and absence of flare manoeuvres before touchdown. The periodic ship motion does have an effect on touchdown; however, it does not affect significantly the glidepath, which is projected assuming the average deck position. The choice of the landing deck size, glideslope, aircraft sink rate and other parameters is made to account for any actual deck position at the moment of touchdown. For example, a steeper glideslope (typically 4°) is used to provide a safe altitude clearance at the deck ramp for its worst possible position (i.e. ship pitched nose down and heaved up). This makes unnecessary to correct the ideal reference trajectory on the fly.
For the UAV shipboard recovery, ship oscillations in high sea cause periodic displacement of the recovery window (the area where capture can be done) several times greater than the size of the window itself. This fact (and also the assumption that the final recovery window position cannot be predicted for a sufficient time ahead) makes it impossible to project an optimal flight path when the final approach starts. Instead, the UAV must constantly track the actual position of the window and approach so that the final miss is minimised. Therefore, it turns out that the UAV recovery problem resembles that of homing guidance rather than typical landing. While stabilisation on a known steady flight path can be done relatively easy with a PID controller, homing guidance to a moving target often requires a more sophisticated control.
Not surprisingly, homing guidance found particularly wide application in ballistic missiles development, hence the accepted terminology owes to this engineering area. In the context of UAV recovery, the UAV moves almost straight towards the 'target' from the beginning (compared to typical missile intercept scenarios) and thus the velocity vector and the line of sight almost coincide.
However, there are two major difficulties that can compromise the effectiveness of Proportional Navigation (PN) for UAV recovery. First, PN laws are known as generating excessive acceleration demands near the target. For a UAV with limited manoeuvrability,

such demands may be prohibitive, especially at low approach airspeeds. On the other hand, the PN guidance strategy does not change during the flight. Several alternative guidance strategies with more favourable acceleration demands exist, e.g. augmented proportional navigation. However, it is unlikely they can sufficiently improve the guidance to an oscillating target such as ship's deck.

2.3 UAV controller structure

The objective is therefore to synthesise such guidance strategy that enables reliable UAV recovery, and to produce a controller that implements the target tracking guidance strategy. The evolutionary design (ED) method applied for this task allows to evolve automatically both the structure and the parameters of the control laws, thus potentially enabling to generate a 'full' controller, which links available measurements directly with the aircraft control inputs (throttle, ailerons, rudder and elevator) and implements both the guidance strategy and flight control (Fig. 3):

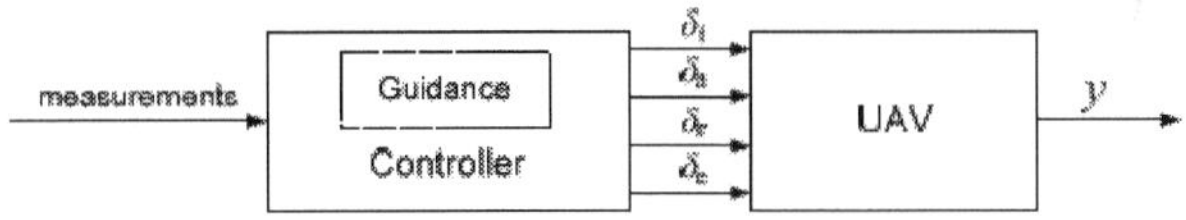

Figure 3. Full controller with embedded guidance strategy

However, this approach, even though appealing at first and requiring minimum initial knowledge, proves to be impractical as the computational demands of the evolutionary algorithms (EAs) soar exponentially with the dimensionality of the problem. It is therefore desirable to reduce complexity of the problem by reducing the number of inputs/outputs and limiting, if appropriate, possible structures of the controllers.

Another difficulty is the evaluation of the controller's performance. In ED, performance or fitness is multiobjective. Therefore, it is highly desirable to decompose this complex task into several simpler problems and to solve them separately.

A natural way of such decomposition is separating the trajectory control (guidance) and flight control. The guidance controller issues commands u_g to the flight controller, which executes these commands by manipulating the control surfaces of the UAV (Fig. 4). These two controllers can be synthesised separately using appropriate fitness evaluation for each case.

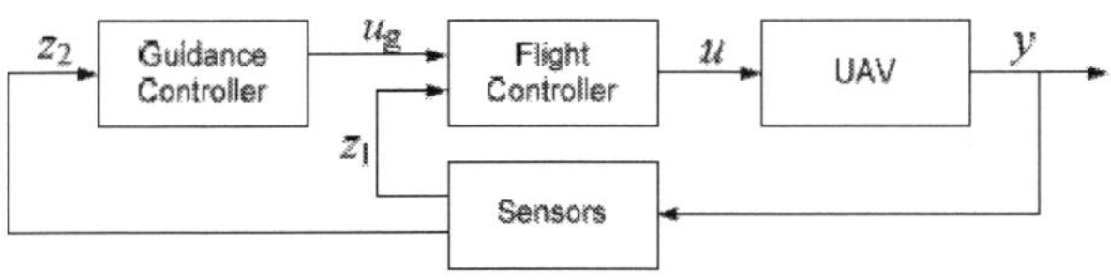

Figure 4. UAV recovery control diagram

2.3.1 Guidance controller

The internal structure of the controller is defined by the automatic evolutionary design based on predefined set of inputs and outputs. It is desirable to keep the number of inputs and outputs to minimum, but without considerably compromising potential performance.

An additional requirement to the outputs u_g is that these signals should be straightforwardly executable by the flight controller. This means that u_g should represent a group of measurable flight parameters such as body accelerations, velocities and Euler angles, which the flight controller can easily track.

The structure of the output part of the guidance controller is as shown in Fig. 5. With this scheme, the guidance laws produce general requests to change trajectory in horizontal and vertical planes. The kinematic converter then recalculates these requests to the form convenient for the flight controller. Both the bank angle and normal body load factor n_y can be relatively easily tracked, with the sensors providing direct measurements of their actual values. At the same time, this approach allows to evolve the horizontal and vertical guidance laws separately, which may be desirable due to different dynamics of the UAV's longitudinal and lateral motion and also due to computational limitations.

Input measurements to the guidance controller should be those relevant to trajectory. First of all, this is all available positioning information, pitch and yaw angles and airspeed. They do not account for steady wind, but still provide substantial information regarding the current 'shape' of trajectory.

The yaw angle fed into the controllers is corrected by the 'reference' yaw ψ_0, which is perpendicular to the arresting wire in the direction of anticipated approach. A zero yaw indicates that the UAV is pointed perpendicularly to the arresting wire (ignoring ship oscillations), which is the ideal condition in the absence of side wind and when the UAV moves along the ideal glidepath. This is similar to rotating the ground reference frame $O_g x_g y_g z_g$ by the correction yaw angle ψ_0. The rotated frame is referred as *approach ground reference frame*.

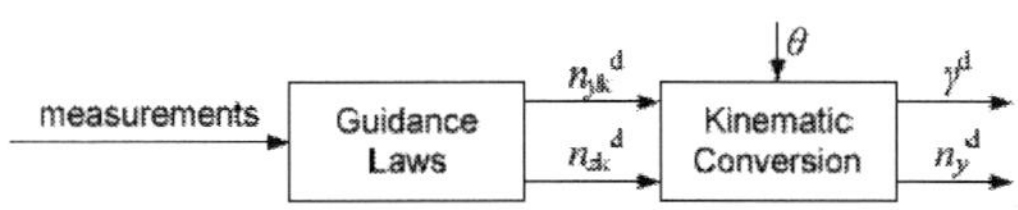

Figure 5. Guidance controller

The derived quantities from raw measurements are the vertical and lateral velocity components with respect to the approach ground reference frame.

Determination of the current UAV position and velocities with respect to the arresting wire is crucial for successful recovery. While previously discussed flight parameters may only help to improve the guidance quality, positioning carries direct responsibility for recovery.

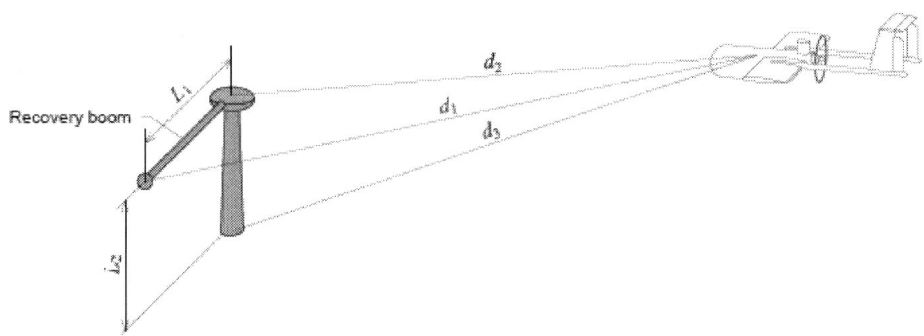

Figure 6. Positioning scheme

The system is based on radio distance metering and provides ten independent raw measurements (Fig. 6): three distances d_1, d_2 and d_3 from the UAV to the radio transmitters located at both ends of the recovery boom which supports the arresting wire and at the base of recovery mast; three rates of change of these distances; distance differences (d_1 - d_2) and (d_3 - d_2); and rates of change of the differences.
The guidance laws evolution process is potentially capable to produce the laws directly from raw measurements, automatically finding necessary relationships between the provided data and the required output.
The target spot elevation is chosen to be a constant, and the value is determined in view of the expected cable sag obtained from simulation of the cable model. For normal approach speed, the cable sag varies between 3.5 and 4.5 m. Accordingly, the target spot elevation is chosen to be h_T = 2 m above the arresting wire.
The recovery procedure, if successful, lasts until the cable hook captures the arresting wire. This happens when the UAV have moved about the full length of the cable *past* the recovery boom. However, position measurements may be unavailable beyond the boom threshold, and even shortly before the crossing the readings may become unreliable. For these reasons, the terminal phase of approach, from the distance about 6–10 m until the capture (or detection of a miss), should be handled separately.
It is possible to disconnect the guidance controller several metres before the recovery boom without affecting the quality of guidance. The allowed error (approximately 2 m in all directions, determined by the lengths of the arresting wire and the cable) should absorb the absence of controlled guidance in the last 0.3 to 1 second (depending on the headwind) in most situations.

2.3.2 Flight controller

Flight controller receives two inputs from the guidance controller: bank angle demand γ^d and normal body load factor demand n_y^d. It should track these inputs as precisely as possible by manipulating four aircraft controls: throttle, ailerons, rudder and elevator.
The available measurements from the onboard sensors are body angular rates ω_x, ω_y, ω_z from rate gyros, Euler angles γ, ψ, θ from strapdown INS, body accelerations n_x, n_y, n_z from the respective accelerometers, airspeed V_a, aerial angles α and β, actual deflection of the control surfaces δ_a, δ_r, δ_e, and engine rotation speed N_{rpm}.
For simplicity of design, the controller is subdivided into independent longitudinal and lateral components. In longitudinal branch, elevator tracks the input signal n_y^d, while throttle is responsible for maintaining a required airspeed. In lateral control, naturally, ailerons track γ^d, while rudder minimises sideforce by keeping n_z near zero.

3. Evolutionary design

The Evolutionary Design (ED) presented in this section, generally, takes no assumptions regarding the system and thus can be used for wide variety of problems, including nonlinear systems with unknown structure. In many parts, this is a novel technique, and the application of ED to the UAV guidance and control problems demonstrates the potential of this design method.

The core of evolutionary design is a specially tailored evolutionary algorithm (EA) which evolves both the structure and parameters of the control laws.

Since the algorithm is used for *creative* work only at the *design* stage, its performance is rather of secondary importance as long as the calculations take a sensible amount of time. The major requirements to automatic design methods are *quality* of the result and *exploration* abilities.

Although the basic framework of an EA is quite simple, there are three key elements that must be prepared before the algorithm can work. They are:

- representation of phenotype (control laws in our case) suitable for genetic operations (genome encoding);
- simulation environment, which enables to implement the control laws within the closed loop system;
- fitness evaluation function, which assesses the performance of given control laws.

These elements, as well as the whole algorithm outline, are addressed below.

Parallel evolution of both the structure and the parameters of a controller can be implemented in a variety of ways. One of the few successfully employed variants is the *block structure controller evolution* (Koza et al., 2000).

In this work the ED algorithm enables to evolve suitable control laws within a reasonable time by utilising gradual evolution with the principle of strong casualty. This means that structure alterations are performed so that the information gained so far in the structure of the control law is preserved. Addition of a new block, though being random, does not cause disruption to the structure. Instead, it adds a new dimension and new potential which may evolve later during numerical optimisation. The principle of strong casualty is often regarded as an important property for the success of continuous evolution (Sendhoff et al., 1997).

The addition of new points or blocks is carried out as a separate dedicated operation (unlike sporadic structure alterations in the sub-tree crossover), and is termed *structure mutation*. Furthermore, in this work structure mutation is performed in a way known as *neutral structure mutation*. That's when the new block should be placed initially with zero coefficient. This will not produce any immediate improvement and may even deteriorate the result slightly because more blocks are used for the same approximation. However, further numerical optimisation should fairly quickly arrive at a better solution. The usefulness of neutral mutations has been demonstrated for the evolution of digital circuits (Van Laarhoven & Aarts, 1987) and aerodynamic shapes (Olhofer et al., 2001).

As a result, the ED algorithm basically represents a numerical EA with the inclusion of structure mutations mechanism.

3.1 Representation of the control laws

Control laws are represented as a combination of static functions and input signals, which are organised as a dynamic structure of *state equations* and *output equations* in form of continuous representation.

The controller being evolved has m inputs, r outputs and n states. The number of inputs and outputs is fixed. The algorithm allows varying number of states; however, in this work, the number of states is also fixed during the evolution. As a result, the controller comprises of n state equations and r output equations:

$$\begin{array}{lcl} \dot{x}_1 = g_1(x,u) & & y_1 = f_1(x,u) \\ \dot{x}_2 = g_2(x,u) & \text{and} & y_2 = f_2(x,u) \\ \vdots & & \vdots \\ \dot{x}_n = g_n(x,u) & & y_n = f_n(x,u) \end{array} \tag{4}$$

where $\boldsymbol{u}$ is size m vector of input signals, $\boldsymbol{x} = [x_1, x_2, \ldots x_n]$ is size n vector of state variables, $y_{1\ldots r}$ are controller outputs. Initial value of all state variables is zero. All $n+r$ equations are built on the same principle and are evolved simultaneously. For structure mutations, a random equation is selected from this pool and mutated.

3.1.1 Representation of input signals

Input signals delivered to each particular controller are directly measured signals as well as the quantities derived from them.

Within each group, inputs are organised in the subgroups of 'compatible' parameters. Compatible parameters are those which have close relationship with each other, have the same dimensions and similarly scaled. The examples of compatible parameters are the pairs (n_x, n_{xg}), (ω_y, $\dot{\psi}$), (V_a, V_{CL}). As a rule, only one of the compatible parameters is needed in a given control law. For this reason, the probability of selection of such parameters for the structure mutation is reduced by grouping them in the subgroups. Each subgroup receives equal chances to be selected. If the selected subgroup consists of more than one input, a single input is then selected with uniform probability.

Therefore, every controller input may be represented by a unique *code* consisting of three indices: the number of group a, the number of subgroup b and the number of item in the subgroup c. The code is designated as u(a,b,c).

3.1.2 Representation of control equations and the structure mutation

Each of the control equations (4) is encoded as described above. To this end, only one single output equation of the form $y = f(\boldsymbol{u})$ will be considered in this section. State variables $\boldsymbol{x}$ are considered as special inputs and have no effect on the encoding.

This is done to speed up the search, which could otherwise be hampered by the multitude of possible operations. It proved to be more effective to include all meaningful quantities derived from source measurements as independent inputs than to implement all the functions and operations which potentially allow to emerge all necessary quantities automatically in the course of evolution.

The encoding should allow a simple way to insert a new parameter in any place of the equation without disrupting its validity and in a way that this insertion initially does not affect the result, thus allowing neutral structure mutations.

Conceptually, the equation is a sum of input signals, in which:

- every input is multiplied by a numeric coefficient or another similarly constructed expression;
- the product of the input and its coefficient (whether numeric or expression) is raised to the power assigned to the input;
- a free (absolute) term is present.

The simplest possible expression is a constant:

$$y = k_0 \tag{5}$$

A linear combination of inputs plus a free term is also a valid expression:

$$y = k_2u_2 + k_1u_1 + k_0 \quad (6)$$

Any numeric constant can be replaced with another expression. An example of a full featured equation is

$$y = ((k_4u_4 + k_3)u_3)-0.5 + k_2u_2 + (k_1u_1)^2 + k_0 \quad (7)$$

This algorithm can be illustrated by encoding the example (7). The respective internal representation of this expression is:

- Equation: $y = ((k_4u_4 + k_3)u_3)-0.5 + k_2u_2 + (k_1u_1)^2 + k_0$
- Expression: { u(3,-0.5) u(4) 1 2 u(2) 3 u(1,2) 4 5 }
- Object parameters: [k_4 k_3 k_2 k_1 k_0]
- Strategy parameters: [s_4 s_3 s_2 s_1 s_0]

This syntax somewhat resembles Polish notation with implicit '+' and '*' operators before each variable. The representation ensures presence of a free term in any sub-expression, such as k_3 and k_0 in the example above.

The algorithm of structure mutation is presented below.

1. Select an input (or a state variable) at random: u(a,b,c).
2. Obtain the initial values of the numeric coefficient and the strategy parameter (initial step size). The initial coefficient k is selected as described above, it can be either 0 or 10^6.
3. Append the object parameters vector with the initial coefficient, and the strategy parameters vector with the initial step size.
4. Form a sub-expression consisting of the selected variable code and the obtained index: { u(a,b,c) n }.
5. With 40% probability, set the insertion point (locus) to 1; otherwise, select a numeric value in the expression at random with equal probability among all numeric values present and set the locus to the index of this value.
6. Insert the sub-expression into the original expression at the chosen locus (*before* the item pointed).

This procedure may produce redundant expressions when the selected variable already exists at the same level. Thus an algebraic simplification procedure is implemented. It parses given expression, recursively collects the factors of each variable encountered and then rebuilds the expression.

3.2 Simulation environment

Fitness evaluation of the controllers is based on the outcome of one or more simulation runs of the models involved with the controller being assessed in the loop.

Simulation environment for this study is constructed in the MATLAB/Simulink software. All the models are implemented as Simulink library blocks (Khantsis, 2006). The models participating in the evolution include:

- The *Ariel* UAV model;
- The atmospheric model;
- The ship model;
- The static cable model.

The sample rate should be kept at minimum, ensuring, however, numerical stability. The fastest and thus the most demanding dynamics in the model is contained in

- sensors (both the lags and noise);
- control actuators;
- turbulence model;
- and potentially, controllers.

These components should be especially carefully examined when adjusting the sample rate. The experimentally determined minimum sample rate for the model is 100 Hz. The model is integrated using the 4th order Runge-Kutta method.

3.3 Fitness evaluation

Fitness evaluation is based on objectives which are individual for each controller and therefore is calculated individually. As discussed above having several specific controllers makes it easier to define their particular objectives and reduces the number of these objectives as compared to one all-purpose autopilot.

Fitness evaluation of a controller can be divided into two main stages. First is the preparation of the sample task and simulation of the model with the controller in the loop. The second stage is analysis of the results obtained from the simulation and evaluation of the fitness as such. These steps are often repeated several times, making the controller perform varying tasks in order to obtain a balanced measure of its performance via multi-task fitness evaluation.

Fitness value may reflect different levels of performance achieved by the controller, including 'hard bounds.' For example, if the controller crashes the UAV, a high penalty score may be assigned.

Other parameters of flight taken into account is control usage (or control effort): it is desirable to keep control usage at minimum. Tracking abilities, if applicable, can be estimated by the weighted sum of the tracking error.

The total fitness value is calculated as the weighted sum of all estimates:

$$F = W_c C_c + W_f C_f + W_e C_e + \dots \tag{8}$$

The exact value of the weighting coefficients W, as well as the number of estimates taken into account, is individual for each controller. As a rule, the weighting coefficients are chosen empirically. It should be noted that the signals are dimensionalised and thus the coefficients may vary significantly. For example, aerodynamic surfaces deflection is measured in degrees and may range from –16 to 16 (for rudder and ailerons); considering 10-fold saturation limits, the peak values may reach ±160. At the same time, throttle range is 0.01 to 1, thus the cost for its usage will be about 30 times as low (for similar control dynamics).

3.4 Evolutionary design algorithm outline

The Evolutionary Design algorithm is implemented as a function with two input arguments: population size and number of generations to run. Other, more specific parameters are initialised within the program. The algorithm framework is as follows.

1. Initialise all parameters. If the evolution is continued, go to step 3.
2. Create initial population.
3. Evaluate fitness of each individual.
4. Save the full state to a temporary file for emergency recovery.
5. If the required number of generations is reached, return.

6. Sort the population according to the fitness of each individual.
7. If elitism is enabled, copy the best scored individual into the new population.
8. Until the new population is filled up:
9. Select the next individual from the sorted list (starting from the elite member).
10. Every (ks)th generation, with probability Ps, perform structure mutation of the selected individual.
11. Reproduce n offspring individuals from the selected (and possibly mutated) individual.
12. Put these n new individuals to the new population.
13. Continue the loop from step 8.
14. Increase the generation counter.
15. Continue from step 3.

Several steps of this algorithm need further clarification.

Initial population is created according to the specified task. By default, it is initialised with the control laws of the form $y = const$ with randomly chosen constants. Most of the controllers, however, are initialised with more meaningful control laws. For example, tracking controllers may be initialised with the laws $y = k_1\varepsilon + k_0$, where ε is the respective error signal and the coefficients k are sampled at random (taking into account default step sizes for the respective signal).

It can be seen that *selection* is performed deterministically. This is the most commonly used way in ES. The populations used in this study were of moderate size, usually 24 to 49 members. Selection pressure is determined by the number of the offspring n of each individual. The smaller n, the lower the selection pressure. For nearly all runs in this work $n = 2$, which means that half of the population is selected. This is a rather mild level of selection pressure.

The parameters determining *structure mutation* occurrence, k_s and P_s, both change during the evolution. As discussed previously, the number of structure mutations should decrease as the complexity of the controllers grows and more time to optimise the coefficients is required. The probability of structure mutation P_s is normally high in the beginning (0.7 to 1.0) and then decrease exponentially to moderate levels (0.4 to 0.6) with the exponent 0.97 to the generation number. Default value of k_s is set according to overall complexity of the controller being evolved. For simpler single-output controllers, as a rule, $k_s = 10$; for more complex controllers $k_s = 20$. However, in the beginning of evolution, k_s is reduced by half until the generation 20 and 100 respectively.

Reproduction is performed simultaneously with mutation, as it is typically done in ES, with the exception that this operation is performed separately for each selected member.

4. Controller synthesis and testing

The UAV control system is synthesised in several steps. First, the flight controller is produced. This requires several stages, since the flight controller is designed separately for longitudinal and lateral channels. When the flight controller is obtained, the guidance control laws are evolved.

Application of the ED algorithm to control laws evolution is fairly straightforward: 1) preparation of the sample task for the controller, 2) execution of the simulation model for the given sample task and 3) analysis of the obtained performance and evaluation of the fitness value. When both the model and fitness evaluation are prepared, the final evolution may be started. Typically, the algorithm is run for 100–200 generations (depending on complexity of the controller being evolved). The convergence and the resulting design is then analysed and the evolution, if necessary, is continued.

4.1 Step 1: PID autothrottle

Initially a simple PID variant of autothrottle is evolved - to ensure a more or less accurate airspeed hold. At the next stage, its evolution is continued in a full form together with the elevator control law. The PID structure of the controller may be ensured by appropriate initialisation of the initial population and by disabling structure mutations. Therefore, the algorithm works as a numerical optimisation procedure. The structure of the autothrottle control law is following:

$$\begin{aligned} \dot{x}_1 &= k_1 x_1 + k_2 \Delta V_a \\ \delta_t &= k_3 x_1 + k_4 \Delta V_a + k_5 \dot{V}_a + k_6 \end{aligned} \tag{9}$$

where $\Delta V_a = V^d - V_a$ is the airspeed error signal, δ_t is the throttle position command and $k_{1...6}$ are the coefficients to be optimised.

A 30-second flight is allocated for performance measurement. Such a long flight is needed because throttle response is quite slow. Since robustness of the PID controller to discrepancies in the UAV model is not topical at this stage (it will be addressed in further evolution), and because structure of the controller is fixed so that irrelevant measurements cannot be attracted, only a single simulation run for each fitness evaluation is performed.

Elevator inputs provide the main source of disturbances for training the autothrottle. A 2.5-degree nose-up step input is executed at time t = 5 s. It produces nearly steady climb without reaching stall angles of attack and saturation on the throttle (except for, possibly, dynamic transition moments).

At t = 14 s, a similar nose-down elevator step is commanded, entering the UAV into glide with small descent angle. In addition, a 3 m/s tailwind gust is imposed at t = 23 s. Generally, manoeuvres and wind disturbances are the most prominent sources of airspeed variations, therefore they are included for autothrottle assessment.

Initial airspeed is set to 23 m/s, i.e. 1 m/s above the required airspeed. This provides an additional small scale disturbance. Initial altitude is 100 m, providing enough elevation to avoid crash for any sensible throttle control.

Algorithm settings are as follows. Since the structure is fixed, large population size N is unnecessary. Population size of 25 members has been used in most runs. N = 13 showed similar results in terms of fitness evaluation demands (red line on the convergence graph in Fig. 7a; adjusted to the population size 25 for comparison). Fitness is evaluated deterministically because all random signals are repeatable from run to run. Elitism is enabled. Fitness is calculated with the following weighting coefficients:

$$F_t = 2000C_e(V_e) + 1000C_c(\delta_t) + 2000C_f(\delta_t) \tag{10}$$

It should be noted that the *commanded* signal δ_t (which enters the actuator) is used for evaluation.

Convergence graphs for three independent runs of the ED algorithm are presented in Fig. 7a. They show the best fitness values for each generation. On average, 2000 to 2500

simulation runs is needed to achieve satisfactory performance. Because this is only an intermediate controller, full optimisation is not necessary at this stage.
The best controller obtained in the experiments is the following:

$$\begin{aligned}\dot{x}_1 &= -2.9966x_1 + 85.57\Delta V_a \\ \delta_t &= 0.012616x_1 + 0.089324\Delta V_a - 0.044004\dot{V}_a + 0.23827\end{aligned} \tag{11}$$

Sample response of this controller is illustrated in Fig. 7b.

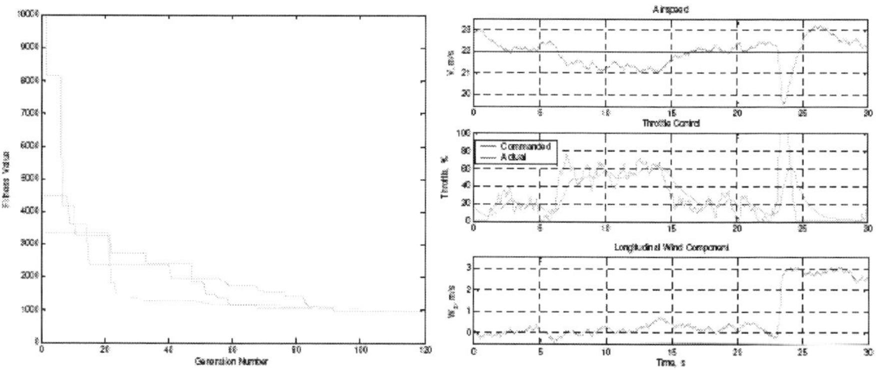

a) b)

Figure 7. PID autothrottle current-best fitness (a) and sample response (b)

4.2 Step 2: Longitudinal control

With a simple autothrottle available, elevator control can be developed to provide tracking of the normal body load factor demand n_y^d.

It is desirable that the sample training task closely reflect the conditions the controller will be subject to in a real guidance task. Unfortunately, this is not possible at the early stages of multistage design process as guidance laws can be evolved only after a capable flight controller has been produced. In the meantime, a replacement pseudoguidance task must be used for the flight controller training.
As a suitable replacement of the guidance law, a simple static altitude hold is used:

$$n_y^d = 0.03\left(H^d - H\right) + \cos\theta \tag{12}$$

A low gain is chosen to ensure guaranteed stability. The $\cos\theta$ term takes into account gravity component, assuming that roll angles are small. However, real guidance in high sea will most likely require a more active control, especially closer to the recovery boom. For this reason, a synthetic 'chirp' signal $c(t)$ is added starting from 10th second of flight. This signal represents a sine wave which frequency increases linearly from f_0 = 0.01 Hz at t_0 = 10 s to f_1 = 0.5 Hz at t_1 = 20 s. For t < 10 s, $c(t)$ = 0, providing 'training' for smooth control. Chirp signal is often used for response analysis of nonlinear systems. Altogether, the reference input signal is

$$n_y^d(t) = 0.03\left(H^d - H(t)\right) + \cos\theta(t) + 0.2c(t) \tag{13}$$

The preset altitude H^d is 5 m above the initial altitude H_0 = 100 m. This causes a small positive step input in the beginning of flight.

Flight time is limited to 20 s, which is close to normal final approach flight time (10–15 s), but allocates more time for better 'training' of slow throttle control. Overall fitness of the controller is calculated as an average of three fitness values obtained for each flight.

Elevator control law is initialised as follows:

$$\begin{aligned} \dot{x}_2 &= 0 \\ \delta_e &= k_1 x_2 + k_2 \Delta n_y + k_3 \end{aligned} \tag{14}$$

where $\Delta n_y = n_y^d - n_y$ is the load factor error and the coefficients $k_{1...3}$ are sampled at random for the initial population. For simplicity, first-order dynamic control law is used. However, when combined with the throttle control, both control laws may include both state variables x_1 and x_2, forming a tightly coupled longitudinal control.

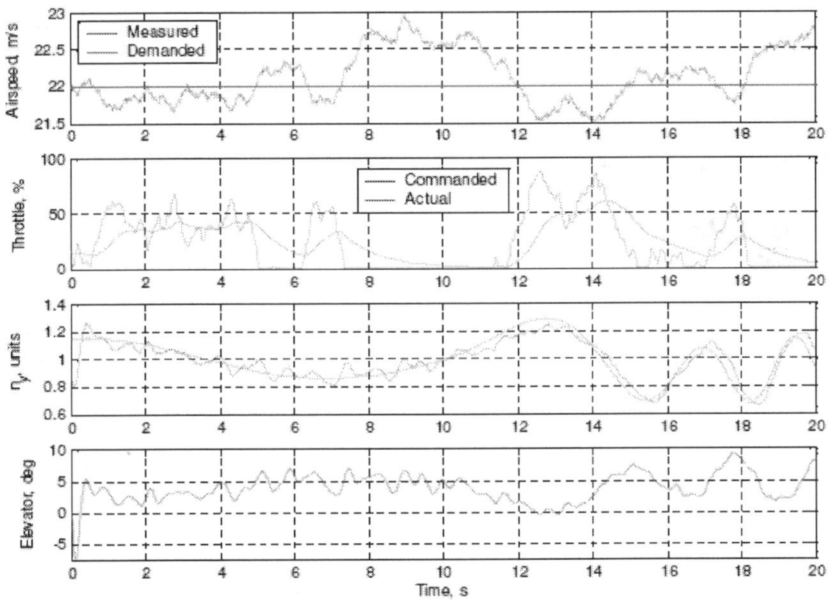

Figure 8. Longitudinal control sample response

Fitness estimation combines both throttle control fitness and elevator control fitness. They are constructed in a similar manner with the following weighting coefficients:

$$F_{te} = F_t + F_e = F_t + 500000C_e(n_y) + 50C_c(\delta_e) + 5000C_f(\delta_e) \tag{15}$$

The weights put slightly more emphasis on elevator control than on throttle control, because the former is more important for guidance. Autothrottle only supplements it to provide safe flight conditions.

Relatively large population size of 49 members has been used in all runs. Initial structure mutation probability P_s was set to 1.0; final value $P_s \to 0.6$. Structure mutations happened every (k_s = 5)th generation for the first 20 generations, then k_s = 10 until 100th generation, then k_s = 20. Satisfactory controllers were obtained after about 100 to 120 generations, which requires approximately 15000 simulation runs. The best performing controller has the following form:

$$\begin{aligned} \dot{x}_1 &= ((-252.611\omega_z - 172.0765)\alpha - 4.411)x_1 + 75.0613\Delta V_a \\ \dot{x}_2 &= -0.77101x_2 - 228.8894\Delta n_y \\ \delta_t &= 0.0011588x_1 + 1.0565\Delta V_a - 0.6989\dot{V}_a + 0.24651 \\ \delta_e &= 41.55940\theta + (-140.8115\alpha)^{0.5} + 0.89016x_2 - 26.1045\Delta n_y + 50.1835\omega_z + 3.871 \end{aligned} \tag{16}$$

Response of this controller to the sample input (in default configuration) is presented in Fig. 8.

4.3 Step 3: Lateral control

Lateral control consists of two channels: ailerons control and rudder control. As a rule, for an aerodynamically stable aircraft such as the *Ariel* UAV, lateral control is fairly simple and is not as vital for flight as longitudinal control. For this reason, both control laws, for ailerons and rudder, are evolved simultaneously in one step.

The set-up is largely similar to that used in the previous step. The just evolved longitudinal control laws (16) are connected to throttle and elevator. Both ailerons and rudder control laws are initialised in a similar manner to (14):

$$\begin{aligned} \dot{x}_3 &= 0 \\ \dot{x}_4 &= 0 \\ \delta_a &= k_{11}x_3 + k_{12}\Delta\gamma + k_{13} \\ \delta_r &= k_{21}x_4 + k_{22}n_z + k_{23} \end{aligned} \tag{17}$$

Fitness is calculated as follows:

$$F_{ar} = F_a + F_r = 50000C_e(\gamma) + C_c(\delta_a) + 2000(\delta_a) + \ldots 200000C_e(n_z) + 10C_c(\delta_r) + 2000C_f(\delta_r) \tag{18}$$

Algorithm settings are the same as for longitudinal control design. Fairly quick convergence (within 70–80 generations) was observed, with slow further progress. Considering that the initial state is zero, state variable x_3 is always zero. The controller is obtained as follows:

$$\begin{aligned} \dot{x}_4 &= -39.2775x_4 - 44.57\omega_x + \left((4.6845x_4)^{0.5} + 69.3478\right)n_z \\ \delta_a &= -26.7812\Delta\gamma + 8.1188\omega_x + 0.57032 \\ \delta_r &= -1.6916x_4 - 1.33n_z + 4.1764\omega_y + 21.2955\beta - 0.65184 \end{aligned} \tag{19}$$

Sample response of this controller is presented in Fig. 9.

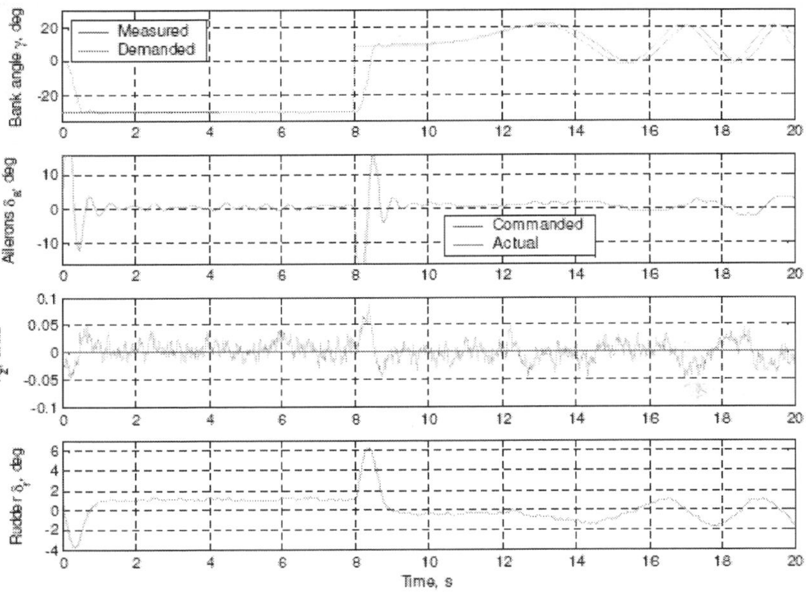

Figure 9. Lateral control sample response

4.4 Step 4: Stall prevention mechanism

Active guidance control may produce high acceleration demands n_y^d, which may lead to exceeding the maximum allowed angle of attack $\alpha_{\max}$ and result in stall.

However, fitness evaluation of the control law with stall prevention would be more complex and time consuming. Separate estimation is required because the objectives are contradictory: adequate limitation of α invariably results in tracking degradation and vice versa.

Though, a separate control law which limits angle of attack is developed. The limiter should work only when dangerous angle of attack is reached, being deactivated in normal conditions.

The control law is evolved in a similar manner to other laws described above. The objective of the control law is to maintain $\alpha = \alpha_{\max}$ by correcting n_y^d as long as this demand is greater than required for safe flight. Also, free term is disregarded in both state and output equations.

Fitness is calculated similarly to all tracking controllers with respect to the command $\alpha = \alpha_{\max}$. Two exceptions are made, however. First, overshoot is penalised twice as heavily as undershoot, because exceeding the maximum angle of attack is considered to be hazardous. Second, since the limiter has no direct control on elevator, the amount of longitudinal oscillations is taken into account as the variance of pitch rate. Altogether, the fitness value is calculated as

$$F_a = 10^6 C_a + 0.1C_c(\delta_e) + 1000C_c(\omega_z) + 1000C_f(\delta_e) \tag{20}$$

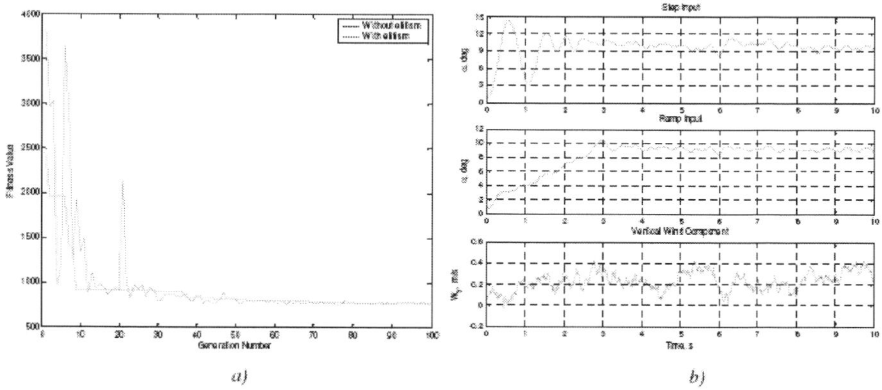

Figure 10. Angle of attack limiter current-best fitness (a) and sample responses (b)

Algorithm settings and initialisations are similar to the case of longitudinal control evolution, except that smaller population of 24 or 25 members is used because only one control law is evolved. Both elitist and non-elitist strategies have been tried with largely similar results. Quick convergence, within 40–50 generations, was observed in all cases, with slow further progress. Examples of current-best convergence are presented in Fig. 10a. A simpler solution obtained in elitist strategy has been selected:

$$\begin{aligned}
&\dot{x}_5 = (93.01\omega_z - 15.0798)(\alpha - \alpha_{\max}) \\
&n_y^{d.corr} = n_y^{d0} - \Delta n_y^d = n_y^{d0} - 0.63208x_5 - 25.246(\alpha - \alpha_{\max}) \\
&K_{\max} = 60
\end{aligned} \tag{21}$$

Where $K_{\max}$ is a damping coefficient and n_y^{d0} is a frozen value for n_y^d. Sample responses of this controller are presented in Fig. 10.

4.5 Step 5: Guidance

At this point, flight controller synthesis is completed and guidance laws can be evolved. Guidance controller comprises two control laws, for the vertical and horizontal load factor demands, n_{yk}^d and n_{zk}^d respectively. This is equivalent to acceleration demands

$$a_{yk}^d = g n_{yk}^d \text{ and } a_{zk}^d = g n_{zk}^d \tag{22}$$

These demands are passed through the kinematic converter to form the flight controller inputs n_y^d and γ^d.

It is important to note that when the controller is evolved in a stochastic environment with many random factors, the evolution progress becomes highly stochastic as well. It is desirable to obtain the fitness value on the basis of multiple simulation runs, reducing thereby sampling errors and making fitness value more deterministic and reliable. Excessive number of simulation runs, however, may slow down the evolution considerably, thus an optimal number should be found.

Second, for the same reason, stagnation of progress of the current best fitness is not a reliable indicator of stagnation of the evolution. Every generation may produce individuals with very good fitness which will not sustain further selections. In some cases, current best fitness may appear stagnated from the very first generation. Average population fitness may be considered as a better measure of evolution progress.

Third, elitism has little sense in a sufficiently stochastic environment and is usually not employed. If it is desirable, for some reason, to use elitist strategy, fitness of the elite solution must be recalculated every generation.

In the guidance task, the random factors include initial state of the UAV; Sea State, atmospheric parameters, and initial state of the ship.

Fitness is calculated as follows:

$$F_g = 40\Delta h_1^2 + 20\Delta z_1^2 + 50|\psi_1| + 25|\gamma_1| + 500C_c\left(n_{yk}^d\right) + 500C_c\left(n_{zk}^d\right) + 100C_f\left(n_{zk}^d\right) \tag{23}$$

where Δh_1 and Δz_1 are vertical and horizontal miss distances, and ψ_1 and γ_1 are final yaw and bank angles. Greater weight for vertical miss than that for horizontal miss is used because vertical miss allowance is smaller (approximately 3-4 m vs. 5 m) and also because vertical miss may result in a crash into the boom if the approach is too low.

Three simulation runs are performed for each fitness evaluation. To provide a more comprehensive estimation of guidance abilities, initial positions of the UAV in these N runs are maximally distributed in space, maintaining, at the same time, a sufficient random component.

Algorithm initialisation is the same as for longitudinal flight control laws evolution, except that elitism is not used and the population size is 48 members. The initial population is sampled with the control laws of the form

$$\begin{aligned} \dot{x}_6 &= 0 \\ \dot{x}_7 &= 0 \\ a_{yk}^d &= k_{11}x_6 + k_{12} \\ a_{zk}^d &= k_{21}x_7 + k_{22} \end{aligned} \tag{24}$$

where all coefficients k are chosen at random. For convenience, the control laws are expressed in terms of accelerations, which are then converted to load factors according to (22). Since these control laws effectively represent 'empty' laws $y = const$ (plus a low-pass output filter with randomly chosen bandwidth), structure mutation is applied to each member of the initial population. Further structure mutations are performed as specified above. With such initialisation, no assumption about possible guidance law is taken.

From the final populations, the best solution is identified by calculating fitness of each member using $N = 25$ simulation runs, taking into account also success rate. The best performing controller is the following (unused state variables are removed) with 100% success rate in the 25 test runs with average fitness 69.52:

$$
\begin{aligned}
\dot{x}_6 &= -3.548\omega_v \\
a_{yk}^d &= 0.916x_6 - 54.3\omega_v - 0.1261 \\
a_{zk}^d &= -107.68\omega_h + 0.0534V_{zT} + 0.4756
\end{aligned} \tag{25}
$$

The other attempted approach is the pure proportional navigation (PPN) law for recovery which was compared with the obtained solutions (Duflos et al., 1999; Siouris & Leros, 1988). The best PN controller have been selected using the fitness values of each member of the final population averaged over 100 simulation runs. It is the following:

$$
\begin{aligned}
a_{yk}^d &= -3.27V_{CLa}\omega_V \\
a_{zk}^d &= -3.18V_{CLa}\omega_h
\end{aligned} \tag{26}
$$

This controller received the fitness 88.85 and showed success rate 95%.

5. Controller testing

Testing is a crucial stage of any controller development process as the designer and end user must be satisfied that the controller meets its performance requirements. In addition, an allowable operational envelope must be obtained. Testing of a developed controller can be performed either within the physical system or in a comprehensive simulation environment. In physical implementation, the controller is working on a true model and hence there are no modelling errors to introduce uncertain effects into the system. In addition, physical testing accurately reflects the operational use of the controller. On the other hand, physical testing involves large time and cost demands for multiple tests. Moreover, the failure of the controller at any stage may lead to a dangerous situation and even to loss of the aircraft.

Testing within a simulation environment has the benefit that many different tests may be applied rapidly with no severe consequences in the event of controller failure. The whole range of operating conditions may be tested. On the downside is that potentially large modelling errors may be introduced, which may bias the results and cause an otherwise excellent controller to fail when implemented physically. In addition, testing situations which cannot occur in practice are possible in the simulation environment.

In this work, two main types of simulation tests are conducted - robustness and performance. Results are presented for robustness test, which is aimed at ensuring the controller has good robustness to modelling uncertainties and to test whether the controller is sensitive to specific perturbations.

5.1 Robustness tests

Testing the controller for robustness to model uncertainty involves perturbing the model and determining the effect upon controller performance. The perturbations can be performed in several ways. Physical quantities such as mass and wing area can be changed directly. Dynamics of the system can be varied by introducing additional dynamic elements and by changing internal variables such as aerodynamic coefficients. For a realistic test, all perturbations should be applied simultaneously to identify the worst case scenario. However, single perturbation tests (the *sensitivity analysis*) allow to analyse the degree of influence of each parameter and help to plan the robustness test more systematically. After single perturbation tests multiple perturbation tests are carried out, where all parameters of the model are perturbed randomly within the identified limits.

5.2 Single perturbation tests

In this type of test, a single model variable is perturbed by a set amount and the effect upon the performance of the controller is determined. Performance evaluation can be measured in a manner similar to fitness evaluation of the guidance controller (equation (23)).

The additional parameters taken into account in performance measurement are impact speed V_{imp} and minimum altitude $H_{\min}$ attained during the approach.

Altogether, the performance cost (PC) is calculated as follows:

$$\begin{aligned} PC = {} & 40\Delta h_1^2 + 20\Delta z_1^2 + 50|\psi_1| + 25|\gamma_1| + 50 f_V + 20 f_H + 10C_c(\delta_a) + 10C_c(\delta_r) + C_c(\delta_e) + \ldots \\ & + 200C_f(\delta_a) + 200C_f(\delta_r) + 200C_f(\delta_e) \end{aligned} \quad (27)$$

where

$$f_V = \begin{cases} V_{imp} - 30, & V_{imp} > 30 \\ 0, & V_{imp} \le 30 \end{cases}, \quad f_H = \begin{cases} 10 - H_{\min}, & H_{\min} > 10 \\ 0, & H_{\min} \le 10 \end{cases}. \quad (28)$$

Impact speed V_{imp} and minimum altitude $H_{\min}$ are measured in m/s and metres respectively. Other designations are as in (23). Unlike fitness evaluation in the flight controller evolution, the commanded control deflections δ_a, δ_r and δ_e are saturated as required for control actuators.

The absolute value of the PC obtained using (27) is not very illustrative for comparison between the results. Smaller values indicate better performance, but the 'ideal' zero value is unreachable because a minimum level of control activity is always present even in very calm environment. For this reason, a *Normalised Performance Cost* (NPC) will be used:

$$NPC = \frac{PC}{PC_{ref}} \quad (29)$$

where PC_{ref} is the reference Performance Cost obtained for the reference (unperturbed) model with the guidance controller being considered. NPC > 1 indicates deterioration of performance. However, the performance with the perturbed model may be better than the reference performance, thus NPC < 1 is also possible.

The environment delivers a great deal of uncertainty. To obtain a reliable estimate of performance, several tens (up to a hundred) of simulation runs in various conditions should be performed at every point. However, this would imply a prohibitively high computational cost. Meanwhile, there is no single 'typical' scenario that could encompass most of the real flight conditions. For these reasons, in this work few different scenarios were used for evaluation of a single Performance Cost to cover most of the possible real world conditions. The range of disturbances and the initial ship phase are chosen to provide a moderately conservative estimation. All random parameters (which include the turbulence time history and the ship initial state) are reproducible between the PC estimations, therefore PC is calculated deterministically. Results presented in this work are for calm environment:

Calm environment. No wind, no turbulence; initial position of the UAV is at the ideal reference point: distance 300 m, elevation 14 m, zero sideways displacement. A small amount of ship motion corresponding to Sea State 2 (SWH)2 is, however, included. This scenario is useful to analyse the performance in benign environment and also to soften the conservative bias to the difficult conditions.

The first test determines the robustness to time delays in the measurement signals.

The tests are carried out for the two controllers selected in Section 4.5. The controller #1 is (26) (pure Proportional Navigation guidance) and the controller #2 is (25). Fig. 11a shows the NPC evaluated for these controllers for varying time delay.

Time delays occur in a physical system due to several factors, which include delays associated with the measurement devices, delays in encoding, decoding and transmitting the signals to the controllers, delays in controller computation and delays in the actuator systems.

The miss distance (averaged over the four scenarios) is shown in Fig. 11b. It can be seen that in terms of guidance accuracy, delays up to 0.07 s for the second controller and up to 0.09 s for the first controller cause little effect. However, considering the NPC value, which also takes into account control activity, sufficient degradation is observable from 0.035 second delay. This is expectable, because delays usually cause oscillations in the closed-loop system.

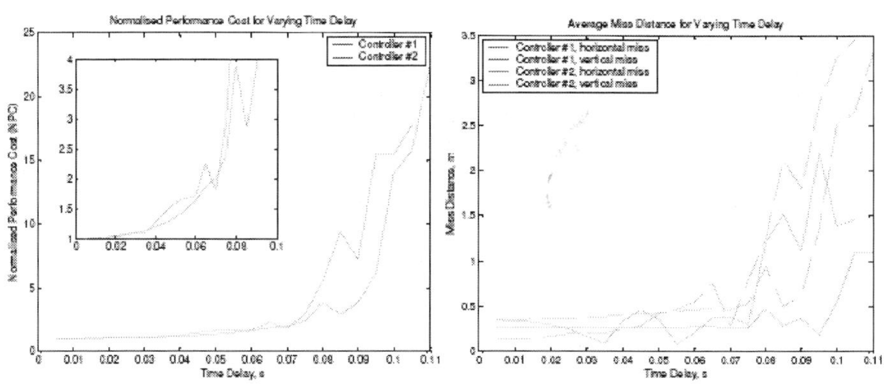

a) b)

Figure 11.10 NPC (a) and miss distance (b) for controllers with time delays

For the selected NPC threshold, the maximum time delay can be defined as 0.065 s for the controller #1 and 0.06 s for the controller #2.

It should be noted, however, that perturbation of any one of the parameters does not provide a comprehensive picture of the effect on performance. For this reason, the effect of perturbation of empty mass m_{empt} will be closely examined as well. Unlike time delay, this perturbation should primarily affect the trajectory rather than the control activity.

m_{empt} is varied from 50% of the reference value (20.45 kg) to 200% linearly with 2% steps. This is done separately for increasing and decreasing the variable until the threshold NPC = 1.8 is crossed, or a failure occurs, or the limit (200% or 50% respectively) is reached.

The parameters corresponding to aircraft geometry and configuration are tested in a similar manner. The range is increased to the scale factors between 0 and 10 (0 to 1000%) with step 0.05. NPC is linearly interpolated at the intermediate points. The allowable perturbations (as factors to the original values) are summarised in Table 1, where * denotes the extreme value tested.

Parameter	Lower limit factor		Upper limit factor	
	Controller 1	Controller 2	Controller 1	Controller 2
Empty mass (m_{empt})	0.50*	0.50*	1.54	1.52
Rolling moment of inertia (I_x)	0.20	0.20	10*	2.43
Yawing moment of inertia (I_y)	0*	0*	10*	10*
Pitching moment of inertia (I_z)	0*	0*	2.17	1.91
xy cross product of inertia (I_{xy})	0*	0*	10*	10*
Wing area (S)	0.74	0.87	1.55	1.60

Table 1. Allowable perturbations of UAV inertial properties and geometry

Fig. 12 demonstraites results for calm environment. Control signals for the worst two cases (with NPC > 3.8) are not shown, they exhibit extremely aggressive oscillations. Dashed cyan and green lines on the trajectory graphs represent the 'unrolled' along the flight path traces of the tips of the recovery boom (lateral position on the top view and vertical position on the side view). The bar on the right hand side illustrates the size of recovery window. The height and position of the window may be slightly different for different trajectories because it is determined by the shape of the arresting cable-hook. However, in most cases this is barely noticeable on the graphs since the difference in flight time and terminal part of trajectory is small. Note that the horizontal and vertical scale on the trajectory graphs is different.

The other parameters evaluated in this research were the perturbations of the power unit parameters, the aircraft aerodynamics parameters and sensor noise (Khantsis, 2006).

Overall, these tests have not exposed any significant robustness problems within the controllers. Large variations in single aircraft parameters caused very few control problems. However, such variations cannot realistically judge the performance of the controllers under simultaneous perturbations of multiple parameters. In order to compare the practical robustness of the aircraft, the following tests are performed.

5.3 Simultaneous multiple perturbation tests

These tests involve simultaneous perturbation of all the aircraft variables by a random amount. This style of testing requires many simulations to ensure adequate coverage of the testing envelope. Normally distributed random scale factors are used for perturbations.

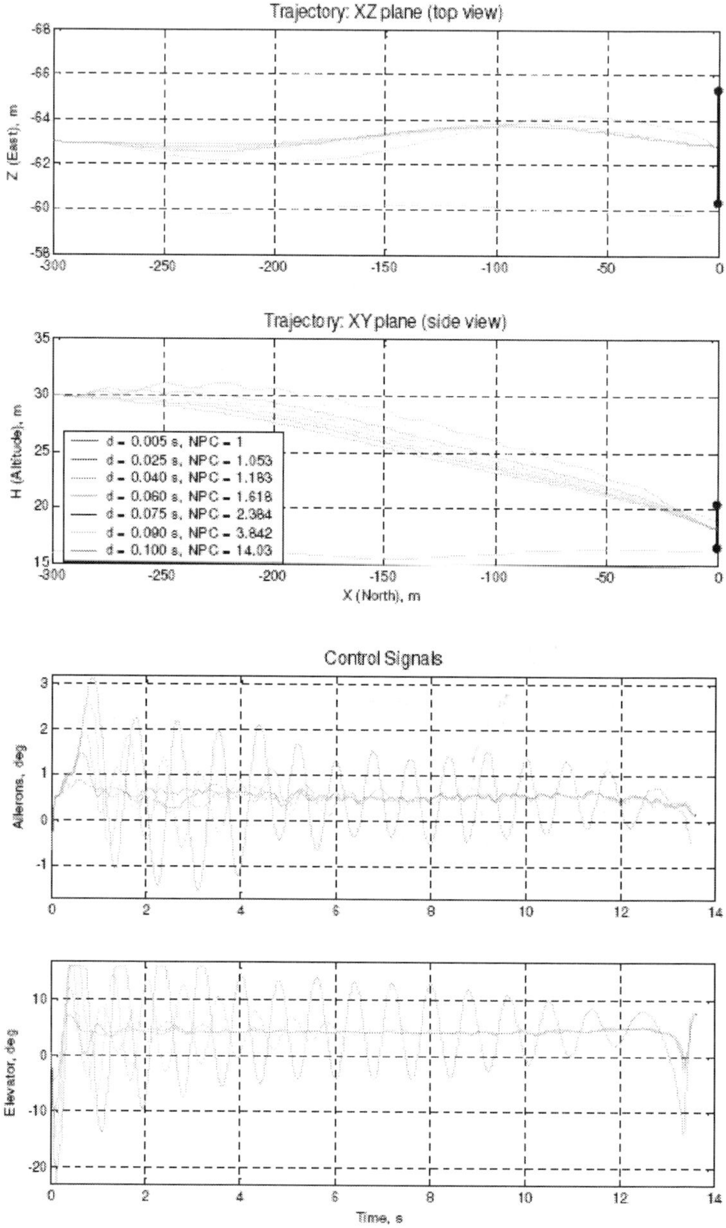

Figure 12. Flight path and control signals with time delays

The chosen standard deviations are approximately one tenth of the maximum allowed perturbations identified in the previous robustness tests.

For each of the two controllers, 1000 tests (4000 simulations considering four scenarios) have been carried out. Note that the perturbations were the same for both controllers in each case. This allows to compare the performance between the controllers in similar conditions. The second controller showed much poorer robustness. 90.4% of test cases showed NPC < 1.8. At the same time, only 69.8% of the tests reported successful recovery. Such a large proportion of cases which acceptable NPC value but without capture (about 20%) indicates that the controller delivers sufficiently inferior guidance strategy when subject to model uncertainty. This could be expected because this controller uses more positioning measurements. The majority of these cases involve a miss in the tailwind scenario, highlighting a potential guidance problem. The number of cases with a crash in one or more scenarios is 2.8%, which is a significant increase over the first controller as well.

From these tests, the Proportional Navigation controller is clearly preferable. However, it must also be tested over a wide range of scenarios.

6. Conclusions

In this chapter, an application of the Evolutionary Design (ED) is demonstrated. The aim of the design was to develop a controller which provides recovery of a fixed-wing UAV onto a ship under the full range of disturbances and uncertainties that are present in the real world environment.

The controller synthesis is a multistage process. However, the approach employed for synthesis of each block is very similar. Evolutionary algorithm is used as a tool to evolve and optimise the control laws. One of the greatest advantages of this methodology is that minimum or no a priori knowledge about the control methods is used, with the synthesis starting from the most basic proportional control or even from 'null' control laws. During the evolution, more complex and capable laws emerge automatically. As the resulting control laws demonstrate, evolution does not tend to produce parsimonious solutions.

The method demonstrating remarkable robustness in terms of convergence indicating that a near optimal solution can be found. In very limited cases, however, it may take too long time for the evolution to discover the core of a potentially optimal solution, and the process does not converge. More often than not, this hints at a poor choice of the algorithm parameters.

The most important and difficult problem in Evolutionary Design is preparation of the fitness evaluation procedure with predefined special intermediate problems. Computational considerations are also of the utmost importance. Robustness of EAs comes at the price of computational cost, with many thousands of fitness evaluations required.

The simulation testing covers the entire operational envelope and highlights several conditions under which recovery is risky. All environmental factors—sea wave, wind speed and turbulence—have been found to have a significant effect upon the probability of success. Combinations of several factors may result in very unfavourable conditions, even if each factor alone may not lead to a failure. For example, winds up to 12 m/s do not affect the recovery in a calm sea, and a severe ship motion corresponding to Sea State 5 also does not represent a serious threat in low winds. At the same time, strong winds in a high Sea State may be hazardous for the aircraft.

Probably the most important consideration in the context of this research is validity of the Evolutionary Design methodology. Whilst it is evident that ED did produce a capable controller that satisfies the original problem statement, several important observations can be made. First of all, in the absence of similar solutions to be compared with, it is unclear how close the result is to the optimum. Considering remarkable robustness of EAs for arriving at the global optimum, it may be believed that the controller is very close to the optimal solution. However, comparing performance of the two generated guidance controllers in different configurations, it may be concluded that there is still room for improvement. The main reason for that is believed to be the limited scope of the testing conditions used in the evolution process, which is mainly due to shortage of computational resources.

On the whole, Evolutionary Design is a useful and powerful tool for complex nonlinear control design. Unlike most other design methodologies, it tries to *solve* the problem at hand automatically, not merely to optimise a given structure. Although ED does not exclude necessity of a thorough testing, it can provide a near optimal solution if the whole range of conditions is taken into account in the fitness evaluation. In principle, no specific knowledge about the system is required, and the controllers can be considered as 'black boxes' whose internals are unimportant. Successful design of the controller for such a challenging task as shipboard recovery demonstrates great potential abilities of this novel technique.

6.1 Limitations of the Evolutionary Design

The first point to note is that ED, as well as any evolutionary method, does not produce inherently robust solutions in terms of both performance and unmodelled uncertainties. As noted above, behaviour of the solutions in the untested during the evolution conditions may be unpredictable. A sufficient coverage of all conditions included in the evolution process invariably requires a large number of simulations, which entails a high computational cost. When the amount of uncertainties in the system and the environment becomes too large, computational cost may become prohibitively high. At the same time, subdivision of the problem into smaller ones and reduction of the number of environmental factors in order to reduce the amount of uncertainties require good understanding of the system. This severely compromises one of the main advantages of ED, which states that little, if any, a priori knowledge of the system is needed.

Another limitation is that evolutionary techniques have little analytical background. Their application is based largely on empirical experience. In particular, most algorithm settings such as population size, selection pressure, probability of genetic operators etc. are adjusted using trial and error method. Whilst EAs are generally robust to these settings, it should be expected that the settings used for controller synthesis in this work may be far from optimal for another application, and their optimisation will take a significant time.

The main limitations of the controller produced in the ED procedure are due to limited validity of the system models available. For a complete control system design, higher fidelity models will be essential.

6.2 Research opportunities on the Evolutionary Design

The ED methodology proved to be easy to apply and extremely suitable for current application. However, the main effort has been directed towards the practical implementation and application of this design method, with little investigation into its general functionality and efficiency.

An investigation into the most effective algorithm settings, such as the population sizes and probabilities of genetic operators, will also be beneficial. Extension and optimisation of the control laws representation may help to expand the area of possible applications as well as improve efficiency of the algorithm. A large room for improvement exists in the computer implementation of the algorithm. Whilst every effort has been made in this work to optimise the internal algorithmic efficiency, significant optimisation of the coding (programming) is possible.

The methodology presented in this thesis can be used to develop a highly optimal controller capable of autonomous recovery of the UAV in adverse conditions. The Evolutionary Design method, which is the core of this methodology, potentially enables to produce controllers for a wide range of control problems. It allows to solve the problems automatically even when little or no knowledge about the controlled system available, which makes it a valuable tool for solving difficult control tasks.

7. References

Astrom, K. J. & Wittenmark B. (1995). A survey of adaptive control applications, *Proceedings of IEEE Conference on Decision and Control* , pp. 649-654, ISBN: 0-7803-2685-7, New Orleans, LA, USA, December 1995

Atherton, D. (1975). *Nonlinear Control Engineering,* Van Nostrand Reinhold Company, ISBN: 0442300174

Blanchini, F. & Sznaier, M. (1994). Rational L1 suboptimal compensators for continuous time systems,. *IEEE Transactions on Automatic Control,* Vol. 39, No 7, page numbers (1487-1492), (July 1994), ISSN: 0018-9286

Bourmistrova, A. (2001). Knowledge based control system design for autonomous flight vehicle, *PhD thesis,* RMIT University: Melbourne, September 2001

Brown, R. G. & Hwang, P. Y. C. (1992). *Introduction to Random Signals and Applied Kalman Filtering.* 2nd ed. John Wiley & Sons, ISBN-10: 0471128392, ISBN-13: 978-0471128397

Bryson, A. J. & Ho Y.-C. (1975). *Applied Optimal Control.* Hemisphere, Washington, DC, 1975

Bu, T., Sznaier, M. & Holmes, M. (1996). A linear matrix inequality approach to synthesizing low order L1 controllers, *Proceedings of IEEE Conference on Decision and Control* proceedings, pp. 1875-1880 vol.2, ISBN: 0-7803-3590-2, Kobe, Japan, December 1996

Bugajski, D. J. & Enns, D. F. (1992). Nonlinear control law with application to high angle-ofattack flight, *Journal of Guidance, Control and Dynamics,* Vol. 15, No 3, (May/June 1992), page numbers (761-767), ISSN 0731-5090

Coley, D. A. (1998). *An Introduction to Genetic Algorithms for Scientists and Engineers.* University of Exeter, ISBN 9810236026, 9789810236021, Published by World Scientific, 1999

Crump, M. R. (2002). The dynamics and control of catapult launching Unmanned Air Vehicles from moving platforms, *PhD thesis,* RMIT University: Melbourne, 1992

Dahleh, M. A. & Pearson, J. B. (1987). L1 optimal feedback compensators for continuous-time systems, *Proceedings of IEEE Transactions Automatic Control*, pp. 889-895, Vol AC-32, ISSN: 0018-9286, October 1987

Dalsamo, M. & Egeland, O. (1995). H∞ control of nonlinear passive systems by output feedback, *Proceedings of IEEE Conference on Decision and Control* proceedings, pp. 351 - 352 vol. 1, ISBN: 0-7803-2685-7, New Orleans, LA, December 1995

DeCarlo, R., Zak, S. & Mathews, G. (1988). Variable structure control of nonlinear multivariable systems: a tutorial. *IEEE*, Vol. 76, No 3, (March 1988), page numbers (212-232), ISSN: 0018-9219

Doyle, J. C., Glover, K., Khargonekar P. P. & Francis B. A. (1989). State-space solutions to standard H_2 and H∞ control problems, *IEEE Transactions on Automatic Control*, Vol. 34, No. 8, (August 1989), page numbers (831-847), ISSN: 0018-9286

Doyle, J. C. & Stein, G. (1981). Multivariable feedback design: concepts for a classical/modern synthesis,. *IEEE Transactions on Automatic Control*, Vol. AC-26, No 1, (February 1981), page numbers (4-16), ISSN: 0018-9286

Duflos, E., Penel, P. & Vanheeghe, P. (1999). 3D guidance law modelling, *IEEE Transactions on Aerospace and Electronic Systems*, Vol. 35, No 1, page numbers (72-83), (January 1999), ISSN: 0018-9251

Graham, D. & McRuler, D. (1971). *Analysis of Nonlinear Control Systems*, Dover, ISBN-10: 0486610144, ISBN-13: 9780486610146

Hyde, R.(1991). The Aplication of Robust Control to VSTOL Aircraft, *PhD thesis*, Cambridge

Kaise, N. & Fujimoto, Y. (1999). Applying the evolutionary neural networks with genetic algorithms to control a rolling inverted pendulum, *Lecture Notes in Computer Science : Simulated Evolution and Learning*, Volume 1585/1999, (January 1999), page numbers (223-230), Springer Berlin / Heidelberg, ISBN 978-3-540-65907-5

Kalman, R. E. (1960). A new approach to linear filtering and prediction problems, *Transaction of the ASME – Journal of Basic Engineering*: p. 35-45, March 1960

Kaminer, I., Khargonekar, P. P. & Robel, G. (1990). Design of a localizer capture and track modes for a lateral autopilot using H∞ synthesis, *IEEE Control Systems Magazine*, Vol. 10, No 4, (June 1990), page numbers (13-21), ISSN: 0272-1708

Khammash, M. & Zou, L., Almquist, J.A.; Van Der Linden, C (1999). Robust aircraft pitch-axis control under weight and center of gravity uncertainty, *Proceedings of the 38th IEEE Conference on Decision and Control*, pp. 1970 - 1975 vol.2, ISBN: 0-7803-5250-5, Phoenix, AZ, December 1999

Khantsis, S., (2006). Control System Design Using Evolutionary Algorithms for Autonomous Shipboard Recovery of Unmanned Aerial Vehicles, *PhD thesis*, RMIT University: Melbourne, 1992

Kim, B. S. & Calise, A. J. (1997). Nonlinear flight control using neural networks,. *Journal of Guidance, Control and Dynamics*, Vol. 20, No 1, page numbers (26-33), (January 1997), ISSN 0731-5090

Koza, J. R., Keane, M. A., Yu, J., Bennett III, F. H. & Mydlowec, W. (2000). Automatic creation of human-competitive programs and controllers by means of genetic programming, *Genetic Programming and Evolvable Machines*, Vol. 1, No 1, page numbers (121-164), (January 2000), ISSN: 1389-2576

Kwakernaak, H. (1993). Robust control and H∞ optimization - tutorial paper, *Automatica*, Vol. 29, No 2, page numbers (255-273), (February 1993), ISSN: 0005-1098

Maciejowski, J. (1989). *Multivariable Feedback Design.* Addison Wesley, ISBN-10: 0201182432, ISBN-13: 978-0201182439, 1989

MathWorks Inc. *Robust Control Toolbox User's Guide.* 2001. Available online at http://www.mathworks.com/access/helpdesk/help/pdf_doc/robust/robust.pdf

Olhofer, M., Jin, Y. & Sendhoff, B. (2001). Adaptive encoding for aerodynamic shape optimization using Evolution Strategies, *Proceedings of Congress on Evolutionary Computation*, pp. 576-583 vol. 1, Seoul, South Korea, 2001, ISBN:1-59593-010-8

Onnen, C., Babuska, R., Kaymak, U., Sousa, J. M., Verbruggen, H. B. & Isermann, R. (1997). Genetic algorithms for optimization in predictive control, *Control Engineering Practice*, Vol. 5, No 10, page numbers (1363-1372), (October 1997), ISSN: 0967-0661

Pashilkar, A. A., Sundararajan, N. & Saratchandran, P. (1999). A fault-tolerant neural aided controller for aircraft auto-landing, *Aerospace Science and Technology*, Vol. 10, No 1, page numbers (49- 61), (January 2006), ISSN: 1270-9638

Postlethwaite, I. & Bates, D. (1999). Robust integrated flight and propulsion controller for the Harrier aircraft, *Journal of Guidance, Control and Dynamics*, Vol. 22, No 2, page numbers (286-290), (February 1999), ISSN 0731-5090

Rugh, W. J. & Shamma, J. S. (2000). Research on gain scheduling,. *Automatica*, Vol. 36, No 10, page numbers (1401- 1425), (October 2000), ISSN: 0005-1098

Sendhoff, B., Kreuz, M. & von Seelen, W. (1997). A condition for the genotype-phenotype mapping: Casualty, *Proceedings of 7th International Conference on Genetic Algorithms*, pp. 73-80, Michigan USA, July 1997, Morgan Kaufmann

Shamma, J. S. & Athans, M. (1992). Gain scheduling: potential hazards and possible remedies, *IEEE Control Systems Magazine*, Vol. 12, No 3, (June 1992), page numbers (101-107), ISSN: 0272-1708

Shue, S. & Agarwal, R. (1999). Design of automatic landing system using mixed H2 / H□□control, *Journal of Guidance, Control and Dynamics*, Vol. 22, No 1, page numbers (103-114), (January 1999), ISSN 0731-5090

Siouris, G. M., Leros, P. (1988). Minimum-time intercept guidance for tactical missiles, *Control Theory and Advanced Technology*, Vol. 4, No 2, page numbers (251-263), (February, 1988), ISSN 0911-0704

Skogestad, S. & Postlethwaite I. (1997). *Multivariable Feedback Control.* John Wiley and Sons, ISBN-10: 0471943304, ISBN-13: 978-0471943303

Spooner, J. T., Maggiore, M., Ordonez ,R. & Passino, K. M. (2002). *Stable Adaptive Control and Estimation for Nonlinear Systems: Neural and Fuzzy Approximator Techniques.* John Wiley and Sons, ISBN-10: 0471415464, ISBN-13: 978-0471415466

Ursem, R. K. (2003). Models for evolutionary algorithms and their applications in system identification and control optimization, *PhD dissertation,* Department of Computer Science, University of Aarhus, Denmark, 2003

van Laarhoven, P. J. M. & Aarts, E. H. L. (1987). *Simulated annealing: theory and applications.* Dordrecht, The Netherlands: D. Reidel, ISBN-10: 9027725136

Wang, Q. & Zalzala, A. M. S. (1996). Genetic control of near time-optimal motion for an industrial robot arm, *Proceedings of IEEE International Conference on Robotics and Automation*, pp. 2592-2597 vol. 3, ISBN: 0-7803-2988-0, Minneapolis, MN, April 1996

Wise, K. & *Nonlinear*, S. J. (1996). H∞ optimal control for agile missiles, *Journal of Guidance, Control and Dynamics*, Vol. 19, No 1, page numbers (157-165), (January 1996), ISSN 0731-5090

Zhou, K. & Doyle, J. C. (1998). *Essentials of Robust Control*. Prentice Hall, ISBN-10: 0135258332, ISBN-13: 978-0135258330

8

Fly-The-Camera Perspective: Control of a Remotely Operated Quadrotor UAV and Camera Unit

DongBin Lee[1], Timothy C. Burg[1], Darren M. Dawson[1] and Günther Dorn[2]

[1]*Clemson University*, [2]*Hochschule Landshut*

[1]*USA*, [2]*Germany*

1. Introduction

This Chapter presents a mission-centric approach to controlling the optical axis of a video camera mounted on a *camera positioner* and fixed to a *quadrotor* remotely operated vehicle. The approach considers that for video collection tasks a single operator should be able to operate the system by *"flying-the-camera"*; that is, collect video data from the perspective that the operator is looking out of and is the pilot of the camera. This will allow the control of the quadrotor and the camera manipulator to be fused into single *robot manipulator* control problem where the camera is positioned using the four degree-of-freedom (DOF) quadrotor and the two DOF camera positioner to provide a full six DOF actuation of the camera view. Design of a closed-loop controller to implement this approach is demonstrated using a *Lyapunov*-type analysis. Computer simulation results are provided to demonstrate the suggested controller.

Historically, the primary driver for UAV reconnaissance capabilities has been military applications; however, we appear to be at the juncture where the cost and capabilities of such systems has become attractive in civilian applications. The success of recent UAV systems has raised expectations for an increased rate of technology development in critical factors such as low-cost, reliable sensors, air-frame construction, more robust and lightweight material, higher energy-density battery technologies, and interfaces that require less operator training. One of the essential technologies is the camera positioner, which includes camera, camera base, and multi-axis servo platform. The potential for UAVs with camera positioners has been well established in many applications as diverse as fire fighting, emergency response, military and civilian surveillance, crop monitoring, and geographical registration. Many research and commercial groups have provided convincing demonstrations of the utility of UAVs in these applications. Most of the commercial systems are equipped with camera positioners as standard equipment; however, the use of the camera is not integrated with the control of the UAV.

The typical structures of a camera positioner include pan-tilt, tilt-roll, or pan/tilt/roll revolute joints or multi-axis gimbals. When considering the actuator of the camera gimbal, rate-gyros or encoders are used to measure the orientations. If the system is small and lightweight, the actuator dynamics can be discounted or neglected in the control of the UAV. Heavier systems, relative to the UAV, may require that the interaction of the camera positioner with the airframe be considered in the control system. In this chapter we demonstrate an approach that couples the positioning of a camera system to the control of the UAV. We believe this approach will serve as the basis for coupling dynamic control of the two systems.

Figure 1. Clemson UAV with Pan-Tilt Camera

1.1 Motivation

A typical camera system has a 2-axis gimbal or pan-tilt attached to the UAV as shown in Fig. 1. This camera positioner, with actuators, is usually fixed to the UAV rigid-body and generally attached to the front of the aerial vehicle where it can be used for surveillance, monitoring, or object targeting. The system is usually open-loop controlled using a controller/ joystick at the ground station. In this case, when the UAV tilts up and down to meet position control objectives of the airframe, the camera will also point up and down as shown in the upper plots of Fig. 2. If left uncompensated, the camera loses the target and fails to meet the surveillance objective. Compensating the camera field-of-view as shown in the lower half of Fig. 2 means that the camera positioner is moved in reaction to the platform motion. In the simplest system, the UAV pilot manually performs this compensation.

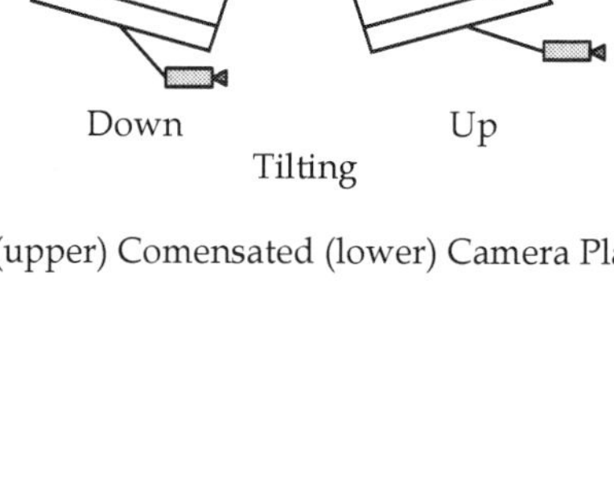

Figure 2. Uncompensated (upper) Comensated (lower) Camera Platform Position for UAV Airframe Tilting Motion

When the navigation or surveillance tasks become complicated, two people may be required to achieve the camera targeting objective that is controlled *independently* from the vehicle: a pilot to navigate the UAV and a camera operator. In automated systems feature tracking software can be used to identify points of interest in the field of view and then generate camera position commands that keep these features in the field of view. It is insightful to consider the actions of the two actors in pilot/camera operator scenario in order to hypothesize a new operational mode. The pilot will work to position the aircraft to avoid obstacles and to put the camera platform, *i.e.*, the aerial vehicle, in a position that then will allow the camera operator to watch the camera feed and move the camera positioner to track a target or survey an area.

An important underlying action on the part of the camera operator that makes this scenario feasible is that the camera operator must compensate for the motions of the UAV that stabilize the camera targeting. Fig. 3 further describes the effect of uncompensated camera platform motion on the camera axis in the upper figures and the bottom figures show a scheme where the camera positioner is used to compensate for the UAV body motion and maintain the camera view. Additionally, there must be communication between the pilot and the camera operator so that the camera platform is correctly positioned or moved to meet the video acquisition objective. More specifically, the camera positioning problem is split between the pilot and the camera operator. Since the operator is not in full control of positioning the camera, she must rely on commands to the pilot to provide movement of the camera platform for motions not included in the camera positioner. For example, if the camera positioner is a simple pan-tilt and the camera operator requires translation of the camera, then a request must be made to the pilot to move the camera platform. The potential shortcomings of this typical operational scenario can be summarized as:

1. multiple skilled technicians are typically required,
2. the camera operator must compensate for the actions of the pilot, and
3. it is not intuitive for a camera operator to split the camera targeting tasks between actions of the camera positioner controlled by the operator and commands to the pilot.

Figure 3. Uncompensated (upper) and Compensated (lower) Camera Platform for Rolling Motion

1.2 Previous Research

Sample research that frames the UAV-camera control problem is now reviewed. The user's complaint of difficulty of locating objects indicates the need for an actuated roll axis as described in Adelstein & Ellis (2000). They created a 3 DOF camera platform, to promote telepresence, by complementing yaw and pitch motions with roll motion in an experiment. The research in Mahony et al. (2002) examined hardware and software to automatically keep the target in the camera's field-of-view and stabilize the image. This work suggested "eye-in-hand" visual servo approach for a robotic manipulator with a camera on the end effector and introduced nonlinear gain into the orientation feedback kinematics to ensure that the target image does not leave the visual field. Jakobsen & Johnson (2005) presented control of a 3-axis Pan-Tilt-Roll camera system using modified servos and optical encoders mounted on a UAV with three different operating modes. Lee et al (2007) presented the modelling of a complete 3 link robotic positioner and suggested a single robotic body system by combining it with a quadrotor model.

The authors, Yoon & Lundberg (2001), presented equations of motion for a two-axis, pan-tilt, gimbal system to simplify the gimbal control and to further illustrate the properties of the configuration. In Sharp et al. (2001), the authors implemented a realtime vision system for a rotorcraft UAV during landing by estimating the vehicle state. A pan-tilt camera vision system is designed to facilitate the landing procedure based on geometric calculations from the camera image. The work in Stolle & Rysdyk (2003) proposed a solution to the problem of the limited range of the camera mechanical positioner. The system attempts to keep a target in the camera field-of-view by applying a circle-based flight path guidance algorithm with a nose-mounted pan-tilt camera. Pieniazek (2003) presented a software-based camera control and stabilization for two degree-of-freedom onboard camera so as to stabilize the image when the aircraft attitude was disturbed by turbulence or attitude changes. Quigley et al. (2005) presented a field-tested mini-UAV gimbal mechanism, flightpath generation algorithm, and a human-UAV interface to enhance target acquisition, localization, and surveillance.

Procerus Technologies produces OnPoint™ targeting system for vision-based target prosecution. The system utilizes feature tracking and video stabilization software, a fixed camera, and no GPS sensor to achieve these goals. Micropilot company manufactures a stabilized pan-tilt-zoom camera system which stabilizes the video camera in yawing or rolling and tilting (elevation) providing a stable image at high zoom. This stabilized gimbal system is mounted on a mechanical interface that can be specified to match the UAV objective with the ground control software HORIZON^{mp}.

1.3 Camera Stabilization and Targeting

A 2-axis camera positioner is shown in Fig. 4 which can be used for surveillance or trajectory-tracking. The problem of providing an intuitive interface with which an operator can move a camera positioner to make a video camera follow a target image appears in many places. The difficulty of moving a system that follows a subject with a video camera was recently addressed in Cooke et al. (2003) where operating a multilink, redundant-joint camera boom for the movie and television industry is described. The interesting result from this work is that an integrated control strategy, using a vision servoing approach to reduce the number of links controlled by the operator, can improve the use of the system. The final result shows an unexperienced operator achieving the same tracking result as an experienced operator; hence, the control strategy has rendered the system more friendly to the operator. The salient point of the control strategy is that there is independent macro- and micro- positioning of the camera - the operator controls the coarse positioning and the vision system controls the fine positioning. The authors suggest that the same approach could be used for other camera platforms; however, it is required that the system have redundant positioning axes. Additionally, an automated vision servoing system may not be desirable for general reconnaissance where the target is not known.

Figure 4 Two-axis Camera Positioner

A different perspective to this basic camera targeting problem was presented in Chitrakaran (2006) and Neff et al. (2007) where the camera platform, a quadrotor UAV, and the camera positioning unit are considered to be controlled concurrently. In this work a controller was developed which simultaneously controls both the quadrotor and the camera positioning unit in a complimentary fashion. Both works show combining the four degrees-of-freedom provided by motion of the quadrotor helicopter with two degrees-of-freedom provided by a camera positioner to provide arbitrary six degree-of-freedom positioning of the on-board video camera. The work in Chitrakaran (2006) is actually directed towards providing an automated means of landing the quadrotor through the vision system but provides an important mathematical framework for analyzing the combined quadrotor/camera system. The work in Neff et al. (2007) builds on Chitrakaran (2006) using a velocity controller and first introduced the "fly-by-camera"concept by combining UAV and 2-axis tilt-roll camera for the combined quadotor/camera system that works from operator commands generated in the camera field-of-view to move both elements, presenting a new interface to the pilot.

The research in Lee et al (2007) is exploited to provide an integrated system combining quadrotor UAV with 3-DOF camera system to present a single robotic unit. The paper designed a complete Pan-Tilt-Roll model which can be used for two camera optical axis; looking forward which can be used for surveillance, tracking, or targeting and looking downward axis for vision-based landing and monitoring the ground situation. In addition, the work suggests a position-based controller to show the upper bounds of position and velocity tracking error which yields Globally Uniformly Ultimately Bounded (SGUUB). The stabilization of the camera comes from this proposed perspective, which will be referred to as the fly-the-camera perspective, where the pilot commands motion from the perspective of the on-board camera - it is as though the pilot is riding on the tip of the camera and commanding movement of the camera ala a six-DOF flying camera. This is subtly different from the traditional remote control approach wherein the pilot processes the camera view and then commands an aircraft motion to create a desired motion of the camera view. The work proposed here exploits this new perspective for fusing vehicle and camera control. In this exposition, a quadrotor UAV model will be combined with a two-DOF camera kinematic model to create a fully actuated camera frame and a positioner controller will be designed.

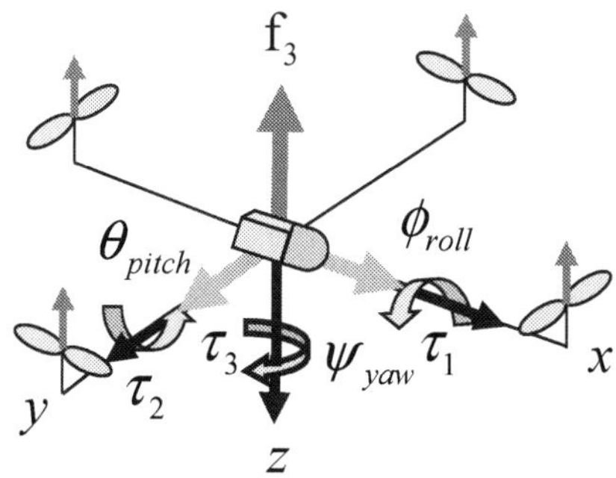

Figure 5. Underactuated Quadrotor Helicopter

2. System Modeling

2.1 Quadrotor UAV

The elements of the quad-rotor unmanned aerial vehicle model are shown in Fig. 5. This system can only directly produce a thrust, f_3, along the z-axis (vertical) and torques about the roll, pitch, and yaw angle axes (τ_1, τ_2, and τ_3); hence, the system is underactuated and the translational motion in the x- and y-directions is indirectly achieved. The quadrotor is assumed to be a rigid body on which thrust and torque act uniformly through the body and that the quadrotor body fixed frame, F, is chosen to coincide with the center of gravity, which implies that it has a diagonal inertia matrix. The kinematic model of a quadrotor expressed in the inertial reference frame, denoted as North-East-Downward (NED), is given by

$$\begin{bmatrix} \dot{p}_F \\ \dot{\Theta}_F \end{bmatrix} = \begin{bmatrix} R_F^I & O_3 \\ O_3 & T_F^I \end{bmatrix} \begin{bmatrix} v^F \\ \omega^F \end{bmatrix} \in R^6 \tag{1}$$

where $v^F(t)$ and $\omega^F = [\omega_x, \omega_y, \omega_y]^T \in R^3$ denote the linear velocity and angular velocity of the quadrotor body-fixed frame, F, with respect to the earth-fixed inertial frame, I, expressed in the body-fixed frame, F. The position and angle, $p_F(t)$ and $\Theta_F(t)$, and velocities $v^F(t)$ and $\omega^F(t)$ are assumed to be measurable. The $\dot{p}_F(t)$ in (1) is the velocity $v^F(t)$ of the quadrotor transformed by the spatial rotation matrix $R_F^I(\Theta) \in SO(3)$ where $\Theta_F = [\phi, \theta, \psi]^T \in R^3$ are roll, pitch, and yaw angles about x-, y-, and z-axes (Fossen, 2002)

$$\dot{R}_F^I = R_F^I S\left(\omega^F\right). \tag{2}$$

$\dot{\Theta}_F(t)$ in (1) represents the angular velocity $\omega^F(t)$ transformed into the inertial frame by the angular orientation matrix $T_F^I(\Theta) \in R^{3x3}$ (Fossen, 2002) which represents the modeling assumption that angular velocity of the quadrotor is calculated directly in lieu of modeling the angular dynamics; that is, $\omega^F(t)$ is considered as the system input. The translational dynamic modeling equation of the quadrotor is based on Chitrakaran (2006) and given by

$$m\dot{v}^F = F^f - mS(\omega^F)v^F + N_1(v^F) + G(R_I^F) \tag{3}$$

where the rotational angular velocity, $\omega^F(t)$, is implemented by control command (see (57)) and (3) contains the gravitational term, $G(R_I^F)$, which is represented in the body-fixed frame as

$$G(R_I^F) = mg(R_I^F)E_3 \in R^3 \tag{4}$$

where $g \in R^1$ denotes gravitational acceleration, $E_3 = [0,0,1]^T$ denotes the unit vector in the coordinates of the inertial frame, $m \in R^1$ is the known mass of the quad-rotor, $N_1(v^F) \in R^3$ represents a bounded, unknown, nonlinear function (*i.e*, aerodynamic damping) and $S(\cdot) \in R^{3x3}$ is a general form of the skew-symmetric matrix as follows

$$S(\xi) = \begin{bmatrix} 0 & -\xi_3 & \xi_2 \\ \xi_3 & 0 & -\xi_1 \\ -\xi_2 & \xi_1 & 0 \end{bmatrix}, \xi = [\xi_1, \xi_2, \xi_3]^T \in R^3. \tag{5}$$

As mentioned at the start of this modeling section, the quadrotor shown in Fig. 5 is underactuated in the sense that it has one translational force along the z-axis. The vector $F^f(t) \in R^3$ refers to the quadrotor translational forces but in reality represents the single translational force which is created by summing the forces generated by the four rotors and is expressed as

$$F^f = B_1 u_1 = \begin{bmatrix} 0 & 0 & u_1 \end{bmatrix}^T \tag{6}$$

where $B_1 = I_3$ is a configuration matrix (actuator dynamics are beyond the scope of this design) and $u_1(t) = f_3(t) \in R^1$. The modeling nomenclature is summarized in 7.3.

2.2 Model Assumptions

The following assumptions are made regarding the system model:

- **A1:** $T_F^I(\Theta)$ are full rank, *i.e.*, $\theta(t) \neq \pm\frac{\pi}{2}$. This will ensure that the orientation angle, θ_a, (defined in (38)) remains within the range $-\pi < \theta_a < \pi$ about the rotation axis $\mu(t)$ and will ensure that $\det(L_\omega)$ exists (see (43)).
- **A2:** Aerodynamic damping term $N_1(v^F)$ in (3) can be replaced by the parameter linearizable form; i.e., $Y_1(v^F)\theta_1 = N_1(v^F)$ where $\left\|Y_1(v^F)\right\| \leq \zeta_0 \left\|v^F\right\|$, ζ_0 is a positive constant, $Y_1(v^F) \in R^{3xn}$ is a known regression matrix, and $\theta_1 \in R^n$ is a known parameter vector. Additionally, $\left\|Y_1(v^F)\theta_1\right\| \leq \zeta_1(\left\|v^F\right\|) \leq \xi_1 \left\|v^F\right\|$ where ζ_1 is a positive function and non-decreasing in $\left\|v^F\right\|$ and $\xi_1 \in R^1$ is a positive constant.

2.3 Camera Positioner

Here we suggest a camera system with two revolute joints to cover forward and downward optical axes as shown in Fig. 6. Considering the integrated system, UAV and camera positioner, the yaw action of the UAV provides the third orientation angle for the camera. To create a general analytic framework for modeling the configurations of a two-link revolute positioner, a 2-axis robot positioner model is proposed. This general two-axis camera positioner will provide the forward kinematics for a tilting-rolling motion and panning-tilting motion where the angles $\theta_{tilt}(t)$ and $\theta_{roll}(t)$ / $\theta_{pan}(t)$ are assumed to be measurable. The Tilt-Roll configuration can be used to compensate for the quadrotor body roll and pitch when the camera is facing forward for tasks such as general navigation or surveillance. On the other hand, the Pan-Tilt camera positioner configuration can be used to compensate for the quadrotor pan and pitch motion when the camera is looking downward for tasks like landing, monitoring, or surveillance.

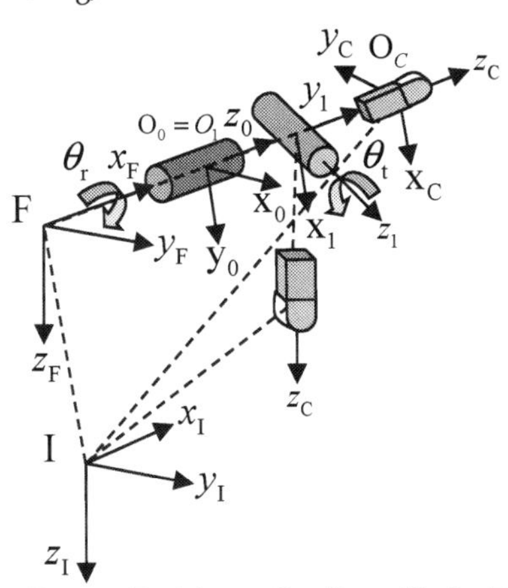

Figure 6. Quadrotor with a Camera Positioner Configurable for Tilt-Roll and Pan-Tilt Operating Modes

The dynamics of the camera unit will be considered negligible. The standard coordinate system definitions are made, specifically, O_0 is used to represent the origin of the coordinate system at the base (B) of the camera, O_1 is the origin of the second link, and O_c is used to represent the camera frame and can be denoted as O_2. Note that the link lengths are assumed to be zero so O_0, O_1, and O_2 are coincident. The camera positioner kinematics will be calculated using Denavit-Hartenberg (D-H) Convention (Spong, 2006). From the convention the rotation matrix R_i^{i-1} from the ith coordinate system to the $(i-1)^{th}$ frame is given

$$R_i^{i-1} = \begin{bmatrix} \cos\theta & \text{-}\sin\theta\cos\alpha & \sin\theta\sin\alpha \\ \sin\theta & \cos\theta\cos\alpha & \text{-}\cos\theta\sin\alpha \\ 0 & \sin\alpha & \cos\alpha \end{bmatrix} \tag{7}$$

where θ is the rotated angle, pan/tilt/roll, and α is the twisted angle between the links.

2.3.1 Case 1: Tilt-Roll Camera Configuration (camera looking forward)

A feasible 2-axis Tilt-Roll camera positioner is pictured in Fig. 7. Note that the matrix O_0 represents a static mounting on the quadrotor as a camera base frame denoted B and is given by

$$R_{0=B}^{F} = R_{\alpha,\frac{\pi}{2}} R_{\theta,\frac{\pi}{2}} = \begin{bmatrix} 0 & 0 & 1 \\ 1 & 0 & 0 \\ 0 & 1 & 0 \end{bmatrix} \tag{8}$$

where θ and α angles are each $+90^\circ$ in order calculated by (7) and $z_0 = \begin{bmatrix} 1 & 0 & 0 \end{bmatrix}^T$ where z_{i-1} (i is the link origin) is the last column of the rotation matrix which will be used in the Jacobian matrix. The Denavit-Hartenberg (D-H) table for the Tilt-Roll configuration when the optical axis is looking forward, as shown in Fig. 7, is in Table I. The camera positioner unit is considered to have coincident rotational links, thus the link lengths are zero (a=0 in Table I).

Link (i)	Offset	Length	Twist	Angle
	D	a	α	θ
1	0	0	$+90^\circ$	$\theta_r + 90^\circ$
2 (forward axis)	0	0	-90°	θ_t

Table 1. D-H for Tilt-Roll Camera Positioner while looking forward optical axis

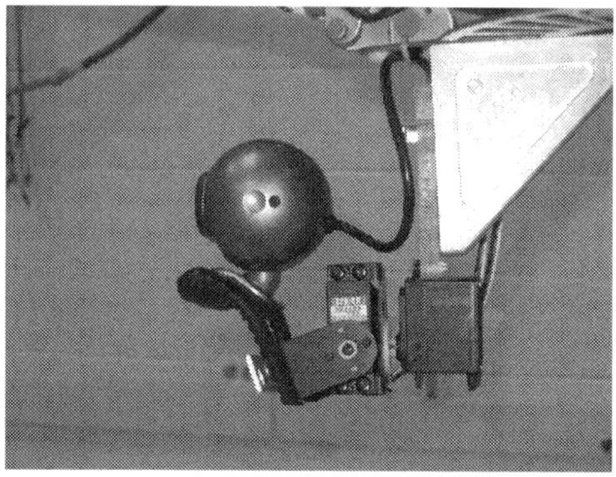

Figure 7. Quadrotor with Tilt-Roll Camera Positioner

The rotation matrices for first and second links are obtained as

$$R_1^B = \begin{bmatrix} -s\theta_r & 0 & c\theta_r \\ c\theta_r & 0 & s\theta_r \\ 0 & 1 & 0 \end{bmatrix} \quad \text{and} \quad R_2^1 = \begin{bmatrix} c\theta_t & 0 & -s\theta_t \\ s\theta_t & 0 & c\theta_t \\ 0 & -1 & 0 \end{bmatrix} \tag{9}$$

where $c\cdot = \cos(\cdot)$, $s\cdot = \sin(\cdot)$ were used, and the rotation matrix from the base frame to first link frame is given

$$R_1^F = R_B^F R_1^B = \begin{bmatrix} 0 & 1 & 0 \\ -s\theta_r & 0 & c\theta_r \\ c\theta_r & 0 & s\theta_r \end{bmatrix} \tag{10}$$

where $z_1 = \begin{bmatrix} 0 & c\theta_r & s\theta_r \end{bmatrix}^T$. The total rotation matrix from the quadrotor through the second link can be obtained as

$$R_{2=C}^F = R_B^F R_1^B R_2^1 = \begin{bmatrix} \sin\theta_t & 0 & \cos\theta_t \\ -\sin\theta_r \cos\theta_t & -\cos\theta_r & \sin\theta_r \sin\theta_t \\ \cos\theta_r \cos\theta_t & -\sin\theta_r & -\cos\theta_r \sin\theta_t \end{bmatrix}. \tag{11}$$

The angular velocity of the camera frame can be written as follows

$$\omega_{BC}^F = R_B^F \omega_{BC}^B = J_C \dot{\theta}_C \tag{12}$$

where the positioner joint angles, $\dot{\theta}_C(t) \in R^3$, are given by

$$\dot{\theta}_C = \begin{bmatrix} \dot{\theta}_r \\ \dot{\theta}_t \end{bmatrix} \text{ and } \theta_C = \begin{bmatrix} \theta_r \\ \theta_t \end{bmatrix}. \quad (13)$$

The Jacobian matrix $J_C(\theta_C) \in R^{3\times 2}$ can be built from the rotation matrices with the final result given by

$$J_C = \begin{bmatrix} z_0 & z_1 \end{bmatrix} = \begin{bmatrix} 1 & 0 \\ 0 & c\theta_r \\ 0 & s\theta_r \end{bmatrix} \quad (14)$$

where the reader is referred to Murray et al. (1994) for the Jacobian matrix of a manipulator.

2.3.2 Case 2: Pan-Tilt Camera Configuration (camera looking downward)

A similar approach to that for case 1, 2.3.1, can be followed to develop the kinematics for the Pan-Tilt configuration when the optical axis is looking downward as shown in Fig. 8.

Figure 8. Quadrotor with Pan-Tilt Camera Positioner

The systematic Denavit-Hartenberg convention is used to obtain the rotation matrices of the Pan-Tilt camera positioner configuration and the detail procedure is shown in Appendix 7.1.

2.4 Position and Orientation of the Camera Positioner

The objective is to control the motion of the camera optical axis. Towards this end, the kinematic relationships in (1) will be extended to include the action of the camera positioning unit and to obtain the position and orientation of the camera. The derivative of the position of the camera in the camera frame, C, with respect to the inertial frame, I, and expressed in the inertial frame, $\dot{p}_C(t) \in R^3$, is defined as

$$\dot{p}_C = R_C^I v^C \quad (15)$$

where $v^C(t) \in R^3$ is the linear velocity in the camera frame, C, referred to inertial frame, I, expressed in the camera frame, C. The rotation matrix R_C^I from the camera frame to the inertial frame can be obtained using the previous equations as

$$R_C^I = R_F^I R_C^F . \tag{16}$$

The velocity $v^C(t)$ can be divided into two components as follows

$$v^C = v^{IF} + v^{FC} = R_F^C v^F \tag{17}$$

where $v^{IF}(t)$ is the UAV velocity expressed in camera frame (C) and $v^{FC}(t) = 0$ since the camera positioning unit only has rotational axes and does not translate from the quadrotor body. Thus, substituting (17) for $v^C(t)$ in (15) yields

$$\dot{p}_C = R_C^I R_F^C v^F = R_F^I v^F . \tag{18}$$

The desired camera position trajectory, $\dot{p}_D(t) \in R^3$, is generated via

$$\dot{p}_D = R_D^I v^D \tag{19}$$

where $v^D(t) \in R^3$ is a desired input velocity vector and the desired rotation matrix expressed in inertial frame, $R_D^I(\Theta, \theta_C, \theta_d) \in R^{3x3}$, can be defined in the following manner

$$R_D^I = R_F^I R_C^F R_D^C \tag{20}$$

where $R_D^C(\theta_d) \in R^{3x3}$ represents a rotation matrix from D to desired frame C as shown in Fig. 9 for the subsequent control development which is the tracking of the camera frame towards the desired camera frame using angle-axis representation (see Appendix 7.2). Utilizing (20), (19) yields

$$\dot{p}_D = R_F^I R_C^F R_D^C v^D . \tag{21}$$

Hence, to quantify the mismatch between the camera (actual) and desired attitudes, we define the rotation matrix $\tilde{R}(\Theta, \theta_C, \theta_d) \in R^{3\times3}$ which can be obtained using measurable rotations and using the angle-axis representation (Spong, 2006) as

$$\tilde{R} = R_C^I (R_D^I)^T . \tag{22}$$

A similar result for the camera angles, $\Theta_C(t)$, in the camera frame from the second equation of (1) is now shown as

$$\dot{\Theta}_C = T_F^I(\Theta)\,\omega_{IC}^F = T_F^I(\Theta)\,R_C^F\,\omega^C \tag{23}$$

where an angular transformation $T_F^I(\Theta)$ is used. Decomposing the angular velocity,

$$\omega^C(t) = \omega^{IF} + \omega^{FB} + \omega^{BC} = R_F^C\omega^F + R_F^C\omega_{FB}^F + R_B^C\omega_{BC}^B, \tag{24}$$

yields

$$\dot{\Theta}_C = T_F^I(\Theta)\,R_C^F(R_F^C\omega^F + R_F^C\omega_{FB}^F + R_B^C\omega_{BC}^B). \tag{25}$$

Then, (25) yields

$$\dot{\Theta}_C = T_F^I(\Theta)\,\omega^F + T_F^I(\Theta)\,J_C\dot{\theta}_c \tag{26}$$

where $R_C^F R_F^C = 1$ is used and $\omega_{FB}^F = 0$ since the camera base is rigidly mounted on the quadrotor frame. Finally, following the same approach the desired camera angle, $\Theta_D(t)$, is obtained from the desired angular velocity of the camera in the quadrotor frame, $\omega^D(t)$, as

$$\dot{\Theta}_D = T_F^I(\Theta)\,R_D^F\omega^D. \tag{27}$$

The changing rate of $R_C^I(\Theta)$ is obtained as follows

$$\dot{R}_C^I = R_C^I S(\omega^C) \quad \text{and} \quad \dot{R}_I^C = S^T(\omega^C)R_I^C = -S(\omega^C)R_I^C. \tag{28}$$

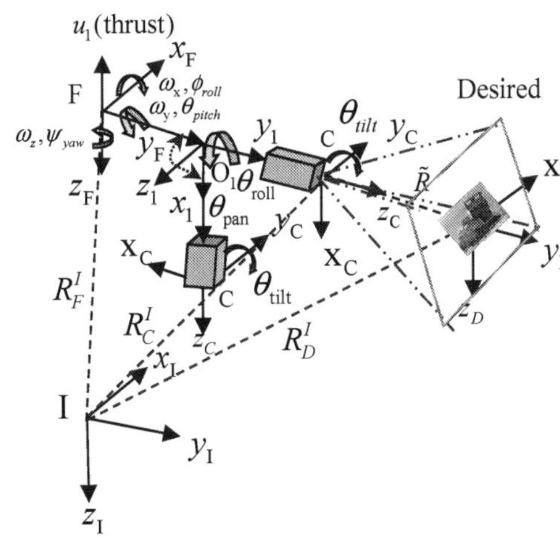

Figure 9. Tracking of the Camera Frame Towards the Desired Camera Frame

3. Control Method

Fig. 10 demonstrates the culmination of the modeling effort to combine the quadrotor and camera positioner to create a means of fully actuating the camera optical axis. In this diagram, it can be seen that the camera is positioned and oriented by using the two camera positioner angles (selected from $\theta_{tilt}(t)$, $\theta_{roll}(t)$, and $\theta_{pan}(t)$, $\theta_{tilt}(t)$, according to the configuration of the camera optical axis), the quadrotor linear force, $F(t)$, and the quadrotor angles $\varphi_{roll}(t)$, $\theta_{pitch}(t)$, and $\psi_{yaw}(t)$ through $\omega_x(t)$, $\omega_y(t)$, and $\omega_z(t)$. In keeping with the fly-the-camera objective, a controller will be designed based on these inputs to move the camera optical axis along a desired trajectory. A control strategy will be proposed to control the camera translational position error, $e_p(t)\in R^3$, and the camera orientation error, $e_\theta(t)\in R^3$.

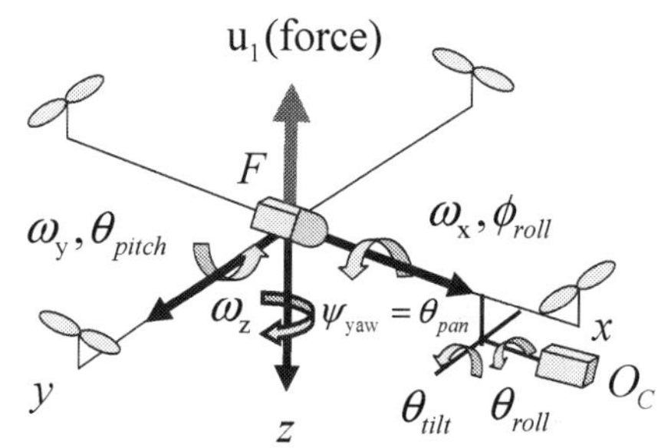

Figure 10. Six-DOF Single Robotic System

3.1 Tracking Error Formulation

The camera translational position error, $e_p(t)\in R^3$, is defined in the camera frame (C) as the transformed difference between the inertial frame based camera position, $p_C(t)$, and the inertial frame based desired camera position, denoted as $p_D(t)\in R^3$, as follows

$$e_p = R_I^C(p_C - p_D). \tag{29}$$

The camera translational position error rate, $\dot{e}_p(t)\in R^3$, is obtained by taking the time derivative of (1) to yield

$$\dot{e}_p = \dot{R}_I^C(p_C - p_D) + R_I^C(\dot{p}_C - \dot{p}_D) \tag{30}$$

where $\dot{p}_C(t)$ and $\dot{p}_D(t)$ were introduced in (18) and (19), respectively. Substituting (28) into the first term in (30) yields

$$\dot{R}_I^C(p_C - p_D) = -S(\omega^C)e_p, \tag{31}$$

and the term $R_I^C\, \dot{p}_C(t)$ in (30) can be rewritten in terms of the quadrotor velocity, $v^F(t)$, as

$$R_I^C\, \dot{p}_C = R_I^C\,(R_C^I v^C) = R_F^C v^F \tag{32}$$

where all translation of the camera is the result of the quadrotor translation. The final term in (30) is rewritten as

$$R_I^C\, \dot{p}_D = R_I^C\,(R_D^I v^D) = R_D^C v^D. \tag{33}$$

Substituting (31), (32), and (33) into (30) yields

$$\dot{e}_p = -S(\omega^C)e_p + R_F^C v^F - R_D^C v^D. \tag{34}$$

To further the controller development, a filtered error, $r(t) \in R^3$, is introduced as

$$r = [r_p^T \quad e_\theta^T]^T \tag{35}$$

where the filtered position error, $r_p(t) \in R^3$, is defined as

$$r_p = k_p e_p + R_F^C v^F + R_F^C \delta, \tag{36}$$

in which $\delta = [0 \quad 0 \quad \delta_3]^T \in R^3$ is a constant design vector. The orientation tracking signal, $e_\theta(t) \in R^3$, is defined in terms of the angle-axis representation of the rotation matrix between the desired and actual camera orientations, $\tilde{R}(\Theta, \theta_C, \theta_d)$, as

$$e_\theta = \theta_a \mu \tag{37}$$

where the scalar angle θ_a is obtained from

$$\theta_a = \cos^{-1}\left(\frac{1}{2}(Tr(\tilde{R}) - 1)\right) \in R^1, \tag{38}$$

in which $Tr(\tilde{R})$ defines the trace of the matrix $\tilde{R}(\Theta, \theta_C, \theta_d)$ and the unit length axis of rotation $\mu \in R^3$ defined as

$$\mu = \frac{1}{2\sin\theta_a}\left[(r_{32} - r_{23}), (r_{13} - r_{31}), (r_{21} - r_{12})\right]^T, \tag{39}$$

for which $\|\mu\|^2 = 1$. In this representation, the rotation angle θ_a is assumed to stay within the range $-\pi < \theta_a < \pi$. Note that the terms on the right-hand side of the definition in (39) come from $\tilde{R}(\Theta, \theta_C, \theta_d)$ as

$$\tilde{R}(\Theta, \theta_C, \theta_d) = \begin{bmatrix} r_{11} & r_{12} & r_{13} \\ r_{21} & r_{22} & r_{23} \\ r_{31} & r_{32} & r_{33} \end{bmatrix}. \tag{40}$$

Substituting the axis-angle representation (38) and (39) into (37) yields

$$e_\theta = \frac{1}{2sinc\{\cos^{-1}(\frac{1}{2}(Tr(\tilde{R})-1))\}} \begin{bmatrix} r_{32} - r_{23} \\ r_{13} - r_{31} \\ r_{21} - r_{12} \end{bmatrix} \tag{41}$$

where $sinc(\theta_a) = \frac{\sin(\theta_a)}{\theta_a}$. The angular error rate can be obtained by taking the time derivative of (37) as

$$\dot{e}_\theta = L_\omega \omega^{CD} \in R^3, \tag{42}$$

and $L_\omega \in R^{3x3}$ can be defined as

$$L_\omega = I_3 - \frac{\theta_a}{2} S(\mu) + (1 - \frac{sinc(\theta_a)}{sinc^2(\frac{\theta_a}{2})}) S^2(\mu), \tag{43}$$

in which $S(\mu)$ is a skew symmetric matrix and details for obtaining

$$S(\mu) = \frac{(\tilde{R} - (\tilde{R})^T)}{2\sin\theta_a}, \tag{44}$$

for obtaining (42) can be found in Malis (1998). The term $\omega^{CD}(t) \in R^3$ introduced in (42) represents the angular velocity of the desired camera frame relative to the actual camera frame as

$$\omega^{CD} = \omega^{CI} + \omega^{ID} = \omega^D - \omega^C = \omega^D - (\omega^C_{IF} + \omega^C_{FC}) \tag{45}$$

where $\omega^{ID}(t) = \omega^D(t)$ and $\omega^{CI}(t) = -\omega^{IC}(t) = -\omega^C(t)$. Substituting (45) along with (12) - (14) into (42) produces

$$\dot{e}_\theta = L_\omega \omega^D - L_\omega R^C_F \omega^F - L_\omega R^C_F J_c \dot{\theta}_c. \tag{46}$$

Taking the time derivative of $r_p(t)$ in (36) yields

$$\dot{r}_p = R_F^C \dot{v}^F + \dot{R}_F^C v^F + k_p \dot{e}_p + \dot{R}_F^C \delta. \tag{47}$$

Substituting from (3) and (34) into (47), utilizing the fact that

$$\dot{R}_F^C(\Theta) = R_F^C S(\omega_{CF}^F) = -R_F^C S(\omega_{FC}^F),$$

and subtracting and adding $R_F^C S(\omega^F)\delta$ yields

$$\dot{r}_p = R_F^C[\frac{1}{m}N_1(v^F) - S(\omega^F)v^F + gR_I^F E_3 + \frac{1}{m}F^f] + R_F^C S(\omega^F)\delta$$
$$+k_p[R_F^C v^F - S(\omega^C)e_p - R_D^C v^D] - R_F^C S(\omega_{FC}^F)v^F - R_F^C\left(S(\omega_{FC}^F) + S(\omega^F)\right)\delta. \tag{48}$$

Combining the angular velocities represented by the second term in the first bracket and the mid-term in the last row of (48) yields

$$-R_F^C S(\omega^F + \omega_{FC}^F)v^F = -R_F^C S(\omega_{IC}^F)v^F. \tag{49}$$

The right-hand side of (49) can be further clarified using $\omega_{IC}^F = R_C^F \omega^C(t)$ and

$$S(\omega_{IC}^F) = R_C^F S(\omega^C) R_F^C, \tag{50}$$

to yield

$$-R_F^C R_C^F S(\omega^C) R_F^C v^F = -S(\omega^C) R_F^C v^F. \tag{51}$$

Combining the angular velocity terms $\omega_{FC}^F(t)$ and $\omega^F(t)$ in the last terms of the last row in (48) with $\omega_{IC}^F(t)$ yields

$$-R_F^C S(\omega^F)\delta - R_F^C S(\omega^{FC})\delta = -R_F^C S(R_C^F \omega^C)\delta = -S(\omega^C) R_F^C \delta. \tag{52}$$

Multiplying both sides of (36) by $S(\omega^C)$ yields

$$-S(\omega^C) r_p = -S(\omega^C)(k_p e_p + R_F^C v^C + R_F^C \delta). \tag{53}$$

Substituting (51) through (53) into (48) yields

$$\dot{r}_p = \frac{R_F^C N_1(v^F)}{m} - S(\omega^C) r_p + gR_I^C E_3 + \frac{R_F^C F^f}{m} + k_p R_F^C v^F - k_p R_D^C v^D - R_F^C S(\delta)\omega^F. \tag{54}$$

By taking the time derivative of $r(t)$ in (35) and substituting (46) and (54) it can be obtained that

$$\dot{r}=\begin{bmatrix}\dfrac{R_F^C F^f}{m}-R_F^C S(\delta)\omega^F \\ -L_\omega R_F^C(\omega^F+J_c\dot{\theta}_c)\end{bmatrix}-\begin{bmatrix}S(\omega^C)r_p \\ O_{3x1}\end{bmatrix}+\begin{bmatrix}\dfrac{R_F^C N_1(v^F)}{m}+k_p(R_F^C v^F-R_D^C v^D)+gR_I^C E_3 \\ -L_\omega\omega^D\end{bmatrix}. \quad (55)$$

Arranging the first term in (55) yields

$$\begin{bmatrix}\dfrac{1}{m}R_F^C F^f-R_F^C S(\delta)\omega^F \\ -L_\omega R_F^C\omega^F-L_\omega R_F^C J_c\dot{\theta}_c\end{bmatrix}=\overline{L}_\omega B\overline{U}=\overline{B}\overline{U} \quad (56)$$

where

$$\overline{L}_\omega=\begin{bmatrix}R_F^C & O_{3x3} \\ O_{3x3} & -L_\omega R_F^C\end{bmatrix}, B=\begin{bmatrix}\dfrac{B_1}{m}, -S(\delta), O_{3x2} \\ O_{3x1}, I_{3x3}, J_c\end{bmatrix}\in R^{6x6}, \overline{U}=\begin{bmatrix}u_1 \\ \omega^F \\ \dot{\theta}_C\end{bmatrix}\in R^6, \quad (57)$$

along with the new term $U(t)$ where $U_1(t)\in R^3$ and $U_2(t)\in R^3$, defined also further clarify the control design procedure.

$$U=\overline{B}\overline{U}=[U_1,U_2]^T\in R^6. \quad (58)$$

Substitution of (57) to (58) into (56) and then substitution of the resulting form of (56) using (55) produces the open-loop filtered error dynamics as follows

$$\dot{r}=\begin{bmatrix}-S(\omega^C)r_p \\ O_{3x1}\end{bmatrix}+\begin{bmatrix}U_1 \\ U_2\end{bmatrix}-\begin{bmatrix}k_p R_D^C v^D \\ L_\omega\omega^D\end{bmatrix}+\begin{bmatrix}\dfrac{R_F^C N_1(v^F)}{m}+k_p R_F^C v^F+gR_I^C E_3 \\ O_{3x1}\end{bmatrix} \quad (59)$$

where it is clear that the control inputs $U_1(t)$ and $U_2(t)$ will be designed to meet the control objective. Note that implementation of the control will require $\overline{U}(t)$ which is obtained from

$$\overline{U}=\overline{B}^{-1}U\in R^6, \quad (60)$$

which requires $\overline{B}^{-1}=B^{-1}(\overline{L}_\omega)^{-1}$ where

$$(\overline{L}_{\omega})^{-1} = \frac{1}{\Delta(\overline{L}_{\omega})}\begin{bmatrix} -L_{\omega}R_F^C & O_3 \\ O_3 & R_F^C \end{bmatrix}, \tag{61}$$

in which $\Delta(\overline{L}_{\omega}) = \left| -R_F^C L_{\omega} R_F^C \right|$ is the determinant of the matrix $\overline{L}_{\omega}(t)$ and $\Delta(\overline{L}_{\omega}) \neq 0$ due to Assumption 1 (A1).

3.2 Lyapunov-based Control Design

The non-negative scalar function $V(t)$ is chosen as

$$V = \frac{1}{2}e_p^T e_p + \frac{1}{2}r^T r.$$

Differentiating yields

$$\dot{V} = e_p^T \dot{e}_p + r_p^T \dot{r}, \tag{62}$$

by substituting (34) and (59) into (62) it can be obtained that

$$\dot{V} = [r_p^T, e_\theta^T]^T [\begin{bmatrix} -S(\omega^C)r_p \\ O_{3x1} \end{bmatrix} + \begin{bmatrix} U_1 \\ U_2 \end{bmatrix} + \begin{bmatrix} \frac{R_F^C N_1(v^F)}{m} + k_p R_F^C v^F + gR_I^C E_3 - k_p R_D^C v^D \\ L_\omega \omega^D \end{bmatrix}] \tag{63}$$

$$+e_p^T[-S(\omega^C)e_p + (r_p - k_p e_p - R_F^C \delta) - v^D]$$

where (36) was utilized. The terms in (63) can be collected to yield

$$\begin{aligned} \dot{V} = [&r_p^T U_1 + \frac{1}{m} r_p^T R_F^C N_1(v^F) + e_p^T r_p + r_p^T k_p R_F^C v^F + r_p^T g R_I^C E_3 \\ &- r_p^T k_p R_D^C v^D - k_p e_p^T e_p - e_p^T R_F^C \delta - e_p^T R_D^C v^D + e_\theta^T U_2 - e_\theta^T L_\omega \omega^D]. \end{aligned} \tag{64}$$

Equation (64) will be utilized to design the control inputs $U_1(t)$ and $U_2(t)$. From the upper equation in (64) we can design $U_1(t)$ to subtract out four terms, add stabilizing feedback, and add robust compensatation for the unknown nonlinear term as follows

$$U_1 = -k_r r_p - \frac{r_p \zeta_1^2\left(\left\|v^F\right\|\right)}{\varepsilon_0} - k_p R_F^C v^F - gR_I^C E_3 - e_p \tag{65}$$

where ε_0 is a positve constant, A2 is used for $\zeta_1(\left\|v^F\right\|)$, and the nonlinear term $\overline{N}_1(t)$ is defined by

$$\bar{N}_1 = \frac{1}{m} R_F^C N_1(v^F). \tag{66}$$

From the lower equation in (59), $U_2(t)$ can be designed as

$$U_2 = L_\omega \omega^D - k_\theta e_\theta. \tag{67}$$

Substituting (65) and (67) into (64) yields

$$\dot{V} = r_p^T(\bar{N}_1 - \frac{r_p \zeta_1^2(\|v^F\|)}{\varepsilon_0}) - k_r r_p^T r_p - r_p^T k_p R_D^C v^D - k_\theta e_\theta^T e_\theta - k_p e_p^T e_p - e_p^T(R_F^C \delta + R_D^C v^D). \tag{68}$$

According to sufficient condition, finally it yields

$$\dot{V} \le -\lambda_2 \|\eta\|^2 + \varepsilon_4. \tag{69}$$

4. Stability Analysis

The closed-loop control law of (65) and (67) ensure that the tracking error is Globally Uniformly Ultimately Bounded (GUUB) in the manner

$$\|\eta\| \le \sqrt{\|\eta(0)\|^2 e^{-2\lambda_2 t} + \frac{\varepsilon_4}{\lambda_2}} \tag{70}$$

where

$$\eta = [r_p^T, e_\theta^T, e_p^T]^T, \tag{71}$$

ε_4 is a positive constant, and λ_2 is a positive constant given by the following form

$$\lambda_2 = \min\left\{(k_r - \frac{\lambda_1}{2}), (k_p - \frac{\lambda_0}{2}), k_\theta\right\} \tag{72}$$

where λ_0, λ_1 are positive constants under the conditions that

$$k_r > \frac{\lambda_1}{2} \quad \text{and} \quad k_{p>} \frac{\lambda_0}{2}. \tag{73}$$

A proof of the theorem can be proven using Lyapunov-type stability analysis (This is beyond the scope).

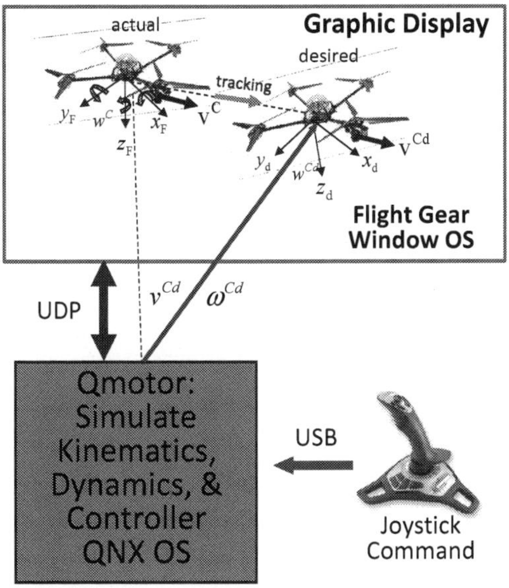

Figure 11. System Overview

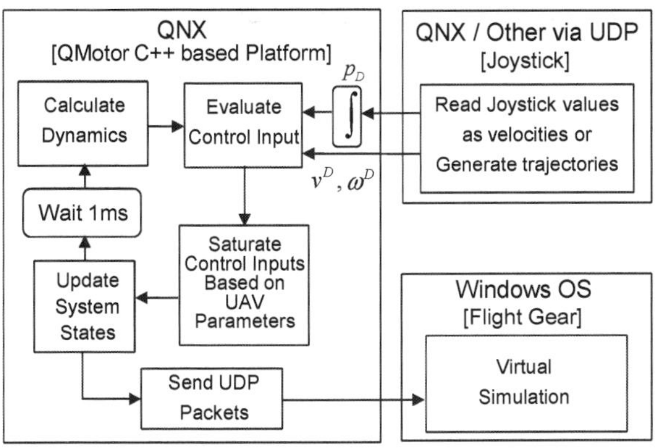

Figure 12. System Interal Outline in Simulation System

5. Simulation

The simulation system overview is given Fig. 11. A two computer system, QNX and Windows, was built to simulate the proposed controller. The QNX software systems are configured to run QNX Real-Time Operating System (RTOS) and host the QMotor (2000) control and simulation package while the Windows XP computer is configured to run and host the FlightGear (v.0.9.10) (2006) open-source flight simulator package. A QMotor program was written to simulate the rigid body kinematics, dynamics, and the proposed control. The output of the dynamics simulation is sent via UDP to set the aircraft/camera position and orientation in the FlightGear virtual world (note that FlightGear is used only as a graphics processor). The desired trajectories are input by the operator using a 6 DOF joystick (Logitech Extreme 3D Pro, 2006). Specifically, the operator indirectly supplies the desired position trajectory used by the controller through monitoring the simulated camera view and using the joystick to command the velocities that move the camera view in the virtual world. A detailed block diagram showing the three inputs on the joystick, labeled as x, y, and twist, are used to generate and control either three translational velocities, $v^D(t)$, or the three angular velocities, $\omega^D(t)$, depending on trigger position. That is, the magnitude and direction of these quantities is derived from the joystick position. The velocity commands are then integrated to produce the desired position trajectory used by the controller. A typical scene from FlightGear (2006) is shown in Fig. 13 where the quadrotor is tilted but the camera view seen by the operator remains level. The quadrotor simulation was developed to approximate the parameters of the DraganFlyer X-Pro (2005). Parameters such as mass (m) and saturation limits for control inputs, and the control gains are chosen to be

$$k_r = k_\theta = diag(1,1,1),\ k_p = diag(1,1,5),\ m = 2.72[kg] \text{ and } g = 9.81[m^3 / kgs^2],$$

$$u_{1\_\max} = 35.586\ [N],\ \theta_{tilt\_\max} = \theta_{roll\_\max} = 4.067\ [Nm].$$

A short timespan of the simulation was captured to demonstrate the operation of the system. The simulation results are shown in Fig. 14 through Fig. 19. The easiest to appreciate operational point is found at t=50 [sec] in Fig. 17 and Fig. 18 where the orientations of the camera and quadrotor rotate in opposite directions measured from the x-, y-, and z-axes to achieve the fly-the-camera perspective. This also can be shown by the torque inputs of UAV roll, the second plot in Fig. 16, and camera roll, the last plot in Fig. 16, which acts in opposite directions as the UAV torque produces the required lateral motion and the camera compensates for the resulting roll. This is also seen in the pitch input in the third plot of Fig. 16 and the camera tilt input in the fourth plot of Fig. 16. Fig. 14 shows the position tracking of the quad-rotor to the desired trajectory and Fig. 15 shows the position errors about the coordinates. The actual quad-rotor trajectory represented by the dotted line follows the desired trajectory represented by the solid line which is commanded to go up at the first time and then, move to forward and again go forward near the end.

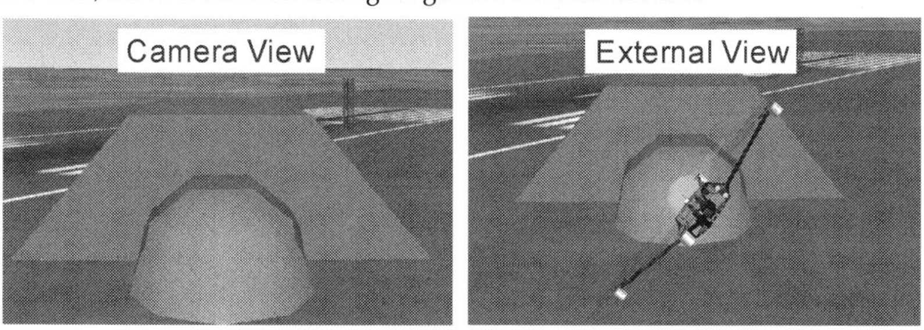

Figure 13. The "fly-the-camera" view used by the operator and an outside view of the quadrotor position

5.1 Simulation Results

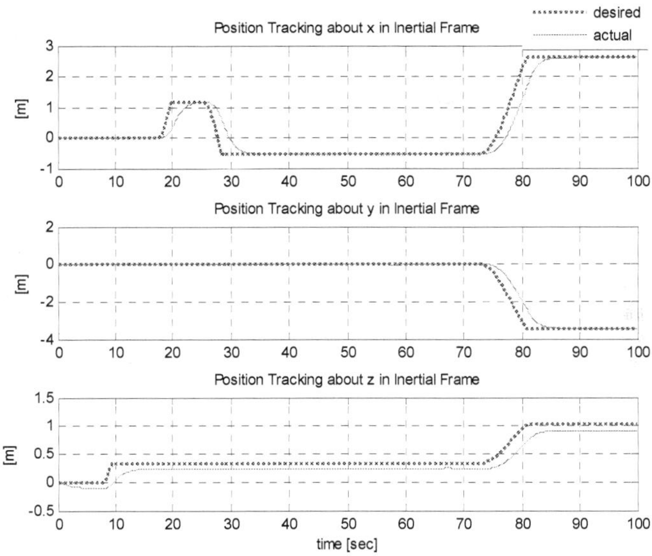

4

Figure 14. Position Tracking

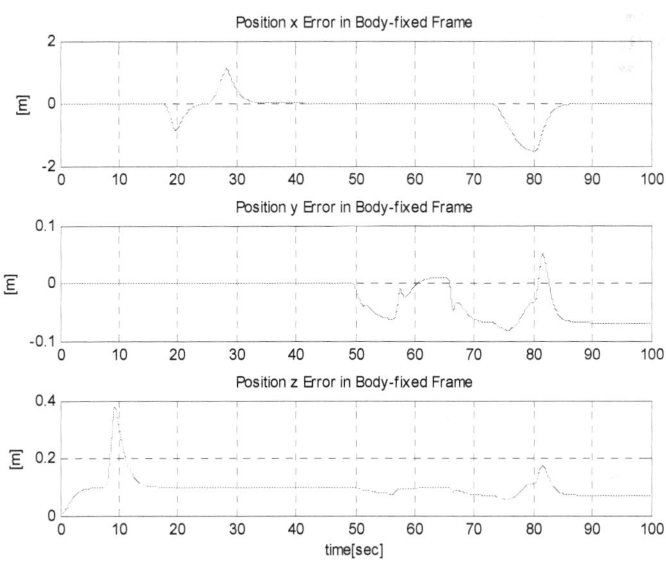

4

Figure 15. Tracking Errors

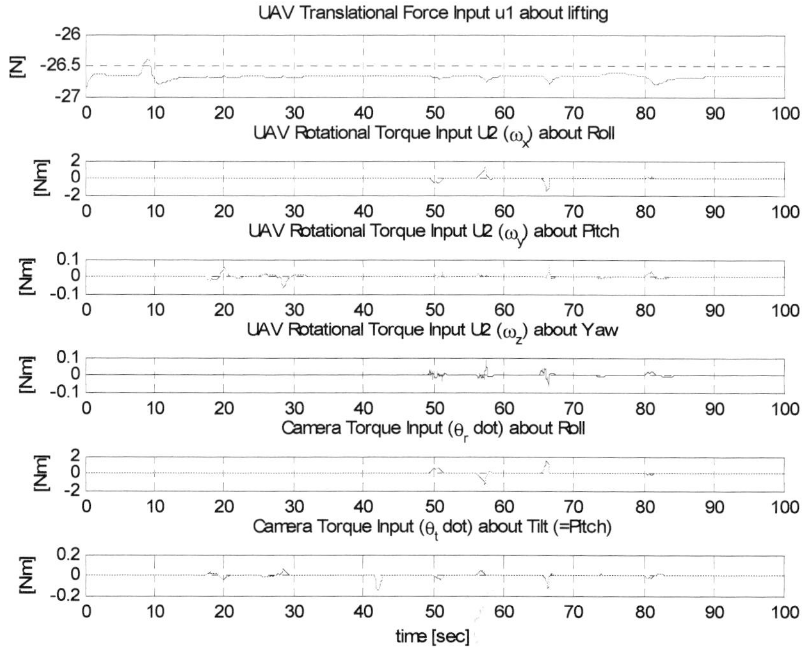

Figure 14. Control Inputs

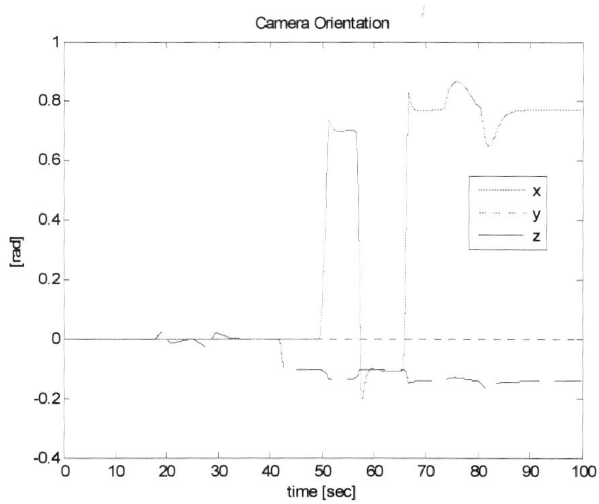

Figure 15. Camera Orientation

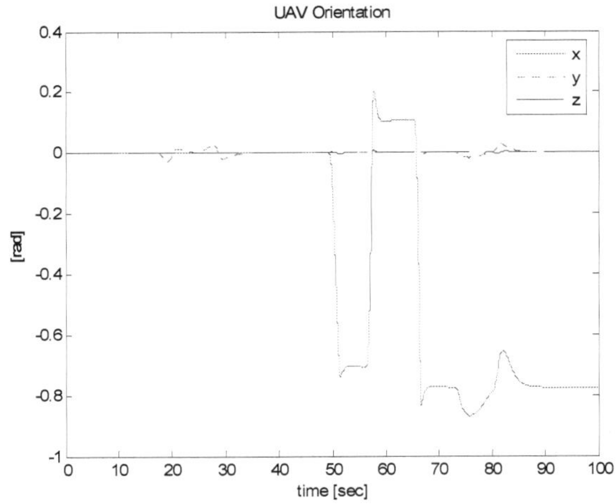

Figure 18. UAV Orientation

6. Conclusion

We have described the fly-the-camera approach to concurrent control of a camera positioner and unmanned aerial vehicle in an operator friendly manner. We detailed the design of a nonlinear controller as one possible embodiment of this philosophy. The fly-the-camera approach considers a single robotic dynamic object which consists of an underactuated aerial vehicle and two complementary camera axes to produce a fully actuated camera targeting platform. The fly-the-camera approach should provide a more intuitive perspective for a single remote pilot to operate the quadrotor vehicle and camera for surveillance, navigation, and landing/monitoring tasks. The approach fuses the often separate tasks of vehicle navigation and camera targeting into a single task where the pilot sees and flies the system as through riding on the camera optical axis. The controller detailed here was shown to provide position and angle based tracking in the form of Globally Uniformly Ultimately Bounded (GUUB) result. The simulation results were shown to demonstrate the proposed system. The development of fly-the-camera perspective system can provide a platform for many areas of commercial, industrial, or research work.

The demonstration of the "fly-the-camera" concept and a control methodology provide the foundation for expanding this approach. Two obvious points would extend the suitability of this work. First, only the kinematic model of the camera has been included. This approximation is only valid when the camera mass and acceleration forces are small relative to the airframe. Second, cognitive loading studies should be performed to evaluate the true potential of this approach.

7. Appendix

7.1 Case 2: Pan-Tilt Camera Configuration (camera looking downward)

The Denavit-Hartenberg (D-H) table for the Pan-Roll configuration when the optical axis is looking downward as shown in Fig. 7 is given in Table II.

Link (i)	Offset	Length	Twist	Angle
	D	a	α	θ
1	0	0	$+90^\circ$	$\theta_p + 90^\circ$
2 (downward axis)	0	0	-90°	$\theta_t - 90^\circ$

Table 2. D-H for Pan-Tilt Camera Positioner while the optical axis is looking forward

The rotation matrices of the Pan-Tilt camera positioning unit for first and second links as shown in Fig. 8 are obtained as

$$R_1^B = \begin{bmatrix} -s\theta_p & 0 & c\theta_p \\ c\theta_p & 0 & s\theta_p \\ 0 & 1 & 0 \end{bmatrix} \quad \text{and} \quad R_2^1 = \begin{bmatrix} s\theta_t & 0 & c\theta_t \\ -c\theta_t & 0 & s\theta_t \\ 0 & -1 & 0 \end{bmatrix} \tag{74}$$

where the first link rotation matrix for panning when the camera looking downward is the same as that of matrix for rolling motion when looking forward as shown in Fig. 6 and 7. Next, we can obtain the matrix from the camera base to first link (also same) but the total rotation matrix by combining all those matrices from quadrotor through the second link to yield

$$R_1^F = \begin{bmatrix} 0 & 1 & 0 \\ -s\theta_p & 0 & c\theta_p \\ c\theta_p & 0 & s\theta_p \end{bmatrix} \quad \text{and} \quad R_{2=C}^F = \begin{bmatrix} -c\theta_t & 0 & c\theta_t \\ -s\theta_t s\theta_p & -c\theta_p & -c\theta_t s\theta_p \\ s\theta_t c\theta_p & -s\theta_p & c\theta_t c\theta_p \end{bmatrix} \tag{75}$$

where $R_1^F = R_B^F R_1^{B=O_0}$ and $R_{2=C}^F = R_1^F R_{2=C}^1$ were used to calculate and the third column in R_1^F is the vector $z_1(t)$ in the Jacobian matrix $J_C(t)$. Thus, the positioner joint angles, $\dot{\theta}_C(t) \in R^3$, are given by

$$J_C = \begin{bmatrix} z_0 & z_1 \end{bmatrix} = \begin{bmatrix} 1 & 0 \\ 0 & c\theta_p \\ 0 & s\theta_p \end{bmatrix}, \quad \dot{\theta}_C = \begin{bmatrix} \dot{\theta}_{p=r} \\ \dot{\theta}_t \end{bmatrix}, \text{ and } \theta_C = \begin{bmatrix} \theta_{p=r} \\ \theta_t \end{bmatrix}. \tag{76}$$

Note that Jacobian matrix of the Pan-Tilt camera manipulator in (76) is the same one used in the Tilt-Roll configuration due to the configuration of the same camera base and the first link. In addition, the pan angle in Pan-Tilt configuration is actually the same of roll used in Tilt-Roll configuration due to the actuation of the second link.

7.2 Rotation Matrix

A rotation matrix, $R_D^C(\theta_d)$ denoted in (20), from D to desired frame C for the tracking of the camera frame to desired frame can be obtained by considering the camera frame with regard to the desired frame as another end link as follows:

Link (i)	Offset	Length	Twist	Angle
	D	a	α	θ
C (looking forward)	0	0	$+90^{\circ}$	$\theta_d + 90^{\circ}$
C (looking downward)	0	0	0°	$\theta_d + 90^{\circ}$

Table 3. D-H Table to find the rotation matrices between the camera and desired frames which yields the following rotation matrix, respectively (for simulation, $\theta_d = 0^{\circ}$)

$$R_D^C = \begin{bmatrix} -s\theta_d & 0 & c\theta_d \\ c\theta_d & 0 & s\theta_d \\ 0 & 1 & 0 \end{bmatrix}_{(forward)} \quad \text{and} \quad R_D^C = \begin{bmatrix} -s\theta_d & -c\theta_d & 0 \\ c\theta_d & -s\theta_d & 0 \\ 0 & 0 & 1 \end{bmatrix}_{(downward)} . \tag{77}$$

7.3 Notation and Nomenclature

R_i^{i-1} Rotation matrix from the origin i to the origin $i-1$ coordinate frame

T_F^I Orientation matrix from the origin of F to the origin of frame I

p_i, $\dot{p}_i$, Camera (C) or UAV (F) position and position rate represented in Inertial frame I - ground-based - quantities denoted using subscript

Θ_i, $\dot{\Theta}_i$ Camera or UAV angle ($[\varphi, \theta, \psi]^T$) and angular rates about roll, pitch, and yaw represented in I frame -ground-based-quantities using subscript

Θ_C Camera angles about pan, tilt, and roll ($[\theta_p, \theta_t, \theta_r]^T$) expressed in the Inertial frame

v^i, ω^i Camera (C), Camera base (B) or UAV linear and angular velocities about x, y, and z-axis which is airborne quantities denoted using superscript

v^{ij}, ω^{ij} Velocities expressed in Camera frame (C) between the origin i and another origin j coordinate frame

v_{jk}^i, ω_{jk}^i UAV or Camera base frame velocities (i) denoting specific quantities between j frame and k frame

8. References

Adelstein, B. D. ; Ellis, S. R. (2000). Rotation and Direction Judgment from Visual Images Head-Slaved in Two and Three Degrees-of-Freedom, *IEEE Transactions on Systems, Man and Cybernetics – Part A: Systems and Humans*, Vol. 30, No. 2, March 2000, pp. 165 - 173.

Chitrakaran, V.; Dawson, D. M.; Kannan, H. (2006). Vision Assisted Autonomous Path Following for Unmanned Aerial Vehicles, in *Proc of 45th IEEE Conference on Decision and Control*, pp. 63 - 68, Dec. 2006.

Cooke, N; Pringle, H. ; Pedersen, H.; Connor, O. (2006). Human Factors of Remotely Operated Vehicles, *ELSVIER JAI*, 2006

DraganFlyer X-Pro (2005): http://www.rctoys.com/ (cited 10/2008).

FlightGear (2006). http://www.flightgear.org/ (cited 10/2008).

Fossen, T. I. (2003). Marine Control Systems : Guidance, Navigation, and Control of Ships, Rigs, and Underwater Vehicles, *Marine Cybernetics*, 2003.

Jakobsen, O. C.; Johnson, E. N. (2005). Control Architecture for a UAV - Mounted Pan/Tilt/Roll Camera Gimbal, *AIAA* 2005-7145, Infotech@Aerospace, Arlington, Virginia, Sep. 2005, pp. 01 - 10.

Lee, D.; Chitrakaran, V.; Burg, T.C.; Dawson, D. M.; Xian, B. (2007). Control of a Remotely Operated Quadrotor Aerial Vehicle and Camera Unit Using a Fly-The-Camera Perspective, in *Proc of 46th IEEE Conference on Decision and Control*, New Orleans, LA, Dec. 2007, pp. 6412-6417.

Logitech Extreme 3D Pro Joystick (2006); http://www.logitech.com

Mahony, R.; Hamel, T.; Chaumette, F. (2002). A Decoupled Image Space Approach to Visual Servo Control of a Robotic Manipulator, in *IEEE Intl Conference on Robotics and Automation*, Washington D. C., May 2002, pp. 3781 - 3786.

Malis, E. (1998) Controbutions a la modelisation et a la commande en asservissement visuel, *Ph.D. Thesis*, University of Rennes I, IRISA, France, 1998.

Micropilot.; http://www.micropilot.com/ (9/2008 cited)

Murray, R. M.; Li, Z.; Sastry S. S. (1994). A Mathematical Introduction to Robotic Manipulator, CRC Press, 1994.

Neff, A.; Lee, D.; Chitrakaran, V.; Dawson, D. M.; Burg, T. C. (2007). Velocity Control of a Quadrotor UAV Fly-By-Camera Interface, in *Proceedings of IEEE Southeast Conference*, pp. 273-278, Richmond, VA, Mar. 2007.

Pieniazek, J. (2003). Software-based Camera Stabilization on Unmanned Aircraft, Aircraft Engineering and Aerospace Technology: *An International Journal*, Vol. 75, No. 6, 2003, pp. 575 - 580.

Procerus Technologies.; http://www.procerusuav.com/ (9/2008 cited)

Qmotor Simulator (2000). A QNX-based C++ Real-time Control Environment, http://www.ece.clemson.edu/crb/research/realtimesoftware/qmotor/index.htm (cited 10/2008).

QNX 6, QNX Neutrino 6.2.1. (2006). QNX Software Systems, Real-Time Control Software. http://www.qnx.com (cited 10/2008)

Quigley, M. ; Goodrich, M. A. ; Griffiths, S. ; Eldredge A. ; Beard, R. W. (2005). Target Acquisition, Localization, and Surveillance Using a Fixed-Wing Mini-UAV and Gimbaled Camera, in *IEEE Intl Conference on Robotics and Automation*, Barcelona, Spain, April 2005, pp. 2611 - 2616.

Sharp, C. S.; Shakernia, O. & Sastry S. S. (2001). A Vision System for Landing an Unmanned Aerial Vehicle, in *IEEE Intl Conference on Robotics and Automation*, Seoul, Korea, May 2001, pp. 1720-1727.

Spong, M. W.; Hutchinson, S.; Vidyasagar M. (2006). Robot Modeling and Control, John Wiley and Sons, Inc:, 2006.

Stanciu, R. & Oh. P. (2007). Human-in-the-Loop Camera Control for a Mechatronic Broadcast Boom, *IEEE/ASME Transactions on Mechatronics*, Vol. 12, No. 1, February 2007.

Stolle, S. & Rysdyk, R. (2003). Flight Path Following Guidance for Unmanned Air Vehicles with Pan-Tilt Camera for Target Observation, in *Proceedings of the AIAA Digital Avionics Systems Conference*, Oct. 2003, pp 01 - 12.

Yoon, S.; Lundberg, J. B. (2001). Equations of Motion for a Two-Axes Gimbal System, *IEEE Transactions on Aerospace and Electronic Systems*, Vol. 37, No. 3, July 2001, pp. 1083 - 1091.

9

A Flight Strategy for Intelligent Aerial Vehicles Learned from Dragonfly

Zheng Hu and Xinyan Deng
University of Delaware
U.S.A.

1. Introduction

Dragonfly is one of the most maneuverable insects and one of the oldest flying species on earth. It is important for human beings to study their flight techniques if we intend to make an insect-like Micro Aerial Vehicle, because their flight performance far exceeds other insects. They can hover, cruise up to 54km/h, turn 180° in three wing beats, fly sideways, glide, and even fly backwards (Alexander, 1984; Appleton, 1974; Whitehouse, 1941). They intercept prey in the air with amazing speed and accuracy. Their thorax are equipped with wing muscles which accounts for 24% (*Aeshna*) of its body weight, compared to 13% of those of the honey bees (Appleton, 1974). Most dragonflies change their wing motion kinematics for different flight modes such as hovering, cruising and turning. Among these kinematic parameters, the most interesting one is the phase difference (γ) between forewing and hindwing. It is defined as the phase angle by which the hindwing leads the forewing. When hovering, dragonflies employ 180° phase difference (anti-phase) (Alexander, 1984; Norberg, 1975; Rüppell, 1989), while 54~100° are used for forward flight (Azuma and Watanabe, 1988; Wang et al., 2003). When accelerating or performing aggressive maneuvers, they use 0° (in-phase) phase difference (Alexander, 1984; Rüppell, 1989; Thomas et al., 2004). Of various phase differences, 270° is rarely observed in dragonfly flight.

Figure 1. Phase difference in dragonfly

The fact that flapping in-phase (0°) appears in situations requiring large acceleration suggests that in-phase might produce higher forces (Alexander, 1984; Rüppell, 1989). The film sequences made by Alexander (Alexander, 1984) showed that in-phase is employed during take-off and sharp turning. It was also found that in a rising flight of a dragonfly, lift was increased during downstroke and drag was increased during upstroke when flying in-phase (Azuma et al., 1985). The conclusion was derived by using the momentum theory and the blade element theory, combined with a numerical method modified from the local circulation method.

However, it was argued that as two wings in tandem are brought closer together, the lift force produced by each wing is reduced (Alexander, 1984). Therefore, forewing and hindwing flapping in-phase would produce less lift because they are closer together than when beating anti-phase. Alexander believed that the reason of dragonflies using in-phase flight may be due to physiological reason as well as the preference of peak forces enhancement at the cost of the mean forces reduction (Alexander, 1984).

Counterstroking (180° or anti-phase) produces uniform flight, whereas flight produced by parallel stroking (0°) is irregular (Rüppell, 1989). This is because inequalities in the aerodynamic effects of the upstroke and downstroke can be compensated to some extent in countertroking. As one pair of wing's upstroke with a steep angle of attack generates strong thrust, the other pair's downstroke with a small angle of attack mainly generates lift. Therefore the net thrust and lift production remains relatively constant during flight due to the alternating force generation on two wings (Rüppell, 1989).

In a recent computational study, (Wang and Russell, 2007) calculated dragonfly's aerodynamic force and power as a function of forewing-hindwing phase difference. They found that anti-phase flapping consumes nearly minimal power while generating sufficient force to balance body weight, and that in-phase motion provides an additional force to accelerate (Wang and Russell, 2007). Furthermore, they proposed an analogy to explain the results by analyzing a model of two cylinders moving in parallel next to each other.

Other computational studies include (Wang and Sun, 2005) and (Huang and Sun, 2007), where they calculated the aerodynamic effects of forewing-hindwing interactions of a specific dragonfly (*Aeshna juncea*) in hover and slow forward flight. They showed that the interaction is detrimental to force generation in almost all cases. At hovering with γ= 180°, the reduction is 8~15%, compared with the force without interaction. The force on hindwing is greatly influenced by the forewing at γ= 180~360°, with the lift coefficient decreased by 20~60%. Furthermore, they proposed a mechanism to explain the effect of forewing on hindwing force reduction: the forewing in each of its downstroke produces a downward "jet" behind it; when the hindwing lags the forewing, it moves into the jet and its effective angle of attack is reduced, resulting in a decrease in its aerodynamic force.

Previous computational studies include (Lan, 1979), where the unsteady quasi-vortex-lattice method was applied to the study of dragonfly aerodynamics, and the results showed that dragonfly can produce high thrust with high efficiency if hindwing leads the forewing by 90°, and that hindwing was able to extract wake energy from the forewing under this condition.

Direct force measurements on tethered dragonflies showed that peak lift increases from approximately 2 to 6.3 times body weight when the animal decreases the phase difference between both flapping wings (Reavis and Luttges, 1988). But maybe this enhancement on peak lift may due to overlap of peak lifts on forewing and hindwing, not necessarily due to wing-wing interaction.

Experimental investigations of the aerodynamic effect of wing-wing interaction was previously performed in (Maybury and Lehmann, 2004), where a pair of robotic wings were vertically stacked to simulate dragonfly hovering flight with horizontal stroke plane. They found that the lift production of the forewing remains approximately constant, while hind wing lift production is reduced to some extent and its maximum value occurs at a phase difference γ= 90°. They attributed the wing-wing interaction to two reasons: LEV destruction and local flow condition (Maybury and Lehmann, 2004). Their results explained the hovering behavior of dragonflies using horizontal stroke plane (*Sympetrum Sanguineum*), while many other dragonfly species employ a 20~70° inclined stroke plane (Alexander, 1984; Norberg, 1975; Rüppell, 1989) and employ an aerodynamic mechanism of "drag based lift generation" (Wang, 2004), which is quite different from that for horizontal stroke plane flight (Wang and Russell, 2007). In this study, we investigate the wing-wing interaction and the underlying mechanism for the inclined stroke plane species.

The effect of the forewing-hindwing interactions in dragonflies has been investigated with some computational and experimental studies, but conclusions are still limited and quite varying. In this study, we constructed a pair of robotic dragonfly wings to investigate the aerodynamic effect of wing-wing interactions in both hovering and forward flight. This apparatus enables us to study inclined stroke plane species such as *Aeshna Juncea* with an inclined stroke plane 60° and varying forward speed. The wing bases are of close proximity to mimic the dragonfly wings. We systematically vary the different phase between the forewing and hindwing to find out why dragonflies apply certain phase differences rather than others in certain flight modes. The conclusion from this research would be valuable for designing a dragonfly-inspired M.A.V (Micro Air Vehicle).

2. A robotic dragonfly model

The Micro Robotics Lab of University of Delaware constructed a pair of dynamically-scaled robotic wing models (flappers) to replicate dragonfly wing motion and measure the instantaneous aerodynamic forces when flapping in a tank filled with mineral oil (Fig. 2).

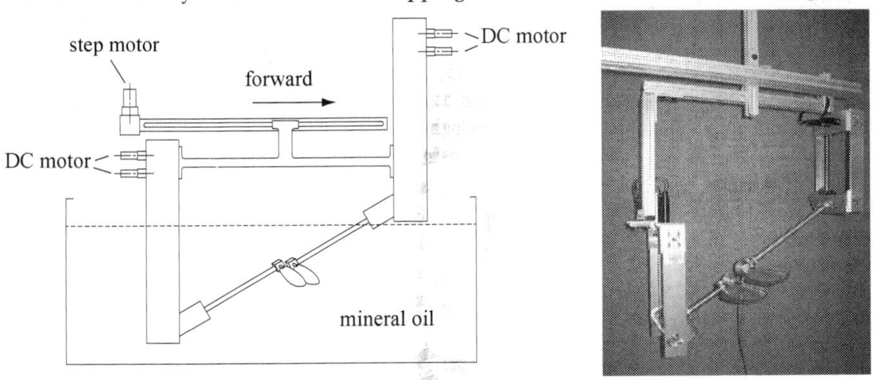

Figure 2. Sketch and image of the experimental setup

For each flapper, a bevel-geared robotic wrist was designed to generate motions of two degrees of freedom (translation and rotation). A set of bevel gears transmits the motion from coaxially driven shafts to the wing holder thus enabling wing translation and rotation (Fig. 3). The two wings are placed with a very close proximity like those of a true dragonfly. In addition, the two flappers are mounted on a linear stage driven by a step motor to achieve forward motion together.

The drive shafts were powered by 16 mm, 0.3 Nm torque DC brush motors (Maxon, Sachseln, Switzerland) equipped with gear heads to reduce speed and magnetic encoders to provide kinematic feedback to ensure motion fidelity. The motors were driven along kinematic patterns provided by a custom MATLAB (Mathworks, Natick, MA) Simulink program with WinCon software (Quanser Consulting, Ontario, Canada). This software provided commands to the real-time control and data acquisition board (Quanser Consulting, Ontario, Canada) communicating with the hardware. We used Proportional-Integral-Derivative (PID) controllers to run the motors with precision of 0.1°. Motion commands from the computer were amplified by analog amplifier units (Advanced Motion Control) which directly controlled the input current received by the motor.

Wings were made from Mylar plastic film with a thickness of 0.25 cm which behaves as a rigid wing in our experiments. A carbon fiber rod was glued on the plastic film to serve as the leading edge. The end of the carbon fiber rod was affixed on to the force sensor. The wings have identical geometry as the dragonfly wings but are four times larger with a length of 19 cm for forewing and 18.5 cm for hindwing. The wing length is calculated as the distance from wing tip to the flapping axis.

The wings along with part of the device were immersed into a tank (46cm × 41cm × 152cm) filled with mineral oil (Kinematic viscosity= 3.4 cSt at 20 oC, density=830 kg/ m3). This overall set-up enabled us to run the wings along a pre-determined dragonfly kinematics while simultaneously measuring the forces on the forewing and the hindwing respectively (Fig. 2).

3. Flight pattern of a typical dragonfly

High-speed photos of the dragonfly (*Aeshna juncea*) in hovering flight were taken by Norberg (Norberg, 1975). The body is held almost horizontal, and the wing stroke plane is tilted 60° relative to the horizontal line. For both forewing and hindwing, the chord is almost horizontal during the downstroke and is close to being vertical during the upstroke (Fig. 4); the stroke frequency is 36 Hz, the stroke amplitude is 60°; the translational angle is from 35° above the horizontal to 25° below for forewing, and is from 45° above to15° below for hindwing; the hindwing leads the forewing in phase by 180°. The mass of the insect is 754mg; forewing length is 4.74 cm; hindwing length is 4.60 cm; the mean chord lengths of the forewing and hindwing are 0.81 cm and 1.12 cm, respectively; the moment of inertial of wing-mass with respect to the fulcrum is 4.54 $g\,cm^{-2}$ for the forewing and 3.77 $g\,cm^{-2}$ for the hindwing (Norberg, 1972).

Due to the low frame rate of the camera used at that time (80 Hz), Norberg's data does not consist of a detailed continuous trajectory of the wing kinematics. Azuma instead, (Azuma et al., 1985) successfully filmed a slow climbing flight of a dragonfly (*Sympetrum Frequens*) with a high speed camera (873 frames per second). He showed that the flapping trajectory can be well represented by a sinusoidal function and the rotation trajectory can be represented by a third harmonic function. Since the two species (*Aeshna juncea and Sympetrum frequens*) share similar values for many kinematic parameters such as translational amplitude, rotational amplitude and stroke plane angle (Azuma et al., 1985; Norberg, 1975), here we assume they also employ similar wing motion trajectories. This assumption of kinematic trajectories can be reasonably applied to other species with highly inclined stroke planes without affecting the main results of this study. We developed a pair of wing kinematic trajectories by matching those for *Sympetrum Frequens* in (Azuma et al., 1985), as shown in Fig. 5.

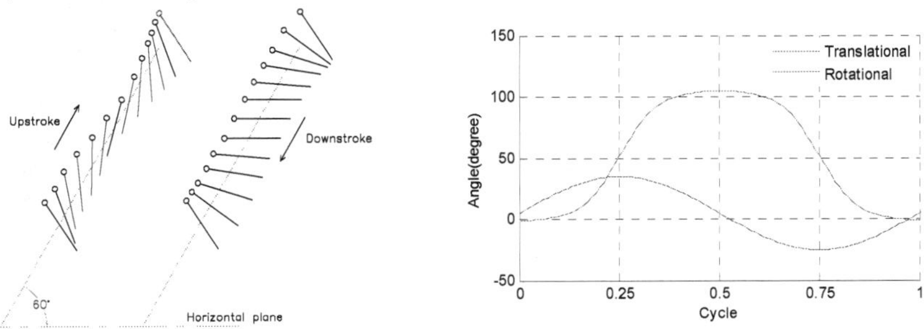

Figure 4. (left) Sketches of the dragonfly wing kineamtics during upstroke and downstroke based on *Aeshna juncea* data (Norberg, 1975). Side views with circles denote the leading edge. Both forewing and hindwing apply; (right) Wing kinematics developed by matching those for *Sympetrum Frequens* from (Azuma et al., 1985) and (Norberg, 1975). Translational angle (blue) and rotation angle (red). This is for forewing only; the kinematics for hindwing is almost the same except a 10° offset for translational angle

4. Force measurements & calculations

For each wing, we measured the instantaneous lift and thrust forces and calculated the average forces. The forces as well as torques on the moving wings can be measured by a six-component force sensor (ATI NANO-17, Apex, NC), with a range of ±12 N for force and ±0.5 Nm for torque along three orthogonal axes. Using appropriate trigonometric conversions, these force measurements were then converted to lift and thrust forces in the earth coordinates. Note that in this study, lift force is defined as the vertical component of the aerodynamics force on the wing and thrust force is the horizontal component in global coordinate frame.

The magnitude of aerodynamic forces acting on an actual dragonfly, F_{fly}, is related to those measured in the robotic model, F_{robot}, according to the following scaling rule (Fry et al., 2005):

$$F_{fly} = F_{robot} \cdot \frac{\rho_{air}}{\rho_{oil}} \cdot \left(\frac{n_{fly}}{n_{robot}}\right)^2 \cdot \left(\frac{r_{fly}}{r_{robot}}\right)^2 \cdot \frac{S_{fly}}{S_{robot}} \cdot \frac{\hat{r}^2_{2fly}}{\hat{r}^2_{2robot}} \tag{1}$$

where ρ is fluid density, n is stroke frequency, r is wing length, S is wing area, and $\hat{r}^2_2$ is the normalized second moment of wing area. $\hat{r}^2_{2fly}$ and $\hat{r}^2_{2robot}$ differ slightly, due to a small null length at the base of the robotic wing required to accommodate the force sensor.

In this study, lift force is defined as the vertical component of the aerodynamics force on the wing; thrust force is the horizontal component. Lift coefficient and thrust coefficient are respectively defined in the following way, similar to (Wang and Sun, 2005):

$$C_l = L/[0.5\rho U^2(S_f + S_h)] \tag{2}$$

$$C_t = T/[0.5\rho U^2(S_f + S_h)] \tag{3}$$

L and T denote the total lift and the total thrust. The reference velocity is calculated to be $U = 2\Phi n r_2 = 2.1\text{ms}^{-1}$; S_f and S_h are the areas of forewing and hindwing, respectively. Accordingly, we define lift coefficient and thrust coefficient for a single wing. For example, for forewing:

$$C_{l,f} = L_f/[0.5\rho U^2(S_f + S_h)] \tag{4}$$

$$C_{t,f} = T_f/[0.5\rho U^2(S_f + S_h)] \tag{5}$$

The total lift coefficient and total thrust coefficient are as follows:

$$C_l = C_{l,f} + C_{l,h} \tag{6}$$

$$C_t = C_{t,f} + C_{t,h} \tag{7}$$

The lift coefficient on a dragonfly should be 2 times C_l, since it has two forewings and two hindwings.

5. Aerodynamic force for anti-phase hovering flight

We first replay the hovering kinematics of dragonfly on the robotic flappers and measured lift and thrust forces on the both wings. Here the forewing and hindwing are anti-phase (180°), which is most commonly observed in dragonfly hovering flight. The results show an average lift force of 0.0711 N on the forewing and 0.082 N on the hindwing during one beat cycle. The average thrust forces are 0.001 N on the forewing and 0.003 N on the hindwing.

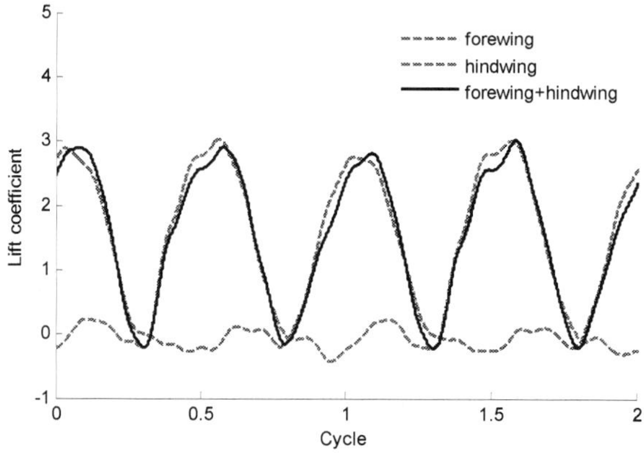

Figure 6. Time trace of lift force coefficients generated when hovering with γ= 180°

We then apply (1) to scale the lift force back to those of a true dragonfly. The resulted average lift force is 187 mg on a forewing and 216 mg on a hindwing. The total average lift on a four-wing dragonfly is therefore (187+216)*2=806 (mg). This result is comparable to the 754 mg body mass measured in (Norberg, 1972). On the other hand, the total average thrust force is almost zero when compared with the average lift force, which follows the hovering condition that the thrust should be zero.

The time traces of instantaneous lift forces are shown in Fig.6. Both forewing and hindwing produce lift peaks during downstrokes. Because the flapping of forewing and hindwing are anti-phase, they generate lift force alternatively during one wingstroke. The average lift force on hindwing is 1.15 times that on forewing, while the area of hindwing is 1.32 times that of forewing. This inconsistency is probably caused by the fact that, compared to forewing, hindwing has more area distributed at positions close to the body. It was proposed that for dragonfly flight, the hindwing acts as power wing which provide more lift force while the forewing is the steering wing (Azuma and Watanabe, 1988; Wang et al., 2003). As we can see, the total lift force is generated alternatively by the forewing and hindwing during one wingbeat. By doing so the insect is able to hover with regular forces which reduce the vibration of the body.

6. Effect of phase difference in hovering flight

We measured the instantaneous forces with the same kinematics of hovering but systematically vary the phase differences from 0° to 360° in steps of 30°. The average lift coefficients of both wings according to varying phase difference are plotted in Fig. 7.

Fig. 7 shows the averaged force coefficients in hover. As the phase difference tends to 0°, forces tend to be higher, and reach their maximum on 0°. The forces get lower values when γ is around 180°. An interesting point is that not all values are below the single wing force, that is, the wing-wing interaction is not always detrimental to force generation. We can conclude here, in range 330° to 30°, the wing-wing interaction enhances the total lift force by up to 6%; in range 150° to 180°, interaction decreases the total lift force by up to 9%. The thrust forces are much smaller than the lift forces, which is reasonable for hovering flight.

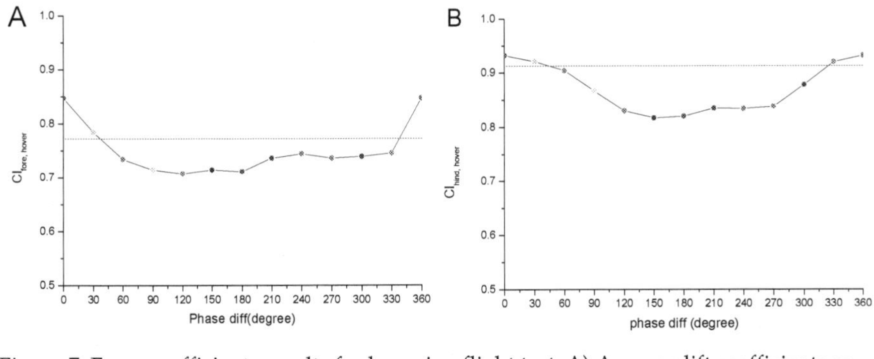

Figure 7. Force coefficients results for hovering flight test. A) Average lift coefficients on forewing; B) average lift coefficients on hindwing; C) total average lift coefficients. The straight lines in each plot indicate the force results without interaction

This part of study supplies a direct proof for the statement that in-phase flight generates larger aerodynamic forces than anti-phase flight, and also larger than the case without wing-wing interaction. This may explain the behavior that dragonfly flies in-phase in case of accelerating or maneuverings that calls for a high force generation.

In order to find out the possible reason that dragonfly uses anti-phase style for hovering mode, we need to compare the time traces of total lift forces generated by $\gamma = 0°$ with those generated by $\gamma=180°$ (Fig. 8). It was noted that anti-phase produces uniform flight, whereas flight produced by in-phase stroking is irregular (Rüppell, 1989). Fig. 8 shows the instantaneous lift force coefficients comparison between in-phase and anti-phase flight. As we can see, in-phase brings larger irregularity in the aerodynamic forces than anti-phase does. Observing from the time trace curve for in-phase hovering, there exists a 1/2 cycle period when lift force is closed to zero, while in another 1/2 cycle the peak value is two times of the peak value for anti-phase flight, because forewing and hindwing peak overlap. This irregularity of instantaneous forces increases the body vibration when hovering, while for anti-phase flight the inequality can be compensated to some extent by evenly distributing the peak forces of forewing and hindwing on the whole cycle. Besides minimizing the force irregularities and keeping body posture stable, anti-phase flight can also save energy that might be lost in body vibration (Wang and Russell, 2007). Moreover, Usherwood and Lehmann showed that dragonfly can also save aerodynamic power during anti-phase hovering (Usherwood and Lehmann, 2008). Thus, it is reasonable that dragonflies would rather lose 15% force production for improving flight stability and vibration suppression as well as power efficiency.

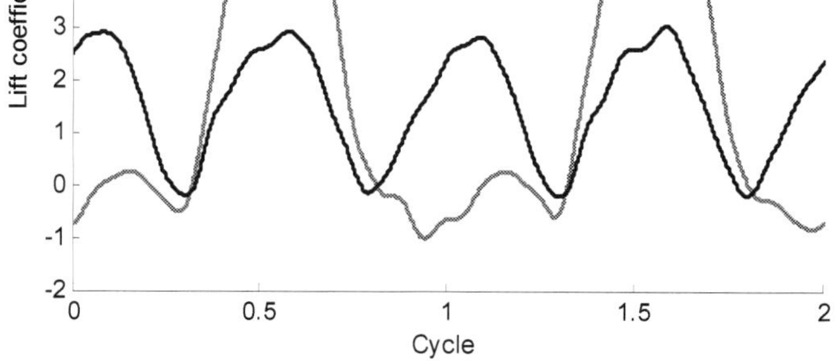

Figure 8. Comparison between total lift forces generated when hovering with $\gamma = 0°$ and $\gamma = 180°$ respectively

7. Effect of phase difference in forward flight

Similarly to the hovering case, we tested the wing aerodynamics by varying phase difference systematically in steps of 30°. The average lift coefficients and average thrust coefficients of both wings are plotted in Fig. 9.

As seen in Fig. 9, during forward flight, interaction patterns for forewing and hindwing differ a lot. Note that the interaction always enhances the forewing's force generation no matter how much γ is, while the hindwing always loses force production because of the interaction. Lift on forewing is enhanced by at most 23% when flapping in-phase and at least 4% when phase difference falls into 120°~330°; Lift on hindwing reaches maximum on 60° and minimum on 270°. Hindwing is subjected to a severe loss on force production up to 38% due to the interaction with forewing. The total force on the two wings does not lose so much force as the hindwing does, due to the considerable enhancement on forewing lift.

Fig. 10 compares the time traces of hindwing lift among the cases of single hindwing, γ=90° and γ=270°. We note that the lift in case of γ=270° was considerably reduced when compared with other cases, and the reduction mainly occurred when hindwing was in the midway of downstroke. This is reasonable because a large portion of lift is produce during this period, since wings reach the highest flapping speed and largest angle of attack around the midway of downstroke.

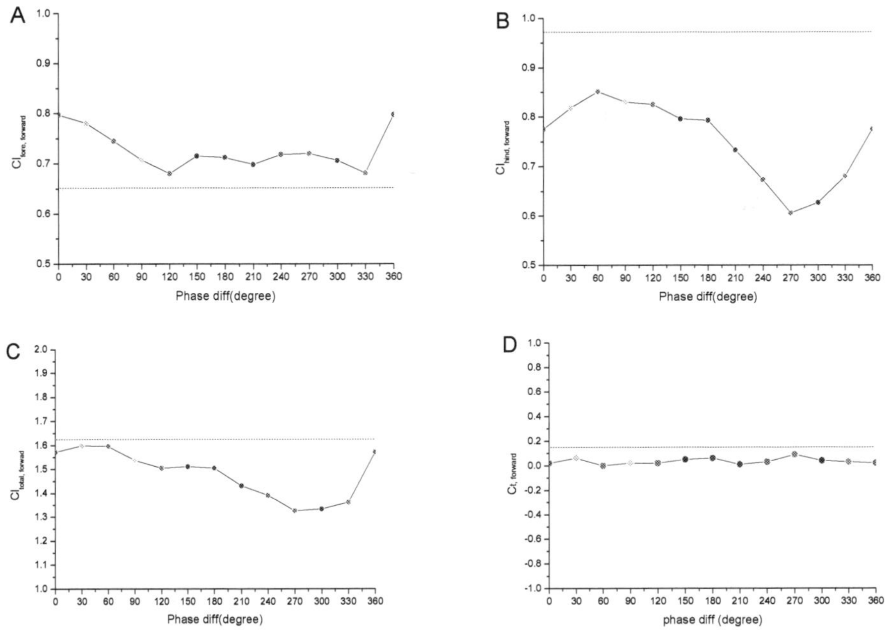

Figure 9. Force coefficients results for forward flight test. A) Average lift coefficients on forewing; B) average lift coefficients on hindwing; C) total average lift coefficients; D) total average thrust coefficients. The straight lines in each plot indicate the force results without interaction

One may explain the reduction in the following way: forewing generates a strong downwash when it finishes a downstroke. If $\gamma = 270°$, that is, forewing leads hindwing by 1/4 cycle, then hindwing is just in the midway of downstroke as forewing finishes downstroke and generates a downwash. In the mean time, because of forward flight, hindwing enters into the downwash area and its lift production decreases due to the downwash effect.

Now we can distinguish the case $\gamma = 90°$ and the case $\gamma = 270°$: the former one can offer dragonfly an 18% higher force than the latter one. On the other hand, 270° offers similar vibration and stability properties as 90°. This may explain why dragonfly never favors the 270° phase difference.

The above results agree qualitatively with the CFD results from (Wang and Sun, 2005) and (Huang and Sun, 2007) to some extent. Their conclusion is that the forewing is only slightly influenced by the wing-wing interaction, but the hindwing lift is greatly reduced by 20~60% during forward flight with a 180°~360° phase difference, compared with that of a single hindwing. In our results, furthermore, there are obvious lift enhancements on the forewing.

For thrust force measurement, (Warkentin and DeLaurier, 2007) conducted a systematic series of wind-tunnel tests on an ornithopter configuration consisting of two sets of symmetrically flapping wings, located one behind the other in tandem. It was discovered that the tandem arrangement can increase thrust for certain relative phase differences and longitudinal spacing between the wing sets. In particular, close spacing on the order of one chord length is generally best, and phase differences of approximately -50° to 50° give the highest thrusts and propulsive efficiencies. Nevertheless, this conclusion does not apply for the dragonfly flight. Instead, the thrust plot above indicates a drop on thrust force caused by interaction, no matter what the phase difference is. This would be reasonable if we notice that the space between forewing and hindwing of dragonfly is much smaller that the one chord spacing between the ornithopter wing sets.

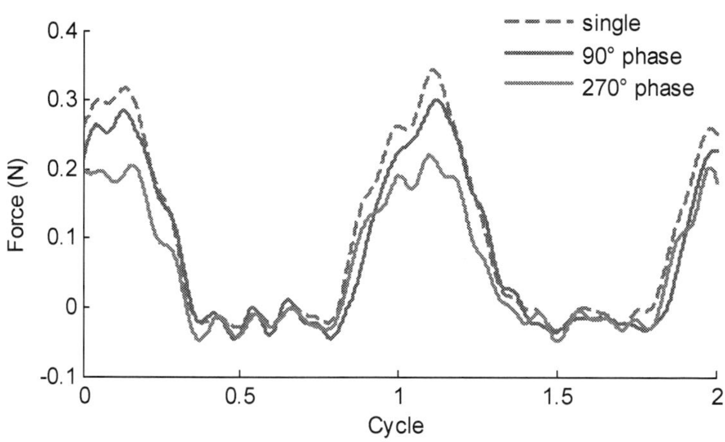

Figure 10. Comparisons of instantaneous hindwing lift in forward flight among the cases of single hindwing, $\gamma = 90°$ and $\gamma = 270°$. Note: the vertical axis indicates the forces generated on the flapper experiment and are unscaled

8. Dragonfly-inspired M.A.V

At last we introduce several dragonfly robots we have developed in our lab. Fig. 11 shows the latest designs. This model has two pairs of wings driven by a double crank rocker mechanism.

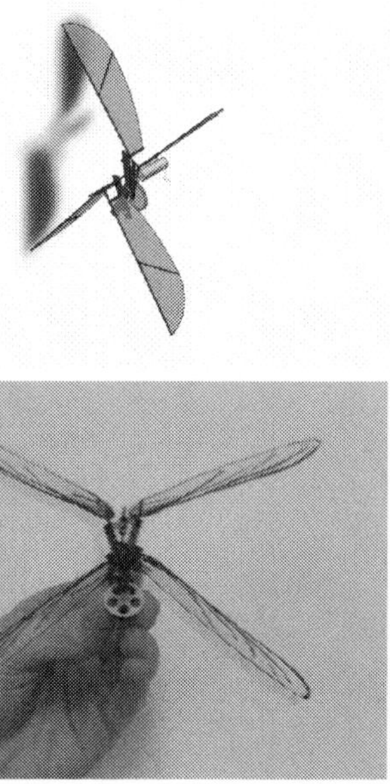

Figure 11. Dragonfly M.A.V prototype

The current prototype weighs 4g including battery and electronics. The following table shows some specifications.

Motor	Battery & Control	Gear
Torque: 44mNm	4V	Maximize Torque
Power: 176mW	0.4A	Ratio: 1:7
16k rpm	Infrared Chip	Precision Molded
Weight: 1.4g	Weight:1.1+1.3g	Weight:0.23g

Table 1. Dragonfly prototype specifications

This prototype was able to generate 28 Hz flapping frequency with four carbon fiber leading edges only and 9 Hz after adding (gluing) the polymer wing onto two leading edges. With all four wings the frequency reduced to 7 Hz. The total weight is 7 grams and the total length is 2 inches.

In order to test different wing performance and property, we have developed a mechanical wing tester. It can be used to study the fatigue cycles of each wing developed and can be also used to visualize the wing kinematics by a high speed camera. Fig.12 shows the wing tester and a sample dragonfly wing made of carbon fiber and polymers. Fig.13 shows the camera images of test wing and Fig.14 shows the frequency plot of different robotic wings under investigation.

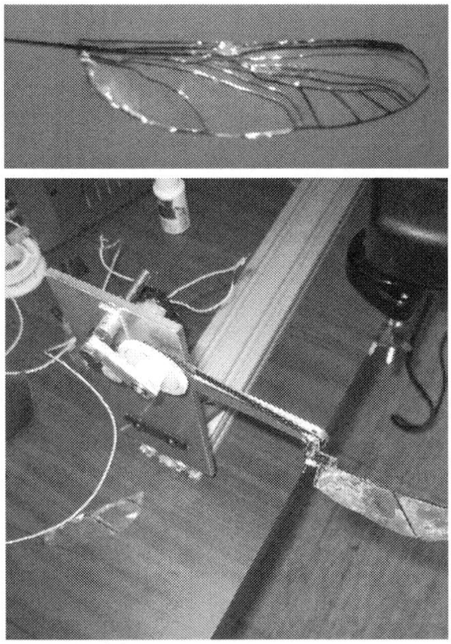

Figure 12. Sample test wing and wing tester

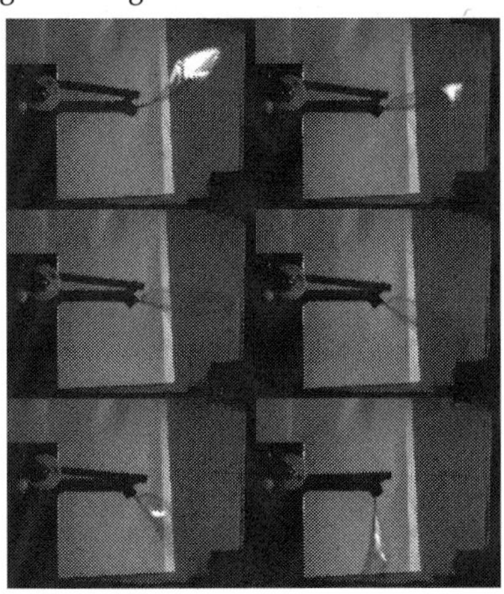

Figure 13. Dragonfly shaped wing flapping at 20Hz

The average size of the wings is bigger compared to real insects under our investigation. Veins are added for added rigidity to generate higher force. With reduced length and size, the frequency becomes higher. This process requires much trial and error as simple differences in wing structure can tear it apart at high frequencies. Dragonflies have a lot of these characteristics naturally build into their wings. Further tests will be aimed to finding out what makes their wings so robust yet so delicate under high frequencies.

9. Conclusion and the future work

The study described here investigates the effect of forewing-hindwing interactions during hovering and forward flight of dragonfly. Overall, wing-wing interaction is detrimental to total lift force generation. Hindwing lift was significantly reduced in forward flight due to the downwash from forewing. In-phase flight generates higher lift than other phase differences, while 270° phase difference generates the lowest lift. In hovering, dragonflies use anti-phase flight which generates a regular lift force for stability and vibration reduction purposes. A prototype for dragonfly-inspired M.A.V has been shown.

Many jobs need to be done in the future aiming to build up a sophisticated M.A.V. The mechanical design should be further improved to achieve a higher flapping frequency and better efficiency, in order to minimize the weight and maximize the lift force. How to make phase difference controllable on a four-wing M.A.V is also challenging the engineers.

10. References

Alexander, D. E. (1984). Unusual phase relationships between the forewings and hindwings in flying dragonflies. *J Exp Biol* 109, 379-383.

Appleton, F. M. (1974). Dragonflies and flight. *Nature Canada* 3(3), 25-29.

Azuma, A., Azuma, S., Watanabe, I. and Furuta, T. (1985). Flight mechanics of a dragonfly. *J Exp Biol* 116, 79-107.

Azuma, A. and Watanabe, T. (1988). Flight performance of a dragonfly. *J Exp Biol* 137, 221-252.

Fry, S. N., Sayaman, R. and Dickinson, M. H. (2005). The aerodynamics of hovering flight in Drosophila. *J Exp Biol* 208, 2303-18.

Huang, H. and Sun, M. (2007). Dragonfly forewing-hindwing interaction at various flight speeds and wing phasing. *AIAA J.* 45, 508-511.

Lan, C. E. (1979). The unsteady quasi-vortex-lattice method with applications to animal propulsion. *J. Fluid Mech.* 93, 747-765.

Maybury, W. J. and Lehmann, F. O. (2004). The fluid dynamics of flight control by kinematic phase lag variation between two robotic insect wings. *J Exp Biol* 207, 4707-26.

Norberg, R. A. (1972). The pterostigma of insect wings and inertial regulator of wing pitch. *J. Comp. Physiol* 81, 9-22.

Norberg, R. A. (1975). Hovering flight of the dragonfly: Aeschna juncea L., kinematics and aerodynamics. *Swimming and Flying in Nature* 2, 763-780.

Reavis, M. A. and Luttges, M. W. (1988). Aerodynamic forces produced by a dragonfly. *AIAA J.* 88-0330, 1-13.

Rüppell, G. (1989). Kinematic analysis of symmetrical flight manoeuvres of odonata. *J Exp Biol* 144, 13-42.

Thomas, A. L., Taylor, G. K., Srygley, R. B., Nudds, R. L. and Bomphrey, R. J. (2004). Dragonfly flight: free-flight and tethered flow visualizations reveal a diverse array of unsteady lift-generating mechanisms, controlled primarily via angle of attack. *J Exp Biol* 207, 4299-323.

Usherwood, J. R. and Lehmann, F. O. (2008). Phasing of dragonfly wings can improve aerodynamic efficiency by removing swirl. *J R Soc Interface*.

Wang, H., Zeng, L., Liu, H. and Yin, C. (2003). Measuring wing kinematics, flight trajectory and body attitude during forward flight and turning maneuvers in dragonflies. *J Exp Biol* 206, 745-57.

Wang, J. K. and Sun, M. (2005). A computational study of the aerodynamics and forewing-hindwing interaction of a model dragonfly in forward flight. *J Exp Biol* 208, 3785-804.

Wang, Z. J. (2004). The role of drag in insect hovering. *J Exp Biol* 207, 4147-55.

Wang, Z. J. and Russell, D. (2007). Effect of forewing and hindwing interactions on aerodynamic forces and power in hovering dragonfly flight. *Physical Review Letters* 99.

Warkentin, J. and DeLaurier, J. (2007). Experimental aerodynamic study of tandem flapping membrane wings. *AIAA J.* 44, 1653-1661.

Whitehouse, F. C. (1941). British Columbia dragonflies (Odonata), with notes on distribution and habits. *The American Midland Naturalist* 26(3), 488.

10

DC Supply System Detector of UAV

Qiongjian Fan, Zhong Yang, Jiang Cui and Chunlin Shen
university
country

1. Introduction

This chapter introduces an embedded smart detector in the DC supply system of an unmanned aerial vehicle (UAV) based on AT89C2051, which can measure load insulation failure in direct current supply system and insure that UAV can operate at steady status. The measuring principles are analyzed first in this chapter and the system design schemes of software and hardware are given out as well. Finally, the measure error has been analyzed and counted. Compare with the Hall instrument, this detector possesses characteristics of low-cost, low-power consumption. And this device has also high measurement precision, and high reliability. Practical experiments show that the sensor system, of which the measurement precision achieves 0.01mA, could meet the demands of detection accuracy, achieve the expected criteria of design, and could be spread.

Unmanned Air Vehicle (UAV) is entering development rapidly since the Middle East wars in last century[1], which is applied to more and more fields, and at the same time a lot of researchers have been engaged in the research. The study of various sensors is always one of key technologies of UAV, this paper mainly engaged in a smart sensor for load insulation failure detection in the DC supply system of UAV.

In the present of DC supply system[2]-[3], Hall instrument is commonly used to detect unbalance status in the load current, which has characteristics of high detection precision, but high expensive price accordingly. So, in one system, simultaneous using two or more hall instruments are very costly and, further, measurement data can not be calculated on-line, and also increase the burden of the detection and computation of the upper PC.

Therefore, a smart sensor based on AT89C2051 is presented in this paper, which can measure load insulation failure in direct current supply system and possess characteristics of low-cost, low-power consumption, and easy to use. This kind of sensor will be pulled-on the positive-negative polar line of each directly current branch circuit. When insulation level of every loop is normal, the current passes through the sensor will have the equal value, and the opposite direction (notice that the current which flow across these coils terminal is defined I_+, the negative current is I_-, and the composed current is ΔI. Then the zero flux condition will be tested in the synthetic magnetic field of direct current

---

Manuscript received October 29, 2005. This work was supported by the Harbin Institute of Technology.
Q.J.Fan. Author, was work in the Aviation University of Air Force, Changchun, 130022 China. She is now study in the College of Automation, Nanjing University of Aeronautics and Astronautics, Nanjing 210016, China (phone: +86-025-84892301; e-mail: fanqiongjian@yahoo.com.cn).

of the sensor, otherwise the output will be nonzero due to $\Delta I = (I_+ - I_-)$. Therefore, by going make a circuit of the output state of the sensor to determine if the loop occur ground fault in direct current supply system. After fault recognition, through detecting data by the sensor, the function that processing fault information and sending alarm signal will be accomplished.

2. Measurement Principle and Hardware Circuit

2.1 Measure Principle

This smart sensor is mainly used to measure unbalance current in each branch circuit, and check circuit if over-current occurs due to overload or else reasons. Coil is the key testing component of this smart sensor, and is also an important unit in the oscillatory circuit. Branch wire that was measured crossed through the center of coil. When synthetic direct current in the loop fluctuate ΔI (magnetic flux produce $\Delta\Phi$ changed, and at the same time coil current produce ΔL_x changed), frequency loop of the oscillatory circuit will change consequent, and then test the variation of frequency Δf to compute current value crossed.

The relational expression can be described as follows in an ideal case:

$$\Phi = \mu \frac{NI}{2R} a^2 \tag{1}$$

Where

μ – core magnetic conductivity;

a – radius of transverse section of endless solenoid core;

R – radius of annular core;

N – turns.

$$L = \frac{N\Phi}{I} \tag{2}$$

$$f = \frac{2R^2C - 3L}{4RLCln(2R^2C / L)} \tag{3}$$

Where R = 5.1 KΩ; C = 1NF.

However, the nonlinear relation exists between coil current and oscillation frequency, which can be detected by conducting experiments, and it is difficult to test current directly. Methods are taken to solve the above problem: firstly, we have a testing experiment and obtain the relation curve between the frequency of the prototype and the corresponding current, and then obtain linear equations of several intervals by piecewise linearization of the curve. Secondly, obtain computing equations best suited for this device by analyzing and comparing the frequency calculated with the data in the table. Finally, actual current value can be calculated using the equation.

2.2 System Design

The whole system architecture of sensor is consisted of several modules: CPU circuit, watchdog circuit, communication circuit, power module and measurement circuit, which is shown in Fig.1.

1) CPU circuit

The CPU chip of smart sensor used is ATMEL AT89C2051 single chip[4], which provides the following features: high performance, ease to use, and high function-price ratio. This chip has 2K flash ROM built-in and be compatible with MCS-51™ Products. The CPU performs the function that records the frequency produced by the oscillatory circuit, and computes the current value according to the relation curve of unbalance current (ΔI) and the oscillatory frequency (Δf). In addition, the CPU is also in charge of the task as communication interface to the Upper PC.

2) Watchdog circuit

MAXIMUM X25045 programmable chip[5] is used in the watchdog circuit, which initialization program is stored in the AT89C2051 chip.

3) Communication circuit

The communication circuit function is mainly performed by MAX485, which is low-power transceiver for RS-485 communication.

Compared with RS-232 communication, the RS-485 communication features high transmission speed and far transmission distance, which is commonly the first-chosen to the lower system. The communication task is controlled by the single chip.

4) Power module

Operating voltages of this system are +5V and +6.5V, which are converted by the voltage regulator chip TL750L05C [6], +6.5V voltage regulator tube and voltage regulator diode from ±8V supply power.

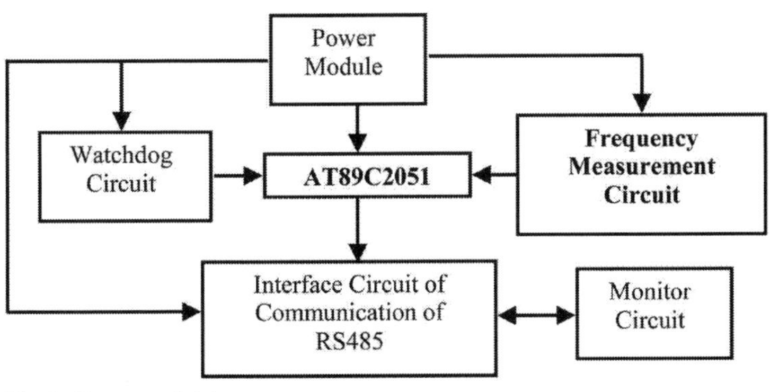

Figure 1. Block diagram of this sensor system

5) Oscillatory circuit

The oscillatory circuit is the key component of frequency measurement circuit, which is consisted of operational amplifier TLE2141CP[7], coils, resistor - condenser oscillatory circuits and high speed diodes, and which operate principle is that the oscillatory frequency of the whole oscillatory circuit will change as the unbalance current flowing across coils center work a change, then the unbalance current value can be calculated by MCU by measuring the varying quantity of the input voltage frequency. Structure diagram is shown in Fig.2 (the direct current resistance value is 6.22Ω in this sensor, and the inductance value is about 100mH as no current flowing through this sensor).

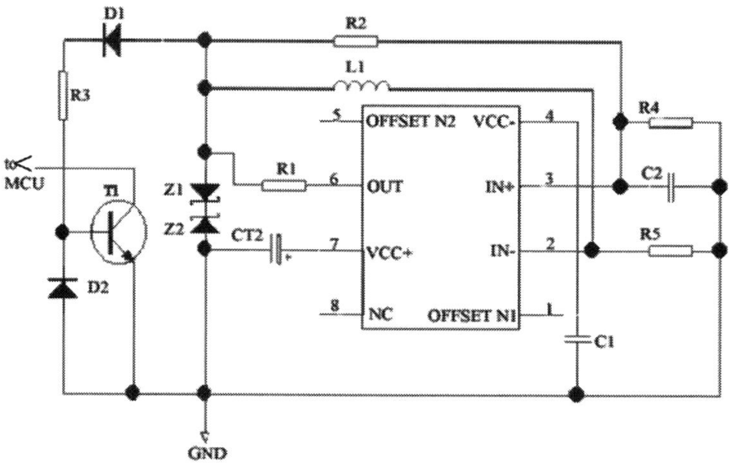

Figure 2. Structure diagram of the oscillatory circuit

3. Software Flowchart and Measurement Methods

3.1 Measure Principle

Software flowchart is shown in Fig.3.

3.2 Introduce of Measurement Methods

Equations (1-3) show that it is complicated to compute the frequency directly, however, in view of frequency range (about 75HZ ~1000HZ), the relation between the value of the machine oscillation cycles tested and current calculated is linear in the smallest possible interval. So, a curve can be obtained by taking the method of measuring oscillation cycle in programs[8], which is shown in Fig.5, and then process piecewise linearization. The lateral coordinate represents machine cycles value tested (the machine cycles value tested obtain by converting from frequency in the table 1, and the vertical ordinate represents current calculated (unit: mA). To improve the measurement precision and extend the measuring range[9], the equation coefficient is verified by taking different approaches, such as multi-byte addition, subtraction, double byte multiplication, multi-byte cycle shift division. As practical situation stand, the precision of the current calculated, which achieves 0.01mA, can satisfy the request of measurement precision.

The oscillation cycle value begins to be tested by the MCU at the T1 time (the waveform shown in Fig.4, which close to the square wave is commutated by the diode, is produced by the Wien bridge oscillatory circuit.). However, the formal measurement sets to work at the T2 time when the wave between the T1~T2 time slot was removed. The T2~T3 time slot is a complete cycle and the data tested will be stored in the MCU after the measurement is completed.

By the same above method, the MCU test 30 groups of cycle data in total and make digital filtering respectively at the same time, then calculate the average value of cycles and look for the interval of current tested by the given data table. Finally, according to the corresponding interpolation computation in the locked intervals, the change value of the current of a cycle will be obtained.

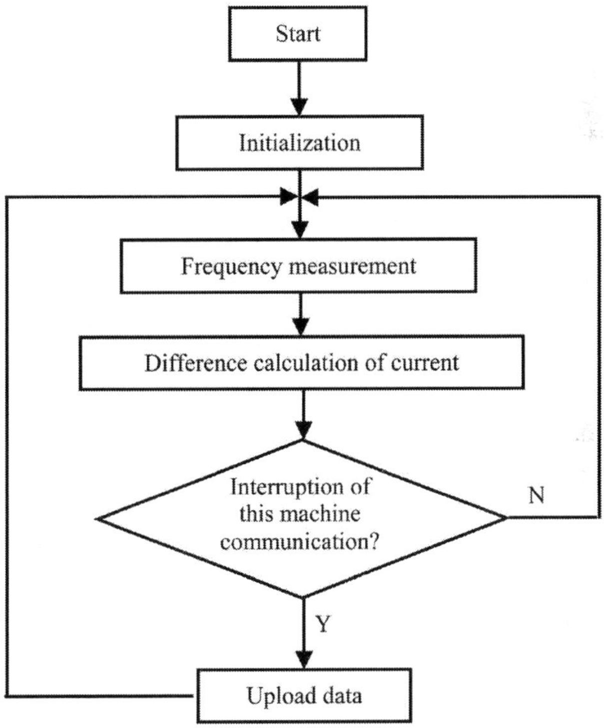

Figure 3. Software flowchart

4. Result Analysis and Methods of Improving Precision

4.1 Result Analysis

During the process of software debugging, part of current values, which are shown in Table 1, are obtained by the actual measurement, and then the subsection curve, shown in Fig.5, will be obtained due to the value. As is shown in Table 1, the current error rate is small, which usually in about 0%-0.8%, therefore the current precision is 0.01mA, which can meet the requirement of the actual project.

Actual current value flowed through coils (mA)	Current value tested by the sensor(mA)	Absolute error rate ×100%
0	0	0%
83.14	83.3	0.2%
96.85	96.98	0.14%
144.69	143.78	0.62%
160.77	161.44	0.45%
192.93	193.63	0.36%
225.08	226.25	0.52%
241.16	246.51	0.59%
305.47	304.05	0.46%
353.7	351.12	0.73%

Table 1. Part of the fault current data Measured by the smart sensor

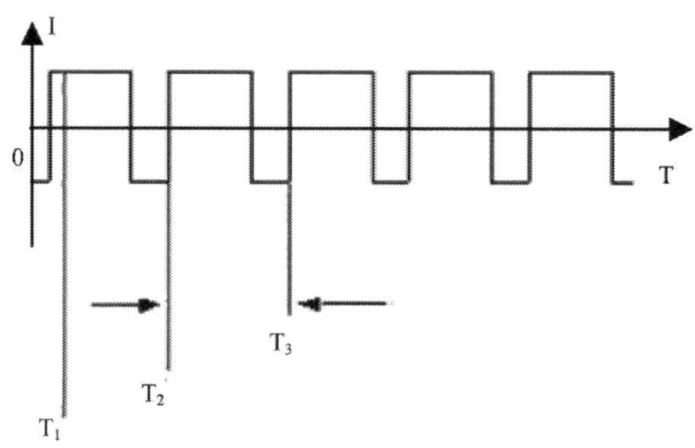

Figure 4. Computing model of a complete cycle

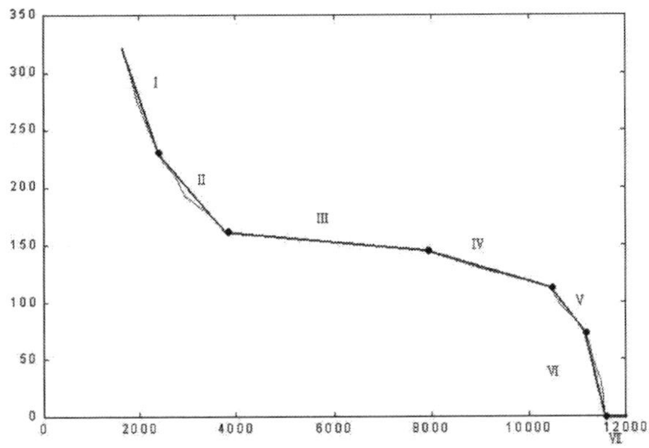

Figure 5. Relation curve between coil current (y-axis) and oscillation cycle (x-axis)

4.2 Methods of Improving Precision

The measurement errors originated from the following sources in this sensor:

1. Relate to the coil sensitivity $\Delta L_X / \Delta I$: The coil inductance changed ΔL_X in direct proportion to the coil current changed ΔI within a certain range, namely $\Delta L_X \propto K \cdot \Delta I$, where K is the proportional coefficient. Coils aging (mainly the heat loss of the coils) would lead to not only decline sensitivity, but also reduce the proportional coefficient K. The resolve method is to replace coils at fixed periods.
2. Relate to the oscillation cycle of coils measured: It would produced errors by measuring the oscillation cycle by calculating the machine cycle, and the maximum frequency value tested is 1KHZ due to the actual measurement, so the biggest error produced (the crystal oscillator of AT89C2051 is 11.0592MHZ) is $(12/11.0592\mu s)/1000\mu s \times 100\% = 0.185\%$ that can satisfy the request of the actual project.
3. Relate to the subsection lineal fitting by the measurement of curve. It can be seen that the effect of the lineal fitting is better within a certain range, however, the lineal fitting errors would be produced due to the little bias existed. The resolve method used is to take the direct interpolation method or increase points measured.

5. Conclusion

This sensor, which has small size, high ratio of performance and price, and high reliability, achieves good results in real operation and measures accuracy. The results show that this design scheme is feasible, could achieve the expected criteria of design and could be spread.

6. References

Major William K Lewis, "UCAV-The Next Generation Air-Superiority Fighter? *School of Advanced Airpowers Studies,* June 2002. Available: http://www.au.af.mil/au/awc/awcgate/saas/lewis.pdf [1]

X.P. Meng, Y. Gao, *Electric Power System Analysis*, Higher Education Press, Beijing, 2004, pp. 100-135. [2]

T.X. Chen, J.D. Duan, T.T. Chen, G. Chen, A new approach to on-line insulation monitoring of electric power cables based on capacitive-current measurement and its criteria, *High Voltage Apparatus*, vol. 40, no. 3, 16, June, 2004, pp. 183-185. [3]

8-bit Microcontroller with 2K Bytes Flash AT89C2051. Available: www.atmel.com/atmel/acrobat/doc0368.pdf [4]

X25045 datasheet. Available: http://tu.cndzz.com/down/soft/2206.htm [5]

TL750L05C datasheet. Available: www.compel.ru/pdf/TI/TL750LXX. [6]

TLE2141CP datasheet. Available: http://www.alldatasheet.com/datasheet-pdf/pdf/TI/TLE2141CP.html [7]

W.J. Dong, The Methods for Software Anti-interference Applied to MCS-51 Monolithic Processor Operation System, *Journal of Luoyang University*, vol. 15, no. 4, Dec. 2000, pp. 56-58. [8]

X.R. Chen, P. Cai, H.Q. Zhou, Several practical methods of frequency measurement based on a single chip computer, *Industrial Instrumentation & Automation*, no.3, 2003, pp. 40-43. [9]

11

Unmanned Aerial Vehicle Formation Flight Using Sliding Mode Disturbance Observers

Dr. Galzi Damien
AUSY
France

1. Introduction

The objective of this chapter is to design a robust controller to achieve outstanding formation flight of unmanned aerial vehicles and to accurately track the desired path while facing unknown disturbances. In other words, the goal is to design the autopilot laws to enforce the desired formation by following the guidance commands. Given a non-linear mathematical model of a quadrotor (X-shaped rotorcraft), a smooth controller composed of three time-scaled nested controllers is developed using sliding mode control driven by sliding mode disturbance observer techniques based on inversion of rotorcraft dynamics and quaternions mathematical representation.

Available research efforts in multiple aircraft formation control (Innocenti et al., 2000) provide us with a variety of control techniques such as the Proportional Integral Derivative (PID) controller, Linear Quadratic Regulator (LQR) feedback controller, Adaptive controller, or control Lyapunov functions. However, these techniques have failed to address the robustness required for the tight formation to real world effects, including external disturbances, environmental aerodynamic uncertainties, end-effectors malfunctions and simulation model inaccuracies. The main limitations of the proposed techniques are the combination of accurate path tracking, adapted guidance laws for extensive formation structure capabilities, and robust formation controllers. The lack of the latter is critical for tight formation keeping in the face of leader maneuvers and unknown disturbances.

Sliding Mode (SM) control is a well-known technique used to achieve robustness and to guarantee the stability of a system (C. Edwards, & S. Spurgeon, 1998). However it presents the inconvenience of control signal high frequency switching which is not appropriate for autopilots. Recent results in Higher Order Sliding Mode (HOSM) control theory (A. Levant, 2003) present the ability to design a continuous and even smooth control, robust to external disturbances, parametric uncertainties, end effectors failures and modeling errors, which is chattering free and provides for enhanced accuracy. Robust sliding mode control using sliding mode disturbance observer techniques allow us to consider tight formation flight where the separation distances between flying aerial vehicles are minimized. The sliding mode control technique provides the unmanned rotorcrafts with bounded, continuous output tracking for the three controllers: translation tracking, attitude tracking and speed controller for the four propellers.

The organization of this chapter is as follow: First the unmanned vehicle is introduced and few guidelines on developing the non-linear model under quaternions representation are proposed Then the formation strategies are developed in a three dimensional environment following triangular formation strategies. Then the controller is developed for a single rotorcraft, considering that, for simplification purposes, all the rotorcrafts have the same characteristics, even though changes in mass and efficiency could easily be taken into account. Finally, simulation results will be presented within the Matlab/Simulink® environment.

2. The vehicle: X-Shaped Rotorcraft

An example of a X-Shaped rotorcraft is the X-4 Flyer®. It is propelled by four independent, fixed-pitch rotors located on the extremities of an 'X' shaped frame. This specific configuration allows a balanced distribution of thrusts around the center of mass of the vehicle. Motion is provided by independently varying the angular speed and thus the thrust of each rotor. Advantages of such a vehicle are numerous, especially for robust formation control, since soft collisions can be tolerated during flight. The major issue of the X-4-Flyer is that it is unstable by design. This makes it entirely inoperable by a human being, without the help of a fast on-board controller and motion sensors.

Figure 1. X-4 Flyer®

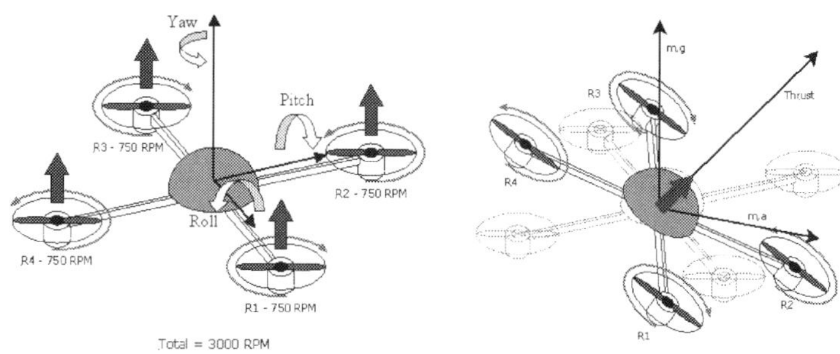

Figure 2. Hover State and Translation

The "X" configuration allows rotations and translations along all three axes, but only four degrees of freedom can be controlled independently because of the coupling between these motions. Like conventional helicopters, pitch and roll rotations are coupled with longitudinal and transversal translations, respectively.

Figure 2 shows examples of rotor speed configurations. In hover state, thrusts produced by the four propellers exactly counteract the effect of gravity. Yaw is controlled by adjusting the torque applied to each motor. The global reaction torque will cause the whole airframe to rotate about its local yaw axis. Pitch and Roll rotations can be achieved by adjusting the difference in rotor speeds within a pair of rotors rotating in the same direction. Translation is achieved by tilting the whole airframe using Pitch/Roll rotations in the direction of the desired acceleration vector, and adjusting the global thrust.

1.1 Mathematical Model

The frames of reference are: a fixed-to-earth frame and a local moving frame attached to the center of mass O of the vehicle. G is assumed to be Galilean. Orientation of frame A relative to frame G is expressed with the rotation matrix, equation (1). The airframe orientation in space is given by the Euler angles: roll (ϕ), pitch (θ) and yaw (ψ) with angular velocities respectively (p, q, r).

$$R_{A/G} = \begin{bmatrix} \cos(\psi)\cos(\theta) & \sin(\phi)\sin(\theta)\cos(\psi) - \cos(\phi)\sin(\psi) & \cos(\phi)\sin(\theta)\cos(\psi) + \sin(\phi)\sin(\psi) \\ \sin(\psi)\cos(\theta) & \sin(\phi)\sin(\theta)\sin(\psi) + \cos(\phi)\sin(\psi) & \cos(\phi)\sin(\theta)\sin(\psi) - \sin(\phi)\cos(\psi) \\ -\sin(\theta) & \cos(\phi)\cos(\theta) & \cos(\phi)\cos(\theta) \end{bmatrix} \quad (1)$$

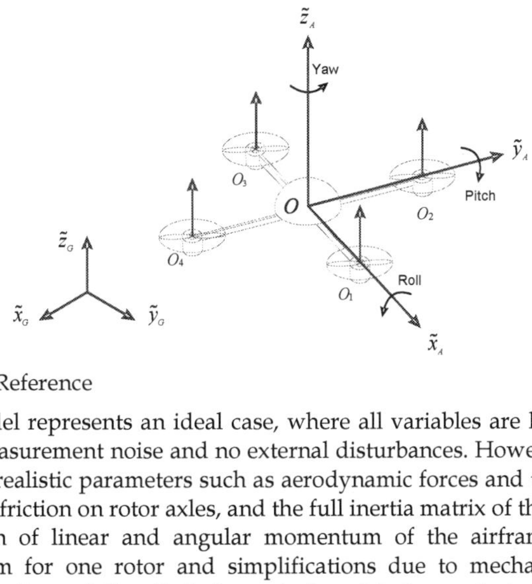

Figure 3. Frame of Reference

The following model represents an ideal case, where all variables are known or measured exactly with no measurement noise and no external disturbances. However, the model does include important realistic parameters such as aerodynamic forces and torques on airframe, propellers, viscous friction on rotor axles, and the full inertia matrix of the airframe.
Using conservation of linear and angular momentum of the airframe, conservation of angular momentum for one rotor and simplifications due to mechanical constraints of Rotor-Airframe coupling, a full realistic dynamical model of a quadrotor is proposed:
Airframe linear acceleration:

$$\begin{pmatrix} \ddot{x} \\ \ddot{y} \\ \ddot{z} \end{pmatrix}_{A/G}^{A} = \dot{\tilde{v}}_{A/G}^{G} = -g \cdot \tilde{z}^{G} + \frac{\rho_{air}}{2m} \cdot [R_{G/A}] \cdot \begin{pmatrix} \sum_{i=1}^{4} \left(c_{rdf_i} \cdot \omega_i \cdot |\omega_i| \cdot \tilde{f}_i^{A} \right) \\ -[C_{ADF}^{A}] \cdot [R_{A/G}] \cdot \tilde{v}_{A/G}^{G} \cdot \left| \tilde{v}_{A/G}^{G\ T} \right| \end{pmatrix} \quad (2)$$

Airframe angular acceleration:

$$\begin{pmatrix} \dot{p} \\ \dot{q} \\ \dot{r} \end{pmatrix}_{A/G}^{A} = \dot{\tilde{\omega}}_{A/G}^{A} = \left[I_{frame/O}^{A}\right]^{-1} \cdot \begin{pmatrix} -sk(\tilde{\omega}_{A/G}^{A}) \cdot \left[\left[I_{frame/O}^{A}\right] \cdot \tilde{\omega}_{A/G}^{A} + \sum_{i=1}^{4}\left(j_{z_i} \cdot \omega_i \cdot \tilde{r}_i^{A}\right)\right] \\ -\sum_{i=1}^{4}\left(\tau_i - c_v \cdot \omega_i\right) \cdot \tilde{r}_i^{A} \\ +\frac{1}{2} \cdot \rho_{air} \cdot \left[-\left[C_{ADF}^{A}\right] \cdot \tilde{\omega}_{A/G}^{A} \cdot \left|\tilde{\omega}_{A/G}^{A\,T}\right| + \sum_{i=1}^{4}\left(c_{rdf_i} \cdot \omega_i \cdot \left|\omega_i\right| \cdot \tilde{f}_i^{A} \times \tilde{L}_i^{A}\right)\right] \end{pmatrix} \tag{3}$$

Rotor angular acceleration:

$$\dot{\omega}_i = \frac{1}{j_{z_i}} \cdot \left(\tau_i - \frac{1}{2} \cdot \rho_{air} \cdot c_{rdt_i} \cdot \omega_i \cdot \left|\omega_i\right| - c_{v_i} \cdot \omega_i\right) \tag{4}$$

Rotation matrix derivative:

$$[\dot{R}_{G/A}] = [R_{G/A}] \cdot sk(\tilde{\omega}_{A/G}^{A}) \tag{5}$$

1.2 Quaternions Representation

A quaternion can be thought of as a complex number with 4 dimensions: 3 imaginary dimensions *i*, *j* and *k*; and one real dimension, all orthogonal to one another. Each of the imaginary dimensions has a unit value of the square root of -1. A quaternion is represented as follows:

$$Q = q_1 \cdot i + q_2 \cdot j + q_3 \cdot k + q_4 \tag{6}$$

or

$$Q = \begin{bmatrix} q_1 \\ q_2 \\ q_3 \\ q_4 \end{bmatrix} \tag{7}$$

When used to describe 3D attitude or orientation, the following rule exists between quaternion coefficients and instantaneous axis ($\tilde{u}$) and angle (θ) of rotation:

$$Q = \begin{bmatrix} q_1 \\ q_2 \\ q_3 \\ q_4 \end{bmatrix} = \begin{bmatrix} u_x \cdot \sin(\theta/2) \\ u_y \cdot \sin(\theta/2) \\ u_z \cdot \sin(\theta/2) \\ \cos(\theta/2) \end{bmatrix} \tag{8}$$

It is possible to rotate vectors in 3D space using quaternion through the same process used with rotation matrices. The following formula shows how to rotate vector $V = [v_x, v_y, v_z]^T$ using rotation axis and angle given by quaternion Q

$$Q_{vrot} = Q \otimes Q_v \otimes Q' \tag{9}$$

$$Q_v = v_x i + v_y j + v_z k + 0 \tag{10}$$

3. Formation Strategies

Formation strategies are developed to manage any size formation according to three specific schemes: "Tracking," "Placement" and "Anti-Collision." The shape of the formation is derived using "triangular" patterns depending on the number of aerial vehicle (AV).

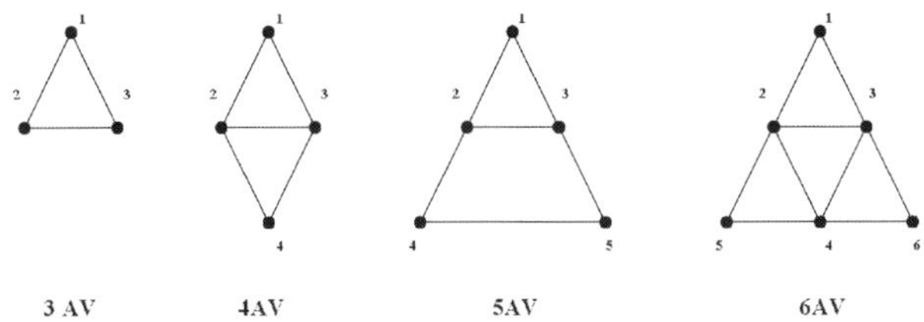

Figure 4. Formations Triangular Patterns

The formation strategies are designed as follows:

- The "Tracking" strategy is specific to the leader of the formation which is in this case the agent at the tip of the triangle. The task of the leader is to track the predefined path.
- The "Placement" strategy consists in positioning an agent in regards to the leader only.
- The "Anti-Collision" strategy consists in positioning an agent in regards to its lateral neighbors (set of safety distances) and to the leader.

Let us consider a formation of three agents in a Cartesian frame centered on the leader (Figure 5). The leader receives the path to follow as a function of time, while the followers receive the distances to keep from each other and to the leader. For an equilateral triangular formation pattern where the distances to keep is d from each vehicle with a vertical separation of h, the formation commands for the followers are:

$$\text{"Placement"}: x_p = x_{leader} - \sqrt{\frac{3}{4}d^2 - h^2}\ ,\ y_p = y_{leader} - \frac{d}{2}\ ,\ z_p = z_{leader} - h$$

$$\text{"Anti-Collision"}: x_A = x_{leader} - \sqrt{\frac{3}{4}d^2 - h^2}\ ,\ y_A = y_P + d\ ,\ z_A = z_{leader} - h$$

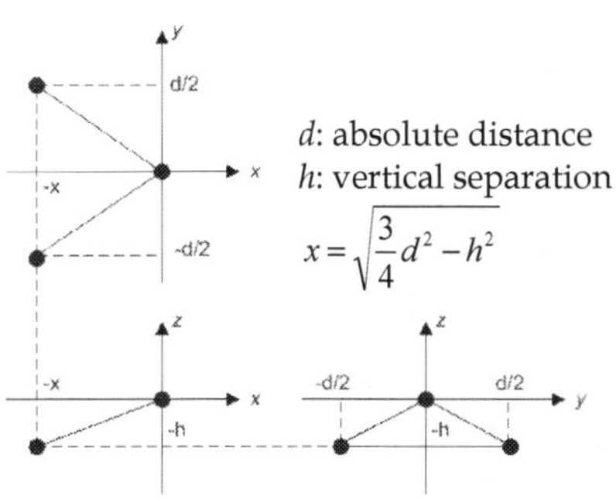

Figure 5. Three Aerial Vehicles 3D Triangular Formation Pattern

4. Controller Architecture

The design of the controller follows the natural time constants of the model, which are ideally separated. A three time scale controller using three nested feedback control loops based on sliding mode driven by sliding mode observer techniques is proposed to include translation tracking controller, attitude tracking controller and speed controller for the four propellers. The control strategy is depicted in figure 6, while the architecture for each controller is depicted in figure 7 including three feedback loops and sensors to measure the real time rotorcraft state.

In loop 1 of figure 7, the translation controller is designed to compute the instantaneous acceleration vector needed to track the reference trajectory by comparing the position commands and the actual measured positions from a Global Positioning System.

Loop 2 works in conjunction with the attitude controller to compute the body torques needed to align the rotorcraft in the direction of the desired acceleration. The desired instantaneous acceleration vector is compared to the angular velocities about the three axis of the airframe measured by gyroscopes. Attitude is derived by integration.

From the desired torque, the desired propeller velocity rates are computed using inverse kinematics (Loop 3). In the loop 4, the propeller controller stabilizes the propeller velocity rates to the commanded value compared to the measured angular rates of each rotor.

The closed-loop time constants of the three outer loop controllers have to be carefully chosen in order to guarantee convergence of each loop. Sliding mode control driven by sliding mode observer techniques is used to provide stabilization of the system dynamics and guarantee robustness; the errors will converge to the origin even in the presence of external and internal disturbances.

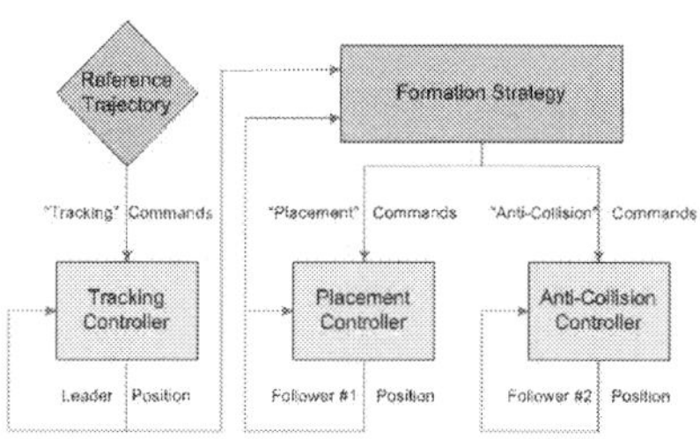

Figure 6. Formation Control Strategy

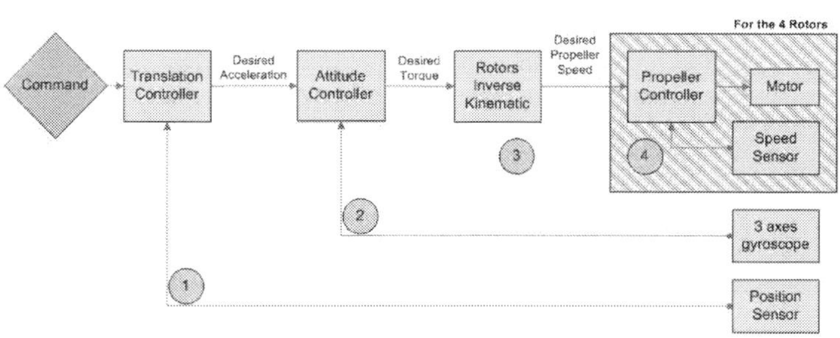

Figure 7. Flight Controller Architecture

5. Sliding Mode

5.1 Introduction to Sliding Mode

Sliding Mode Control is a control technique that allows the user to drive a system to a desired state, referred to as sliding surface, and consists of two stages: the reaching phase and the sliding phase as shown in figure 8. The reaching phase focuses on driving the system state to the sliding surface. The sliding phase consists of bringing the sliding variable to zero and is characterized by high-frequency, discontinuous switching of the control.

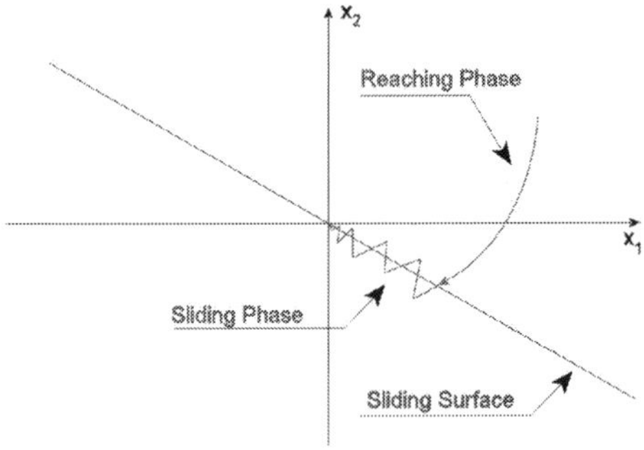

Figure 8. Sliding Mode Motion

The sliding surface is to be designed so that the system exhibits the desired behavior. The control is to be designed to assure that the system reaches the sliding surface and stays on it thereafter

Output tracking problems consider Multi Input Multi Output (MIMO) system such as the following

$$\begin{cases} \dot{x} = f(x,t) + G(x,t)u \\ y = h(x,t) \end{cases} \tag{11}$$

where $x \in \Re^n$, $f(x,t) \in \Re^n$, $y \in \Re^m$, $u \in \Re^m$ and $h(x,t) = [h_1, h_2, ..., h_m]^T \in \Re^m$, $G(x,t) = [g_1, g_2, ..., g_m]^T \in \Re^{n\times m}$ are analytic vector and matrix functions.

The system can be expressed in normal form assuming the system is completely feedback linearizable (Isidori, 1995) in a reasonable compact domain $x \in \Gamma(x)$. $L_f^{r_i-1}h_i$ and $L_{g_i}\left(L_f^{r_i-1}h_i\right)$ $\forall i = \overline{1,m}$ are corresponding Lie derivatives, $|E(x,t)| \neq 0$, $\forall x \in \Gamma$ and $\bar{r} = [r_1, r_2, ..., r_m]^T$ is a vector relative degree.

$$\begin{bmatrix} y_1^{(r_1)} \\ y_2^{(r_2)} \\ \cdots \\ y_m^{(r_m)} \end{bmatrix} = \begin{bmatrix} L_f^{r_1}h_1(x,t) \\ L_f^{r_2}h_2(x,t) \\ \cdots \\ L_f^{r_m}h_m(x,t) \end{bmatrix} + E(x,t)u \tag{12}$$

$$E(x,t)=\begin{bmatrix} L_{g_1}\left(L_f^{r_1-1}h_1\right) & L_{g_2}\left(L_f^{r_1-1}h_1\right) & \dots & L_{g_m}\left(L_f^{r_1-1}h_1\right) \\ L_{g_1}\left(L_f^{r_2-1}h_2\right) & L_{g_2}\left(L_f^{r_2-1}h_2\right) & \dots & L_{g_m}\left(L_f^{r_2-1}h_2\right) \\ \dots & \dots & \dots & \dots \\ L_{g_1}\left(L_f^{r_m-1}h_m\right) & L_{g_2}\left(L_f^{r_m-1}h_m\right) & \dots & L_{g_m}\left(L_f^{r_m-1}h_m\right) \end{bmatrix} \quad (13)$$

The goal of output tracking techniques is to derive the control u to drive the system output y to the desired trajectory y_c (command). In other words, the goal is to drive the tracking error $e = y - y_c$ to zero in a finite time. The design problem is in choosing a sliding variable σ that will drive the tracking error to zero and maintain it thereafter.
Choose the following sliding variable

$$\sigma_i = e_i^{(r_i-1)} + c_{i,r_i-2}e_i^{(r_i-2)} + \dots + c_{i,1}e_i^{(1)} + c_{i,0}e_i + c_{i,-1}e_{i,-1} \quad (14)$$

where $\sigma = [\sigma_1, \sigma_2, \dots, \sigma_m]^T$, $e_i = y_{i,c} - y_i$, $e_i^{(k)} = \dfrac{d^k e_i}{dt^k}$, $e_{i,-1} = \int e_i dt$ and $c_{i,k} > 0 \in \Re$.
Taking the derivatives on both sides,

$$\dot{\sigma}_i = e_i^{(r_i)} + c_{i,r_i-2}e_i^{(r_i-1)} + \dots + c_{i,1}e_i^{(2)} + c_{i,0}e_i^{(1)} + c_{i,-1}e_i \quad (15)$$

From equation (12),

$$y_i^{(r_i)} = L_f^{r_i}h_i(x,t) + (Eu)_i \quad (16)$$

Introducing $\tilde{u}_i = (Eu)_i$,

$$e_i^{(r_i)} = y_{i,c}^{(r_i)} - L_f^{r_i}h_i(x,t) - \tilde{u}_i \quad (17)$$

and substituting (17) into (15),

$$\dot{\sigma}_i = y_{i,c}^{(r_i)} - L_f^{r_i}h_i(x,t) - \tilde{u}_i + c_{i,r_i-2}e_i^{(r_i-1)} + \dots + c_{i,1}e_i^{(2)} + c_{i,0}e_i^{(1)} + c_{i,-1} \quad (18)$$

Let us use the following notation

$$\begin{cases} \Psi_i = y_{i,c}^{(r_i)} - L_f^{r_i}h_i(x,t) + c_{i,r_i-2}e_i^{(r_i-1)} + \dots + c_{i,1}e_i^{(2)} + c_{i,0}e_i^{(1)} + c_{i,-1} \\ \Psi_i = \psi_i^0 + \Delta\psi_i \end{cases} \quad (19)$$

Where ψ_i^0 represents the known part of Ψ_i and $\Delta\psi_i$ represents unknown bounded disturbances such that $\|\Delta\psi_i\| \le L_i$. Introducing (19) into (18),

$$\dot{\sigma}_i = \psi_i^0 + \Delta\psi_i - \tilde{u}_i \quad (20)$$

Let us design the control as

$$\tilde{u}_i = \tilde{u}_{i,0} + \tilde{u}_{i,1} \quad (21)$$

where $\tilde{u}_{i,0}$ is designed to compensate for the known terms

$$\tilde{u}_{i,0} = \psi_i^0 \tag{22}$$

Therefore, becomes,

$$\dot{\sigma}_i = \Delta\psi_i - \tilde{u}_{i,1} \tag{23}$$

The control problem is to design the control $\tilde{u}_{i,1}$ that will drive $\dot{\sigma}_i$ to zero.

5.2 Traditional Sliding Mode Control

Traditional sliding mode control is based on the analysis of Lyapunov functions to insure asymptotic stability with finite time convergence. Consider the following Lyapunov function

$$V_i = \frac{1}{2}\sigma_i^2 \tag{24}$$

Taking the time derivative, we obtain

$$\dot{V}_i = \dot{\sigma}_i \sigma_i \tag{25}$$

From Lyapunov theory, we derive the following finite-time asymptotic convergence condition

$$\dot{V}_i \leq -\rho_i \sqrt{2V_i} \tag{26}$$

with $\rho_i > 0,\ \rho_i \in \Re$, which guarantees a reaching time of $t_{r,i} \leq \dfrac{\sqrt{2V_i(0)}}{\rho_i}$.

Substituting (24) and (25) into (26) we obtain

$$\sigma_i \dot{\sigma}_i \leq -\rho_i \sqrt{\sigma_i^2} \tag{27}$$

$$\sigma_i \dot{\sigma}_i \leq -\rho_i |\sigma_i| \tag{28}$$

From (23), we also have

$$\sigma_i \dot{\sigma}_i = \sigma_i \left(\Delta\psi_i - \tilde{u}_{i,1}\right) \tag{29}$$

$$\sigma_i \dot{\sigma}_i = \sigma_i \Delta\psi_i - \sigma_i \tilde{u}_{i,1} \tag{30}$$

which can be rewritten as

$$\sigma_i \dot{\sigma}_i \leq |\sigma_i||\Delta\psi_i| - \sigma_i \tilde{u}_{i,1} \tag{31}$$

The disturbances being bounded, $\|\psi_i\| \leq L_i$, (31) becomes

$$\sigma_i \dot{\sigma}_i \leq L_i |\sigma_i| - \sigma_i \tilde{u}_{i,1} \tag{32}$$

or, using the definition of the sign function

$$\sigma_i \dot{\sigma}_i \leq |\sigma_i| \left[L_i - \tilde{u}_{i,1} sign(\sigma_i) \right] \tag{33}$$

The control $\tilde{u}_{i,1}$ has to satisfy the sliding mode existence condition expressed in (28). Therefore, from (33) and (28) we obtain

$$L_i - \tilde{u}_{i,1} sign(\sigma_i) = -\rho_i \tag{34}$$

$$\tilde{u}_{i,1} sign(\sigma_i) = \rho_i + L_i \tag{35}$$

Therefore,

$$u_{i,1} = (\rho_i + L_i) sign(\sigma_i) \tag{36}$$

The final sliding mode control is expressed using (22), (36) and recalling that $\tilde{u}_i = (Eu)_i$

$$u = E^{-1} \left[\psi^0 + (\rho + L) \; sign(\sigma) \right] \tag{37}$$

This traditional sliding mode control will stabilize the sliding variable on the sliding surface but is characterized by high frequency switching. Further developments in higher order sliding mode control and sliding mode control driven by sliding mode observer are employed to reduce the switching motions and accurately estimate the disturbances.

5.3 Sliding Mode Driven by Sliding Mode Disturbance Observer

The main idea of this technique is to design a sliding mode based observer to accurately estimate the unknown disturbance in order to compensate for it.
Introduce the auxiliary sliding variables S_i and Z_i such that

$$\begin{cases} s_i = \sigma_i + z_i \\ \dot{z}_i = \tilde{u}_{i,1} - v_i \end{cases} \tag{38}$$

Therefore, the dynamic of the auxiliary sliding variable is

$$\dot{s}_i = \left(\Delta\psi_i - \tilde{u}_{i,1} \right) + \left(\tilde{u}_{i,1} - v_i \right) \tag{39}$$

$$\dot{s}_i = \Delta\psi_i - v_i \tag{40}$$

As seen previously in (36), the control that will drive S_i to zero and maintain it thereafter is

$$v_i = \left(\rho_i + L_i \right) sign(s_i) \tag{41}$$

To avoid the high frequency switching motion of the control, a Low-Pass filter is introduced

$$\hat{v}_i = \frac{1}{1+\beta_i s} v_i \tag{42}$$

Where β_i is the time constant to be chosen according to design specifications. Therefore, when the auxiliary sliding variable S_i stabilizes at zero, (26) becomes

$$\Delta\psi_i - \hat{v}_i = 0 \tag{43}$$

As a result, the control $\hat{v}_i$ is an exact estimate of the unknown disturbances $\Delta\psi_i$.
Let us design the control $\tilde{u}_{i,1}$ to stabilize σ_i to zero in (23) of the form:

$$\tilde{u}_{i,1} = \hat{v}_i + K_{0,i}\sigma_i \tag{44}$$

Where $K_{0,i} > 0$ is chosen to ensure fast reaching of the sliding surface by σ, as soon as the disturbance is estimated and compensated for (i.e., the convergence of σ_i cannot be faster than that of S_i). Therefore, the dynamic of the sliding variable is

$$\dot{\sigma}_i = -K_{0,i}\sigma_i \tag{45}$$

which satisfies the sliding mode existence condition: $\sigma_i\dot{\sigma}_i < 0$.
The final sliding mode control driven by sliding mode observer is

$$u = E^{-1}[\psi^0 + \hat{v} + K_0\sigma] \tag{46}$$

6. Controller Design

6.1 Translation Controller

The following equation expresses the airframe linear position, velocity and acceleration errors, respectively

$$\begin{aligned} \tilde{\varepsilon}_p &= \tilde{p}_c - \tilde{p} \\ \tilde{\varepsilon}_v &= \dot{\tilde{\varepsilon}}_p = \dot{\tilde{p}}_c - \dot{\tilde{p}} \\ \tilde{\varepsilon}_\gamma &= \ddot{\tilde{\varepsilon}}_p = \ddot{\tilde{p}}_c - \ddot{\tilde{p}} \end{aligned} \tag{47}$$

where $\ddot{\tilde{p}}_c, \dot{\tilde{p}}_c, \tilde{p}_c$ are given by the desired trajectory commands in position, velocity and acceleration, respectively, and $\dot{\tilde{p}}, \tilde{p}$ are measured or estimated in real time by on-board sensors.
Following sliding mode observer techniques, a sliding variable of second order is designed

$$\tilde{\sigma}_p = k_{v1}\dot{\tilde{\varepsilon}}_p + k_{p1}\tilde{\varepsilon}_p + k_{i1}\int \tilde{\varepsilon}_p dt \tag{48}$$

A third order pole placement is used to compute values for the coefficients k_{v1}, k_{p1} and k_{i1}. In this case, a triangular left half plane pattern pole placement is used to compute values for the coefficients, using the desired response time for position tracking.
The dynamics of the sliding variable can be expressed as

$$\dot{\tilde{\sigma}}_p = \tilde{\psi}_p + \Delta\tilde{\psi}_p - k_{v1}\tilde{\gamma} \tag{49}$$

where $\tilde{\psi}_p$ represents the known terms,

$$\tilde{\psi}_p = k_{v1}(\ddot{\tilde{p}}_c + \tilde{g}) + k_{p1}\dot{\tilde{\varepsilon}}_p + k_{i1}\tilde{\varepsilon}_p \tag{50}$$

And $\tilde{\psi}_p$ the unknown bounded disturbances, $|\Delta\tilde{\psi}_p| \le L_p$, where L_p is a positive constant. Therefore, the control can be expressed in a sum of control terms compensating respectively for the known and unknown terms

$$\tilde{\gamma} = (\tilde{u}_{0p} + \tilde{u}_{1p}) / k_{v1} \tag{51}$$

The disturbance observer is designed considering the following auxiliary sliding variable

$$\tilde{\psi}_p = k_{v1}(\ddot{\tilde{p}}_c + \tilde{g}) + k_{p1}\dot{\tilde{\varepsilon}}_p + k_{i1}\tilde{\varepsilon}_p \tag{52}$$

$$\begin{cases} \tilde{s}_p = \tilde{\sigma}_p + \tilde{z}_p \\ \dot{\tilde{z}}_p = \tilde{u}_{1p} - \tilde{v}_p \end{cases} \tag{53}$$

The resulting dynamics are

$$\dot{\tilde{s}}_p = \Delta\tilde{\psi}_p - \tilde{v}_p \tag{54}$$

The dynamics of the auxiliary variable are completely compensated with a control of the form

$$\tilde{v}_p = \rho_p \, sign(\tilde{s}_p), \; \rho_p >> L_p \tag{55}$$

The high frequency switching is filtered out using the equivalent control

$$\hat{\tilde{v}}_p = \frac{\tilde{v}_p}{\tau_p s + 1} \tag{56}$$

where τ_p is the filter time constant to be tuned.
Finally, the resulting designs of the control terms are

$$\begin{aligned} \tilde{u}_{0p} &= \tilde{\psi}_p \\ \tilde{u}_{1p} &= \hat{\tilde{v}}_p + K_p\tilde{\sigma}_p \end{aligned} \tag{57}$$

where $K_p > 0$ assures exponential convergence of the sliding variable to zero.

6.2 Attitude Controller

Applying inverse rotation kinematics, the desired instantaneous axis and angle of rotation are defined by computing the cross product between current thrust force vector $\widetilde{F}$ and the desired acceleration vector $\widetilde{\gamma}$ normalized to unit length.

$$\widetilde{u}_d = \frac{\widetilde{F}}{\|\widetilde{F}\|} \times \frac{\widetilde{\gamma}}{\|\widetilde{\gamma}\|} \tag{58}$$

The direction of unit vector $\widetilde{u}_d$ gives the desired axis of rotation. Therefore, the actual attitude error angle is

$$\widetilde{\varepsilon}_\theta = \cos^{-1}\left(\frac{\widetilde{F}}{\|\widetilde{F}\|} \times \frac{\widetilde{\gamma}}{\|\widetilde{\gamma}\|} \right) \tag{59}$$

Equations (60) express the attitude, angular velocity and angular acceleration errors, respectively

$$\begin{aligned} \widetilde{\varepsilon}_\theta &= \widetilde{\theta}_c - \widetilde{\theta} = \cos^{-1}\left(\frac{\widetilde{F}}{\|\widetilde{F}\|} \times \frac{\widetilde{\gamma}}{\|\widetilde{\gamma}_d\|} \right) \cdot \frac{\widetilde{u}_u}{\|\widetilde{u}_d\|} \\ \widetilde{\varepsilon}_w &= \dot{\widetilde{\varepsilon}}_\theta = \widetilde{\omega}_c - \widetilde{\omega} = -\widetilde{\omega} \\ \dot{\widetilde{\varepsilon}}_w &= \ddot{\widetilde{\varepsilon}}_\theta = \dot{\widetilde{\omega}}_c - \dot{\widetilde{\omega}} = -\dot{\widetilde{\omega}} \end{aligned} \tag{60}$$

Following similar sliding mode observer techniques as presented in Section 2.D, a sliding variable of second order is designed

$$\widetilde{\sigma}_\theta = k_{v2}\dot{\widetilde{\varepsilon}}_\theta + k_{p2}\widetilde{\varepsilon}_\theta + k_{i2}\int \widetilde{\varepsilon}_\theta dt \tag{61}$$

A third order pole placement is used to compute values for the coefficients k_{v2}, k_{p2} and k_{i2}. In this case, a Butterworth optimal pole placement as presented in previous work [45] is used to compute values for the coefficients, using the desired response time for attitude tracking.

Equation (62) expresses the dynamics of the sliding variable

$$\dot{\widetilde{\sigma}}_\theta = \widetilde{\psi}_\theta + \Delta\widetilde{\psi}_\theta - k_{v2}\dot{\widetilde{\omega}} \tag{62}$$

where $\widetilde{\psi}_\theta$ represents the known terms,

$$\widetilde{\psi}_\theta = k_{p2}\dot{\widetilde{\varepsilon}}_\theta + k_{i2}\widetilde{\varepsilon}_\theta \tag{63}$$

and $\Delta\widetilde{\psi}_\theta$ the unknown bounded disturbances, $|\Delta\widetilde{\psi}_\theta| < L_\theta$, where L_θ is a positive constant.

Once again, the control is expressed in terms of a sum of control terms compensating respectively for the known and unknown terms

$$\dot{\widetilde{\omega}} = (\widetilde{u}_{0\theta} + \widetilde{u}_{1\theta}) / k_{v2} \tag{64}$$

The disturbance observer is designed considering the following auxiliary sliding variable

$$\begin{cases} \tilde{s}_\theta = \tilde{\sigma}_\theta + \tilde{z}_\theta \\ \dot{\tilde{z}}_\theta = \tilde{u}_{1\theta} - \tilde{v}_\theta \end{cases} \tag{65}$$

The resulting dynamics are

$$\dot{\tilde{s}}_\theta = \Delta\tilde{\psi}_\theta - \tilde{v}_\theta \tag{66}$$

The dynamics of the auxiliary variable are completely compensated for with control of the form

$$\tilde{v}_\theta = \rho_\theta \, sign(\tilde{s}_\theta),\ \rho_\theta >> L_\theta \tag{67}$$

The high frequency switching is filtered out using the equivalent control

$$\hat{\tilde{v}}_\theta = \frac{\tilde{v}_\theta}{\tau_\theta s + 1} \tag{68}$$

where τ_θ is the filter time constant to be tuned. Finally, the resulting design of the control terms is

$$\begin{aligned} \tilde{u}_{0\theta} &= \tilde{\psi}_\theta \\ \tilde{u}_{1\theta} &= \hat{\tilde{v}}_\theta + K_\theta \tilde{\sigma}_\theta \end{aligned} \tag{69}$$

where $K_\theta > 0$ assure exponential convergence of the sliding variable to zero.

6.3 Propeller Controller

The speed of each rotor $(w_i, i = 1...4)$ is computed using inverse dynamics as presented in and compared to the desired speed $(w_{id}, i = 1...4)$ corresponding to the desired linear $(\tilde{\gamma})$ and angular $(\dot{\tilde{\omega}})$ accelerations. The necessary torques to apply to each electric motor are then computed in order to achieve these desired speeds.
The angular velocity error of each rotor is defined as

$$\begin{aligned} \varepsilon_{\omega i} &= \omega_{id} - \omega_i \\ \dot{\varepsilon}_{\omega i} &= \dot{\omega}_{id} - \dot{\omega}_i = -\dot{\omega}_i \end{aligned} \tag{70}$$

Using the same sliding mode observer techniques as for the previous two controllers, the sliding variable

$$\sigma_{\omega i} = k_{p3}\varepsilon_{\omega i} + k_{i3}\int \varepsilon_{\omega i}\,dt \tag{71}$$

A classical second order placement is used, with the desired response time, to define k_{p3} and k_{i3}. The dynamics of the sliding variable are expressed as

$$\dot{\sigma}_{\omega i} = \psi_{\omega i} + \Delta\psi_{\omega i} - k_{p3}\dot{\omega}_i \tag{72}$$

where $\psi_{\omega i}$ represents the known terms

$$\psi_{\omega i} = k_{i3}\varepsilon_{\omega i} \tag{73}$$

and $\Delta\psi_{\omega i}$ the unknown bounded disturbances, $\Delta\psi_{\omega i} < L_{\omega i}$, where $L_{\omega i}$ is a positive constant. Once again, the control is expressed in terms of a sum of control terms compensating respectively for the known and unknown terms

$$\dot{\omega}_i = (u_{0\omega i} + u_{1\omega i}) / k_{p3} \tag{74}$$

The disturbance observer is designed considering the auxiliary sliding variable

$$\begin{cases} s_{\omega i} = \sigma_{\omega i} + z_{\omega i} \\ \dot{z}_{\omega i} = u_{1\omega i} - v_{\omega i} \end{cases} \tag{75}$$

The resulting dynamics are

$$\dot{s}_{\omega i} = \Delta\psi_{\omega i} - v_{\omega i} \tag{76}$$

The dynamics of the auxiliary variable are completely compensated for with control of the form

$$v_{\omega i} = \rho_{\omega i}\, sign(s_{\omega i}),\ \rho_{\omega i} >> L_{\omega i} \tag{77}$$

The high frequency switching is filtered out using the equivalent control

$$\hat{v}_{\omega i} = \frac{v_{\omega i}}{\tau_{\omega i} s + 1} \tag{78}$$

where τ_{wi} is the filter time constant to be tuned. Finally, the resulting design of the control terms is

$$\begin{aligned} u_{0\omega i} &= \psi_{\omega i} \\ u_{1\omega i} &= \hat{v}_{\omega i} + K_{\omega i}\sigma_{\omega i} \end{aligned} \tag{79}$$

where $K_{wi} > 0$ assures exponential convergence of the sliding variable to zero.

7. Simulation Results

The following figures are the simulation results for the formation of three X-4 Flyer flying in a close formation in the face of simulated wind gusts acting upon each rotorcraft (modeled as offset on the velocity vector in frame G). The formation is to fly in a triangular pattern with an absolute distance of three meters (d = 3m) between each rotorcraft and a vertical separation of one meter to the leader. The trajectory to follow is a sinusoid at a constant elevation of 2 meters and a constant longitudinal velocity of 2m/s. Note that the first follower will stay at the initial position for one second following take off of the leader and then follow the "Placement" commands. The second follower will stay at the initial position for two seconds following the take off of the leader (or one second following the take off of the first follower) and then follow the "Anti-Collision" commands. The simulation time is 10sec with a computing frequency of a 1000Hz.

7.1 Leader "Path Tracking"

The following figures represent the "Path Tracking" simulation results for the leader of the formation. The characteristics of the X-4 Flyer, such as position, velocity, angular velocity and propeller angular velocity are illustrated, as well as the control torques and the HOSM controller variables.

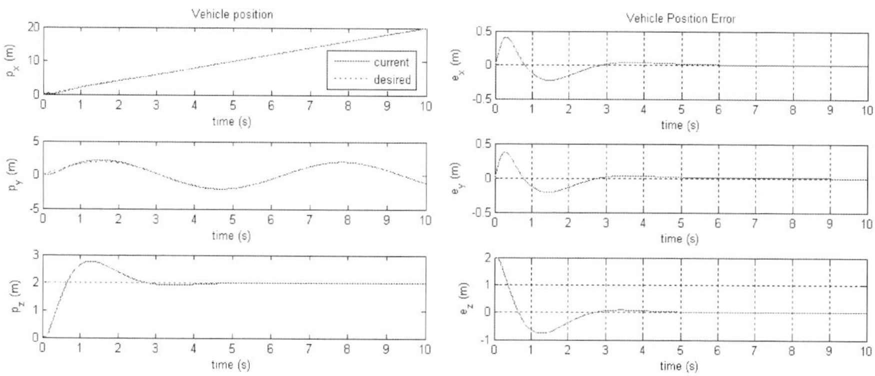

Figure 9. Leader Position and Position Error versus Time

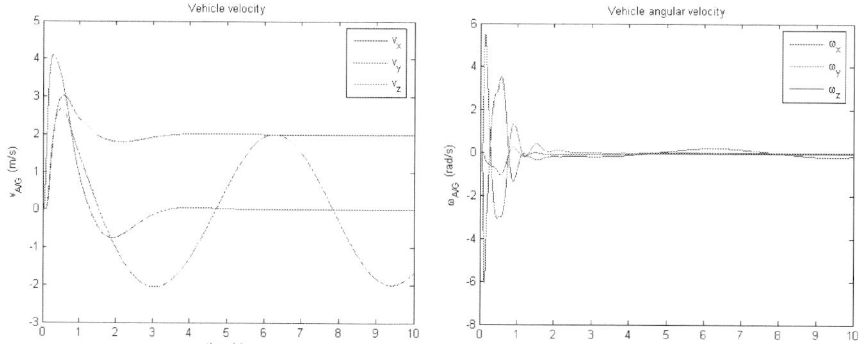

Figure 10. Leader Velocity and Angular Velocity versus Time

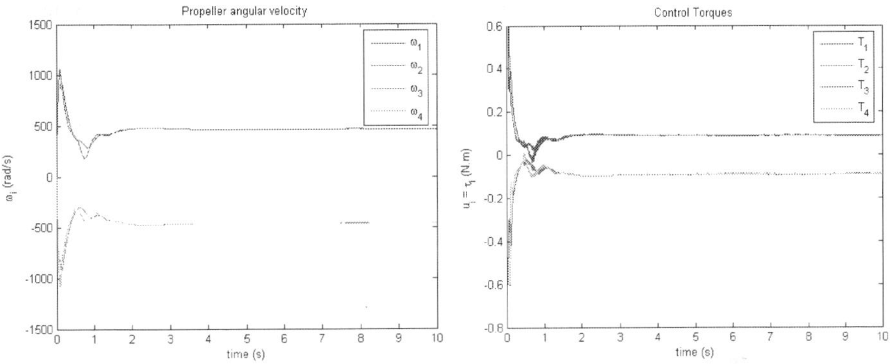

Figure 11. Leader Propeller Angular Velocity and Control Torques versus Time

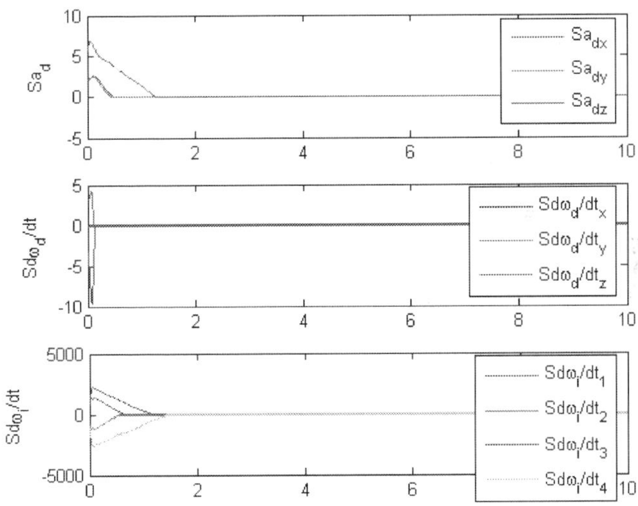

Figure 12. Leader Auxiliary Sliding Variables versus Time

The leader achieves perfect tracking of the desired path in a smooth and robust manner. The position errors in 3 reach zero in finite time. The unknown disturbances are estimated and completely compensated for using robust HOSM techniques.

7.2 First Follower "Placement"

The following figures represent the "Placement" simulation results for the first follower. The characteristics of the X-4 Flyer, such as position, velocity, angular velocity and propeller angular velocity are illustrated, as well as the control torques and the HOSM controller variables.

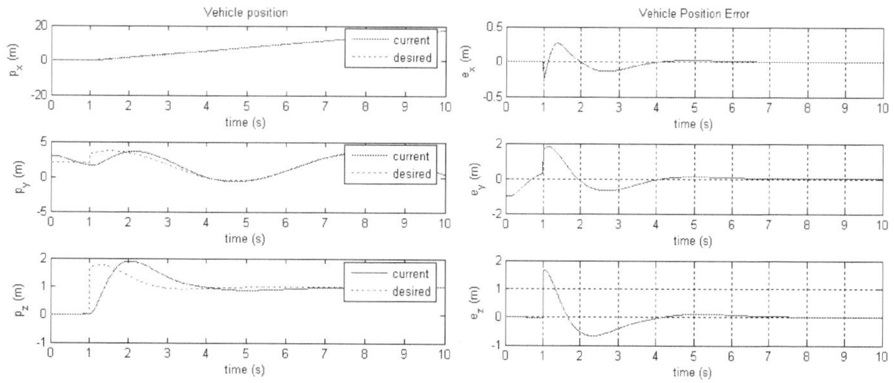

Figure 13. First Follower Position and Position Error versus Time

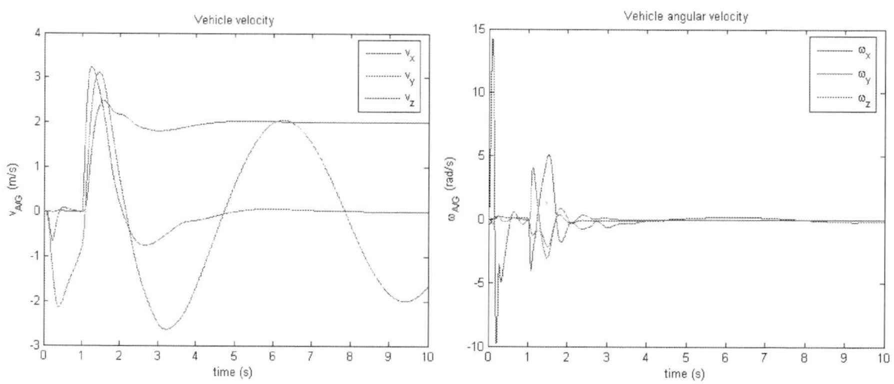

Figure 14. First Follower Velocity and Angular Velocity versus Time

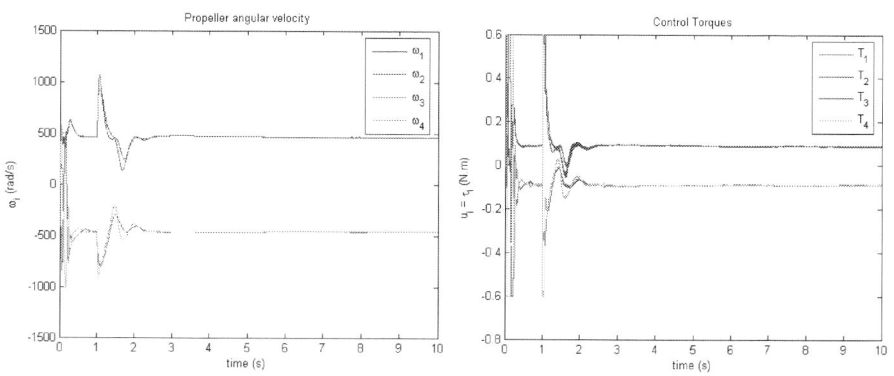

Figure 15. First Follower Propeller Angular Velocity and Control Torques versus Time

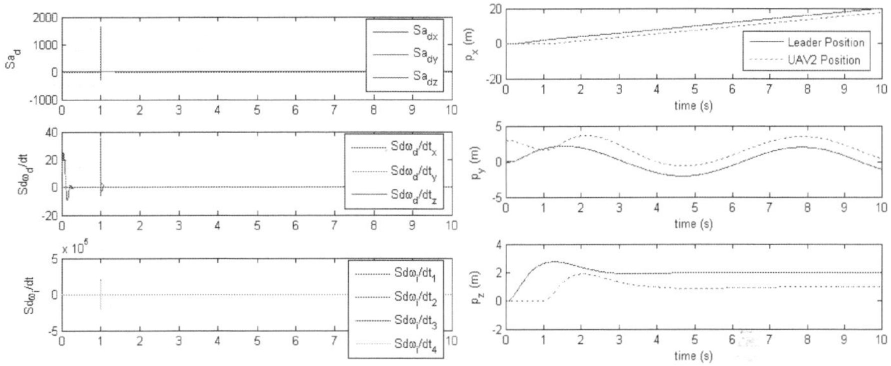

Figure 16. First Follower Auxiliary Sliding Variables and Separation versus Time

After staying at the initial position for the first second, the rotorcraft initiates the formation following the "Placement" commands. The rotorcraft achieves perfect positioning after a smooth transient, the position errors reach zero in finite time. The disturbances are estimated and fully compensated for, via robust HOSM.

7.3 Second Follower "Anti-Collision"

The following figures represent the "Anti-Collision" simulation results for the second follower. The characteristics of the X-4 Flyer, such as position, velocity, angular velocity and propeller angular velocity are illustrated, as well as the control torques and the HOSM controller variables.

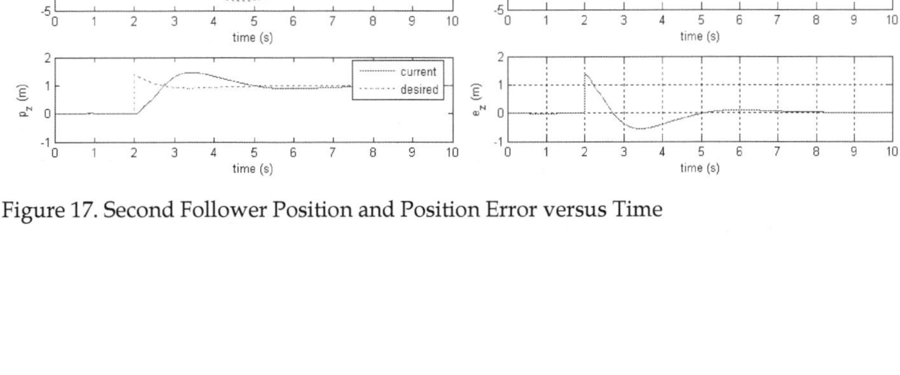

Figure 17. Second Follower Position and Position Error versus Time

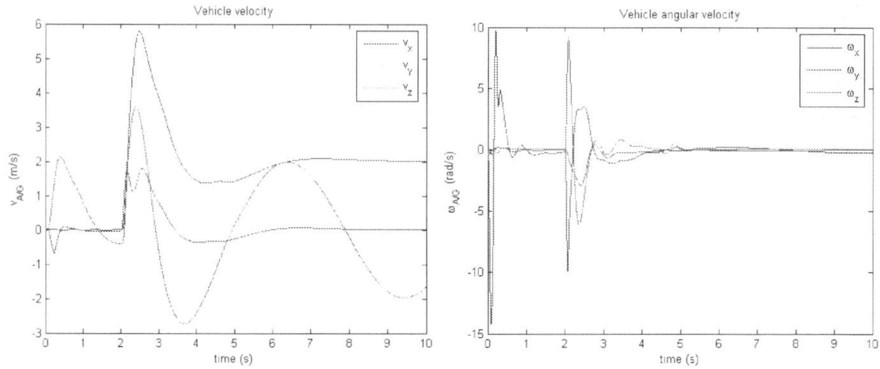

Figure 18. Second Follower Velocity and Angular Velocity versus Time

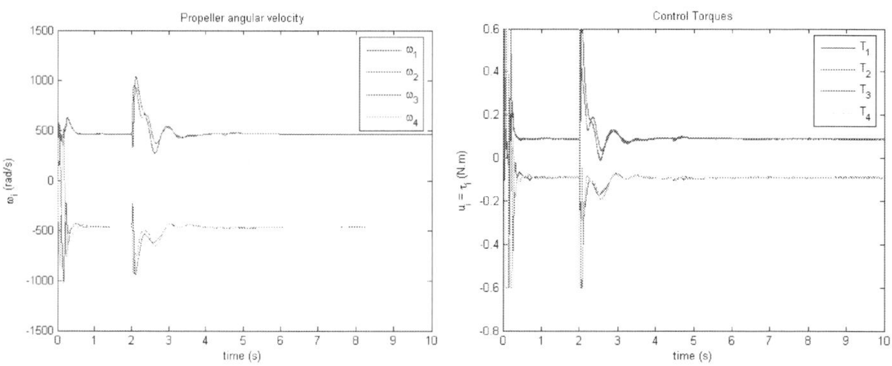

Figure 19. Second Follower Propeller Angular Velocity and Control Torques versus Time

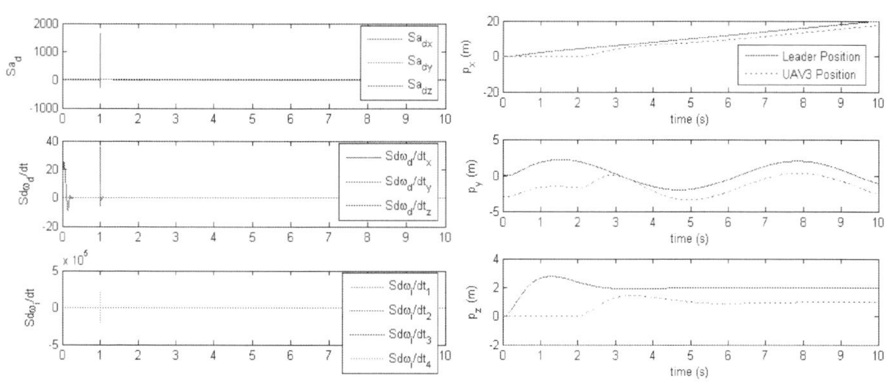

Figure 20. Second Follower Auxiliary Sliding Variables and Separation versus Time

Finally, the second follower initiates the "Anti-Collision" strategy after two seconds of delay at initial condition. The rotorcraft achieves perfect positioning after a short transient. The rotorcraft has to "catch up" with the rest of the formation and, therefore, must utilize all available resources to rapidly reach the desired position, which can be seen by the saturation of the torques . Once again, the disturbances are estimated and fully compensated for via robust HOSM.

7.4 Formation tracking

The rotorcrafts initiate and keep the desired formation on the right path after a certain transient. The absolute distance between rotorcrafts reaches exactly the desired separation. The desired triangular formation following the prescribed sinusoidal path is represented in the following figures.

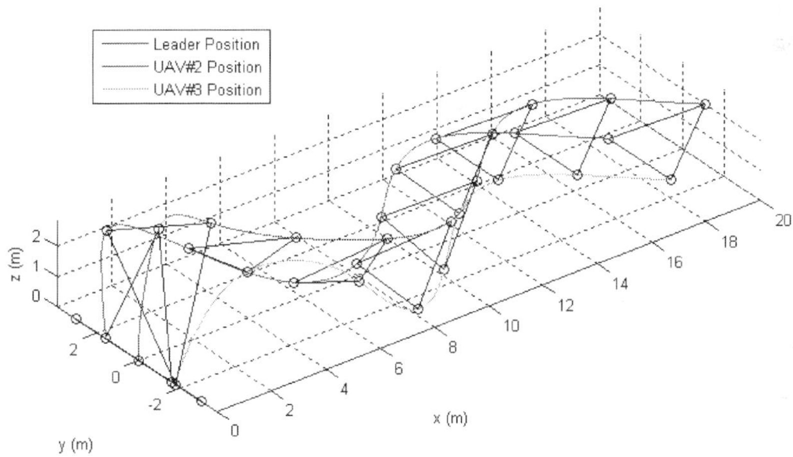

Figure 21. Three X-4 Flyer Formation Flight

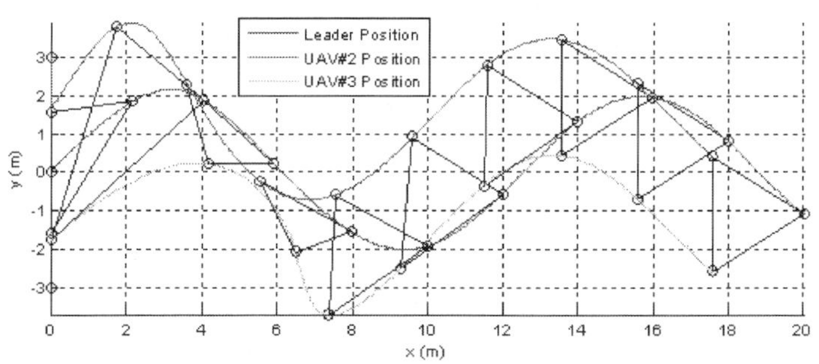

Figure 22. Three X-4 Flyer Formation Flight X-Y Projection

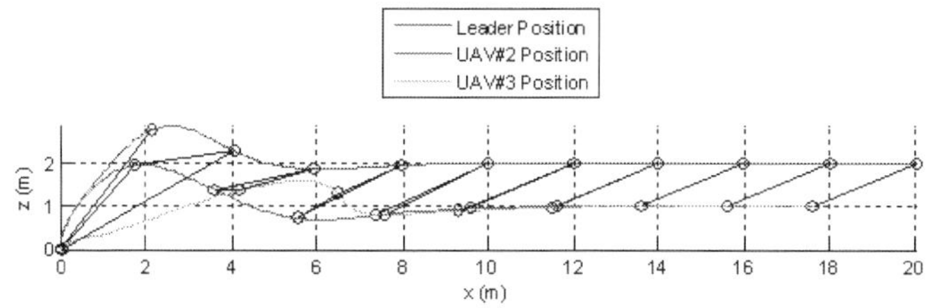

Figure 23. Three X-4 Flyer Formation Flight X-Z Projection

7.5 Actuator Malfunction

One of the propeller of the first follower is not working properly; 4 seconds into the simulation, the propeller loses efficiency and the motor torque is dropped by 50%. In order to keep the rotorcraft balanced, the propeller angular velocity must be restored as fast as possible by adjusting the applied torque. 32 shows that the controller doubles the appropriate motor torques to compensate for the loss of efficiency quickly, restoring the angular velocity. The flying formation is not affected by the actuator malfunction, thus proving the robustness of the proposed HOSM techniques.

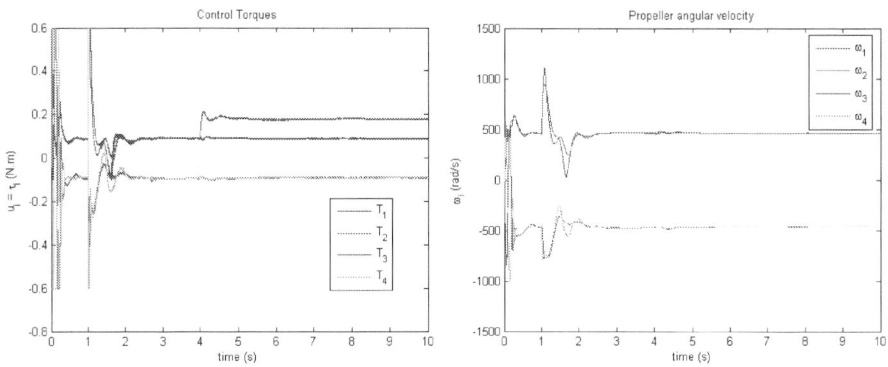

Figure 24. Compensating Torques and Restored Angular Velocity

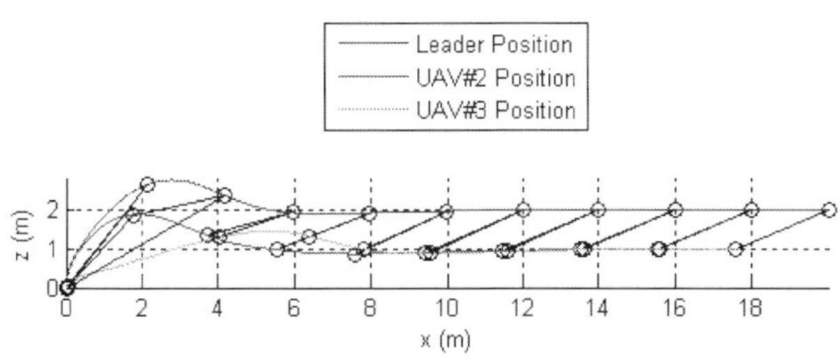

Figure 25. Formation Flying with Actuator Malfunction X-Z projection

8. Conclusion

In this chapter, formation strategies are applied to the close formation flight of three X-Shaped rotorcrafts in a full 3D environment. Robust formation tracking is achieved via sliding mode driven by sliding mode disturbance observer techniques. Simulation results show that SMDO controllers exhibit smooth and continuous controls in the presence of unknown bounded disturbances and actuator malfunction. The controls provide finite time convergence of the tracking errors to zero after a short transient corresponding to the initialization of the formation. Perfect close formation is achieved in the face of leader maneuvers and unknown disturbances, which proves the effectiveness of the proposed techniques.

Possible extensions of the research include the implementation of actual formation flight scenarios with leader reassignment and formation reconfiguration. Also, an embedded collision avoidance algorithm could be added to each aerial vehicle in the formation. Results using potential field techniques are interesting and available in the literature.

9. List of Symbols

Symbol	Description
$\tilde{x}_A, \tilde{y}_A, \tilde{z}_A$	Unit vectors respectively along X_A, Y_A and Z_A axes of frame A
$\tilde{V}^A$	Vector V expressed in frame A
$\dot{\tilde{V}}^A$	Derivative of vector V expressed in frame A
$[M^A]$	Matrix M expressed in frame A
$\tilde{F}^A_{B_1 \to B_2}$	Force vector exerted from body B_1 to body B_2 expressed in frame A
$\tilde{\tau}^A_{B_1 \to B_2}$	Torque vector exerted from body B_1 to body B_2 expressed in frame A
$sk(\tilde{V}^A)$	Skew matrix of vector V expressed in frame A (used for cross product)
(ϕ, θ, ψ)	Euler angles
(p, q, r)	Angular velocities
$[R_{G/A}]$	Rotation matrix to go from frame A to frame G
D	Diameter of propeller (m)
$\left[I^A_{frame/O}\right]$	Inertia matrix of airframe expressed in frame A at point O ($kg.m^2$)
$\left[I^A_{rotor/O_i}\right]$	Inertia matrix of rotor 'i' expressed in frame A at point O_i ($kg.m^2$)
$\tilde{v}^G_{A/G}$	Linear velocity vector of frame A relative to frame G (ground) ($m.s^{-1}$)
$\tilde{\omega}^A_{A/G} = (p, q, r)^T$	Angular velocity vector of frame A relative to frame G ($rad.s^{-1}$)
$\tilde{\omega}^A_{R_i/A}$	Angular velocity vector of rotor 'i' relative to frame A ($rad.s^{-1}$)
ω_i	Angular velocity of rotor 'i' with respect to frame A ($rad.s^{-1}$)
$\tilde{r}^A_i$	Axis of rotation (unit vector) of rotor 'i' expressed in frame A.
$\tilde{f}^A_i$	Direction of thrust produced by rotor 'i' expressed in frame A.
τ_i	Torque produced by motor 'i' on rotor 'i' ($N.m = kg.m^2.s^{-2}$)

F_i	Thrust force produced by rotor 'i' expressed in frame A (N = kg.m.s^{-2})
$\tilde{L}_i^A$	Vector $\overrightarrow{O_iO}$ between rotor and center of mass expressed in frame A (m)
$\tilde{\tau}_{F_i}^A$	Torque produced by rotor 'i' on airframe expressed in frame A (N.m)
$[C_{ADF}^A]$	Drag force coefficient matrix of airframe expressed in frame A (m^2)
$[C_{ADT}^A]$	Drag torque coefficient matrix of airframe expressed in frame A (m^5)
$[C_{RDT_i}^A]$	Drag torque coefficient matrix of rotor 'i' expressed in frame A (m^5)
$[C_{V_i}^A]$	Viscous coefficient matrix of rotor 'i' expressed in frame A (kg.m^2.s^{-1})
c_{rdf_i}	Drag torque coefficient of rotor 'i' along its axis of rotation (m^5)
c_{v_i}	Viscous friction coefficient of rotor 'i' (kg.m^2.s^{-1})
$\tilde{p}_c$, $\dot{\tilde{p}}_c$, $\ddot{\tilde{p}}_c$	Commanded trajectory position, velocity and acceleration
$\tilde{\varepsilon}_p$, $\tilde{\varepsilon}_v$, $\tilde{\varepsilon}_\gamma$	Position, velocity and acceleration error vectors respectively
k_{v1}, k_{p1}, k_{i1}	Position controller feedback coefficients
k_{v2}, k_{p2}, k_{i2}	Attitude controller feedback coefficients
k_{p3}, k_{i3}	Propeller controller feedback coefficients
$\tilde{F}_i$	Thrust force produced by each rotor expressed in frame A (N)
$\tilde{F}$	Current total thrust force expressed in frame G (N)

10. References

C. Edwards, & S. Spurgeon, Sliding Mode Control, Taylor & Francis, Bristol, PA, 1998.

M. Innocenti, L. Pollini, & F. Guilietti, Autonomous Formation Flight, *IEEE Controls System Magazine*, Vol. 20, No. 6, 2000, pp. 34-44.

A. Isidori, Nonlinear Control Systems, Springer-Verlag, London, 3rd Ed., 1995.

A. Levant, Higher-Order Sliding Modes, Differentiation and Output-Feedback Control, *International journal of Control*, 76 (9/10), pp. 924-941, 2003.

12

Autonomous Formation Flight – Design and Experiments

Yu Gu, Giampiero Campa, Brad Seanor,
Srikanth Gururajan and Marcello R. Napolitano
West Virginia University
U.S.A.

1. Introduction

Formation flight has long been performed by many species of birds for its social and aerodynamic benefits. The traditional "V" shape formation flown by birds not only helped communication between individuals, but also decreased the induced drag for each trailing bird, and thus reduced the energy required for flying (Weimerskirch et al. 2001). The benefits of formation flight have also been evaluated for manned aircraft. However, due to the high level of risk, human-piloted close formations are rarely sustained for a long enough time to fully appreciate the aerodynamic benefits. Therefore, reliable autonomous formation control can be an attractive capability for both human-piloted aircraft and Unmanned Aerial Vehicles (UAVs).

The formation control problem has been extensively discussed in recent years with numerous applications with ground mobile robots, aircraft systems, and space vehicles. In their survey paper, (Scharf et al. 2004) classified the spacecraft formation flight control algorithms into five architectures:

- *"Multiple-Input Multiple-Output, in which the formation is treated as a single multiple-input, multiple-output plan;*
- *Leader/Follower, in which individual spacecraft controllers are connected hierarchically;*
- *Virtual Structure, in which spacecraft are treated as rigid bodies embedded in an overall virtual structure;*
- *Cyclic, in which individual spacecraft controllers are connected nonhierarchically;*
- *Behavioral, in which multiple controllers for achieving different (and possibly competing) objectives are combined."*

Similar classifications can be extended to the formation control of other types of vehicles. However, due to the complexity and non-linearity associated with the aircraft dynamics, the 'leader-follower' approach was by far the most popular method for aircraft formation flight control. The advantage of the 'leader-follower' approach lies in its conceptual simplicity, where the formation flight problem is reduced to a set of tracking problems that can be analyzed and solved using standard control techniques.

In the early 1990s, a series of publications from D'Azzo and his colleagues (Dargan et al. 1992) (Buzogany et al. 1993) (Reyna et al. 1994) (Veth et al. 1995) outlined the foundation for the control of the 'leader-follower' formation flight using compensation-type controllers.

Since then, a variety of control techniques has been evaluated includeing optimal control (McCammish et al. 1996) (Dogan et al. 2005), adaptive control (Boskovic & Mehra 2003), fuzzy control (Li et al. 2005), robust control (Li et al. 2006), feedback linearization (Singh et al. 2000) (Venkataramanan & Dogan 2003), and sliding mode (Schumacher & Singh 2000). (Allen et al. 2002) performed a string stability analysis of an autonomous formation for measuring of how position errors propagate from one vehicle to another in a cascaded system. (Giulietti et al. 2000) simulated the scenario with the presence of a failure of one of the nodes, such as the loss of an aircraft.

Experimental studies for evaluating the aerodynamic effects of formation flight and for validating formation control laws have been conducted with both wind tunnel experiments (Gingras 1999) (Fowler & D'Andrea 2003) (Kutay et al. 2005) and flight-testing (Napolitano 2005) (Gu et al. 2006) (Lavretsky 2002) (Hanson et al. 2002) (How et al. 2004) (Johnson et al. 2004). Related to flight-testing efforts, the NASA Dryden Flight Research Center Autonomous Formation Flight (AFF) project in 2001 demonstrated formation control in the lateral and vertical channels with a pair of F/A-18 aircraft (Hanson et al. 2002). (How et al. 2004) at MIT performed a 2-aircraft formation flight using timing control. (Johnson et al. 2004) at Georgia Tech have been performing a series of vision-based formation flight since 2004.

This chapter presents the research effort leading to the flight demonstration of autonomous formation flight using three YF-22 research aircraft designed, built, instrumented, and tested at West Virginia University (WVU). The rest of the chapter is organized as follows. Section two presents the formation geometry and formation controller design. Section three describes the test-bed aircraft and the avionics system. The identification of the linear and nonlinear aircraft mathematical model is provided in Section four. Section five describes the formation flight simulation environment and the on-board software. Different flight-testing phases and final experimental results are presented in Section six. A brief conclusion is then provided in the final section.

2. Formation Controller Design

In the selected 'leader-follower' formation flight configuration, a Radio Control (R/C) pilot manually controls the 'leader' throughout the flight. Each 'follower' executes the formation control laws to maintain a pre-defined position and orientation with respect to the 'leader'. A main objective of the controller design is to maintain the formation geometry under maneuvered flight conditions. In addition, a minimum amount of information exchange is also desired between the 'leader' and 'follower' aircraft.

Since the trajectory dynamics generally have a much larger time constant than the attitude dynamics, the formation controller can be designed with an inner/outer-loop structure. In this configuration, the outer-loop controller minimizes the lateral, forward, and vertical distance error while the inner-loop controller performs disturbance attenuation and attitude tracking. The definition of formation geometry and designs of the inner-loop and outer-loop controllers are described next.

2.1 Formation Geometry

Consider that the flight path typically lies in a horizontal plane, the formation flight control can be simplified as two decoupled horizontal and vertical tracking problems. The general formation geometry is shown in Fig. 1. For navigation purposes, the position and velocity of both 'leader' and 'follower' aircraft are expressed with respect to a pre-defined Local Tangent Plane (LTP) and are measured by the on-board GPS receivers.

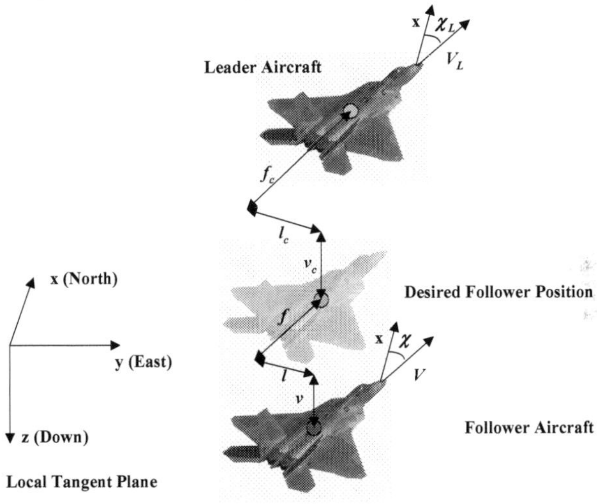

Figure 1. Formation Geometry

A. Horizontal Geometry

The horizontal formation geometric parameters include the desired forward clearance f_c and the desired lateral clearance l_c, as defined in Fig. 1. The forward distance error f and lateral distance error l can be calculated using the following relationship:

$$\begin{bmatrix} l \\ f \end{bmatrix} = \begin{bmatrix} \sin(\chi_L) & -\cos(\chi_L) \\ \cos(\chi_L) & \sin(\chi_L) \end{bmatrix} \begin{bmatrix} x_L - x \\ y_L - y \end{bmatrix} - \begin{bmatrix} l_c \\ f_c \end{bmatrix} \tag{1}$$

, where χ is the aircraft azimuth angle, x and y are the aircraft position along the x and y axis, and a subscript 'L' is used for all 'leader' parameters. A trigonometric expression for the azimuth angle is given as:

$$\sin(\chi) = \frac{V_y}{\sqrt{V_x^2 + V_y^2}} \tag{2}$$

, where V_x and V_y are the projections of the velocity along the x and y axes of the LTP. Equation (1) transforms the position error from an LTP reference frame to a reference frame oriented along the velocity of the 'leader'. In addition, the derivatives of f and l can be described as:

$$\begin{bmatrix} \dot{\ell} \\ \dot{f} \end{bmatrix} = \begin{bmatrix} V_{xy} \sin(\chi - \chi_L) \\ V_{Lxy} - V_{xy} \cos(\chi - \chi_L) \end{bmatrix} + \Omega_L \begin{bmatrix} f \\ -\ell \end{bmatrix} \tag{3}$$

, where Ω is the aircraft angular turn rate and V_{xy} is the aircraft velocity in the horizontal plane. It is obvious that the lateral and forward distance controls are coupled.

B. Vertical Geometry

The vertical distance error v can be obtained as:

$$v = z_L - z - v_c \tag{4}$$

, where v_c is the desired vertical clearance and z is the aircraft position along the z axis. The vertical geometry is also shown in Fig. 1.

2.2 Outer-Loop Controller

The outer-loop controller is designed separately for the decoupled vertical and horizontal formation geometries. The vertical controller is a simple linear altitude tracker providing the desired pitch angle θ_d to be followed by the inner-loop controller

$$\theta_d = K_v v + K_{vs}\dot{v} \tag{5}$$

, where K is the feedback gain to be selected later.

The outer-loop horizontal controller calculates the desired throttle position δ_T and the desired roll angle command ϕ_d for the inner-loop control laws to follow:

$$\begin{bmatrix} \delta_T \\ \phi_d \end{bmatrix} = f(\chi - \chi_L, \begin{bmatrix} f \\ l \end{bmatrix}) \tag{6}$$

, where *f()* is a nonlinear function to be determined next. The design of the horizontal controller is based on a Non-Linear Dynamic Inversion (NLDI) approach, which algebraically transforms the nonlinear system into a linear one (Isidori 1995) (Slotine & Li 1991) (Calise & Rysdyk 1998) so that standard linear control techniques can be applied. The use of nonlinear technique provides an effective way for controlling the aircraft under a wide range of maneuvering conditions.

In this specific problem, to minimize the forward and lateral distance errors f and l, the desired bank angle ϕ_d and throttle command δ_T are used as outer-loop control inputs. From equation (3), the second derivatives of f and l can be calculated as:

$$\begin{bmatrix} \ddot{\ell} \\ \ddot{f} \end{bmatrix} = \begin{bmatrix} \sin(\chi - \chi_L) \\ -\cos(\chi - \chi_L) \end{bmatrix} \dot{V}_{xy} + \begin{bmatrix} \cos(\chi - \chi_L) \\ \sin(\chi - \chi_L) \end{bmatrix} V_{xy}(\Omega - \Omega_L) + \begin{bmatrix} f \\ -\ell \end{bmatrix} \dot{\Omega}_L + \begin{bmatrix} \dot{f} \\ -\dot{\ell} \end{bmatrix} \Omega_L \tag{7}$$

To establish the relationship between this equation and control inputs $[\delta_T, \phi_d]$, let's first look at the aircraft forward translational acceleration equation (Stevens & Lewis 1992):

$$\begin{aligned} \dot{V} &= \frac{1}{m}(Y\sin\beta - D\cos\beta + T\cos\alpha\cos\beta) - g\sin\gamma \\ &= \left(\frac{\cos\alpha\cos\beta}{m}\right)T - \left(\frac{\bar{q}S(C_D\cos\beta - C_Y\sin\beta)}{m} + g\sin\gamma\right) \triangleq \omega_1 T - \omega_2 \end{aligned} \tag{8}$$

, where D, Y, and T are the drag, side force, and thrust,, respectively, m is the aircraft mass, g is the acceleration due to gravity, α, β, and γ are the angle of attack, angle of sideslip, and flight path angle, respectively, with $\cos\gamma = V_{xy}/V$, $\bar{q}$ is the dynamic pressure, S is the wing area, and C_D and C_Y are the dimensionless drag and side force coefficients.

The projection of $\dot{V}$ onto the level plan is given by (assuming a quasi steady state condition with $\dot{\gamma}=0$):

$$\dot{V}_{xy} = \dot{V}\cos\gamma = T\omega_1\cos\gamma - \omega_2\cos\gamma = \frac{V_{xy}}{V}\omega_1 T - \frac{V_{xy}}{V}\omega_2 = \frac{V_{xy}}{V}\omega_1(T_b + K_T\delta_T) - \frac{V_{xy}}{V}\omega_2 \tag{9}$$

, where T_b and K_T are constants to be provided by the aircraft propulsion model. Assuming a coordinated turn condition for both the 'leader' and 'follower' aircraft:

$$\Omega = \dot{\chi} \cong \dot{\psi} \cong \frac{g}{V}\tan\phi \tag{10}$$

, where ψ is the aircraft heading angle and ϕ is the roll angle. Also assuming a steady wings-level or steady turning flight condition for the 'leader':

$$\dot{\Omega}_L = 0 \tag{11}$$

equation (7) becomes

$$\begin{bmatrix} \ddot{\ell} \\ \ddot{f} \end{bmatrix} = \begin{bmatrix} V_{xy}\cos(\chi-\chi_L) & \frac{V_{xy}}{V}\omega_1\sin(\chi-\chi_L) \\ V_{xy}\sin(\chi-\chi_L) & -\frac{V_{xy}}{V}\omega_1\cos(\chi-\chi_L) \end{bmatrix} \begin{bmatrix} \frac{g}{V}\tan(\phi_d) \\ T_b + K_T\delta_T \end{bmatrix} + \frac{V_{xy}}{V}\omega_2 \begin{bmatrix} -\sin(\chi-\chi_L) \\ \cos(\chi-\chi_L) \end{bmatrix} - \Omega_L V_{xy} \begin{bmatrix} \cos(\chi-\chi_L) \\ \sin(\chi-\chi_L) \end{bmatrix} + \Omega_L \begin{bmatrix} \dot{f} \\ -\dot{\ell} \end{bmatrix} \tag{12}$$

Since the (2 x 2) matrix relating inputs and second derivatives of the output from (12) is invertible, the resulting inversed relationship is given by:

$$\begin{bmatrix} \frac{g}{V}\tan(\phi_d) \\ T_b + K_T\delta_T \end{bmatrix} = \frac{1}{V_{xy}} \begin{bmatrix} \cos(\chi-\chi_L) & \sin(\chi-\chi_L) \\ \frac{V}{\omega_1}\sin(\chi-\chi_L) & -\frac{V}{\omega_1}\cos(\chi-\chi_L) \end{bmatrix} \begin{bmatrix} \ddot{\ell}_d \\ \ddot{f}_d \end{bmatrix} + \begin{bmatrix} \Omega_L \\ \frac{\omega_2}{\omega_1} \end{bmatrix} + \begin{bmatrix} \dot{\ell}\sin(\chi-\chi_L) - \dot{f}\cos(\chi-\chi_L) \\ -\frac{V}{\omega_1}\dot{\ell}\cos(\chi-\chi_L) - \frac{V}{\omega_1}\dot{f}\sin(\chi-\chi_L) \end{bmatrix} \frac{\Omega_L}{V_{xy}} \tag{13}$$

By imposing $\alpha = \alpha_0$, $\beta = 0$, the lateral NLDI control law is:

$$\phi_d = \arctan\left\{ \frac{1}{g\cos\gamma}\left[\ddot{\ell}_d\cos(\chi-\chi_L) + \ddot{f}_d\sin(\chi-\chi_L)\right] + \frac{V}{g}\Omega_L + \left[\dot{\ell}\sin(\chi-\chi_L) - \dot{f}\cos(\chi-\chi_L)\right]\frac{\Omega_L}{g\cos\gamma} \right\} \tag{14}$$

and the forward control law is:

$$\begin{aligned}\delta_T =& \frac{m}{K_T \cos\gamma}\left[\ddot{\ell}_d \sin(\chi-\chi_L) - \ddot{f}_d \cos(\chi-\chi_L)\right] \\ &+ \frac{1}{K_T}\left(\frac{1}{2}\rho_0 V^2 S\left(C_{D0} + C_{D\alpha}\alpha_0\right) + m\sin\gamma - T_b\right) \\ &- \frac{m}{K_T \cos\gamma}\Omega_L\left[\dot{\ell}\cos(\chi-\chi_L) + \dot{f}\sin(\chi-\chi_L)\right]\end{aligned} \tag{15}$$

The application of the control inputs (14) and (15) to the system described by (12) cancels the non-linearities, leading to the linear relationship:

$$\begin{bmatrix}\ddot{\ell} \\ \ddot{f}\end{bmatrix} = \begin{bmatrix}\ddot{\ell}_d \\ \ddot{f}_d\end{bmatrix} \tag{16}$$

, which can be controlled with compensator-type linear control laws:

$$\begin{aligned}\ddot{\ell}_d &= -K_{\ell s}\dot{\ell} - K_\ell \ell \\ \ddot{f}_d &= -K_{fs}\dot{f} - K_f f\end{aligned} \tag{17}$$

2.3 Inner-Loop Controller

The objective of the inner-loop controller design is to achieve desirable disturbance attenuation and tracking capabilities while maintaining a reasonable stability margin and damping ratio. The longitudinal inner-loop control law tracks the desired pitch angle, as supplied by the outer-loop controller, using the following relationship:

$$i_H = K_q q + K_\theta(\theta - \theta_d) \tag{18}$$

, where i_H is the aircraft stabilator deflection, q is the roll rate, and θ is the roll angle.
The lateral-directional inner-loop control laws track a desired bank angle, supplied by the outer-loop controller, while augmenting the lateral-directional stability of the aircraft:

$$\delta_A = K_p p + K_\phi(\phi - \phi_d) \tag{19}$$

$$\delta_R(s) = K_r \frac{s}{s+\omega_0} r \tag{20}$$

where δ_A and δ_R are the aileron and rudder deflections, p and r are the pitch and yaw rates, ϕ is the pitch angle, and ω_0 is the washout filter constant to be selected.

3. Test-Bed Development

3.1 Aircraft Platform

A set of three YF-22 research aircraft were designed and developed for the formation flight experiments. Although these aircraft feature similarities with the Lockheed's YF-22 aircraft, they are not dynamically scaled models. Instead, the research team focused on designing aircraft that could provide desirable handling quality and payload capacity. The WVU YF-22 fleet is shown in Fig. 2. Additional information about the design and manufacturing of the WVU YF-22 research aircraft is available at (Napolitano 2005).

Figure 2. WVU YF-22 Research Aircraft (Formation Flight Fleet)

An overview of the aircraft specifications is provided in Table 1.

Wingspan	1.96 m
Length	3.05 m with probe
Height	0.61 m
Wing Area	1.37 m^2
Weight	23 Kg
Fuel Capability	3.5 L
Maximum Flight Duration	12 minutes
Cruise Airspeed	42 m/s
Takeoff Speed	30 m/s
Engine / Thrust	RAM1000 / 125 N
T/W Ratio (fully fueled)	0.55
W/S Ratio (fully fueled)	16.5 Kg/ m^2

Table 1. Specifications of the test-bed aircraft

During takeoff/landing and part of the flight, the aircraft is under manual control with a 10-channel Pulse Code Modulation (PCM) R/C system. The ground pilot has control of the aircraft primary control surfaces (stabilators, ailerons, and rudders), secondary control surfaces (flaps), engine throttle, brakes, and a 'controller switch' to activate/deactivate the on-board autonomous flight control.

The turbine propulsion system generates up to 125 N of thrust. An Electronic Control Unit (ECU) monitors the exhaust gas temperature and engine compressor pressure, and, in turn, controls the turbine RPM by regulating the fuel supplied to the turbine. The fuel consumption is rated at approximately 0.35 liter/minute for a maximum RPM (127,000) setting. Throughout the flight experiment, with exception of the takeoff phase, the throttle setting is typically within the (½ - ¾) range, with fuel consumption in the range of (0.15-0.3) liter/minute.

3.2 Avionics System

The avionics system is designed as a modulated system to meet the requirements of a wide range of research topics including formation flight control, fault-tolerant flight control, and vision-based navigation. Each 'follower' aircraft equips a complete set of avionics capable of data acquisition, communication, and flight control. It receives the 'leader' position information at a 50Hz update rate through a 900 Hz RF modem. The 'leader' avionics is a stripped down version of the 'follower' avionics with main objectives as data acquisition and communication. Fig. 3 shows the formation configuration and capabilities of the 'leader' and 'follower' avionics systems.

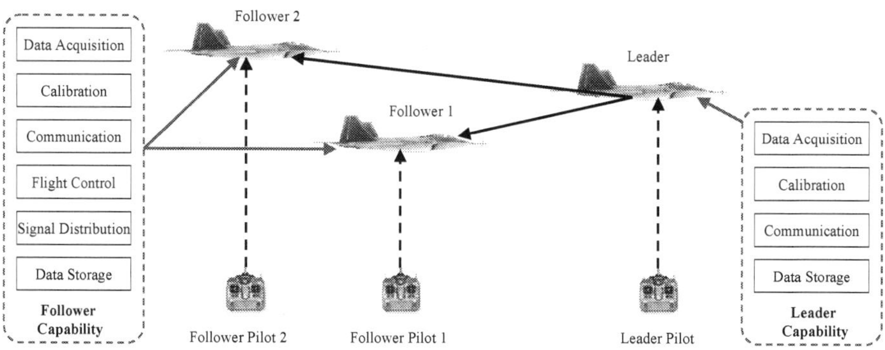

Figure 3. Formation Configuration and Capabilities

A view of the installed 'follower' avionics system is shown in Fig. 4. In general, the avionics receives pilot commands, monitors aircraft states, performs data communication, generates formation control commands, and distributes control signals to primary control surfaces and the propulsion system. A description of major avionics sub-systems is provided next.

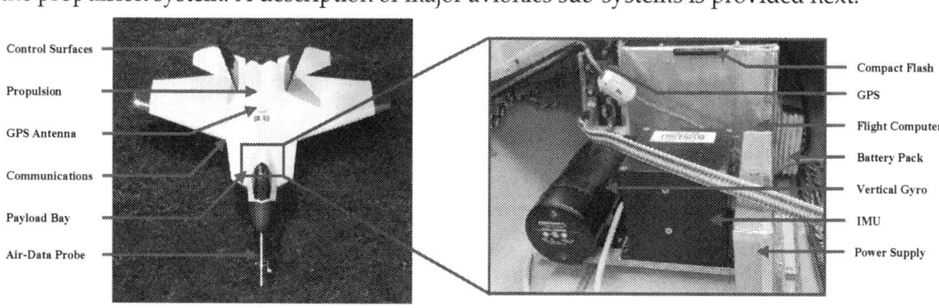

Figure 4. 'Follower' Aircraft and Avionics System

A. Flight Computer

The flight computer is based on a PC-104 format computer stack with a 300Mhz CPU module, a 32-channel 16-bit data acquisition module, a power supply/communication module, and an IDE compact flash adapter. In addition, a set of custom Printed Circuit

Board (PCB) was developed for interfacing sensor components, generating Pulse-Width Modulation (PWM) control signals, and distributing signals to each control actuator. The PC-104 format is selected because of its compact size and expandability. An 8 MB compact flash card stores the operation system, the flight control software, and the collected flight data. A 14.8v 3300mAh Li-Poly battery pack can power the avionics system for more than an hour, providing sufficient ground testing and flight mission time.

B. Sensor Suite

Flight data is collected and calibrated on-board for both real-time control and post-flight analysis. The sensor suite include a SpaceAge mini air-data probe, two SenSym pressure sensors, a Crossbow IMU400 Inertial Measurement Unit (IMU), a Goodrich VG34 vertical gyro, a Novatel OEM4 GPS receiver, a thermistor, and eight potentiometers measuring primary control surfaces deflections (stabilators, ailerons, rudders) and flow angles (α, β). A digital video camera is also installed on one of the 'followers' for flight documentation. A total of 22 analog channels are measured with a 16-bit resolution. The sampling rate was initially set at 100 Hz for data acquisition flights and later reduced to 50 Hz for matching the control command update rate (limited by the R/C system). Consider the aircraft short period mode of 7.7 rad/sec (1.2 Hz), a 50 Hz sampling rate provides a substantial amount of oversampling.

The analog signals measured on-board include absolute pressure (0-103.5 kPa), dynamic pressure (0-6.9 kPa), angle of attack (±25°), sideslip angle (±25°), air temperature (-10-70°C), roll angle (±90°), pitch angle (±60°), 3-axis accelerations (±10g), 3-axis angular rates (±200°/sec), 6-channel primary control surfaces deflections (±15°), and several avionics health indicators. A GPS receiver provides direct measurements of the aircraft 3-axis position and velocity with respect to an Earth-Centered-Earth-Fixed (ECEF) Cartesian coordinate system. These measurements are then transformed into a LTP used by the formation controller. The GPS measurement is updated at a rate of 20 Hz, providing a substantial advantage over the low-cost 1Hz GPS system.

C. Control Signal Distribution System

A Control Signal Distribution System (CSDS) is designed to give the 'follower' pilot the freedom to switch between manual and autonomous modes at any time during the flight. A block diagram for the CSDS is shown in Fig. 5.

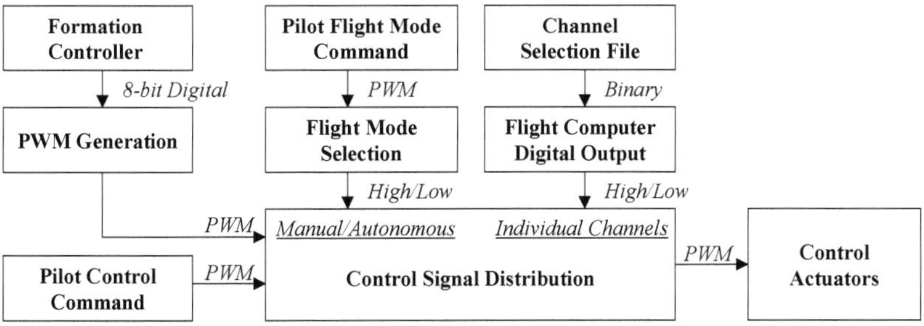

Figure 5. Control Signal Distribution System

During the autonomous mode, the flight computer can have control of all or a subset of six control channels including the left stabilator, right stabilator, left aileron, right aileron, dual rudders, and engine throttle. Two switching mechanisms are designed to ensure the safety of the aircraft - 'Hardware Switching' and 'Software Switching'. 'Hardware Switching' allows the pilot to switch back to manual control instantly under any circumstance. In the case of avionics power loss, the manual control is engaged automatically. 'Software Switching' gives the flight computer the flexibility of controlling any combination of the aircraft's primary control surfaces and propulsion with pre-programmed selections. The 'Software Switching' is implemented through a synthesis of both hardware and software modules. Specifically, the on-board software reads pre-determined channel selection information from a log file during the initialization stage of the execution. Once the 'controller switch' is activated, the software sends out the channel selection signal through the digital output port of the data acquisition card. This signal is then passed to a controller board to select the pilot/on-board control. By using this feature, individual components of the flight control system can be tested independently. This, in turn, increases the flexibility and improves the safety of the flight-testing operation.

F. Electro-Magnetic Interference

Electro-Magnetic Interference (EMI) can pose significant threats to the safety of the aircraft. This is especially true for small UAVs, where a variety of electronic components are confined within a limited space. The most vulnerable part of the avionics system is often the R/C link between the ground pilot and the aircraft, which directly affects the safety of the aircraft and ground crew. Being close in distance to several interference sources such as the CPU, vertical gyro, RF modem, and any connection cable acting as an antenna, the range of the R/C system can be severely reduced. Since prevention is known to be the best strategy against EMI, special care is incorporated into the selection of the 'commercial-off-the-shelf' products as well as the design and installation of the customized components. Specifically, low pass filters are designed for the power system; all power and signal cables are shielded and properly grounded; and aluminum enclosures are developed and sealed with copper or aluminum tape to shield the hardware components. Once the avionics system is integrated within the airframe, ferrite chokes are installed along selected cables based on the noise level measured with a spectrum analyzer. Nevertheless, although detailed lab EMI testing has been proven important, because of the unpredictable nature of the EMI issue, strict R/C ground range test procedures are followed before each takeoff to ensure the safety of the flight operation.

4. Modeling and Parameter Identification

The availability of an accurate mathematical model of the test-bed is critical for the selection of formation control parameters and the development of a high-fidelity simulation environment. The modeling process is mainly based on the empirical data collected through both ground tests and flight-testing experiments.

4.1 Identification of the Aircraft Linear Mathematical Model

The decoupled linear aircraft model is determined through a Parameter IDentification (PID) effort. A series of initial test flights are performed to collect data used for the identification process. Typical pilot-injected maneuvers, including stabilator doublets, aileron doublets, rudder doublets, and aileron/rudder doublets, are performed with various magnitudes to excite the aircraft longitudinal and lateral-directional dynamics. Fig. 7 represents a typical aileron/rudder doublets maneuver, where a rudder doublet is performed immediately after an aileron doublet.

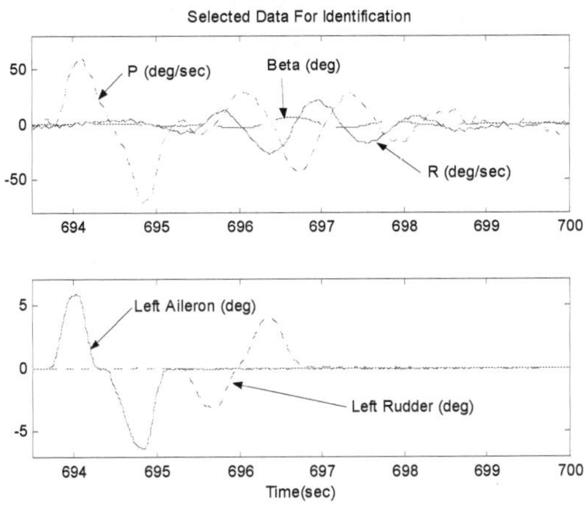

Figure 6. Flight Data for Linear Model Identification

The identification of the linear model is performed using a 3-step process. First, after a detailed examination of the flight data, two data segments with the best quality for each class of maneuvers are selected. Next, a subspace-based identification method (Ljung 1999) is used to perform the parameter identification with one set of data. Finally, the identified linear model is validated through comparing the simulated aircraft response with the remaining unused data set. This identification process is repeated until a satisfactory agreement is achieved. Following the identification study, the estimated linear longitudinal and lateral-directional aerodynamic model in continuous time are found to be:

$$\begin{bmatrix} \dot{V}_T \\ \dot{\alpha} \\ \dot{q} \\ \dot{\theta} \end{bmatrix} = \begin{bmatrix} -0.284 & -23.096 & 0 & -0.171 \\ 0 & -4.117 & 0.778 & 0 \\ 0 & -33.884 & -3.573 & 0 \\ 0 & 0 & 1 & 0 \end{bmatrix} \begin{bmatrix} V_T \\ \alpha \\ q \\ \theta \end{bmatrix} + \begin{bmatrix} 20.168 \\ 0.544 \\ -39.085 \\ 0 \end{bmatrix} i_H \tag{21}$$

$$\begin{bmatrix} \dot{\beta} \\ \dot{p} \\ \dot{r} \\ \dot{\phi} \end{bmatrix} = \begin{bmatrix} 0.430 & 0.094 & -1.030 & 0.237 \\ -67.334 & -7.949 & 5.640 & 0 \\ 20.533 & -0.655 & -1.996 & 0 \\ 0 & 1 & 0 & 0 \end{bmatrix} \begin{bmatrix} \beta \\ p \\ r \\ \phi \end{bmatrix} + \begin{bmatrix} 0.272 & -0.771 \\ -101.845 & 33.474 \\ -6.261 & -24.363 \\ 0 & 0 \end{bmatrix} \begin{bmatrix} \delta_A \\ \delta_R \end{bmatrix} \tag{22}$$

, where V_T is the true airspeed. This model represents the aircraft in a steady and level flight at V_T= 42 m/s, H= 310 m above the sea level, at trimmed condition with α = 3 deg, with inputs i_H= -1°, δ_A = δ_R =0° and a thrust force along the x body axis of the aircraft T = 54.62 N. The decoupled linear model is used later for the formation controller design.

4.2 Identification of the Non-Linear Mathematical Model

A more detailed non-linear mathematical model is identified for the development of a formation flight simulator. The identification process for a non-linear dynamic system relies on detailed knowledge of the system dynamics along with the application of minimization algorithms (Maine & Iliff 1986). In general, the non-linear model of an aircraft system can be described using the following general form (Stevens & Lewis 1992), (Roskam 1995):

$$\begin{aligned} \dot{x} &= f(x,\delta,G,F_A(x,\delta),M_A(x,\delta)); \\ y &= g(x,\delta,G,F_A(x,\delta),M_A(x,\delta)); \end{aligned} \tag{23}$$

where x is the state vector, y is the output vector, δ is the input vector, G is a vector of geometric parameters and inertia coefficients, and F_A and M_A are aerodynamic forces and moments acting on the aircraft. The functions f and g are known as analytic functions modeling the dynamics of a rigid-body system. The aerodynamic forces and moments are expressed using the aerodynamic coefficients (Roskam 1995), including drag coefficient C_D, side force coefficient C_Y, lift coefficient C_L, rolling moment coefficient C_l, pitching moment coefficient C_m, and yawing moment coefficient C_n:

$$F_A = \bar{q}S \begin{bmatrix} C_D(x,\delta) \\ C_Y(x,\delta) \\ C_L(x,\delta) \end{bmatrix}, \quad M_A = \bar{q}S \begin{bmatrix} bC_l(x,\delta) \\ \bar{c}C_m(x,\delta) \\ bC_n(x,\delta) \end{bmatrix} \tag{24}$$

The moments of inertia of the aircraft are experimentally evaluated with a 'swing pendulum' experimental set-up (Soule & Miller 1934), as shown in Fig. 7

Figure 7 Experimental Setup for Measuring Aircraft Moments of Inertia

The product of inertia Ixz could not be evaluated using the pendulum-based method. Thus, the remaining issue is to determine Ixz along with the values of the aerodynamic derivatives of the aircraft. The relationship from the coefficients of the linear models (21) and (22) to the values of the aerodynamic derivatives and geometric-inertial parameters are known (Stevens & Lewis 1992). After inverting these relationships and using the experimental values of the geometric and inertial parameters, initial values for each of the aerodynamic stability derivatives are calculated. A Sequential Quadratic Programming (SQP) technique (Hock & Schittowski 1983) is then used to iteratively minimize the Root Mean Square (RMS) of the difference between the actual and simulated aircraft outputs [Campa et al. 2007]. The resulting non-linear mathematical model is given by:

Geometric and inertial:

$\bar{c}$ = 0.76 m, b = 1.96 m, S = 1.37 m^2

I_{xx} = 1.61 Kg m^2, I_{yy} = 7.51 Kg m^2, I_{zz} = 7.18 Kg m^2, I_{xz} = -0.24 Kg m^2

M = 20.64 Kg, T = 54.62 N

Longitudinal aerodynamic derivatives:

C_{D0} = 0.0085, C_{Da} = 0.5079, C_{Dq} = 0.0000, C_{DiH} = -0.0339

C_{L0} = -0.0492, C_{La} = 3.2580, C_{Lq} = -0.0006, C_{LiH} = 0.1898

C_{m0} = 0.0226, C_{ma} = -0.4739, C_{mq} = -3.4490, C_{miH} = -0.3644

Lateral-Directional aerodynamic derivatives:

C_{Y0} = 0.0156, $C_{Y\beta}$ = 0.2725, C_{Yp} = 1.2151,

C_{Yr} = -1.1618, $C_{Y\delta A}$ = 0.1836, $C_{Y\delta R}$ = -0.4592

C_{l0} = -0.0011, $C_{l\beta}$ = -0.0380, C_{lp} = -0.2134,

C_{lr} = 0.1147, $C_{l\delta A}$ = -0.0559, $C_{l\delta R}$ = 0.0141

C_{n0} = -0.0006, $C_{n\beta}$ = 0.0361, C_{np} = -0.1513,

C_{nr} = -0.1958, $C_{n\delta A}$ = -0.0358, $C_{n\delta R}$ = -0.0555

where $\bar{c}$ is the mean aerodynamic chord, b is the wing span, S is the wing area, and m is the aircraft mass with a 60% fuel capacity.

A final validation of the non-linear model is then conducted using the validation flight data set, as it was performed for the linear mathematical model. Figure 8 shows a substantial agreement between the measured and the simulated data with the non-linear model.

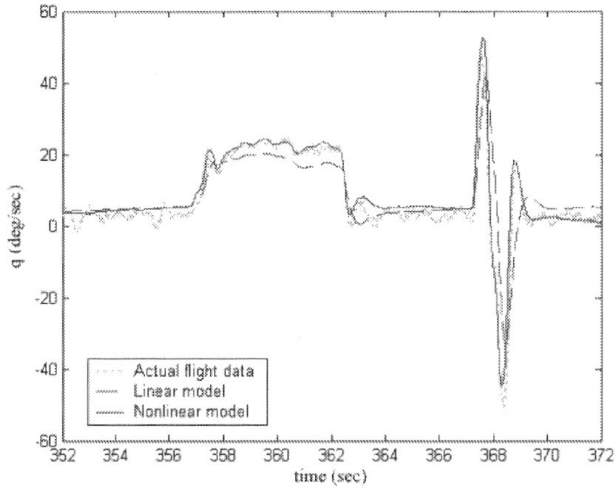

Figure 8. Linear and Non-linear Model Simulations Compared to Actual Flight Data

4.3 Engine and Actuator Models

The engine mathematical model is defined as the transfer function from the throttle command to the actual engine thrust output. The evaluation of this model is important as the jet propulsion system has a substantially lower bandwidth compared with rest of the control system. Fig. 9 provides a photo and a schematic drawing of the experimental set-up used for the identification process.

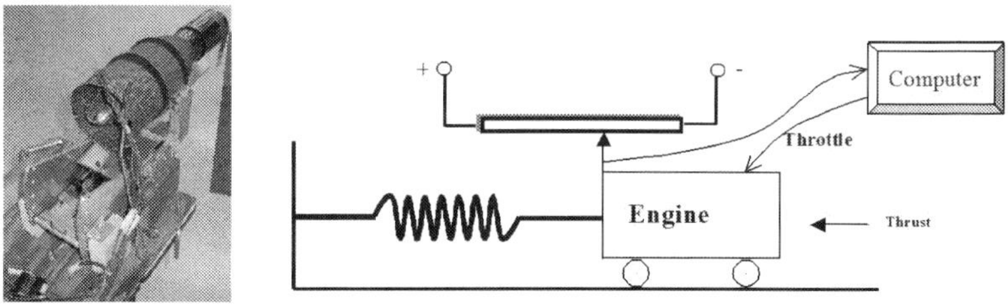

Figure 9. Engine Ground Test Setup and Schematic

The turbine is mounted on a customized engine test stand where the motion is limited to be only along the thrust force (x) direction. The thrust is then measured by reading the displacement of a linear potentiometer. The throttle control is based on 8-bit PWM signal generated by the computer with a throttle range between 0 and 255. During the test, a sequence of throttle commands is sent to the turbine and the corresponding thrust is measured with the data acquisition system, as shown in Fig. 10.

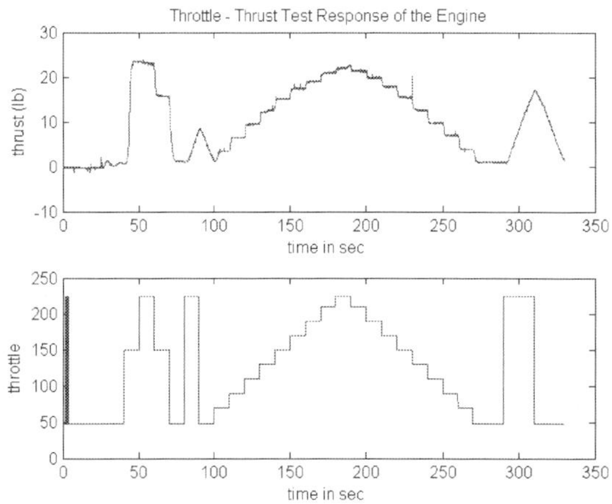

Figure 10. Throttle Thrust Response in Test Time Sequence

The first step of the engine model identification is to identify the static gain of the engine response from the throttle position to the thrust output. For simplicity purposes, a linear fitting is used. The linearized input-output relationship under steady-state condition is found to be:

$$T\,(N) = T_b + K_T \delta_T \tag{25}$$

, with $K_T = 0.624$ and $T_b = -25.86$.

To quantify the transient response of the engine dynamics, a standard prediction error method is applied to selected data segments where the throttle input consists of a series of step-like signals, as shown in Fig. 10. The identification result shows that the engine dynamic response can be approximated with a 1st order system and a pure time delay:

$$G_T(s) = \frac{T(s) - T_b}{\delta_T(s)} = \frac{K_T}{1 + \tau_T s} e^{-\tau_d s} \tag{26}$$

, with $\tau_T = 0.25\,\text{sec}$, and $\tau_d = 0.26\,\text{sec}$. As indicated by the values of τ_T and τ_d, the low bandwidth of the turbine propulsion system poses a fundamental limitation of the achievable formation flight performance under maneuvered flight conditions.

Digital R/C servos are used as actuators for the aircraft primary control surfaces. The actuator dynamics is defined as the transfer function from the 8-bit digital command to the actuator's actual position. During ground experiments, a set of step inputs is sent to the actuator. Both the control command and aircraft surface deflection are then recorded. The procedure is repeated for all six actuators on each of the primary control surfaces. From data analysis it is found that the actuator model could be approximated by the following transfer function:

$$G_{Act}(s) = \frac{1}{1 + \tau_a s} e^{-\tau_{ad} s} \tag{27}$$

where the actuator time constant τ_a and the time delay constant τ_{ad} were identified to be 0.04 sec and 0.02 sec respectively.

5. Controller Implementations and Simulation

5.1 Controller Parameters

Once a complete set of aircraft mathematical model is available, controller parameters are designed based on the classic root-locus method. The time delays in the engine model (26) and actuator model (27) are replaced by 1st order Pade approximations to facilitate the controller design. The final selections of controller parameters are listed in Table 2.

Inner-Loop Controller			Outer-Loop Controller		
Longitudinal	Lateral	Directional	Forward	Lateral	Vertical
$K_q = 0.12$	$K_p = 0.04$	$K_r = 0.16$	$K_f = 0.24$	$K_\ell = 0.20$	$K_v = 3.23$
$K_\theta = 0.50$	$K_\phi = 0.35$	$\omega_0 = 1.80$	$K_{fs} = 2.06$	$K_{\ell s} = 0.89$	$K_{vs} = 1.76$

Table 2. Formation Controller Parameters

5.2 Simulation Environment

A Simulink®-based formation flight simulation environment is developed using the mathematical model and the formation control laws described in previous sections. This environment provides a platform for validating and refining the formation control laws prior to performing the actual flight tests. The simulation schemes are interfaced with the Matlab® Virtual Reality Toolbox (VRT), where objects and events of a virtual world can be driven by signals from the simulation. The collected flight data can also be played back 'side-by-side' with the simulated aircraft response. This provides an important tool for validating the accuracy of the identified nonlinear aircraft model. In addition, the ability for VRT to visualize the entire formation flight operation, especially with the freedom of selecting different viewpoints, provides a substantial amount of intuition during the controller design and flight planning process. Figure 11 shows the formation flight simulation environment.

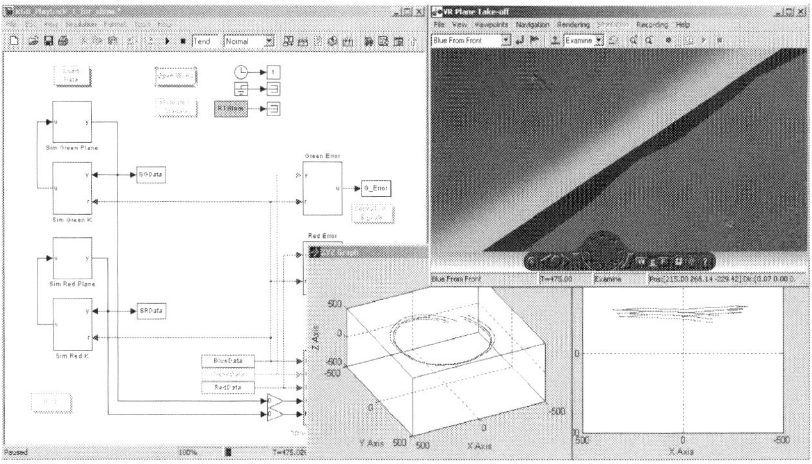

Figure 11. Formation Flight Simulation Environment

5.3 Robustness Assessment

The robustness of the formation controller is investigated with a Monte Carlo method, where a series of simulation studies is performed to evaluate the degradation of the close-loop stability and tracking performance caused by the measurement noise and modeling error. The following two formation configurations are analyzed:

Configuration #1: $$l_c = -20m, \quad f_c = 20m, \quad v_c = 20m \tag{28}$$

Configuration #2: $$l_c = 20m, \quad f_c = 20m, \quad v_c = -20m \tag{29}$$

These two configurations are later used during the flight-testing program. To simulate the effects of the measurement error/noise, a set of random noise is applied on all inputs of the formation controller. Specifically, random values following Gaussian distributions with zero means are added to the simulation parameters using the following standard deviation values:

- 2 deg/sec for angular rates (p, q, r);
- 2 deg for Euler angles (θ, ϕ);
- 4 m for horizontal position components (x,y);
- 8 m for vertical position component (z);
- 2 m/sec for horizontal velocity components (V_x, Vy);
- 4 m/s for vertical velocity component (V_z).

These values are substantially higher than the typical measurement noise observed in the actual flight data. Simulation studies reveal that the average tracking error increased by 6% and 20% for configurations #1 and #2, respectively, compared to the ideal conditions without measurement error.

An assessment of the closed-loop stability with the existence of multiplicative modeling uncertainties is also performed with a 2-step process. First, a ±10% variation is applied on each of the 30 longitudinal and lateral-directional aerodynamic derivatives, one at a time. Simulation studies show no unstable conditions for all configurations with the three most sensitive coefficients found to be $C_{D\alpha}$, $C_{L\alpha}$, and $C_{m\alpha}$. The second step is to vary the value of these three coefficients along with seven additional parameters by ±5% and perform a simulation for each possible combination. The selected parameters are C_{D0}, C_{miH}, C_{Y0}, $C_{l\beta}$, C_{lp}, $C_{l\delta A}$, $C_{n\beta}$, $C_{n\delta R}$. Therefore, a batch set of simulations (2048 total) by varying combinations of parameter changes are performed using both formation configuration #1 and #2. Again, no unstable conditions are observed in this analysis. The worst-case degradation of tracking performance is found to be 1.98 m for the lateral distance error, 1.05 m for the forward distance error, and 4.41 m for the vertical distance error. Overall, the simulation result indicates that the designed formation controller has adequate robustness characteristics with respect to modeling errors.

5.4 On-Board Software

The formation controller software module, once validated through simulation studies, is integrated with other software components to perform real-time data acquisition, communication, and control. The on-board software is implemented as a Simulink scheme with each component written in C-language as a Matlab 'S-function'. An executable file is compiled using Matlab Real-Time Workshop (RTW) as a real-time extended DOS target for flight test experiments. The modulated software design provides flexibility for quick on-site reconfigurations to meet various flight-testing objectives.

The main tasks for 'leader' on-board software is to perform data acquisition, communication, and data storage. The 'follower' software also executes the formation control laws, selects the operational mode of the aircraft, decides primary control channels to be controlled on-board, and calibrates the flight control commands. Figure 12 shows a sample of the 'follower' on-board software scheme.

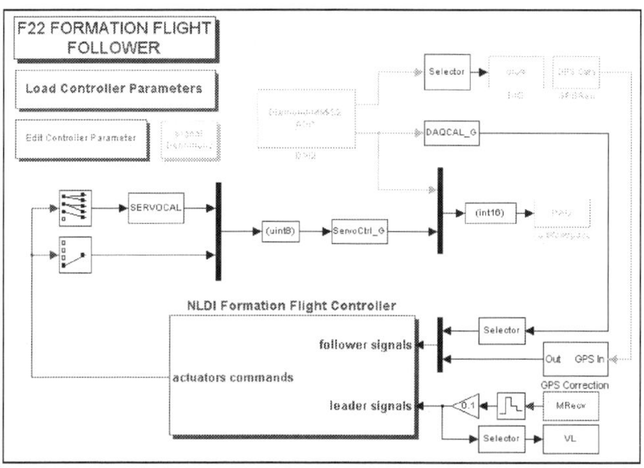

Figure12. Simulink Diagram of the 'Follower' Aircraft Software

6. Flight-Testing Of Formation Control Laws

Flight-testing is the most realistic step for controller validating and by far the most risky one. Despite the fact that the use of UAV can greatly reduce the risk associated with control system validation, careful planning is still of paramount importance for ensuring the safe operation and help identifying potential problems. This section provides an overview of different flight-testing phases and the outcomes of the autonomous formation flight experiments.

6.1 Flight Testing Phases

The flight-testing program is divided into six major phases with increasing complexity and risk level:

A. Flight for Assessment of Handling Qualities

This initial phase is for evaluating the handling qualities and dynamic characteristics of the test-bed aircraft. After a few satisfactory test flights, 'artificial' payloads of incremental weight are installed to test the structural integrity and the handling qualities under a full payload configuration.

B. Data Acquisition Flights

The avionics system is installed and flight data is collected for the PID analysis. A set of dedicated PID maneuvers is performed throughout multiple flights to excite the aircraft dynamics. Typical PID maneuvers include stablator doublets, aileron-rudder doublets, and a range of engine throttle inputs.

C. Inner-Loop Controller Validation

The stability and tracking performance of the designed inner-loop linear controller is validated during this phase. Both the longitudinal and lateral-directional inner-loop control laws are tested. The flight control hardware is also validated during this phase.

D. Outer-Loop Controller Sub-System Validation

Individual sub-systems of the outer-loop controller are tested. Experiments are performed to test the altitude-hold, heading-hold, and velocity-hold control and their combinations. Sample flight data in Fig. 13 shows the result from a heading-hold control experiment.

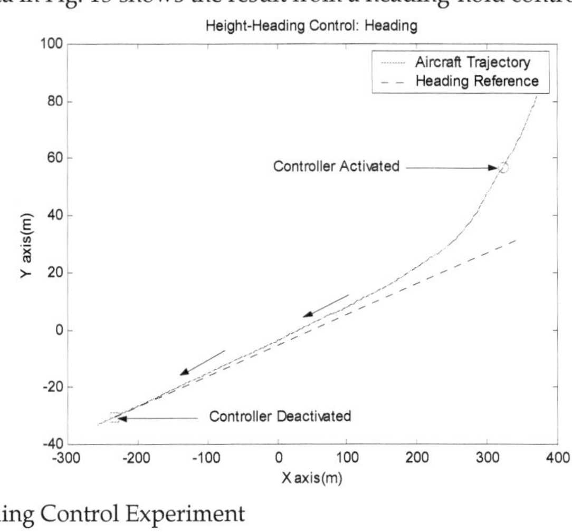

Figure 13. Heading Control Experiment

E. 'Virtual Leader' Flights

A 'Virtual Leader' (VL) approach is implemented as an alternative method for testing the formation controller without the risk and logistic issues associated with a full-blown multiple-aircraft experiment. The VL experiment consists of a single aircraft tracking a previously recorded flight trajectory. This trajectory is initially stored on-board the 'follower' aircraft and later moved to a ground station to test the performance of the communication link. The VL flights are proven to be invaluable for the validation and fine-tuning of formation control laws. A total of 12 VL flights are performed using various formation parameters. Fig. 14 is a sample flight data demonstrating the ability for the formation controller to reduce a large initial error and maintain the formation flight.

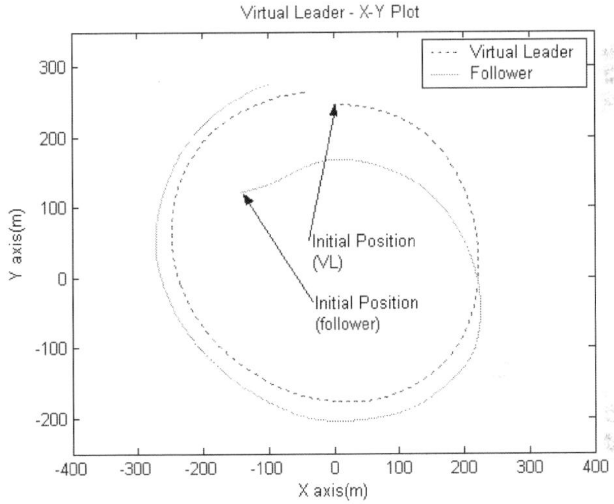

Figure 14. 'Virtual leader' Test - X-Y Plane

F. Multiple Aircraft Formation Flights

After various formation geometries and initial conditions are explored with the VL experiments, the flight-testing program proceeds to the multiple aircraft testing. A total of four 2-aircraft formation flight experiments are performed along with one 3-aircraft formation demonstration.

6.2 Three-Aircraft Formation Flight Experiment

The procedure for the 3-aircraft formation experiment is the following. The 'blue' aircraft, acting as the 'leader', takes off first while the 'red' aircraft ('follower #1') takes off approximately 35 seconds later. After the 'red' aircraft reaches a pre-defined 'rendezvous' area behind the 'leader', the ground pilot engages the on-board formation control. Once the 2-aircraft formation is stabilized for approximately 50 seconds, the 'green' aircraft ('follower #2') takes off and approaches a 'rendezvous' area behind the 2-aircraft already in formation. After the 'green' pilot engages the autonomous control, the trajectory of the 3-aircraft formation is solely controlled by the 'leader' R/C pilot. Fig. 15 shows a ground photo of the 3-aircraft formation experiment.

Figure 15. 3-Aircraft Formation Flight Test

The pre-selected formation geometries include configuration #1 (Equation 28) for the 'red' aircraft and configuration #2 (Equation 29) for the 'green' aircraft. Fig. 16 represents a 40-second portion of flight trajectory during the formation flight. The 3-aircraft formation configuration is engaged for approximately 275 seconds. Fig. 17 shows the aircraft altitude during the formation flight.

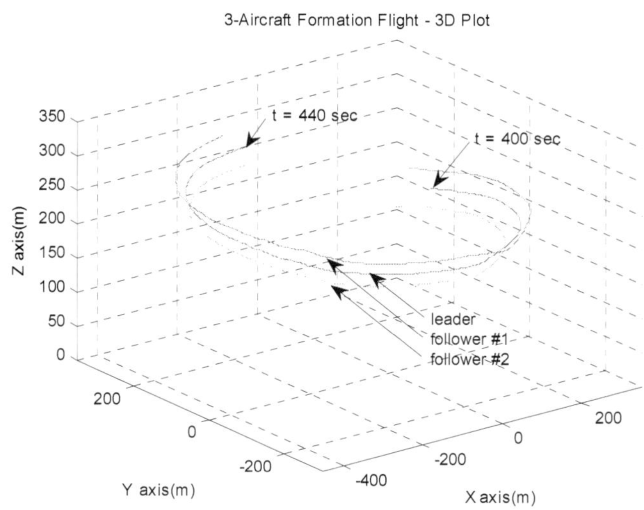

Figure 16. 3- Aircraft Formation Test - 3D Trajectory
(Blue='leader', Red=Outside 'follower', Green=Inside 'follower')

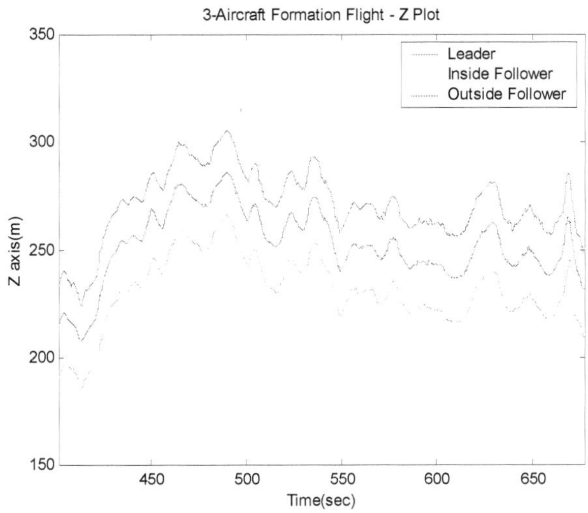

Figure 17. 3-Aircraft Formation Test - Altitude

The mean and standard deviation of the steady state tracking error for the flight test are shown in Table 3. The simulation results calculated with the same 'leader' trajectory are also supplied for comparison purposes.

3-aircraft formation Experiment		f_c (m)	l_c (m)	v_c (m)	Forward Distance Error (m)		Lateral Distance Error (m)		Vertical Distance Error (m)	
					Mean	Std. Dev	Mean	Std. Dev	Mean	Std. Dev
Green Aircraft	Flight Data	20	-20	20	-2.49	2.46	13.24	3.45	1.15	1.04
	Simulation	20	-20	20	-3.59	2.50	14.31	3.30	1.34	0.71
Red Aircraft	Flight Data	20	20	-20	27.28	3.73	-2.59	2.29	1.15	0.95
	Simulation	20	20	-20	25.30	3.82	-0.46	1.98	1.19	0.66

Table 3. 3-Aircraft Formation Test – Error Analysis

The 3-aircraft formation experiment validates the overall design of the formation controller, test-bed aircraft, and on-board avionics system. The statistical analysis shows that the 'outside' aircraft - 'follower #2' - achieves desirable lateral tracking performance but with a larger forward tracking error. On the contrary, the 'inside' aircraft shows desirable forward tracking and a slightly degraded lateral tracking performance. Both 'follower' aircraft exhibits excellent tracking performance for the vertical channel. Overall, the standard deviation for all of the tracking errors are found to be relatively small, with a maximum value of 3.73 m, showing a smooth trajectory following performance. In addition, a substantial agreement between the simulation result and actual flight data is noticed, indicating an accurate nonlinear mathematical model of the aircraft.

7. Conclusion

This chapter summarizes the results of an effort towards demonstrating closed-loop formation flight using research UAVs. A 'leader-follower' strategy is followed during the formation controller design. A two-time-scale approach is used with a nonlinear outer-loop and linear inner-loop controller. The flight-testing program was conducted over three flight seasons (2002 through 2004) with approximately 100 flight sessions. The incremental flight-testing phases validates the overall design of the formation control laws and the performance of the test-bed aircraft and avionics systems. The application of a 'virtual leader' technique proves to be an invaluable and safe approach for an initial testing of the formation control laws. During the final flight sessions, a total of five formation flight experiments are successfully performed, including four 2-aircraft formations and one 3-aircraft formation. Flight data confirms satisfactory performance for the designed 'leader-follower' type formation control laws.

8. References

Weimerskirch, H.; Martin, J.; Clerquin, Y.; Alexandre, P. & Jiraskova, S. (2001). Energy savings in flight formation. *Nature*. Vol. 413, pp. 697-698.

Scharf, D. P.; Hadaegh, F. Y. & Ploen, S. R. (2004). A survey of spacecraft formation flying guidance and control (part II): control. *American Control Conference*, Boston, MA, June 2004, pp. 2976–2985.

Dargan, J.L.; Patcher, M. & D'Azzo, J.J. (1992). Automatic formation flight control. *Proc. AIAA Guidance, Navigation and Control Conf.*. Hilton Head, SC, Aug. 1992, pp. 838-857.

Buzogany, L.E.; Patcher, M. & D'Azzo, J.J. (1993). Automated control of aircraft in formation flight. *Proc. AIAA Guidance, Navigation and Control Conf.*. Monterey, CA, Aug. 1993, pp. 1349-1370.

Veth, M.; Pachter, M. & D'Azzo, J.J. (1995). Energy preserving formation flight control. AIAA-1995-335, *Aerospace Sciences Meeting and Exhibit*, 33rd, Reno, NV, Jan 9-12, 1995

Reyna, V.P.; Pachter, M. & D'Azzo, J.J. (1994). Formation Flight. Control Automation. *Proc. AIAA Guidance, Navigation, and Control Conference*, pp 1379-1404,

McCammish, S.; Pachter, M.; D'Azzo, J. J. & Reyna, V. (1996) Optimal Formation Flight Control. *AIAA Guidance, Navigation, and Control Conference*, San Diego, CA, Jul. 1996.

Dogan, A.; Sato, S.; & Blake W. (2005). Flight Control and Simulation for Aerial Refueling. AIAA paper 2005-6264, *Proc. AIAA Guidance, Navigation, and Control Conference*, San Francisco, CA, Aug 15-18, 2005

Boskovic, J.D. & Mehra, R.K. (2003). An adaptive reconfigurable formation flight control design, *Proc. American Control Conference*. June 2003 Vol 1, pp. 284-289

Li, Y.; Li, B.; Sun, Z. & Song, Y.D. (2005). Fuzzy technique based close formation flight control. *Industrial Electronics Society*, 2005. IECON 2005. *32nd Annual Conference of IEEE*, Nov. 2005

Li, B.; Liao, X.H.; Sun, Z.; Li, Y.H. & Song, Y.D. (2006). Robust Autopilot for Close Formation Flight of Multi-UAVs, System Theory. *Proceeding of the Thrity-Eighth Southeastern Symposium*, March 2006, pp. 258- 262

Singh, S.N.; Chandler, P.; Schumacher, C.; Banda, S. & Pachter M. (2000). Adaptive feedback linearizing nonlinear close formation control of UAVs, *Proc. American Control Conference*, 2000. vol.2, pp.854-858

Venkataramanan, S. & Dogan, A. (2003). Nonlinear Control for Reconfiguration of UAV Formation. AIAA paper 2003-5725, *Proc. AIAA Guidance, Navigation, and Control Conference*, Austin, TX, Aug. 2003

Schumacher, C.J. & Singh, S.N. (2000). Nonlinear Control of Multiple UAVs in Close-Coupled Formation Flight. *AIAA Paper* 2000-4373, Aug. 2000

Allen, M.J.; Ryan, J.; Hanson, C.E. & Parle, J.F. (2002). String Stability of a Linear Formation Flight Control System. *AIAA Paper* 2002-4756, Aug. 2002

Giulietti, F.; Pollini, L. & Innocenti, M. (2000). Autonomous formation flight. *IEEE Control Systems Magazine*, Vol. 20, No. 6, pp. 34-44, Dec. 2000

Gingras, R.D. (1999). Experimental Investigation of a Multi-Aircraft Formation, *AIAA. Paper*-99-3143, 1999

Fowler, J. M. & D'Andrea, R. (2003). A formation flight experiment: Constructing a test-bed for research in control of interconnected systems. *Control Systems Magazine*, 23(5) pp.35-43, 2003

Kutay, A.T.; Fowler, J.M.; Calise, A.J. & D'Andrea, R. (2005). Distributed Adaptive Output Feedback Control Design and Application to a Formation Flight Experiment. *AIAA Guidance, Navigation and Control Conference*, August 2005

Napolitano, M.R. (2005). Development of formation flight control algorithms using 3 YF-22 flying models. *AFOSR Report* A994434, Apr. 2005, Available: (http://www.stormingmedia.us/99/9944/A994434.html)

Gu, Y.; Seanor, B.; Campa, G.; Napolitano, M.R.; Rowe, L.; Gururajan, S.; Perhinschi, M.G. & Wan, S. (2006). Design and Flight Testing Evaluation of Formation Control Laws. *IEEE Transactions on Control Systems Technology*, November 2006

Lavretsky, E. (2002). F/A-18 autonomous formation flight control system design. AIAA Guidance, Navigation and Control Conference, *AIAA Paper* 2002-4757, Monterey, CA, Aug. 2002.

Hanson, C.E.; Ryan, J.; Allen, M.J. & Jacobson, S.R. (2002). An Overview of Flight Test Results for a Formation Flight Autopilot. *AIAA Paper* 2002-4755, Aug. 2002

How, J. P.; King, E. & Kuwata, Y. (2004). Flight Demonstrations of Cooperative Control for UAV Teams. *Proc. AIAA 3rd Unmanned Unlimited Technical Conference, Workshop and Exhibit*, Sept. 2004. AIAA-2004-6490

Johnson, E.N.; Calise, A.J.; Sattigeri, R. & Watanabe, Y. (2004) Approaches to Vision-Based Formation Control. *Proc. IEEE Conference on Decision and Control*, December 2004.

Isidori, A. (1995). *Nonlinear Control Systems*, Springer-Verlag, London, Third edition, 1995.

Slotine, J.-J. E. & Li, W. (1991). Applied Nonlinear Control, Prentice Hall, New Jersey, 1991.

Calise, A.J. & Rysdyk, R.T. (1998). Nonlinear adaptive flight control using neural networks. *IEEE Control Systems Magazine*, Vol. 18, No. 6, pp. 14-25, Dec. 1998

Stevens, B. & Lewis, F. (1992). *Aircraft Control and Simulation*, John Wiley & Sons, NY, 1992.

Ljung, L. (1999). *System Identification: Theory for the User*, 2nd Ed., PTR Prentice Hall, Upper Saddle River, Englewood Cliffs, NJ, 1999.

Maine, R.E. & Iliff, K.W. (1986). Identification of dynamic systems: theory and formulation. *NASA RF 1168*, June 1986.

Roskam, J. (1995) *Airplane Flight Dynamics and Automatic Flight controls*, DARcorporation, KS, 1995.

Soule, H.A. & Miller, M. P. (1934). The experimental determination of the moments of inertia of airplanes. *NACA Report 467*, 1934. Available: (http://naca.larc.nasa.gov/reports/1934/).

Hock, W. & Schittowski, K. (1983). A Comparative performance evaluation of 27 nonlinear programming codes. *Computing*, Vol. 30, p. 335, 1983.

Campa, G.; Gu, Y.; Seanor, B.; Napolitano, M. R.; Pollini, L. & Fravolini, M. L. (2007) Design And Flight Testing Of Nonlinear Formation Control Laws. *Control Engineering Practice*, pp 1077-1092, Vol. 15, Issue 9, September 2007.

13

Vibration-induced PM Noise in Oscillators and its Suppression

Archita Hati, Craig Nelson and David Howe
National Institute of Standards and Technology
USA

1. Introduction

High-precision oscillators have significant applications in modern communication and navigation systems, radars, and sensors mounted in unmanned aerial vehicles, helicopters, missiles, and other dynamic platforms. These systems must provide their required performance even when subject to mild to severe dynamic environmental conditions. Oscillators often can provide sufficiently low intrinsic phase modulation (PM) noise to satisfy particular system requirements when in a static environment. However, these oscillators are sensitive to acceleration that can be in the form of steady acceleration, vibration, shock, or acoustic pickup. In most applications the acceleration experienced by an oscillator is in the form of vibration, which can introduce mechanical deformations that deteriorate the oscillator's otherwise low PM noise (Filler, 1988; Vig et al., 1992; Howe et al., 2005). This degrades the performance of the entire electronic system that depends on this oscillator's low phase noise. For example, when radars and sensors mounted on helicopters are subjected to severe low- and medium-frequency vibration environments, the vibration noise induced into the system's reference oscillator translates to blurring of targets and possibly false detection.

This sensitivity to vibration originates most commonly from phase fluctuations within the oscillator's positive-feedback loop, due usually to the physical deformations in the frequency determining element, the resonator. Factors that lead to high acceleration sensitivity of the resonator include nonlinear or sensitive mechanical coupling effects and lack of mechanical symmetry that serve to cancel frequency changes in the resonator. Vibration also causes mechanical deformations in non-frequency-determining electronic components that then cause phase fluctuations (Steinberg, 2000). Because these fluctuations are inside the oscillator feedback loop and are integrated according to Leeson's model (Leeson, 1966), they can become large at Fourier, or offset, frequencies close to carrier frequency. An oscillator's sensitivity to vibration is characterized traditionally by acceleration sensitivity, which is the normalized frequency change per g (1 g is the acceleration of gravity near the earth's surface, approximately 9.8 m/s^2). Typically, frequency shifts in oscillators are on the order of 10^{-8} to 10^{-10} per g, primarily because of the physical deformations.

Vibration-induced noise can be suppressed by physical means and further by electronic means if a suitably low-cost way of measuring and correcting the vibration-induced noise from an oscillator is implemented. Passive mechanical isolation systems consist of elastic and damping materials that translate vibration energy to different frequencies where they are less troublesome and/or damped (Renoult et al., 1989). Active mechanical systems use accelerometers and mechanical actuators to measure and cancel motion induced by the vibration. Hybrid active-passive systems allow higher degrees of vibration isolation to be achieved, but such systems are not easily miniaturized, are somewhat complex, and are power-consuming (Weglein, 1989). In principle, atom-based frequency-determining elements such as those used in atomic frequency standards have extremely low acceleration sensitivity (Thieme et al., 2004). However, the state-selection, RF interrogation, and detection electronics are more complex than in oscillators, and the corresponding large volume of atomic standards make them equally vulnerable to mechanical deformation under vibration. Some method of suppressing induced frequency shifts is often required to even approach 10^{-10} per g (Kwon & Hahn, 1983). More compact atomic standards allow for simpler mechanical vibration isolation to be incorporated (Riley, 1992).

Strategies for electronically reducing acceleration sensitivity have traditionally relied on accurately detecting this vibration with sensors (Healy et al., 1983) and even using the resonator itself as a vibration sensor (Watts et al., 1988). Suppression at one vibration frequency along one axis in quartz oscillators by electronic means has been explored with success (Rosati & Filler, 1981). More recently, significant advances have been made in which this electronic vibration suppression is effective over a wide range of vibration frequencies from a few hertz to 200 Hz. This is accomplished by fabricating high-Q quartz resonators in which the "cross" g-sensitivities of the three orthogonal axes are decoupled to a high degree (Bloch et al., 2006).

This chapter is intended to introduce the subject of vibration-induced PM noise by discussing the method of characterizing acceleration sensitivity and reporting such characterization on a sample of devices operating at microwave frequencies. Schemes for reducing vibration-induced noise are also discussed.

2. Defining Acceleration Sensitivity

If the vibration frequency from mechanical shock or other external processes is f_v, the vibration-induced phase fluctuations cause the carrier frequency to deviates from its nominal frequency, f_0, by an amount $\pm\ \Delta f$, at a rate of f_v. Spurious sidebands, a highly undesirable type of noise in many applications, will appear at $f_0 \pm f_v$. The red curve in Figure 1 shows the PM noise of one test oscillator that is subjected to 100 Hz vibration along one axis. Note that the intrinsic random electronic noise is degraded by additional noise due to this vibration (shown as the noise pedestal on both sides of an ideal carrier signal). Also, the blue curve indicates that as the vibration increases, so do the sidebands, eventually exceeding the carrier power.

Low acceleration sensitivity at one frequency such as 100 Hz does not necessarily mean that phase noise due to acoustic and structure-borne vibration is suppressed. While vibration-induced noise modulation on an oscillator may be proportional to overall acceleration sensitivity, the proportionality as a function of f_v can be complicated in the range of audio frequencies of concern here (from a few Hertz to 2 kHz). Resonator deformations that affect its center frequency depend on designs of mounting, elastic properties of materials, acoustic resonances, sound and vibration isolation, orientation, etc. Therefore, suppression of only "dc or time-independent" acceleration sensitivity due to what is commonly called 2 g-tipover (Vig et al., 1992) or steady acceleration has limitations and is insufficient to solve the larger problem of "ac or time dependent" acceleration sensitivity due to vibration. Acceleration sensitivity and vibration sensitivity are often used interchangeably for time-dependent accelerations. The acceleration sensitivity is characterized more fully as a function of f_v, as discussed next.

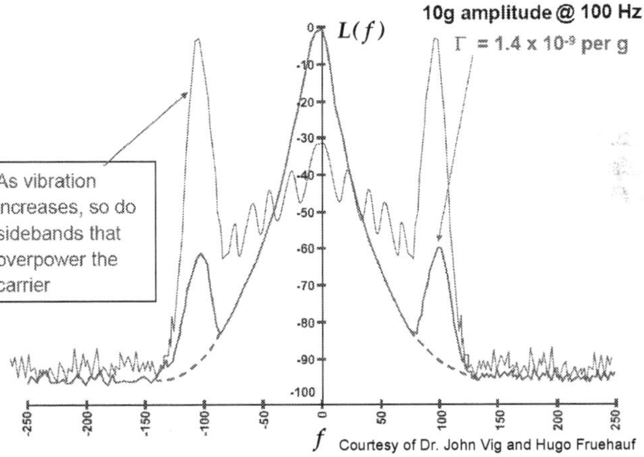

Figure 1. Phase noise of an oscillator that is subjected to vibration at f_v = 100 Hz. f is the offset frequency from the carrier

Acceleration sensitivity of an oscillator is explained in detail by Filler [Filler, 1988]. When an oscillator is subjected to acceleration, its resonant frequency shifts. The peak frequency shift Δf_{peak}, which is proportional to magnitude of the acceleration and dependent on the direction of acceleration, is given by a peak fractional-frequency change y_{peak} as

$$y_{peak} = \frac{\Delta f_{peak}}{f_0} = \vec{\Gamma} \bullet \vec{a}, \qquad (1)$$

where f_0 is the frequency of the oscillator with no acceleration, $\vec{\Gamma}$ is the acceleration sensitivity vector and $\vec{a}$ is the peak applied acceleration vector. The magnitude of acceleration is expressed in units of g. When the direction of applied acceleration is parallel to the axis of acceleration sensitivity vector, it will have the greatest effect on Δf_{peak}.

By definition, $S_y(f)$ is the power spectral density of root-mean-square (rms) fractional-frequency change, y_{rms} (Sullivan et al., 1990), and is given by

$$S_y(f) = \frac{|y_{rms}|^2}{BW} = \left|\frac{\Delta f_{rms}}{f_0}\right|^2 \frac{1}{BW} = \left|\frac{\Delta f_{peak}}{\sqrt{2} f_0}\right|^2 \frac{1}{BW} \quad [1/\text{Hz}], \qquad (2)$$

where f is the offset, or Fourier, frequency away from the carrier and BW is the bandwidth of the spectral-density measurement. Also, $S_y(f)$ is related to power spectral density of phase fluctuations as

$$S_{\phi}(f) = S_y(f)\left(\frac{f_0}{f}\right)^2 \text{.....[rad}^2\text{/Hz]}, \tag{3}$$

and the single sideband PM noise $L(f)$ is defined as

$$L(f) \equiv \frac{1}{2} S_{\phi}(f) \quad \text{[dBc/Hz]}, \tag{4}$$

where dBc/Hz is dB below the carrier in a 1 Hz bandwidth. Substituting $f = f_v$ the vibration frequency, and normalizing to a 1 Hz bandwidth, $L(f_v)$ can be related to acceleration sensitivity for a small modulation index as

$$\begin{aligned} L(f_v) &= \frac{1}{2}\left(\frac{\Delta f_{peak}}{\sqrt{2} f_0}\right)^2 \left(\frac{f_0}{f_v}\right)^2 \\ &= \left(\frac{\vec{\Gamma} \bullet \vec{a}}{2 f_v} f_0\right)^2. \end{aligned} \tag{5}$$

Or, when expressed in dB,

$$L(f_v) = 20 \log\left(\frac{\vec{\Gamma} \bullet \vec{a}}{2 f_v} f_0\right). \tag{6}$$

Equation 6 may be rearranged to obtain

$$\Gamma_i = \frac{2 f_v}{a_i f_0} 10^{\left(\frac{L(f_v)}{20}\right)}, \tag{7}$$

where Γ_i is the component of acceleration sensitivity vector in the i ($i = x, y$ and z) direction. For a sinusoidal vibration, $|\vec{a}|$ is the peak applied vibration level in units of g, and $L(f_v)$ is expressed in units of dBc. In most cases, vibration experienced by an oscillator is random instead of sinusoidal. Under random vibration the power is randomly distributed over a range of frequencies, phases, and amplitudes, and the acceleration is represented by its power spectral density (PSD). For random vibration, $|\vec{a}| = \sqrt{2PSD}$, and its unit is g/√Hz. Also, for a random vibration, $L(f_v)$ is expressed in units of dBc/Hz. The sum of acceleration sensitivity squared in all three axes gives the total acceleration sensitivity, or gamma ($\vec{\Gamma}$), and its magnitude is defined as

$$\left|\vec{\Gamma}\right| = \sqrt{\Gamma_x^2 + \Gamma_y^2 + \Gamma_z^2}\,. \tag{8}$$

$\vec{\Gamma}$ of an oscillator can be calculated from equation 8 once the PM noise of the oscillator is measured for all three axes. Noteworthy to this discussion, the sidebands generated by oscillators under vibration are a more serious issue, as the signal frequency increases due either to frequency multiplication or direct frequency generation at higher frequency. Systems are in place that require ultralow PM noise from reference oscillators operating in the range of 6 to 18 GHz. The vibration-induced PM noise of an oscillator with frequency f_0 upon frequency multiplication by a factor of N is given by

$$L(f_v) = 20\log\left[\frac{\vec{\Gamma}\bullet\vec{a}}{2f_v}(Nf_0)\right]. \tag{9}$$

The vibration frequency f_v is unaffected because it is an external influence. It is clear from equation 9 that, given a nominal $\left|\vec{\Gamma}\right| \sim 1 \times 10^{-9}/\text{g}$, the level of vibration sidebands in phase-noise plots of $L(f)$ can become excessively large at X-band and higher ranges. For example, a 10 MHz oscillator with a vibration sensitivity of $1 \times 10^{-9}/\text{g}$ when experiencing an acceleration of 5 g produces a sideband level of -72 dBc at f_v = 100Hz. For the same $\left|\vec{\Gamma}\right|$ and under identical vibration conditions, a 10 GHz oscillator will produce a sideband of -12 dBc, a factor of 20 log (N =1000) higher. Often the sidebands are larger than the carrier, and there are also conditions where the carrier disappears and all of the power appears in the sidebands. This seriously affects or even prohibits the use of microwave systems that employ phase-locked loops, because large sidebands due to vibration cause large phase excursions and unlock the loops (Filler, 1988; Wallin et al., 2003).

3. Measurement Techniques

In order to measure the acceleration sensitivity of different microwave oscillators and components, the device needs to be characterized while subjected to vibration. The equipment needed to vibrate the device consists of a mechanical actuator or "shaker," its associated power amplifier, an accelerometer, and computer control system. The computer uses the accelerometer to sense the vibration of the actuator and generates the desired vibration profile using closed-loop feedback. The single axis actuator used for these tests has the capability to vibrate either in a random vibration pattern or in various sinusoidal patterns, including continuous wave and swept. When the actuator vibrates in one axis, the cross-axis leakage is low, as shown in Figure 3. These data are taken with a 3-axis accelerometer mounted on the actuator.

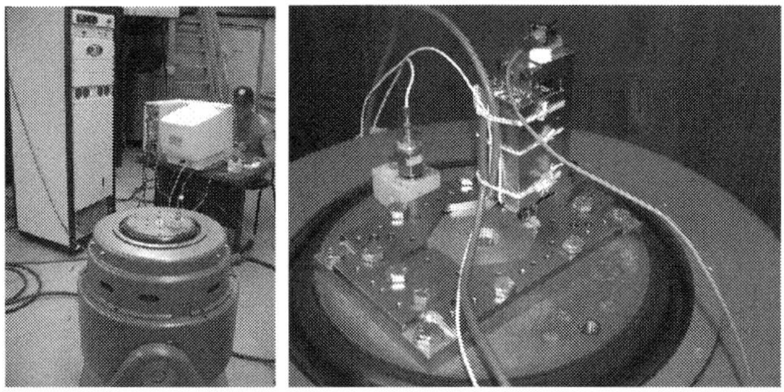

Figure 2. The picture on the left shows the vibration actuator (shown in blue); the amplifier for driving the actuator (the vertical rack-mount system); and the controlling computer. The picture on the right is a device under test (DUT) mounted on the actuator

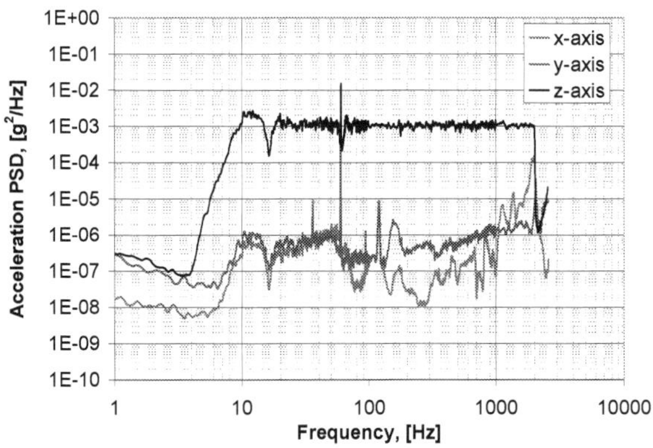

Figure 3. Plot showing the acceleration power spectral density (PSD) along x, y and z axes when vibration is along the z-axis, showing that cross-axis leakage is small. A random vibration profile of acceleration PSD = 1.0 mg²/Hz (rms) is used for 10 Hz $\leq f_v \leq$ 2000 Hz; f_v is the vibration frequency

For all the vibration tests discussed later, both sinusoidal and random vibration testing are chosen. First, a random vibration pattern is used, vibrating at frequencies between 10 and 2000 Hz, followed by a sinusoidal vibration at 10, 20, 30, 50, 70, 90, 100, 200, 300, 500, 700, 900, 1000, and 2000 Hz. This frequency range is chosen because it is the full range for the available vibration table, adequately covering smaller ranges associated with most applications (Section 5).

3.1 Experimental Setup for Residual PM Noise Measurement

Residual noise is the noise that is added to a signal by its passage through a two-port device. Figure 4 shows the block diagram of a PM noise measurement system used to measure the residual noise of a two-port or a non-oscillatory device such as a bandpass filter or amplifier as well as a cable and connector under vibration (Walls & Ferre-Pikal, 1999). The output power of a reference oscillator is split into two paths. One path is used to drive the device under test (DUT), and the other path is connected to a delay line. The delay is chosen so that the delay introduced in one path is equal to the delay in the other path. A phase shifter is used to set phase quadrature or 90-degrees between two paths, and the resulting signals are fed to a double-balanced mixer, acting as a phase detector. The baseband signal at the output of the phase detector is amplified and measured on a fast Fourier transform (FFT) analyzer. The output voltage $V_0(t)$ is given by

$$V_0(t) = k_d G \Delta\phi(t) \quad \text{for } \Delta\phi(t) << 1, \tag{10}$$

where k_d is the mixer sensitivity, G is the gain of the baseband IF amplifier, and $\Delta\phi(t)$ is the difference in phase fluctuations between two inputs to the phase detector. The PM noise is obtained from

$$S_\phi(f) = \frac{PSD\left[V_0(t)\right]}{\left(k_d G\right)^2}. \tag{11}$$

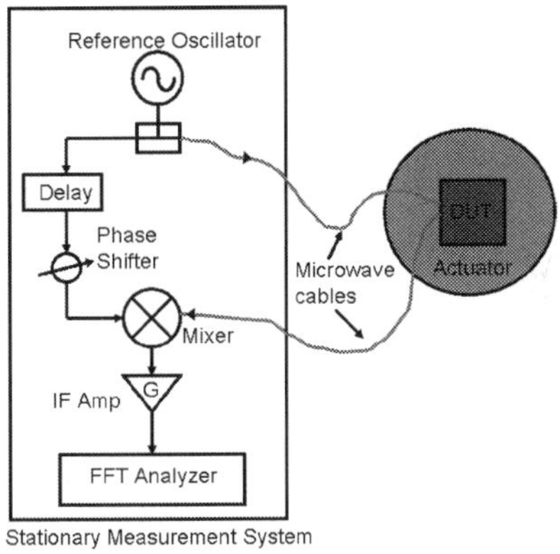

Figure 4. Block diagram of an experimental setup for residual PM noise measurement of components under vibration. DUT — Device Under Test; IF Amp — Intermediate Frequency Amplifier

Because the delays in the two signal paths are equal, the PM noise from the reference oscillator is equal and correlated in each path and thus cancels. At the output of the mixer, the noise from the vibrating DUT and connecting cables appears because it is not correlated between the two inputs of the mixer. A low noise phase detector and IF amplifier are chosen for this measurement and their noise contributions are much lower than the dominating vibration-induced noise of DUT and cables.

In order to accurately measure the vibration sensitivity of a DUT, it is very important to know the vibration sensitivity noise floor first. For the noise floor measurement, the DUT is replaced with an appropriate length of rigid coaxial cable. Compared to all other experimental components, it is the microwave cables, blue in color (Figure 4) and connected between the measurement system and the DUT mounted on the actuator that generally set the vibration sensitivity noise floor. When the DUT is under vibration, the cables flex between the vibrating and stationary (measurement system) reference frames. The flexure of the cables causes the relative position between the different segments of the outer conductor, the dielectric, and the inner conductor of the cable to vary and thus changes the electrical characteristics of the cables. The main challenge is to obtain a reproducible low noise floor at close to carrier offset frequencies. The noise floor is very dependent on the configuration and tension of cables the running between the vibrating and stationary reference frames; a slight change may cause the noise to vary anywhere from 10 to 30 dB, as shown in Figure 5.

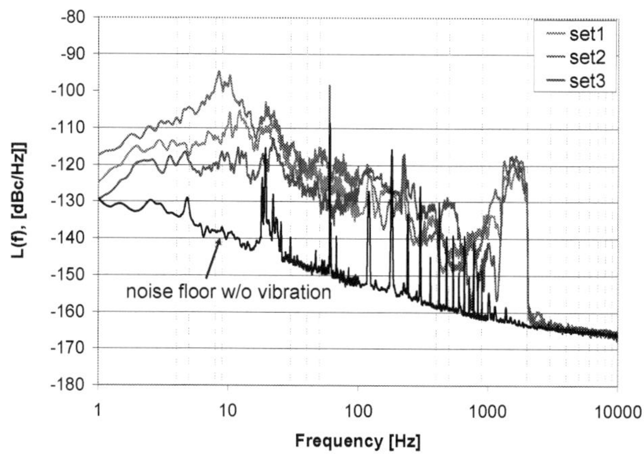

Figure 5. Residual PM noise floor of the measurement system measured with and without vibration. A random vibration profile of acceleration PSD = 1.0 mg²/Hz (rms) is used for 10 Hz ≤ f_v ≤ 2000 Hz. It shows the variation in the close to carrier noise for three different sets of measurement cable configurations. The bottom curve shows the noise floor measured under no vibration. Narrow spurs are power line EMI pick-up and should be ignored

Sometimes the noise floor is so high that it is impossible to accurately characterize low-vibration-sensitive components. In order to measure the acceleration sensitivity of a component accurately, the following precautionary measures should be taken:

- Rigidly mount the DUT on the vibration table to avoid any mechanical resonance inside the frequency range of interest.
- Experiment with different amount of cable slack or tension between the stationary and vibrating reference frames to obtain the best noise floor.
- Properly secure the cables to minimize flexing and strain due to vibration. It is also important to properly secure the power leads for the DUT.
- Reduce the acoustic noise and external vibration in the test area.
- The vibration actuator often has cooling fans; prevent this airflow from disturbing the cables.
- No other components except the DUT and accelerometer should be mounted on the vibration table.

- If possible, use 1 to 3 dB attenuators at the connector interfaces to minimize the effect of voltage-standing-wave-ratio (VSWR) induced mechanical and multipath phase fluctuations.
- Check the noise floor in between the measurements by replacing the DUT with a short cable.

3.2. Experimental Setup for Absolute PM Noise Measurement

Absolute noise is the noise measured on a one-port device such as an oscillator. Several measurement techniques can be used to measure the PM noise and thus the acceleration sensitivity of an oscillator. The most commonly used techniques are the heterodyne (two-oscillator) PM noise measurement system and the homodyne (delay-line discriminator) measurement system (Lance et al., 1984; Sullivan et al., 1990; Walls & Ferre-Pikal, 1999). In this section, these measurement techniques are briefly described. The block diagram of a single-channel heterodyne (two-oscillator) PM noise measurement system is shown in Figure 6. In this system, the signals from the DUT mounted on the vibration actuator and a stationary reference oscillator of the same frequency are fed into a double-balanced mixer. A lowpass filter (LPF) is used after the mixer to filter the higher-order harmonics, and a phase-locked loop (PLL) is used to lock the reference frequency to the DUT frequency and to maintain quadrature between the two input signals to the mixer. The output voltage of the mixer is proportional to the difference between the phase fluctuations of the two sources. This signal is amplified, and its power spectral density is measured on a FFT analyzer.

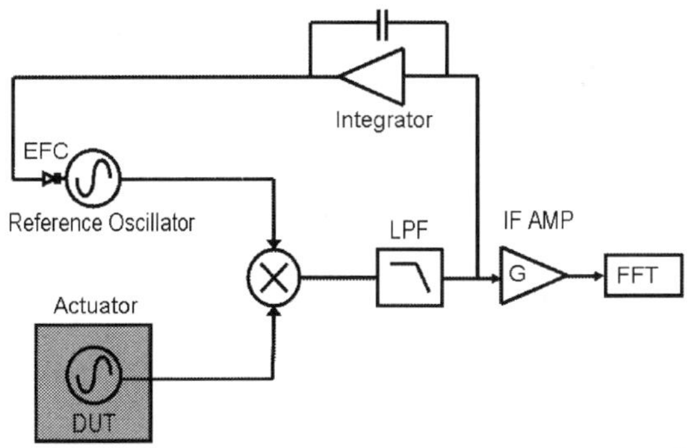

Figure 6. Block diagram for a heterodyne PM noise measurement system. DUT- Device Under Test; LPF – Low Pass Filter; EFC – Electrical Frequency Control; IF AMP – Intermediate Frequency Amplifier

The advantage of the heterodyne measurement system is its high sensitivity; however, there are notable disadvantages. It requires a reference oscillator of the same frequency, a PLL, and calibration of the mixer sensitivity (k_d). Among these, the main measurement uncertainty comes from the PLL. In order to measure the vibration induced PM noise at close to carrier offset frequencies, a low loop bandwidth is required. Furthermore, when the DUT experiences a large level of vibration, it may be difficult to keep the DUT and reference oscillator phase locked due to large phase excursions. This situation requires the PLL bandwidth to be increased, which complicates the calibration inside the PLL bandwidth.

A PM noise measurement system that needs neither a second source nor a phase-locked loop is a delay-line discriminator system, shown in Figure 7. As discussed in Section 3.1, in a residual PM noise measurement of a device, it is important to keep the delay in both signal paths as equal as possible so that the source noise is correlated and cancels at the phase detector. However, in a delay-line discriminator technique, introduction of a long delay in one path intentionally un-correlates the noise by increasing the delay time so that one can measure the PM noise of the source. The differential delay introduced in one path creates a frequency dependent phase shift between the two inputs of the phase detector. This frequency dependent phase shift converts frequency fluctuations of the source into differential phase fluctuations between the inputs of the phase detector. When these two signals are adjusted for phase quadrature, the phase detector converts the phase fluctuations into their voltage equivalent for measurement and analysis. The voltage at the output of the IF amplifier is given by

$$V_0(t) \cong (2\pi\nu_0 k_d \tau) G y_{rms}(t) = \nu_0 k_\nu G y_{rms}(t), \tag{12}$$

where ν_0, τ, k_d ,G, $y_{rms}(t)$, and k_ν are respectively the carrier frequency, the time delay introduced by the delay line, the mixer sensitivity to phase fluctuations, the voltage gain of the baseband IF amplifier, the rms fractional frequency fluctuation and mixer sensitivity to frequency fluctuations. Equation 12 indicates that the output voltage of the mixer is proportional to the frequency fluctuation in the source; thus, this system measures the frequency modulated (FM) noise of the DUT. The FM and PM noises can be calculated from the following relations respectively,

$$\left.\begin{aligned} S_y(f) &= \frac{PSD[V_0(t)]}{(\nu_0 k_\nu G)^2} \\ S_\phi(f) &= \left(\frac{1}{f^2}\right)^2 \frac{PSD[V_0(t)]}{(k_\nu G)^2} \end{aligned}\right\} \text{ for } f \ll 1/\tau. \tag{13}$$

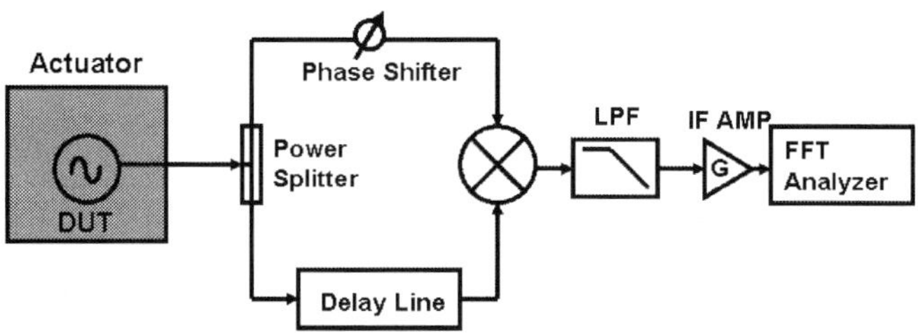

Figure 7. Block diagram of a measurement system that uses a delay-line to measure PM noise of an oscillator under vibration

The close to carrier noise floor of a delay line discriminator measurement is inversely proportional to the time delay, τ. By increasing the time delay, the noise floor can be lowered; however, longer delay causes higher insertion loss of the RF signal thus reducing measurement sensitivity, which effectively increases the noise floor. This problem can be overcome by use of a low loss delay, such as a photonic delay line (Rubiola et al., 2005). The long fiber delay provides a low-loss carrier of RF signals. A 2 km fiber typically provides 0.4 dB loss and about 10 μs delay whereas insertion loss due to a 25 m semi-rigid cable at 10 GHz is approximately 20 dB with a delay of about 100 ns. Using a 10 μs photonic delay instead of a 100 ns RF delay, almost 40 dB of improvement in the noise floor can be achieved.

There is another technique known as direct digital PM noise measurement (Grove et al., 2004), which requires two sources similar to a heterodyne technique. However, like a delay line discriminator technique it needs no phase-locked loop. This PM noise measurement system uses fast analog-to-digital converters to digitize the input RF signal and performs all down-conversion and phase detection functions by digital signal processing. It provides a low noise floor, require no calibration of k_d or k_v, and can compare oscillators at different frequencies. High dynamic range with no phase-locked loop involved means the PM noise of a vibrating noisy source can be measured accurately for offset frequencies very close to the carrier. Figure 8 shows the setup used to measure acceleration sensitivity of different microwave oscillators discussed in Section 5.2 using this measurement system. The oscillator under test is mounted on a vibration table with the output going to one input of a mixer. This output is then mixed with a very-low-PM-noise oscillator to generate a beat frequency anywhere between 1 MHz to 30 MHz for the most straightforward digitizers. A low-noise 10 MHz quartz crystal oscillator serves nicely as the digitizer's reference clock.

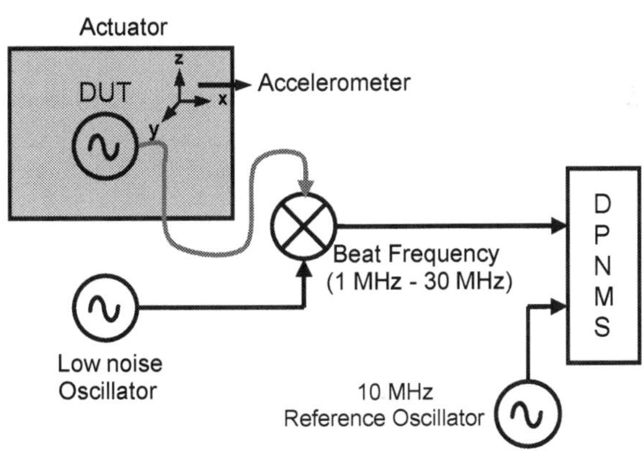

Figure 8. Block diagram of an experimental setup for measuring acceleration sensitivity of an oscillator. DPNMS — Direct digital PM noise measurement system

4. Results - Acceleration Sensitivity of Non-oscillatory Devices

Most microwave components are sensitive to vibrations to some extent. Among them, microwave cables, bandpass filters, mechanical phase shifters and connectors are the most sensitive. Therefore, it is very important to select vibration-tolerant components in order to build a system with low vibration sensitivity. The local oscillator is one of the prime performance-limiting components in an RF receiver. However, with increasing demand for low acceleration sensitivity of oscillators in various applications, the sensitivity to vibration of these non-oscillatory components will soon be the limiting factor and thus cannot be ignored (Driscoll & Donovan, 2007). In this section, the performance of some of the non-oscillatory microwave and optical components under vibration is discussed. The components used for test are commercially available and shown in Figure 9.

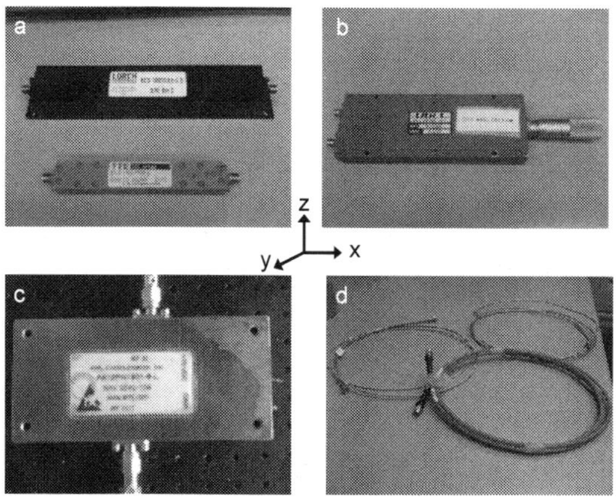

Figure 9. Picture of the non-oscillatory components used for vibration test. (a) Bandpass filter, (b) Mechanical phase shifter, (c) Amplifier and (d) Microwave cables. Arbitrary x and y axes are chosen in the plane of the page, and the z-axis is normal to the page

4.1 Microwave Components

First, two 10 GHz bandpass cavity filters of different quality factors (Q) are tested under random vibration. A random vibration profile of acceleration PSD 1.0 mg^2/Hz (rms) for offset frequencies 10 Hz to 2000 Hz is used. Figure 10(a) shows the PM noise floor of the measurement system as well as the PM noise of the filters under vibration. A comparison of acceleration sensitivity of these filters in rad/g is shown in Figure 10(b). The result is obtained by converting $L(f)$ to $S_\phi(f)$ using equation 4 and dividing it by $|\vec{a}| = \sqrt{2PSD}$. The result shows that the filter with higher Q (narrow bandwidth) is more sensitive to vibration. This is due to the fact that the transfer function phase of a band pass filter has its steepest slope at the center frequency. Any vibration that modulates the resonant structure of the filter also modulates the center frequency and thus the phase shift through the filter. The phase slope is proportional to the filter Q; this causes the high-Q filter to be more sensitive to small mechanical distortions under vibration. Figure 10(b) also indicates that the microwave cables used for the measurement set a noise floor that provides an acceleration sensitivity of 10^{-4} to 10^{-6} rad/g. For perspective, this means that in order to measure a 10 GHz oscillator with an acceleration sensitivity of 10^{-12}/g experiencing a random noise of 0.05 g^2/Hz (rms) between 10 Hz $\leq f_v \leq$ 1000 Hz, the cable should have vibration sensitivity better than 7 × 10^{-4} rad/g and 7 × 10^{-6} at offset frequencies 10 Hz and 1000 Hz, respectively.

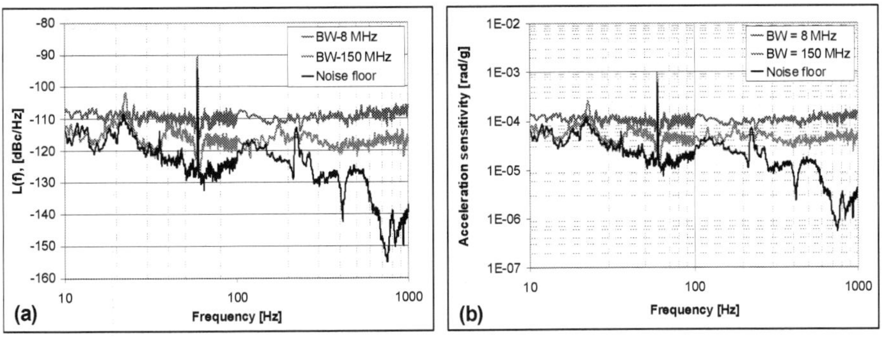

Figure 10. (a) PM noise of two 10 GHz bandpass cavity filters under vibration. A random vibration profile of acceleration PSD = 1.0 mg²/Hz (rms) is used for 10 Hz ≤ f_v ≤ 2000 Hz. The bottom curve shows the PM noise floor set by flexing of cables under vibration.
(b) Acceleration sensitivity of the same two 10 GHz bandpass cavity filters. BW is the bandwidth of band pass filter

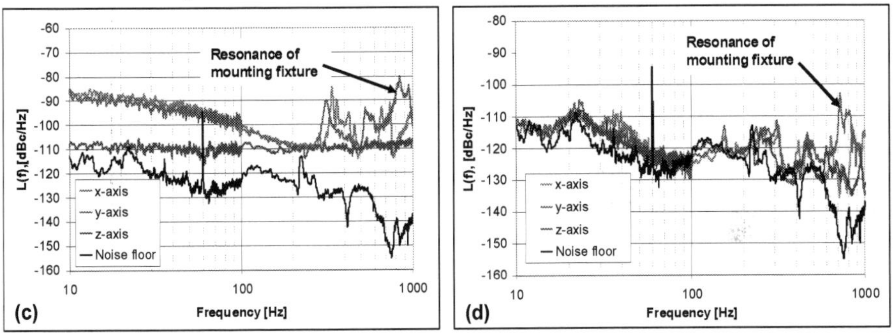

Figure 10. (c) PM noise of a 10 GHz high-Q band pass filter along x, y and z axes. (d) PM noise of a mechanical phase shifter along x, y and z axes. A random vibration profile of acceleration PSD = 1.0 mg²/Hz (rms) is used for 10 Hz ≤ f_v ≤ 2000 Hz

The PM noise of a high-Q bandpass filter and a mechanical phase shifter is measured along all three axes. See Figure 9 for the orientation of the axes. Figure 10(c) shows the PM noise of a bandpass filter under vibration along all three axes. Close to the carrier, the filter is more sensitive to vibration along the x- and y-axis than along the z-axis. It is also important to note that sometimes the mechanical resonances of the fixture used to mount the component on the actuator can appear. This effect is visible for the bandpass filter above 300 Hz. Therefore, it is important to either minimize this effect, if possible, or separate it from the results of actual vibration induced noise. In the case of the phase shifter, the noise is limited mostly by the noise floor of the measurement system, as shown in Figure 10(d).

Next, three different types of cables, viz. flexible, hand formable and semi-rigid cables, each 12 feet long, are chosen for the noise floor measurement. The PM noise floors set by each of these cables under vibration are comparable, similar to that shown in Figure 5. Finally, the PM noises of a few amplifiers at 10 GHz are also measured under vibration (Hati et al. 2007). The acceleration sensitivity of these components is much lower than the acceleration sensitivity noise floor provided by the cables. As a result, an accurate measurement is not possible. However, it can be concluded from the experimental results that the acceleration sensitivity of the phase shifter and amplifier under test is no greater than 10^{-4} rad/g.

4.2 Optical Components

Optical fibers are widely used in fiber-optic communication, ring-laser gyroscopes, optoelectronic oscillators (OEOs), and delay line discriminators, among a number of other applications (Yao & Maleki, 1996; Römisch et al., 2000; Kadiwar & Giles, 1989; Rubiola et al., 2005; Minacian, 2006). Mechanical distortions due to vibration induce phase fluctuations in the fiber. In this section, the effect of vibration on the phase fluctuations in the optical fiber wound on a spool is discussed. Letting l be the length of the fiber (assumed to be uniformly wound on a cylindrical spool) and v_g the group velocity of amplitude modulated electromagnetic waves down the fiber; then the group delay through the fiber is $\tau_d = l/v_g$. In fiber, stretching can occur either as a result of temperature changes of the spool on which the fiber is wound or as a result of axial vibrations that accelerate, and hence, deform the spool. The fiber length change (δl) due to deformation of spool, which is mounted at the bottom, is given by (Ashby et al., 2007)

$$\frac{\delta l}{l} = \frac{\nu \rho a(t) h}{2E}, \tag{14}$$

where ν, ρ, $a(t)$, h and E are, respectively, Poisson's ratio of the spool material, spool density, time-dependent acceleration, height of the spool, and Young's modulus (describing the spool's stiffness). Equation 14 shows that the stiffer the spool, the smaller the length change, and the denser the spool, the larger the length change, because the force on the fiber is larger. It also shows a linear dependence on the time-dependent acceleration. Further, a change in length due to stretching will also result in stresses within the fiber that can change the optical properties of the fiber, resulting in a change in the group velocity through its stress-optic coefficients. When these changes are small, the fractional frequency shift is a superposition of these two contributions:

$$\frac{\delta f}{f_0} \approx \frac{1}{v_g}\left(\frac{\delta v_g}{\delta\left(\frac{\delta l}{l}\right)} - v_g\right)\frac{\delta l}{l}. \tag{15}$$

The changes in the group velocity for a stretched fiber are quite small, and we can neglect them in a first approximation so that the fractional frequency shift due to vibration-induced fiber length changes can be adequately approximated by

$$\frac{\delta f}{f_0} \approx -\frac{\delta l}{l}. \tag{16}$$

Thus, when used in an optoelectronic oscillator, which is described in Section 5.2, the main effect on a desired frequency comes from the change of length itself rather than from the change in group velocity. The small time-delay fluctuations due to fiber length fluctuations inside the loop directly map to larger frequency fluctuations at the output of the oscillator (Leeson, 1966), which is principally why suppressing fiber vibration sensitivity is so important.

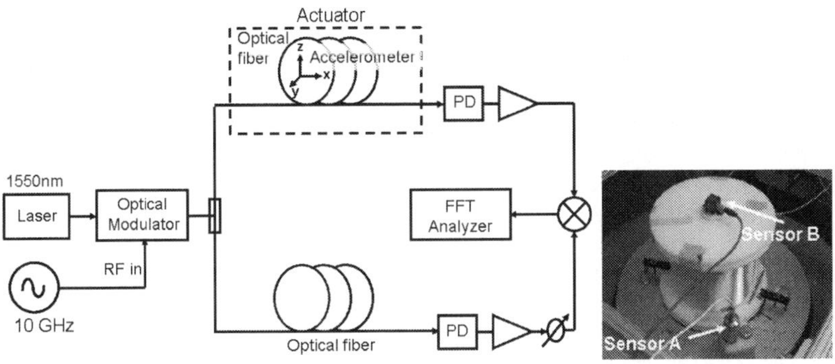

a) b)

Figure 11. (a) Experimental setup to study vibration-induced PM noise in a fiber delay line mounted on a vibration table. (b) Optical fiber of length 3 km wound on a ceramic spool and mounted on a vibration table. Orientation as shown presents the vibration (acceleration) along the z-axis of spool

Figure 11(a) shows the setup used to measure the effect of vibration on a fiber delay line using a residual PM noise measurement. It consists of a 1550 nm communications-grade laser whose output is sent into an optical modulator that is amplitude-modulated by a 10 GHz RF signal. This RF-modulated optical signal is then split into two signal paths, each composed of a 3 km length of single mode fiber (SMF-28) wound on a cylindrical spool, a photo-detector, an RF amplifier, and a phase shifter. As discussed earlier, a mechanical phase shifter is used to set phase quadrature between the two signals so that the output voltage noise of the mixer is proportional to the PM noise. This signal is then measured on a FFT analyzer. Because the delays in the two signal paths are equal, this technique cancels the PM noise introduced by the laser, optical modulator, and the reference oscillator.

In order to test the vibration properties of the fiber, one of the 3 km fibers wound on a stiff cylindrical ceramic spool is secured to a vibration table. The cylindrical spool is approximately 11.5 cm in diameter with a length between end caps of 10 cm. An accelerometer mounted to the actuator, as shown in Figure 11(b) (sensor A), provides closed-loop feedback to a computer for control of the vibration profile. Sensor B is a triaxial accelerometer that is used to sense vibrations affecting the fiber spool. Its z-axis is aligned with the axis of the spool, and both the x and y axes are in the radial directions. For this test, the z-axis sensitivity is the subject of consideration because the vibration sensitivity of a fiber-on-spool is greatest along this axis (Ashby et al. 2007; Huang et al. 2000). The spool is subjected to a random vibration of acceleration PSD equal to 0.5 mg^2/Hz (rms) along the z-axis, and the PM noise is measured. In Figure 12, the bottom curve is the noise floor, and the topmost curve is the same measurement of PM noise under vibration. The dotted brown line shows that the acceleration sensitivity of this fiber is 2.5×10^{-3} rad/g, which is calculated for $L(f)$ = -85 dBc/Hz. The resonance at 600 Hz is a characteristic of the hollow cylindrical spool geometry. The ceramic spool is chosen due to its stiffness because it is less sensitive to vibration than other materials such as plastics (Taylor et al., 2008). The measurement of acceleration sensitivity of other optical components such as optical connectors, fiber laser, optical modulator, and photo-detector are also important; however, the largest sensitivity to vibration is usually from the fiber.

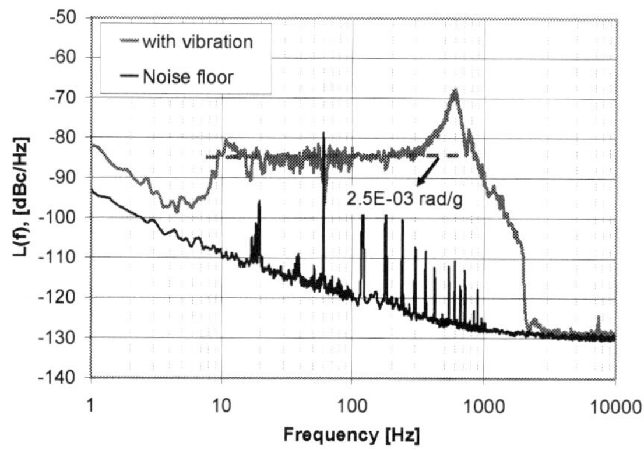

Figure 12. The residual PM noise of a 3 km fiber under random vibration along the z-axis is shown. A random vibration profile of acceleration PSD of 0.5 mg^2/Hz (rms) is used for 10 Hz $\leq f_v \leq$ 2000 Hz. The dotted brown curve shows the acceleration sensitivity corresponding to $L(f)$ = -85 dBc/Hz. The bottom curve is the noise floor measured under no vibration

5. Results - Acceleration Sensitivity of Different Classes of Oscillators

Low-noise oscillators must operate under "real world" vibration/acceleration conditions. This is crucial for high-speed communications systems, advanced surveillance systems, weapons detection, and many other applications. The PM noise requirement for each application is different, and a single oscillator cannot meet the PM noise requirement over a broad range of offset frequencies. For example, a good low noise 10 MHz quartz oscillator typically has a PM noise L(10 Hz) = -130 dBc/Hz and a white PM noise level of -170 dBc/Hz. The PM noise of this 10 MHz signal at 10 Hz offset, when multiplied to generate a signal at 10 GHz, becomes -70 dBc/Hz, which is reasonably low and suitable for a 10 GHz radar to detect slow-moving objects (Fruehauf, 2007). However, the white PM noise after ideal multiplication becomes -110 dBc/Hz and is too high to detect fast-moving targets such as supersonic aircraft. There are several X-band oscillators topologies such as a YIG (Yttrium, Iron and Garnet) oscillator, dielectric resonator oscillator (DRO), surface transverse wave (STW) oscillator, sapphire loaded cavity oscillator, or a simple air-dielectric cavity resonator oscillator that can provide very low phase noise at higher offset frequencies. A combination of any of these oscillators with a quartz oscillator can achieve a low PM noise performance over a broad frequency range.

Oscillators that are capable of satisfying specific system requirements in quiet environments are readily available. However, in the vibrating environments of airborne platforms, the PM noise of oscillators degrades significantly. Table 1 and Figure 13 show the typical level of acceleration for certain operating environments (Mancini, 2004). Acceleration levels at the oscillator depend on how and where the oscillator is mounted. Platform resonances can greatly amplify the acceleration levels.

Environment	Acceleration typical levels, in units of g
Tractor-trailer (3-80 Hz)	0.2 peak
Armored personnel carrier	0.5 to 3 rms
Ship — calm seas	0.02 to 0.1 peak
Ship — rough seas	0.8 peak
Propeller aircraft	0.3 to 5 rms
Helicopter*	0.1 to 7 rms
Missile — boost phase	15 peak
Jet aircraft*	0.02 to 2 rms

Table 1. Typical level of acceleration for certain operating environments. * See Figure 13

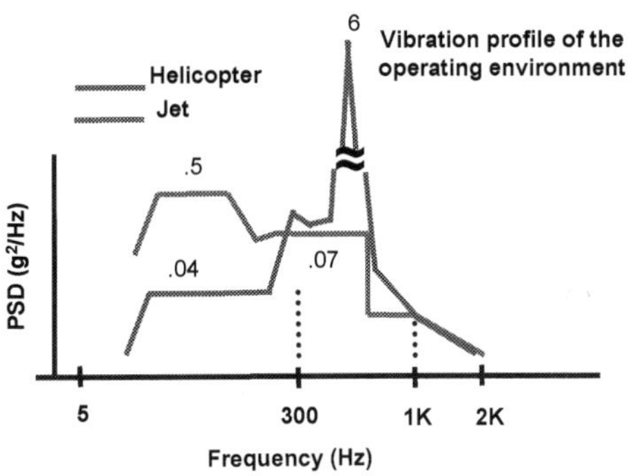

Figure 13. Typical aircraft random vibration profile

Section 5 is primarily focused on studying the effect of vibration on the phase noise of electronic and opto-electronic oscillators at 10 GHz and also comparing their acceleration sensitivity. The acceleration sensitivity of a quartz oscillator is not discussed here because enormous amounts of work have already been done and the data are readily available in the literature (Vig, 1992; Driscoll, 1993; Filler, 1983; Kosinski, 2000). A good quartz oscillator typically has an acceleration sensitivity of $1 \times 10^{-9}/g$, and the lowest known acceleration sensitivity for a quartz oscillator is $2 \times 10^{-11}/g$ for 10 Hz $\leq f_v \leq$ 200 Hz (Bloch et al., 2006).

5.1 Electronic Oscillators

Figure 14 shows three different types of oscillators chosen for the vibration test, viz., a DRO at 10 GHz, a silicon germanium (SiGe) amplifier-based STW oscillator at 2.5 GHz (Hay, et al., 2004), and a TE_{023} mode air-dielectric ceramic-cavity resonator oscillator (ACCRO) at 10 GHz (Hati et al., 2006). A STW oscillator is chosen because these are the best performing oscillators in the frequency range 1 to 3 GHz. Below 1 GHz surface acoustic wave (SAW) oscillators perform well (Parker & Montress, 1988), and above 3 GHz DROs provide the best compromise between performance and cost. There are several other commercially available oscillators above 3 GHz that have extremely low phase noise but are large, specialized and expensive by comparison. At first, the PM noise of these oscillators is measured with no vibration; the PM noise plots are shown in Figure 15.

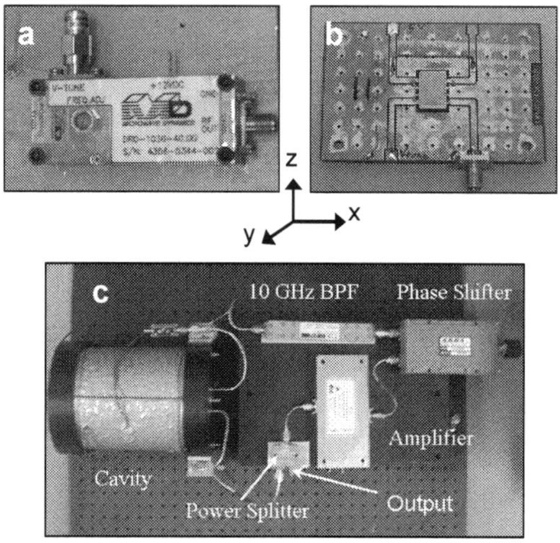

Figure 14. Pictures of three different types of oscillators used for vibration test. (a) DRO, (b) STW oscillator, (c) ACCRO. Arbitrary x and y axes are chosen in the plane of the page, and the z-axis is normal to the page

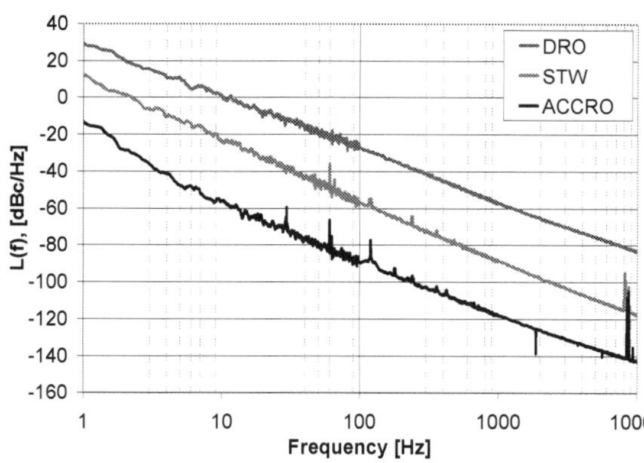

Figure 15. PM noise of three different oscillators at 10 GHz without vibration. For straight comparison, the PM noise of 2.5 GHz STW oscillator is normalized to 10 GHz

A commercial DRO at 10 GHz is subjected to a random vibration along three axes independently. The degradation in PM noise due to vibration in the z-axis is shown in Figure 16(a). The effect of random vibration in the x and y axes is not noticeable because the PM noise of the stationary DRO is much higher than the noise induced by random vibration.

In order to measure the acceleration sensitivity in all three axes, the oscillator is subjected to sinusoidal vibration with higher g-levels at different frequencies. The results are shown in Figure 16(b). This particular DRO is more sensitive to vibration along the z-axis than along the other two axes. Further, the acceleration sensitivities of two DROs of comparable size and weight but different Q are compared. Figure 17(a) and 17(b) respectively show the PM noise and acceleration sensitivity of these DROs. The DRO with higher Q is more sensitive to vibration for the same reason as discussed in Section 4.1 for a high-Q bandpass filter.

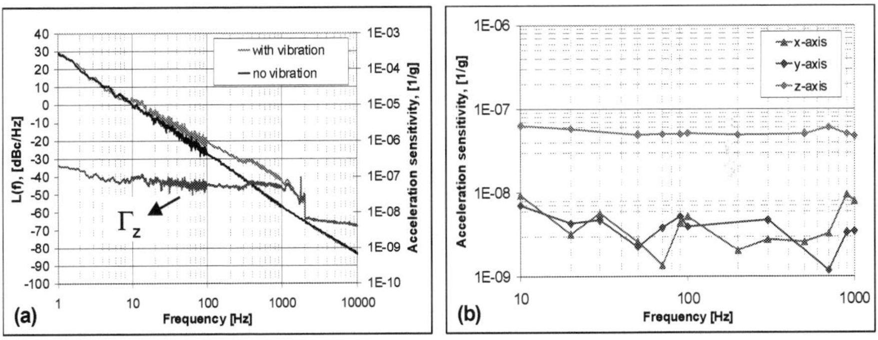

Figure 16. (a) PM noise of the DRO with and without vibration along the z-axis. The brown curve indicates the z-axis acceleration sensitivity of this particular DRO. (b) x, y, and z-axis gamma. An acceleration PSD of 0.5 mg^2/Hz (rms) is used for random vibration (Fig. 16a) and a peak acceleration of 1 g is used for sinusoidal vibration (Fig. 16b)

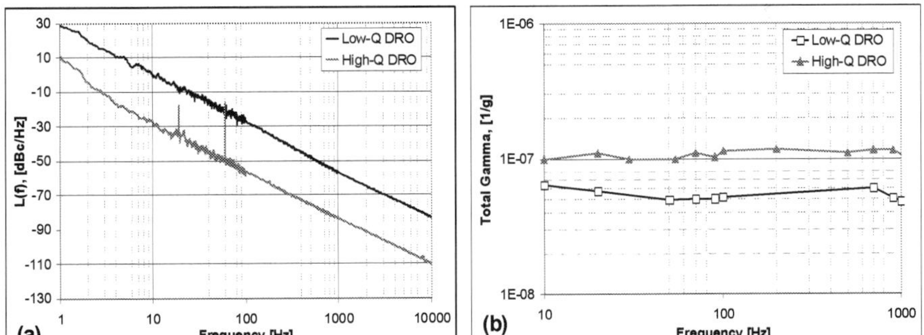

Figure 17. (a) PM noise of two DROs of different Q without vibration. (b) Plot of total gamma for the DROs. An acceleration PSD = 0.5 mg^2/Hz (rms) is used for 10 Hz $\leq f_v \leq$ 2000 Hz. The lower PM noise oscillator has correspondingly higher acceleration sensitivity

Vibration-sensitivity experiments are also performed for a STW oscillator in all three axes, and the results are shown in Figure 17(c). For the ACCRO, the vibration measurement is confined to a single axis normal to the mounting plate, as shown in Figure 10. Figure 17(d) shows the z-axis acceleration sensitivity of three oscillators; the acceleration sensitivity of the STW oscillator is two orders of magnitude lower than that of the DRO.

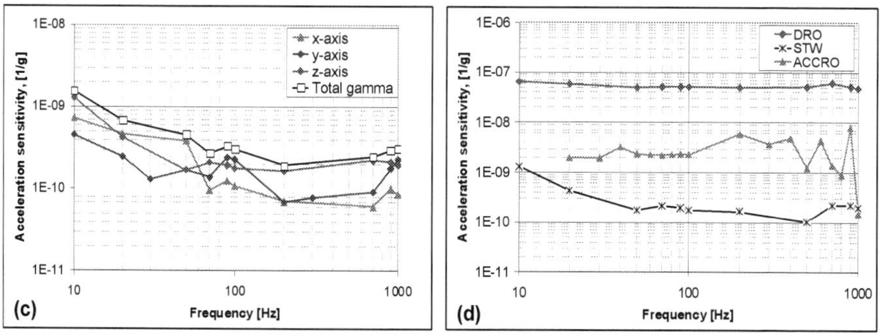

Figure 17. (c) x, y and z-axis acceleration sensitivity of a STW oscillator. (d) Comparison of z-axis acceleration sensitivity of different oscillators. A peak acceleration of 1 g is used

5.2 Optoelectronic Oscillator

Low-noise, microwave-frequency oscillators are key components of systems that require high spectral purity. An optoelectronic oscillator (OEO) is an example that has emerged as a low-noise source in recent years (Yao & Maleki, 1996; Römisch et al., 2000; Eliyahu et al., 2008). The high spectral purity signal of an OEO is achieved by using a long optical fiber that provides a very high quality factor (Q). However, the close to carrier spectral purity of an OEO is degraded mostly by environmental sensitivities, one being the vibration-induced phase fluctuations in its optical fiber (Howe et al., 2007; Hati et al., 2008).

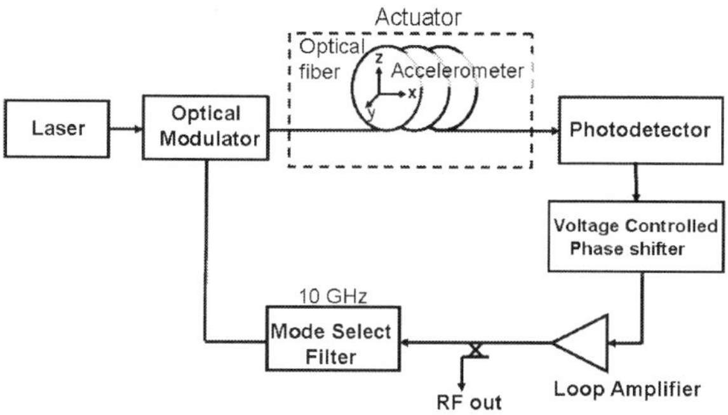

Figure 18. Block diagram of an optoelectronic oscillator

In a typical optoelectronic device as shown in Figure 18, light from a laser passes through an electro-optic amplitude modulator, the output of which is fed to a long optical fiber and detected with a photodetector. The output of the photodetector is then amplified, filtered, and fed back to the modulator port, which amplitude-modulates the laser light. When loop gain is greater than 1, this configuration leads to self-sustained oscillations. A typical OEO gives a large number of modes with frequencies given by (Yao & Maleki, 1996; Römisch et al., 2000)

$$f_0 = \frac{(K + 1/2)}{\tau_d}. \tag{17}$$

where K is an integer whose value is selected by the filter and $\tau_d = l/v_g = nl/c$ is the group delay through the fiber with index of refraction n and length l and c is the velocity of light. The quality factor of an OEO is proportional to length l and is given by $Q = \pi\tau_d f_0 = \pi nlf_0/c$. An OEO at 10 GHz is designed by using the same 3 km (SMF-28) fiber as mentioned in Section 4.2 to study the effect of vibration on the overall PM noise performance of the oscillator. The PM noise of this OEO subjected to random vibration is shown in Figure 19. The PM noise under vibration degrades almost 30 to 40 dB from its normal stationary PM noise performance. This is due to the fact that the phase perturbations due to vibration in the fiber, which is the most vibration sensitive component in the loop, translate to frequency fluctuations inside the OEO's resonator bandwidth. The z-axis acceleration sensitivity of this OEO is approximately $5 \times 10^{-9}/g$, as shown by the brown plot in Figure 19.

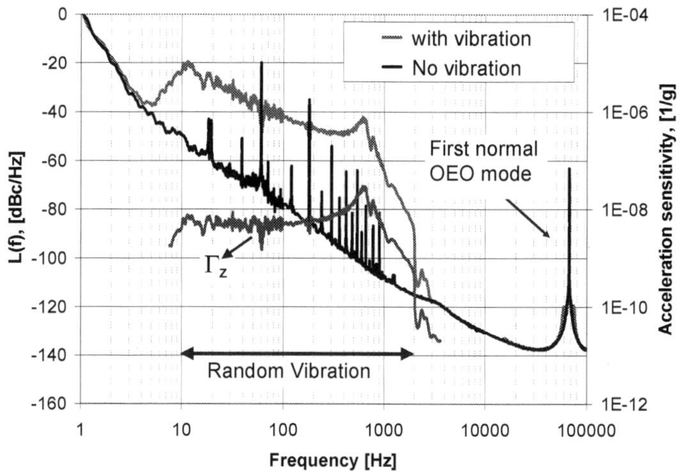

Figure 19. Plot comparing the PM noise of an optoelectronic oscillator with and without vibration. A random vibration profile of acceleration PSD = 0.5 mg²/Hz (rms) is used for 10 Hz $\leq f_v \leq$ 2000 Hz. The brown curve shows the z-axis acceleration sensitivity of this OEO, which is approximately $5 \times 10^{-9}/g$. The first mode for a 3 km fiber, which is approximately 67 kHz from the carrier, is also shown

6. Acceleration Sensitivity Reduction

In this section, a few methods of reducing vibration-induced noise from vibration-sensitive components are discussed. The most common approach for reducing vibration-induced PM noise is to select low-vibration-sensitive materials. A comparison is made for two air-dielectric cavity oscillators at 10 GHz, one cavity made of aluminum and another one made of ceramic. The two cavities are chosen so that they have comparable volume, almost identical loaded Q's of 22,000 (TE_{023} mode) and insertion losses of 6 dB. All the other components of the two oscillators are identical. These two oscillators are tested for different sinusoidal vibration frequencies. The acceleration sensitivity of the ceramic cavity oscillator is found to be almost one sixth that of the aluminum cavity oscillator, as shown in Figure 20. This is because ceramic is stiffer than aluminum and thus less sensitive to vibration.

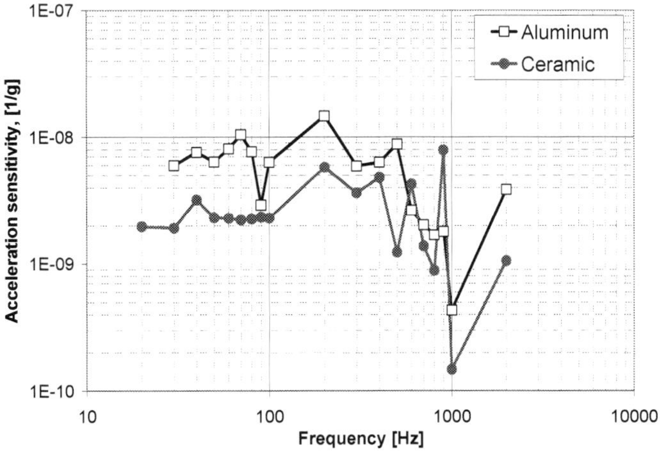

Figure 20. Comparison of acceleration sensitivity of aluminum and ceramic air-dielectric cavity resonator oscillators

It is also worth noting the results of a few tests of passive mechanical dampers and isolators on different test oscillators. The most common approach to reduce vibration-induced PM noise is to select vibration isolators of low natural frequency. Small stranded wire rope isolators and urethane shock mounts, as shown in Figure 21(a), provide excellent damping and omnidirectional isolation. When size is critical, these small shock absorbers are often incorporated inside the oscillator package to improve the PM noise performance of the oscillator in high vibration induced environments. Further, when space is not an issue, an external vibration isolation platform can be used. In this case, the whole system including the oscillator can be mounted on this vibration-free platform to achieve the precision needed for the particular application. There are several commercially available vibration isolation platforms with extremely good performance and very low natural resonance frequencies, less than 1 Hz. A significant improvement in the acceleration sensitivity of the DRO and STW oscillators is observed when tested under vibration with these passive dampers as shown in Figure 21(b).

Finally, an active electronic vibration cancellation technique can be sometimes used to reduce the vibration sensitivity. A 3 km long optical fiber and an optoelectronic oscillator are chosen to illustrate this scheme. Work on control of environmental noise in optical fiber has previously been implemented in systems where either a portion of the system undergoes vibration or a stable reference is available to measure the vibration-induced noise (Foreman et al., 2007). It is very important to assess the degree to which vibration induced phase fluctuations $\phi_v(t)$ can be correlated with and predicted by an accelerometer so that it becomes possible to electronically cancel the effect of vibration-induced noise in the fiber (Hati et al., 2008).

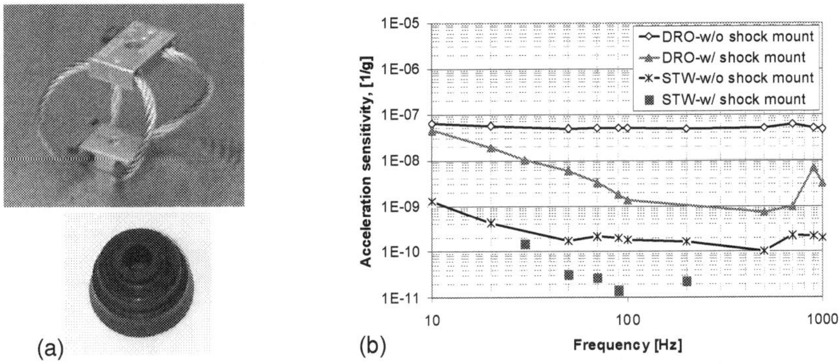

Figure 21. (a) Wire-rope and urethane passive shock mounts. (b) Improvement in acceleration sensitivity resulting from use of shock mounts

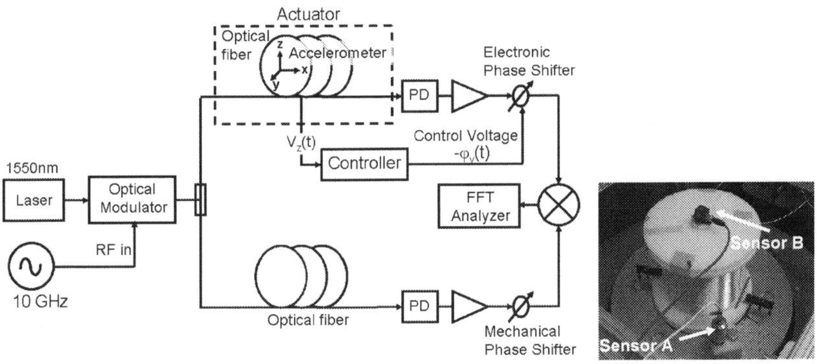

Figure 22. Experimental setup to study vibration-induced PM noise and its suppression in a fiber delay line mounted on a vibration table. PD is a photodetector that converts 10 GHz optical modulation to RF

The 3 km optical fiber (SMF-28) wound on a ceramic spool is subjected to vibration. While the fiber is under vibration, an estimate of the opposite phase of the $\phi_v(t)$ signal is generated based on vibration sensors, in this case, a z-axis accelerometer (sensor B) mounted on the top of the spool. The z-axis is the most sensitive axis, by an order of magnitude or more compared to the x and y axes (Ashby et al. 2007, Huang et al. 2000). An electronic phase shifter, as shown in Figure 22, corrects the phase perturbations of the demodulated 10 GHz signal sensed by the accelerometer by feed-forward correction. Figure 23(a) shows preliminary results and proof-of-concept of active noise control applied to the ceramic spool of optical fiber. The bottom curve is the noise floor, and the topmost curve is the same measurement of PM noise while the spool is subjected to a random vibration. The middle curve is the residual PM noise with the noise control on. Particularly noteworthy is that the residual PM noise through the spool of fiber is reduced by 15 dB to 25 dB. For this experiment, a flat frequency response is used for the feed-forward phase correction. The acceleration sensitivity of 3 km fiber with and without feed-forward cancellation is also shown in Figure 23(b).

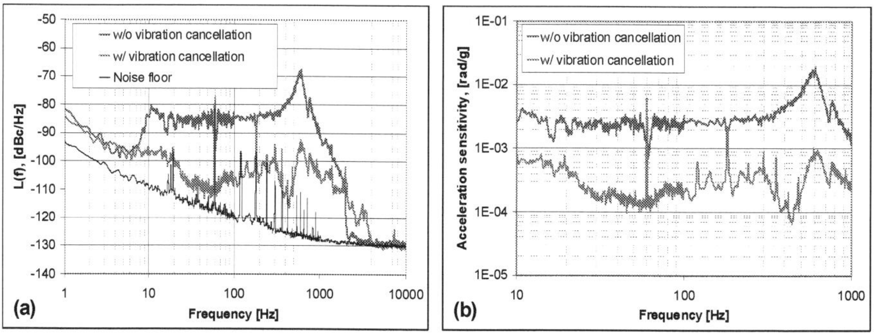

Figure 23. (a) Plot comparing the residual PM noise of fiber under random vibration with and without feed-forward cancellation. (b) Acceleration sensitivity of a 3 km fiber with and without feed-forward cancellation. A random vibration profile of acceleration PSD = 0.5 mg^2/Hz (rms) is used for 10 Hz $\leq f_v \leq$ 2000 Hz

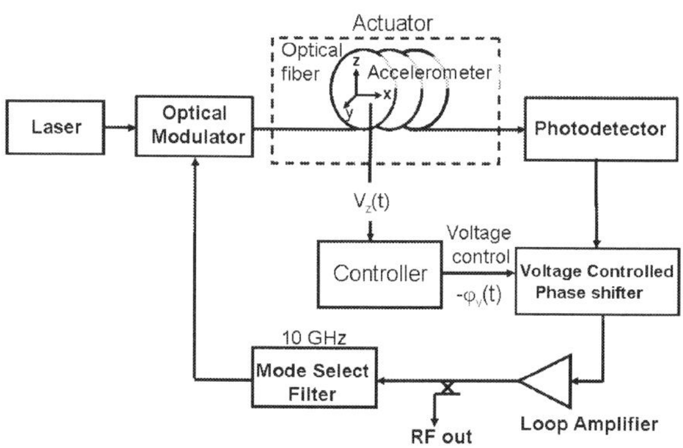

Figure 24. Block diagram of an OEO with active vibration-induced noise control

The same technique is implemented in an OEO to cancel the vibration-induced noise. In an OEO under vibration, an accelerometer signal is used to accurately estimate the complex conjugate of the vibration-induced phase modulation, as depicted in Figure 24. This estimate modulates the oscillator's output frequency by virtue of Leeson's model (Leeson, 1966) in such a way as to correct the induced in-loop phase perturbations. The PM noise of the 10 GHz OEO with and without vibration, as well as with vibration cancellation, is shown in Figure 25(a). Finally, the z-axis acceleration sensitivity of the OEO at 10 GHz is calculated by use of equation 7 and Figure 25(a) and is shown in Figure 25(b). There is an improvement by almost an order of magnitude in acceleration sensitivity over the full range of vibration frequencies tested. A flat frequency response is used for the feed-forward phase correction in this case. However, a custom-tailored frequency response can be used to achieve better cancellation at different vibration frequencies. In order to verify this, the fiber spool is subjected to a sinusoidal acceleration of 1 g at 10 Hz, 50 Hz, 100 Hz, 200 Hz and 1 kHz. The bottom curve of Figure 25(b) shows that further improvement in z-axis sensitivity is achieved by optimizing the phase and amplitude of the feed-forward phase correction at these specific sinusoidal frequencies.

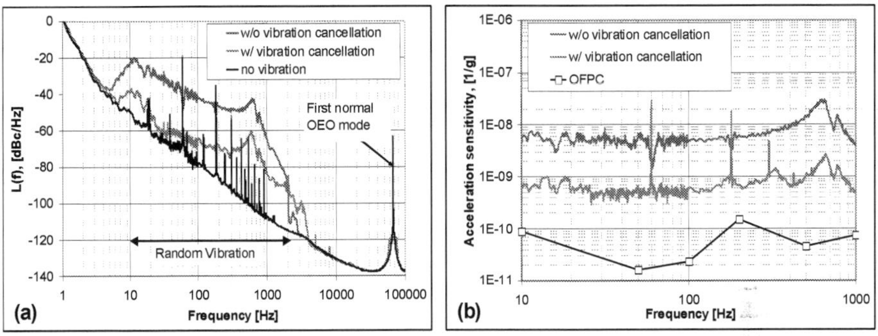

Figure 25. (a) PM noise of a 10 GHz OEO with and without vibration cancellation. A random vibration profile of acceleration PSD of 0.5 mg²/Hz (rms) is used for 10 Hz $\leq f_v \leq$ 2000 Hz. (b) Plot of z-axis acceleration sensitivity with and without vibration cancellation. The bottom curve corresponds to a single-frequency sinusoidal peak acceleration of 1 g. For each sinusoidal frequency, an optimized feed-forward phase correction (OFPC) is used

7. Conclusion

Structure-borne vibration is routine for many applications, causing an increase in PM noise of oscillators that disables many systems. Therefore, it is very important to select components that show low phase changes under vibration in order to build a system with low vibration sensitivity. In this chapter, the acceleration sensitivity and its relation to PM noise of an oscillator is derived. Different techniques for measuring the PM noise level of components under vibration are also discussed. The acceleration sensitivity of several state-of-the-art microwave and optical components are measured. Results show that high-Q components are generally more sensitive to vibration. Finally, different passive and active techniques to suppress or cancel vibration-induced noise in these oscillators are also discussed. Results show that proper use of passive and active cancellation schemes can improve the acceleration sensitivity of an oscillator by several orders of magnitude.

8. Acknowledgement

The authors thank Neil Ashby, Jeff Jargon and Jennifer Taylor for useful discussion and valuable suggestions.

9. References

Ashby, N.; Howe, D. A.; Taylor, J.; Hati, A. & Nelson, C. (2007). Optical fiber vibration and acceleration model, *2007 IEEE International Frequency Control Symposium Jointly with the 21st European Frequency and Time Forum*, pp. 547-551, Geneva, Switzerland, 29 May-1 June, 2007

Bloch, M; Mancini, O. & Stone, C. (2005). Method for achieving highly reproducible acceleration insensitive quartz crystal oscillators, *U.S. Patent 07106143*, 2006

Driscoll, M.M. (1993). Reduction of Quartz Oscillator flicker-of-frequency and white phase noise (floor) levels of acceleration sensitivity via use of multiple resonators. *IEEE Transactions on Ultrasonics, Ferroelectrics, and Frequency Control,* Vol. 40, No.4, pp. 427-430, July 1993

Driscoll, M.M. & Donovan, J.B. (2007). Vibration-Induced Phase Noise: It Isn't Just About the Oscillator, *Proceedings of 2007 IEEE International Frequency Control Symposium Jointly with the 21st European Frequency and Time Forum,* pp. 535-540, Geneva, Switzerland, 29 May–1 June, 2007

Eliyahu, D.; Seidel, D. & Maleki, L. (2008). RF amplitude and phase-noise reduction of an optical link and an opto-electronic oscillator. *IEEE Transactions on Microwave Theory and Techniques,* Vol. 56, No. 2, pp. 449-456, February 2008

Filler, R.L.; Kosinski, J.A. & Vig, J.R. (1983). Further Studies on the Acceleration Sensitivity of Quartz Resonators, *Proceedings of 37th Annual Symposium on Frequency Control,* pp. 265 - 271, 1983

Filler, R.L. (1988). The acceleration sensitivity of quartz crystal oscillators: A review. *IEEE Transactions on Ultrasonics, Ferroelectrics, and Frequency Control,* Vol. 35, No. 3, pp. 297-305, May 1988

Foreman, S.M; Ludlow, A.D; Miranda, M.H.G.; Stalnaker, J. E.; Diddams, S.A. & Ye, J. (2007). Coherent optical phase transfer over a 32-km fiber with 1 s instability at 10^{-17}. *Physical Review Letter,* Vol. 99, pp. 153601 (4), October 2007

Fruehauf, H. (2007). "g"-Compensated, Miniature, High Performance Quartz Crystal Oscillators, http://www.freqelec.com/tech_lit.html, April 2007.

Grove, J.; Hein, J.; Retta, J.; Schweiger, P.; Solbrig, W. & Stein, S.R. (2004). Direct-digital phase-noise measurement, *Proceedings of the 2004 IEEE International Frequency Control Symposium and Exposition,* pp. 287 - 291, 23-27, August 2004

Hati, A.; Howe, D.A. & Nelson, C.W. (2006). Comparison of AM noise in commercial amplifiers and oscillators at X-band, *Proceedings of 2006 IEEE International Frequency Control Symposium,* pp. 740-744, June 2006

Hati, A.; Nelson, C.W; Howe, D.A.; Ashby, N.; Taylor, J., Hudek, K.M.; Hay, C.; Seidel, D.J. & Eliyahu, D. (2007). Vibration sensitivity of microwave components, *Proceedings of 2007 IEEE International Frequency Control Symposium Jointly with the 21st European Frequency and Time Forum,* pp. 541-546, Geneva, Switzerland, 29 May–1 June, 2007

Hati, A.; Nelson, C.W.; Taylor, J.; Ashby, N. & Howe, D.A. (2008). Cancellation of Vibration-Induced Phase Noise in Optical Fibers. *IEEE Photonics Technology Letters, Accepted for future publication,* 2008

Hay, C.E.; Harrell, M.E. & Kansy, R.J. (2004). 2.4 and 2.5 GHz miniature, low-noise oscillators using surface transverse wave resonators and a SiGe sustaining amplifier, *Proceedings of the 2004 IEEE International Frequency Control Symposium and Exposition,* pp. 174 - 179, 23-27 August 2004

Healy, D.J.; Hahn, H. & Powell, S. (1983). A measurement technique for determination of frequency vs. acceleration characteristics of quartz crystal units, *Proceedings of the 37th Annual Symposium on Frequency Control,* pp. 284-289, 1983

Howe, D.A; Lanfranchi, J.; Cutsinger, L.; Hati, A. & Nelson, C.W. (2005). Vibration-induced PM noise in oscillators and measurements of correlation with vibration sensors, *Proceedings of the 2005 IEEE International Frequency Control Symposium and Exposition,* pp. 494-498, 29-31 August, 2005

Howe, D.A.; Hati, A.; Nelson, C.W.; Taylor, J. & Ashby, N. (2007). Active vibration-induced PM noise control in optical fibers: preliminary studies, *2007 IEEE International Frequency Control Symposium Jointly with the 21st European Frequency and Time Forum*, pp. 552-556, Geneva, Switzerland, 29 May–1 June, 2007

Huang, S.; Tu, M.; Yao, S. & Maleki, L. (2000). A turnkey optoelectronic oscillator with low acceleration sensitivity, *Proceedings of the 2000 IEEE/EIA International Frequency Control Symposium and Exhibition*, pp. 269-279, 2000

Kadiwar R.K & Giles, I.P. (1989). Optical fibre Brillouin ring laser gyroscope. *Electronics Letters*, Vol. 25, No. 25, pp. 1729 - 1731, 1989

Kosinski, J. A. (2000), Theory and design of crystal oscillators immune to acceleration: Present state of the art, *Proceedings of the 2000 IEEE/EIA International Frequency Control Symposium and Exhibition*, pp. 260-268, 2000.

Kwon, T.M. & Hahn, T. (1983). Improved vibration performance in passive atomic frequency standards by servo-loop control, *Proceedings of the 37th Annual Symposium on frequency control*, pp. 28-20, 1983

Lance, A.L.; Seal, W.D. & Labaar, F. (1984). Phase noise and AM noise measurements in the frequency domain. *Infrared and Millimeter Waves*, Vol. 11, pp. 239-289, 1984

Leeson, D.B. (1966). A simple model of feed back oscillator noise spectrum, *Proceeding of the IEEE*, Vol. 54, No. 2, pp. 329-330, 1966

Mancini, O. (2004). Tutorial on Precision Frequency Generation, *Microwave Theory & Techniques Society of the IEEE Long Island Section*, 18 March 2004. http://www.ieee.li/pdf/viewgraphs_freq.pdf

Minasian, R. (2006). Photonic signal processing of microwave signals. *IEEE Transactions on Microwave Theory and Technology*, Vol. 54, No. 2, pp. 832–846, February 2006

Parker, T.E. & Montress, G.K. (1988). Precision surface-acoustic-wave (SAW) oscillators. *IEEE Transactions on Ultrasonics, Ferroelectrics, and Frequency Control*, Vol. 35, No. 3, pp. 342 - 364, May 1988

Renoult, P.; Girardet, E. & Bidart, L. (1989). Mechanical and acoustic effects in low phase noise piezoelectric oscillators, *Proceeding of the 43rd Annual Symposium on Frequency Control*, pp. 439-446, 31 May-2 June, 1989

Riley, W. (1992). The Physics of the environmental sensitivity of rubidium gas cell atomic Frequency standards. *IEEE Transactions on Ultrasonics, Ferroelectrics, and Frequency Control*, Vol. 39, pp. 232-240, 1992

Römisch, S.; Kitching, J.; Ferre-Pikal, E.S.; Hollberg L. & Walls, F.L. (2000). Performance evaluation of an optoelectronic oscillator. *IEEE Transactions on Ultrasonics, Ferroelectrics, and Frequency Control*, Vol. 47, pp. 1159-1165, September 2000

Rosati V.J. & Filler, R. L. (1981). Reduction of the effects of vibration on SC-cut quartz crystal oscillators, *Proceedings of the 35th Annual Symposium on Frequency Control*, pp. 117-121, 1981

Rubiola, E.; Salik, E.; Huang, S.; Yu, N. & Maleki, L. (2005). Photonic delay technique for PM noise measurement of microwave oscillators. *Journal Optical Society of America B: Optical Physics*, Vol. 22, No.5, pp. 987-997, May 2005

Steinberg, D.S. (2000). *Vibration Analysis for Electronic Equipment*, John Wiley & Sons, Inc. 9780471376859, NY, USA

Sullivan, D.B; Allan, D.W.; Howe, D.A. & Walls F. L. (1990). Characterization of clocks and oscillators. *National Institute of Standards and Technology Technical Note 1337*, Section A-6, March 1990

Taylor, J.; Nelson, C.W.; Hati, A.; Ashby, N. & Howe, D.A. (2008). Vibration-induced PM noise measurements of a rigid optical fiber spool, *Proceedings of the 2008 IEEE International Frequency Control Symposium*, pp. 807-810, Hawaii, May 2008

Thieme, B., Zoschg, D. & Baister, G. (2004). Space worthy electronics package for the 35kg space active hydrogen maser on ACES, *18th European Frequency and Time Forum*, April 2004

Vig, J.R.; Audoin, C.; Cutler, L.S.; Driscoll, M.M.; EerNisse, E.P.; Filler, R.L.; Garvey, R.M.; Riley, W.L.; Smythe, R.C. & Weglein, R.D. (1992). Acceleration, vibration and shock effects-IEEE standards project P1193, *Proceedings of the 46rd Annual. Symposium on Frequency Control*, pp. 763-781, 1992

Wallin, T.; Josefsson, L. & Lofter, B. (2003). Phase noise performance of sapphire microwave oscillators in airborne radar systems, *GigaHertz 2003. Proceedings from the Seventh Symposium*, Linköping, Sweden, November 4–5, 2003

Walls, F.L. & Ferre-Pikal, E.S. (1999). Measurement of frequency, phase noise and amplitude noise. *Wiley Encyclopedia of Electrical and Electronics Engineering*, Vol. 12, pp. 459-473, June 1999

Watts, M.H.; EerNisse, E.P.; Ward, R.W. & Wiggins, R.B. (1988). Technique for measuring the acceleration sensitivity of SC-cut quartz resonators, *Proceedings of the 42rd Annual Symposium on Frequency Control*, pp. 442-446, 1988

Weglein, R.D. (1989). The vibration-induced phase noise of a visco-elastically supported crystal resonator, *Proceedings of the 43rd Annual Symposium on Frequency Control*, pp. 433-438, 1989

Yao, X.S. & Maleki, L. (1996). Optoelectronic microwave oscillator. *Journal Optical Society of America B*, Vol. 13, No.8, pp. 1725-1735, 1996

14

Neural Network Control and Wireless Sensor Network-based Localization of Quadrotor UAV Formations

Travis Dierks and S. Jagannathan
Missouri University of Science and Technology
United States of America

1. Introduction

In recent years, quadrotor helicopters have become a popular unmanned aerial vehicle (UAV) platform, and their control has been undertaken by many researchers (Dierks & Jagannathan, 2008). However, a team of UAV's working together is often more effective than a single UAV in scenarios like surveillance, search and rescue, and perimeter security. Therefore, the formation control of UAV's has been proposed in the literature.

Saffarian and Fahimi present a modified leader-follower framework and propose a model predictive nonlinear control algorithm to achieve the formation (Saffarian & Fahimi, 2008). Although the approach is verified via numerical simulations, proof of convergence and stability is not provided. In the work of Fierro et al., cylindrical coordinates and contributions from wheeled mobile robot formation control (Desai et al., 1998) are considered in the development of a leader-follower based formation control scheme for aircrafts whereas the complete dynamics are assumed to be known (Fierro et al., 2001). The work by Gu et al. proposes a solution to the leader-follower formation control problem involving a linear inner loop and nonlinear outer-loop control structure, and experimental results are provided (Gu et al., 2006). The associated drawbacks are the need for a dynamic model and the measured position and velocity of the leader has to be communicated to its followers. Xie et al. present two nonlinear robust formation controllers for UAV's where the UAV's are assumed to be flying at a constant altitude. The first approach assumes that the velocities and accelerations of the leader UAV are known while the second approach relaxes this assumption (Xie et al., 2005). In both the designs, the dynamics of the UAV's are assumed to be available. Then, Galzi and Shtessel propose a robust formation controller based on higher order sliding mode controllers in the presence of bounded disturbances (Galzi & Shtessel, 2006).

In this work, we propose a new leader-follower formation control framework for quadrotor UAV's based on spherical coordinates where the desired position of a follower UAV is specified using a desired separation, s_d, and a desired- angle of incidence, α_d and bearing, β_d. Then, a new control law for leader-follower formation control is derived using neural networks (NN) to learn the complete dynamics of the UAV online, including unmodeled dynamics like aerodynamic friction in the presence of bounded disturbances. Although a quadrotor UAV is underactuated, a novel NN virtual control input scheme for leader follower formation control is proposed which allows all six degrees of freedom of the UAV to be controlled using only four control inputs. Finally, we extend a graph theory-based scheme for discovery, localization and cooperative control. Discovery allows the UAV's to form into an ad hoc mobile sensor network whereas localization allows each UAV to estimate its position and orientation relative to its neighbors and hence the formation shape. This chapter is organized as follows. First, in Section 2, the leader-follower formation control problem for UAV's is introduced, and required background information is presented. Then, the NN control law is developed for the follower UAV's as well as the formation leader, and the stability of the overall formation is presented in Section 3. In Section 4, the localization and routing scheme is introduced for UAV formation control while Section 5 presents numerical simulations, and Section 6 provides some concluding remarks.

2. Background

2.1 Quadrotor UAV Dynamics

Consider a quadrotor UAV with six DOF defined in the inertial coordinate frame , E^a, as $[x, y, z, \phi, \theta, \psi]^T \in E^a$ where $\rho = [x, y, z]^T \in E^a$ are the position coordinates of the UAV and $\Theta = [\phi, \theta, \psi]^T \in E^a$ describe its orientation referred to as roll, pitch, and yaw, respectively. The translational and angular velocities are expressed in the body fixed frame attached to the center of mass of the UAV, E^b, and the dynamics of the UAV in the body fixed frame can be written as (Dierks & Jagannathan, 2008)

$$M\begin{bmatrix} \dot{v} \\ \dot{\omega} \end{bmatrix} = \bar{S}(\omega)\begin{bmatrix} v \\ \omega \end{bmatrix} + \begin{bmatrix} N_1(v) \\ N_2(\omega) \end{bmatrix} + \begin{bmatrix} G(R) \\ 0_{3x1} \end{bmatrix} + U + \tau_d \tag{1}$$

where $U = \begin{bmatrix} 0 & 0 & u_1 & u_2^T \end{bmatrix}^T \in \Re^6$,

$$M = \begin{bmatrix} mI_{3x3} & 0_{3x3} \\ 0_{3x3} & J \end{bmatrix} \in \Re^{6x6}, \quad \bar{S}(\omega) = \begin{bmatrix} -mS(\omega) & 0_{3x3} \\ 0_{3x3} & S(J\omega) \end{bmatrix} \in \Re^{6x6}$$

and m is a positive scalar that represents the total mass of the UAV, $J \in \Re^{3x3}$ represents the positive definite inertia matrix, $v(t) = [v_{xb}, v_{yb}, v_{zb}]^T \in \Re^3$ represents the translational velocity, $\omega(t) = \begin{bmatrix} \omega_{xb}, \omega_{yb}, \omega_{zb} \end{bmatrix}^T \in \Re^3$ represents the angular velocity, $N_i(\bullet) \in \Re^{3x1}, i = 1, 2$, are the nonlinear aerodynamic effects, $u_1 \in \Re^1$ provides the thrust along the z-direction, $u_2 \in \Re^3$ provides the rotational torques, $\tau_d = [\tau_{d1}^T, \tau_{d2}^T]^T \in \Re^6$ and $\tau_{di} \in \Re^3$, $i = 1, 2$ represents unknown, but bounded disturbances such that $\|\tau_d\| < \tau_M$ for all time t, with τ_M being a known positive constant, $I_{nxn} \in \Re^{nxn}$ is an nxn identity matrix, and $0_{mxl} \in \Re^{mxl}$ represents an mxl matrix of all zeros. Furthermore, $G(R) \in \Re^3$ represents the gravity vector defined as $G(R) = mgR^T(\Theta)E_z$ where $E_z = [0,0,1]^T$ is a unit vector in the inertial coordinate frame, $g = 9.81 m/s^2$, and $S(\bullet) \in \Re^{3x3}$ is the general form of a skew symmetric matrix defined as in (Dierks & Jagannathan, 2008). It is important to highlight $w^T S(\gamma)w = 0$ for any vector $w \in \Re^3$, and this property is commonly referred to as the *skew symmetric property* (Lewis et al., 1999).

The matrix $R(\Theta) \in \Re^{3x3}$ is the translational rotation matrix which is used to relate a vector in the body fixed frame to the inertial coordinate frame defined as (Dierks & Jagannathan, 2008)

$$R(\Theta) = R = \begin{bmatrix} c_\theta c_\psi & s_\phi s_\theta c_\psi - c_\phi s_\psi & c_\phi s_\theta c_\psi + s_\phi s_\psi \\ c_\theta s_\psi & s_\phi s_\theta s_\psi + c_\phi c_\psi & c_\phi s_\theta s_\psi - s_\phi c_\psi \\ -s_\theta & s_\phi c_\theta & c_\phi c_\theta \end{bmatrix} \tag{2}$$

where the abbreviations $s_{(\bullet)}$ and $c_{(\bullet)}$ have been used for $\sin(\bullet)$ and $\cos(\bullet)$, respectively. It is important to note that $R^{-1} = R^T$, $\dot{R} = RS(\omega)$ and $\dot{R}^T = -S(\omega)R^T$. It is also necessary to define a rotational transformation matrix from the fixed body to the inertial coordinate frame as (Dierks & Jagannathan, 2008)

$$T(\Theta) = T = \begin{bmatrix} 1 & s_\phi t_\theta & c_\phi t_\theta \\ 0 & c_\phi & -s_\phi \\ 0 & s_\phi / c_\theta & c_\phi / c_\theta \end{bmatrix} \tag{3}$$

where the abbreviation $t_{(\bullet)}$ has been used for $\tan(\bullet)$. The transformation matrices R and T are nonsingular as long as $-(\pi/2) < \phi < (\pi/2)$, $-(\pi/2) < \theta < (\pi/2)$ and $-\pi \le \psi \le \pi$. These regions will be assumed throughout the development of this work, and will be referred to as the *stable operation regions* of the UAV. Under these flight conditions, it is observed that $\|R\|_F = R_{\max}$ and $\|T\|_F < T_{\max}$ for known constants $R_{\max}$ and $T_{\max}$ (Neff et al., 2007).

Finally, the kinematics of the UAV can be written as

$$\begin{aligned} \dot{\rho} &= Rv \\ \dot{\Theta} &= T\omega \end{aligned} \tag{4}$$

2.2 Neural Networks

In this work, two-layer NN's are considered consisting of one layer of randomly assigned constant weights $V_N \in \Re^{axL}$ in the first layer and one layer of tunable weights $W_N \in \Re^{Lxb}$ in the second with a inputs, b outputs, and L hidden neurons. A compromise is made here between tuning the number of layered weights with computational complexity. The *universal approximation property* for NN's (Lewis et al., 1999) states that for any smooth function $f_N(x_N)$, there exists a NN such that $f_N(x_N) = W_N^T \sigma(V_N^T x_N) + \varepsilon_N$ where ε_N is the bounded NN functional approximation error such that $\|\varepsilon_N\| < \varepsilon_M$, for a known constant ε_M

and $\sigma(\cdot):\Re^a \rightarrow \Re^L$ is the activation function in the hidden layers. It has been shown that by randomly selecting the input layer weights V_N, the activation function $\sigma(\bar{x}_N)=\sigma(V_N^T x_N)$ forms a stochastic basis, and thus the approximation property holds for all inputs, $x_N \in \Re^a$, in the compact set S. The sigmoid activation function is considered here. Furthermore, on any compact subset of $\Re^n$, the target NN weights are bounded by a known positive value, W_M, such that $\|W_N\|_F \le W_M$. For complete details of the NN and its properties, see (Lewis et al., 1999).

2.3 Three Dimensional Leader-Follower Formation Control

Throughout the development, the follower UAV's will be denoted with a subscript 'j' while the formation leader will be denoted by the subscript 'i'. To begin the development, an alternate reference frame is defined by rotating the inertial coordinate frame about the z-axis by the yaw angle of follower j, ψ_j, and denoted by E_j^a. In order to relate a vector in E^a to E_j^a, the transformation matrix is given by

$$R_{aj}=\begin{bmatrix}\cos\psi_j & \sin\psi_j & 0\\ -\sin\psi_j & \cos\psi_j & 0\\ 0 & 0 & 1\end{bmatrix}, \tag{5}$$

where $R_{aj}^T = R_{aj}^{-1}$.

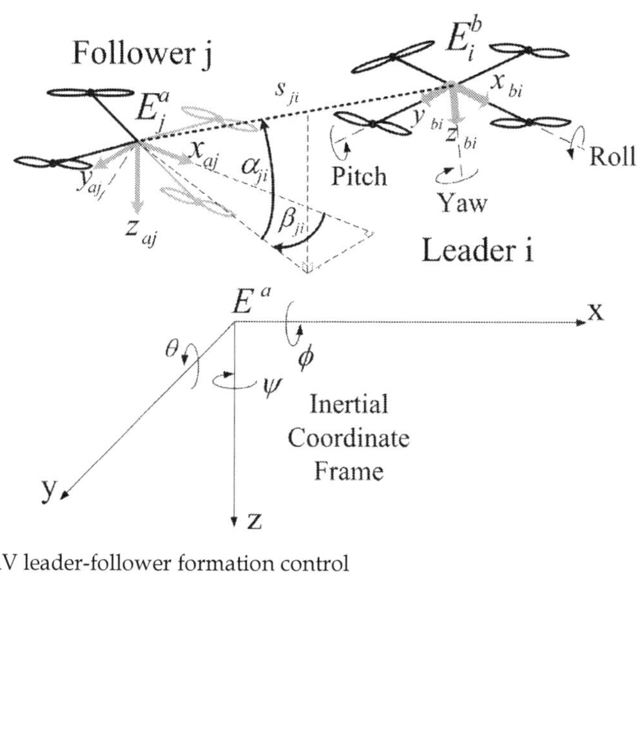

Figure 1. UAV leader-follower formation control

The objective of the proposed leader-follower formation control approach is for the follower UAV to maintain a desired separation, s_{jid}, at a desired angle of incidence, $\alpha_{jid} \in E_j^a$, and bearing, $\beta_{jid} \in E_j^a$, with respect to its leader. The incidence angle is measured from the $x_{aj}-y_{aj}$ plane of follower *j* while the bearing angle is measured from the positive x_{aj}-axis as shown in Figure 1. It is important to observe that each quantity is defined relative to the follower *j* instead of the leader *i* (Fierro et al., 2001), (Desai et al., 1998). Additionally, in order to specify a unique configuration of follower *j* with respect to its leader, the desired yaw of follower *j* is selected to be the yaw angle of leader *i*, $\psi_i \in E^a$ as in (Saffarian & Fahimi, 2008). Using this approach, the measured separation between follower *j* and leader *i* is written as

$$\rho_i - \rho_j = R_{aj}^T s_{ji} \Xi_{ji}, \tag{6}$$

where

$$\Xi_{ji} = \begin{bmatrix} \cos\alpha_{ji} \cos\beta_{ji} \\ \cos\alpha_{ji} \sin\beta_{ji} \\ \sin\alpha_{ji} \end{bmatrix}. \tag{7}$$

Thus, to solve the leader-follower formation control problem in the proposed framework, a control velocity must be derived to ensure

$$\left.\begin{array}{l} \lim_{t\to\infty}(s_{jid} - s_{ji}) = 0,\ \lim_{t\to\infty}(\beta_{jid} - \beta_{ji}) = 0, \\ \lim_{t\to\infty}(\alpha_{jid} - \alpha_{ji}) = 0,\ \lim_{t\to\infty}(\psi_{jd} - \psi_j) = 0 \end{array}\right\}. \tag{8}$$

Throughout the development, s_{jid}, α_{jid} and β_{jid} will be taken as constants, while the constant total mass, m_j, is assumed to be known. Additionally, it will be assumed that reliable communication between the leader and its followers is available, and the leader communicates its measured orientation, Θ_i, and its *desired* states, $\psi_{id}, \dot{\psi}_{id}, \ddot{\psi}_{id}$, $v_{id}, \dot{v}_{id}$. This is a far less stringent assumption than assuming the leader communicates all of its *measured* states to its followers (Gu et al., 2006). Additionally, future work will relax this assumption. In the following section, contributions from single UAV control will be considered and extended to the leader-follower formation control of UAV's.

3. Leader-Follower Formation Tracking Control

In single UAV control literature, the overall control objective UAV *j* is often to track a desired trajectory, $\rho_{jd} = [x_{jd}, y_{jd}, z_{jd}]^T$, and a desired yaw ψ_{jd} while maintaining a stable flight configuration (Dierks & Jagannathan, 2008). The velocity v_{jzb} is directly controllable with the thrust input. However, in order to control the translational velocities v_{jxb} and v_{jyb}, the pitch and roll must be controlled, respectively, thus redirecting the thrust. With these objectives in mind, the frameworks for single UAV control are extended to UAV formation control as follows.

3.1 Follower UAV Control Law

Given a leader i subject to the dynamics and kinematics (1) and (4), respectively, define a reference trajectory at a desired separation s_{jid}, at a desired angle of incidence, α_{jid}, and bearing, β_{jid} for follower j given by

$$\rho_{jd} = \rho_i - R_{ajd}^T s_{jid} \Xi_{jid} \tag{9}$$

where R_{ajd} is defined as in (5) and written in terms of ψ_{jd}, and Ξ_{jid} is written in terms of the desired angle of incidence and bearing, $\alpha_{jid}, \beta_{jid}$, respectively, similarly to (7). Next, using (6) and (9), define the position tracking error as

$$e_{j\rho} = \rho_{jd} - \rho_j = R_{aj}^T s_{ji} \Xi_{ji} - R_{ajd}^T s_{jid} \Xi_{jid} \in E^a \tag{10}$$

which can be measured using local sensor information. To form the position tracking error dynamics, it is convenient to rewrite (10) as $e_{j\rho} = \rho_i - \rho_j - R_{ajd}^T s_{jid} \Xi_{jid}$ revealing

$$\dot{e}_{j\rho} = R_i v_i - R_j v_j - \dot{R}_{ajd}^T s_{jid} \Xi_{jid} . \tag{11}$$

Next, select the desired translational velocity of follower j to stabilize (11)

$$v_{jd} = [v_{jdx} \; v_{jdy} \; v_{jdz}]^T = R_j^T \left(R_i v_{id} - \dot{R}_{ajd}^T s_{jid} \Xi_{jid} + K_{j\rho} e_{j\rho}\right) \in E^b \tag{12}$$

where $K_{j\rho} = diag\{k_{j\rho x}, k_{j\rho y}, k_{j\rho z}\} \in \Re^{3x3}$ is a diagonal positive definite design matrix of positive design constants and v_{id} is the desired translational velocity of leader i. Next, the translational velocity tracking error system is defined as

$$e_{jv} = \begin{bmatrix} e_{jvx} \\ e_{jvy} \\ e_{jvz} \end{bmatrix} = \begin{bmatrix} v_{jdx} \\ v_{jdy} \\ v_{jdz} \end{bmatrix} - \begin{bmatrix} v_{jxb} \\ v_{jyb} \\ v_{jzb} \end{bmatrix} = v_{jd} - v_j . \tag{13}$$

Applying (12) to (11) while observing $v_j = v_{jd} - e_{jv}$ and similarly $e_{iv} = v_{id} - v_i$, reveals the closed loop position error dynamics to be rewritten as

$$\dot{e}_{j\rho} = -K_{j\rho} e_{j\rho} + R_j e_{jv} - R_i e_{iv} . \tag{14}$$

Next, the translational velocity tracking error dynamics are developed. Differentiating (13), observing

$$\dot{v}_{jd} = -S(\omega_j)v_{jd} + R_j^T\left(R_i S(\omega_i)v_{id} + R_i\dot{v}_{id} - \ddot{R}_{ajd}^T s_{jid}\Xi_{jid}\right) + R_j^T K_{j\rho}\left(R_i v_i - R_j v_j - \dot{R}_{ajd}^T s_{jid}\Xi_{jid}\right),$$

substituting the translational velocity dynamics in (1), and adding and subtracting $R_j^T(K_{j\rho}(R_i v_{id} + R_j v_{jd}))$ reveals

$$\begin{aligned}\dot{e}_{jv} = \dot{v}_{jd} - \dot{v}_j = -N_{j1}(v_j)/m_j - S(\omega_j)e_{jv} - G(R_j)/m_j - u_{j1}E_{jz}/m_j - \bar{\tau}_{jd1} \\ + R_j^T\left(R_i S(\omega_i)v_{id} + R_i\dot{v}_{id} - \ddot{R}_{ajd}^T s_{jid}\Xi_{jid} + K_{j\rho}\left(R_j e_{jv} - K_{j\rho}e_{j\rho}\right)\right) - R_j^T K_{j\rho}(R_i e_{iv} - R_j e_{jv})\end{aligned} \quad (15)$$

Next, we rewrite (2) in terms of the *scaled* desired orientation vector, $\bar{\Theta}_{jd} = [\bar{\theta}_{jd}\ \bar{\phi}_{jd}\ \psi_{jd}]^T$ where $\bar{\theta}_{jd} = \pi\theta_{jd}/(2\theta_{d\max})$, $\bar{\phi}_{jd} = \pi\phi_{jd}/(2\phi_{d\max})$, and $\theta_{d\max} \in (0, \pi/2)$ and $\phi_{d\max} \in (0, \pi/2)$ are the maximum desired roll and pitch, respectively, define $R_{jd} = R_j(\bar{\Theta}_{jd})$, and add and subtract $G(R_{jd})/m_j$ and $R_{jd}^T\Lambda_j$ with $\Lambda_j = R_i\dot{v}_{id} - \ddot{R}_{ajd}^T s_{jid}\Xi_{jid} + K_{j\rho}R_j e_{jv} - K_{j\rho}e_{j\rho}$ to $\dot{e}_{jv}$ to yield

$$\dot{e}_{jv} = -G(R_{jd})/m_j + R_{jd}^T\Lambda_j + A_{jc1}f_{cj1}(x_{cj1}) - u_{j1}E_{jz}/m_j - K_{j\rho}R_i e_{iv} - \bar{\tau}_{jd1} \quad (16)$$

where $A_{jc1} = diag\{\cos(\bar{\theta}_{jd}), \cos(\bar{\phi}_{jd}), 1\} \in \Re^{3x3}$ and

$$\begin{aligned}f_{cj1}(x_{cj1}) = A_{jc1}^{-1}\left(G(R_{jd})/m_j - G(R_j)/m_j + (R_j^T - R_{jd}^T)\Lambda_j\right) + \\ A_{jc1}^{-1}\left(K_{j\rho}R_j e_{jv} - S(\omega_j)e_{jv} - N_{j1}(v_j)/m_j + R_j^T R_i S(\omega_i)v_{id} + R_j^T K_{j\rho}(1 - K_{j\rho})e_{j\rho}\right)\end{aligned} \quad (17)$$

is an unknown function which can be rewritten as $f_{jc1}(x_{jc1}) = \left[f_{jc11}\ f_{jc12}\ f_{jc13}\right]^T \in \Re^3$. In the forthcoming development, the approximation properties of NN will be utilized to estimate the unknown function $f_{jc1}(x_{jc1})$ by bounded ideal weights W_{jc1}^T, V_{jc1}^T such that $\|W_{jc1}\|_F \le W_{Mc1}$ for an unknown constant W_{Mc1}, and written as $f_{jc1}(x_{jc1}) = W_{jc1}^T\sigma(V_{jc1}^T x_{jc1}) + \varepsilon_{jc1}$ where $\varepsilon_{jc1} \le \varepsilon_{Mc1}$ is the bounded NN approximation error where ε_{Mc1} is a known constant. The NN estimate of f_{jc1} is written as $\hat{f}_{jc1} = \hat{W}_{jc1}^T\sigma\left(V_{jc1}^T\hat{x}_{jc1}\right) = \hat{W}_{jc1}^T\hat{\sigma}_{jc1} = [\hat{W}_{jc11}^T\hat{\sigma}_{jc1}\ \hat{W}_{jc12}^T\hat{\sigma}_{jc1}\ \hat{W}_{jc13}^T\hat{\sigma}_{jc1}]^T$ where $\hat{W}_{jc1}^T$ is the NN estimate of W_{jc1}^T, $\hat{W}_{jc1i}^T, i = 1,2,3$ is the ith row of $\hat{W}_{jc1}^T$, and $\hat{x}_{jc1}$ is the NN input defined as

$$\hat{x}_{jc1} = [1\ \Theta_j^T\ \Theta_i^T\ \Lambda_j^T\ v_{jd}^T\ v_{id}^T\ \dot{v}_{id}^T\ \psi_{jd}\ \dot{\psi}_{jd}\ \ddot{\psi}_{jd}\ \omega_j^T\ v_j^T\ e_{jv}^T\ e_{j\rho}^T]^T.$$

Note that $\hat{x}_{jc1}$ is an estimate of x_{jc1} since the follower does not know ω_i. However, Θ_i is directly related to ω_i; therefore, it is included instead.

Remark 1: In the development of (16), the scaled desired orientation vector was utilized as a design tool to specify the desired pitch and roll angles. If the un-scaled desired orientation

vector was used instead, the maximum desired pitch and roll would remain within the stable operating regions. However, it is desirable to saturate the desired pitch and roll before they reach the boundaries of the stable operating region.
Next, the virtual control inputs θ_{jd} and ϕ_{jd} are identified to control the translational velocities v_{jxb} and v_{jyb}, respectively. The key step in the development is identifying the *desired* closed loop velocity tracking error dynamics. For convenience, the *desired* translational velocity closed loop system is selected as

$$\dot{e}_{jv} = -K_{jv}e_{jv} - \bar{\tau}_{jd1} - K_{j\rho}R_i e_{iv} \tag{18}$$

where $K_{jv} = diag\{k_{jv1}\cos(\bar{\theta}_{jd}), k_{jv2}\cos(\bar{\phi}_{jd}), k_{v3}\}$ is a diagonal positive definite design matrix with each $k_{vi} > 0$, $i = 1,2,3$, and $\bar{\tau}_{jd1} = \tau_{jd1}/m_j$. In the following development, it will be shown that $\theta_d \in (-\pi/2, \pi/2)$ and $\phi_d \in (-\pi/2, \pi/2)$; therefore, it is clear that $K_v > 0$. Then, equating (16) and (18) while considering only the first two velocity error states reveals

$$-g\begin{bmatrix} -s_{\bar{\theta}jd} \\ c_{\bar{\theta}jd}s_{\bar{\phi}jd} \end{bmatrix} + \begin{bmatrix} c_{\bar{\theta}jd}(k_{jv1}e_{jvx} + f_{jc11}) \\ c_{\bar{\phi}jd}(k_{jv2}e_{jvy} + f_{jc12}) \end{bmatrix} + \begin{bmatrix} c_{\bar{\theta}jd}c_{\psi jd} & c_{\bar{\theta}jd}s_{\psi jd} & -s_{\bar{\theta}jd} \\ s_{\bar{\phi}jd}s_{\bar{\theta}jd}c_{\psi jd} - c_{\bar{\phi}jd}s_{\psi jd} & s_{\bar{\phi}jd}s_{\bar{\theta}jd}s_{\psi jd} + c_{\bar{\phi}jd}c_{\psi jd} & s_{\bar{\phi}jd}c_{\bar{\theta}jd} \end{bmatrix} \begin{bmatrix} \Lambda_{j1} \\ \Lambda_{j2} \\ \Lambda_{j3} \end{bmatrix} = \begin{bmatrix} 0 \\ 0 \end{bmatrix} \tag{19}$$

where $\Lambda_j = [\Lambda_{j1} \ \ \Lambda_{j2} \ \ \Lambda_{j3}]^T$ was utilized. Then, applying basic math operations, the first line of (19) can be solved for the desired pitch θ_{jd} while the second line reveals the desired roll ϕ_{jd}. Using the NN estimates, $\hat{f}_{cj1}$, The desired pitch θ_{jd} can be written as

$$\theta_{jd} = \frac{2\theta_{\max}}{\pi} a\tan\left(\frac{N_{\theta jd}}{D_{\theta jd}}\right) \tag{20}$$

where $N_{\theta jd} = c_{\psi jd}\Lambda_{j1} + s_{\psi jd}\Lambda_{j2} + k_{jv1}e_{jvx} + \hat{f}_{jc11}$ and $D_{\theta jd} = \Lambda_{j3} - g$. Similarly, the desired roll angle, ϕ_{jd}, is found to be

$$\phi_{jd} = \frac{2\phi_{\max}}{\pi} a\tan\left(\frac{N_{\phi jd}}{D_{\phi jd}}\right) \tag{21}$$

where $N_{\phi jd} = s_{\psi jd}\Lambda_{j1} - c_{\psi jd}\Lambda_{j2} - k_{jv2}e_{jvy} + \hat{f}_{jc12}$ and $D_{\phi jd} = c_{\bar{\theta}jd}(\Lambda_{j3} - g) + s_{\bar{\theta}jd}c_{\psi jd}\Lambda_{j1} + s_{\bar{\theta}jd}s_{\psi jd}\Lambda_{j2}$.

Remark 2: The expressions for the desired pitch and roll in (20) and (21) lend themselves very well to the control of a quadrotor UAV. The expressions will always produce desired values in the *stable operation regions* of the UAV. It is observed that $a\tan(\bullet)$ approaches $\pm\pi/2$ as its argument increases. Thus, introducing the scaling factors in $\bar{\theta}_{jd}$ and $\bar{\phi}_{jd}$ results in $\theta_{jd} \in (-\theta_{\max}, \theta_{\max})$ and $\phi_{jd} \in (-\phi_{\max}, \phi_{\max})$, and the aggressiveness of the UAV's maneuvers can be managed.

Now that the desired orientation has been found, next define the attitude tracking error as

$$e_{j\Theta} = \Theta_{jd} - \Theta_j \in E^a \tag{22}$$

where the dynamics are found using (4) to be $\dot{e}_{j\Theta} = \dot{\Theta}_{jd} - T_j\omega_j$. In order to drive the orientation errors (22) to zero, the desired angular velocity, ω_{jd}, is selected as

$$\omega_{jd} = T_j^{-1}(\dot{\Theta}_{jd} + K_{j\Theta}e_{j\Theta}) \tag{23}$$

where $K_{j\Theta} = diag\{k_{j\Theta 1}, k_{j\Theta 2}, k_{j\Theta 3}\} \in \Re^{3x3}$ is a diagonal positive definite design matrix all with positive design constants. Define the angular velocity tracking error as

$$e_{j\omega} = \omega_{jd} - \omega_j \tag{24}$$

and observing $\omega_j = \omega_{jd} - e_{j\omega}$, the closed loop orientation tracking error system can be written as

$$\dot{e}_{j\Theta} = -K_{j\Theta}e_{j\Theta} + T_j e_{j\omega} \tag{25}$$

Examining (23), calculation of the desired angular velocity requires knowledge of $\dot{\Theta}_{jd}$; however, $\dot{\Theta}_{jd}$ is not known in view of the fact $\dot{\Lambda}_j$ and $\dot{\hat{f}}_{jc1}$ are not available. Further, development of u_{j2} in the following section will reveal $\dot{\omega}_{jd}$ is required which in turn implies $\ddot{\Lambda}_j$ and $\ddot{\hat{f}}_{jc1}$ must be known. Since these requirements are not practical, the *universal approximation property* of NN is invoked to estimate ω_{jd} and $\dot{\omega}_{jd}$ (Dierks and Jagannathan, 2008).

To aid in the NN virtual control development, the desired orientation, $\Theta_{jd} \in E^a$, is reconsidered in the fixed body frame, E^b, using the relation $\dot{\Theta}^b_{jd} = T_j^{-1}\dot{\Theta}_{jd}$. Rearranging (23), the dynamics of the proposed virtual controller when the all dynamics are known are revealed to be

$$\begin{aligned}\dot{\Theta}^b_{jd} &= \omega_{jd} - T_j^{-1}K_{j\Theta}e_{j\Theta} \\ \dot{\omega}_{jd} &= \dot{T}_j^{-1}(\dot{\Theta}_{jd} + K_{j\Theta}e_{j\Theta}) + T_j^{-1}(\ddot{\Theta}_{jd} + K_{j\Theta}\dot{e}_{j\Theta})\end{aligned} \tag{26}$$

For convenience, we define a change of variable as $\Omega_d = \omega_d - T^{-1}K_\Theta e_\Theta$, and the dynamics (26) become

$$\begin{aligned}\dot{\Theta}^b_{jd} &= \Omega_{jd} \\ \dot{\Omega}_{jd} &= \dot{T}_j^{-1}\dot{\Theta}_{jd} + T_j^{-1}\ddot{\Theta}_{jd} = f_{j\Omega}(x_{j\Omega}) = f_{j\Omega}\end{aligned} \tag{27}$$

Defining the estimates of Θ^b_{jd} and Ω_{jd} to be $\hat{\Theta}^b_{jd}$ and $\hat{\Omega}_{jd}$, respectively, and the estimation error $\tilde{\Theta}^b_{jd} = \Theta^b_{jd} - \hat{\Theta}^b_{jd}$, the dynamics of the proposed NN virtual control inputs become

$$\begin{aligned} \dot{\hat{\Theta}}^b_{jd} &= \hat{\Omega}_{jd} + K_{j\Omega 1}\tilde{\Theta}^b_{jd} \\ \dot{\hat{\Omega}}_{jd} &= \hat{f}_{j\Omega} + K_{j\Omega 2}\tilde{\Theta}^b_{jd} \end{aligned} \tag{28}$$

where $K_{j\Omega 1}$ and $K_{j\Omega 2}$ are positive constants. The estimate $\hat{\omega}_{jd}$ is then written as

$$\hat{\omega}_{jd} = \hat{\Omega}_{jd} + K_{j\Omega 3}\tilde{\Theta}^b_{jd} + T_j^{-1}K_{j\Theta}e_{j\Theta} \tag{29}$$

where $K_{j\Omega 3}$ is a positive constant.

In (28), *universal approximation property* of NN has been utilized to estimate the unknown function $f_{j\Omega}(x_{j\Omega})$ by bounded ideal weights $W^T_{j\Omega}, V^T_{j\Omega}$ such that $\|W_{j\Omega}\|_F \le W_{M\Omega}$ for a known constant $W_{M\Omega}$, and written as $f_{j\Omega}(x_{j\Omega}) = W^T_{j\Omega}\sigma(V^T_{j\Omega}x_{j\Omega}) + \varepsilon_{j\Omega}$ where $\varepsilon_{j\Omega}$ is the bounded NN approximation error such that $\|\varepsilon_{j\Omega}\| \le \varepsilon_{\Omega M}$ for a known constant $\varepsilon_{\Omega M}$. The NN estimate of $f_{j\Omega}$ is written as $\hat{f}_{j\Omega} = \hat{W}^T_{j\Omega}\sigma(V^T_{j\Omega}\hat{x}_{j\Omega}) = \hat{W}^T_{j\Omega}\hat{\sigma}_{j\Omega}$ where $\hat{W}^T_{j\Omega}$ is the NN estimate of $W^T_{j\Omega}$ and $\hat{x}_{j\Omega}$ is the NN input written in terms of the virtual control estimates, desired trajectory, and the UAV velocity. The NN input is chosen to take the form of $\hat{x}_{\Omega} = [1\ \Lambda_j^T\ (\Theta^b_{jd})^T\ \hat{\Omega}^T_{jd}\ v_j^T\ \omega_j^T]^T$.

Observing $\tilde{\omega}_{jd} = \omega_{jd} - \hat{\omega}_{jd} = \tilde{\Omega}_{jd} - K_{j\Omega 3}\tilde{\Theta}^b_{jd}$, subtracting (28) from (27) and adding and subtracting $W^T_{j\Omega}\hat{\sigma}_{j\Omega}$, the virtual controller estimation error dynamics are found to be

$$\begin{aligned} \dot{\tilde{\Theta}}^b_{jd} &= \tilde{\omega}_{jd} - (K_{j\Omega 1} - K_{j\Omega 3})\tilde{\Theta}^b_{jd} \\ \dot{\tilde{\Omega}}_{jd} &= \tilde{f}_{j\Omega} - K_{j\Omega 2}\tilde{\Theta}^b_{jd} + \xi_{j\Omega} \end{aligned} \tag{30}$$

where $\tilde{\Omega}_{jd} = \Omega_{jd} - \hat{\Omega}_{jd}$, $\tilde{f}_{j\Omega} = \tilde{W}^T_{j\Omega}\hat{\sigma}_{j\Omega}$, $\tilde{W}^T_{j\Omega} = W^T_{j\Omega} - \hat{W}^T_{j\Omega}$, $\xi_{j\Omega} = \varepsilon_{j\Omega} + W^T_{j\Omega}\tilde{\sigma}_{j\Omega}$, and $\tilde{\sigma}_{j\Omega} = \sigma_{j\Omega} - \hat{\sigma}_{j\Omega}$. Furthermore, $\|\xi_{j\Omega}\| \le \xi_{\Omega M}$ with $\xi_{\Omega M} = \varepsilon_{\Omega M} + 2W_{M\Omega}\sqrt{N_\Omega}$ a computable constant with N_Ω the constant number of hidden layer neurons in the virtual control NN. Similarly, the estimation error dynamics of (29) are found to be

$$\dot{\tilde{\omega}}_{jd} = -K_{j\Omega 3}\tilde{\omega}_{jd} + \tilde{f}_{j\Omega} - \bar{K}_{j\Omega}\tilde{\Theta}^b_{jd} + \xi_{j\Omega} \tag{31}$$

where $\bar{K}_{j\Omega} = K_{j\Omega 2} - K_{j\Omega 3}(K_{j\Omega 1} - K_{j\Omega 3})$. Examination of (30) and (31) reveals $\tilde{\Theta}^b_{jd}, \tilde{\omega}_{jd}$, and $\tilde{f}_{j\Omega}$ to be equilibrium points of the estimation error dynamics when $\|\xi_{j\Omega}\| = 0$.

To this point, the desired translational velocity for follower j has been identified to ensure the leader-follower objective (8) is achieved. Then, the desired pitch and roll were derived to drive $v_{jxb} \rightarrow v_{jdx}$ and $v_{jyb} \rightarrow v_{jdy}$, respectively. Then, the desired angular velocity was found to ensure $\Theta_j \rightarrow \Theta_{jd}$. What remains is to identify the UAV thrust to guarantee $v_{jzb} \rightarrow v_{jdz}$ and rotational torque vector to ensure $\omega_j \rightarrow \omega_{jd}$. First, the thrust is derived.

Consider again the translational velocity tracking error dynamics (16), as well as the *desired* velocity tracking error dynamics (18). Equating (16) and (18) and manipulating the third error state, the required thrust is found to be

$$\begin{aligned} u_{j1} &= m_j c_{\bar{\theta}jd} c_{\bar{\phi}jd} (\Lambda_{j3} - g) + m_j (c_{\bar{\phi}jd} s_{\bar{\theta}jd} c_{\psi jd} + s_{\bar{\phi}jd} s_{\psi jd}) \Lambda_{j1} \\ &+ m_j (c_{\bar{\phi}jd} s_{\bar{\theta}jd} s_{\psi jd} - s_{\bar{\phi}jd} c_{\psi jd}) \Lambda_{j2} + m_j k_{jvz} e_{vj3} + m_j \hat{f}_{jc13} \end{aligned} \tag{32}$$

where $\hat{f}_{jc13}$ is the NN estimate in (17) previously defined. Substituting the desired pitch (20), roll (21), and the thrust (32) into the translational velocity tracking error dynamics (16) yields

$$\dot{e}_{jv} = -K_{jv} e_{jv} + A_{jc1}(W_{jc1}^T \sigma_{jc1} + \varepsilon_{jc}) - A_{jc1} \hat{W}_{jc1}^T \hat{\sigma}_{jc1} - K_{jp} R_i e_{iv} - \bar{\tau}_{jd1},$$

and adding and subtracting $A_{jc1} W_{jc1}^T \hat{\sigma}_{jc1}^T$ reveals

$$\dot{e}_{jv} = -K_{jv} e_{jv} + A_{jc1} \tilde{W}_{jc1}^T \hat{\sigma}_{jc1} - K_{jp} R_i e_{iv} + \xi_{jc1} \tag{33}$$

with $\xi_{jc1} = A_{jc1} W_{jc1}^T \tilde{\sigma}_{jc1}^T + A_{jc1} \varepsilon_{jc1} - \bar{\tau}_{jd1}$, $\tilde{W}_{jc1} = W_{jc1} - \hat{W}_{jc1}$, and $\tilde{\sigma}_{jc1} = \sigma_{jc1} - \hat{\sigma}_{jc1}$. Further, $\|A_{jc1}\|_F = A_{c1\max}$ for a known constant $A_{c1\max}$, and $\|\xi_{jc1}\| \le \xi_{Mc1}$ for a computable constant $\xi_{Mc1} = A_{c1\max} \varepsilon_{Mc1} + 2 A_{c1\max} W_{Mc1} \sqrt{N_c} + \tau_M / m_j$.

Next, the rotational torque vector, u_{j2}, will be addressed. First, multiply the angular velocity tracking error (24) by the inertial matrix J_j, take the first derivative with respect to time, and substitute the UAV dynamics (1) to reveal

$$J_j \dot{e}_{j\omega} = f_{jc2}(x_{jc2}) - u_{j2} - \tau_{jd2} \tag{34}$$

with $f_{jc2}(x_{jc2}) = J_j \dot{\omega}_{jd} - S(J_j \omega_j)\omega_j - N_{j2}(\omega_j)$. Examining $f_{jc2}(x_{jc2})$, it is clear that the function is nonlinear and contains unknown terms; therefore, the *universal approximation property* of NN is utilized to estimate the function $f_{jc2}(x_{jc2})$ by bounded ideal weights W_{jc2}^T, V_{jc2}^T such that $\|W_{jc2}\|_F \le W_{Mc2}$ for a known constant W_{Mc2} and written as $f_{jc2}(x_{jc2}) = W_{jc2}^T \sigma(V_{jc2}^T x_{jc2}) + \varepsilon_{jc2}$ where ε_{jc2} is the bounded NN functional reconstruction error such that $\|\varepsilon_{jc2}\| \le \varepsilon_{Mc2}$ for a known constant ε_{Mc2}. The NN estimate of f_{jc2} is given by

$\hat{f}_{jc2}=\hat{W}^T_{jc2}\sigma(V^T_{jc2}\hat{x}_{jc2})=\hat{W}^T_{jc2}\hat{\sigma}_{jc2}$ where $\hat{W}^T_{jc2}$ is the NN estimate of W^T_{jc2} and $\hat{x}_{jc2}=[1\ \omega^T_j\ \dot{\hat{\Omega}}^T_{jd}\ \widetilde{\Theta}^{b\ T}_{jd}\ e^T_{j\Theta}]^T$ is the input to the NN written in terms of the virtual controller estimates. By the construction of the virtual controller, $\dot{\hat{\omega}}_{jd}$ is not directly available; therefore, observing (29), the terms $\dot{\hat{\Omega}}^T_{jd}$, $\widetilde{\Theta}^{b\ T}_{jd}$, and $e^T_{j\Theta}$ have been included instead.

Using the NN estimate $\hat{f}_{jc2}$ and the estimated desired angular velocity tracking error $\hat{e}_{j\omega}=\hat{\omega}_{jd}-\omega_j$, the rotational torque control input is written as

$$u_{j2}=\hat{f}_{jc2}+K_{j\omega}\hat{e}_{j\omega}, \tag{35}$$

and substituting the control input (35) into the angular velocity dynamics (34) as well as adding and subtracting $W^T_{jc2}\hat{\sigma}_{jc}$, the closed loop dynamics become

$$J_j\dot{e}_{j\omega}=-K_{j\omega}e_{j\omega}+\widetilde{W}^T_{jc2}\hat{\sigma}_{jc2}+K_{j\omega}\widetilde{\omega}_{jd}+\xi_{jc2}\ , \tag{36}$$

where $\widetilde{W}^T_{jc2}=W^T_{jc2}-\hat{W}^T_{jc2}$, $\xi_{jc2}=\varepsilon_{jc2}+W^T_{jc2}\widetilde{\sigma}_{jc}-\tau_{jd2}$, and $\widetilde{\sigma}_{jc2}=\sigma_{jc2}-\hat{\sigma}_{jc2}$. Further, $\|\xi_{jc2}\|\le\xi_{Mc2}$ for a computable constant $\xi_{Mc2}=\varepsilon_{Mc2}+2W_{Mc2}\sqrt{N_{c2}}+\tau_{dM}$ where N_{c2} is the number of hidden layer neurons.

As a final step, we define $\widetilde{W}_{jc}=[\widetilde{W}_{jc1}\ 0;\ 0\ \widetilde{W}_{jc2}]$ and $\hat{\sigma}_{jc}=[\hat{\sigma}^T_{jc1}\ \ \hat{\sigma}^T_{jc2}]^T$ so that a single NN can be utilized with N_c hidden layer neurons to represent $\hat{f}_{jc}=[\hat{f}^T_{jc1}\ \ \hat{f}^T_{jc2}]^T\in\Re^6$. In the following theorem, the stability of the follower *j* is shown while considering $e_{iv}=0$. In other words, the position, orientation, and velocity tracking errors are considered along with the estimation errors of the virtual controller and the NN weight estimation errors of each NN for follower *j* while ignoring the interconnection errors between the leader and its followers. This assumption will be relaxed in the following section.

Theorem 3.1.1: (Follower UAV System Stability) Given the dynamic system of follower *j* in the form of (1), let the desired translational velocity for follower *j* to track be defined by (12) with the desired pitch and roll defined by (20) and (21), respectively. Let the NN virtual controller be defined by (28) and (29), respectively, with the NN update law given by

$$\dot{\hat{W}}_{j\Omega}=F_{j\Omega}\hat{\sigma}_{j\Omega}\left(\widetilde{\Theta}^b_{jd}\right)^T-\kappa_{j\Omega}F_{j\Omega}\hat{W}_{j\Omega}, \tag{37}$$

where $F_{j\Omega}=F^T_{j\Omega}>0$ and $\kappa_{j\Omega}>0$ are design parameters. Let the dynamic NN controller for follower *j* be defined by (32) and (35), respectively, with the NN update given by

$$\dot{\hat{W}}_{jc}=F_{jc}\hat{\sigma}_{jc}\left(A_{jc}\hat{e}_{jS}\right)^T-\kappa_{jc}F_{jc}\hat{W}_{jc}, \tag{38}$$

where $A_{jc}=[A_{jc1}\ 0_{3x3};0_{3x3}\ I_{3x3}]\in\Re^{6x6}$, $\hat{e}_{jS}=\begin{bmatrix}e_{jv}^T & \hat{e}_{j\omega}^T\end{bmatrix}^T$, $F_{jc}=F_{jc}^T>0$ and $\kappa_{jc}>0$ are constant design parameters. Then there exists positive design constants $K_{j\Omega1}, K_{j\Omega2}, K_{j\Omega3}$, and positive definite design matrices $K_{j\rho}, K_{j\Theta}, K_{jv}, K_{j\omega}$, such that the virtual controller estimation errors $\tilde{\Theta}_{jd}^b$, $\tilde{\omega}_{jd}$ and the virtual control NN weight estimation errors, $\tilde{W}_{j\Omega}$, the position, orientation, and translational and angular velocity tracking errors, $e_{j\rho}, e_{j\Theta}, e_{jv}, e_{j\omega}$, respectively, and the dynamic controller NN weight estimation errors, $\tilde{W}_{jc}$, are all *SGUUB*.

Proof: Consider the following positive definite Lyapunov candidate

$$V_j=V_{j\Omega}+V_{jc}, \tag{39}$$

where

$$V_{j\Omega}=\frac{1}{2}\tilde{\Theta}_{jd}^{b\ T}\overline{K}_{j\Omega}\tilde{\Theta}_{jd}^b+\frac{1}{2}\tilde{\omega}_{jd}^{\ T}\tilde{\omega}_{jd}+\frac{1}{2}tr\{\tilde{W}_{j\Omega}^T F_{j\Omega}^{-1}\tilde{W}_{j\Omega}\}$$

$$V_{jc}=\frac{1}{2}e_{j\rho}^T e_{j\rho}+\frac{1}{2}e_{j\Theta}^T e_{j\Theta}+\frac{1}{2}e_{jv}^T e_{jv}+\frac{1}{2}e_{j\omega}^T J_j e_{j\omega}+\frac{1}{2}tr\{\tilde{W}_{jc}^T F_{jc}^{-1}\tilde{W}_{jc}\}$$

whose first derivative with respect to time is given by $\dot{V}_j=\dot{V}_{j\Omega}+\dot{V}_{jc}$. Considering first $\dot{V}_{j\Omega}$, and substituting the closed loop virtual control estimation error dynamics (30) and (31) as well as the NN tuning law (37) , reveals

$$\dot{V}_{j\Omega}=-\overline{K}_{j\Omega2}\tilde{\Theta}_{jd}^{b\ T}\tilde{\Theta}_{jd}^b-K_{j\Omega3}\tilde{\omega}_{jd}^{\ T}\tilde{\omega}_{jd}+\tilde{\omega}_{jd}^{\ T}\xi_{j\Omega}+tr\left\{\tilde{W}_{j\Omega}^T\left(\kappa_{j\Omega}\hat{W}_{j\Omega}-\hat{\sigma}_{j\Omega}\left(\tilde{\Theta}_{jd}^b\right)^T+\hat{\sigma}_{j\Omega}\tilde{\omega}_{jd}^{\ T}\right)\right\}$$

where $\overline{K}_{j\Omega2}=(K_{j\Omega1}-K_{j\Omega3})(K_{j\Omega2}-K_{j\Omega3}(K_{j\Omega1}-K_{j\Omega3}))$ and $\overline{K}_{j\Omega2}>0$ provided $K_{j\Omega1}>K_{j\Omega3}$ and $K_{j\Omega2}>K_{j\Omega3}(K_{j\Omega1}-K_{j\Omega3})$. Observing $\|\hat{\sigma}_{j\Omega}\|\le\sqrt{N_{j\Omega}}$, $\|W_{\Omega j}\|_F\le W_{M\Omega}$ for a known constant, $W_{M\Omega}$, and $tr\{\tilde{W}_{j\Omega}^T(W_{j\Omega}-\tilde{W}_{j\Omega})\}\le\|\tilde{W}_{j\Omega}\|_F W_{M\Omega}-\|\tilde{W}_{j\Omega}\|_F^2$, $\dot{V}_{j\Omega}$ can then be rewritten as

$\dot{V}_{\Omega}\le-\overline{K}_{j\Omega2}\|\tilde{\Theta}_{jd}^b\|^2-K_{j\Omega3}\|\tilde{\omega}_{jd}\|^2-\kappa_{j\Omega}\|\tilde{W}_{j\Omega}\|_F^2+\|\tilde{\omega}_{jd}\|\xi_{\Omega M}+\|\tilde{\Theta}_{jd}^b\|\|\tilde{W}_{j\Omega}\|_F\sqrt{N_{j\Omega}}+\|\tilde{\omega}_{jd}\|\|\tilde{W}_{j\Omega}\|_F\sqrt{N_{j\Omega}}+\kappa_{j\Omega}\|\tilde{W}_{j\Omega}\|_F W_{M\Omega}$.

Now, completing the squares with respect to $\|\tilde{W}_{j\Omega}\|_F$, $\|\tilde{\Theta}_{jd}^b\|$, and $\|\tilde{\omega}_{jd}\|$, an upper bound for $\dot{V}_{j\Omega}$ is found to be

$$\dot{V}_{j\Omega}\le-\left(\overline{K}_{j\Omega2}-\frac{N_{j\Omega}}{\kappa_{j\Omega}}\right)\|\tilde{\Theta}_{jd}^b\|^2-\left(\frac{K_{j\Omega3}}{2}-\frac{N_{j\Omega}}{\kappa_{j\Omega}}\right)\|\tilde{\omega}_{jd}\|^2-\frac{\kappa_{j\Omega}}{4}\|\tilde{W}_{j\Omega}\|_F^2+\eta_{j\Omega} \tag{40}$$

where $\eta_{j\Omega}=\kappa_{j\Omega}W_{M\Omega}^2+\xi_{\Omega M}^2/(2K_{j\Omega3})$. Next, considering $\dot{V}_{jc}$ and substituting the closed loop kinematics (14) and (25), dynamics (33) and (36), and NN tuning law (38) while considering $e_{iv}=0$ reveals

$$\dot{V}_{jc} = -e_{j\rho}^T K_{j\rho} e_{j\rho} - e_{j\Theta}^T K_{j\Theta} e_{j\Theta} - e_{jv}^T K_{jv} e_{jv} - e_{j\omega}^T K_{j\omega} e_{j\omega} + e_{j\rho}^T R_j e_{jv} + e_{j\Theta}^T T_j e_{j\omega} + e_{jv}^T \xi_{jc1} + e_{j\omega}^T K_{j\omega} \tilde{\omega}_{jd} + e_{j\omega}^T \xi_{jc2}$$
$$+ \kappa_{jc} tr\{\tilde{W}_{jc}^T (W_{jc} - \tilde{W}_{jc})\} + tr\{\tilde{W}_{jc}^T \hat{\sigma}_{jc2} (e_{j\omega}^T - \hat{e}_{j\omega}^T)\}$$

Then, observing $\tilde{\omega}_{jd} = e_{j\omega} - \hat{e}_{j\omega}$ and completing the squares with respect to $e_{j\rho}, e_{j\Theta}, e_{jv}, e_{j\omega}$ and $\tilde{W}_{jc}$, and upper bound for $\dot{V}_{jc}$ is found to be

$$\dot{V}_{jc} \le -\frac{K_{j\rho\min}}{2}\|e_{j\rho}\|^2 - \frac{K_{j\Theta\min}}{2}\|e_{j\Theta}\|^2 - \left(\frac{K_{jv\min}}{2} - \frac{R_{\max}^2}{2K_{\rho\min}}\right)\|e_{jv}\|^2 - \frac{\kappa_{jc}}{3}\|\tilde{W}_{jc}\|_F^2 - \left(\frac{K_{j\omega\min}}{3} - \frac{T_{\max}^2}{2K_{j\Theta\min}}\right)\|e_{j\omega}\|^2$$
$$+ \frac{3K_{j\omega\min}}{4}\|\tilde{\omega}_{jd}\|^2 + \frac{3N_{jc}}{4\kappa_{jc}}\|\tilde{\omega}_{jd}\|^2 + \eta_{jc} \quad (41)$$

where $K_{j\rho\min}, K_{j\Theta\min}, K_{jv\min}$, and $K_{j\omega\min}$ are the minimum singular values of $K_{j\rho}$, $K_{j\Theta}$, K_{jv}, and $K_{j\omega}$, respectively, and $\eta_{jc} = \xi_{c1M}^2/(2K_{jv\min}) + \xi_{c2M}^2/(2K_{j\omega\min}) + 3W_{Mc}\kappa_{jc}/4$. Now, combining (40) and (41), an upper bound for $\dot{V}_j$ is written as

$$\dot{V}_j \le -\left(\bar{K}_{j\Omega} - \frac{N_{j\Omega}}{\kappa_{j\Omega}}\right)\|\tilde{\Theta}_{jd}^b\|^2 - \left(\frac{K_{j\Omega 3}}{2} - \frac{N_{j\Omega}}{\kappa_{j\Omega}} - \frac{3K_{j\omega\min}}{4} - \frac{3N_{jc}}{4\kappa_{jc}}\right)\|\tilde{\omega}_{jd}\|^2 - \frac{K_{j\rho\min}}{2}\|e_{j\rho}\|^2 - \frac{K_{j\Theta\min}}{2}\|e_{j\Theta}\|^2$$
$$-\left(\frac{K_{jv\min}}{2} - \frac{R_{\max}^2}{2K_{\rho\min}}\right)\|e_{jv}\|^2 - \left(\frac{K_{j\omega\min}}{3} - \frac{T_{\max}^2}{2K_{j\Theta\min}}\right)\|e_{j\omega}\|^2 - \frac{\kappa_{j\Omega}}{4}\|\tilde{W}_{j\Omega}\|_F^2 - \frac{\kappa_{jc}}{3}\|\tilde{W}_{jc}\|_F^2 + \eta_{j\Omega} + \eta_{jc} \quad (42)$$

Finally, (42) is less than zero provided

$$\bar{K}_{j\Omega 2} > \frac{N_{j\Omega}}{\kappa_{j\Omega}}, \quad K_{j\Omega 3} > \frac{2N_{j\Omega}}{\kappa_{j\Omega}} + \frac{3K_{j\omega\min}}{2} + \frac{3N_{jc}}{2\kappa_{jc}}, \quad K_{jv\min} > \frac{R_{\max}^2}{K_{\rho\min}}, \quad K_{j\omega\min} > \frac{3T_{\max}^2}{2K_{j\Theta\min}} \quad (43)$$

and the following inequalities hold:

$$\|\tilde{\Theta}_{jd}^b\| > \sqrt{\frac{\eta_{j\Omega} + \eta_{jc}}{\bar{K}_{j\Omega 2} - N_{j\Omega}/\kappa_{j\Omega}}} \quad or \quad \|\tilde{W}_{jc}\|_F > \sqrt{\frac{3(\eta_{j\Omega} + \eta_{jc})}{\kappa_{jc}}} \quad or \quad \|e_{jv}\| > \sqrt{\frac{2(\eta_{j\Omega} + \eta_{jc})}{K_{jv\min} - R_{\max}^2/K_{\rho\min}}} \quad (44)$$
$$or \quad \|e_{j\rho}\| > \sqrt{\frac{2(\eta_{j\Omega} + \eta_{jc})}{K_{j\rho\min}}} \quad or \quad \|e_{j\Theta}\| > \sqrt{\frac{2(\eta_{j\Omega} + \eta_{jc})}{K_{j\Theta\min}}} \quad or \quad \|\tilde{W}_{j\Omega}\|_F > \sqrt{\frac{4(\eta_{j\Omega} + \eta_{jc})}{\kappa_{j\Omega}}}$$
$$or \quad \|e_{j\omega}\| > \sqrt{\frac{\eta_{j\Omega} + \eta_{jc}}{K_{j\omega\min}/3 - T_{\max}^2/(2K_{j\Theta\min})}} \quad or \quad \|\tilde{\omega}_{jd}\| > \sqrt{\frac{\eta_{j\Omega} + \eta_{jc}}{\frac{K_{j\Omega 3}}{2} - \frac{N_{j\Omega}}{\kappa_{j\Omega}} - \frac{3K_{j\omega\min}}{4} - \frac{3N_{jc}}{4\kappa_{jc}}}}$$

Therefore, it can be concluded using standard extensions of Lyapunov theory (Lewis et al., 1999) that $\dot{V}_j$ is less than zero outside of a compact set, revealing the virtual controller estimation errors, $\tilde{\Theta}_{jd}^b, \tilde{\omega}_{jd}$, and the NN weight estimation errors, $\tilde{W}_{j\Omega}$, the position, orientation, and translational and angular velocity tracking errors, $e_{j\rho}, e_{j\Theta}, e_{jv}, e_{j\omega}$, respectively, and the dynamic controller NN weight estimation errors, $\tilde{W}_{jc}$, are all *SGUUB*.

3.2 Formation Leader Control Law

The dynamics and kinematics for the formation leader are defined similarly to (1) and (4), respectively. In our previous work (Dierks and Jagannathan, 2008), an output feedback control law for a single quadrotor UAV was designed to ensure the robot tracks a desired path, $\rho_{id} = [x_{id}, y_{id}, z_{id}]^T$, and desired yaw angle, ψ_{id}. Using a similarly approach to (10)-(14), the state feedback control velocity for leader i is given by (Dierks and Jagannathan, 2008)

$$v_{id} = [v_{idx}\ v_{idy}\ v_{idz}]^T = R_i^T(\dot{\rho}_{id} + K_{i\rho}e_{i\rho}) \in E^b \tag{45}$$

The closed loop position tracking error then takes the form of

$$\dot{e}_{i\rho} = -K_{i\rho}e_{i\rho} + R_i e_{iv} \tag{46}$$

Then, using steps similar to (15)-(21), the desired pitch and roll angles are given by

$$\theta_{id} = \frac{2\theta_{\max}}{\pi} a\tan\left(\frac{N_{i\theta d}}{D_{i\theta d}}\right) \tag{47}$$

where $N_{i\theta d} = c_{i\psi d}(\ddot{x}_{id} + k_{i\rho x}\dot{x}_{id} - v_{iR1}) + s_{i\psi d}(\ddot{y}_{id} + k_{i\rho y}\dot{y}_{id} - v_{iR2}) + k_{iv1}e_{vx} + \hat{f}_{ic11}$ and $D_{i\theta d} = \ddot{z}_{id} + k_{i\rho z}\dot{z}_{id} - v_{iR3} - g$ and

$$\phi_{id} = \frac{2\phi_{\max}}{\pi} a\tan\left(\frac{N_{i\phi d}}{D_{i\phi d}}\right) \tag{48}$$

where $N_{i\phi d} = s_{i\psi d}(\ddot{x}_{id} + k_{i\rho x}\dot{x}_{id} - v_{iR1}) - c_{i\psi d}(\ddot{y}_{id} + k_{i\rho y}\dot{y}_{id} - v_{iR2}) - k_{iv2}e_{vy} + \hat{f}_{ic12}$, $D_{\phi d} = c_{\bar{\theta}di}(\ddot{z}_{id} + k_{i\rho z}\dot{z}_{id} - v_{iR3} - g) + s_{\bar{\theta}id}c_{\psi id}(\ddot{x}_{id} + k_{i\rho x}\dot{x}_{id} - v_{iR1}) + s_{\bar{\theta}di}s_{\psi id}(\ddot{y}_{id} + k_{i\rho y}\dot{y}_{id} - v_{iR2})$ with $v_{iR} = [v_{iR1}\ v_{iR2}\ v_{iR3}]^T = K_{i\rho}R_i v_i$ and $\hat{f}_{ic1} = [\hat{f}_{ic11}\ \hat{f}_{ic12}\ \hat{f}_{ic13}]^T \in \Re^3$ is a NN estimate of the unknown function $f_{ic1}(x_{ic1})$ (Dierks and Jagannathan, 2008). The desired angular velocity as well as the NN virtual controller for the formation leader is defined similarly to (23) and (28) and (29), respectively, and finally, the thrust and rotation torque vector are found to be

$$\begin{aligned} u_{i1} &= m_i c_{\bar{\theta}id}c_{\bar{\phi}id}(\ddot{z}_{id} + k_{i\rho z}\dot{z}_{id} - v_{iR3} - g) + m_i(c_{\bar{\phi}id}s_{\bar{\theta}id}s_{\psi id} - s_{\bar{\phi}id}c_{\psi id})(\ddot{y}_{id} + k_{i\rho y}\dot{y}_{id} - v_{iR2}) + m_i k_{iv3}e_{ivz} \\ &+ m_i(c_{\bar{\phi}id}s_{\bar{\theta}id}c_{\psi id} + s_{\bar{\phi}id}s_{\psi id})(\ddot{x}_{id} + k_{i\rho x}\dot{x}_{id} - v_{iR1}) + m_i\hat{f}_{ic13} \end{aligned} \tag{49}$$

$$u_{i2} = \hat{f}_{ic2} + K_{i\omega}\hat{e}_{i\omega} \tag{50}$$

where $\hat{f}_{ic2} \in \Re^3$ is a NN estimate of the unknown function $f_{ic2}(x_{ic2})$ and $\hat{e}_{i\omega} = \hat{\omega}_{id} - \omega_i$. The closed loop orientation, virtual control, and velocity tracking error dynamics for the

formation leader are found to take a form similar to (25), (30) and (31), and (33) and (36), respectively (Dierks and Jagannathan, 2008).
Next, the stability of the entire formation is considered in the following theorem while considering the interconnection errors between the leader and its followers.

3.3 Quadrotor UAV Formation Stability

Before proceeding, it is convenient to define the following augmented error systems consisting of the position and translational velocity tracking errors of leader i and N follower UAV's as

$$e_\rho = \left[e_{i\rho}^T \; e_{j\rho}^T \Big|_{j=1} \cdots e_{j\rho}^T \Big|_{j=N} \right]^T \in \Re^{3(N+1)}$$

$$e_v = \left[e_{iv}^T \; e_{jv}^T \Big|_{j=1} \cdots e_{jv}^T \Big|_{j=N} \right]^T \in \Re^{3(N+1)}.$$

Next, the transformation matrix (2) is augmented as

$$R_F = diag\left\{ R_i, R_j \Big|_{j=1}, \ldots, R_j \Big|_{j=N} \right\} \in \Re^{3(N+1)x3(N+1)} \tag{51}$$

while the NN weights for the translational velocity error system are augmented as

$$\hat{W}_{c1} = diag\left\{ \hat{W}_{ic1}, \hat{W}_{jc1} \Big|_{j=1}, \ldots, \hat{W}_{jc1} \Big|_{j=N} \right\} \in \Re^{(N \cdot N_{jc1} + N_{ic1})x3(N+1)}$$

$$\hat{\sigma}_{c1} = \left[\hat{\sigma}_{ic1}^T \; \hat{\sigma}_{jc1}^T \Big|_{j=1}, \ldots, \hat{\sigma}_{jc1}^T \Big|_{j=N} \right]^T \in \Re^{(N \cdot N_{jc1} + N_{ic1})}$$

Now, using the augmented variables above, the augmented closed loop position and translational velocity error dynamics for the entire formation are written as

$$\dot{e}_\rho = -K_\rho e_\rho + (I - G_F) R_F \, e_v \tag{52}$$

$$\dot{e}_v = -K_v e_v + A_{cF} \tilde{W}_{c1}^T \hat{\sigma}_{c1} - K_\rho G_F R_F e_v + \xi_{c1} \tag{53}$$

where $A_{cF} = diag\left\{ A_{ic}, A_{jc} \Big|_{j=1}, \ldots, A_{jc} \Big|_{j=N} \right\}$ with A_{ic} defined similarly to A_{jc} in terms of $\overline{\Theta}_{id}$, $K_\rho = diag\left\{ K_{i\rho}, K_{j\rho} \Big|_{j=1}, \ldots, K_{j\rho} \Big|_{j=N} \right\}$, $K_v = diag\left\{ K_{iv}, K_{jv} \Big|_{j=1}, \ldots, K_{jv} \Big|_{j=N} \right\}$, G_F is a constant matrix relating to the formation interconnection errors defined as

$$G_F = [0 \; 0; \; F_T \;\; 0] \in \Re^{(N+1)x(N+1)} \tag{54}$$

and $F_T \in \Re^{NxN}$ is constant and dependent on the specific formation topology. For instance, in a string formation where each follower follows the UAV directly in front of it, follower *1* tracks leader *i*, follower *2* tracks follower *1*, etc., and F_T becomes the identity matrix.

In a similar manner, we define augmented error systems for the virtual controller, orientation, and angular velocity tracking systems as

$$\tilde{\Theta}_d^b = \left[(\tilde{\Theta}_{id}^b)^T, (\tilde{\Theta}_{jd}^b)^T\Big|_{j=1} ...(\tilde{\Theta}_{jd}^b)^T\Big|_{j=N}\right]^T \in \Re^{3(N+1)},\ \tilde{\omega}_d = \left[\tilde{\omega}_{id}^T, \tilde{\omega}_{jd}^T\Big|_{j=1} ...\tilde{\omega}_{jd}^T\Big|_{j=N}\right]^T \in \Re^{3(N+1)},$$

$$e_\Theta = \left[e_{i\Theta}^T\ e_{j\Theta}^T\Big|_{j=1} ....\ e_{j\Theta}^T\Big|_{j=N}\right]^T \in \Re^{3(N+1)},\ e_\omega = \left[e_{i\omega}^T\ e_{j\omega}^T\Big|_{j=1} ....\ e_{j\omega}^T\Big|_{j=N}\right]^T \in \Re^{3(N+1)}, \tag{55}$$

respectively. It is straight forward to verify that the error dynamics of the augmented variables (55) takes the form of (30), (31), (25), and (36), respectively, but written in terms of the augmented variables (55).

Theorem 3.3.1: (UAV Formation Stability) Given the leader-follower criterion of (8) with 1 leader and N followers, let the hypotheses of *Theorem 3.1.1* hold. Let the virtual control system for the leader i be defined similarly to (28) and (29) with the virtual control NN update law defined similarly to (37). Let control velocity and desire pitch and roll for the leader be given by (45), (47), and (48), respectively, along with the thrust and rotation torque vector defined by (49) and (50), respectively, and let the control NN update law be defined identically to (38). Then, the position, orientation, and velocity tracking errors, the virtual control estimation errors, and the NN weights for each NN for the entire formation are all *SGUUB*.

Proof: Consider the following positive definite Lyapunov candidate

$$V = V_\Omega + V_c, \tag{56}$$

where

$$V_\Omega = \frac{1}{2}\tilde{\Theta}_d^{b\,T}\bar{K}_\Omega\tilde{\Theta}_d^b + \frac{1}{2}\tilde{\omega}_d^T\tilde{\omega}_d + \frac{1}{2}tr\{\tilde{W}_\Omega^T F_\Omega^{-1}\tilde{W}_\Omega\}$$

$$V_c = \frac{1}{2}e_\rho^T e_\rho + \frac{1}{2}e_\Theta^T e_\Theta + \frac{1}{2}e_v^T e_v + \frac{1}{2}e_\omega^T J e_\omega + \frac{1}{2}tr\{\tilde{W}_c^T F_c^{-1}\tilde{W}_c\}$$

with $\bar{K}_\Omega = diag\{\bar{K}_{i\Omega}I, \bar{K}_{j\Omega}I\big|_{j=1} ..\bar{K}_{j\Omega}I\big|_{j=N}\}$, $\tilde{W}_\Omega = diag\{\tilde{W}_{i\Omega}, \tilde{W}_{j\Omega}\big|_{j=1} ..\tilde{W}_{j\Omega}\big|_{j=N}\}$, $J = diag\{J_i, J_j\big|_{j=1} ..J_j\big|_{j=N}\}$, and $\tilde{W}_c = diag\{\tilde{W}_{ic}, \tilde{W}_{jc}\big|_{j=1} ..\tilde{W}_{jc}\big|_{j=N}\}$. The first derivative of (56) with respect to time is given by $\dot{V} = \dot{V}_\Omega + \dot{V}_c$, and performing similar steps as those used in (40)-(41) reveals

$$\dot{V}_\Omega \le -\left(\bar{K}_{\Omega 2\min} - \frac{N_\Omega}{\kappa_{\Omega\min}}\right)\left\|\tilde{\Theta}_d^b\right\|^2 - \left(\frac{K_{\Omega 3\min}}{2} - \frac{N_\Omega}{\kappa_{\Omega\min}}\right)\left\|\tilde{\omega}_d\right\|^2 - \frac{\kappa_\Omega}{4}\left\|\tilde{W}_\Omega\right\|_F^2 + \eta_\Omega, \tag{57}$$

$$\dot{V}_c \le -\frac{K_{\rho\min}}{2}\left\|e_\rho\right\|^2 - \frac{K_{\Theta\min}}{2}\left\|e_\Theta\right\|^2 - \left(\frac{K_{v\min}}{2} - \frac{\eta_1^2}{2K_{\rho\min}} - \eta_2\right)\left\|e_{jv}\right\|^2 - \frac{\kappa_{c\min}}{3}\left\|\tilde{W}_c\right\|_F^2 - \left(\frac{K_{\omega\min}}{3} - \frac{(N+1)T_{\max}^2}{2K_{\Theta\min}}\right)\left\|e_\omega\right\|^2, \tag{58}$$

$$+\frac{3K_{\omega\min}}{4}\left\|\tilde{\omega}_d\right\|^2 + \frac{3N_c}{4\kappa_{c\min}}\left\|\tilde{\omega}_d\right\|^2 + \eta_c$$

where $\overline{K}_{\Omega 2\min}$ is the minimum singular value of $\overline{K}_{2\Omega}$, $\kappa_{\Omega\min}$ is the minimum singular value of $\kappa_{\Omega} = diag\{\kappa_{i\Omega} I,\ \kappa_{j\Omega} I\big|_{j=1} \ldots \kappa_{j\Omega} I\big|_{j=N}\}$ with I being the identity matrix, $K_{\Omega 3\min}$ is the minimum singular value of $K_{\Omega 3} = diag\{K_{i\Omega 3} I,\ K_{j\Omega} 3I\big|_{j=1} \ldots K_{j\Omega 3} I\big|_{j=N}\}$, N_{Ω} is the number of hidden layer neurons in the augmented virtual control system, and η_{Ω} is a computable constant based on $\eta_{i\Omega}$ and $\eta_{j\Omega}, j = 1,\ldots N$. Similarly, $K_{\rho\min}$, $K_{\Theta\min}$, $K_{v\min}$, $K_{\omega\min}$, and $\kappa_{c\min}$ are the minimum singular values of the augmented gain matrices $K_{\rho}, K_{\Theta}, K_{v}, K_{\omega}$, and κ_c respectively, where $\eta_1 = R_{F\max}\sqrt{1+2N}$, $\eta_2 = K_{\rho\min} R_{F\max}\sqrt{N}$ are a known computable constants and η_c is a computable constant dependent on η_{ic} and $\eta_{jc}, j = 1,\ldots N$. Now, using (57) and (58), an upper bound for $\dot{V}$ is found to be

$$\dot{V}_{\Omega} \le -\left(\overline{K}_{\Omega 2\min} - \frac{N_{\Omega}}{\kappa_{\Omega\min}}\right)\left\|\widetilde{\Theta}_d^b\right\|^2 - \left(\frac{K_{\Omega 3\min}}{2} - \frac{N_{\Omega}}{\kappa_{\Omega\min}} - \frac{3K_{\omega\min}}{4} - \frac{3N_c}{4\kappa_{c\min}}\right)\left\|\widetilde{\omega}_d\right\|^2 - \frac{\kappa_{\Omega}}{4}\left\|\widetilde{W}_{\Omega}\right\|_F^2 - \frac{\kappa_{c\min}}{3}\left\|\widetilde{W}_c\right\|_F^2 \tag{59}$$
$$-\frac{K_{\rho\min}}{2}\left\|e_{\rho}\right\|^2 - \frac{K_{\Theta\min}}{2}\left\|e_{\Theta}\right\|^2 - \left(\frac{K_{v\min}}{2} - \frac{\eta_1^2}{2K_{\rho\min}} - \eta_2\right)\left\|e_{jv}\right\|^2 - \left(\frac{K_{\omega\min}}{3} - \frac{(N+1)T_{\max}^2}{2K_{\Theta\min}}\right)\left\|e_{\omega}\right\|^2 + \eta_c + \eta_{\Omega}$$

Finally, (59) is less than zero provided

$$\overline{K}_{\Omega 2\min} > \frac{N_{\Omega}}{\kappa_{\Omega\min}},\quad K_{\Omega 3\min} > \frac{2N_{\Omega}}{\kappa_{\Omega\min}} + \frac{3K_{\omega\min}}{2} + \frac{3N_c}{2\kappa_{c\min}},\quad K_{v\min} > \frac{\eta_1^2}{K_{\rho\min}} + 2\eta_2,\quad K_{\omega\min} > \frac{3(N+1)T_{\max}^2}{2K_{\Theta\min}} \tag{60}$$

and the following inequalities hold:

$$\left\|\widetilde{\Theta}_d^b\right\| > \sqrt{\frac{\eta_{\Omega}+\eta_c}{\overline{K}_{\Omega 2\min} - N_{\Omega}/\kappa_{\Omega\min}}} \quad or \quad \left\|\widetilde{W}_c\right\|_F > \sqrt{\frac{3(\eta_{\Omega}+\eta_{jc})}{\kappa_{c\min}}} \quad or \quad \left\|e_{jv}\right\| > \sqrt{\frac{2(\eta_{\Omega}+\eta_c)}{K_{v\min} - \eta_1^2/K_{\rho\min} - 2\eta_2}} \tag{61}$$
$$or \quad \left\|e_{\rho}\right\| > \sqrt{\frac{2(\eta_{\Omega}+\eta_c)}{K_{\rho\min}}} \quad or \quad \left\|e_{\Theta}\right\| > \sqrt{\frac{2(\eta_{\Omega}+\eta_c)}{K_{\Theta\min}}} \quad or \quad \left\|\widetilde{W}_{\Omega}\right\|_F > \sqrt{\frac{4(\eta_{\Omega}+\eta_c)}{\kappa_{\Omega\min}}}$$
$$or \quad \left\|e_{\omega}\right\| > \sqrt{\frac{\eta_{\Omega}+\eta_c}{K_{\omega\min}/3 - (N+1)T_{\max}^2/(2K_{\Theta\min})}} \quad or \quad \left\|\widetilde{\omega}_d\right\| > \sqrt{\frac{\eta_{\Omega}+\eta_c}{\frac{K_{\Omega 3\min}}{2} - \frac{N_{\Omega}}{\kappa_{\Omega\min}} - \frac{3K_{\omega\min}}{4} - \frac{3N_c}{4\kappa_{c\min}}}}$$

Therefore, it can be concluded using standard extensions of Lyapunov theory (Lewis et al., 1999) that $\dot{V}$ is less than zero outside of a compact set, revealing the position, orientation, and velocity tracking errors, the virtual control estimation errors, and the NN weights for each NN for the entire formation are all *SGUUB*.

Remark 3: The conclusions of *Theorem 3.3.1* are independent of any specific formation topology, and the Lyapunov candidate (56) represents the most general form required show the stability of the entire formation. Examining (60) and (61), the minimum controller gains and error bounds are observed to increase with the number of follower UAV's, N. These results are not surprising since increasing number of UAV's increases the sources of errors which can be propagated throughout the formation.

Remark 4: Once a specific formation as been decided and the form of F_T is set, the results of *Theorem 3.3.1* can be reformulated more precisely. For this case, the stability of the formation is proven using the sum of the individual Lyapunov candidates of each UAV as opposed to using the augmented error systems (51)-(55).

4. Optimized energy-delay sub-network routing (OEDSR) protocol for UAV Localization and Discovery

In the previous section, a group of UAV's was modeled as a nonlinear interconnected system. We have shown that the basic formation is stable and each follower achieves its separation, angle of incidence, and bearing relative to its leader with a bounded error. The controller assignments for the UAV's can be represented as a graph where a directed edge from the leader to the followers denotes a controller for the followers while the leader is trying to track a desired trajectory. The shape vector consists of separations and orientations which in turn determines the relative positions of the follower UAV's with respect to its leader.

Then, a group of N UAV's is built on two networks: a physical network that captures the constraints on the dynamics of the lead UAV and a sensing and communication network, preferably wireless, that describes information flow, sensing and computational aspects across the group. The design of the graph is based on the task in hand. In this graph, nodes and edges represent UAV's and control policies, respectively. Any such graph can be described by its adjacency matrix (Das et al., 2002).

In order to solve the leader-follower criterion (8), ad hoc networks are formed between the leader(s) and the follower(s), and the position of each UAV in the formation must be determined on-line. This network is dependent upon the sensing and communication aspects. As a first step, a leader is elected similar to the case of multi-robot formations (Das et al., 2002) followed by the discovery process in which the sensory information and physical networks are used to establish a wireless network. The outcome of the leader election process must be communicated to the followers in order to construct an appropriate shape. To complete the leader-follower formation control task (8), the controllers developed in the previous section require only a single-hop protocol; however, a multi-hop architecture is useful to relay information throughout the entire formation like the outcome of the leader election process, changing tasks, changing formations, as well alerting the UAV's of approaching moving obstacles that appear in a sudden manner.

The optimal energy-delay sub-network routing (OEDSR) protocol (Jagannathan, 2007) allows the UAV's to communicate information throughout the formation wirelessly using a multi-hop manner where each UAV in the formation is treated as a hop. The energy-delay routing protocol can guarantee information transfer while minimizing energy and delay for real-time control purposes even for mobile ad hoc networks such as the case of UAV formation flying.

We envision four steps to establish the wireless ad hoc network. As mentioned earlier, leader election process is the first step. The discovery process is used as the second step where sensory information and the physical network are used to establish a spanning tree. Since this is a multi-hop routing protocol, the communication network is created on-demand unlike in the literature where a spanning tree is utilized. This on-demand nature would allow the UAV's to be silent when they are not being used in communication and generate a communication path when required. The silent aspect will reduce any inference to others. Once a formation becomes stable, then a tree can be constructed until the shape changes. The third step will be assignment of the controllers online to each UAV based on the location of the UAV. Using the wireless network, localization is used to combine local sensory information along with information obtained via routing from other UAV's in order to calculate relative positions and orientations. Alternatively, range sensors provide relative separations, angles of incidence, and bearings. Finally cooperative control allows the graph obtained from the network to be refined. Using this graph theoretic formulation, a group is modeled by a tuple $\Gamma = (\vartheta, \mathrm{P}, \mathrm{H})$ where ϑ is the reference trajectory of the robot, P represents the shape vectors describing the relative positions of each vehicle with respect to the formation reference frame (leader), and H is the control policy represented as a graph where nodes represent UAV and edges represent the control assignments. Next, we describe the OEDSR routing protocol where each UAV will be referred to as a "node."

In OEDSR, sub-networks are formed around a group of nodes due to an activity, and nodes wake up in the sub-networks while the nodes elsewhere in the network are in sleep mode. An appropriate percentage of nodes in the sub-network are elected as cluster heads (CHs) based on a metric composed of available energy and relative location to an event (Jagannathan, 2007) in each sub-network. Once the CHs are identified and the nodes are clustered relative to the distance from the CHs, the routing towards the formation leader (FL) is initiated. First, the CH checks if the FL is within the communication range. In such case, the data is sent directly to the FL. Otherwise, the data from the CHs in the sub-network are sent over a multi-hop route to the FL. The proposed routing algorithm is fully distributed since it requires only local information for constructing routes, and is proactive adapting to changes in the network. The FL is assumed to have sufficient power supply, allowing a high power beacon from the FL to be sent such that all the nodes in the network have knowledge of the distance to the FL. It is assumed that all UAV's in the network can calculate or measure the relative distance to the FL at any time instant using the formation graph information or local sensory data. Though the OEDSR protocol borrows the idea of an energy-delay metric from OEDR (Jagannathan, 2007), selection of relay nodes (RN) does not maximize the number of two hop neighbors. Here, any UAV can be selected as a RN, and the selection of a relay node is set to maximize the link cost factor which includes distance from the FL to the RN.

4.1 Selection of an Optimum Relay-Node-Based link cost factor

Knowing the distance information at each node will allow the UAV to calculate the Link Cost Factor (LCF). The link cost factor from a given node to the next hop node 'k' is given by (62) where D_k represent the delay that will be incurred to reach the next hop node in range, the distance between the next hop node to the FL is denoted by Δx_k, and the remaining energy, E_k, at the next hop node are used in calculation of the link cost as

$$LCF_k = \frac{E_k}{D_k \cdot \Delta x_k} \tag{62}$$

In equation (62), checking the remaining energy at the next hop node increases network lifetime; the distance to the FL from the next hop node reduces the number of hops and end-to-end delay; and the delay incurred to reach the next hop node minimizes any channel problems. When multiple RNs are available for routing of the information, the optimal RN is selected based on the highest LCF. These clearly show that the proposed OEDSR protocol is an on demand routing protocol. For detailed discussion of OEDSR refer to (Jagannathan, 2007). The route selection process is illustrated through the following example. This represents sensor data collected by a follower UAV for the task at hand that is to be transmitted to the FL.

4.2 Routing Algorithm through an Example

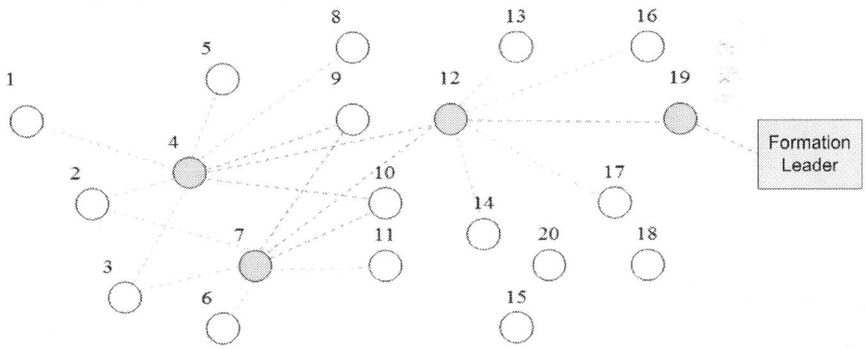

Figure 2. Relay node selection

Consider the formation topology shown in Figure 2. The link cost factors are taken into consideration to route data to the FL. The following steps are implemented to route data using the OEDSR protocol:

1. Start with an empty relay list for source UAV n: *Relay(n)={ }*. Here UAV n_4 and n_7 are CHs.
2. First, CH n_4 checks with which nodes it is in range with. In this case, CH n_4 is in range with nodes n_1, n_2, n_3, n_5, n_8, n_9, n_{12}, and n_{10}.
3. The nodes n_1, n_2, and n_3 are eliminated as potential RNs because the distance from them to the FL is greater than the distance from CH n_4 to the FL.
4. Now, all the nodes that are in range with CH n_4 transmit RESPONSE packets and CH n_4 makes a list of possible RNs, which in this case are n_5, n_8, n_9, n_{12}, and n_{10}.
5. CH n_4 sends this list to CH n_7. CH n_7 checks if it is range with any of the nodes in the list.
6. Nodes n_9, n_{10}, and n_{12} are the nodes that are in range with both CH n_4 and n_7. They are selected as the potential common RNs.
7. The link cost factors for n_9, n_{10}, and n_{12} are calculated.
8. The node with the maximum value of LCF is selected as the RN and assigned to *Relay(n)*. In this case, *Relay(n)={n12}*.
9. Now UAV n_{12} checks if it is in direct range with the FL, and if it is, then it directly routes the information to the FL.

10. Otherwise, n_{12} is assigned as the RN, and all the nodes that are in range with node n_{12} and whose distance to the FL is less than its distance to the FL are taken into consideration. Therefore, UAV's n_{13}, n_{16}, n_{19}, and n_{17} are taken into consideration.
11. The LCF is calculated for n_{13}, n_{16}, n_{19}, n_{14}, and n_{17}. The node with the maximum LCF is selected as the next RN. In this case $Relay(n) = \{n19\}$.
12. Next the RN n_{19} checks if it is in range with the FL. If it is, then it directly routes the information to the FL. In this case, n_{19} is in direct range, so the information is sent to the FL directly.

4.3 Optimality Analysis for OEDSR

To prove that the proposed route created by OEDSR protocol is optimal in all cases, it is essential to show it analytically.

Assumption 1: It is assumed that all UAV's in the network can calculate or measure the relative distance to the FL at any time instant using the formation graph information or local sensory data.

Theorem 4.3.1: The link cost factor-based routing generates viable RNs to the FL.

Proof: Consider the following two cases

Case I: When the CHs are one hop away from the FL, the CH selects the FL directly. In this case, there is only one path from the CH to the FL. Hence, OEDSR algorithm does not need to be used.

Case II: When the CHs have more than one node to relay information, the OEDSR algorithm selection criteria are taken into account. In Figure 3, there are two CHs, CH_1 and CH_2. Each CH sends signals to all the other nodes in the network that are in range. Here, *CH1* first sends out signals to *n1, n3, n4, and n5* and makes a list of information about the potential RN. The list is then forwarded to *CH2*. *CH2* checks if it is in range with any of the nodes in the list. Here, *n4* and *n5*are selected as potential common RNs. A single node must be selected from both *n4* and *n5* based on the OEDSR link cost factor. The cost to reach *n* from *CH* is given by (2). So based on the OEDSR link cost factor, *n4* is selected as the RN for the first hop. Next, *n4* sends signals to all the nodes it has in range, and selects a node as RN using the link cost factor. The same procedure is carried on till the data is sent to the FL.

Lemma 4.3.2: The intermediate UAV's on the optimal path are selected as RNs by the previous nodes on the path.

Proof: A UAV is selected as a RN only if it has the highest link cost factor and is in range with the previous node on the path. Since OEDSR maximizes the link cost factor, intermediate nodes that satisfy the metric on the optimal path are selected as RNs.

Lemma 4.3.3: A UAV can correctly compute the optimal path (with lower end to end delay and maximum available energy) for the entire network topology.

Proof: When selecting the candidate RNs to the CHs, it is ensured that the distance form the candidate RN to the FL is less than the distance from the CH to the FL. When calculating the link cost factor, available energy is divided by distance and average end-to-end delay to ensure that the selected nodes are in range with the CHs and close to the FL. This helps minimize the number of multi-point RNs in the network.

Theorem 4.3.4: OEDSR protocol results in an optimal route (the path with the maximum energy, minimum average end-to-end delay and minimum distance from the FL) between the CHs and any source destination.

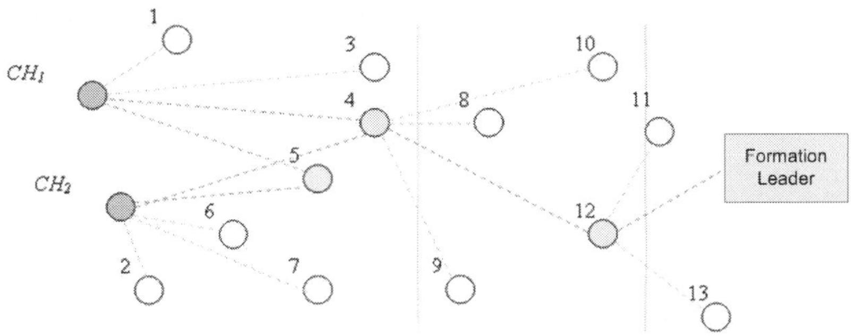

Figure 3. Link cost calculation

5. Simulation Results

A wedge formation of three identical quadrotor UAV's is now considered in MATLAB with the formation leader located at the apex of the wedge. In the simulation, follower 1 should track the leader at a desired separation $s_{jid} = 2\ m$, desired angle of incidence $\alpha_{jid} = 0\ (rad)$, and desired bearing $\beta_{jid} = \pi/3\ (rad)$ while follower 2 tracks the leader at a desired separation $s_{jid} = 2\ m$ desired angle of incidence, $\alpha_{jid} = -\pi/10(rad)$, and desired bearing $\beta_{jid} = -\pi/3\ (rad)$. The desired yaw angle for the leader and thus the formation is selected to be $\psi_d = \sin(0.3\pi\, t)$. The inertial parameters of the UAV's are taken to be as $m = 0.9\,kg$ and $J = diag\{0.32, 0.42, 0.63\}\ kg\ m^2$, and aerodynamic friction is modeled as in (Dierks and Jagannathan, 2008). The parameters outlined in Section 3.3 are communicated from the leader to its followers using single hop communication whereas results for OEDSR in a mobile environment can be found in (Jagannathan, 2007).

Each NN employs 5 hidden layer neurons, and for the leader and each follower, the control gains are selected to be, $K_{\Omega 1} = 23,\ K_{\Omega 2} = 80,\ K_{\Omega 3} = 20$, $K_\rho = diag\{10, 10, 30\}$, $k_{v1} = 10, k_{v2} = 10, k_{v3} = 30$, $K_\Theta = diag\{30, 30, 30\}$, and $K_\omega = diag\{25, 25, 25\}$. The NN parameters are selected as, $F_\Omega = 10, \kappa_\Omega = 1$, and $F_c = 10, \kappa_c = 0.1$, and the maximum desired pitch and roll values are both selected as $2\pi / 5$.

Figure 4 displays the quadrotor UAV formation trajectories while Figures 5-7 show the kinematic and dynamic tracking errors for the leader and its followers. Examining the trajectories in Figure 4, it is important to recall that the bearing angle, β_{ji}, is measured in the inertial reference frame of the follower rotated about its yaw angle. Examining the tracking errors for the leader and its followers in Figures 5-7, it is clear that all states track their desired values with small bounded errors as the results of *Theorem 3.3.1* suggest. Initially, errors are observed in each state for each UAV, but these errors quickly vanish as the virtual control NN and the NN in the actual control law learns the nonlinear UAV dynamics. Additionally, the tracking performance of the underactuated states v_x and v_y implies that the desired pitch and roll, respectively, as well as the desired angular velocities generated by the virtual control system are satisfactory.

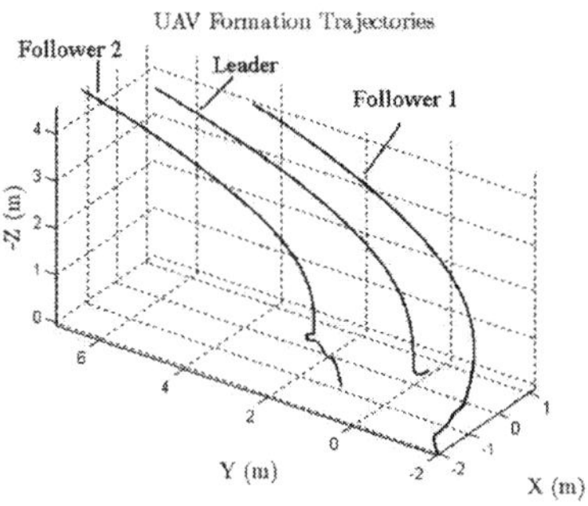

Figure 4. Quadrotor UAV formation trajectories

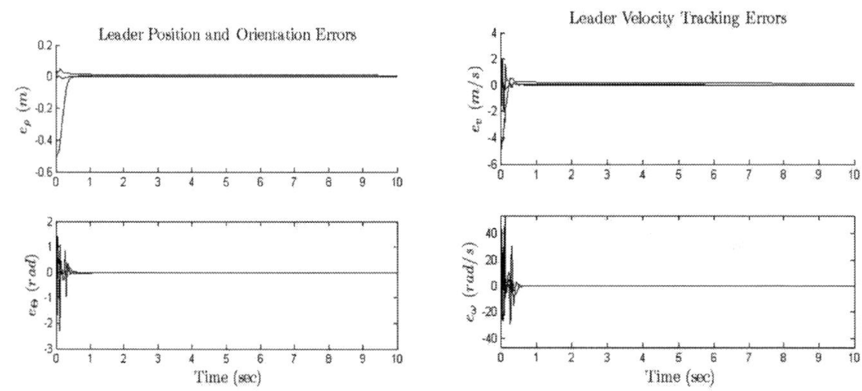

Figure 5. Tracking errors for the formation leader

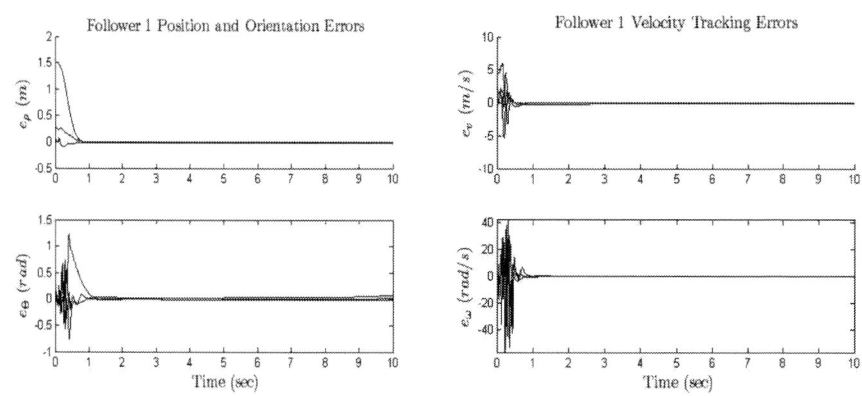

Figure 6. Tracking errors for follower 1

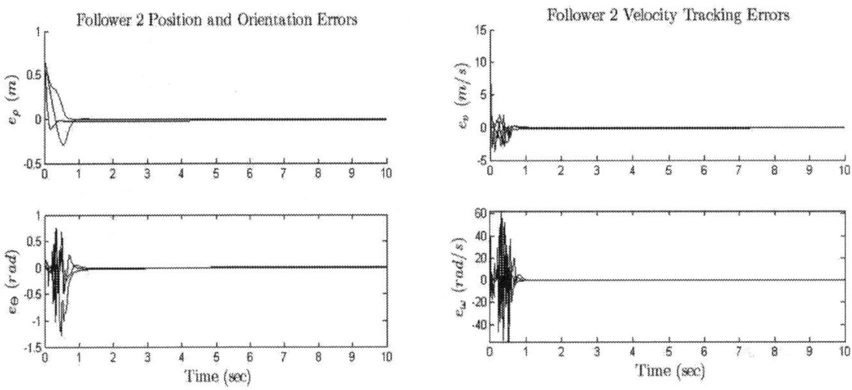

Figure 7. Tracking errors for follower 2

6. Conclusions

A new framework for quadrotor UAV leader-follower formation control was presented along with a novel NN formation control law which allows each follower to track its leader without the knowledge of dynamics. All six DOF are successfully tracked using only four control inputs while in the presence of unmodeled dynamics and bounded disturbances. Additionally, a discovery and localization scheme based on graph theory and ad hoc networks was presented which guarantees optimal use of the UAV's communication links. Lyapunov analysis guarantees *SGUUB* of the entire formation, and numerical results confirm the theoretical conjectures.

7. References

Das, A.; Spletzer, J.; Kumar, V.; & Taylor, C. (2002). Ad hoc networks for localization and control of mobile robots, *Proceedings of IEEE Conference on Decision and Control,* pp. 2978-2983, Philadelphia, PA, December 2002

Desai, J.; Ostrowski, J. P.; & Kumar, V. (1998). Controlling formations of multiple mobile robots, *Proceedings of IEEE International Conference on Robotics and Automation,* pp. 2964-2869, Leuven, Belgium, May 1998

Dierks, T. & Jagannathan, S. (2008). Neural network output feedback control of a quadrotor UAV, *Proceedings of IEEE Conference on Decision and Control,* To Appear, Cancun Mexico, December 2008

Fierro, R.; Belta, C.; Desai, J.; & Kumar, V. (2001). On controlling aircraft formations, *Proceedings of IEEE Conference on Decision and Control,* pp. 1065-1079, Orlando, Florida, December 2001

Galzi, D. & Shtessel, Y. (2006). UAV formations control using high order sliding modes, *Proceedings of IEEE American Control Conference,* pp. 4249-4254, Minneapolis, Minnesota, June 2006

Gu, Y.; Seanor, B.; Campa, G.; Napolitano M.; Rowe, L.; Gururajan, S.; & Wan, S. (2006). Design and flight testing evaluation of formation control laws. *IEEE Transactions on Control Systems Technology*, Vol. 14, No. 6, (November 2006) page numbers (1105-1112)

Jagannathan, S. (2007). *Wireless Ad Hoc and Sensor Networks*, Taylor & Francis, ISBN 0-8247-2675-8, Boca Raton, FL

Lewis, F.L.; Jagannathan, S.; & Yesilderek, A. (1999). *Neural Network Control of Robot Manipulators and Nonlinear Systems*, Taylor & Francis, ISBN 0-7484-0596-8, Philadelphia, PA

Neff, A.E.; DongBin, L.; Chitrakaran, V.K.; Dawson, D.M.; & Burg, T.C. (2007). Velocity control for a quad-rotor uav fly-by-camera interface, *Proceedings of the IEEE Southeastern Conference*, pp. 273-278, Richmond, VA, March 2007

Saffarian, M. & Fahimi, F. (2008). Control of helicopters' formation using non-iterative nonlinear model predictive approach, *Proceedings of IEEE American Control Conference*, pp. 3707-3712, Seattle, Washington, June 2008

Xie, F.; Zhang, X.; Fierro, R; & Motter, M. (2005). Autopilot based nonlinear UAV formation controller with extremum-seeking, *Proceedings of IEEE Conference on Decision and Control*, pp 4933-4938, Seville, Spain, December 2005

15

Asymmetric Hovering Flapping Flight: a Computational Study

Jardin Thierry, Farcy Alain and David Laurent
LEA, University of Poitiers, CNRS, ENSMA
France

1. Introduction

In the early 90's, Micro Air Vehicles (MAV's) appeared as a possible solution for missions of reconnaissance in constrained environments. The American Defense Advanced Research Projects Agency (DARPA) initiated workshops on the concept and, in 1997, raised funds to conduct a multi-year programme whose objective was to develop a low cost, high autonomy unmanned aircraft, with a largest dimension limited to 15 centimeters (6 inches). In terms of aerodynamics, this typical size specification places the corresponding airflow in the range of low Reynolds number flows, typically between 10^2 and 10^4. Several prototypes were tested, based on the conventional fixed and rotary wing concepts. However, at such low Reynolds numbers and notably supported by the researches carried on the analysis of insects' flight, the flapping wing concept appeared as an alternative answer, suggesting enhanced aerodynamic performances (lift, efficiency), flight agility, capability to hover coupled with a low noise generation. The latter arises from the complex wing motion defined by superimposed translating (downstroke and upstroke) and rotating (supination and pronation) motions.
Pioneer works relying on the aerodynamics of flapping wings were proposed by biologists whose objective was to evaluate the amount of lift generated by insects. After several attempts based on the quasi-steady approach, it was admitted that unsteady aerodynamic mechanisms are essential to keep an insect aloft, especially while hovering (Jensen, 1956; Weis-Fogh, 1973; Ellington 1984). Precisely, three phenomena may be distinguished:

1. The presence of a leading edge vortex (LEV or dynamic stall mechanism) during the translating phases, acting as a low pressure suction region on the extrados of the wing. Due to its importance in aeronautics, this phenomenon was extensively studied experimentally (Walker, 1931) and analytically (Polhamus, 1971) before its evidence was demonstrated and analysed in the context of flapping wings (Maxworthy, 1979; Dickinson & Götz, 1993).
2. The Kramer effect, assimilated to the supplementary air circulation resulting from the combined translating and rotating motions (Kramer, 1932; Bennett, 1970; Dickinson et al, 1999).
3. The wake capture mechanism, occurring as the wing encounters and interacts with its own wake generated during previous phases (Dickinson, 1994; Dickinson et al, 1999).

One should keep in mind that the spatial and temporal behaviours of such unsteady phenomena highly depend on the wing kinematics. Thus, when analysing the flow dynamics generated by a flapping wing, it is essential to clearly precise the flight configuration studied. Basically, two main approaches may be distinguished. The first one is the forward-flapping flight configuration for which the wing flaps into a head wind. The resulting aerodynamic force can be divided into two components: the forward (horizontal) and the upward (vertical) components which must be sufficiently strong to counter the body drag force and the weight respectively. The second one, referred to as the hovering-flapping flight configuration, is in fact an extreme mode of flight where the head wind velocity is zero. The resulting mean aerodynamic force is here strictly vertical, the average horizontal force over a flapping period being null. In this particular case, the unsteady effects are preponderant relatively to the quasi steady effects. Considering hovering flapping flight leads to further differentiation whether the wing flaps along a horizontal stroke plane (symmetric or normal hovering) or an inclined stroke plane (asymmetric hovering). The former case has been under much consideration since it appeared to be the most common configuration observed in the world of insects. Specifically, the influence of various kinematic parameters (e.g. angle of attack, position of the centre of rotation, Reynolds number etc) was experimentally (Sane & Dickinson, 2001; Sane & Dickinson, 2002; Kurtulus, 2005) and numerically (Wu & Sun, 2004; Kurtulus, 2005) analysed, giving satisfying agreement. On the other hand, asymmetric hovering studies seemed restricted to biologic configurations (reproducing the wing kinematics of the dragonfly) and very few works reported parametrical results (Wang, 2004) dedicated to MAVs improvement. Yet, introducing asymmetry in hovering flapping flight is an appealing approach as the lifting force results from the combination of both lift and drag, presupposing enhanced aerodynamic efficiency.

In this chapter, two dimensional numerical computations are used to evaluate the flow dynamics and the resulting aerodynamic loads experienced by a wing undergoing asymmetric hovering flapping motions at Reynolds 1000. On the contrary to previous works, the present parametrical analysis focuses on the aerodynamic performances of non-biological configurations for application to Micro Air Vehicles. Reference to a larger context is ensured by comparing asymmetric configurations with the widely studied symmetric configurations.

2. Flapping kinematics and parameters

Normal (or symmetric) flapping motions are characterized by strictly opposed downstroke and upstroke wing kinematics. As a consequence, the drag (aerodynamic force component collinear to the stroke plane) generated during downstroke counteracts the drag generated during upstroke. If the stroke plane is set as the horizontal, the drag may be assimilated to the horizontal force component whose mean value over a period is hence null, ensuring the hovering condition. In this study, asymmetry is introduced by fixing different downstroke and upstroke angles of attack. The resulting asymmetric motions exhibit dissimilar downstroke and upstroke drag such that setting the stroke plane as the horizontal no longer ensures the hovering condition. Thus, the latter should be tilted (angle β), resulting in a combined action of both lift and drag as the effective lifting (or vertical) force.

A set of symmetric and asymmetric cases is analysed for a two dimensional NACA0012 profile. The wing kinematics result from the combination of translating and rotating motions as shown in figure 1. Basically, the flapping motions may be decomposed into different regions whether the wing is translating at constant speed and fixed angle of attack (region T) or is submitted to both varying translation speed and rotating motion (regions R). Region T and regions R are respectively 4 and 1 chord long, so that the wing travels along a total course of 6 chords during one stroke. The rotation is applied around a spanwise axis located ¼ chord away from the leading edge. The constant wing velocity V_0 reached during the pure translation phases (region T) is calculated with respect to the Reynolds number such that:

$$Re = \frac{cV_0}{\upsilon} \tag{1}$$

where c is the chord of the NACA0012 profile and υ the kinematic viscosity of air. Note that the Reynolds number represents the adimensional ratio between inertial and viscous forces and is here fixed to 1000 such that the flow is considered to be laminar. The flapping period T is defined as:

$$T = \frac{4c}{|V_0|}(2 + \frac{\pi}{2}) \tag{2}$$

leading to a flapping frequency of approximately 10 Hz. The varying parameters are chosen to be the downstroke and upstroke angles of attack (α_d, α_u), the difference between the two reflecting the asymmetry of the motion. In order to ensure the continuity of the translating and rotating accelerations as necessary to numerically solve the Navier-Stokes equations, the instantaneous velocities follow 4th order polynomial motion laws as displayed in figure 1.

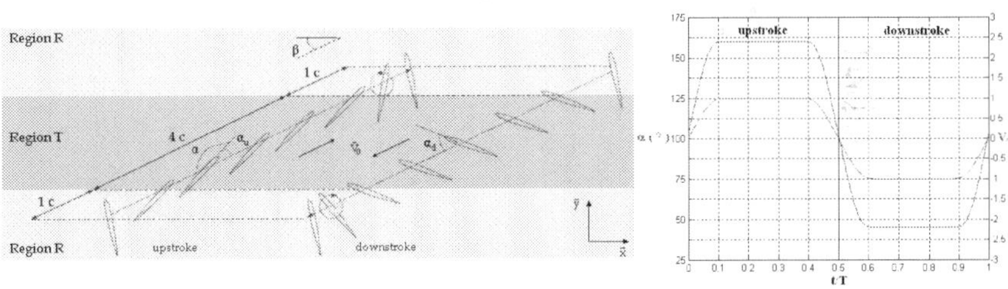

Figure 1. Schematization of the flapping kinematics and time evolution of the angle of attack and translating velocity (α_d=45°, α_u=20°)

3. Investigation tools

3.1 Numerical solver

The aerodynamic flow established in the vicinity of the airfoil is computed using a finite element method. The two dimensional time-dependent Navier-Stokes equations (equations (3) and (4) in Cartesian tensor notation for general compressible and incompressible flows) are directly solved (DNS) in the fixed inertial reference frame through a moving mesh method, assuming laminar incompressible flow.

$$\frac{\partial \rho}{\partial t} + \frac{\partial}{\partial x_j}(\rho u_j) = 0 \tag{3}$$

$$\frac{\partial \rho u_i}{\partial t} + \frac{\partial}{\partial x_j}(\rho u_j u_i - \tau_{ij}) = -\frac{\partial p}{\partial x_i} \quad (4)$$

with the constitutive relation for a laminar Newtonian fluid:

$$\tau_{ij} = 2\mu s_{ij} - \frac{2}{3}\mu \frac{\partial u_k}{\partial x_k}\delta_{ij} \quad (5)$$

$$s_{ij} = \frac{1}{2}\left(\frac{\partial u_i}{\partial x_j} + \frac{\partial u_j}{\partial x_i}\right) \quad (6)$$

Symbols are listed in table 1 below.

t	time
x_i	Cartesian coordinate
u_i	Fluid velocity component
p	Piezometric pressure
ρ	density
μ	Molecular dynamic viscosity
τ_{ij}	Stress tensor components
s_{ij}	Rate of strain tensor
δ_{ij}	Kronecker delta

Table 1. Definition of the Navier-Stokes equations symbols

The Navier-Stokes equations are solved on a non-conformal O-type grid as shown in figure 2. The latter is composed of 57500 cells, forming a refined rectangular domain close to the profile and a coarser circular domain at the far field. The instantaneous airfoil velocity is computed by means of user-defined subroutines and implemented through a no-slip boundary condition. The pressure is fixed and assumed to be the standard air pressure at the far field, located 15 chords away from the airfoil centre of rotation. The velocities at the corresponding cell faces are linked to the local pressure gradients, allowing the local flow to be inwards or outwards. For a two-dimensional configuration, symmetric boundary conditions are applied on the front and back side of the grid. The spatial and temporal discretisations are performed using an upwind differencing scheme and a PISO scheme respectively. The PISO approach consists of a special implicit scheme, based on the fully implicit Euler scheme and explicit deferred correctors, which results in a formal accuracy lying between first and second order. Setting 1000 time steps per flapping period leads to a maximum and mean Courant number (CFL) of approximately 10 and 0.5 respectively. Note that the results exposed in the present paper concern the 7th flapping period for which the flow is ensured to be periodical.

Higher resolution and larger computational domain tests (keeping a similar CFL) demonstrated that the solution is mesh independent.

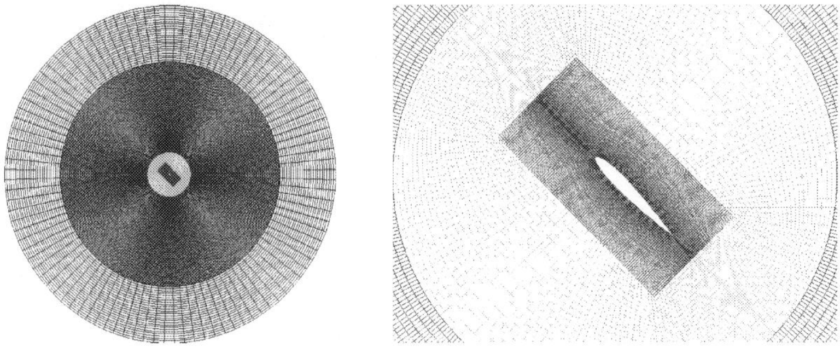

Figure 2. General and zoomed view of the computational grid

3.2 Vortex characterization

The numerical flow solver returns the velocity and the pressure flow fields in the whole computational domain at each time step. In order to characterize the flow dynamics, it is convenient to derive specific criteria which facilitate the vortex cores identification. The present analysis focuses on the λ_2 criteria (Jeong & Hussain, 1995), whose definition for a two-dimensional flow in the xy plane of velocity components uv is:

$$\lambda_2 = \frac{\partial u}{\partial y}\frac{\partial v}{\partial x} - \frac{\partial u}{\partial x}\frac{\partial v}{\partial y} \tag{7}$$

On the contrary to the vorticity, the λ_2 criterion differentiates the strain regions from the vortex core regions. Precisely, λ_2 is the second highest eigenvalue of the tensor Ω^2+S^2, where S and Ω are respectively the symmetric and antisymmetric components of the velocity gradient tensor. Negative values of λ_2 put into evidence local minima of pressure usually illustrating vortex cores.

3.3 Aerodynamic loads

The presence of vortical structures in the vicinity of the flapping airfoil highly influences the generation of aerodynamic loads. As will be exposed in the following section, vortex cores are usually assimilated to low pressure regions which may act as lift enhancers if located on the extrados of the airfoil. As a consequence, the understanding of flapping airfoil performances imposes to establish a correlation between the vortex behaviours and the resulting aerodynamic force F. The latter is inferred by integrating both pressure and viscous stresses along the wing surface and may be decomposed into the lift F_l and drag F_d components (respectively orthogonal and collinear to the stroke plane) or the vertical F_y and horizontal F_x components, linked by the following relations:

$$F_x = \pm F_d \cos\beta - F_l \sin\beta \tag{8}$$

$$F_y = F_l \cos\beta \pm F_d \sin\beta \tag{9}$$

The basic two-dimensional aerodynamic coefficients are obtained by adimensionalizing the previous components using the constant translating velocity V_0 and the chord of the airfoil c:

$$C_i = \frac{2F_i}{\rho c V_0^2} \tag{10}$$

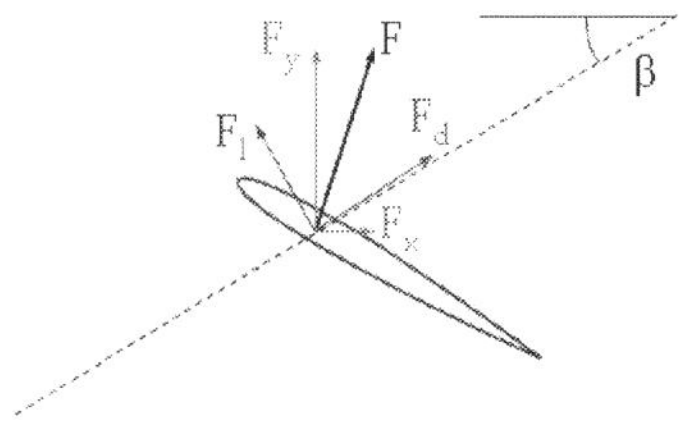

Figure 3. Relation between the resulting aerodynamic force F and the various components F_l, F_d, F_y and F_x

As a first approach, it is of interest to evaluate the mean values of these coefficients by averaging them over a flapping period T:

$$\overline{C_i} = \frac{1}{T}\int_0^T C_i(t)dt \tag{11}$$

Further coefficients are introduced to evaluate aerodynamic efficiency, power consumption and quality factor:

$$\overline{C_{eff}} = \frac{\overline{C_y}}{\overline{C_d}} \tag{12}$$

$$\overline{C_p} = \frac{1}{T}\int_0^T C_d(t)\frac{V(t)^3}{V_0^3}dt \tag{13}$$

$$\overline{C_q} = \frac{\overline{C_y}^{\frac{3}{2}}}{2\overline{C_p}}\sqrt{\frac{c}{a}} \tag{14}$$

where V(t) is the instantaneous airfoil velocity. The efficiency ratio represents the ratio between the effective lifting force necessary for flight and the drag force countering the wing motion, acting as a source of energy consumption. The power ratio may be assimilated to the power consumption resulting from the presence of this drag force. Finally, the Froude theory leads to the definition of a quality coefficient which estimates the relative importance of the lifting force and the power consumption. Note that in this case "a" stands for the total flapping amplitude.

4. Results and discussion

4.1 Mean analysis

The aerodynamic performances inherent to the different kinematics are evaluated using the mean coefficients described in the previous section. Figure 4 represents the mean drag, vertical (or lifting) force, efficiency ratio, power and quality coefficients for all the configurations tested. The latter is regrouped according to the value of the downstroke angle of attack α_d consecutively fixed to 30°, 45° and 60°. In this way, each graph exhibits the effect of desymmetrization on the global aerodynamic performances of flapping airfoils. Precisely, the right hand symbols correspond to symmetric cases, the asymmetry being increased with decreasing upstroke angle of attack.

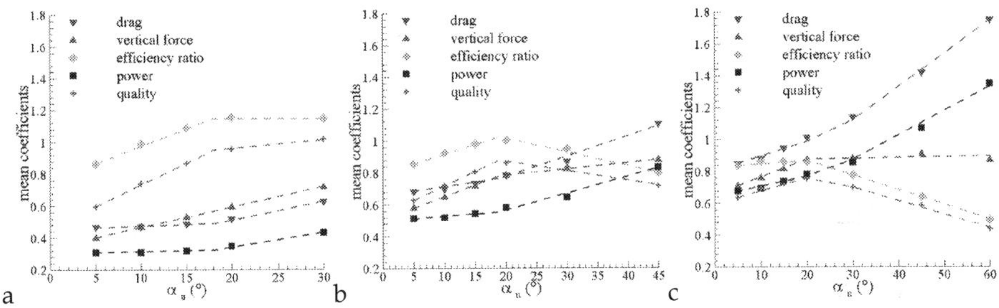

Figure 4. Mean drag, vertical force, efficiency ratio, power and quality ($5^*\overline{C_q}$) coefficients in function of α_u for the α_d =30° (a), α_d =45° (b) and α_d =60° (c) configurations

The overall comparison demonstrates that the mean drag and vertical force coefficients, hence the global aerodynamic force F, increase with the downstroke/upstroke angle of attack, as denoted in previous works (Sane & Dickinson, 2001; Wu & Sun, 2004; Kurtulus, 2005). This remark is illustrated by both the enhancement of the drag and vertical force tendency curves from one graph to another (influence of α_d) and the latter's positive slope observed within a graph (influence of α_u).

In order to characterize the effect of asymmetry on aerodynamic performances, it is of interest to further analyse the behaviour of these tendency curves within a graph. Figure 4.a shows a linear increase of the vertical force coefficient with the upstroke angle of attack. This augmentation is similar for the drag coefficient when α_u is set above 18°, but less pronounced for low α_u values. The direct consequence resides in a clear change of the efficiency ratio tendency curve near α_u=18°. Comparable changes in evolution are further put into evidence for both drag and vertical force coefficients in figure 4.b. In this particular case, the behaviours of the drag and the vertical force seem opposed, resulting in a maximum efficiency ratio of approximately 1.05 at $\alpha_u \approx 19°$. Besides these changes, Figure 4.c exhibits a quasi constant vertical force coefficient and a sharp increase in drag with increasing upstroke angle of attack. Thus, the efficiency ratio is maintained to approximately 0.85 until $\alpha_u \approx 20°$, after which it is strongly affected and decreases to 0.498 for the normal hovering case α_d=60°, α_u=60°. As a conclusion, the common characteristic of all groups (α_d=30°, α_d=45° and α_d=60°) resides in the presence of a critical upstroke angle of attack for which the tendency curve representing the efficiency ratio as a function of α_u changes behaviour and reaches a strong (maximum in some cases) value. If the cause for the existence of this critical angle of attack may arise from trivial trigonometric considerations, it is also probable that the occurence of non-linear unsteady flow structures strongly affect aerodynamic performances from one configuration to another, and may participate in the changes of behaviour previously exposed.

Furthermore, in all cases, the drag and power coefficients are characterized by comparable tendencies, still shifted by a value which increases with the upstroke angle of attack. The effect on the quality coefficient is such that the latter demonstrates analogous changes around $\alpha_u \approx 20°$.

α_d	α_u	$\overline{C_d}$	$\overline{C_y}$	$\overline{C_{eff}}$	$\overline{C_p}$	$\overline{C_q}$	β
30	5	0.4651	0.4018	0.864	0.310	0.119	34
30	10	0.4698	0.4647	0.989	0.309	0.148	24
30	15	0.4863	0.5300	1.090	0.321	0.173	17
30	20	0.5185	0.5966	1.151	0.348	0.191	10
30	30	0.6275	0.7190	1.146	0.433	0.203	0
45	5	0.6807	0.5802	0.852	0.512	0.125	51
45	10	0.6983	0.6444	0.923	0.520	0.144	41
45	15	0.7314	0.7149	0.977	0.543	0.161	33
45	20	0.7823	0.7826	1.000	0.580	0.172	25
45	30	0.8685	0.8217	0.946	0.644	0.167	14
45	45	1.1038	0.8801	0.797	0.832	0.143	0
60	5	0.8477	0.7080	0.835	0.676	0.127	64
60	10	0.8805	0.7598	0.863	0.695	0.138	54
60	15	0.9435	0.8134	0.862	0.735	0.144	45
60	20	1.0129	0.8706	0.860	0.782	0.150	38
60	30	1.1367	0.8811	0.775	0.855	0.140	25
60	45	1.4182	0.9087	0.641	1.070	0.117	14
60	60	1.7509	0.8713	0.498	1.350	0.087	0

Table 2. Aerodynamic coefficients survey

4.2 Flow unsteadiness

Focusing on the spatial and temporal evolution of the vortical structures governing the flow dynamics appears as an essential step in the comprehension of the asymmetric flapping flight aerodynamic performances. In order to explain the different tendencies in the generation of lifting force and in the behaviour of the efficiency ratio from one configuration to another, the adimensional λ_2^* (figure 6) and pressure C_p (figure 7) contours are displayed for the symmetric configuration α_d=45°, α_u=45° and the asymmetric configuration α_d=45°, α_u=20°. These flow properties are correlated with the time-dependent lift and drag coefficients (figure 5), putting into evidence the force generating mechanisms. The two configurations show comparable downstroke kinematics, hence comparable quasi-steady aerodynamic forces, but different upstroke kinematics. As a consequence, differences observed in the flow dynamics and in the generation of forces during the downstroke phase directly arise from the previous strokes history, especially through the wake capture phenomenon. Note that the downstroke kinematics of both cases cannot be qualified as identical since the rotation velocities are not strictly equal.

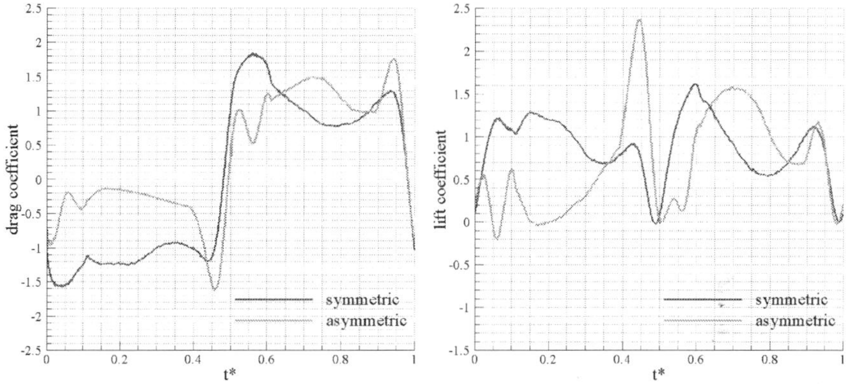

Figure 5. Drag and lift coefficient histories of the symmetric α_d=45°, α_u=45° and asymmetric α_d=45°, α_u=20° configurations

All flow properties and aerodynamic coefficients are represented at time t*, where t* corresponds to the time elapsed since the beginning of the flapping cycle, adimensionalized by the period T.

At t*=0, the airfoil begins the upstroke phase. Notwithstanding the zero translating velocity which characterizes the flapping kinematics at this instant, the drag coefficients reach a strong value (≈-1) for both configurations. Indeed, two counter-rotating vortical structures dominate the flow in the vicinity of the profile (figure 6 - t*=0), inducing the presence of a fluid jet directed towards the intrados and which acts as a drag enhancer. As previously expressed, this interaction between the airfoil and its own wake is typical of the flapping wings aerodynamics.

Between t*=0 and t*=0.1, the airfoil accelerates while still rotating to reach a constant translating speed and fixed angle of attack. The production of drag and lift here derives from the complex combination of inertial effects and vortex behaviours. Hence, significant differences can be observed between both symmetric and asymmetric configurations. On the one hand, the aerodynamic force resulting from the symmetric motion seems to primarily arise from the accelerating inertial effect, leading to the presence of a strong pressure region covering most of the airfoil intrados. On the other hand, the relative position of the airfoil and the previously shed vortices characterizing the asymmetric configuration induces a competition between strong (leading edge) and low (trailing edge) intrados pressure regions (figure 7 - t*=T/12). Therefore, at t*=0.06, the asymmetric configuration lift and drag decrease sharply whereas the symmetric ones are maintained to relatively strong values.

At t*=0.1, both cases demonstrate the formation of a LEV acting as a low pressure suction region on the extrados of the airfoil. Note that the latter is promoted by the wing/wake interaction as shown in previous studies (Birch & Dickinson, 2003). The LEV is essential when considering the symmetric motion since it dominates the vortical flow field during most of the upstroke phase. Thus, absolute values of lift and drag above 1 are reached. On the contrary, the LEV generated in the asymmetric case has a limited lifetime such that its influence is only perceptible through the occurrence of a narrow lift peak. Note that this vortical structure does not significantly affect the drag since the upstroke angle of attack is set to a low value (20°), i.e. the airfoil surface projection on the axis perpendicular to the stroke plane is weak. Moreover, one can remark that the lifetime is partially shortened by the wake capture phenomenon which leads to the LEV shedding (figure 6 - t*=T/12). Consequently, the asymmetric upstroke phase leads to a partially attached flow supported by the presence of a fluid downwash engendered by the previous strokes.

The time evolutions of the drag and lift coefficients roughly respond to the spatial and temporal evolution of the upstroke LEV until t*=0.4. At this instant, the airfoil starts to rotate while decelerating. Figure 5 shows the presence of drag and lift peaks, particularly pronounced in the asymmetric configuration for which the angle of attack changes from α_u=20° to α_d=45° with a rotating speed superior to the one encountered in the symmetric configuration. Such lift enhancements arise from the supplementary circulation generated by the combined rotating and translating motions (Kramer effect). Furthermore, whereas the rotation strengthens the Trailing Edge Vortex (TEV) formed during the latter part of the downstroke phase in the symmetric case, it is the cause for the formation of a Rotating Trailing Edge Vortex (RTEV) in the asymmetric case (figure 6 - t*=5T/12).

At t*=0.5, as observed at t*=0, the drag coefficient resulting from the symmetric motion exhibits a strong value linked to the presence of a fluid jet generated by the two shed counter rotating vortical structures. The flow behaviour induced by the asymmetric motion considerably differs since no strong separated flow yielded to the presence of a vortex dipole (figure 6 - t*=6T/12). Hence, any fluid jet tends to increase the drag coefficient which is here quasi null.

Between t*=0.5 and t*=0.6, the airfoil accelerates in the opposite direction to initiate the downstroke phase. Despite the symmetric aspect of the normal hovering case, the presence of highly unsteady non-linear structures leads to non-strictly equivalent upstroke and downstroke aerodynamic forces, as noticed by other authors on biologic cases (Van Oudheusden et al, 2008). However, global variations of upstroke and downstroke coefficients strongly resemble as arising from comparable flow behaviour. The symmetric motion engenders a rapid augmentation of lift and drag through the accelerating inertial effects. The latter is supported and maintained by the formation of a downstroke LEV on the airfoil extrados (figure 6 - t*=7T/12). On the contrary, the influence of wake capture is reduced in the asymmetric case. Indeed, besides the action of the RTEV assimilated to a low pressure region under the intrados of the airfoil, the previously shed vorticity is not sufficient to accelerate the formation of a downstroke LEV. Consequently, the drag and lift are maintained to low values until t*=0.6. The interesting aspect here resides in the fact that the LEV develops smoothly, staying closely attached to the extrados. Thus, from t*=0.6, despite the fact that the wing kinematics are identical in both symmetric and asymmetric configurations (i.e. identical quasi-steady forces), the asymmetric motion leads to stronger aerodynamic coefficients. This observation is evidence that the wake capture phenomenon not only affects the early stages but most of the stroke resulting forces.

From t*=0.9, the wake of both configurations still exhibit different behaviour hence different drag productions as the airfoil jointly rotates and decelerates. However, the lift coefficients tend to similar time evolutions since the inertial forces conducting the Kramer effect are quasi equivalent in both configurations.

4.3 Effects of desymetrization

The correlation between the fluid dynamics and the loads experienced by the flapping airfoil for two specific symmetric and asymmetric configurations (respectively α_d=45°, α_u=45° and α_d=45°, α_u=20°) demonstrated that reducing the upstroke wake by fixing a low upstroke angle of attack leads 1) to a reduction of the aerodynamic coefficients immediately after stroke reversal and 2) to an enhancement of the latter during most of the downstroke translating phase. This characteristic partially results from the absence of significant upstroke leading edge separation, hence significant wake capture, inducing the generation of a closely attached LEV on the extrados of the airfoil. The analysis of the unsteady drag and lift coefficients obtained for all configurations tested support this observation (figure 8). Note that each graph of figure 8 regroups the flapping kinematics parameterized with a common downstroke angle of attack such that it effectively illustrates the effect of desymetrization on the production of lift and drag during the downstroke phase (the quasi steady downstroke forces being equivalent). Nevertheless, analysing the lift and drag coefficients averaged over the downstroke phase shows that both effects seem to inhibit each other. For instance, the mean drag and lift coefficients obtained during the downstroke phase of the α_d=45°, α_u=45° and the α_d=45°, α_u=20° configurations are respectively 1.09, 0.87 and 1.11, 0.89.

Besides the changes observed in the fluid dynamics, asymmetric hovering motions benefit from the combined action of both lift and drag as the effective lifting force (Wang, 2004). Indeed, for such motions, the hovering condition imposes the stroke plane to be inclined such that the drag has a vertical component and thus provides a lifting force rather than exclusively acting as a power consumer. According to the β angles calculated and listed in table 1, the resulting horizontal and vertical unsteady components are deduced from the drag and lift components and represented in figure 9. In all cases, increasing the difference α_d-α_u, i.e. the stroke plane angle, considerably enhance the resulting downstroke vertical force. For instance, the mean lifting force coefficient obtained during the downstroke phase of the α_d=45°, α_u=45° and the α_d=45°, α_u=20° configurations are respectively 0.87 and 1.28 which corresponds to an increase of approximately 47%. However, the upstroke phase appears as a harmful phase and may sometimes lead to a negative vertical force component as exhibited when the upstroke angle of attack is set to 10°. Note that such negative values can be obtained with still non-negligible angle of attack since the presence of a fluid downwash tends to reduce the effective angle of attack. Consequently, as dedicated to MAVs application, asymmetric kinematics with an upstroke angle of attack fixed below 15° should be avoided. Moreover, the quality of the resulting aerodynamic performances resides in the relative importance of the upstroke and downstroke phases.

4.4 Improving aerodynamic performances

Previous sections showed that comparatively to symmetric flapping motions, asymmetric flapping motions may lead to sustained efficiency ratio coupled with a consistent lifting force when the upstroke angle of attack is lowered, weakening the wake capture phenomenon and implying the contribution of drag to the vertical force component. Furthermore, it obviously appeared that, the upstroke phase being harmful, the relative importance of the upstroke and downstroke phases should be deeply considered. In this brief section, we suggest a way to ameliorate the aerodynamic performances of asymmetric motions by increasing the upstroke translating velocity, i.e. reducing the relative importance

of the upstroke phase. This approach is consistent since the inherent characteristics of the previous cases are not altered, the flow dynamics not depending on the Reynolds number in this range (Wu & Sun, 2004; Kurtulus, 2005).
Results are obtained for velocity ratios (defined as the ratio r_v between the upstroke and the downstroke translating velocity) of 1, 1.2 and 1.4. The aerodynamic coefficients are determined by adimensionalizing the upstroke and downstroke forces by means of the upstroke and the downstroke translating velocities respectively. Tables 3 and 4 illustrate the effect of increasing the upstroke velocity on two chosen asymmetric cases. It is shown that the lifting force and the efficiency can be increased in a prompter way than the power consumption such that the quality coefficient can be significantly enhanced. Indeed, one might observe that the influence on the power consumption is not monotonous (increasing between r_v=1 and r_v=1.2, decreasing between r_v=1.2 and r_v=1.4).

	$r_V = 1$	$r_V = 1.2$	$r_V = 1.4$
$\overline{C_d}$	0.7314	0.7497 (+2.5%)	0.7733 (+5.7%)
$\overline{C_y}$	0.7149	0.7825 (+9.5%)	0.8308 (+16.2%)
$\overline{C_{eff}}$	0.977	1.044 (+6.9%)	1.074 (+9.9%)
$\overline{C_p}$	0.543	0.604 (+11.2%)	0.583 (+7.4%)
$\overline{C_q}$	0.161	0.165 (+2.5%)	0.187 (+16.1%)

Table 3. Effect of velocity ratio on aerodynamic coefficients for the asymmetric configuration α_d=45°, α_u=15°

	$r_V = 1$	$r_V = 1.2$	$r_V = 1.4$
$\overline{C_d}$	0.7823	0.7848 (+0.3%)	0.8144 (+4.1%)
$\overline{C_y}$	0.7826	0.8212 (+4.9%)	0.8498 (+8.6%)
$\overline{C_{eff}}$	1.000	1.046 (+4.6%)	1.043 (+4.3%)
$\overline{C_p}$	0.580	0.628 (+8.3%)	0.599 (+3.3%)
$\overline{C_q}$	0.172	0.171 (-0.6%)	0.189 (+9.9%)

Table 4. Effect of velocity ratio on aerodynamic coefficients for the asymmetric configuration α_d=45°, α_u=20°

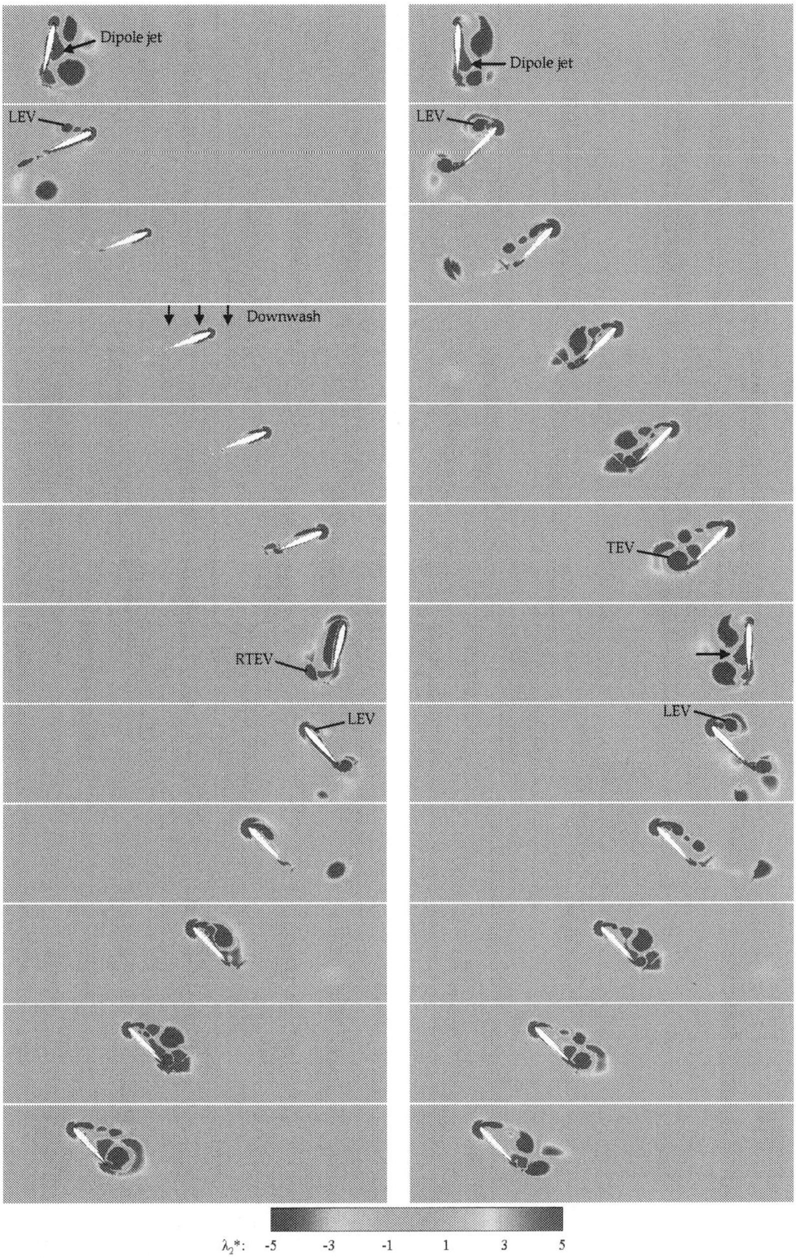

Figure 6. λ_2^* distribution at t*=nT/12 (n ∈ [0;11]) from top to bottom for the symmetric (right) and asymmetric (left) configurations

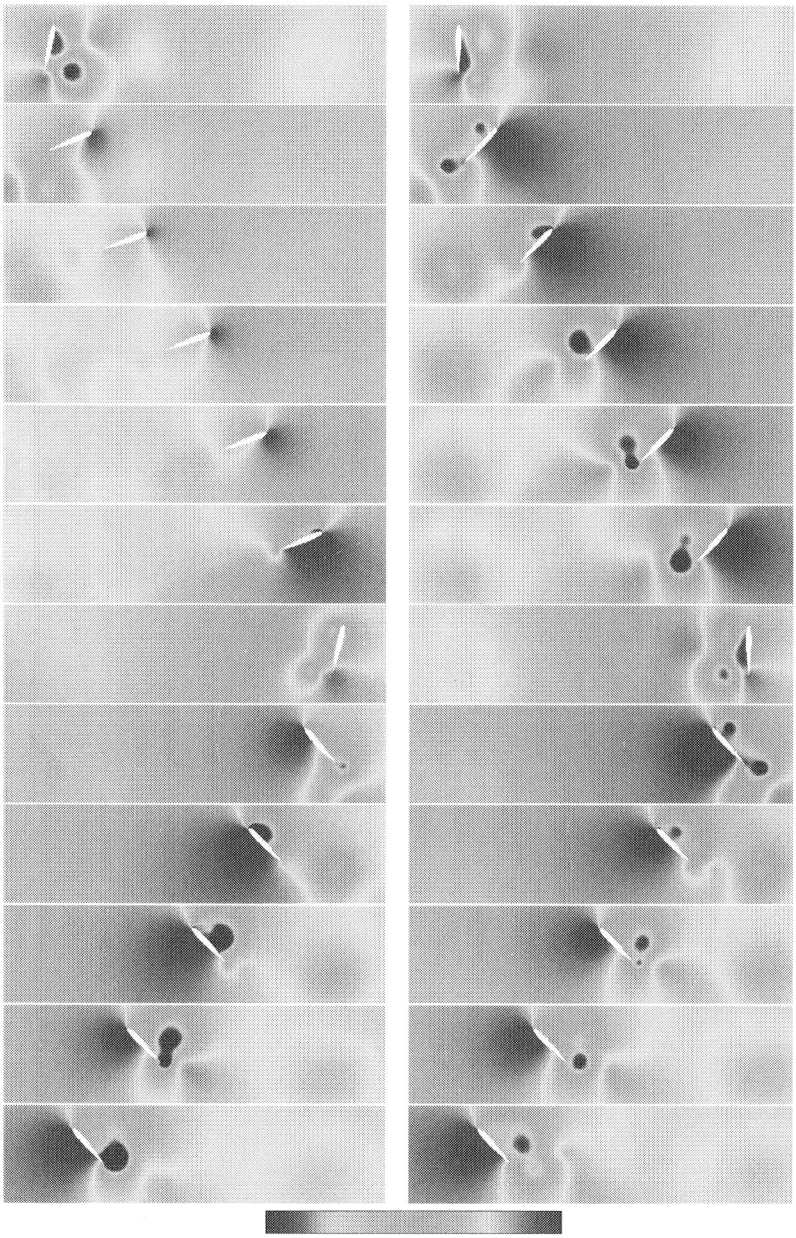

Figure 7. Adimensional pressure distribution at t*=nT/12 (n ∈ [0;11]) from top to bottom for the symmetric (right) and asymmetric (left) configurations

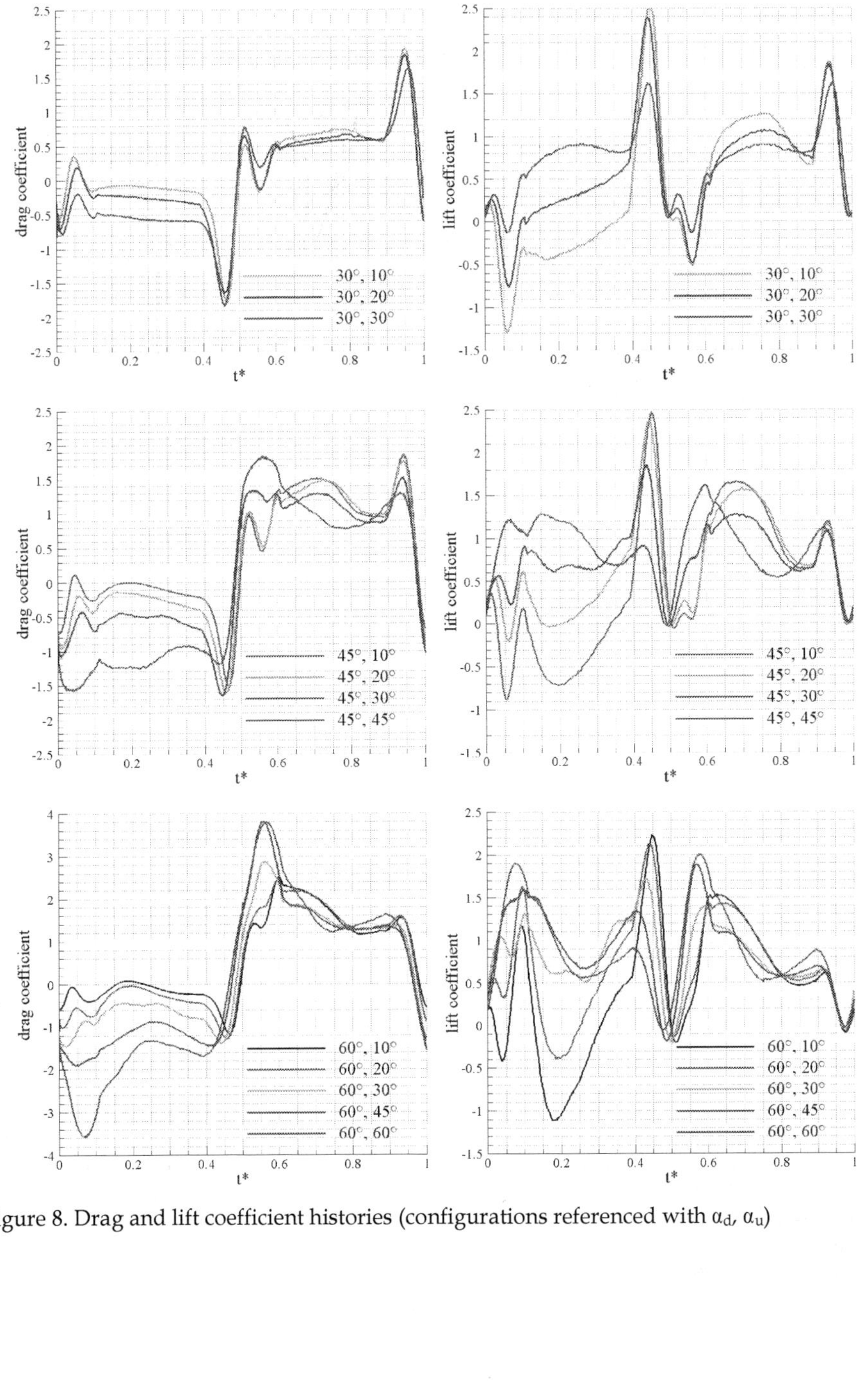

Figure 8. Drag and lift coefficient histories (configurations referenced with α_d, α_u)

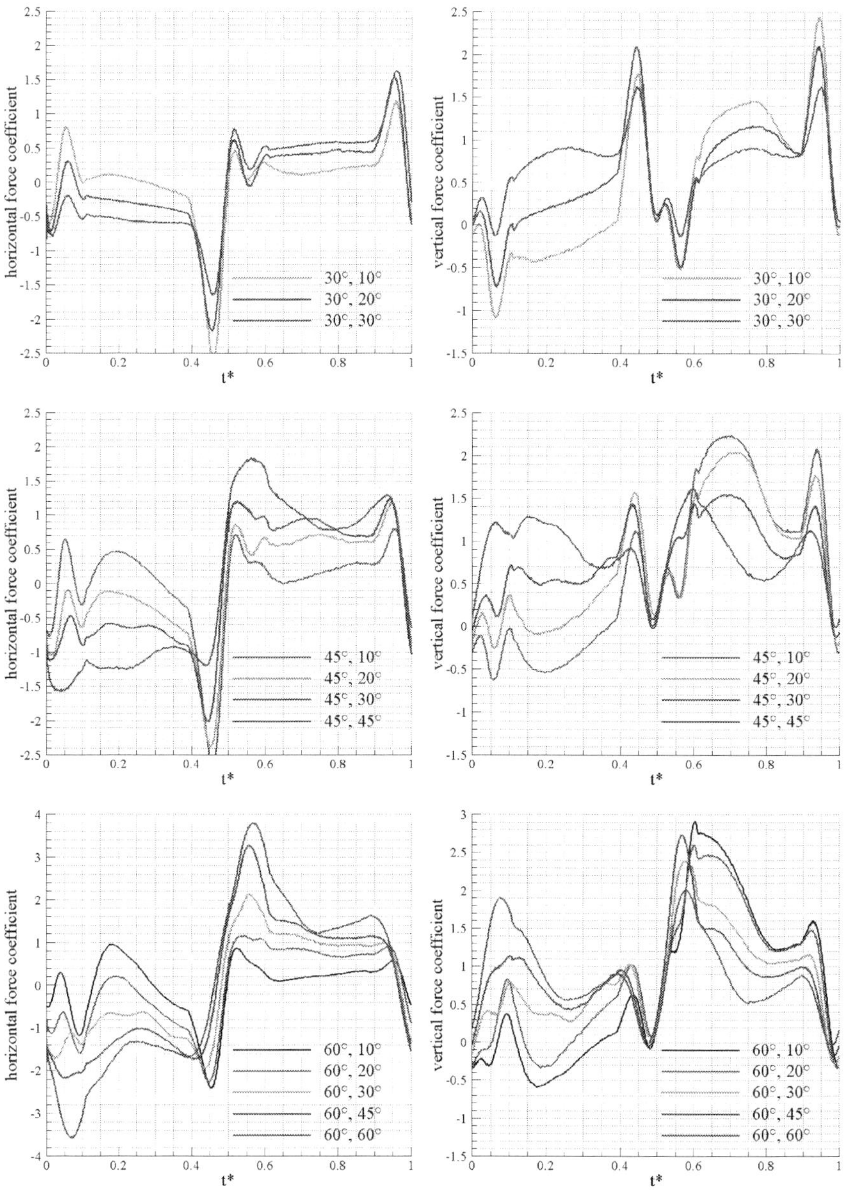

Figure 9. Horizontal and vertical force coefficient histories (configurations referenced with α_d, α_u)

5. Conclusion

Micro Air Vehicles are characterized by limited dimensions (15 cm) which places the corresponding aerodynamic flow in the range of low Reynolds number flows (10^2-10^4). At such Reynolds numbers, the flapping wing concept appears as an alternative solution to the conventional fixed and rotary wings, presupposing enhanced aerodynamic performances. Previous works relied on the analysis of normal (symmetric) hovering configurations as being the most common motion kinematics observed in the world of insects. In this study, parameterized asymmetric flapping motions at Reynolds 1000 are investigated by means of two-dimensional DNS calculations and compared to symmetric flapping motions.

In a first step, the mean analysis of the resulting aerodynamic coefficients demonstrate that introducing asymmetry (by differentiating the upstroke angle of attack to the downstroke one) generally lowers the aerodynamic force experienced by the airfoil while enhancing both efficiency and quality coefficients. Furthermore, when the latter is displayed as a function of the upstroke angle of attack, two distinct tendency behaviours are denoted on both sides of $\alpha_u \approx 20°$.

In a second step, the flow unsteadiness is analysed by means of the λ_2 and the pressure contours. The symmetric case parameterized with $\alpha_d = \alpha_u = 45°$ exhibits a strong leading edge separation during the upstroke translating phase, implying the occurrence of significant wing/wake interactions at stroke reversal. On the contrary, the asymmetric case is parameterized with $\alpha_d = 45°$, $\alpha_u = 20°$ such that the upstroke flow is attached, resulting in a reduced wake capture phenomenon. The direct consequences are 1) the absence of lift peak at stroke reversal and 2) the presence of a closely attached Leading Edge Vortex inducing enhanced downstroke lift. Such observations suggest that the behaviours observed on mean coefficients might arise from the presence or not of strong separation during upstroke. Moreover, the hovering condition applied on asymmetric motions imposes the stroke plane to be inclined such that part of the drag provides a lifting force rather than exclusively consuming power. Hence, the benefit of a closely attached LEV producing both lift and drag is further increased.

Despite the generation of enhanced downstroke lifting force, asymmetric configurations are characterized by a harmful upstroke phase. This aspect partially engenders a decrease in mean resulting aerodynamic forces. However, in order to improve global performances, it is of interest to shorten the latter, weakening the relative importance of upstroke comparatively to downstroke.

Consequently, this chapter brings further insight into the aerodynamics of asymmetric flapping motions and provides interesting perspectives for the development of high efficiency/quality Micro Air Vehicles.

6. References

Bennett, L. (1970). Insect flight: lift and the rate of change of incidence. *Science*, 167, 177-179

Birch, J.M., Dickinson, M.H. (2003). The influence of wing-wake interactions on the production of aerodynamic forces in flapping flight. *Journal of Experimental Biology*, 206, 2257-2272

Bos, F.M., Lentink, D., Van Oudheusden, B.W., Bijl, H. (2008). Influence of wing kinematics on aerodynamic performance in hovering insect flight. *Journal of Fluid Mechanics*, 594, 341-368

Dickinson, M.H., Götz, K.G. (1993). Unsteady aerodynamic performance of model wings at low Reynolds numbers. *Journal of Experimental Biology*, 174, 45-64

Dickinson, M.H., Lehmann, F.O., Sane, S.P. (1999). Wing rotation and the aerodynamic basis of insect flight. *Science*, 284, 1954-1960

Ellington, C.P. (1984). The aerodynamics of hovering insect flight I–V. *Philosophical Transactions of the Royal Society. Series B: Biological Sciences*, 305, 1122, 1-181

Jensen, M. (1956). Biology and physics of locust flight. III. The aerodynamics of locust flight. *Philosophical Transactions of the Royal Society. Series B: Biological Sciences*, 239, 667, 511-552

Jeong, J., Hussain, F. (1995). On the identification of a vortex. *Journal of Fluid Mechanics*, 285, 69-94

Kramer, M. (1932). Die zunahme des maximalauftriebes von tragflugeln bei plotzlicher anstellwinkelvergrosserung. *Z Flugtech Motorluftschiff*, 23, 185-189

Kurtulus, D.F. (2005). Numerical and experimental analysis of flapping motion in hover. Application to Micro Air-Vehicles. *Phd Thesis*, University of Poitiers, Laboratoire d'Etudes Aérodynamiques

Maxworthy, T. (1979). Experiments on the Weis-Fogh mechanism of lift generation by insects in hovering flight Part 1. Dynamics of the 'fling'. *Journal of Fluid Mechanics*, 93, 47-63

Polhamus, E.C. (1971). Predictions of vortex-lift characteristics by a leading-edge suction analogy. *Journal of Aircraft*, 8, 4, 193 -199

Sane, S.P., Dickinson, M.H. (2001). The control of flight force by a flapping wing: lift and drag production. *Journal of Experimental Biology*, 204, 2607-2626

Sane, S.P. , Dickinson, M.H. (2002). The aerodynamic effects of wing rotation and a revised quasi-steady model of flapping flight. *Journal of Experimental Biology*, 205, 1087-1096

Walker, P.B. (1931). Growth of circulation about a wing and an apparatus for measuring fluid motion. *ARC report*

Wang, Z.J. (2004). The role of drag in insect hovering. *Journal of Experimental Biology*, 207, 4147-4155

Weis-Fogh, T. (1973). Quick estimates of flight fitness in hovering animals, including novel mechanism for lift production. *Journal of Experimental Biology*, 59, 169-230

Wu, J., Sun, M. (2004). Unsteady aerodynamic forces of a flapping wing. *Journal of Experimental Biology*, 207, 1137-1150

16

UAV Path Planning in Search Operations

Farzad Kamrani and Rassul Ayani
Royal Institute of Technology (KTH)
Sweden

1. Introduction

An Unmanned Aerial Vehicle (UAV) is a powered pilotless aircraft, which is controlled remotely or autonomously. UAVs are currently employed in many military roles and a number of civilian applications. Some 32 nations are developing or manufacturing more than 250 models of UAVs and 41 countries operate some 80 types of UAVs (U. S. Department of Defense, 2005). By all accounts utilization of UAVs in military and civilian application is expanding both in the short term and long term.

The two basic approaches to implementing unmanned flight, remote control and autonomy, rely predominantly on remote data communication and microprocessor technologies (U. S. Department of Defense, 2005). Advances in these technologies, which have grown exponentially since introduction, have dramatically improved the capabilities of the UAVs to address more complicated tasks. Increasing availability of low-cost computational power will stretch the boundary of what is possible to accomplish with less oversight of human operators, a feature generally called autonomy.

In many civil and military flight missions the aircrafts freedom of action and movement is restricted and the path is predefined. Given the path, the control task of the aircraft is to generate the trajectory, i.e. to determine required control manoeuvres to steer the aircraft from one point to another. However, in some flight missions the path is not predefined but should dynamically be determined during the mission, e.g. in military surveillance, search & rescue missions, fire detection, and disaster operations. In this type of scenarios the goal of the UAV is to find the precise location of a searched object in an area of responsibility. Usually some uncertain a priori information about the initial location of the object is available. Since during the search operation this information may be modified due to new reports from other sources or the UAV's observation, the path of the UAV can not be determined before starting the mission. In this chapter, we try to address this problem and introduce a framework for autonomous and dynamic UAV path planning in search operations.

The rest of this chapter is organized as follows: section 2 describes the problem in general and the instance of the problem that we solve, section 3 presents history and related work, section 4 provides an overall description of the proposed solution, section 5 and 6 explains Sequential Monte Carlo (SMC) methods and how we have applied it, in section 7 simulation of a test case scenario and the obtained results are presented, and section 8 summarizes the chapter.

2. Problem Definition

In surveillance or search and rescue missions an area of responsibility is assigned to a UAV with the task to find a target object. Usually the area is large and the detection time to find the object is the critical parameter that should be minimized. If no information about the target and the area of responsibility is available, then the only strategy is to exhaustively and uniformly search the area. However, in real life usually some information is available that justifies that the search effort is not evenly distributed over the entire area. An example of such situation is when some uncertain information about the former location of the target is available. This information may be combined with assumptions about the velocity and movement of the target to yield a time-dependent probability of the location of the target. Another situation is when sensor observations or report from other sources exclude some parts of the area. Furthermore, if the geographical map of the area of responsibility is available, one could use this information to concentrate search efforts on parts of the area, where the target is more likely to be found. Usually management of these fragments of information is performed by human operators, especially those with high experience in the field. The overall aim of the operator is to increase the utilization of the UAV resources by conducting the search operation in a manner that areas with higher probability of finding the object are prioritized and/or searched more thoroughly. When during the mission new information becomes available, it is required to repeat the procedure and modify the path if necessary. There are two major drawbacks with this approach. Firstly, since both the target and the sensor (UAV) are mobile, it is not always a trivial task to determine high probability areas and find the appropriate path. Analysis of this information may be beyond the capacity of a human brain. Secondly, due to the time-critical nature of these missions, it is not feasible to assign this task to an operator. Valuable time may be lost before information is processed by the human operator and it may be impossible to fulfil the time requirements, specially, where the information changes frequently or its volume is very high.

A more efficient approach is to automate the path planning process and integrate the reasoning about the locations of the target into the autonomous control system. In order to make this possible all available information has to be conveyed to the UAV and the autonomous control system should dynamically plan and modify the route.

In this work a simulation based method is introduced to address UAV path planning in search and surveillance missions, where some uncertain a priori information about the target and environment is available. Although the suggested framework is applicable in more general contexts, we have implemented and tested the method for a scenario in which a UAV searches for a mobile target moving on a known road network.

3. Related Work

Work on modern search theory began in the US Navy's Antisubmarine Warfare Operations Research Group (ASWORG) in 1942 (Morse, 1982; Stone, 1989). Bernard Osgood Koopman is credited with publishing the first scientific work on the subject in 1946, Search and Screening, which was classified for ten years before it was published (Stone, 1989). He defined many of the basic search concepts and provided the probabilistic model of the optimal search for a stationary target. However, developing algorithms for optimal search plan for moving targets started in the early 1970s and when computer technology became more available. The next step in developing search planners was to consider the dynamic nature of the search process. Computer Assisted Search Planning (CASP), developed for US Coast Guard in the 1970s by Richardson is a pioneer software system for dynamic planning of search for ships and people lost at sea (Richardson & Discenza, 1980; Stone, 1983). CASP employed Monte Carlo methods to obtain the target distribution using a multi-scenario approach. The scenarios were specified by choosing three scenario types and the required parameter values for each scenario. A grid of cells was used to build a probability map from the target distribution, where each cell had a detection probability associated with it. A search plan was developed based on the probability map. Feedbacks from the search were incorporated in the probability map for future search plans, if the first search effort did not succeed. The shortage of computer power and display technique did not allow CASP to be a truly dynamic tool operating in real-time aboard aircraft. Advances in computer technology provided the possibility of developing more feasible tools, Search and Localization Tactical decision aid (SALT) was a prototype air-antisubmarine search planner system for real-time use aboard aircraft (Stone, 1989).

The problem of searching for a lost target at sea by a single autonomous sensor platform (UAV) is discussed by (Bourgault et al., 2003a). In this paper the target may be static or mobile but not evading. The paper presents a Bayesian approach to the problem and the feasibility of the method is investigated using a high fidelity UAV simulator. Bayesian analysis is a way to recursively combine the motion model of the target and the sensor measurements to calculate the updated probability distribution of the target. Time is discretized in time steps of equal length and the distribution is calculated numerically. The search algorithm chooses a strategy that minimizes the expected time to find the target or alternatively maximizes the cumulative probability of finding the target given a restricted amount of time. The paper chooses one-step lookahead, i.e. the time horizon used for optimization is one time step. Because of this myopic planning, the UAV fails to detect the target if it is outside its sensor range. A decentralized Bayesian approach is suggested to solve the same problem by coordinating multiple autonomous sensor platforms (UAVs) in (Bourgault et al., 2003b; Bourgault et al., 2004). The coordinated solution is claimed to be more efficient, however, the simulations in these papers suffer from the short time horizon as well, i.e. one-step lookahead.

The problem of path planning of UAVs in search and surveillance missions (sensor platform steering) can be considered as a sensor resource management problem and is investigated by the Information fusion community as well. Sensor management is formally described as the process of coordinating the usage of a set of sensors or measurement devices in a dynamic, uncertain environment to improve the performance of data fusion and ultimately that of perception. A brief but highly insightful review of the multi-sensor management is presented by (Xiong & Svensson, 2003). The idea of simulating the target's future movements and choosing sensor control parameters to maximize a utility function is described in (Ahlberg et al., 2004). Given a situation X_0, all possible future situations X that are consistent with the positions in X_0 at time $t=0$ are generated. For each of these X's, the utility of each sensor control scheme S is calculated by simulating observations of X using scheme S. The S whose average over all X is "best" is then chosen. However, to overcome the computation complexity, the set of possible sensor schemes is kept relatively small. Simulation-based planning for allocation of sensor resources is also discussed by (Svenson & Mårtenson, 2006). For each considered sensor allocation, the possible future paths of the target are partitioned into equivalence classes. Two futures are considered equivalent with respect to a given sensor allocation if they would give rise to the same set of observations. This partitioning to equivalence classes decreases the computational complexity of the problem and speeds up the calculation process.

4. Our Approach

The approach employed in this work to address the path planning problem is simulation-based. In short, this approach can be described as a method that uses simulation to approximate the future state of the target and tests alternative paths against the estimated future by running what-if simulations. These what-if simulations are conducted continuously during the mission (on-line). Utilizing information, even when it is incomplete or uncertain, is essential in constructing efficient search strategies and a system that uses all pieces of information in general performs better compared to systems not considering this information. In order to utilize this information, modeling and simulation techniques are used, which have shown to be a feasible tool handling complex and "difficult-to-analyze" systems.

The on-line simulation method for path planning in a search mission can be described as the following. The mission length is divided by a sequence of *time check points*, $\{t_0, t_1, \ldots\}$ where t_0 is the start time of the mission. At time t_0 the UAV chooses a default (random) path. At each time check point $t_k \in \{t_0, t_1, \ldots, t_n\}$, a set of simulations are started. In each simulation the state of the target for time $t \geq t_{k+1}$ and the effect of choosing an alternative UAV path for time $t \geq t_{k+1}$ are estimated. These simulations are completed during the time period $[t_k, t_{k+1}]$ and the results of these simulations are compared to choose the most appropriate path. At time t_{k+1} the chosen path is applied and a new set of simulations are started. Observations and other received information continuously modify the estimation of the target, but this updated model is employed when the UAV reaches the next time check point. That is observations obtained in time period $[t_k, t_{k+1}]$ affect simulations conducted in period $[t_{k+1}, t_{k+2}]$ which determine the path of the UAV after time t_{k+2}.

Apart from difficulties in constructing an on-line simulation system in general, some other problems should be addressed before this method can be employed in UAV path planning. Given the state of a system, the aim of on-line simulations is to predict the future state of the system and choose a course of action that is most beneficial for the system. In a surveillance mission, the state of the system (target) is not available. Indeed, the objective of the on-line simulation in this case is to optimize the process of acquiring information. The sensor data, before the target is detected, consists mostly of "negative" information i.e. lack of sensor measurement where it was (with some probability) expected (Koch, 2004). This information should be utilized to modify our estimation of the target's location. The process of drawing conclusions from sensor data is a problem studied by the information fusion community. One powerful estimation technique used in information fusion is Sequential Monte Carlo (SMC) methods also known as Particle Filtering which is based on point mass (or particle) representation of probability densities (Arulampalam et al., 2002).

In tracking applications, SMC is an on-line simulation process, which runs in parallel with the data collection process. In on-line UAV path planning we use SMC methods to estimate the current state of the target. This estimation (particle set) which is our only picture of the reality and is updated continuously is employed in "what-if" simulations to determine how the UAV should move to collect new data as effectively as possible. This path planning algorithm consists of two parts. The first part is a main loop running in real-time in which information is collected and our picture of the state of the system is updated. The other part is a set of "what-if" simulations that are initiated and executed periodically and after reaching time check points. These simulations run faster than real-time and are executed concurrently. Comparing these simulation outputs determines the most appropriate course of action. In the two next sections we describe SMC methods briefly and explain how it is applied in the path planning framework.

5. Sequential Monte Carlo Methods

In order to analyze a dynamic system using a sequence of noisy measurements, at least two models are required: First, a transition model which describes how the system changes over time and second, a sensor model which relates the noisy measurements to the state (Arulampalam et al., 2002). Usually these models are available in probabilistic form and since measurements are assumed to be available at discrete times, a discrete-time approach is convenient. In this approach the transition model, $p(x_k \mid x_{k-1})$, gives the conditional probability of the state x_k given x_{k-1}. The sensor model, $p(z_k \mid x_k)$, gives the conditional probability of observation z_k, given the state x_k. We are usually interested in the conditional state of the system, given the sequence of observations $z_{1:k} = \{z_1, z_2, \dots, z_k\}$, i.e. $p(x_k \mid z_{1:k})$. In general, this conditional probability density function, may be obtained recursively in two stages, prediction and update. The prediction is calculated before the last observation z_k is available

$$
\begin{aligned}
p(x_k \mid z_{1:k-1}) &= \\
\int p(x_k \mid x_{k-1}, z_{1:k-1}) p(x_{k-1} \mid z_{1:k-1}) dx_{k-1} &= \\
\int p(x_k \mid x_{k-1}) p(x_{k-1} \mid z_{1:k-1}) dx_{k-1}&.
\end{aligned} \tag{1}
$$

The first equality follows from $p(x_k) = \int p(x_k \mid x_{k-1}) p(x_{k-1}) dx_{k-1}$ and the second equality is a result of the fact that the process is Markovian, i.e. given the current state, old observations have no effect on the future state (Arulampalam et al., 2002).
In the update stage the conditional probability $p(x_k \mid z_{1:k})$ is calculated using the prediction result when the latest observation z_k becomes available via Bayes' rule:

$$
p(x_k \mid z_{1:k}) = \frac{p(z_k \mid x_k) p(x_k \mid z_{1:k-1})}{p(z_k \mid z_{1:k-1})} \tag{2}
$$

where the denominator is calculated using

$$
p(z_k \mid z_{1:k-1}) = \int p(z_k \mid x_k) p(x_x \mid z_{1:k-1}) dx_k . \tag{3}
$$

If the transition model and the sensor model are linear and the process noise has a Gaussian distribution, which is a rather restrictive constraint, these calculations can be performed analytically by using Kalman Filter, otherwise some approximate method such as Particle Filtering (SMC methods) should be used (Arulampalam et al., 2002).

SMC methods are a set of simulation-based methods, which have been shown to be an appropriate tool for estimating the state of a non-linear system using a sequence of noisy measurements. Intuitively, SMC methods are simulations of how the state changes according to the transition model, and filtering the result using the sensor model. Since the system changes over time, this process is repeated in parallel with the real system when new data is received. Even if new observations are not available the prediction stage still can be used to predict the future state of the system. The procedure would be the same with the exception that since future measurements are not known yet, the update stage is not performed.

In SMC methods the probability density function of the state of the system in each time-step k is represented as a set of n points x_k^i in the state-space and corresponding weights w_k^i, i.e. $p_k^i = \{(x_k^i, w_k^i)\}_{i=0}^n$ where p_k^i is particle number *i* in time $t = k$.

The simulation begins with sampling S_0, a set of n particles, from the a prior distribution $p(x_0)$, such that

$$S_0 = \{(x_0^i, w_0^i)\}_{i=1}^n, w_0^i = \frac{1}{n} \tag{4}$$

and number of particles in each interval [a, b] is in proportion to $\int p(x_0)dx_0$. At each iteration, particles in the set S_{k-1} are updated using the transition model, i.e. by sampling from

$$p(x_k^i \mid x_{k-1}^i) \tag{5}$$

and when observations arrive the weights are recalculated using

$$w_k^i \propto w_{k-1}^i p(z_k \mid x_k^i). \tag{6}$$

Particles are resampled periodically considering their weights, i.e. they will be sampled with replacement in proportion to their weights and weights are set to $w_k^i = 1/n$. This step is necessary to replicate particles with large weights and eliminate particles with low weights and avoid degeneracy of the algorithm (Arulampalam et al., 2002).

6. Applying SMC methods to UAV Search

One natural application area of SMC methods is target tracking and surveillance. The transition model is then derived from properties of the target, terrain characteristics and other forehand information available about the target. The sensor model depends on the characteristics of the sensors and the signature of the target. Many examples of applying

SMC methods in surveillance are provided in (Doucet et al., 2001; Ristic et al., 2004). Examples of applying the methods in terrain-aided tracking are found in (Ristic et al., 2004). To demonstrate how the SMC methods work in practice, we present briefly how we have implemented them in the suggested framework. Here we assume that a single target is moving on a known road network. Some uncertain information about the initial location of the target, an approximation of its velocity and some assumption about its goal are available. The SMC methods include the following four stages: sampling, prediction, update, and resampling.

6.1 Sampling

The simulation starts with sampling $S_0 = \{(x_0^i, w_0^i)\}_{i=1}^N$ particles randomly from the a priori information $p(x_0)$, such that the number of particles on each (small) road segment is proportional to the probability of existence of the target on that road segment. Each particle is assigned a velocity randomly sampled from the distribution of the target's velocity. The weights of all particles are set equally to $1/N$.

6.2 Prediction

At iteration k, particles in the set S_{k-1} are propagated forward, that is the new state of the particles are calculated using their current location, velocity and a process noise based on the transition model, $p(x_k \mid x_{k-1})$. Since the motions of the particles (vehicle) are constrained by a known road network, their state can be specified by the vector $x_k = [r_k, d_k, v_k]$, where r_k is the current road segment, d_k is the distance the particle has moved on road r_k and v_k is the instantaneous speed of the particle.

6.3 Update

As described in the previous section, the sensor model is generally described by the probabilistic model $p(z_k \mid x_k)$, where x_k is the state of the system, and z_k is the observation at time $t = k$. The dimensions of the state z_k are usually, but not necessarily, less than the dimensions of x_k, since the system is not completely observable. We choose to distinguish between the part of the system-state which is not under our control, i.e. the state of the target x_k and the state of the UAV (sensor), y_k. Hence the probabilistic sensor model would be $p(z_k \mid x_k, y_k)$. We assume that the video interpretation task is solved by some means, i.e. we have a sensor that analyzes the incoming video at a constant rate and alarm with some probability if any target is observed. That is $z_k \in [ALARM, \neg ALARM]$.

Inspired by (Lichtenauer et al., 2004) we suggest the following model for sensor observations at a standard height.

$$p(ALARM \mid x_k, y_k) = \begin{cases} p_d & d \le \delta_{in} \\ p_d - \dfrac{(p_d - p_f)(d - \delta_{in})}{\delta_{out} - \delta_{in}} & \delta_{in} \le d \le \delta_{out} \\ p_f & d \ge \delta_{out} \end{cases} \tag{7}$$

In this model given the position of the target x_k and the position of the sensor y_k, the probability that the sensor indicates an ALARM, is calculated using four sensor constants. These characteristics are: detection probability (p_d), false alarm probability (p_f), inner detection range (δ_{in}), and outer detection limit (δ_{out}), see Figure 1.

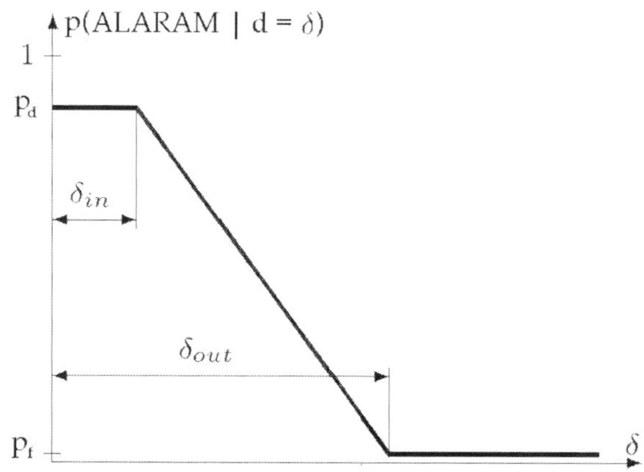

Figure 1. A graphic representation of the sensor model

After the propagation the weights of the particles are modified depending on the sensor model and current sensor observation. A sensor signal in a point increases the importance (weights) of the particles near that point. On the contrary, lack of sensor signals decreases the weights of the particles which are near the sensor. For instance if we have a perfect sensor and the UAV flies over a road segment and no sensor signal is supplied, the weights of all particles in that road segment are set to zero. After modifying the weights, they are normalized by dividing by the sum of the weights.

6.4 Resampling

A common problem with SMC methods is the degeneracy phenomenon, which refers to the fact that after many iterations, the number of particles with negligible weight increases and a large computational effort is devoted to updating particles, whose contribution to $p(x_k | z_k)$ is almost zero. One method to reduce the effect of degeneracy is to resample particles, i.e. to eliminate particles that have small weights and concentrate on particles with large weights. The resampling algorithm generates a new set of particles by sampling (with replacement) N times from the cumulative distribution function of weights of particles. The weights of these new particles are assigned the value 1/N.

7. Implementation and Test of a Scenario

The suggested path planning algorithm consists of two loops, the main control loop that includes the UAV and interacts with it at each time check point to modify the path of the

UAV to the *best known path*, and a simulation loop that estimates a picture of the reality, i.e. the probability density function of position of the target. At each time check point this picture of the reality is used in a series of what-if simulations with different possible paths of the UAV. The path that decreases the amount of uncertainty about the future is considered to be a "better" path.

Information entropy (Mackay, 2005), which is a measure of uncertainty associated with a discrete random variable $X \in \{x_1, x_2, ..., x_n\}$ is defined as $H(X) = -\sum_{i=1}^{n} p(x_i) \log_2 p(x_i)$.

H(X) takes only non-negative values, where $H(X) = 0$ indicates no uncertainty and larger values correspond to higher uncertainty. We suggest the expectation of the information entropy, $E[H(X)]$, as an objective function for comparing candidate UAV paths. In each step the path that decreases the expectation of the information entropy is chosen. SMC methods, which estimate the location of the target with a set of particles, provide an appropriate mechanism to estimate the expectation of the information entropy. Consider the UAV, being at point A in Figure 2(a), is facing the decision of whether to choose path ABC or ADE. The current estimate of the location of the target is shown by particles on these road segments.

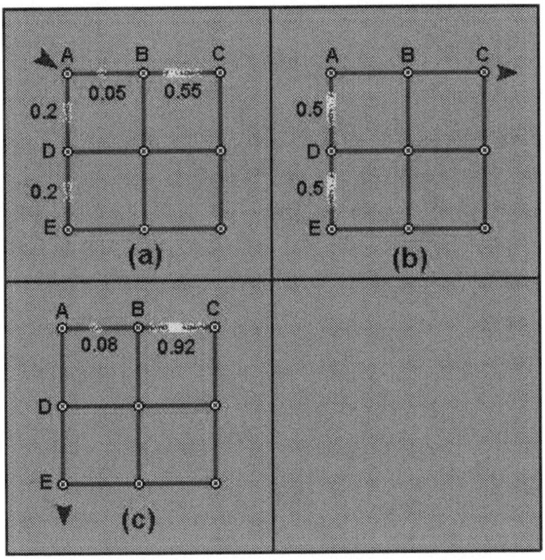

Figure 2. The impact of choosing different UAV paths on distribution of particles

Despite the fact that the total probability of finding the target on road ABC is 0.6, the most favourable path for the UAV is to choose ADE. Comparing Figures 2(b) and 2(c), shows that expectation of the information entropy by choosing path ABC is much more than choosing path ADE. See equations 8 and 9.

$$E[H(path = ABC)] = 0.p_{ABC} + (1 - p_{ABC})(-p_{AD} \log_2 p_{AD} - p_{DE} \log_2 p_{DE}) = 0.4(-0.5 \log_2 0.5 - 0.5 \log_2 0.5) = 0.4 \tag{8}$$

$$E[H(path = ADE)] = 0.p_{ADE} + (1 - p_{ADE})(-p_{AB}\log_2 p_{AB} - p_{BC}\log_2 p_{BC}) = \\ 0.6(-0.08\log_2 0.08 - 0.92\log_2 0.92) = 0.24 \quad (9)$$

To evaluate the performance of the suggested method, a test scenario is designed and simulations are performed using a special purpose simulation tool, called S2-simulator, introduced in (Kamrani et al., 2006). The tool contains a "real-world" simulator including a two dimensional terrain, a target object, and a UAV that can employ different search methods, one of which is on-line simulation method as described here. The on-line simulation method employs the simulation of this "real-world" to search for the target. Clearly, the information about the location of the target in the "real-world" is not available for the UAV. However, for simplicity we assume that some part of UAV's perception from reality is exact, e.g. the map of the terrain is accurate. Terrain is modelled as a two-dimensional landscape and includes two basic elements, nodes and road segments used to build a road network.

The geography of the test scenario, as shown in Figure 3, consists of a regular road network of perpendicular crossroads. Each road segment is 15 km long and the 60 road segments make an area of responsibility that covers a square of size 75 km by 75 km. For convenience a 2D coordinate system that has the origin located at the upper left-most node, with x values increasing to the right and y values increasing downwards is introduced.

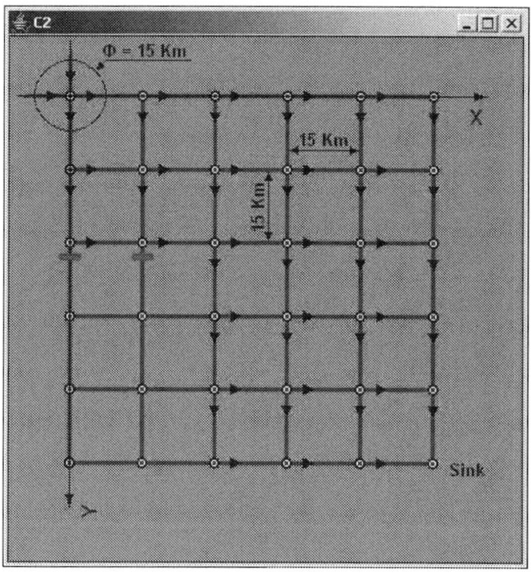

Figure 3. The target starts at the upper left-most node moving towards the sink

The target is initially located at the upper left-most node at origin. At this node and all other nodes the target moves either to east or toward south with equal probability, if any of these options are available. Hence, after passing 10 nodes and traversing 150 km the target reaches the lower right-most node and stops there. Considering these directions, the road network can be modelled as a directed acyclic graph with a source at origin and the sink located at the node in (75 km, 75 km). During the mission the road network may change as a result of the presence of friendly units, e.g. tanks may block certain roads. For example, in the scenario in Figure 3, two vertical road segments are blocked, which is shown by red bars in the figure. As seen in the figure, blocking a road segment does not affect only the blocked road segment but may change the distribution of particles in other areas. The target has a predefined path, which is unknown to the UAV. If a road in the path of the target is blocked the target chooses an alternative path. If directions to east and south are both blocked, the target stops. The average velocity of the target is 20 m/s with a velocity noise, thus it reaches its goal after approximately 7500 seconds. This velocity is unknown to the UAV.

The UAV starts its mission from a point in the III quadrant on the line y = x with a distance u_0 from the origin. Velocity of the UAV is 100 m/s. A large distance between the initial location of the UAV and the area of responsibility ensures that the target has a lead over the UAV. For example if u_0 = 180 km, the UAV reaches the area of responsibility after 1/2 hour.

The information available to the UAV system consists of the approximate initial location and velocity of the target, i.e. it is known that the target starts from a point uniformly distributed on the roads passing origin having a maximum distance of 7.5 km and has an average velocity between 15 and 25 m/s. It is as well known, that the target chooses one of the unblocked outgoing roads (if more than one) to south or east and there is no reason to believe that the target prefers one of these outgoing roads.

In the on-line simulation every 50 seconds (time check points), a new set of 60 alternative paths are compared by simulation. The length of these simulations is 1200 seconds. These simulations are run with the maximum possible speed, which depends on the computational power of the host computer. Simulations of the real world are run by time factor 5, i.e. 5 simulated seconds are equal to 1 (wall-clock) second. All simulations are run 10 times and the presented values are the average of these results.

Figure 4 illustrates the detection time of the target as a function of the distance of the UAV to the origin. Road network and the target's paths are also depicted in the same figure for 3 different cases. Blocked road segments are marked by red bars. Blocking of road segments occurs always during the mission, but before the target reaches the blocking points.

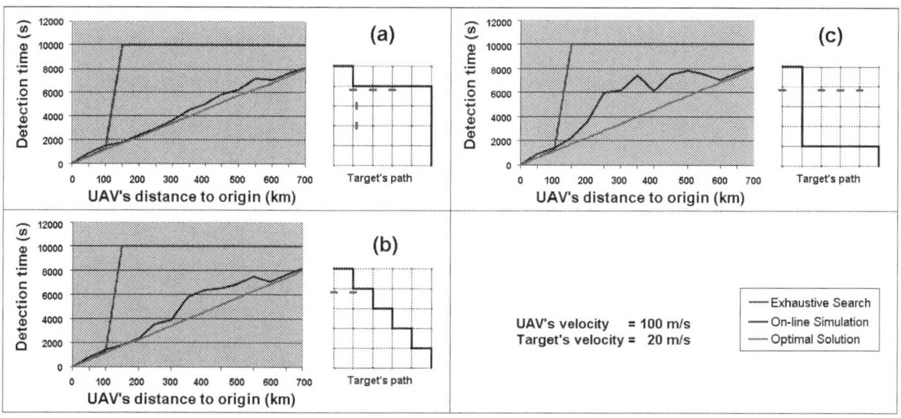

Figure 4. Detection time as a function of UAV's distance to the origin in online simulation compared with exhaustive search and the optimal solution

The values on the horizontal axis show the distance u_0 of the UAV with the origin, beginning from 0 to a maximum of 700 km. The vertical axis shows the average time for detection of the target. If the UAV fails to detect the target in 10000 seconds, the search is stopped. In each case the detection time for an exhaustive search method is also shown. In the exhaustive search method the UAV searches the entire road network indiscriminately starting from the road segments closest to origin and moving to the segments that are further from the origin.

Furthermore, a lower bound for the optimal value of "detecting" the target is also drawn in Figure 4. By optimal, we mean the time the UAV needs to meet the target if the path and the velocity of the target are completely known to the UAV. As expected, the detection time increases by increasing the distance of the UAV to the origin. However, the detection times for on-line simulation method remain significantly under 10000 seconds for all distances less than 700Km, while the exhaustive search method fails to detect the target when the UAV starts from a distance larger than 150 km. Moreover, comparing the result of the on-line simulation with the "optimal" solution in Figures 4-a, 4-b and 4-c, indicates that in some intervals the detection time approaches the optimal value, i.e. the time needed to reach the target if the UAV has all information about the target and the path.

8. Summary and Conclusion

In this chapter, we investigated the problem of autonomous UAV path planning in search or surveillance mission, when some a priori information about the target and the environment is available. A search operation that utilizes the available uncertain information about the initial location of the target, terrain data, and reasonable assumptions about the target movement can in average perform better than a uniform search that does not incorporate this information. We introduced a simulation-based framework for utilizing uncertain information in path planning. Search operations are generally dynamic and should be modified during the mission due to sensor observations, changes in the environment, and reports from other sources, hence an on-line simulation method was suggested. This method fuses continuously all available information using Sequential Monte Carlo methods to yield an updated picture of the probability density of the target's location. This estimation is used periodically to run a set of what-if simulations to determine which UAV path is most promising. From a set of different UAV paths the one that decreases the uncertainty about the location of the target is preferable. Hence, the expectation of information entropy is used as a measure for comparing different courses of action of the UAV.

The suggested framework was applied to a test case scenario involving a single UAV searching for a single target moving on a known road network. The performance of the method was tested by simulation, which indicated that the on-line path planning has generally a high performance. The result obtained by the on-line simulation method was compared with an exhaustive search, where the UAV searched the entire road network indiscriminately. The on-line simulation method showed significantly higher performance and detected the target in a considerably shorter time. Furthermore, the performance of the method was compared with the detection time when the UAV had the exact information about the initial location of the target, its velocity, and its path (minimum detection time). Comparison with this value indicated that the on-line simulation method in many cases achieved a "near" optimal performance in the studied scenario.

9. References

Ahlberg, S. & Hörling, P. & Jöred, K. & Mårtenson, C. & Neider, G. & Schubert, J. & Sidenbladh, H. & Svenson, P. & Svensson, P. & Undén, K. & Walter, J. (2004). The IFD03 information fusion demonstrator. *Proceedings of the Seventh International Conference on Information Fusion*, Vol. 2, pp. 936–943, June 2004

Arulampalam, S. & Maskell, S. & Gordon, N. & Clapp, T. (2002). A tutorial on Particle Filters for on-line non-linear/non-Gaussian Bayesian tracking. *IEEE Transactions on Signal Processing*, Vol. 50, No. 2, pp. 174–188, February 2002

Bourgault, F. & Furukawa T. & Durrant-Whyte, H.F. (2003a). Optimal Search for a Lost Target in a Bayesian World. *Proceedings of the 4th International Conference on Field and Service Robotics (FSR'03)*, volume 24, pp. 209–222, Lake Yamanaka, Japan, July 2003

Bourgault, F. & Furukawa, T & Durrant-Whyte, H.F. (2003b). Coordinated Decentralized Search for a Lost Target in a Bayesian World. *Proceedings of the 2003 IEEE/RSJ International Conference on Intelligent Robots and Systems*, Vol. 1, pp. 48–53, Las Vegas, Nevada, October 2003

Bourgault, F. & Göktogan, A. & Furukawa, T. & Durrant-Whyte H.F. (2004). Coordinated Search for a Lost Target in a Bayesian World. *Advanced Robotics*, Vol. 18, No. 10, pp. 979–1000

Doucet, Arnaud & de Freitas, Nando & Gordon, Neil. (2001). *Sequential Monte Carlo Methods in Practice*. Springer Verlag, ISBN 0-387-95146-0

Kamrani, Farzad & Lozano, Marianela Garcia & Ayani, Rassul. (2006). Path planning for UAVs using symbiotic simulation. *Proceedings of the 20th annual European Simulation and Modelling Conference, ESM'2006*, pp. 215–238, Toulouse, France, October 2006

Koch, Wolfgang. (2004). On 'negative' information in tracking and sensor data fusion: Discussion of selected examples. *Proceedings of the Seventh International Conference on Information Fusion*, Vol. 1, pp. 91–98, June 2004

Lichtenauer, J. & Reinders, M. & Hendriks, E. (2004). Influence of the Observation Likelihood Function on Particle Filtering Performance in Tracking Applications. *Proceedings of the 6th International Conference on Automatic Face and Gesture Recognition*. pp. 767- 772, May 2004

Mackay, David J. C. (2005). *Information Theory, Inference, and Learning Algorithms*. Cambridge University Press, pp. 22-46

Morse, Philip M. (1982). In Memoriam: Bernard Osgood Koopman, 1900-1981. *Operations Research*, Vol. 30, No. 3, May - June 1982, pp. 417–427

Richardson, Henry R. & Discenza, Joseph H. (1980). The United States Coast Guard Computer-Assisted Search Planning System (CASP), *Naval Research Logistics Quarterly*, Vol. 27, No. 4, March - April 1980, pp. 659–680

Ristic, Branko & Arulampalam, Sanjeev & Gordon, Neil. (2004). *Beyond the Kalman Filter: Particle Filters for Tracking Applications*. Artech House Series Publishers, ISBN 1-58053-631-x

Stone, Lawrence D. (1983). The Process of Search Planning: Current Approaches and Continuing Problems. *Operations Research*, Vol. 31, No. 2, March – April 1983, pp. 207–233

Stone, Lawrence D. (1989). What's Happened in Search Theory since the 1975 Lanchester Prize? *Operations Research*, Vol. 37, No. 3, May - June 1989, pp. 501–506

Svenson, P. & Mårtenson, C. (2006). SB-Plan: Simulation-based support for resource allocation and mission planning. *Proceedings of the Conference on Civil and Military Readiness (CIMI 2006)*, Enköping, Sweden, May 16–18, 2006

U. S. Department of Defense. (2005). *Unmanned Aircraft Systems (UAS) Roadmap, 2005-2030,* August 2005, pp. 47-52

Xiong, N. & Svensson, P. (2003). Multi-sensor management for information fusion: issues and approaches. *Information Fusion,* Vol. 3, No. 2, pp. 163–186

17

Optimal Circular Flight of Multiple UAVs for Target Tracking in Urban Areas[1]

Jongrae Kim[1] and Yoonsoo Kim[2]

[1]*Department of Aerospace Engineering, University of Glasgow*

[2] *Department of Mechanical and Mechatronic Engineering, University of Stellenbosch*

[1]*U.K.*, [2]*South Africa*

1. Abstract

This work is an extension of our previous result in which a novel single-target tracking algorithm for fixed-wing UAVs (Unmanned Air Vehicles) was proposed. Our previous algorithm firstly finds the centre of a circular flight path, r_c, over the interested ground target which maximises the total chance of keeping the target inside the camera field of view of UAVs, J, while the UAVs fly along the circular path. All the UAVs keep their maximum allowed altitude and fly along the same circle centred at r_c with the possible minimum turn radius of UAVs. As discussed in [1,4], these circular flights are highly recommended for various target tracking applications especially in urban areas, as for each UAV the maximum altitude flight ensures the maximum visibility and the minimum radius turn keeps the minimum distance to the target at the maximum altitude.

Assuming a known probability distribution for the target location, one can quantify J, which is incurred by the travel of a single UAV along an arbitrary circle, using line-of-sight vectors. From this observation, (the centre of) an optimal circle $\mathcal{C}^*$ among numerous feasible ones can be obtained by a gradient-based search combined with random sampling, as suggested in [1]. This optimal circle $\mathcal{C}^*$ is then used by the other UAVs jointly tracking the same target. As the introduction of multiple UAVs may minimise J further, the optimal spacing between the UAVs can be naturally considered. In [1], a typical line search method is suggested for this optimal spacing problem. However, as one can easily expect, the computational complexity of this search method may undesirably increase as the number of UAVs increases.

The present work suggests a remedy for this seemingly complex optimal spacing problem. Instead of depending on time-consuming search techniques, we develop the following algorithm, which is computationally much more efficient. Firstly, We calculate the distribution $J(x)$, where x is an element of $\mathcal{C}^*$, which is the chance of capturing the target by one camera along $\mathcal{C}^*$. Secondly, based on the distribution function, $J(x)$, find separation angles between UAVs such that the target can be always tracked by at least one UAV with a guaranteed probabilistic measure. Here, the guaranteed probabilistic measure is chosen by taking into account practical constraints, e.g. required tracking accuracy and UAVs'

[1] This work has been supported by Subcommittee B in the University of Stellenbosch, South Africa

minimum and maximum speeds. Our proposed spacing scheme and its guaranteed performance are demonstrated via numerical simulations.
Key words: target tracking; optimal circular flight; multiple UAVs

2. Introduction and Problem Statement

Motivated by our previous novel formulation for the STTP (Single-Target Tracking Problem) using multiple UAVs (Unmanned Air Vehicles) in [1], we extend our previous ideas of handling the STTP using multiple UAVs in this work. The STTP to be considered here can be stated as follows:

Problem 2.1 *Given a ground-based target and fixed-wing UAVs equipped with a camera, find an optimal strategy in a known urban environment such that the target is kept inside at least one camera's field-of-view.*

Due to the importance and enormous applications of this problem, it has attracted a great deal of attention from researchers and has been studied in various directions, e.g. target identification or classification, fault-tolerant target tracking, multi-sensor target tracking, target position estimation. See [1] for a detailed literature review. However, it is our observation that most of the existing works are mainly focused on sensory data processing or fusion. In this work, we are more interested in *path planning* for target trackers, especially in densely populated urban domains. As far as our knowledge is concerned, there are very few works in this research area. We note that [2,3] directly address the present STTP. In particular, [3] poses the STTP as a stochastic optimisation and then proposes a real-time genetic algorithm which gradually improves an initially guessed solution.

In this present work, we develop a deterministic algorithm which is computationally much more efficient than the existing algorithms. As suggested in [1, 4], the main idea here begins with using circular motions of the fixed-wing UAVs tracking the same target. These UAVs' circular motions are highly recommended for various target tracking applications especially in urban areas, as for each UAV the maximum altitude ensures the maximum visibility and the minimum radius turn keeps the minimum distance to the target at the maximum altitude. More precisely, we first find a position, r_c, over the interested target (location), r_{tar}, which maximises the total chance of keeping the target inside the camera field of view, J, during the circular motions of the UAVs around r_c. Here, all the UAVs tracking r_{tar} keep their maximum allowed altitude $\bar{h}_{uav}$ and fly along the same circle $\mathcal{C}^*$, centred at r_c with the UAVs' minimum turn radius $\underline{r}_{uav}$. We assume that the probability distribution $p(t)$ of the target location at time t is known. Note that, as the UAVs are a fixed-wing type in our present discussion, the following practical constraints are part of the STTP:

$$0 < \underline{\omega} \leq \omega_i \leq \overline{\omega} \quad \text{for} \quad i = 1, 2, \ldots, n, \tag{1}$$

where ω_i is the turning rate (magnitude) of the ith UAV, and $\underline{\omega}$ and $\overline{\omega}$ are the minimum and the maximum turning rates of UAVs, respectively. In contrast with [1], we here use ω_i as a control variable to solve the STTP problem. Note that the minimum radius turn is a function of angular velocity and bank angle. Therefore, in order to turn with the minimum radius but different speed, the angular speed must be increased or decreased by thrust, and bank angle must be adjusted appropriately so that the minimum turn radius can be maintained.

In §3.1, we first summarise how to find r_c and C^*. For more details, we direct readers to the reference [1]. Once the UAVs' flying path C^* is determined, it is natural to question about how to place more than one UAV on C^*. To answer this question, our present work is focused on the following problem:

Problem 2.2 *Determine spacing Φ between the UAVs flying along C^* such that at least one UAV keeps the target inside its camera's field-of-view all the time.*

This UAV spacing problem is mathematically formulated in §3.2 and two spacing schemes are proposed thereafter. Finally, numerical examples will demonstrate the efficacy of the proposed spacing schemes in §4.

3. UAV Path Planning for STTP

3.1 Optimal circle over target

In this section, we briefly summarise a method to find the centre r_c of an optimal circular path, along which UAVs fly and track a target at r_{tar}.[2]

Suppose t_c is the current time and the optimal circular path is updated every t_{com} seconds. Then, assuming a known probability density function $p(r_{tar})$ of the target location during the time interval $[t_c, t_c + t_{com}]$ (or at $t_c + t_{com}/2$), one can quantify J incurred by any UAV's travel (360-degree turn) along a circular path $\mathcal{C}$ using line-of-sight vectors, i.e.

$$J(\mathcal{C}) = \int_{x \in \mathcal{C}} \left[\int_{y \in \mathcal{S}(x)} p(y)\, dy \right] dx, \tag{2}$$

where $\mathcal{S}(x)$ is the set of all the ground locations within the UAV's camera's field-of-view at $x \in \mathcal{C}$. Thus, the following optimisation problem is of our interest:

$$\mathcal{C}^* = \underset{\mathcal{C}}{\arg\max}\ J(\mathcal{C}).$$

A sub-optimal $\mathcal{C}^*$ (centred at r_c with a radius of $\underline{r}_{uav}$) may be obtained by a gradient-based search (starting with a circle whose centre is precisely over the most probable target location) combined with a random sampling scheme. Note that this search is not much computationally involved as long as a moderate number of samples is used to approximate the integration (2).

3.2 Spacing Strategy I

Once $\mathcal{C}^*$ is calculated, we allow more UAVs to join $\mathcal{C}^*$ for tracking the same target. As the introduction of more UAVs on $\mathcal{C}^*$ may decrease the chance of losing the target further, we are now interested in finding spacing between the UAVs so that the target is always within at least one UAV's camera's field-of-view.

Suppose the first UAV is placed on $\mathcal{C}^*$ at $x_1 = r_c + [\underline{r}_{uav} \cos\theta_1, \underline{r}_{uav} \sin\theta_1]$[3] and other $n-1$ UAVs are placed on $\mathcal{C}^*$ according to the following sequence of spacing,

$$\Phi = \{\Phi_1, \Phi_2, \ldots, \Phi_{n-1}\},$$

[2] See [1] for more details.

[3] The z coordinate of x_1 is the maximum altitude that the first UAV can assume, and is dropped in the expressions hereafter.

where Φ_i defines the angle measured from the ith UAV to the (i+l)th UAV counterclockwise with respect to r_c. That is, $\Phi_i = \theta_{i+1} - \theta_i$ for $i = 1, 2, \ldots, n-1$ and $\Phi_n = \theta_1 - \theta_n$. We then consider the following new cost function J(x), where $\mathbf{x} = [x_1, x_2, \ldots, x_n] \subset \mathcal{C}^*$,

$$J(\mathbf{x}) = \max_{i \in \{1,2,\ldots,n\}} \int_{y \in \mathcal{S}_i(x_i)} p(y)\, dy, \tag{3}$$

where $\mathcal{S}_i(x_i)$ is the set of all the ground locations within the ith UAV's camera's field-of-view at x_i. As x can be written in terms of angles, we also call (3) $J(\Theta)$, where $\Theta = \{\theta_1, \theta_2, \ldots, \theta_n\}$ and

$$x_i = r_c + [\underline{r}_{uav} \cos\theta_i, \underline{r}_{uav} \sin\theta_i].$$

for $i = 1, 2, \ldots, n$.

As implied before, the direct optimisation to find Θ may be computationally involved as the number of UAVs on $\mathcal{C}^*$ increases. For this reason, we here slightly modify the problem statement as follows:

Problem 3.1 *Find Θ such that the target is kept inside at least one UAV's camera's field-of-view with a fixed probability or probabilistic measure μ, i.e.*

$$J(\Theta) \geq \mu > 0 \tag{4}$$

as UAVs fly along $\mathcal{C}^*$.

To this end, we first consider $J(\Theta)$ when $n = 1$, i.e $\Theta = \theta_1 \stackrel{\text{def}}{=} \theta$. For a fixed probability distribution for the target location $p(r_{tar})$ during the time interval $[t_c, t_c + t_{com}]$, one can compute $J(\theta)$ for each $\theta \in [0, 2\pi]$ or $x_1 \stackrel{\text{def}}{=} x \in \mathcal{C}^*$. Figure 1 depicts one period of a typical $J(\theta)$. Note that, as shown in Figure 1, $\theta_{\tau j}$, which belong to $[0, 2\pi)$ for $j = 1, 2, \ldots, n_\tau$, are the angles for which the sign of $J(\theta) - \mu$ changes. The corresponding positions to $\theta_{\tau j}$ are called *transition points*.[4]

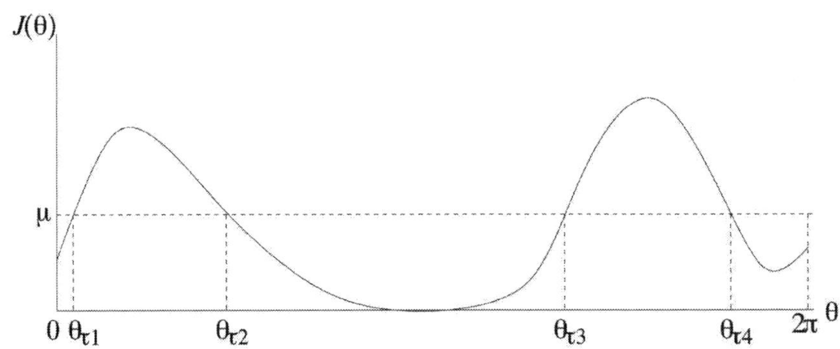

Figure 1. A typical example of $J(\theta)$ for $\theta \in [0, 2\pi]$

[4] We assume that $n_\tau \geq 1$. If $n_\tau = 0$, i.e. no transition point exists, (4) is trivially satisfied.

Our first spacing strategy is given in Table 1.[5] Clearly, this strategy guarantees the existence of at least one UAV flying along a *positive section* of C^* in which $J(\theta) \geq \mu$. Indeed, it guarantees an even better bound on average.

(1)	Place the ith UAV at $x_i = r_c + [r_{uav} \cos\theta_{\tau k}, r_{uav} \sin\theta_{\tau k}]$, where $k = \text{mod}(i, n_\tau)$. That is, $k = i$ if n is less than or equal to n_τ; otherwise, ith-UAV, where $i > n_\tau$, is placed at one of the already occupied transition points.
(2)	UAV's angular speeds are chosen such that the times to travel between any two consecutive transition points are all equal. UAVs keep constant angular speeds during their flights between two consecutive transition points.

Table 1. Spacing Strategy I

Theorem 3.1 *Spacing Strategy I given in Table 1 guarantees (4). Furthermore, during UAVs' 360-degree turns along C^*,*

$$J(\Theta) \geq \mu + \frac{2}{n_\tau |\overline{\mathcal{I}^+}|} \int_{\theta \in \mathcal{I}^+} [J(\theta) - \mu]\, d\theta, \tag{5}$$

on average, where $\mathcal{I}^+$ is the set of intervals in which $J(\theta) \geq \mu$, and $|\overline{\mathcal{I}^+}|$ is the length of the largest interval in $\mathcal{I}^+$.

Proof: As $J(\Theta) > \mu$ is straightforward to prove, we prove the second part only. Suppose

$$\mathcal{I}^+ = \{[\theta_{\tau 1}, \theta_{\tau 2}], [\theta_{\tau 3}, \theta_{\tau 4}], \ldots\} \tag{6}$$

and the interval lengths in $\mathcal{I}^+$ are l_1, l_3, . . ., respectively. In view of the second part of Spacing Strategy I, we set T to the time to travel between any two consecutive transition points. Then, for the example profile of $J(\theta) = J(\theta_1)$ as shown in Figure 2, if two UAVs start at $\theta_{\tau 1}$ and $\theta_{\tau 2}$ at $t = 0$ and turn 360 degrees along the circle, the corresponding $J(\Theta)$ in the time-domain can be calculated as shown in Figure 3. As seen in the figures, each positive part (peak) of $J(\theta) - \mu$ in the θ-domain appears twice with its stretched, shrunk or modified shape in the time-domain, depending on the two UAVs' locations.

More precisely, when t belongs to the interval $[(k - 1)T, kT)$ for some $k = 1, 2, \ldots, n_\tau$, the cost is given by

$$J(\Theta) = \mu + \max_{i \in \{1,2,\ldots,n\}} [J(\theta_i) - \mu] = \mu + \max_{i \in \{1,2,\ldots,n\}} [J(\theta_{0k} + \omega_{ik} t) - \mu],$$

where ω_{ik} is the constant angular velocity of the ith UAV during the given time interval, and for $i = 1,2, \ldots ,n$

$$\begin{aligned} \theta_i &= \frac{\theta_{\tau k}(t_{ik} + T) - \theta_{\tau(k+1)} t_{ik}}{T} + \omega_{ik} t \\ &= k\theta_{\tau k} - (k-1)\theta_{\tau(k+1)} + \omega_{ik} t \overset{\text{def}}{=} \theta_{0k} + \omega_{ik} t, \end{aligned}$$

where t_{ik} is the time when the ith UAV is at θ_{ik}.

[5] We here do not consider collision avoidance issues between UAVs.

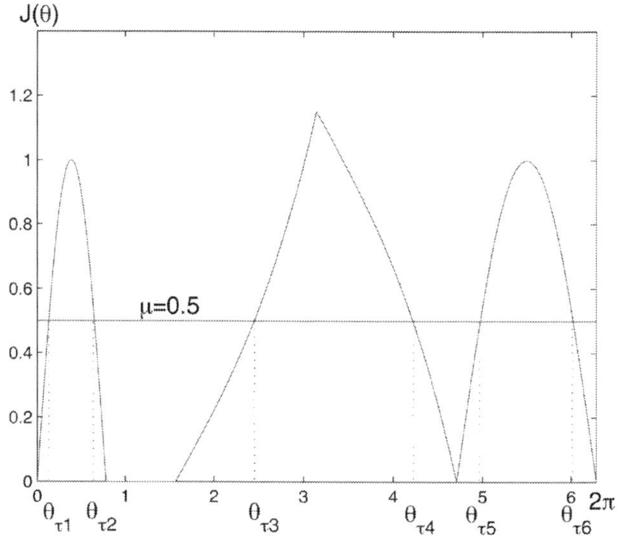

Figure 2. An example profile of $J(\theta)$ for $\theta \in [0, 2\pi]$: $\theta_{\tau j}$ (j = 1,2,..., 6) are transition points

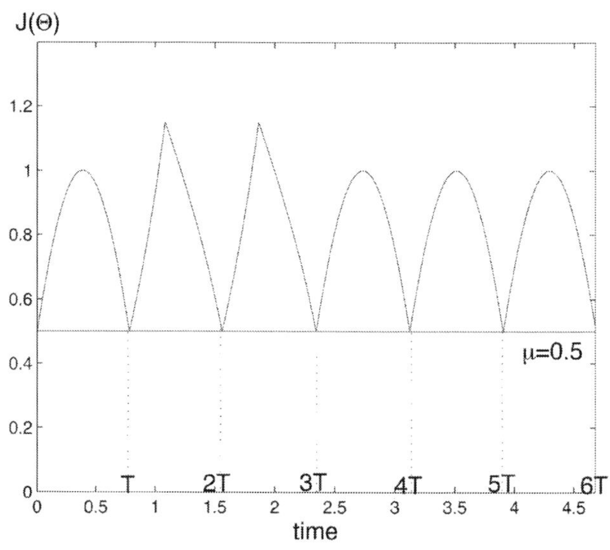

Figure 3. $J(\Theta)$ for the $J(\theta)$ shown in Figure 2: T is the time for UAVs to travel between any two consecutive transition points

Therefore, the mean value of $J(\theta)$ during UAVs' 360-degree turns can be expressed as

$$\mu + \frac{1}{n_\tau T}\sum_{k=1}^{n_\tau}\int_{(k-1)T}^{kT}\left\{\max_{i\in\{1,2,\ldots,n\}}[J(\theta_{0k}+\omega_{ik}t)-\mu]\right\}dt$$

$$\geq \mu + \frac{2}{n_\tau T}\sum_{k=1}^{n_\tau/2}\int_{2(k-1)T}^{(2k-1)T}[J(\theta_{0k}+\omega_{1k}t)-\mu]\,dt.$$

Note that n_τ is always an even number when $J(\theta)$ is continuous. Using the constant angular velocity assumption between any two consecutive transition points, one can easily check that for each $k = 1, 2, \ldots, n_\tau/2$,

$$\frac{1}{T}\int_{2(k-1)T}^{(2k-1)T}[J(\theta_{0k}+\omega_{1k}t)-\mu]\,dt = \frac{1}{l_{2k-1}}\int_{\theta_{\tau(2k-1)}}^{\theta_{\tau(2k)}}[J(\theta_1)-\mu]\,d\theta_1$$

$$\geq \frac{1}{|\mathcal{I}^+|}\int_{\theta_{\tau(2k-1)}}^{\theta_{\tau(2k)}}[J(\theta_1)-\mu]\,d\theta_1.$$

Thus, the claimed inequality follows. ■

This theorem allows us to approximately calculate average $J(\Theta)$ (guaranteed average *performance level)* once $\mathcal{C}^*$ is given, regardless of the time T to travel between any two consecutive transition points. It is obvious that one may wish to choose μ, so that the performance level in (2.5) becomes as high as possible, but this may result in transition points between which the interval is too small. Clearly, such a small interval makes UAVs flight infeasible. Therefore, μ must be determined based on various flight constraints such as (1). In the next section, we modify the first spacing strategy to accommodate those flight constraints.

3.3 Spacing Strategy II

In this section, we take into account flight constraints on UAV's travel speed and time along $\mathcal{C}^*$. In view of (1.1), μ must be chosen such that

$$\underline{\omega}T \leq l_i \leq \overline{\omega}T \quad \forall i,$$

where $l_i = \theta_{\tau(i+1)} - \theta_{\tau i}$, as defined in (6).

These flight constraints immediately imply that μ cannot be arbitrarily large. In this regard, we propose the second spacing strategy which now involves a search for a proper μ. The second strategy starts with setting μ to the average $J(\theta)$ over $0 \leq \theta \leq 2\pi$, i.e. $\int_0^{2\pi} J(\theta)\,d\theta/2\pi$, and check if there exist(s) an interval(s) $[\chi_{2j-1}, \chi_{2j}]$ $(j = 1, 2, \ldots, n_\chi) \subset \mathcal{I}_i^+$ in (6) such that

$$\begin{aligned}
&\underline{\omega}T \leq \chi_{2j} - \chi_{2j-1} \leq \overline{\omega}T \quad \forall j = 1, 2, \ldots, n_\chi,\\
&\underline{\omega}T \leq \frac{\chi_{2j+1} - \chi_{2j}}{n-1} \leq \overline{\omega}T \quad \forall j = 1, 2, \ldots, n_\chi - 1 \text{ (if } n_\chi > 1),\\
&\underline{\omega}T \leq \frac{2\pi - (\chi_{2n_\chi} - \chi_1)}{n-1} \leq \overline{\omega}T,
\end{aligned} \tag{7}$$

are satisfied. If there exist such intervals, one can increase μ and repeat the same step above. Otherwise, one should decrease μ until such intervals are found and the search is terminated.

Once the search is finished with a proper μ and its corresponding $[\chi_{2j-1}, \chi_{2j}]$, UAVs are placed on $\mathcal{C}^*$ at

$$x_1 = r_c + [\underline{r}_{uav} \cos \chi_1, \underline{r}_{uav} \sin \chi_1] \tag{8}$$

and if $n_\chi > 1$,

$$\begin{aligned} x_i = r_c \quad + \quad & \left\{ \underline{r}_{uav} \cos \left[\chi_2 + \frac{i-2}{n-1} (\chi_3 - \chi_2) \right], \right. \\ & \left. \underline{r}_{uav} \sin \left[\chi_2 + \frac{i-2}{n-1} (\chi_3 - \chi_2) \right] \right\}, \end{aligned} \tag{9}$$

otherwise

$$\begin{aligned} x_i = r_c \quad + \quad & \left\{ \underline{r}_{uav} \cos \left[\chi_2 + \frac{i-2}{n-1} (2\pi + \chi_1 - \chi_2) \right], \right. \\ & \left. \underline{r}_{uav} \sin \left[\chi_2 + \frac{i-2}{n-1} (2\pi + \chi_1 - \chi_2) \right] \right\} \end{aligned} \tag{10}$$

for $i = 2, 3, \ldots, n$.

The conditions in (7) ensure that at least one UAV flies over one of the intervals in $\mathcal{I}^+$, and so $J(\theta) \geq \mu$. In fact, one can easily show that the bound obtained in Proposition 2.1 still holds.

(1) Maximise μ subject to (7).
(2) Place UAVs at the locations specified by (8), (9) and (10).

Table 2. Spacing Strategy II

Theorem 3.2 *Spacing Strategy II guarantees (4) and (5).*

However, finding such intervals may not be a simple task, as $\mathcal{I}^+$ is likely to be non-convex (a union of disjoint intervals). In this regard, assuming that there is a small number of intervals in $\mathcal{I}^+$, one can go through every possible combination of intervals in J^+ to find feasible $[\chi_{2j-1}, \chi_{2j}]$ (j = 1,...,n_χ).[6] In fact, we start the search with the case of n_χ = 1. For example, consider an interval $[\theta_{\tau 1}, \theta_{\tau 2}] \subset \mathcal{I}^+$ and solve the following linear program for χ_1, χ_2 and T: minimise T subject to

$$\begin{aligned} & \theta_{\tau 1} \leq \chi_1 \leq \theta_{\tau 2}, \quad \theta_{\tau 1} \leq \chi_2 \leq \theta_{\tau 2}, \quad \underline{\omega} T \leq \chi_2 - \chi_1 \leq \overline{\omega} T, \\ & \underline{\omega} T \leq \frac{2\pi - (\chi_2 - \chi_1)}{n-1} \leq \overline{\omega} T. \end{aligned}$$

[6] MILP (Mixed Integer Linear Programming) could be used for this purpose.

If no feasible solution is found for each interval in $\mathcal{I}^+$, n_χ is now increased to 2 and two intervals are chosen from $\mathcal{I}^+$.[7] For example, we choose $[\theta_{\tau 1}, \theta_{\tau 2}]$, $[\theta_{\tau 3}, \theta_{\tau 4}] \in \mathcal{I}_i^+$, and solve the following linear program for $\chi_1, \chi_2, \chi_3, \chi_4$ and T: minimise T subject to

$$\theta_{\tau 1} \leq \chi_1 \leq \theta_{\tau 2}, \quad \theta_{\tau 1} \leq \chi_2 \leq \theta_{\tau 2}, \quad \theta_{\tau 3} \leq \chi_3 \leq \theta_{\tau 4}, \quad \theta_{\tau 3} \leq \chi_4 \leq \theta_{\tau 4},$$
$$\underline{\omega} T \leq \chi_2 - \chi_1 \leq \overline{\omega} T, \quad \underline{\omega} T \leq \chi_4 - \chi_3 \leq \overline{\omega} T,$$
$$\underline{\omega} T \leq \frac{\chi_3 - \chi_2}{n-1} \leq \overline{\omega} T, \quad \underline{\omega} T \leq \frac{2\pi - (\chi_4 - \chi_1)}{n-1} \leq \overline{\omega} T.$$

When this trial is still not successful with different two intervals in $\mathcal{I}^+$, n_χ is increased (up to $n_\tau/2$) and gives rise to similar linear programs as above. If no feasible solution is found for each n_χ, μ must be decreased to relax the constraints.[8]

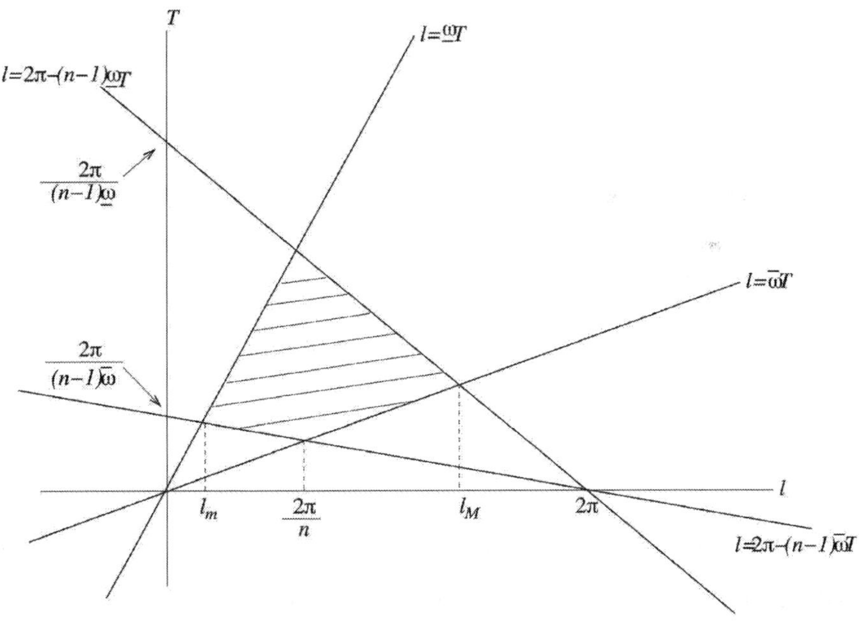

Figure 4. The polygon (dashed area) formed by the constraints when $n_\chi = 1$

[7] When time is critical, one may not want to increase n_χ, but to decrease μ and keep considering the case of $n_\chi = 1$ with new $\mathcal{I}^+$.

[8] It is not exactly true that less μ implies less stringent constraints. However, it is true that decreasing μ eventually yields a feasible solution.

As a matter of fact, the case of $n_\chi = 1$ allows a more efficient way of obtaining the solution. First note that the constraints in the associated linear program can be written as

$$\underline{\omega}T \leq l \leq \overline{\omega}T \quad \text{and} \quad \underline{\omega}T \leq \frac{2\pi - l}{n-1} \leq \overline{\omega}T,$$

where $l = \chi_2 - \chi_1$. These two constraints in l and T form a polygon in two-dimensional space, as shown in Figure 4. From the figure, one can read that the minimum and maximum l possible in the polygon are

$$l_m = \frac{2\pi}{1+(n-1)(\overline{\omega}/\underline{\omega})} \quad \text{and} \quad l_M = \frac{2\pi}{1+(n-1)(\underline{\omega}/\overline{\omega})}.$$

Therefore, in order for some μ to be feasible, the corresponding $\mathcal{I}^+$ must contain an interval whose length is in between l_m and l_M. If there exists such an interval(s) in $\mathcal{I}^+$ whose lengths are $l_1, l_2, \ldots$, one can then read the corresponding $T_1, T_2, \ldots$ from the figure, i.e.

$$T_j = \begin{cases} (2\pi - l_j)/(n\overline{\omega} - \overline{\omega}) & \text{if} \quad l_m \leq l_j \leq 2\pi/n, \\ l_j/\overline{\omega} & \text{if} \quad 2\pi/n < l_j \leq l_M. \end{cases}$$

Thus, the solution $[\chi_1, \chi_2]$ is the interval in $\mathcal{I}^+$ whose length l^* is such that T_j is minimised over all j. One can repeat finding $[\chi_1, \chi_2]$ with an increased μ to improve the solution.

4. Numerical Tests

To demonstrate the performance of the suggested algorithm (Spacing Strategy II), we randomly generate urban-area maps and model a ground target as a random jump process. As the employed strategy directs, each UAV flies along a designed circular path (consisting of several segments which are defined by separation angle(s)) at a different speed, but at a constant speed along each segment. In our tests, UAVs' minimum turn radius is assumed to be equal to 200 m, and their flight altitude is fixed to 100 m. The corner velocity, i.e. the velocity for a minimum radius turn at a maximum rate, is 30 m/s (approximately 100 km/h). Therefore, the maximum angular rate ($\overline{\omega}$) is given by 0.15 rad/s. The minimum angular velocity ($\underline{\omega}$) is assumed to be equal to the half of the maximum rate, i.e. 0.075 rad/s. The number of UAVs is two and they are of the same type.

One test example is shown in Figure 5. In the figure, two UAVs fly along the optimised circular path (thick solid line) in a counterclockwise direction, with an initial separation of 130 degrees. The starting position of each UAV is indicated by a triangle. The ground moving target (denoted by a red star) jumps to a random place every 1 s. Note that the target's operational area is densely populated by buildings, and thus tracking the target is very challenging. The simulation shows that, as expected, the two UAVs swap their starting positions and reach there at the same time by changing their angular velocities. Figure 6 shows $J(\theta)$ (dotted line; before introducing two UAVs), $J(\Theta)$ (solid line; after introducing two UAVs with the separation angle suggested by our spacing strategy), and the corresponding lower bound μ. The actual cost slightly violates the lower bound at 27 s. This violation is caused by the discretisation of $J(\theta)$ when numerically obtaining separation angles. A finer discretisation may alleviate this problem, but this slight violation does not make any significant physical difference in tracking the target. Figure 7 shows when the target is captured by the cameras. During one complete 360-degree turn along the circular path, the joint visibility is guaranteed during more than 69% of the total flight time. That is, the target is outside the two cameras' field-of-views for about 16s and the longest continuous time during which the target is lost is about 6 s.

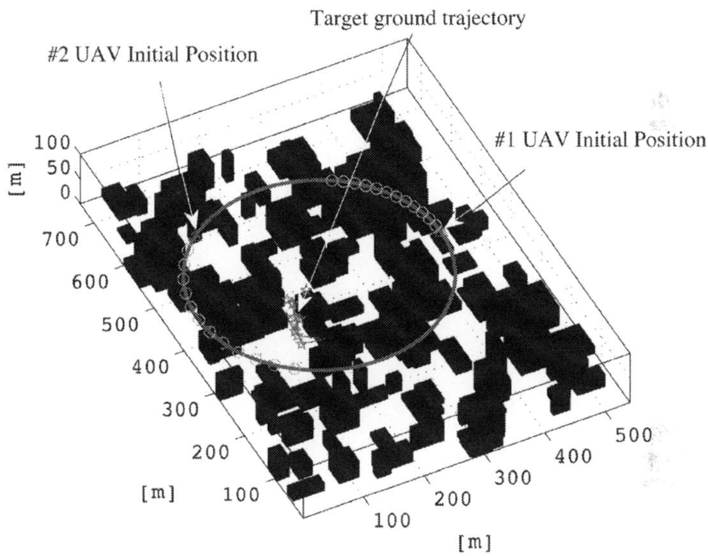

Figure 5. Two UAVs tracking a ground moving target in a dense urban area

A 1000-Monte-Carlo random simulation is also performed. This simulation is done using a personal computer (Windows-XP Professional version 2002, Intel Core 2 Quad CPU, 2.66GHz, 2.75GB of RAM, MATLAB 7.5 (R2007b)). The mean time spent to calculate separation angles for each case is 0.4963 s. In Figure 8, our proposed spacing strategy is compared with a simple separation strategy in which two UAVs are initially separated by 180 degrees. Figure 8 depicts, for both strategies, the number of cases versus the per cent of visible time during one 360-degree turn along the circular path associated with each case. The mean visibility is around 56% for both strategies. However, the variance corresponding to our spacing strategy is slightly better (smaller) than the simple one: our spacing strategy yields about 23% and the simple one about 27%. Also note that the two distributions look quite different: the distribution for the simple strategy is almost uniform, whereas the one for our proposed spacing strategy is skewed towards the right. This implies that our separation strategy handles difficult scenarios much better than the simple one.

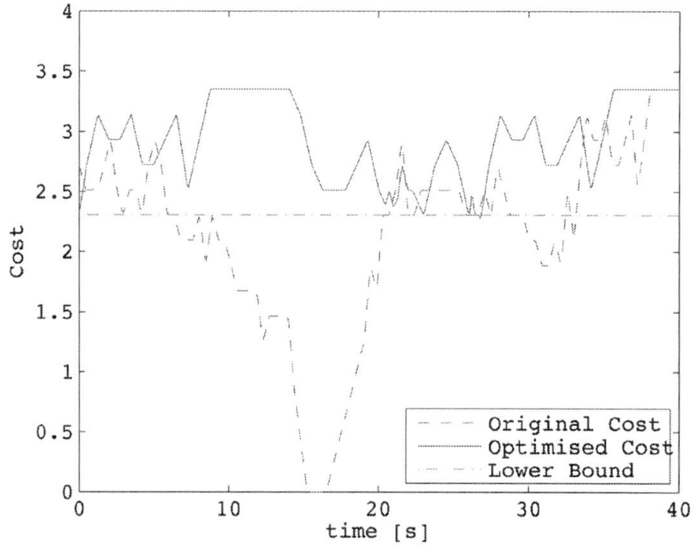

Figure 6. Original cost $J(\theta)$, optimised cost $J(\Theta)$ and lower bound

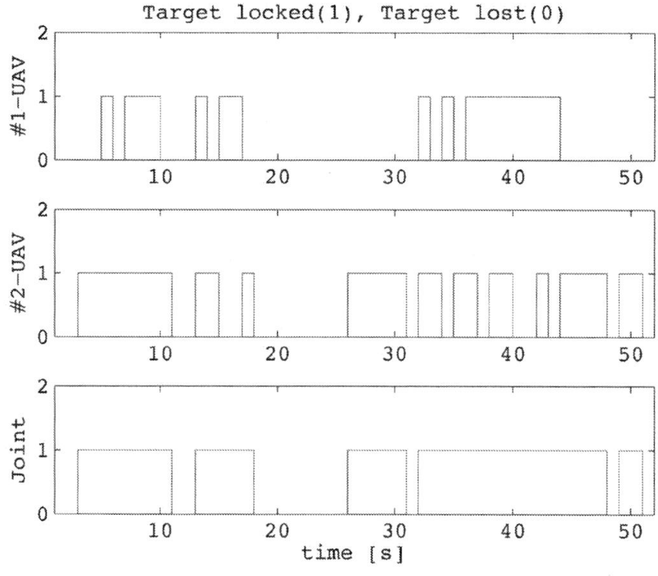

Figure 7. Target visibility during a 360-degree turn along the circular path shown in Figure 5

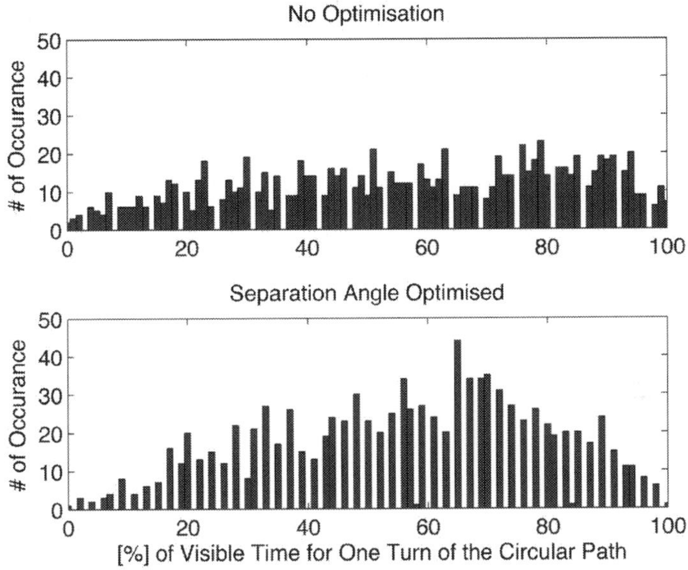

Figure 8. Target visibility distribution for 1000 random cases during a 360-degree turn along the circular path associated with each case

5. Conclusion

In this work, we extended our previous result by suggesting two spacing strategies which allow more than one UAV to jointly track a target with a guaranteed probabilistic bound. As opposed to typical line search methods, the proposed spacing strategies are independent of the number of UAVs tracking the same target in terms of computational complexity, and so easy to implement on UAVs especially with low computational power. However, there are a few points that need to be addressed in our future work. For example, suppose that a target moves beyond the area covered by a circular path $\mathcal{C}_1$, and so a new circular path $\mathcal{C}_2$ and the corresponding spacing are calculated. Then, one needs to consider a transition strategy which allows UAVs to move from $\mathcal{C}_1$ to $\mathcal{C}_2$ while still guaranteeing a probabilistic bound. Needless to say, extensions to the multiple target tracking case should be useful as well.

6. References

J. Kim and Y. Kim. Moving ground target tracking in dense obstacle areas using UAVs, In *Proceedings of the 17th IFAC World Congress,* July 2008. [1]

Mark A. Peot, Thomas W. Altshuler, Arlen Breiholz, Richard A. Bueker, Kenneth W. Fertig, Aaron T. Hawkins and Sudhakar Reddy. Planning sensing actions for UAVs in urban domains, In *Proceedings of the SPIE,* October 2005. [2]

Vitaly Shaferman and Tal Shima. Unmanned aerial vehicles cooperative tracking of moving ground target in urban environments, *Journal of Guidance, Control, and Dynamics,* Vol. 31, No. 5, September-October 2008, pp. 1360-1371. [3]

F. Rafi, S. Khan, K. Shafiq and M. Shah. Autonomous target following by unmanned aerial vehicles, In *Proceedings of the SPIE,* May 2006. [4]

18

Stiffness-Force Feedback in UAV Tele-operation

T. M. Lam, M. Mulder and M. M. van Paassen
Delft University of Technology
The Netherlands

1. Introduction

In the tele-operation of an uninhabited aerial vehicle (UAV), the operator and vehicle are physically separated. The limited sources of information of the environment, e.g., lack of a complete outside view, sounds, vehicle motions and vibrations, often lead to poor situation awareness. Tele-operation usually involves the use of a visual interface on a ground station, providing a navigation display and an outside visual generated by a camera mounted onboard the vehicle. The visual information, however, is usually not sufficient due to for instance a limited field of view, leading to an often not very efficient and sometimes even unsafe tele-operation (Elhaij et al., 2001; McCarley & Wickens, 2005).

Previous research efforts indicated that a haptic interface, using force feedback via a haptic control device, can be used to complement the visual interface. Haptic feedback provides information about the environment through the sense of touch (Lam et al., 2004; Lam et al., 2007). Indeed, the multi-sensory haptic interface improved operator performance and significantly reduced the number of collisions, leading to an overall highly increased level of safety (Lam et al., 2007). Operator control activity and workload, however, increased as well. This may be attributed to the incompatibility between the haptic interface and the operator's intentions, based on her or his internal representation of the environment.

High operator control activity and workload particularly occur in situations where the UAV is surrounded by many obstacles. Although force feedback would indeed help the operator to avoid a collision with one obstacle, through deflecting the control device away from the direction of the obstacle, it may also direct the UAV towards another obstacle located at the other side. Whereas the haptic interface does not know about the other closely-located obstacle, the operator might already have located it from the visual interface, and would prefer to adjust the direction of motion only "just enough" to avoid collision.

The reported incompatibilities between the motions of the haptic device and the operator's intentions may have been caused by the particular implementation of the haptic interface, i.e., through applying force feedback alone. Here, an external force offset will actively deflect the haptic control device, in such a way that the vehicle moves away from the direction of an obstacle. The offset force, a function of the relative position and velocity of the UAV with respect to the obstacle, still exists even after the operator releases her hand from the control device and may cause the control device to deflect to the other side of the zero deflection. For control of a UAV helicopter, the focus of our study, the control deflection represents a velocity command. Therefore, the force feedback sometimes does not allow the operator to "rest" and follow the motions of the control device, but forces the

operator to counteract them to avoid collision with another obstacle. This all contributes to high operator control activity and workload.
This chapter introduces two alternative implementations of haptic feedback: stiffness feedback and stiffness-force feedback. These alternatives aim to relieve the operator from continuously counteracting the repulsive forces from the haptic control device. The study involves both a theoretical discussion on the differences between the haptic feedback alternatives, as well as a pilot-in-the-loop experimental evaluation.

2. Haptic feedback

The use of haptic feedback through a control device allows the operator to perceive tactile cues through the sense of touch. The tactile information can represent various physical or mechanical properties, such as temperature level and relative distance (Elhaij et al., 2001). For collision avoidance in the tele-operation of a moving vehicle, it would be compatible with the operator's internal representation when the tactile cues would represent "repulsive forces" exerted from obstacles located near-by in the environment. These "repulsive forces" can be generated by an artificial force field (AFF) (Boschloo et al., 2004; Krogh, 1984).
The purpose of the artificial force field is to transform the UAV position and velocity relative to an obstacle, to a certain "value" that can be fed back to a haptic control device, yielding an impedance on the operator's control deflections. Fig. 1 shows a schematic representation of haptic feedback from a haptic control device, in our study a side-stick, to the human operator.

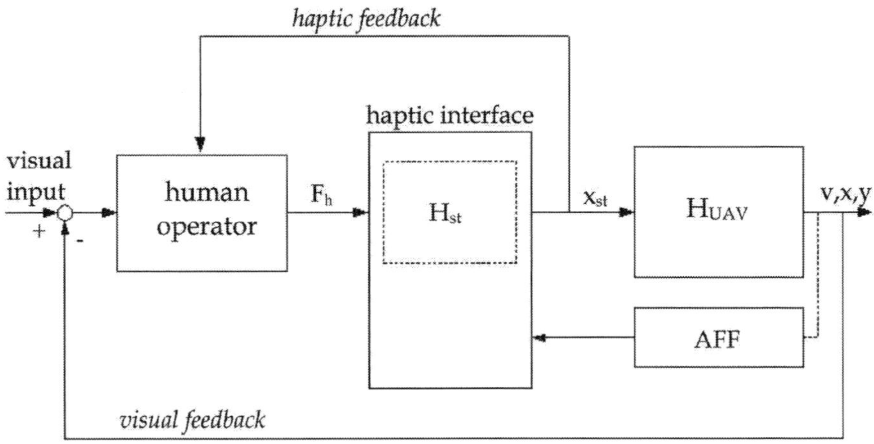

Figure 1. Schematic representation of a human-in-the-loop system with visual and haptic feedback

The impedance of the haptic control device consists of the stick mass-spring-damper dynamics H_{st}, depicted in Eq. (1), and the extra impedance from the output of the AFF.

$$H_{st}(s) = \frac{1}{ms^2 + bs + k}, \qquad (1)$$

with m, b and k the stick's mass, damping coefficient and spring constant, respectively. The generated value of the AFF can be fed back in many different ways to the haptic device. In the following sections, three alternatives will be discussed.

2.1 Force feedback

In force feedback, a force offset F_f is applied to the control manipulator to guide the operator, see Fig. 2. The total force that the operator perceives is the sum of the reaction force from the stick dynamics, F_{st}, and the external force offset. Assume that the stick is displaced to a certain position, x_{st}, then the force exerted by the operator, i.e., the force on the hand, F_h, is written as:

$$F_h(x_{st},i) = F_{st}(x_{st}) + F_f(i), \quad (2)$$

$$F_h(x_{st},i) = kx_{st} + F_f(i), \quad (3)$$

with $F_{st}(x_{st})$ and $F_f(i)$ the reaction force from the control device as function of the stick deflection x_{st}, and the external force offset as function of haptic feedback information i, respectively. Note that, with haptic feedback, the stick deflection x_{st} acts as the input to the system to be controlled, in our case the UAV helicopter.

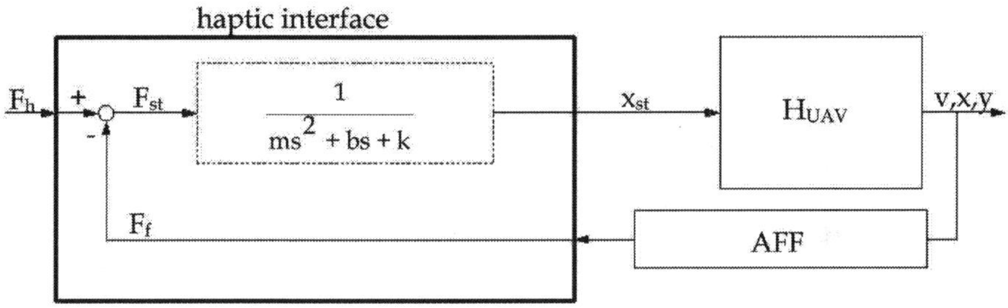

Figure 2. Schematic representation of the force feedback algorithm

Fig. 3 shows that due to the force offset, the stick will have a non-zero neutral position (A), i.e., the position where the stick is in equilibrium in the absence of external forces. When the stick is released, $F_h(x_{st}, i)=0$, repulsive forces can still exist and the stick indeed *actively deflects* away from the direction with a possible collision: $x_{st}=-F_f(i)/k$. This active deflection with hands-off can be considered as an "autonomous collision avoidance" function. In fact, the force feedback can be regarded to yield a "commanded" stick deflection that the operator should follow as good as possible. That is, when yielding to the forces applied on the hand, the operator would deflect the stick in a way that satisfies the collision avoidance function. Because a control device is generally limited in its deflections, there is also a limit to the amplitude of the force feedback, i.e., a natural constraint to the force feedback gain.

In previous studies, Lam et al. found that the amplitudes of stick motions due to neutral position changes may contribute to workload (Lam et al., 2007; Lam et al., 2008). In particular when flying through a narrow corridor, or when moving along multiple smaller and closely-spaced obstacles, these force offsets may vary continuously. The operator is not able to follow the force feedback accurately and overshoots and control oscillations occur.

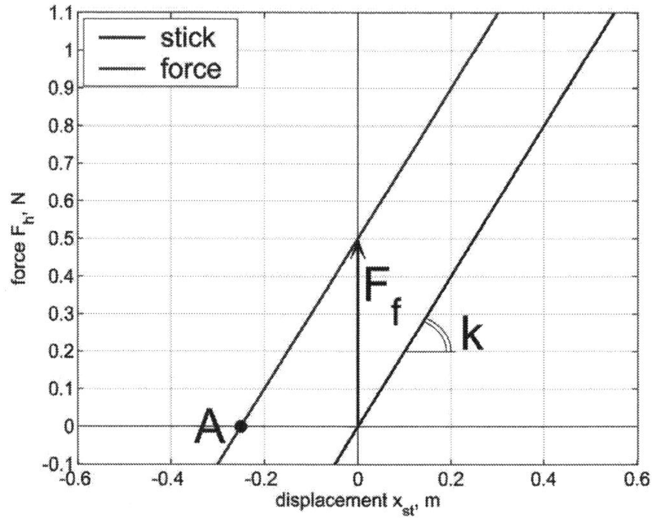

Figure 3. Force-displacement relation of a force feedback system

A possible solution would be to reduce the force offset magnitude, i.e., to decrease the gain of the force feedback. This might result, however, in repulsive forces that are too small in magnitude, in particular for operators who adopt a high neuromuscular stiffness. It would therefore significantly reduce the effectiveness of the collision avoidance command and limits the haptic presentation of information regarding a potential collision. Tuning the force feedback magnitude needs a compromise between, on the one hand, the effectiveness of the haptic interface and, on the other hand, the workload it imposes on the operator.

2.2 Stiffness feedback

Stiffness feedback involves addition of an extra spring load, $k_s(i)$ as a function of haptic feedback information i, to the nominal stick dynamics' spring constant k, see Fig. 4. Instead of having a force offset, the stick becomes stiffer when in the presence of an obstacle, that is, the extra stiffness provides an impedance, resulting in an extra force that depends on the deflection of the stick by the operator. When the stick is released it will *not actively deflect away* from a possible collision, as is the case when using force feedback, but just returns to the neutral position. Hence, stiffness feedback alone is not suitable for implementing autonomous collision avoidance. The total force that an operator perceives in this situation can be written as:

$$F_h(x_{st},i) = F_{st}(x_{st}) + F_s(x_{st},i), \tag{4}$$

$$F_h(x_{st},i) = kx_{st} + k_s(i)\, x_{st}. \tag{5}$$

with $F_s(x_{st}, i)$ the force due to an extra spring load $k_s(i)$. Fig. 5 shows that the slope of the force-excursion relation increases due to the extra spring load. The line rotates around the origin, and therefore a zero displacement leads to zero repulsive force.

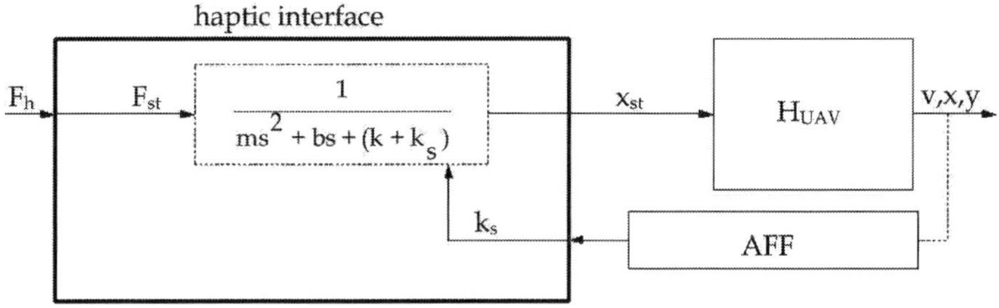

Figure 4. Schematic representation of the stiffness feedback algorithm

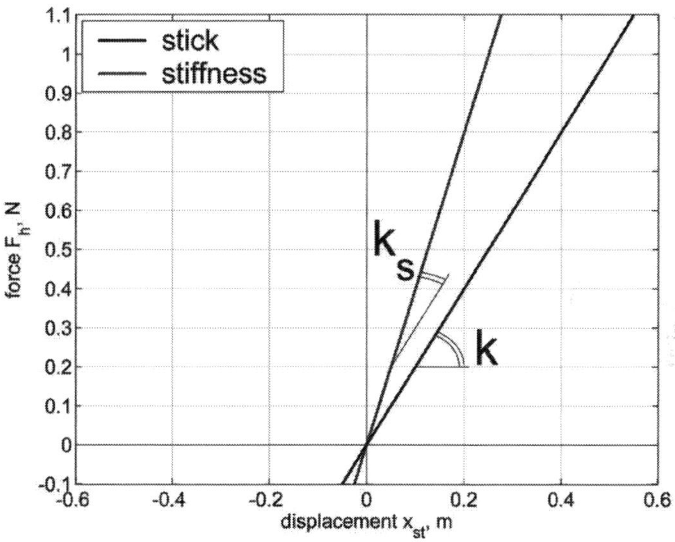

Figure 5. Force-displacement relation of a stiffness feedback system

Since small displacements result in relatively small repulsive forces, human operators may not perceive sufficient information when using only small stick displacements. With a zero stick displacement the operator will not be provided with any haptic feedback. When in this case the UAV would have moved very close to the obstacle, like in the case of drift due to wind, the stiffness feedback would have become so large that the stick cannot be moved at all. Providing only stiffness feedback is therefore undesirable. The haptic device should be capable of actively deflecting the stick away from a possible collision, preferably with only small overshoot from the zero displacement in order to not introduce control problems. It could be helpful, however, when the haptic device would be able to provide large resistance when the stick is deflected in the "wrong" direction, i.e., with high risk of collision. Hence, when combining the force feedback with stiffness feedback these two properties of haptic feedback can be achieved. This combination will be discussed below.

2.3 Stiffness-force feedback

The lack of active repulsive deflections of the control device with stiffness feedback can be resolved by combining it with force feedback, see Fig. 6. In this combination, referred to as "stiffness-force" feedback, the total exerted force by the human operator is defined as:

$$F_h(x_{st},i) = F_{st}(x_{st}) + F_s(x_{st},i) + F_f(i), \tag{6}$$

$$F_h(x_{st},i) = kx_{st} + k_s(i)\, x_{st} + F_f(i). \tag{7}$$

Fig. 7 shows the force-excursion relation of stiffness-force feedback. A property of this configuration is that due to the increase in stiffness, the desired offset of the neutral position, (A), commanded by the force feedback will decrease, which results in (B). It can also be seen, however, that in the hands-off case ($F_h(x_{st}, i)=0$) the stick deflection x_{st} becomes equal to $-F_f(i)/(k+k_s(i))$, i.e., the stiffness feedback actually *reduces* the effects of the force feedback.

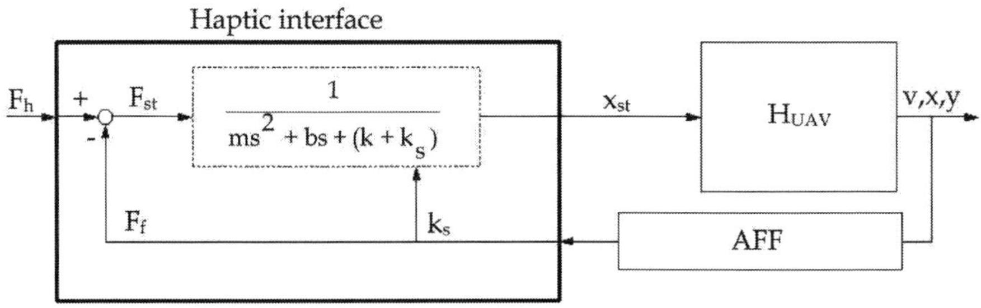

Figure 6. Schematic representation of the stiffness-force feedback algorithm

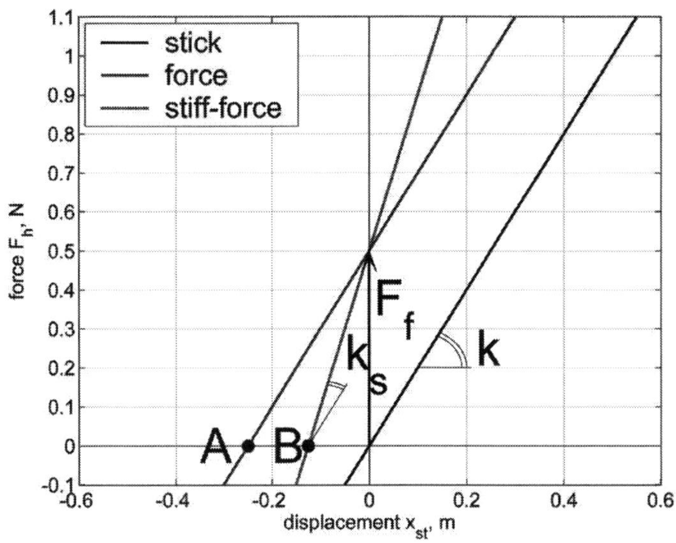

Figure 7. Force-displacement relation of a stiffness-force feedback system

2.4 Stability analysis

As mentioned in the introduction, force feedback may cause the stick to overshoot the neutral position. Operators sometimes need to counteract the active motions, particularly when obstacles are present at the other side. This indicates that active stick motion, a property of force feedback, could be a source for undesired oscillations. This section will consider the stability of force feedback, stiffness feedback and stiffness-force feedback at the hand of computer simulations, using the closed-loop system from Fig. 1. The human operator model contained a neuromuscular model for the lateral direction from (Lam et al., 2005). The stick dynamics were described as:

$$H_{st}(s) = \frac{1}{0.015s^2 + 0.3s + 3}. \tag{8}$$

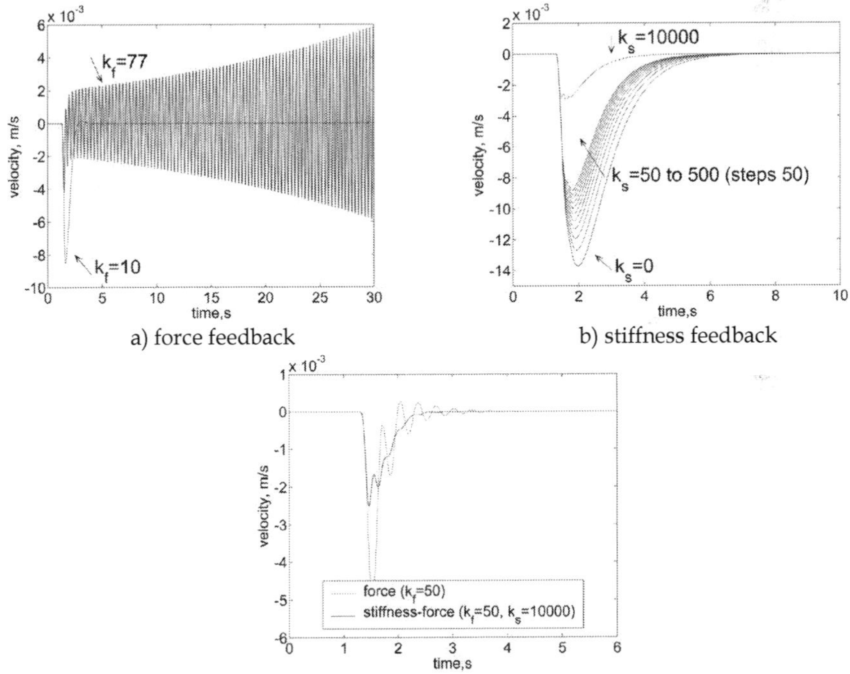

Figure 8. Step responses of the different haptic feedback configurations

The UAV helicopter was assumed to be control-augmented, with dynamics equivalent to a first order relation between stick deflection and UAV velocity (Voorsluijs et al., 2004):

$$H_{UAV}(s) = \frac{2}{s+2}. \tag{9}$$

The function of the AFF was to generate a value, based on relative position and velocity toward an obstacle, which could be used as input to the haptic device. Since this theoretical investigation only focuses on the behavior of the closed-loop system response to the different feedback of the AFF signal, the actual (non-linear) algorithm of AFF was left out and replaced by a gain for direct feedback of the velocity. For force feedback and stiffness feedback, the gains were k_f and k_s, respectively. It was assumed that the AFF primarily depended on the velocity.

Figs. 8(a) and 8(b) show step responses of a force feedback and stiffness feedback for increasing k_f and k_s gains, respectively. Fig. 8(c) shows the influence of stiffness feedback on a step response with stiffness-force feedback. It can be seen that a step response with force feedback becomes unstable for large force gain $k_f \geq 77$, whereas the stiffness feedback will not result in instability. For large stiffness gains k_s the stick dynamics will become "infinitely" stiff and no deflection will occur at all. The better stability and no change of neutral position suggests that stiffness feedback would result in less workload and control activity. Additionally, stiffness feedback has a reducing effect on force feedback, as was already illustrated in Fig. 7.

3. Tuning of the stiffness feedback

Both force feedback and stiffness feedback will cause the human operator to exert a larger force to keep the same deflection. However, the two configurations have a different contribution to the repulsive force. The force feedback is directly mapped as a repulsive force, whereas the stiffness feedback is only an increase of the spring constant. Hence, when a same value from the AFF is used to generate external force or external spring constant, the latter will result in a smaller repulsive force, which also depends on the stick deflection.

Because the side stick in the experiment uses moments and angles instead of forces and displacements, we will refer to moments and angles from this point. When considering a hand that is exerting a certain moment M_h on the haptic device the following holds:

$$M_h(\theta_{st}, i) = M_{st}(\theta_{st}) + M_{ext}(\theta_{st}, i), \tag{10}$$

$$M_h(\theta_{st}, i) = \underbrace{I\ddot{\theta}_{st} + B\dot{\theta}_{st} + K\theta_{st}}_{\text{stick dynamics}} + \underbrace{M_f(i) + M_s(\theta_{st}, i)}_{\substack{\text{haptic}\\ \text{dynamics}}}, \tag{11}$$

with I, B, and K the stick moment of inertia, damping coefficient, and spring constant, respectively. θ_{st} is the stick rotation, $M_{st}(\theta_{st})$ is the reaction moment from the stick dynamics, and $M_{ext}(i)$ is the repulsive moment from the haptic device. $M_{ext}(i)$ may consist of a moment offset $M_f(i)$ from force feedback and/or $M_s(i)$ from the stiffness feedback, due to extra spring constant.

To compare force feedback and stiffness feedback, it is important to tune the stiffness feedback in such a way that both systems will generate approximately the same level of repulsive force. The strategy adopted in this chapter is to tune the maximum repulsive moment based on the so-called Parametric Risk Field (PRF), an AFF adopted from (Boschloo et al., 2004).

For a maximum risk value of 1 (the limit of the Parametric Risk Field) and a force gain k_f=1.5, the maximum repulsive moment ($M_{ext}(\theta_{st},i)$) the force feedback can generate is 1.5 Nm. For tuning the stiffness gain, the maximum repulsive moment provided by stiffness feedback should also be 1.5 Nm. Consider the contribution of stiffness to the haptic dynamics:

$$M_s(\theta_{st},i) = K_s(i)\theta_{st}\,. \tag{12}$$

The maximum repulsive moment from stiffness feedback depends on the extra spring constant $K_s(i)$ and the maximum stick deflection, which was $\theta_{st(long)}$ = 0.35 rad and $\theta_{st(lat)}$ = 0.4 rad in the longitudinal and lateral direction, respectively. Hence, for $M_{ext}(\theta_{st},i)$ = $M_s(\theta_{st},i)$ = 1.5 Nm the stiffness gain for the longitudinal and lateral direction is:

$$K_{s(long)} = \frac{1.5}{0.35} = 4.29 \quad \text{Nm/rad}\,, \tag{13}$$

$$K_{s(lat)} = \frac{1.5}{0.40} = 3.75 \quad \text{Nm/rad}\,. \tag{14}$$

Pure stiffness feedback, however, has some drawbacks as mentioned in Section 2.2. First, the repulsive force generated by the stiffness feedback only equals the force feedback when the stick is maximally deflected. Second, for small deflections, small repulsive forces are generated that are very hard to sense. Third, in case of drifting toward an obstacle due to wind the operator would not feel any feedback (assuming there is no controller, rejecting the drift), because the stick is not deflected into the direction of the obstacle. But the stiffness would also be very high, making it difficult for the operator to deflect the stick away from the direction of a potential collision.

It is, therefore, recommended to use stiffness-force feedback with a small part of force feedback in order to provide some feedback when small or no deflections are given, while the UAV is in a close vicinity of an obstacle. Through trial and error, it was decided to use 20% and 40% of the force offset in the longitudinal and lateral direction, respectively. Now, for a maximal deflection, the total external moment can exceed 1.5 Nm due to the partial addition of the force feedback. Since operators are able to release the stick, resulting in a rapid decrease of the repulsive force to avoid large overshoot, it is unlikely that the higher maximum external moment would lead to larger workload and control activity.

4. Experiment

A human-in-the-loop experiment was conducted to investigate the effects of the various force feedback and stiffness-force feedback alternatives on human operator performance, collision avoidance, control activity and workload.

4.1 Subjects and instructions

Eight subjects with no flight experience participated in the experiment. Their main task was to fly from waypoint to waypoint as accurately as possible in an obstacle-laden environment. The experiment simulated a reconnaissance task in a hazardous environment. The waypoints were represented by smoke plumes, strategically positioned near particular obstacles in the environment. Note that subjects were not provided with information whether or not, or when a collision occurred. After each run, a subject was asked to rate the workload she or he experienced using the NASA TLX rating scale (Hart & Staveland, 1988).

4.2 Apparatus

The experiment was conducted in a fixed-base simulator. Subjects were seated on an aircraft chair in front of an 18 inch screen, presenting a navigation display (Fig. 9(a)), and looking forward towards a large white wall, on which the outside view from a hypothetical onboard camera was projected (Fig. 9(b)).

a) 2-dimensional navigation display b) 3-dimensional camera display

Figure 9. Experimental navigation and camera displays

On the right side of the aircraft chair, an electro-hydraulic side stick was used as the haptic control device. The mass-spring-damper stick dynamics were simulated with the inertia $I = 0.01$ kgm^2, damping coefficient $B = 0.2$ Nms/rad and spring constant $K = 2$ Nm/rad. The UAV was simulated by a control-augmented helicopter model, Eq. (9), with a maximum velocity of 5 m/s and maximum acceleration of 1 m/s^2. The altitude was kept constant by the control augmentation.

4.3 Independent variables

Two independent variables were used in the experiment: three levels of haptic feedback (HF) and six levels of subtask (ST).
The three haptic feedback conditions were:

Haptic off (HFoff): Subjects would only feel the side stick simulated mass-spring-damper dynamics;

Force (HF1): Subjects would feel external forces due to force offset generated by the PRF (Boschloo et al., 2004); and

Stiffness-force (HF2): Subjects would feel external forces, due to external spring constant and partial force offset, also based on the PRF.

The six subtasks are listed below, with the item number corresponding to the subtask number. The subtasks are illustrated in Fig. 10, with an arrow showing the flight direction and the red stars indicating the waypoint location, i.e., the smoke plumes. Note that a secondary purpose of the smoke plumes was to reduce the visibility of the obstacle boundaries.

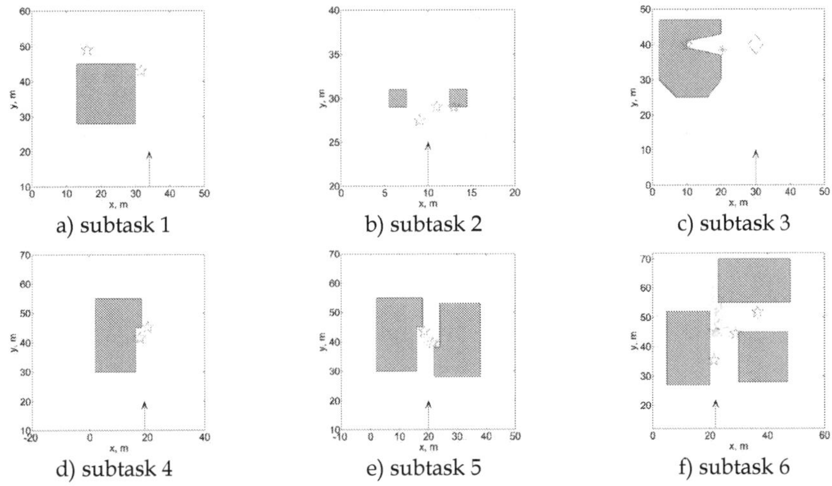

Figure 10. Top-down view of the six experimental subtasks

Subtask 1 In this subtask, the helicopter had to make a 90 degrees turn around a building, Fig. 10(a). Before the turn, the UAV had to approach the first waypoint, forcing the UAV to fly closely to the corner. The second waypoint was located after the corner in order to force subjects to make a turn as sharp as possible.

Subtask 2 In this subtask, the helicopter was to fly through a narrow gate of which the posts served as two closely-spaced, small obstacles. Fig. 10(b) shows a cross section of the gate. The smoke plume positioned between the posts was meant to reduce the visibility of the posts, making it more difficult to fly through the gate, rather than that it should serve as a waypoint.

Subtask 3 This subtask demanded a special task during hover. Once the helicopter had reached the (blue) diamond, it should hover backward toward the building until the operator could see a certain stop sign that was below the flight altitude and fixed in the world (blue asterisk), Fig. 10(c). Here, the camera would not point in the direction of motion and it was expected that haptic feedback would become very useful. Smoke on top of the building was placed to reduce the visibility of the edges of the building, leading to difficult estimation of the building edges, rather than to serve as waypoint.

Subtask 4 This subtask consisted of a building with a discrete change in the shape of the wall. The smoke was located before the discrete change of the wall and forced the UAV to approach the wall followed by an escape maneuver to avoid collision with the extension of the wall, Fig. 10(d). The second smoke plume was placed in order to force the UAV to stay close to the building during the escape maneuver.

Subtask 5 In this subtask, two buildings with discrete changes in opposite directions might lead to oscillatory behaviour in the stick and might cause considerable control difficulties. The first location of the smoke plume would force the UAV to make a sharp turn, whereas the second smoke plume would force the UAV to make an escape maneuver, Fig. 10(e).

Subtask 6 In this subtask, the turn radius with haptic feedback would be limited due to the obstacles in front and at the left side. It was expected that this subtask would lead to control difficulties, when approaching with high speed. The first smoke plume would force the UAV to approach the side of a building, whereas the second smoke plume would force the UAV to make a quick turn to fly closely along the corner of the building. The third smoke plume would force the UAV to approach another wall after the turn, Fig. 10(f).

4.4 Trajectory

In general, the trajectories (scenarios) were the same as was used in (Lam et al., 2007). The difference was the absence of a reference path in this experiment. For navigation, the subtask areas had a darker background colour on the navigation display (Fig. 9(a)).
The trajectories consisted of three sectors, which were in turn clusters of the six subtasks in a different order. To prevent boredom, 6 different trajectories were designed. Each trajectory contained a different order of the sectors. Each trajectory was flown once for each condition. Hence, each subject had to fly 6x3=18 runs; each run took approximately 5 min.

4.5 Dependent measure

The performance of collision avoidance was expressed by the number of collisions (cnt). The level of safety was expressed by the minimum allowable distance toward an obstacle (D_{min}). The performance of tele-operation was expressed by the minimum distance from the smoke location (D_{smoke}). Speed-related measures were expressed by the average of the total speed ($\bar{v}$) and the standard deviation of the total speed (σ_v).

The standard deviations of the total stick deflection $\sigma_{\delta tot}$ and the total exerted moment by the hand (σ_{Mh}) represented control activity. Haptic activity was represented by the standard deviation of the total external moment by the haptic device (σ_{Mext}). Workload was measured using the NASA TLX rating scale, after each measurement run.

4.6 Procedure

Each subject flew 6 runs for each haptic configuration. Before the actual experiment, subjects got the opportunity to get familiar with the three haptic configurations by training runs. After each experiment run, subjects were asked to rate their workload using the NASA TLX rating scale. After the experiment, subjects were asked to complete a questionnaire, regarding their experience with the three haptic feedback configurations.

4.7 Results and discussion

The main results are discussed in this section. A full-factorial ANOVA was applied. The means and 95% confidence intervals of the dependent measures are shown in Fig. 12.

4.7.1 Collisions

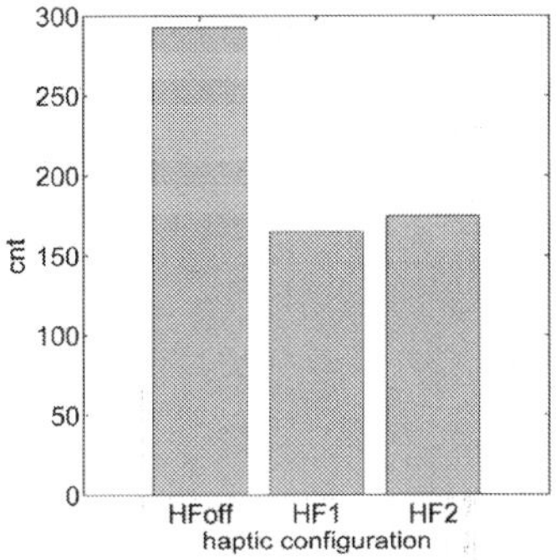

Figure 11. Total number of collisions (data of all subjects)

The total number of collisions, illustrated in Fig. 11, was highest when flying without haptic feedback, which was expected. The absence of haptic feedback, HFoff, led to a lack of situation awareness, especially when the visual information was reduced due to the smoke plumes. The force feedback, HF1, resulted in slightly less collisions than the stiffness-force feedback, HF2.

4.7.2 Approach performance

Performance expressed in terms of the distance from the smoke plumes (the waypoints) is shown in Fig. 12(a). Subtasks 2 and 3 were not included, because here the smoke plumes were placed with the objective to reduce the visual information of the obstacle boundaries, rather than to serve as a waypoint. Fig. 12(a) shows that HF1 resulted in the largest distance from the waypoints, whereas no haptic feedback resulted in the smallest distance from the waypoints, a highly-significant effect (HF: $F_{2,14} = 8.129$, $p \leq 0.01$). A post-hoc analysis revealed that the stiffness-force feedback lies in-between the force feedback and no haptic feedback conditions (Student-Newman Keuls (SNK), $\alpha = 0.05$). Regarding the effect of subtask, in subtasks 1 and 4, subjects were able to fly closely to the smoke. In subtasks 5 and 6, with objects surrounding the vehicle, subjects were not able to closely approach the waypoints, due to the difficult escape maneuver the close approach would cause afterwards, a highly-significant effect (ST: $F_{3,21} = 10.024$, $p \leq 0.01$).

4.7.3 Control activity

Fig. 12(b) indicates no significant effect of HF on the stick deflections. In subtask 3, subjects had to fly backward and forward, resulting in the largest standard deviation of the stick deflection. In subtask 2, subjects only had to fly straight through a passage, resulting in the smallest standard deviation of the stick deflection, a highly-significant effect (ST: $F_{5,35} = 29.343$, $p \leq 0.01$). Fig. 12(c) shows that without haptic feedback the standard deviation of the measured moment was significantly smallest (HF: $F_{2,14} = 8.173$, $p \leq 0.01$). Post-hoc analysis (SNK, $\alpha = 0.05$) revealed that the difference between force and stiffness-force feedback was not significant. Similar to the stick deflection, subtask 3 and subtask 2 resulted in the largest and smallest variation of the measured moment, respectively (ST: $F_{5,35} = 33.592$, $p \leq 0.01$).

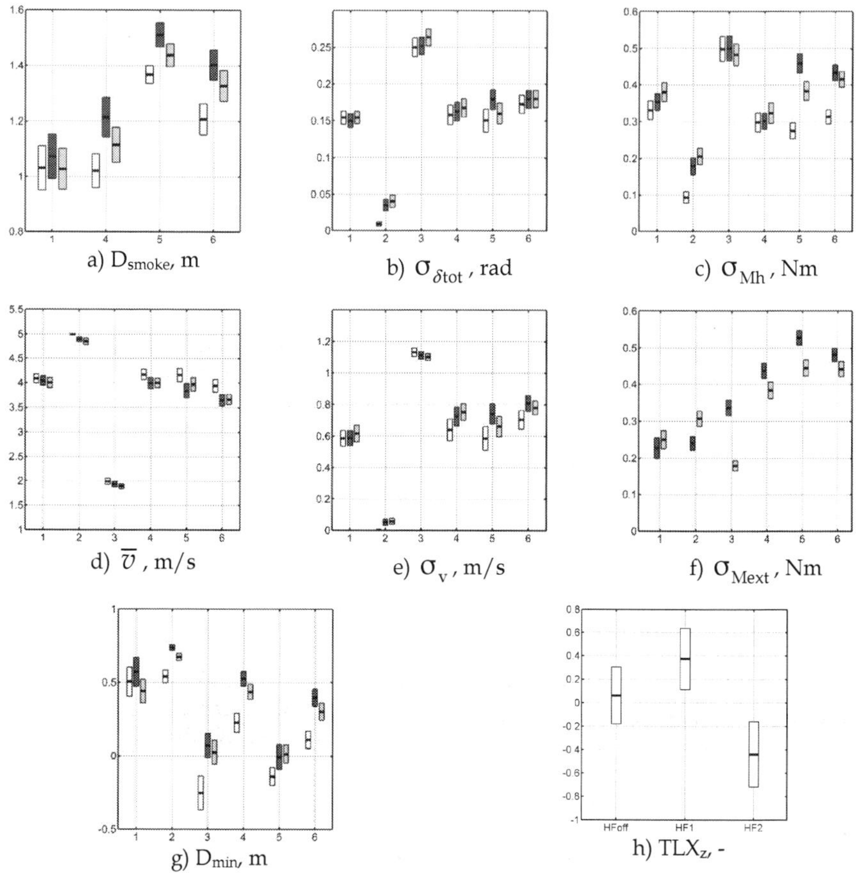

Figure 12. The means and 95% confidence intervals of the main dependent measures. The numbers 1 to 6 on the horizontal axis correspond to the subtask numbers. In all subfigures except (h), the white, dark gray, and light gray bars represent the haptic feedback conditions HFoff, HF1, and HF2, respectively

4.7.4 Haptic activity

Fig. 12(f) indicates that the standard deviation of the external moment was highly-significantly larger for the force feedback condition (HF: $F_{1,7} = 39.700$, $p \leq 0.01$). Obviously, the condition without haptic feedback was not considered. Subtasks 4, 5 and 6, all demanding a difficult approach and escape maneuver, resulted in large variations of the external moment (ST: $F_{5,35} = 28.255$, $p \leq 0.01$). In subtask 2, the external moment was larger for the stiffness-force feedback; this may be due to the partial addition of force feedback as discussed in the theoretical analysis section.

4.7.5 Speed

The average of the UAV velocity, a measure of tele-operation efficiency, was highly-significantly larger without haptic feedback, due to the lack of the repulsive forces in the opposite flight direction (HF: $F_{2,14} = 7.338$, $p \leq 0.01$), Fig. 12(d). Post-hoc analysis revealed that the difference between HF1 and HF2 was not significant (SNK, $\alpha = 0.05$). In subtask 2 (short and relatively easy) and 3 (difficult), the average speed was highly-significantly largest and smallest, respectively (ST: $F_{5,35} = 59.978$, $p \leq 0.01$). Fig. 12(e) shows that subtasks 2 and 3 resulted in a highly-significantly smallest and largest standard deviation of the total speed, respectively (ST: $F_{5,35} = 24.635$, $p \leq 0.01$).

4.7.6 Minimum distance

Fig. 12(g) shows that HFoff resulted in the smallest distance, whereas HF1 resulted in the largest distance, a highly-significant effect (HF: $F_{2,14} = 24.209$, $p \leq 0.01$). Subtasks 3 and 5 resulted in the smallest distance from a building, a highly-significant effect (ST: $F_{5,35} = 19.058$, $p \leq 0.01$). Note that a negative distance was due to collisions and the smallest distance due to HFoff corresponded to the largest amount of collisions as discussed in Section 4.7.1.

4.7.7 Workload

Fig. 12(h) shows the TLX z-scores and indicates that stiffness-force feedback, as expected, resulted in the lowest workload, a significant effect (HF: $F_{2,14} = 5.888$, $p = 0.014$).

5. Conclusions and recommendations

It can be concluded that, when considering the approach performance and workload, stiffness-force feedback outperforms force feedback with equivalent control activity, level of safety and UAV velocities. The improved approach performance and lower workload can be attributed to the relatively lower repulsive force activity with stiffness-force feedback.
The additional stiffness changes the dynamics of the haptic interface, which influences the interactions with the operator. A stiffer side stick would make it more difficult for the operator to overrule the force feedback. Hence, stiffness feedback in combination with force feedback will not only add extra repulsive force, but it may also serve as a haptic feedback design parameter to establish the level of operator autonomy. In order to achieve this, however, the reducing effect of stiffness feedback on force feedback should be eliminated. The first steps towards this new concept, provisionally called "force-stiffness feedback" have proven to be successful (Abbink & Mulder, 2008).

6. References

Abbink, D. A. & Mulder, M. (2008), Exploring the Dimensions of Haptic Feedback Support in Manual Control, *Journal of Computing and Information Science Engineering*, accepted

Boschloo, H. W.; Lam, T. M.; Mulder, M. & Van Paassen, M. M. (2004), Collision Avoidance for a Remotely-Operated Helicopter Using Haptic Feedback, *Proceedings of the IEEE International Conference on Systems, Man and Cybernetics*, The Hague, The Netherlands, 10-13 October, pp. 229-235

Elhaij, I.; Xi, N.; Fung, W. K.; Liu, Y. H.; Li, W. J.; Kaga, T. & Fukuda, T., (2001). Haptic Information in Internet-Based Tele-operation, *IEEE/ASME Transactions on Mechatronics,* Vol. 6, No. 3, pp. 295-304

Hart, S. G. & Staveland, L. E. (1988), Development of NASA-TLX (Task Load Index): Results of Empirical and Theoretical Research, Hancock, *Human Mental Workload,* P. A. and Meshkati, N., pp. 139-183, Elsevier Science Publishers (North-Holland)

Krogh, B. H. (1984), A Generalized Potential Field Approach to Obstacle Avoidance Control, Society *of Manufacturing Engineers Technical Paper,* No. MS84-484.

Lam, T. M.; Boschloo, H. W.; Mulder, M.; Van Paassen, M. M. & Van der Helm, F. C. T. (2004), Effect of Haptic Feedback in a Trajectory Following Task with an Unmanned Aerial Vehicle, *Proceedings of IEEE International Conference on Systems, Man and Cybernetics,* The Hague, The Netherlands, 10-13 October, pp. 2500-2506

Lam, T. M.; Mulder, M.; Van Paassen, M. M. & Van der Helm, F. C. T. (2005), Tele-operating a UAV Using Haptics - Modeling the Neuromuscular System, *Proceedings of the IEEE International Conference on Systems, Man and Cybernetics,* Hawaii, USA, 10-12 October, pp. 2695-2700

Lam, T. M.; Mulder, M. & Van Paassen, M. M. (2007), Haptic Interface for UAV Collision Avoidance, *International Journal of Aviation Psychology,* Vol. 17, No. 2, pp. 167-195

Lam, T. M.; Mulder, M. & Van Paassen, M. M. (2008), Haptic Feedback in UAV Tele-operation with Time Delay, *AIAA Journal of Guidance, Control, and Dynamics,* Vol. 31, No. 6, pp. 1728-1739

McCarley, J. S. & Wickens, C. D. (2005), Human Factors Implications of UAVs in the National Airspace, *Technical report Aviation Human Factors Division,* AHFD-05-05/FAA-05-01, pp. 1-5, Savoy, Illinois

Voorsluijs, G. M.; Bennani, S. & Scherer, C. W. (2004), Linear and Parameter-dependent Robust Control Techniques Applied to a Helicopter UAV, *Proceedings of the 38-th AIAA Guidance, Navigation and Control Conference,* Providence (RI), August 16-19, AIAA paper 2004-4909.

19

Objectively Optimized Earth Observing Systems

David John Lary
Joint Center for Earth Systems Technology (JCET) UMBC, NASA/GSFC
United States

1. Introduction

In this chapter we discuss an approach to making Earth Observing Systems autonomous by integrating the situational awareness provided by our theoretical and observational knowledge of the system being observed. Our theoretical understanding of the system is encapsulated in a deterministic model. The observational knowledge is integrated via a data assimilation system. We will start by surveying what has been done to date, then outline our contribution to the field, and finally suggest some future areas of development.

2. Current Observing Systems and Autonomy

Modern observing systems are typically composed of a variety of components, or assets, that are interconnected to form a 'Sensor Web' (Botts et al., 2007). The communications fabric of the Sensor Web is a vital component that links the sensors and control systems and allows coordinated data transmission and instrument control. Effective autonomous control of individual sensors and/or the overall Sensor Web is highly desirable, maybe even indispensable, if the observing system is to be able to respond in real time to the evolving state of the environment.

The current Earth observing capability depends primarily on spacecraft missions and ground-based networks to provide the critical on-going observations necessary for improved understanding of the Earth system. Aircraft missions have played an important role in process studies but are often limited to relatively short-duration flights. However, for over a decade, it has been recognized that autonomous aerial observations can make an important contribution to Earth observing systems (Coronado et al., 1998; Sandford et al., 2004). The ability of the observing system to respond in real time to an evolving situation can, in some cases, be critical. An example would be robotic reconnaissance operations in extreme environments that are potentially hazardous (such as volcanic gas sampling, mine fields, battlefield environments, enemy occupied territories, terrorist infiltrated environments, or areas that have been exposed to biochemical agents or radiation) (Caltabiano et al., 2005; Chen et al., 2006; Fink et al., 2007). In other cases autonomy is necessary as the vehicles are operating in inaccessible locations, such as remote planetary atmospheres (Ippolito et al., 2005; Young et al., 2005; Morrow et al., 2006).

Observing platform autonomy is a relatively new field (Davis et al., 1992; Schiller et al., 1993). The nature of the autonomy depends in part on the type of asset. For space based orbital assets the autonomy could be directing the real time instrument pointing, or instrument mode (e.g. zoom in, global survey, step and stare, etc). For sub-orbital assets such as aircraft or unmanned aerial vehicles (UAVs) the autonomy is generally much more developed and typically includes automated route planning for a single UAV (Sullivan et al., 2004; Kunchev et al., 2006; Ceccarelli et al., 2007; Pehlivanoglu et al., 2007; Pehlivanoglu and Hacioglu, 2007) and coordinated route planning for a fleet of UAVs (Bellur et al., 2002; Dargar et al., 2002; Mahler and Prasanth, 2002; McInnes, 2003; Zelinski et al., 2003; Gilmore and Garbarino, 2004; Hope et al., 2004; King et al., 2004; Maddula et al., 2004; Chen et al., 2005; Jin et al., 2006; King et al., 2006; Sinha et al., 2006; Smith and Nguyen, 2006c;b;a; Fink et al., 2007; Lamont et al., 2007; Smith and Nguyen, 2007; Yuan et al., 2007). When we are dealing with a fleet of distributed autonomous systems there can also be an element of decentralization, so that in addition to concerted action, each member may exhibit some degree of individual autonomy (Richards and How, 2004; Smith and Nguyen, 2005). The UAV targets may be a set of fixed locations or mobile targets (Rafi et al., 2006).

Autonomy can even be used to improve mission-level functional reliability through better system self-awareness and adaptive mission planning (Valenti et al., 2007). The flight duration of UAVs is continually improving; some UAVs are now capable of continuous flight over many days (Noth et al., 2006), for such vehicles autonomy is again a benefit. Presently, the UAVs come in all shapes and sizes, from palm top micro UAVs, to helicopters, to giant strategic UAVs that can loiter over targets for extended periods of time (Natarajan, 2001). Unmanned aerial vehicles also include blimps (Elfes et al., 2001; Elfes et al., 2003; Hygounenc et al., 2004). A blimp is a small airship that has no metal framework and collapses when deflated (Beji and Abichou, 2005). In some cases the blimp can only control its vertical position by ascent or descent, with the surrounding wind field controlling its horizontal location. For conventional balloon borne observations the autonomy could be in directing the optimal time and location for launch.

Autonomous direction is not limited to orbital and sub-orbital platforms; they can include unmanned ground vehicles (UGV) (Kamsickas and Ward, 2003; Kelly et al., 2006; Li and Cassandras, 2006), unattended ground sensors (UGS) (Roberts et al., 2003), and subsea and surface (UUV and USV) vehicles (Davis et al., 1992; Hansen et al., 2006; Steinberg, 2006) operating together with minimal human oversight.

UAVs (aircraft or helicopter) are advanced and complex robotics platforms used for a variety of tasks. For example, UAVs can be used in applications for environmental monitoring (weather and/or pollution), traffic monitoring (Dudziak, 1998; Chung and He, 2007), surveillance (Persson, 2002; Ryan et al., 2007), intelligence gathering (Stottler et al., 2007), terrain mapping (Templeton et al., 2007), emergency services assistance (Ryan and Hedrick, 2005; Beard et al., 2006; Onosato et al., 2006; Snarski et al., 2006; Fabiani et al., 2007), studying the movement of agricultural threat agents, pollen, plant pathogens, and other biological particles (Wang et al., 2007), crop condition (Herwitz et al., 2004; Johnson et al., 2004; Furfaro et al., 2007), photogrammetry, and surveying (Schmale et al., 2008).

UAV autonomy is accomplished by a complex interconnection of systems related to a wide range of topics, e.g., flight low level control, navigation and task-based planning, elaboration of sensor signals, software architecture for reactive behaviors, communication (Zingaretti et al., 2008). In some applications the UAVs will be working in a cooperative network with other autonomous agents, such as UGVs (Unmanned Ground Vehicles) and UGSs (Chandler et al., 2002; Murphey and O'Neal, 2002; Roberts et al., 2003; Zingaretti et al., 2008) to accomplish specific tasks that may, or may not, have been defined a priori.

The autonomy has to be based on objective criteria that can be evaluated in real time via methodologies such as optimization, information theory, computer vision, mixed integer linear programming, fuzzy logic, genetic algorithms, artificial cognition, geographic information systems (GIS), set partition theory, vehicle health status, and/or threat assessments (Atkins et al., 1998; Chandler and Pachter, 1998; Sinopoli et al., 2001; Chandler et al., 2002; Levin et al., 2002; Cruz et al., 2003; Flint et al., 2003; Miller and Larsen, 2003; Nikolos et al., 2003; Lee et al., 2004; Ertl and Schulte, 2005; Kamal et al., 2005; KrishnamurthyGopalan et al., 2005; Vachtsevanos et al., 2005; Pettersson and Doherty, 2006; Pongpunwattana et al., 2006; Skoglar et al., 2006; Yang et al., 2006; Smith, 2007; Smith and Nguyen, 2007). Autonomy is often implemented using multi-layer hierarchical control architectures (Boskovic et al., 2002). The autonomous operation can include features such as the Cognitive Emotion Layer (CEL) that uses dynamical emotional response mechanisms to model response to continuous stimuli and provides adaptive decision making and control capabilities for the exploration platform (Ippolito et al., 2005). A given control system may provide more than one autonomous mode, for example, (Sasiadek and Duleba, 2000) outline an approach where the motion planning problem is split into two stages: a decision mode and a trace mode. In the decision mode, the vehicle selects its current direction of motion on the basis of the current value of a performance index. In the trace mode vehicle traces boundary and edges of obstacles using its on-board sensors.

Alternatively, an operator can issue the commands remotely. For example, an operator in a manned aircraft can issue mission level commands to an autonomous aircraft in real time (Sandewall et al., 2003; Schouwenaars et al., 2006). The open source movement has had a radical impact on the software Architecture for Autonomy (Broten et al., 2006) open source frameworks can reduce the cost and risk of systems engineering.

Hybrid UAVs also exist, for example, the McDonnell Douglas Bird Dog, where a semi-autonomous UAV has been coupled with a manned ground vehicle (Kolding and Pouliot, 1997). In addition to making observations and delivering payloads UAVs can facilitate communications by acting as routing nodes (Francl, 2000). This is a role that may be come increasingly important.

3. Situational Awareness via Theory and Observation

A valuable addition can be made to the studies described above, a situational awareness that comes from directly coupling our theoretical and observational understanding of the specific system being observed. The disciplines of numerical weather prediction and oceanography have long benefited from coupling a theoretical understanding as encapsulated by a deterministic model with a wide variety of observations through the mathematical formalism of data assimilation (Pielke et al., 1992; Swinbank and O'Neill, 1994; Kalnay et al., 1996). The use of data assimilation has dramatically improved forecast accuracy (Simmons and Hollingsworth, 2002).

3.1 Data Assimilation

Data assimilation is the term given to recursive Bayesian estimation in the geosciences (Wikipedia, 2008a). Data assimilation proceeds by analysis cycles. In each analysis cycle, observations of the current (and possibly, past) state of a system are combined with the results from a deterministic mathematical model (the forecast) to produce an analysis, which is considered as 'the best' estimate of the current state of the system. This is called the analysis step. Essentially, the analysis step tries to balance the uncertainty in the data and in the forecast. The model is then advanced in time and its result becomes the forecast in the next analysis cycle.

The analysis and forecasts are best thought of as probability distributions. The analysis step is an application of the Bayes theorem. Advancing the probability distribution in time would be done exactly in the general case by the Fokker-Planck equation (Wikipedia, 2008b), but that is unrealistically expensive, so various approximations operating on simplified representations of the probability distributions are used instead. If the probability distributions are normal, they can be represented by their mean and covariance, which gives rise to the Kalman filter (Kalman, 1962).

More recently the use of data assimilation has also been extended to cover chemically reactive atmospheric systems relevant to issues such as ozone depletion and air quality (Fisher and Lary, 1995; Elbern et al., 1997; Menard et al., 2000; Lary et al., 2003).

It is current practice for today's forecasts for weather, air-quality, fire likelihood etc. to be generated on a fixed time schedule. However, new technologies and improved numerical model physics are enabling on demand forecasts in response to particular events (Plale et al., 2006). These forecasts ingest global and regional data in real time and can consume large computational resources. It is now timely to couple these concepts and take the next step of creating a coupled prediction and observation system. Such a coupled system can provide improved forecast accuracy and efficient collaborative adaptive sensing to make best use of the suite of observational assets. This synergistic approach offers distinct benefits to the respective user communities, and when used together, promise a paradigm shift in atmospheric science research.

As a growing volume of data is now freely available, it is a powerful approach where the incorporation of existing data into the analysis can be used to guide the optimal collection of new data. Large volumes of data are routinely provided by governmental organizations including in situ observations, geospatial data sets, remote sensing products, and simulation model output. Increasingly these data are being made available as web-services. Web services offer a means for sharing data more openly by providing a standard protocol for machine-to-machine communication. An example of the use of web-services in this manner is the machine accessible interface for the National Water Information System (NWIS) (Goodall et al., 2008), an online repository of historical and real-time stream-flow, water-quality, and ground water level observations maintained by the United States Geological Survey (USGS). These services provide a middle-layer of abstraction between the NWIS database and hydrologic analysis systems, allowing such analysis systems to proxy the NWIS server for on-demand data access. The services are intentionally designed to be generic and applicable to other hydrologic databases, in order to provide interoperability between disparate data sources.

3.2 Autonomous Observation for Atmospheric Chemistry

Our research effort has been focused on incorporating such situational awareness provided by coupling theory and observation. Our specific case study has been creating a software infrastructure for an autonomous observing system for atmospheric chemistry. The system works on generic principles that should be easily transferable to other scenarios. The observing system consists of many elements, orbital (several satellite instruments), sub-orbital (aircraft, UAVs and balloons), and ground-based assets. The satellite assets include those with fixed pointing for which autonomy is not relevant and those with flexible pointing and multiple modes for which autonomy are most useful. The software infrastructure needs to be aware of all the observational elements and their capabilities at any given time *and* to be aware of the state of the system we are observing and the precision with which we know its current state.

Such an autonomous Objectively Optimized Observation Direction System is of great utility for NASA's observation and exploration objectives. In particular, it is useful to have a fleet of smart assets that can be reconfigured based on the changing needs of science and technology. Our project integrates a modeling and assimilation system within a Sensor Web. This enables a two-way interaction between the modeling/assimilation system and the sensing system to enhance Sensor Web performance and usage. Our aim is to develop a key technology currently missing in Sensor Web implementations, an autonomous Objectively Optimized Observation Direction System (OOODS). The OOODS will objectively optimize the workflow management of a set of assets, i.e. plan, monitor, and control the resources. This workflow management will involve a high level of situation awareness.

An important concept in such a system is the direct two-way feedback between the distributed Sensor Web and an autonomous OOODS (involving a numerical prediction and assimilation system). The project concentrates on the principles of how such an OOODS would operate, its architecture, and development as a plug-in for a Sensor Web simulator/controller. Two scenarios have been chosen to show the relevance of the technology to both current and future objectives of NASA Earth science. The project leverages technology that has already won five NASA awards and is currently being used in several NASA Aura validation and earth system science projects.

The goal of the Sensor Web OOODS described is to employ an objectively optimized data acquisition strategy for integrated Earth observation that is responsive to environmental events for both application and scientific purposes. Such an OOODS Sensor Web can achieve science objectives beyond the abilities of a single platform by specifically directing observations of high scientific value. This will be achieved by using objective criteria for the observation direction enabling a high degree of collaboration among the sensing and analysis assets. Once such a framework is in place, what the objective criteria are can readily be changed for other applications, depending on the desired science or application goals.

A Sensor Web with such a goal oriented OOODS can be of particular use for decision support systems. For example, in the scenarios we have chosen here, an obvious application would be for air quality decision support systems and in responding to a large hazardous release of atmospheric contaminants.

3.3 Key Principles

Metrics of what we do not know (state vector uncertainty) are used to define what we need to measure and the required mode, time and location of the observations, i.e. to define in real time the observing system targets. Data assimilation provides us with the state vector uncertainty. Below we will consider in detail how we construct the state vector uncertainty. Metrics of how important it is to know this information (information content) are used to assign a priority to each observation. The metrics are passed in real time to the Sensor Web observation scheduler to implement the observation plan for the next observing cycle. The same system could also be used to reduce the cost and development time in an Observation

Sensitivity Simulation Experiment (OSSE) mode for the optimum development of the next generation of observing systems.

4. Implementation

Now that we have considered the principles behind the autonomy let us move to the details of the two key components that we have used. The first component is a Sensor Web simulator. The second is our data assimilation system. Figure 1 shows an engineering diagram for our system.

4.1 Sensor Web Simulator

The Sensor Web simulator describes the capabilities and locations of each sensor as a function of time, whether they are orbital, sub-orbital, or ground based. The simulator has been implemented using Analytical Graphics Incorporated (AGI) Satellite Tool Kit (STK). STK makes it easy to analyze and visualize optimal solutions for complex space scenarios (www.agi.com). STK performs complex analysis of land, sea, air, and space assets, and shares results in one integrated solution. The STK functionality enables us to calculate the position and orientation of each asset, to evaluate inter-visibility times, and to determine quality of dynamic spatial relationships.

STK can be used for real time monitoring of operations with live feeds from sensors or for mission concept design and visualization. The Sensor Web data assimilation system feeds multivariate objective measures to an STK Matlab plug-in (e.g. information content for observation priority and system uncertainty for target definition). The target definition and priority are then passed to the STK Scheduler plug-in (our smart scheduler) that is aware of each assets capability and then constructs an optimal observation schedule for each asset in the Sensor Web. The STK Scheduler is a fully integrated add-on module for STK that provides a powerful scheduling and planning application to mission designers and operations engineers alike. Users can define tasks and related resource requirements, request schedule solutions, and analyze the results through a user-friendly graphical user interface (GUI) or via the Application Program Interface (API). The STK Scheduler is powered by a scheduling engine from Optwise Corporation that finds better solutions in a shorter amount of time than traditional heuristic algorithms. The global search algorithm within this engine is based on neural network technology that not only outperforms traditional scheduling engines, but can find solutions to larger and more complex problems. STK Scheduler allows system planners to maximize the value of limited resources. Examples of other uses of STK for UAV systems in real-time operations and post-mission analysis can be found online at http://www.agi.com/solutions/specializedAreas/uav/.

Some complementary work in building Earth Observing simulators for meteorology is also underway (Higgins et al., 2005; Kalb et al., 2005; Halem et al., 2007; Seablom et al., 2007). The end product of these studies will be a "sensor web simulator" (SWS) that would objectively quantify the scientific return of a fully functional model-driven meteorological sensor web.

4.2 Chemical Assimilation System

The chemical modeling and assimilation system provides situational awareness by supplying the time evolution of uncertainty and information content metrics that are used to tell us what we need to observe and the priority we should give to the observations.

State Vector Uncertainty for Target Definition

We use the state vector uncertainty to define our targets. As the state vector uncertainty is a product of the modeling and assimilation system let us consider the mathematical formalism that we use.

In the case considered here the model consists of a system of coupled ordinary differential equations (ODEs). Each ODE describes the time evolution at a given location of a chemical constituent. In our standard configuration we consider fifty-eight atmospheric constituents cast in Lagrangian coordinates. The fifty-eight constituents give a comprehensive description of the reactive oxygen, hydrogen, nitrogen, chlorine, bromine and methane chemistry of the atmosphere. We use the extensively validated AutoChem model described (Fisher and Lary, 1995; Lary et al., 1995; Lary, 1996). The model is explicit and uses an adaptive time-step, error monitoring time integration scheme for stiff systems of equations (Stoer and Bulirsch, 2002; Press, 2007). AutoChem was the first model to ever have the facility to perform 4D variational data assimilation (4D-Var) (Fisher and Lary, 1995) and now also includes a Kalman filter (Khattatov et al., 1999; Lary et al., 2003). AutoChem can also autonomously provide analyses of the results from the Sensor Web assets to the users. An example of the analysis AutoChem can provide can be found at www.CDACentral.info.

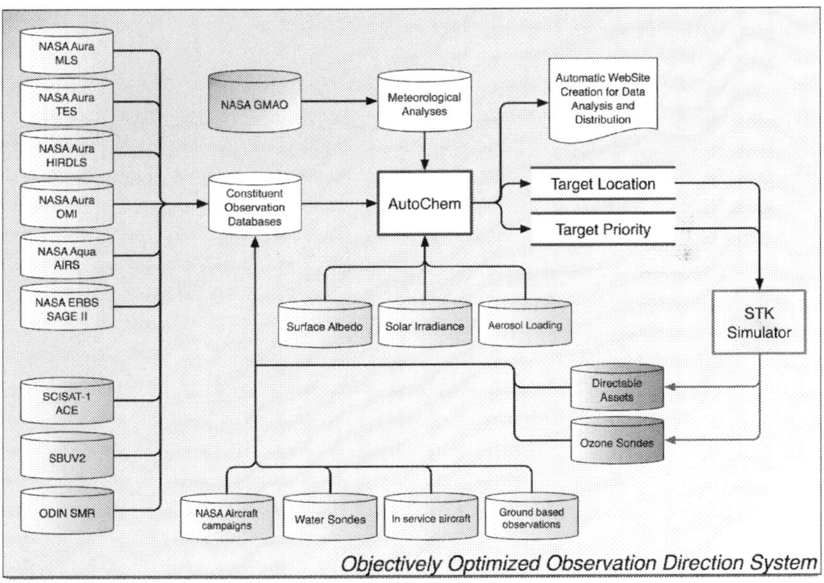

Figure 1. Modeling and Assimilation System Engineering Diagram

AutoChem is NASA release software constituting an automatic code generation, symbolic differentiator, analysis, documentation, and web site creation tool for atmospheric chemical modeling and data assimilation. Figure 1 is a schematic of the system and how the AutoChem plug-in fits into the Sensor Web. AutoChem is a suite of Fortran90 and Matlab programs. The code written by AutoChem is in Fortran90, the documentation is written in LaTeX and then converted to PDF files, the websites are written in HTML and JavaScript.

Given databases of reactions (unimolecular, bimolecular, trimolecular, photolysis, heterogeneous and cosmic ray ionization) AutoChem automatically selects the reactions involved in a user-defined constituent list. AutoChem can use online XML reaction databases maintained by NIST. AutoChem then constructs the time derivatives and symbolically differentiates them to give the Jacobian, and then symbolically differentiates the Jacobian to give the Hessian. All the Fortran90 code and ancillary files for the forward model, its adjoint and a full Kalman filter are automatically generated. The code and documentation generation is extremely fast, typically taking less than a second. When the code generation is complete all that is required is compilation with the supplied makefile. AutoChem is very flexible and has already been used in a wide variety of kinetic applications. It has also been adapted for use within the Earth Science Modeling Framework (ESMF). It could also be used for a variety of other applications such as combustion modeling, interstellar chemistry and modeling of metabolism. AutoChem is also enabled for the machine learning of ODEs to accelerate the solution of the stiff ODEs.

The chemical Kalman filter allows the optimal combination of model simulations and measurements taking into account their respective uncertainties. Consider a model of a physical system represented by operator (generally nonlinear) $\boldsymbol{M}$, and let vector $\boldsymbol{x}$ with dimension N_x be a set of input parameters for the model. These input parameters are used to predict the state of the system, vector $\boldsymbol{y}$ with dimension N_y:

$$y = M(x) \tag{1}$$

Assume that vector $\boldsymbol{x}$ represents the state of a time-dependent numerical photochemical model, i.e., concentrations of modeled species at model grid points in the atmosphere. In the case of a box model that includes N species, the dimension of vector $\boldsymbol{x}$ would be N. We will now limit the discussion to the case when $\boldsymbol{M}$ is used to predict the state of the system at some future time from past state estimates. Formally, in this case

$$x = x_t, y = x_t + \Delta t \tag{2}$$

$$\text{and } x_t + \Delta t = M(t, x_t) \tag{3}$$

Let vector $\boldsymbol{y}_o$ contain observations of the state. Usually, the dimension of $\boldsymbol{y}_o$ is less than N_y, the dimension of the model space, since not all model species are usually observed. The connection between $\boldsymbol{y}_o$ and $\boldsymbol{y}$ can be established through the so-called observational operator $\boldsymbol{H}$:

$$y_o = H(y) \tag{4}$$

Combining the above two equations, we get

$$y_o = H(M(x)) \tag{5}$$

We now assume that that the probability density functions associated with $\boldsymbol{x}$ and $\boldsymbol{y}$ can be satisfactory approximated by Gaussian functions:

$$\text{PDF}(y) \approx \exp\left(-\frac{(x-x_t)^T C^{-1} (x-x_t)}{2}\right) \tag{6}$$

where $\mathbf{x_t}$ is the true value of $\mathbf{x}$ and $\mathbf{C}$ is the corresponding error covariance matrix. Its diagonal elements are the uncertainties (standard deviations) of x, and the off-diagonal elements represent correlation between uncertainties of different elements of vector x. The covariance matrix $\mathbf{C}$ is defined as

$$C = \left\langle (x - x_t)(x - x_t)^T \right\rangle \tag{7}$$

where angle brackets represent averaging over all available realizations of x.
For most practical applications we need to introduce the linear approximation. In the linear approximation we assume that for small perturbations of the parameter vector Δx the following is approximately true:

$$M(x + \Delta x) = M(x) + L\Delta x \tag{8}$$

Formally, L is a derivative of M with respect to x:

$$L = \frac{dM}{dx} \tag{9}$$

For small variations of x one can show that the evolution of error covariance matrix C_t is given by:

$$C_{t+\Delta t} = LC_t L^T + Q \tag{10}$$

Matrix Q is the error covariance matrix introduced to take into account uncertainties of the model calculations. The Kalman filter equations are

$$x_t + \Delta t = M(t, x_t) \tag{11}$$

$$C_t + \Delta t = LC_t L^T + Q \tag{12}$$

$$\hat{x}_t = x_t + C_t H^T (HC_t H^T + O)^{-1} (y_o - Hx_t) \tag{13}$$

$$\hat{C}_t = C_t + C_t H^T (HC_t H^T + O)^{-1} HC_t \tag{14}$$

At the end of each analysis period the model value (x_t) and the corresponding observation (y_o) are ‚mixed' (see (13)) with weights inversely proportional to their respective errors to produce the analysis, $\hat{x}_t$. Then the model is integrated forward in time starting from the obtained analysis. Once an observation has been incorporated in the model, the analysis error covariance is updated to reflect this (see (14)). In the absence of observations, the model state is updated using (5), while evolution of the error covariance is obtained from the linearized model equations as in (12).
If no observations are available, then

$$\hat{x}_t = x_t \tag{15}$$

$$\text{and } \hat{C}_t = C_t \tag{16}$$

The diagonal elements of C_t are our state vector uncertainty providing us with an objective measure of "what we don't know" and are used to define our targets.

Information Content for Target Priority

A basic postulate of information theory is that information can be treated like a measurable physical quantity, such as density or mass. Consequently we can construct an optimization method for use in observing systems where there is an objective optimization for information content. This allows the best return of information for a given investment in measuring systems. Based on information content and level of uncertainty we can create a dynamic observation control system that adapts what measurements are made, where they are made, and when they are made, in an online fashion to maximize the information content, minimize the uncertainty in characterizing the system's state vector.

As a dynamic system evolves with time not all of the state variables within the state vector contain equal amounts of information (information content), and not all state variables are known to the same precision. It is therefore clearly desirable that the observations made of a system both contain the maximum information content possible with a given observing platform capability *and* allow the systems state to be characterized with a minimum uncertainty. Information content is a broad term that could be quantified in any number of ways depending on the system or problem being studied. Therefore, although we describe some specific measures of information content for the atmosphere's chemical system, these measures could easily be substituted with alternative measures that may be more suitable depending on the given objectives of an investigation. Although we describe a specific example here from atmospheric chemistry, the principle is clearly more general. It is suggested that the concepts of information content and uncertainty could be used generically to determine: what should be measured, when and where. Thus providing a cost effective strategy for using resources and minimizing the data storage required in characterizing a system with a given level of precision.

It is valuable to have objective measures of chemical information content, for example, to assist in the best use of our earth observing systems. This is particularly true as many recently launched and soon to be launched remote sensing instruments have high spectral resolution. The high resolution often enables individual spectral lines to be resolved. Consequently we are about to experience a dramatic increase in the total capacity with which to observe the atmosphere. However, this will not be accompanied by the corresponding increase in human resources to process (retrieve) the data and produce constituent profiles. So any assistance in directing our attention will be useful.

Information theory has been widely applied by communication engineers and some of its concepts have found application in psychology and linguistics. The ideas of information content and uncertainty are closely related and their application to atmospheric composition have been rather lacking to date. There are many conceivable measures of chemical information content that could be used. The most appropriate will depend on the purpose of a given investigation. Claude Shannon (Shannon and Weaver, 1949; Shannon, 1997) provided a framework for engineers to approach problems related to transmitting information content through communications channels. In chemical systems we would like to ask what are the key chemical players?

Ranking the Key Chemical Players

To make best use of any observing system it is useful to construct a ranked list of variables/constituents that characterizes their chemical information content. This list is obviously a function of the question asked as well as time and location. Such an index could be based on answering the question in going from time t to time $t+\Delta t$ what are the key chemical players?

Let vector $\boldsymbol{x}$ represent the state of a time-dependent photochemical box model, i.e., concentrations of all modeled species in a parcel of air at a given instant. In the case of a box model that includes N species, the dimension of vector $\boldsymbol{x}$ would be N. The photochemical box model $\boldsymbol{M}$ describes the transformation of vector $\boldsymbol{x}$ from time t to time $t+\Delta t$. Formally,

$$x_{(t+\Delta t)} = M(t, x_t) \tag{17}$$

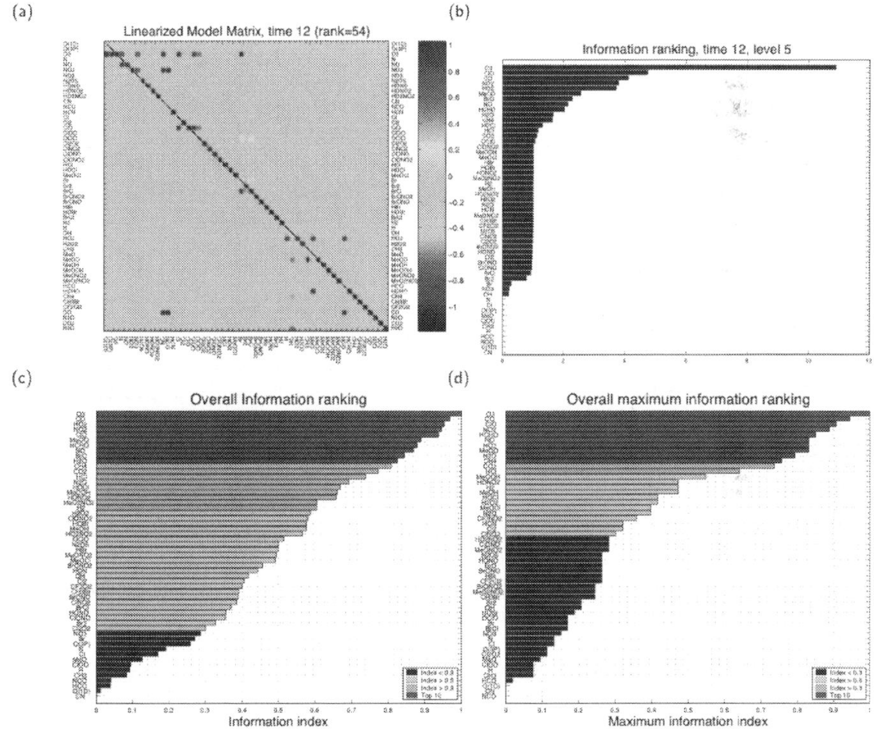

Figure 2 (a) shows the linearized model matrix for a local solar time 12:15 at a potential temperature of 426 K (≈18 km) on 30 March 1992 at 38°S. (b) shows the chemical information content index, I_c. (c) shows the information index $TotI_c$. (d) shows the corresponding index based on the maximum value of elements of M. In (c) and (d) the values have been normalized by the maximum value of $TotI_c$ so that the values range from zero to one.

An example of the model matrix is shown in Figure 2 (a). The figure shows the linearized model matrix for a local solar time 12:15 at a potential temperature of 426 K (an altitude of approximately 18 km) on 30 March 1992 at 38°S. The photochemical model was carefully initialized from a Kalman filter chemical data assimilation analyses of measurements made by the Atmospheric Trace Molecule Spectroscopy Experiment (ATMOS) instrument (Atlas-

1). This model simulation simultaneously agreed with ATMOS Atlas-1 observations of O_3, NO, NO_2, N_2O_5, HNO_3, HO_2NO_2, HCN, $ClONO_2$, HCl, H_2O, CO, CO_2, CH_4, and N_2O to within the observation and model uncertainties. The analyses and its reliability was described in detail in (Lary et al., 2003).

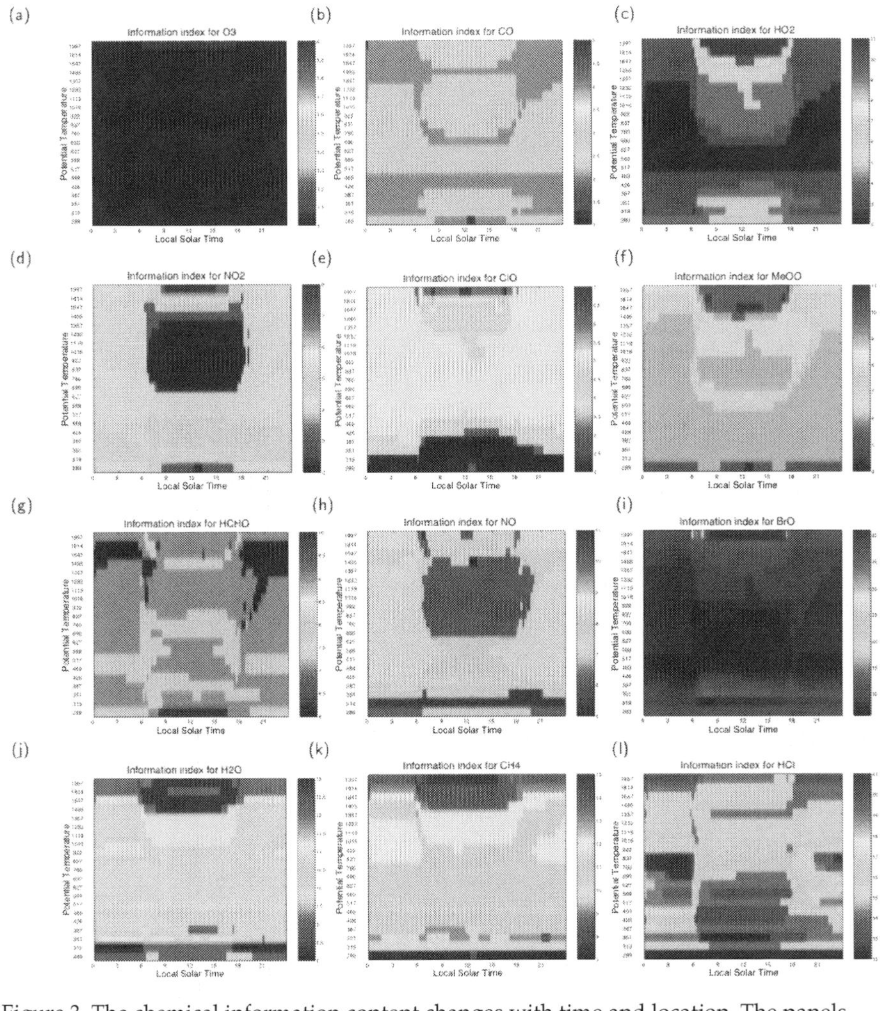

Figure 3. The chemical information content changes with time and location. The panels show some examples of how the information index changes with time (at 15 minute intervals) and location in a vertical profile at 38°S

It can be seen that some rows of $\boldsymbol{M}$ (an example is in Figure 2 (a)) such as the one for O_3 have many more significant elements than others, i.e. highlighting the fact that it is currently a key player in the dynamically evolving system. We can use this to construct an information content index, I_c.

$$I_c = \sum_i M(i,j) \tag{18}$$

where *i* and *j* refer to the rows and columns of ***M***. An example of this information content index, I_c, is shown in Figure 2 (b) corresponding to the model matrix, ***M***, shown in Figure 2 (a). As one would expect O_3 comes out at the top of the list. The order of this list is a rather strong function of time and location.

If we now sum the index I_c at each grid point in a vertical profile and at each time throughout a diurnal cycle we get a composite index of the species role throughout the day.

$$TotI_c = \sum I_c \tag{19}$$

An example of this is shown in Figure 2 (c). The values have been normalized by the maximum value of $TotI_c$ so that the values range from zero to one. One being the greatest information content, zero being the least. For some species their importance is shorter lived (i.e. not so persistent, they are important for just a short while, for these species using a definition based on the maximum value of elements of ***M*** is more suitable. An example of this is shown in Figure 2 (d).

The ranking of species provided is sensible, when sorted in order of maximum $TotI_c$ the top 10 species are O_3, HO_2, NO_2, ClO, CH_3OO, HCHO, NO, BrO, and H_2O. We see key representatives of carbon, nitrogen, hydrogen, and bromine species reminding us how coupled atmospheric chemistry is. It is interesting that two of the top ten are peroxy radicals, number three is HO_2 and number six is CH_3OO. Figure 3 shows how the chemical information content changes with time and location for ten of the species with the highest information content. The panels show some examples of how the information index changes with time (at 15 minute intervals) and location for a vertical profile at 38°S.

It is interesting to note that both the uncertainty used to define the targets and the information content used to define the priorities can be a strong function of altitude. This emphasizes that real time diagnosis of these metrics is more advantageous that a priori prescription.

A Hybrid Approach

We noted earlier that an autonomous system could have multiple modes of operation. So it is quite conceivable that the metrics used above for target selection and priority based on uncertainty and information content could be used in different ways. For example, during the validation campaign of a new instrument we may want to target regions where we know the state of the system with the highest precision for our validation. In this case we would use targets defined by the minima in our state vector uncertainty. Conversely, during routine operation we would like the observing system to be adaptively reducing the total uncertainty, so would use targets defined on the maxima in our state vector uncertainty.

Taking this one step further, it may be of use to have feature recognition as part of the targeting. For example, we may be focusing on ship tracks, or jet streaks in the weather systems. In this case we can combine the approaches by choosing, for example, the ship tracks where we have the highest information content and uncertainty.

5. Future Directions

As data volumes increase communication bandwidth becomes an increasingly important consideration. This can be made part of the multi-objective optimization. It is conceivable that the distributed sensor web also becomes part of the communications network for the system.

6. Conclusion

We have described how situational awareness can be integrated into an autonomous observing system by incorporating our theoretical and observational knowledge of the system. Metrics of what we do not know (state vector uncertainty) can be used to define what we need to measure and the required mode, time and location of the observations. The mathematical formalism of data assimilation provides us with the state vector uncertainty. Metrics of how important it is to know this information (information content) are used to assign a priority to each observation. A basic postulate of information theory is that information can be treated like a measurable physical quantity, such as density or mass. Consequently we can construct an optimization method for use in observing systems where there is an objective optimization for information content. This allows the best return of information for a given investment in measuring systems. Based on information content and level of uncertainty we can create a dynamic observation control system that adapts what measurements are made, where they are made, and when they are made, in an online fashion to maximize the information content, minimize the uncertainty in characterizing the system's state vector. The metrics are passed in real time to the Sensor Web observation scheduler to implement the observation plan for the next observing cycle.

7. References

Atkins, E.M., Miller, R.H., Van Pelt, T., Shaw, K.D., Ribbens, W.B., Washbaugh, P.D. & Bernstein, D.S. (1998) Solus: An autonomous aircraft for flight control and trajectory planning research. *Proceedings of the 1998 American Control Conference, Vols 1-6*, 689-693.

Beard, R.W., Mclain, T.W., Nelson, D.B., Kingston, D. & Johanson, D. (2006) Decentralized cooperative aerial-surveillance using fixed-wing miniature UAVs. *Proceedings of the IEEE*, 94, 1306-1324.

Beji, L. & Abichou, A. (2005) Tracking control of trim trajectories of a blimp for ascent and descent flight manoeuvres. *International Journal of Control*, 78, 706-719.

Bellur, B.R., Lewis, M.G. & Templin, F.L. (2002) Tactical information operations for autonomous teams of unmanned aerial vehicles (UAVs). *2002 IEEE Aerospace Conference Proceedings, Vols 1-7*, 2741-2756.

Boskovic, J.D., Prasanth, R. & Mehra, R.K. (2002) A Multi-Layer control architecture for unmanned aerial vehicles. *Proceedings of the 2002 American Control Conference, Vols 1-6*, 1825-1830.

Botts, M., Percivall, G., Reed, C. & Davidson, J. (2007) OGC Sensor Web Enablement: Overview and High Level Architecture. Version 3 ed.

Broten, G.S., Monckton, S.P., Collier, J. & Giesbrecht, J. (2006) Architecture for autonomy - art. no. 62300H.

Caltabiano, D., Muscato, G., Orlando, A., Federico, C., Giudice, G. & Guerrieri, S. (2005) Architecture of a UAV for volcanic gas sampling. *Etfa 2005: 10th IEEE International Conference on Emerging Technologies and Factory Automation, Vol 1, Pts 1 and 2, Proceedings*, 739-744.

Ceccarelli, N., Enright, J.J., Frazzoli, E., Rasmussen, S.J. & Schumacher, C.J. (2007) Micro UAV path planning for reconnaissance in wind. *2007 American Control Conference, Vols 1-13*, 2059-2064.

Chandler, P.R. & Pachter, M. (1998) Research issues in autonomous control of tactical UAVs. *Proceedings of the 1998 American Control Conference, Vols 1-6*, 394-398.

Chandler, P.R., Pachter, M., Swaroop, D., Fowler, J.M., Howlett, J.K., Rasmussen, S., Schumacher, C. & Nygard, K. (2002) Complexity in UAV cooperative control. *Proceedings of the 2002 American Control Conference, Vols 1-6*, 1831-1836.

Chen, Y.L., Peot, M., Lee, J., Sundareswaran, V. & Altshuler, T. (2005) Autonomous Collaborative Mission Systems (ACMS) for multi-UAV missions. *Defense Transformation and Network-Centric Systems*, 5820, 152-159.

Chen, Y.L., Peot, M., Lee, J., Sundareswaran, V. & Altshuler, T. (2006) Autonomous collaborative behaviors for multi-UAV missions - art. no. 62490L. *Defense Transformation and Network-Centric Systems*, 6249, L2490-L2490.

Chung, Y.C. & He, Z.H. (2007) Low-complexity and reliable moving objects detection and tracking for aerial video surveillance with small UAVs. *2007 IEEE International Symposium on Circuits and Systems, Vols 1-11*, 2670-2673.

Coronado, P.L., Stetina, F. & Jacob, D. (1998) New technologies to support NASA's mission to planet earth satellite remote sensing product validation: The use of an Unmanned Autopiloted Vehicle (UAV) as a platform to conduct remote sensing. *Robotic and Semi-Robotic Ground Vehicle Technology*, 3366, 38-49.

Cruz, J.B., Chen, G.S., Garagic, D., Tan, X.H., Li, D.X., Shen, D., Wei, M. & Wang, X. (2003) Team dynamics and tactics for mission planning. *42nd IEEE Conference on Decision and Control, Vols 1-6, Proceedings*, 3579-3584.

Dargar, A., Kamel, A., Christensen, G. & Nygard, K. (2002) An agent-based framework for UAV collaboration. *Intelligent Systems*, 54-59.

Davis, R.E., Webb, D.C., Regier, L.A. & Dufour, J. (1992) The Autonomous Lagrangian Circulation Explorer (Alace). *Journal of Atmospheric and Oceanic Technology*, 9, 264-285.

Dudziak, M.J. (1998) Coordinated traffic incident management using the I-Net embedded sensor architecture. *Mobile Robots Xiii and Intelligent Transportation Systems*, 3525, 404-416.

Elbern, H., Schmidt, H. & Ebel, A. (1997) Variational data assimilation for tropospheric chemistry modeling. *Journal of Geophysical Research-Atmospheres*, 102, 15967-15985.

Elfes, A., Bueno, S.S., Bergerman, M., De Paiva, E.C., Ramos, J.G. & Azinheira, J.R. (2003) Robotic airships for exploration of planetary bodies with an atmosphere: Autonomy challenges. *Autonomous Robots*, 14, 147-164.

Elfes, A., Carvalho, J.R.H., Bergerman, M. & Bueno, S.S. (2001) Towards a perception and sensor fusion architecture for a robotic airship. *Sensor Fusion and Decentralized Control in Robotic Systems Iv*, 4571, 65-75.

Ertl, C. & Schulte, A. (2005) Enabling autonomous UAV co-operation by onboard artificial cognition. *Foundations of Augmented Cogniton, Vol 11*, 1165-1174.

Fabiani, P., Fuertes, V., Piquereau, A., Mampey, R. & Teichteil-Konigsbuch, F. (2007) Autonomous flight and navigation of VTOL UAVs: from autonomy demonstrations to out-of-sight flights. *Aerospace Science and Technology,* 11, 183-193.

Fink, W., George, T. & Tarbell, M.A. (2007) Tier-scalable reconnaissance: The challenge of sensor optimization, sensor deployment, sensor fusion, and sensor interoperability - art. no. 655611. *Micro (Mems) and Nanotechnologies for Defense and Security,* 6556, 55611-55611.

Fisher, M. & Lary, D.J. (1995) Lagrangian 4-Dimensional Variational Data Assimilation of Chemical-Species. *Quarterly Journal of the Royal Meteorological Society,* 121, 1681-1704.

Flint, M., Fernandez-Gaucherand, E. & Polycarpou, M. (2003) Cooperative control for UAV's searching risky environments for targets. *42nd IEEE Conference on Decision and Control, Vols 1-6, Proceedings,* 3567-3572.

Francl, M. (2000) UAVs as communications routing nodes. *Airborne Reconnaissance Xxiv,* 4127, 40-45.

Furfaro, R., Ganapol, B.D., Johnson, L.F. & Herwitz, S.R. (2007) Neural network algorithm for coffee ripeness evaluation using airborne images. *Applied Engineering in Agriculture,* 23, 379-387.

Gilmore, J.F. & Garbarino, J.E. (2004) UAV team behaviors in operational scenarios. *Unattended/Unmanned Ground, Ocean, and Air Sensor Technologies and Applications Vi,* 5417, 197-205.

Goodall, J.L., Horsburgh, J.S., Whiteaker, T.L., Maidment, D.R. & Zaslavsky, I. (2008) A first approach to web services for the National Water Information System. *Environmental Modelling & Software,* 23, 404-411.

Halem, M., Patwardhan, A., Dornbush, S., Seablom, M. & Yesha, Y. (2007) Sensor web design studies for realtime dynamic congestion pricing. *Fifth Annual Ieee International Conference on Pervasive Computing and Communications Workshops, Proceedings,* 413-418.

Hansen, E., Huntsberger, T. & Elkins, L. (2006) Autonomous maritime navigation developing autonomy skill sets for USVs - art. no. 62300U.

Herwitz, S.R., Johnson, L.F., Dunagan, S.E., Higgins, R.G., Sullivan, D.V., Zheng, J., Lobitz, B.M., Leung, J.G., Gallmeyer, B.A., Aoyagi, M., Slye, R.E. & Brass, J.A. (2004) Imaging from an unmanned aerial vehicle: agricultural surveillance and decision support. *Computers and Electronics in Agriculture,* 44, 49-61.

Higgins, G., Kalb, M., Mahoney, R., Lutz, R., Mauk, R., Seablom, M. & Talabac, S. (2005) Architecture vision and technologies for post-NPOESS weather prediction system: Two-way interactive observing and modeling. Part II, Use case scenario. *Enabling Sensor and Platform Technologies for Spaceborne Remote Sensing,* 5659, 242-252.

Hope, T., Marston, R. & Richards, D. (2004) Decision support for decision superiority: Control strategies for multiple UAVs. *Human Performance, Situation Awareness and Automation: Current Research and Trends, Vol 1,* 290-294.

Hygounenc, E., Jung, I.K., Soueres, P. & Lacroix, S. (2004) The autonomous blimp project of LAAS-CNRS: Achievements in flight control and terrain mapping. *International Journal of Robotics Research,* 23, 473-511.

Ippolito, C., Pisanich, G. & Young, L.A. (2005) Cognitive emotion layer architecture for intelligent UAV planning, behavior and control. *2005 IEEE Aerospace Conference, Vols 1-4*, 2964-2979.

Jin, Y., Liao, Y., Minai, A.A. & Polycarpou, M.M. (2006) Balancing search and target response in cooperative unmanned aerial vehicle (UAV) teams. *IEEE Transactions on Systems Man and Cybernetics Part B-Cybernetics*, 36, 571-587.

Johnson, L.F., Herwitz, S.R., Lobitz, B.M. & Dunagan, S.E. (2004) Feasibility of monitoring coffee field ripeness with airborne multispectral imagery. *Applied Engineering in Agriculture*, 20, 845-849.

Kalb, M., Higgins, G., Mahoney, R., Lutz, R., Mauk, R., Seablom, M. & Talabac, S. (2005) Architecture vision and technologies for post-NPOESS weather prediction system: Two-way interactive observing and modeling. *Enabling Sensor and Platform Technologies for Spaceborne Remote Sensing*, 5659, 221-232.

Kalman, R.E. (1962) Canonical Structure of Linear Dynamical Systems. *Proceedings of the National Academy of Sciences of the United States of America*, 48, 596-&.

Kalnay, E., Kanamitsu, M., Kistler, R., Collins, W., Deaven, D., Gandin, L., Iredell, M., Saha, S., White, G., Woollen, J., Zhu, Y., Chelliah, M., Ebisuzaki, W., Higgins, W., Janowiak, J., Mo, K.C., Ropelewski, C., Wang, J., Leetmaa, A., Reynolds, R., Jenne, R. & Joseph, D. (1996) The NCEP/NCAR 40-year reanalysis project. *Bulletin of the American Meteorological Society*, 77, 437-471.

Kamal, W.A., Gu, D.W. & Postlethwaite, I. (2005) Real time trajectory planning for UAVs using MILP. *2005 44th IEEE Conference on Decision and Control & European Control Conference, Vols 1-8*, 3381-3386.

Kamsickas, G.M. & Ward, J.N. (2003) Developing UGVs for the FCS program. *Unmanned Ground Vehicle Technology V*, 5083, 277-284.

Kelly, A., Stentz, A., Amidi, O., Bode, M., Bradley, D., Diaz-Calderon, A., Happold, M., Herman, H., Mandelbaum, R., Pilarski, T., Rander, P., Thayer, S., Vallidis, N. & Warner, R. (2006) Toward reliable off road autonomous vehicles operating in challenging environments. *International Journal of Robotics Research*, 25, 449-483.

Khattatov, B.V., Gille, J.C., Lyjak, L.V., Brasseur, G.P., Dvortsov, V.L., Roche, A.E. & Waters, J.W. (1999) Assimilation of photochemically active species and a case analysis of UARS data. *Journal of Geophysical Research-Atmospheres*, 104, 18715-18737.

King, E., Kuwata, Y., Alighanbari, M., Bertuccelli, L. & How, J. (2004) Coordination and control experiments on a multi-vehicle testbed. *Proceedings of the 2004 American Control Conference, Vols 1-6*, 5315-5320.

King, E., Kuwata, Y. & How, J.P. (2006) Experimental demonstration of coordinated control for multi-vehicle teams. *International Journal of Systems Science*, 37, 385-398.

Kolding, J. & Pouliot, M. (1997) Bird dog: Coupling the Longbow Apache(TM) attack helicopter with an unmanned aerial vehicle. *American Helicopter Society - 53rd Annual Forum Proceedings, Vols 1 and 2*, 281-287.

Krishnamurthygopalan, A., Davari, A. & Manish, A. (2005) Optimal path planning for an unmanned aerial vehicle. *Proceedings of the Thirty-Seventh Southeastern Symposium on System Theory*, 258-261.

Kunchev, V., Jain, L., Ivancevic, V. & Finn, A. (2006) Path planning and obstacle avoidance for autonomous mobile robots: A review. *Knowledge-Based Intelligent Information and Engineering Systems, Pt 2, Proceedings,* 4252, 537-544.

Lamont, G.B., Slear, J.N. & Melendez, K. (2007) UAV swarm mission planning and routing using multi-objective evolutionary algorithms. *2007 IEEE Symposium on Computational Intelligence in Multi-Criteria Decision Making,* 10-20.

Lary, D.J. (1996) Gas phase atmospheric bromine photochemistry. *Journal of Geophysical Research-Atmospheres,* 101, 1505-1516.

Lary, D.J., Chipperfield, M.P. & Toumi, R. (1995) The Potential Impact of the Reaction Oh+ClO->HCl+O2 on Polar Ozone Photochemistry. *Journal of Atmospheric Chemistry,* 21, 61-79.

Lary, D.J., Khattatov, B. & Mussa, H.Y. (2003) Chemical data assimilation: A case study of solar occultation data from the ATLAS 1 mission of the Atmospheric Trace Molecule Spectroscopy Experiment (ATMOS). *Journal of Geophysical Research-Atmospheres,* 108.

Lee, D.J., Beard, R.W., Merrell, P.C. & Zhan, P.C. (2004) See and avoidance behaviors for autonomous navigation. *Mobile Robots Xvii,* 5609, 23-34.

Levin, E., Kupiec, S., Forrester, T., Debacker, A. & Jannson, T. (2002) GIS-based UAV real-time path planning and navigation. *Sensors, and Command, Control, Communications and Intelligence (C31) Technologies for Homeland Defense and Law Enforcement,* 4708, 296-303.

Li, W. & Cassandras, C.G. (2006) Centralized and distributed cooperative Receding Horizon control of autonomous vehicle missions. *Mathematical and Computer Modelling,* 43, 1208-1228.

Maddula, T., Minai, A.A. & Polycarpou, M.M. (2004) Multi-target assignment and path planning for groups of UAVs. *Recent Developments in Cooperative Control and Optimization,* 3, 261-272.

Mahler, R.P. & Prasanth, R.K. (2002) Technologies for unified collection and control of UCAVs. *Signal Processing, Sensor Fusion, and Target Recognition Xi,* 4729, 90-101.

Mcinnes, C.R. (2003) Velocity field path-planning for single and multiple unmanned aerial vehicles. *Aeronautical Journal,* 107, 419-426.

Menard, R., Cohn, S.E., Chang, L.P. & Lyster, P.M. (2000) Assimilation of stratospheric chemical tracer observations using a Kalman filter. Part I: Formulation. *Monthly Weather Review,* 128, 2654-2671.

Miller, R.H. & Larsen, M.L. (2003) Optimal fault detection and isolation filters for flight vehicle performance monitoring. *2003 IEEE Aerospace Conference Proceedings, Vols 1-8,* 3197-3203.

Morrow, M.T., Woolsey, C.A. & Hagerman, G.M. (2006) Exploring titan with autonomous, buoyancy driven gliders. *Jbis-Journal of the British Interplanetary Society,* 59, 27-34.

Murphey, R.A. & O'neal, J.K. (2002) A cooperative control testbed architecture for smart loitering weapons. *Proceedings of the Fifth International Conference on Information Fusion, Vol I,* 694-699.

Natarajan, G. (2001) Ground control stations for unmanned air vehicles. *Defence Science Journal,* 51, 229-237.

Nikolos, I.K., Valavanis, K.P., Tsourveloudis, N.C. & Kostaras, A.N. (2003) Evolutionary algorithm based offline/online path planner for UAV navigation. *IEEE Transactions on Systems Man and Cybernetics Part B-Cybernetics,* 33, 898-912.

Noth, A., Engel, W. & Siegwart, R. (2006) Design of an ultra-lightweight autonomous solar airplane for continuous flight. *Field and Service Robotics,* 25, 441-452.

Onosato, M., Takemura, F., Nonami, K., Kawabata, K., Miura, K. & Nakanishi, H. (2006) Aerial robots for quick information gathering in USAR. *2006 Sice-Icase International Joint Conference, Vols 1-13,* 1592-1595.

Pehlivanoglu, Y.V., Baysal, O. & Hacioglu, A. (2007) Path planning for autonomous UAV via vibrational genetic algorithm. *Aircraft Engineering and Aerospace Technology,* 79, 352-359.

Pehlivanoglu, Y.V. & Hacioglu, A. (2007) Vibrational genetic algorithm based path planner for autonomous UAV in spatial data based environments. *2007 3rd International Conference on Recent Advances in Space Technologies, Vols 1 and 2,* 573-578.

Persson, M. (2002) Visual-servoing based tracking for an UAV in a 3D simulation environment. *Acquisition, Tracking, and Pointing Xvi,* 4714, 65-75.

Pettersson, P.O. & Doherty, P. (2006) Probabilistic roadmap based path planning for an autonomous unmanned helicopter. *Journal of Intelligent & Fuzzy Systems,* 17, 395-405.

Pielke, R.A., Cotton, W.R., Walko, R.L., Tremback, C.J., Lyons, W.A., Grasso, L.D., Nicholls, M.E., Moran, M.D., Wesley, D.A., Lee, T.J. & Copeland, J.H. (1992) A Comprehensive Meteorological Modeling System - Rams. *Meteorology and Atmospheric Physics,* 49, 69-91.

Plale, B., Gannon, D., Brotzge, J., Droegemeier, K., Kurose, J., Mclaughlin, D., Wilhelmson, R., Graves, S., Ramamurthy, M., Clark, R.D., Yalda, S., Reed, D.A., Joseph, E. & Chandrasekar, V. (2006) CASA and LEAD: Adaptive cyberinfrastructure for real-time multiscale weather forecasting. *Computer,* 39, 56-+.

Pongpunwattana, A., Wise, R., Rysdyk, R. & Kang, A.J. (2006) Multi-vehicle cooperative control flight test. *2006 IEEE/AIAA 25th Digital Avionics Systems Conference, Vols 1-3,* 781-791.

Press, W.H. (2007) *Numerical recipes : the art of scientific computing,* Cambridge, UK ; New York, Cambridge University Press.

Rafi, F., Khan, S., Shafiq, K. & Shah, M. (2006) Autonomous target following by unmanned aerial vehicles - art. no. 623010.

Richards, A. & How, J. (2004) A decentralized algorithm for robust constrained model predictive control. *Proceedings of the 2004 American Control Conference, Vols 1-6,* 4261-4266.

Roberts, R.S., Kent, C.A., Cunningham, C.T. & Jones, E.D. (2003) UAV cooperation architectures for persistent sensing. *Sensors, and Command, Control, Communications, and Intelligence (C3i) Technologies for Homeland Defense and Law Enforcement Ii,* 5071, 306-314.

Ryan, A. & Hedrick, J.K. (2005) A mode-switching path planner for UAV-assisted search and rescue. *2005 44th IEEE Conference on Decision and Control & European Control Conference, Vols 1-8,* 1471-1476.

Ryan, A., Tisdale, J., Godwin, M., Coatta, D., Nguyen, D., Spry, S., Sengupta, R. & Hedrick, J.K. (2007) Decentralized control of unmanned aerial vehicle collaborative sensing missions. *2007 American Control Conference, Vols 1-13*, 1564-1569.

Sandewall, E., Doherty, P., Lemon, O. & Peters, S. (2003) Words at the right time: Real-time dialogues with the WITAS unmanned aerial vehicle. *Ki 2003: Advances in Artificial Intelligence*, 2821, 52-63.

Sandford, S.P., Harrison, F.W., Langford, J., Johnson, J.W., Qualls, G. & Emmitt, D. (2004) Autonomous aerial observations to extend and complement the Earth Observing System: A science driven, systems oriented approach. *Remote Sensing Applications of the Global Positioning System*, 5661, 142-159.

Sasiadek, J.Z. & Duleba, I. (2000) 3D local trajectory planner for UAV. *Journal of Intelligent & Robotic Systems*, 29, 191-210.

Schiller, I., Luciano, J.S. & Draper, J.S. (1993) Flock Autonomy for Unmanned Vehicles. *Mobile Robots Vii*, 1831, 45-51.

Schmale, D.G., Dingus, B.R. & Reinholtz, C. (2008) Development and application of an autonomous unmanned aerial vehicle for precise aerobiological sampling above agricultural fields. *Journal of Field Robotics*, 25, 133-147.

Schouwenaars, T., Valenti, M., Feron, E., How, J. & Roche, E. (2006) Linear programming and language processing for human/unmanned-aerial-vehicle team missions. *Journal of Guidance Control and Dynamics*, 29, 303-313.

Seablom, M.S., Talabac, S.J., Higgins, G.J. & Womack, B.T. (2007) Simulation for the design of next-generation global earth observing systems - art. no. 668413. *Atmospheric and Environmental Remote Sensing Data Processing and Utilization Iii: Readiness for Geoss*, 6684, 68413-68413.

Shannon, C.E. (1997) The mathematical theory of communication (Reprinted). *M D Computing*, 14, 306-317.

Shannon, C.E. & Weaver, W. (1949) *The mathematical theory of communication*, Urbana,, University of Illinois Press.

Simmons, A.J. & Hollingsworth, A. (2002) Some aspects of the improvement in skill of numerical weather prediction. *Quarterly Journal of the Royal Meteorological Society*, 128, 647-677.

Sinha, A., Kirubarajan, T. & Bar-Shalom, Y. (2006) Autonomous search, tracking and classification by multiple cooperative UAVs - art. no. 623508. *Signal Processing, Sensor Fusion, and Target Recognition Xv*, 6235, 23508-23508.

Sinopoli, B., Micheli, M., Donato, G. & Koo, T.J. (2001) Vision based navigation for an unmanned aerial vehicle. *2001 IEEE International Conference on Robotics and Automation, Vols I-Iv, Proceedings*, 1757-1764.

Skoglar, P., Nygards, J. & Ulvklo, M. (2006) Concurrent path and sensor planning for a UAV - Towards an information based approach incorporating models of environment and sensor. *2006 IEEE/Rsj International Conference on Intelligent Robots and Systems, Vols 1-12*, 2436-2442.

Smith, J.F. (2007) Fuzzy logic planning and control for a team of UAVS. *Procedings of the 11th Iasted International Conference on Artificial Intelligence and Soft Computing*, 286-294.

Smith, J.F. & Nguyen, T.H. (2005) Distributed autonomous systems: resource management, planning, and control algorithms. *Signal Processing, Sensor Fusion, and Target Recognition Xiv,* 5809, 65-76.

Smith, J.F. & Nguyen, T.H. (2006a) Fuzzy logic based resource manager for a team of UAVs. *Nafips 2006 - 2006 Annual Meeting of the North American Fuzzy Information Processing Society, Vols 1 and 2,* 484-491.

Smith, J.F. & Nguyen, T.H. (2006b) Fuzzy logic based UAV allocation and coordination. *Icinco 2006: Proceedings of the Third International Conference on Informatics in Control, Automation and Robotics,* 9-18.

Smith, J.F. & Nguyen, T.H. (2006c) Resource manager for an autonomous coordinated team of UAVs - art. no. 62350C. *Signal Processing, Sensor Fusion, and Target Recognition Xv,* 6235, C2350-C2350.

Smith, J.F. & Nguyen, T.H. (2007) Fuzzy decision trees for planning and autonomous control of a coordinated team of UAVs - art. no. 656708. *Signal Processing, Sensor Fusion, and Target Recognition Xvi,* 6567, 56708-56708.

Snarski, S., Scheibner, K., Shaw, S., Roberts, R., Larow, A., Breitfeller, E., Lupo, J., Neilson, D., Judge, B. & Forrenc, J. (2006) Autonomous UAV-based mapping of large-scale urban firefights - art. no. 620905. *Airborne Intelligence, Surveillance, Reconnaissance (ISR) Systems and Applications Iii,* 6209, 20905-20905.

Steinberg, M. (2006) Intelligent autonomy for unmanned naval vehicles - art. no. 623013.

Stoer, J. & Bulirsch, R. (2002) *Introduction to numerical analysis,* New York, Springer.

Stottler, R., Ball, B. & Richards, R. (2007) Intelligent Surface Threat Identification System (ISTIS). *2007 IEEE Aerospace Conference, Vols 1-9,* 2028-2040.

Sullivan, D., Totah, J., Wegener, S., Enomoto, F., Frost, C., Kaneshige, J. & Frank, J. (2004) Intelligent mission management for uninhabited aerial vehicles. *Remote Sensing Applications of the Global Positioning System,* 5661, 121-131.

Swinbank, R. & O'neill, A. (1994) A Stratosphere Troposphere Data Assimilation System. *Monthly Weather Review,* 122, 686-702.

Templeton, T., Shim, D.H., Geyer, C. & Sastry, S.S. (2007) Autonomous vision-based landing and terrain mapping using an MPC-controlled unmanned rotorcraft. *Proceedings of the 2007 IEEE International Conference on Robotics and Automation, Vols 1-10,* 1349-1356.

Vachtsevanos, G., Tang, L., Drozeski, G. & Gutierrez, L. (2005) From mission planning to flight control of unmanned aerial vehicles: Strategies and implementation tools. *Annual Reviews in Control,* 29, 101-115.

Valenti, M., Bethke, B., How, J.R., De Farias, D.P. & Vian, J. (2007) Embedding health management into mission tasking for UAV teams. *2007 American Control Conference, Vols 1-13,* 3486-3492.

Wang, J., Patel, V., Woolsey, C.A., Hovakimyan, N. & Schmale, D. (2007) L-1 adaptive control of a UAV for aerobiological sampling. *2007 American Control Conference, Vols 1-13,* 5897-5902.

Wikipedia (2008a) Data assimilation --- Wikipedia{,} The Free Encyclopedia.

Wikipedia (2008b) Fokker Planck equation --- Wikipedia{,} The Free Encyclopedia.

Yang, Y.Y., Zhou, R. & Chen, Z.J. (2006) Autonomous trajectory planning for UAV based on threat assessments and improved Voronoi graphics - art. no. 63584J.

Young, L.A., Pisanich, G., Ippolito, C. & Alena, R. (2005) Aerial vehicle surveys of other planetary atmospheres and surfaces: imaging, remote-sensing, and autonomy technology requirements. *Real-Time Imaging Ix,* 5671, 183-199.

Yuan, H.L., Gottesman, V., Falash, M., Qu, Z.H., Pollak, E. & Chunyu, J.M. (2007) Cooperative formation flying in autonomous unmanned air systems with application to training. *Advances in Cooperative Control and Optimization,* 369, 203-219.

Zelinski, S., Koo, T.J. & Sastry, S. (2003) Hybrid system design for formations autonomous vehicles. *42nd IEEE Conference on Decision and Control, Vols 1-6, Proceedings*, 1-6.

Zingaretti, P., Mancini, A., Frontoni, E., Monteriu, A. & Longhi, S. (2008) Autonomous helicopter for surveillance and security. *DETC2007: Proceedings of the ASME International Design Engineering Technology Conference and Computers and Information in Engineering Conference,* 4, 227-234.

20

Performance Evaluation of an Unmanned Airborne Vehicle Multi-Agent System

Zhaotong Lian[1] and Abhijit Deshmukh[2]
[1]*Faculty of Business Administration, University of Macau,*
[2]*Department of Industrial and Systems Engineering, Texas A&M University*
[1]*China,* [2]*USA*

1. Introduction

Consider an unmanned airborne vehicle (UAV) multi-agent system. A UAV agent is aware of the destination or goal to be achieved, its own quantitative or qualitative, of encountering enemy defenses in the region. Each agent plans its moves in order to maximize the chances of reaching the target before the required task completion time (see Fig. 1). The plans are developed based on the negotiations between different UAVs in the region with the overall goal in mind. The model is actually motivated by another large research project related to multi-agent systems. The information about enemy defenses can be communicated between UAVs and they can negotiate about the paths to be taken based on their resources, such as fuel, load, available time to complete the task and the information about the threat. In this system, we can also model the behavior of enemy defenses as independent agents, with known or unknown strategies. Each enemy defense site or gun has a probability of destroying a UAV in a neighborhood. The UAVs have an expectation of the location of enemy defenses, which is further refined as more information becomes available during the flight or from other UAVs. To successfully achieve the goal with a high probability, the UAVs need to select a good plan based on coordination and negotiation between each other.
One paper dealing with this model is Atkins et al. (Atkins et al., 1996), which considered an agent capable of safe, fully-automated aircraft flight control from takeoff through landing. To build and execute plans that yield a high probability of successfully reaching the specified goals, the authors used state probabilities to guide a planner along highly-probable goal paths instead of low-probability states. Some probabilistic planning algorithms are also developed by the other researchers. Kushmerick et al. (Kushmerick et al., 1994) concentrate on probabilistic properties of actions that may be controlled by the agent, not external events. Events can occur over time without explicit provocation by the agent, and are generally less predictable than state changes due to actions. Atkins et al. (Atkins et al., 1996) presented a method by which local state probabilities are estimated from action delays and temporally-dependent event probabilities, then used to select highly probable goal paths and remove improbable states. The authors implemented these algorithms in the Cooperative Intelligent Real-time Control Architecture (CIRCA). CIRCA combines an AI planner, scheduler performance for controlling complex real-world systems (Musliner et al., 1995). CIRCA's planner is based on the philosophy that building a plan to handle all world

states (Schoppers, 1987}) is unrealistic due to the possibility of exponential planner execution time (Ginsberg, 1989), so it uses heuristics to limit state expansion and minimizes its set of selected actions by requiring only one goal path and guaranteeing failure avoidance along all other paths.

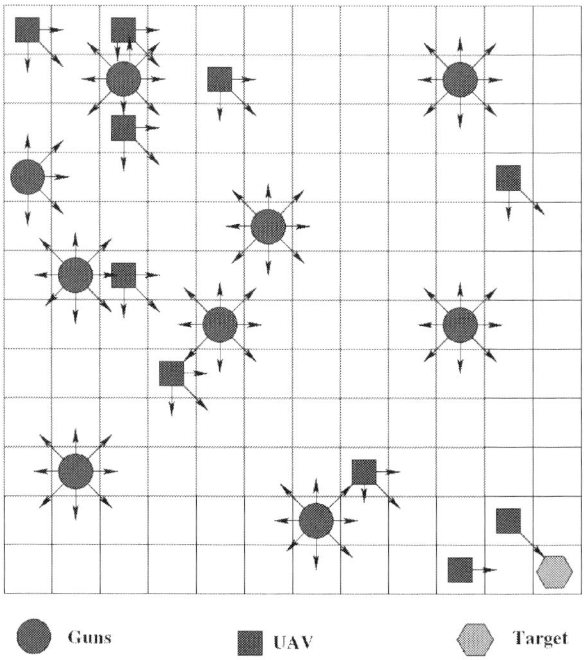

Figure 1. Unmanned aircraft system

McLain et al. (McLain et al., 2000) considered two or more UAVs, a single target in a known location, battle area divided into low threat and high threat regions by a threat boundary, and threats that `pop up' along the threat boundary. The objective is to have the UAVs arrive at the target simultaneously, in a way that maximizes the survivability of the entire team of UAVs. The approach the authors used is to decentralize the computational solution of the optimization problem by allowing each UAV to compute its own trajectory that is optimal with respect to the needs of the team. The challenge is determine what information must be communicated among team members to give them an awareness of the situation of the other team members so that each may calculate solutions that are optimal from a team perspective.

An important methodology using in this paper is Markov decision process (MDP) based approach. An important aspect of the MDP model is that it provides the basis for algorithms that probably find optimal policies given a stochastic model of the environment and a goal. The most widely used algorithms for solving MDPs are iterative methods. One of the best known of these algorithms is due to Howard (Howard, 1960), and is known as *policy iteration*, with which, some large size MDPs (Meuleau et al., 1998; Givan et al., 1997; Littman et al., 1995) can be solved approximately by replacing the transition probability with stationary probability.

MDP models play an important role in current AI research on planning (Dean et al., 1993; Sutton, 1990) and learning (Barto et al., 1991; Watkins & Dayan, 1992). As an extension of the MDP model, partially observable Markov decision processes (POMDP) were developed within the context of operation research (Monahan, 1982; Lovejoy, 1991; Kaelbling et al., 1998). The POMDP model provides an elegant solution to the problem of acting in partially observable domains, treating actions that affect the environment and actions that only affect the agent's state of information uniformly.

Xuan et al. (Xuan et al., 1999) considered the communication in multi-agent MDPs. Assume that each agent only observes part of the global system state. Although agents do have the ability to communicate with each other, it is usually unrealistic for the agents to communicate their local state information to all agents at all times, because communication actions are associated with a certain cost. Yet, communication is crucial for the agents to coordinate properly. Therefore, the optimal policy for each agent must balance the amount of communication such that the information is sufficient for proper coordination but the cost for communication does not outweigh the expected gain.

In this paper, we assume that there are multiple guns and UAVs in the lattice. The UAVs and guns can move to the neighboring sites at each discrete time step. To avoid the attacks from the guns, the UAVs need to figure out the optimal path to successfully reach the target with a high probability. However, a UAV cannot directly observe the local states of other UAVs, which are dynamic information. Instead, a UAV has a choice of performing a communication between two moving actions. The purpose of the communication for one UAV is to know the current local state of the other UAVs, i.e., the location and the status (dead or alive). By using the traditional MDP approach, we conduct an analytical model when there are one or two UAVs on the lattice. We extend it to a multi-UAV model by developing a heuristic algorithm.

The remainder of this paper is organized as follows. In Section 2, we derive the probability transition matrix of guns by formulate the action of guns as a Markov process. When there are only one or two UAVs in the lattice, we analyze the model as an MDP. In Section 3, we conduct extensive numerical computations. We develop an algorithm to derive the moving directions for the multi-UAV case. A sample path technique is used to calculate the probability that reaching the target is successful. Finally in Section 4, we conclude with the summary of results and suggestions for this model and the future research.

2. MDP Models

2.1 The probability transition matrix of guns

In this subsection, we discuss the action of the guns in the lattice. We assume that the size of lattice is $m_1 \times m_2$. Let $A = \{(i,j): 0 \le i \le m_1 - 1, 0 \le j \le m_2 - 1\}$ be the set of all sites in the lattice.

Each site $\mathbf{a} \in A$ is associated the number of guns $\delta_{\mathbf{a}}^t$ which can assume $q+1$ different values $(\delta_{\mathbf{a}} = 0, 1, \cdots, q)$ at time t. A complete set $\{\delta_{\mathbf{a}}^t, \mathbf{a} \in A, t \ge 0\}$ of lattice variables specifies a configuration of the gun system.

Since guns move to their neighbors randomly in each step without depending on their past positions, we can derive the probability transition matrix of guns by constructing a Markov chain. When the lattice is large, however, the size of the state space becomes so big that the computation of the transition probabilities is complicated. Fortunately, the number of guns in a certain site only depends on the previous states of this site and its neighbors. we can directly derive the probability of having guns in a certain site by using some recursive formulae.

We assume that each gun has 9 possible directions to move including the current site. We denote the set of the directions by using a two-dimensional vector set $\Phi = \{(k,h): k,h = -1,0,1\}$ (see Fig. 2).

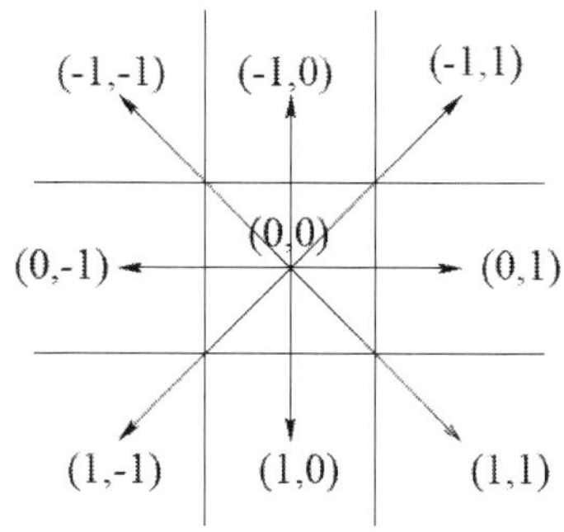

Figure 2. Walking directions of the UAVs and guns

In practice, there would not be too many guns located at one site at the same time. In order to attack UAVs more effectively, we assume that the guns negotiate with each other if there is more than one guns located in the same site. That is, they would not go to the same direction in the next step. To handle the model more easily, we restrict that there are at most 9 guns at each site.

If there is only one gun in a site, to simplify the model, we assume that the gun moves to any direction with the same probability of 1/9 including the case that the gun doesn't move at all. Obviously the probability that a gun moves to any direction is $pr(k,h,N) = N/9$ if there are N guns in the site $(N \le 9)$.

Denote $\rho^{(t)}(\mathbf{a}, n)$ as the probability that there are n guns in site $\mathbf{a}$ at time t, where $\mathbf{a} = (a_1, a_2)$. Suppose we know $\rho^{(0)}(\mathbf{a}, n)$, $n \le 9$. Let's see how to calculate $\rho^{(t)}(\mathbf{a}, n)$ when $t \ge 1$ by using recursive equations.

Suppose there are j guns at site $(a_1 + k, a_2 + h)$ at time $t-1$, then there exists one gun moving to site $\mathbf{a}$ with probability $j/9$. Therefore, the probability that there exists one gun moving from site $(a_1 + k, a_2 + h)$ to site (a_1, a_2) is $\sum_{j=1}^{9} j\rho^{(t-1)}((a_1 + k, a_2 + h), j)/9$.

There will be n guns in site $\mathbf{a}$ if there are $n_{k,h}$ guns moving from site $(a_1 + k, a_2 + h)$, where $\sum_{k,h=-1,0,1} n_{k,h} = n$ and $n_{k,h} = 0$ or 1. We define

$$B(n_{k,h}) = \begin{cases} \sum_{j=1}^{9} j\rho^{(t-1)}((a_1+k, a_2+h), j)/9, & n_{k,h}=1 \\ 1-\sum_{j=1}^{9} j\rho^{(t-1)}((a_1+k, a_2+h), j)/9, & n_{k,h}=0 \end{cases} \tag{1}$$

It is easy to see that we have the following recursive equations:

$$\rho^{(t)}(\mathbf{a}, n) = \sum_{\substack{\sum n_{k,h}=n \\ k,h=-1,0,1}} \prod_{k,h=-1,0,1} B(n_{k,n}) \tag{2}$$

and the stationary probability distribution $\pi(\mathbf{a}, n)$ exists:

$$\lim_{t\to\infty} \rho^{(t)}(\mathbf{a}, n) = \pi(\mathbf{a}, n) \tag{3}$$

That means, there exists a number T_0, when $t > T_0$, $\forall \mathbf{a}$ and $n \geq 0$,

$$\rho^{(t)}(\mathbf{a}, n) \approx \rho^{(T_0)}(\mathbf{a}, n) \tag{4}$$

2.2 A general MDP model on UAVs

Since UAVs are agent-based, they determine their paths independently although they have the same global objective which is to maximize the successful probability of at least one UAV reaching the target. A UAV wouldn't know the status of other's unless they communicate with each other. We assume that there are totally N UAVs, $X_0, \cdots, X_{N-1}$ in the lattice, where N is a finite number. Let m_t be a 2^N dimensional vector standing for all communication actions of UAVs. We use 1 and 0 to represent communicating or not between two UAVs. Then m_t is a combination of 0 and 1. Let $x_i^t(m_{t-1})$ be the action of UAV X_i at time t, $i = 0, \cdots, N-1$. Let δ^t be the state of the guns at time t. Since the values of $(x_0^{t+1}, \cdots, x_{N-1}^{t+1})$, and δ^{t+1} only depend on the values of $(x_0^t, \cdots, x_{N-1}^t)$, m_t and δ^t, $\{x_0^t, \cdots, x_{N-1}^t, m_t, \delta^t, t = 0, 1, \cdots, T\}$ consists a Markov decision process in a finite horizon T.

We can define a very simple global reward function r: any move is free and receives no reward. If one of UAVs reaches the target in t time units, the terminal reward is $\beta^t R$, where R is the probability of reaching the target and $0 \leq \beta \leq 1$ is a time discount factor. If at time T none of the UAVs reach the target, the terminal reward is 0. Each communication costs a constant c.

Denote $M^{(t)}(x_0^s(m_{s-1}), \cdots, x_{N-1}^s(m_{s-1}), 1 \leq s \leq t)$ be the probability of reaching the target within t time units, $t \leq T$. Then the objective is

$$\max_{t} \max_{m_{s-1}, x_0^s, \cdots, x_{N-1}^s, 1 \le s \le t} \left\{ \beta^t M^{(t)}\left(x_0^s, \cdots, x_{N-1}^s, 1 \le s \le t\right) - c \sum_{i=0}^{t} m_i \mathbf{e} \right\} \quad (5)$$

where $\mathbf{e}$ is a column vector in which all elements are 1.

Unfortunately, calculating optimal decision for (5) is not going to be computationally feasible since the combination of the decision policy is huge. Thus, we seek to reduce the size of the policy by defining approximation policies using heuristic approaches. We consider two folds. Firstly, we consider there are only two UAVs in the lattice. UAVs can communicate at every stage regardless of the history. Global states are known to both UAVs at all times, and thus we can regard it as a centralized problem where global states are observable. Once we know how to deal with the model of one or two UAVs, we can develop a scheme to handle the multi-UAV model. We will analyze it later.

Now let's consider the model with two UAVs. As a byproduct, we will see that the model with a single UAV is a special case of the model with two UAVs.

First of all, let's introduce the concept of the distance between two site $\mathbf{a} = (a_1, a_2)$ and $\mathbf{b} = (b_1, b_2)$ which is

$$S(\mathbf{a}, \mathbf{b}) = \max\{|a_1 - b_1|, |a_2 - b_2|\}. \quad (6)$$

We say a vector $\mathbf{b}$ is a neighbor of vector $\mathbf{a}$, if $S(\mathbf{a}, \mathbf{b}) \le 1$. We denote the set of vector $\mathbf{a}$'s neighbors by $\mathcal{D}(a) = \{\mathbf{b} : S(\mathbf{a}, \mathbf{b}) \le 1\}$.

Denote $V^{(t)}(\mathbf{x}, \mathbf{y})$ as the maximum probability the UAVs successfully reach the target within t time units when both UAVs are alive and their locations are $\mathbf{x}, \mathbf{y}$ at time 0 in the centralized sense. Denote $U^{(t)}(\mathbf{a})$ as the maximum reaching probability of a UAV within time t when there is only one UAV located at site $\mathbf{a}$ left in the lattice. Obviously, $U^{(t)}(\mathbf{a}) = 1$ if $\mathbf{a} = \mathbf{g}$ and $V^{(t)}(\mathbf{x}, \mathbf{y}) = 1$ if $\mathbf{x} = \mathbf{g}$ or $\mathbf{y} = \mathbf{g}$, where $\mathbf{g}$ is the position of the target location.

We can derive the rest $U^{(t)}(\mathbf{a})$, $V^{(t)}(\mathbf{x}, \mathbf{y})$ and the optimal path by the following recursive equations

$$V^{(0)}(\mathbf{x}, \mathbf{y}) = \begin{cases} 1, & \mathbf{x} = \mathbf{g}, \text{ or } \mathbf{y} = \mathbf{g}, \\ 0, & \text{otherwise}, \end{cases} \quad (7)$$

$$U^{(0)}(\mathbf{a}) = \begin{cases} 1, & \mathbf{a} = \mathbf{g}, \\ 0, & \mathbf{a} \ne \mathbf{g}, \end{cases} \quad (8)$$

$$U^{(t)}(\mathbf{a}) = \max_{\bar{\mathbf{a}} \in \mathrm{D}(\mathbf{a})} \left\{\rho^{(T-t)}(\bar{\mathbf{a}}, 0) U^{(t-1)}(\bar{\mathbf{a}})\right\} \quad (9)$$

and

$$V^{(t)}(\mathbf{x}, \mathbf{y}) = \max_{\bar{\mathbf{x}} \in \mathrm{D}(\mathbf{x}), \bar{\mathbf{y}} \in \mathrm{D}(\mathbf{y})} \left\{ \begin{array}{l} \rho^{(T-t)}(\bar{\mathbf{x}}, 0) V^{(t-1)}(\bar{\mathbf{x}}, \bar{\mathbf{y}}) \\ \quad + \rho^{(T-t)}(\bar{\mathbf{x}}, 0)\left(1 - \rho^{(T-t)}(\bar{\mathbf{y}}, 0)\right) U^{(t-1)}(\bar{\mathbf{x}}) \\ \quad + \rho^{(T-t)}(\bar{\mathbf{y}}, 0)\left(1 - \rho^{(T-t)}(\bar{\mathbf{x}}, 0)\right) U^{(t-1)}(\bar{\mathbf{y}}) \end{array} \right\}, \quad 1 \le t \le T \quad (10)$$

where T is the lifetime of the UAV at time 0.

We denote the algorithm based on the above formula as **Double-MDP**. Usually, we have to use the classical dynamic programming to solve the above MDP problem. Note that only the first step is optimal because we assume that the UAVs communicate all the time. Once finish the first step, they have to figure out the status of each other (dead or alive) again. Obviously, the game is over if both UAVs die. If both are still alive, we can repeat the above MDP to obtain the optimal paths. If only one UAV is alive, we will use (9) to obtain the optimal path for this UAV.

Specially, when there is only one UAV in the lattice at the beginning, we can calculate the optimal path and the maximum successful probability only by using (8) and (9). We call the algorithm based on one UAV a **Single-MDP**. Even there are more than one UAV in the lattice, we still call it Single-MDP based approach if only they find their moving directions independently according to their own local objectives.

Intuitively, the UAVs affect each other only when they are close, for instance, when they are neighbors. Hence we can develop an heuristic algorithm in which, UAVs communicate with each other only when they are neighbors. Since the agent based UAVs know each other at the beginning, and they use the same Double-MDP approaches, they should know when they are neighbors at time 0 or at the time of their last communication.

Furthermore, we can improve the above algorithm by extend the definition of neighbor for UAVs, in which, the UAVs communicate with each other only when they are neighbors.

Definition 1. *We say two UAVs located in* $\mathbf{a}$ *and* $\mathbf{b}$ *are neighbors if their distance* $S(\mathbf{a},\mathbf{b}) \le d$, *where* d *is a non-negative integer.*

When $d = 0$, the UAVs communicate when they are in the same site;

When $d = 1$, the UAVs communicate when they are in the same site or they are 'real' neighbors;

When $d > m$ where $m \times m$ is the size of the lattice, the UAVs communicate with each other in all steps.

To evaluate the above algorithm and choose a suitable d, we have to calculate successful probabilities for each d. Unfortunately, it is hard to obtain the successful probabilities analytically. What we are going to do is to calculate the successful times by combining the MDP and sample path technique. We generate the gun's status (guns exist or not) on the location of the UAVs according to the transition probability of guns. If there exist guns in the location of a UAV, the UAV is killed. If both of UAVs die before the reach the target, we say the UAVs fail, otherwise, we say the UAVs success.

2.3 Negotiation between UAVs

Besides Double-MDP approach, we can also consider that two UAVs derive their moving directions by using Single-MDP. They may negotiate with each other when they are neighbors by changing directions. The basic idea is let the UAVs negotiate when they have the same optimal direction in the next step. We call this Single-MDP based approach **Nego-MDP**. Since Nego-MDP is also a one-dimensional MDP, it should be much faster than Double-MDP to obtain the numerical results. Assume that both UAVs A and B have the same first choice $\mathbf{c}$ at time t. The successful probabilities are $A.p(\mathbf{c})$ and $B.p(\mathbf{c})$ respectively. A and B have the second choices $\mathbf{a}$ and $\mathbf{b}$ (see Figure 3). The successful probabilities are $A.p(\mathbf{a})$ and $B.p(\mathbf{b})$ respectively. Let $\rho^{(t+1)}(\mathbf{x},0)$ be the probability of having no gun in the position $\mathbf{x}$ at time $t+1$. We define the following probabilities.

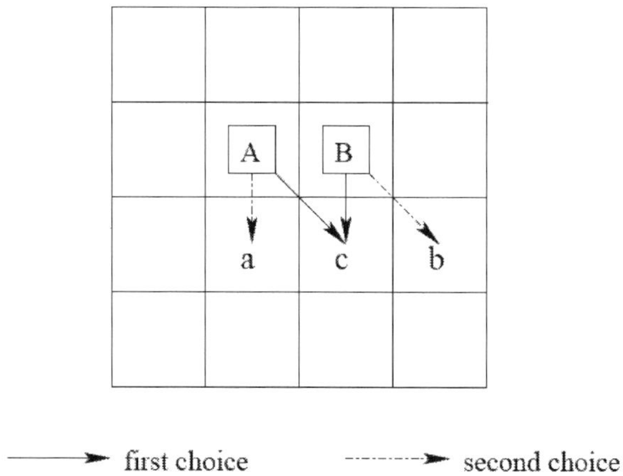

Figure 3. Negotiation Analysis

P_1: the successful probability of both A and B going to site $\mathbf{c}$ at the time $t+1$.

P_2: the successful probability of A going to site $\mathbf{c}$ and B going to site $\mathbf{b}$ at time $t+1$.

P_3: the successful probability of A going to site $\mathbf{a}$ and B going to site $\mathbf{c}$ at time $t+1$.

We have

$$P_1 = \left(V_A^{(T_A-t)}(\mathbf{c}) + V_B^{(T_B-t)}(\mathbf{c}) - V_A^{(T_A-t)}(\mathbf{c})V_B^{(T_B-t)}(\mathbf{c})\right)\rho^{(t+1)}(\mathbf{c},0) \tag{11}$$

$$P_2 = V_A^{(T_A-t-1)}(\mathbf{c}) + V_B^{(T_B-t-1)}(\mathbf{b}) - V_A^{(T_A-t-1)}(\mathbf{c})V_B^{(T_B-t-1)}(\mathbf{b}) \tag{12}$$

$$P_3 = V_A^{(T_A-t-1)}(\mathbf{a}) + V_B^{(T_B-t-1)}(\mathbf{c}) - V_A^{(T_A-t-1)}(\mathbf{a})V_B^{(T_B-t-1)}(\mathbf{c}) \tag{13}$$

where $V_A^{(T_A-t)}(\mathbf{a})$ stands for the successful probability for aircraft A starting from site $\mathbf{a}$ at time t when the initial gas is T_A.

Comparing among these three probabilities, we choose the corresponding activity when the successful probability is maximum (see Fig. 4). That is,

Case 1: P_1 is the maximum, both A and B go to site $\mathbf{c}$.

Case 2: P_2 is the maximum, A goes to $\mathbf{c}$, B goes to site $\mathbf{b}$.

Case 3: P_3 is the maximum, A goes to $\mathbf{a}$ and B goes to site $\mathbf{c}$.

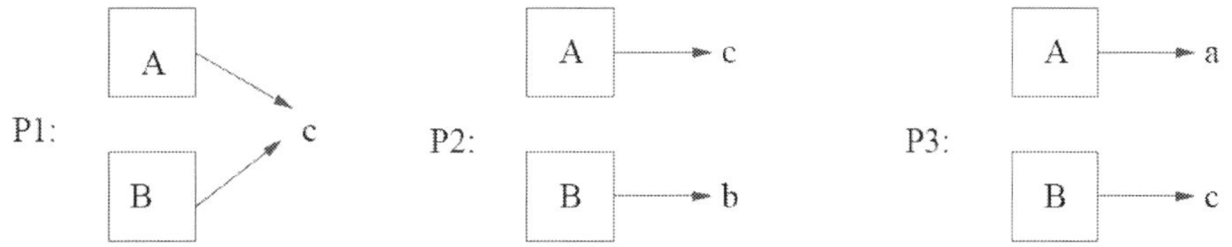

Figure 4. Negotiation Analysis

Once we have the results of single UAV and double UAV models, we can apply the Single-MDP, Nego-MDP and Double-MDP to the multi-UAV model numerically. We will discuss that in detail in the next section.

3. Numerical Analysis

In the section, we analyze the UAV model numerically. We firstly discuss the communication issue in the case with two UAVs. Then we compare the successful probabilities among the Single-MDP, Nego-MDP and Double-MDP approach in the multi-UAV model.

In the following examples, we assume that the size of the lattice is $m \times m$, where $m = 15$, and the target is located at $\mathbf{g} = (m-1, m-1)$. The amount of the gas in each UAV is 20 units. We assume each unit time the UAV needs to spend a unit gas. The probability $pr(k,h) = 1/9, \forall k, h = -1, 0, 1$.

A C++ program is written to calculate the successful probabilities and the optimal path. First of all, we calculate $\rho^{(t)}(\mathbf{a}, 0)$ and save them as an array p0[x][y][t]. We then calculate $U^{(t)}(\mathbf{a})$ and $V^{(t)}(\mathbf{x}, \mathbf{y})$ by using (7), (8), (9) and (10), and save them as arrays $U[x][y][t]$ and $V[x_1][y_1][x_2][y_2][t]$ respectively.

3.1 Double-MDP in the two-UAV model

In this subsection, we assume that there are only two UAVs in the lattice. By using Double-MDP algorithm, we are going to detect how the successful probabilities are different between communication or non-communication.

The UAVs start from the sites $(0, posi)$ and $(posi, 0)$, where $posi \le m-1$. Since we assume that the UAVs know each other at the beginning, they can calculate the whole moving path for both UAVs individually because they are using the same algorithm. When they are neighbors, they will communicate with each other to see the status of the other UAV (dead or alive). We generate the number of guns at a certain site according to the probability of having guns at this site to determine whether the UAVs are dead or not. The game is over when all UAVs die or at least one UAV reach the target. By using the sample path technique, we can simulate the whole procedure. Below is the pseudo-code to describe how the UAVs find the path to reach the target.

```
-------------------------------------------------------------------
gas=20; t=0; \\initializing
While(at least one UAV is alive){
  if(both UAVs are alive){
     Obtain the optimal directions for both UAVs according to
     the Double-MDP and move one step;
     Generate the status of guns on the locations of both UAVs;
  }
  else{
     Obtain the optimal direction for the alive UAV according
     to the Single-MDP and move one step;
     Generate the status of guns on the location of the UAV;
  }
   if(At least one UAV reaches the target){
     break;
  }
   else{
      gas++; t++;
  }
}
-------------------------------------------------------------------
```

gun rate = 0.09									
d \ posi	0	1	2	3	4	5	6	7	8
0	0.568	0.543	0.543	0.357	0.370	0.397	0.427	0.455	0.469
1	0.586	0.562	0.562	0.525	0.370	0.397	0.427	0.455	0.469
2	0.586	0.562	0.562	0.529	0.370	0.397	0.427	0.455	0.469
3	0.586	0.562	0.562	0.529	0.370	0.397	0.427	0.455	0.469
4	0.586	0.562	0.562	0.529	0.370	0.397	0.427	0.455	0.469
gun rate = 0.12									
0	0.481	0.455	0.455	0.427	0.227	0.215	0.218	0.233	0.257
1	0.500	0.476	0.476	0.443	0.227	0.215	0.218	0.233	0.257
2	0.500	0.476	0.476	0.447	0.227	0.215	0.218	0.233	0.257
3	0.500	0.476	0.476	0.447	0.227	0.215	0.218	0.233	0.257
4	0.500	0.476	0.476	0.447	0.237	0.218	0.218	0.234	0.256
gun rate = 0.15									
0	0.425	0.400	0.400	0.360	0.176	0.164	0.169	0.181	0.200
1	0.446	0.421	0.421	0.375	0.176	0.164	0.169	0.181	0.200
2	0.446	0.421	0.421	0.379	0.176	0.164	0.169	0.181	0.200
3	0.446	0.421	0.421	0.379	0.176	0.164	0.170	0.182	0.200
4	0.446	0.421	0.421	0.379	0.176	0.165	0.171	0.182	0.201

Table 1. Successful probabilities for symmetric UAVs

We consider two different cases for guns at time t: symmetric and asymmetric cases. In the symmetric case, we assume $p0[x][y][0] = 0.09, 0.12, or\ 0.15$ for all (x, y) where $p0[x][y][0]$ is the probability that there is no gun at site (x, y) at time 0. We call these probabilities **gun rates**. In the asymmetric case, we assume $gun\ rate = 0.03, 0.06, or\ 0.09$ only when $x > y$. $p0[x][y][0] = 0.0$ for the other (x, y). We consider the neighbor meter $d = 0, 1, 2, 3\ or\ 4$. At time 0, the UAVs are located at $(0, posi)$ and $(posi, 0)$, where $posi = 0, 1, \cdots, 8$. Each simulation, we take 30 different seeds, and run 500 replications for each seed, then we calculate the averages and the coefficients of the successful probability. We found that the coefficients are about 5% which is acceptable. Table 1 and 2 are the numerical results.

From the Table 1 and Table 2, we see that the maximum differences of successful probabilities between non-communication ($d = 0$) and communication ($d > 0$) are not significantly different. And only when both UAVs start from the same site (0, 0), the differences reach 2%. The conclusion is that it is not necessary to communicate with each other to know whether another UAV is still alive or not. Specially when they are not neighbors.

gun rate = 0.03									
$d \backslash$ posi	0	1	2	3	4	5	6	7	8
0	0.984	0.984	0.985	0.985	0.985	0.985	0.985	0.978	0.982
1	0.984	0.985	0.985	0.985	0.985	0.985	0.985	0.978	0.982
2	0.985	0.985	0.986	0.986	0.986	0.986	0.986	0.978	0.983
3	0.985	0.986	0.986	0.986	0.986	0.986	0.986	0.980	0.984
4	0.985	0.986	0.986	0.986	0.986	0.986	0.986	0.980	0.984
gun rate = 0.06									
$d \backslash$ posi	0	1	2	3	4	5	6	7	8
0	0.896	0.914	0.919	0.924	0.921	0.913	0.913	0.900	0.929
1	0.897	0.916	0.919	0.924	0.921	0.913	0.913	0.900	0.929
2	0.897	0.915	0.921	0.926	0.921	0.913	0.913	0.900	0.931
3	0.896	0.916	0.921	0.927	0.924	0.916	0.916	0.900	0.932
4	0.896	0.916	0.921	0.927	0.925	0.918	0.918	0.904	0.932
gun rate = 0.09									
$d \backslash$ posi	0	1	2	3	4	5	6	7	8
0	0.795	0.808	0.823	0.823	0.806	0.786	0.786	0.786	0.810
1	0.797	0.811	0.823	0.823	0.806	0.786	0.786	0.786	0.810
2	0.798	0.812	0.826	0.823	0.806	0.786	0.786	0.786	0.810
3	0.797	0.813	0.827	0.830	0.813	0.786	0.786	0.786	0.810
4	0.797	0.813	0.827	0.830	0.814	0.796	0.796	0.796	0.817

Table 2. Successful probabilities for symmetric UAVs

3.2 Multi-UAV

Based on the conclusion of the case with two UAVs, we can develop a scheme to deal with the case of more than two UAVs. We know that the UAVs affect each other only when they get close. As an intuitive algorithm, we let the UAVs group themselves two by two dynamically based on their distance at the beginning.

In the following examples, we consider the symmetric case. We assume $gun\ rate = 0.02, 0.04, \cdots, 0.18$. We also symmetrically launch the UAVs from the sites $(0, posi)$ and $(posi, 0)$, $0 \le posi \le 14$ depending on the number of UAVs. There is at most only one UAV each site. For example, if there is only one UAV, we launch it from (0, 0); if there are 5 UAVs, we launch them from (2, 0), (1, 0), (0, 0), (0, 1) and (0, 2) etc. We assume that the number of UAVs are $1, 3, 5, \cdots, 2m$-1 so that the UAVs are launched symmetrically. For example, if we have 5 UAVs, we group them as three groups: {(2, 0), (1, 0)}, {(0, 0), (0, 1)} and {(0, 2)}. Since all UAVs use the same algorithm in each experiment and they know each other at the beginning, they can figure out the moving directions of all UAVs individually without communication. When one or more UAVs reach the target, we say the UAVs success. Below is the algorithm based on the Double-MDP. The algorithm based on the Nego-MDP is similar except that we need to replace Double-MDP with Nego-MDP in step III.

I. Let $gas = 20$ and $t = 0$;

II. Group UAVs two by two;

III. Obtain the optimal paths for the alive UAV according to the Single-MDP or Double-MDP and move one step;

IV. Generate the status of guns on the locations of UAVs and see if the UAVs will be attacked by the guns;

V. If at least one UAV reaches the target, the UAV is successful and stop; otherwise, let $gas = gas - 1$, $t = t + 1$ and go back to III;

Comparing with the Double-MDP or Nego-MDP algorithm, the Single-MDP algorithm is simpler, in which, all UAVs figure out their moving directions independently based on the Single-MDP algorithm till one of UAVs reaches the target.

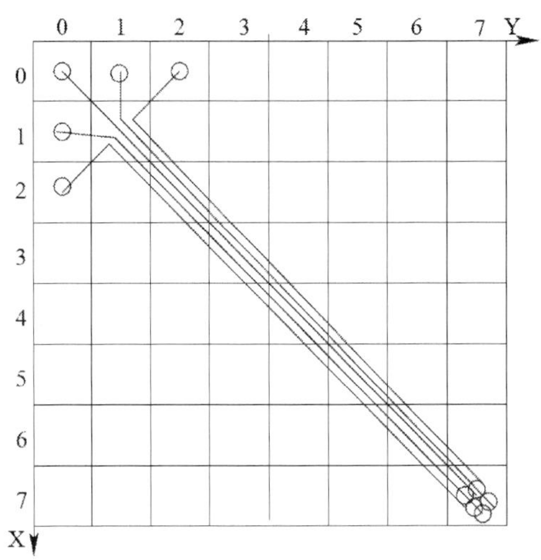

	1	2	3	4	5
0	2,0	1,0	0,0	0,1	0,2
1	1,1	1,1	1,1	1,1	1,1
2	2,2	2,2	2,2	2,2	2,2
3	3,3	3,3	3,3	3,3	3,3
4	4,4	4,4	4,4	4,4	4,4
5	5,5	5,5	5,5	5,5	5,5
6	6,6	6,6	6,6	6,6	6,6
7	7,7	7,7	7,7	7,7	7,7

Figure 5. A sample of moving paths for Single-MDP

Fig. 5, 6 and 7 are the sample paths based on the Single-MDP, Nego-MDP and Double-MDP respectively when the size of lattice is 8×8 and the gas of the UAV is 15 units. We assume there are 5 UAVs and $gun\ rate = 0.10$. In Fig. 5, since the UAVs independently make decision based on their own objective, they have the same local optimal path: (1, 1), (2, 2), (3, 3), (4, 4), (5, 5), (6, 6), (7, 7). In Fig. 6 and 7, UAVs cooperate two by two. UAV 1 and UAV 2 have the different moving paths, so do UAV 3 and UAV4, but UAV 5 still acts independently. On the other hand, Group 1 (UAV 1, 2) and Group 2 (UAV 3, 4) are independent, they have the similar paths. Obviously, if we increase the group size, or let all UAVs cooperate with each other, the successful probability can still be improved. But it will increase the complexity of the algorithm.

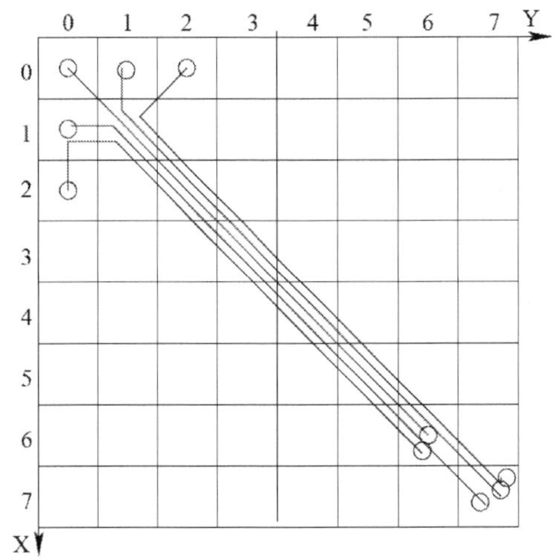

	1	2	3	4	5
0	2,0	1,0	0,0	0,1	0,2
1	1,0	1,1	0,0	1,1	1,1
2	1,1	2,2	1,1	2,2	2,2
3	2,2	3,3	2,2	3,3	3,3
4	3,3	4,4	3,3	4,4	4,4
5	4,4	5,5	4,4	5,5	5,5
6	5,5	6,6	5,5	6,6	6,6
7	6,6	7,7	6,6	7,7	7,7

Figure 6. A sample of moving paths for Nego-MDP

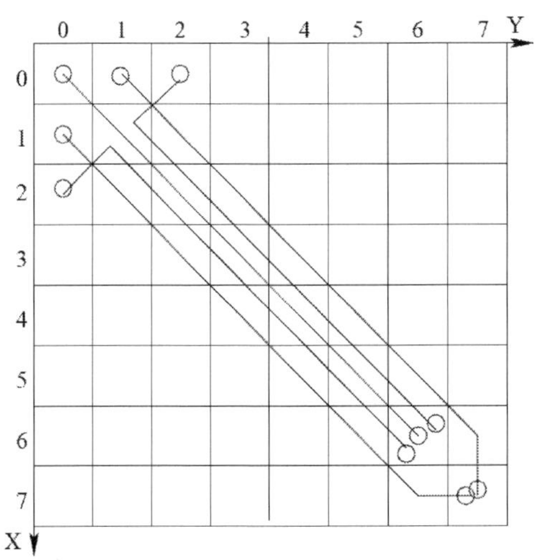

	1	2	3	4	5
0	2,0	1,0	0,0	0,1	0,2
1	1,1	2,1	1,1	1,2	1,1
2	2,2	3,2	2,2	2,3	2,2
3	3,3	4,3	3,3	3,4	3,3
4	4,4	5,4	4,4	4,5	4,4
5	5,5	6,5	5,5	5,6	5,5
6	6,6	7,6	6,6	6,7	6,6
7	6,6	7,7	6,6	7,7	7,7

Figure 7. A sample of moving paths for Double-MDP

Fig. 8 and 9 also show the sample paths based on the Single-MDP and Nego-MDP for the asymmetric case when the gun rate=0.65 in a site (i,j) if $i < j$, and the gun rate= 0.55 if $i \geq j$. We can see the paths are very different between the Single-MDP and Nego-MDP.

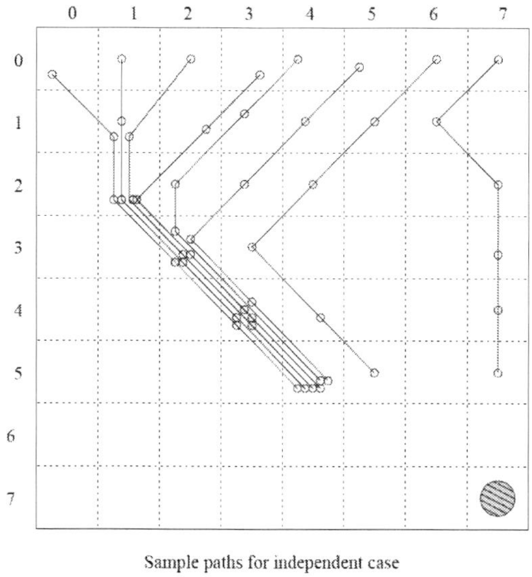

Figure 8. Sample paths for Single-MDP (asymmetric guns)

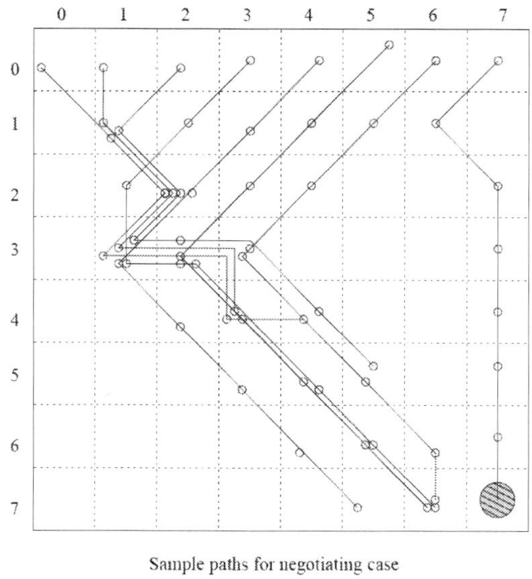

Figure 9. Sample paths for Nego-MDP (asymmetric guns)

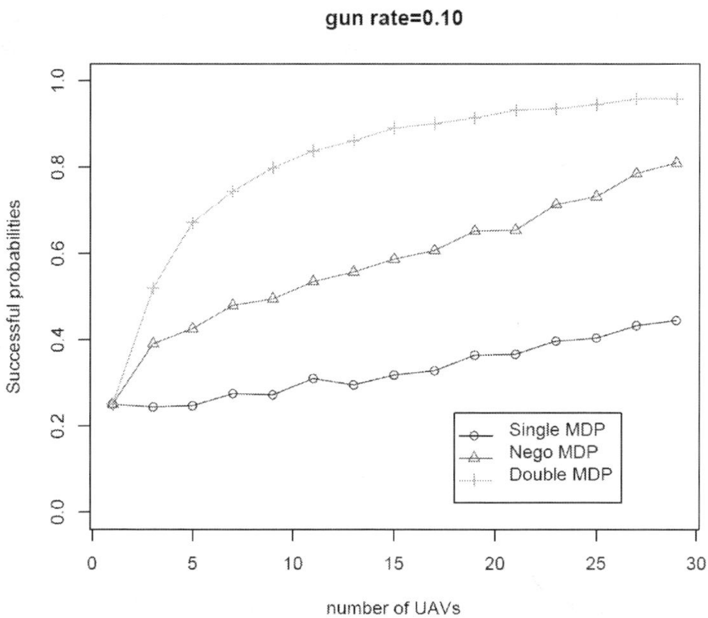

Figure 10. Successful probabilities for different numbers of UAVs

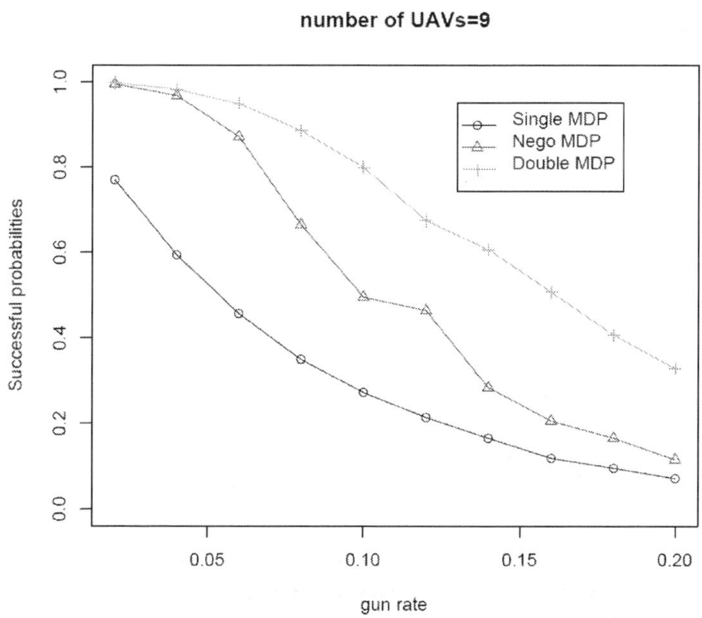

Figure 11. Probabilities of successfully reaching the target for different gun rates

Fig. 10 shows the relation between the successful probability and the number of UAVs. Fig. 11} shows the relation between the gun rate and the probability of successfully reaching the target. Obviously, for all three algorithms, the probabilities of reaching the target are increasing with the number of UAVs, and decreasing with the gun rate. We found that the Double-MDP algorithm is the best algorithm. Nego-MDP is much better than the Single-MDP algorithm. The average probability difference between Double-MDP and Nego-MDP are about 0.2. From Fig. 11, however, we see both Double-MDP and Nego-MDP reach 0.95 when the gun rate is less than 0.05, while the Single-MDP only reach about 0.7. Since Nego-MDP is a one-dimensional MDP based on the Single-MDP algorithm, it is much faster than Double-MDP which is two-dimension MDP (see Fig. 12 and 13). When the gun rate is smaller, Nego-MDP is good enough to use. When the gun rate is larger, however, the successful probability of Nego-MDP is close to the successful probability of Single-MDP. In this case, we recommend to use double MDP. From Fig. 11, we can see that the successful probability for the Double-MDP is increasing concave function of the UAV launching rate, which means that the successful probabilities would not increase significantly when the launching rate is large. This result tells us that it is not necessary to launch so many UAVs in order to reach the target with a certain probability.

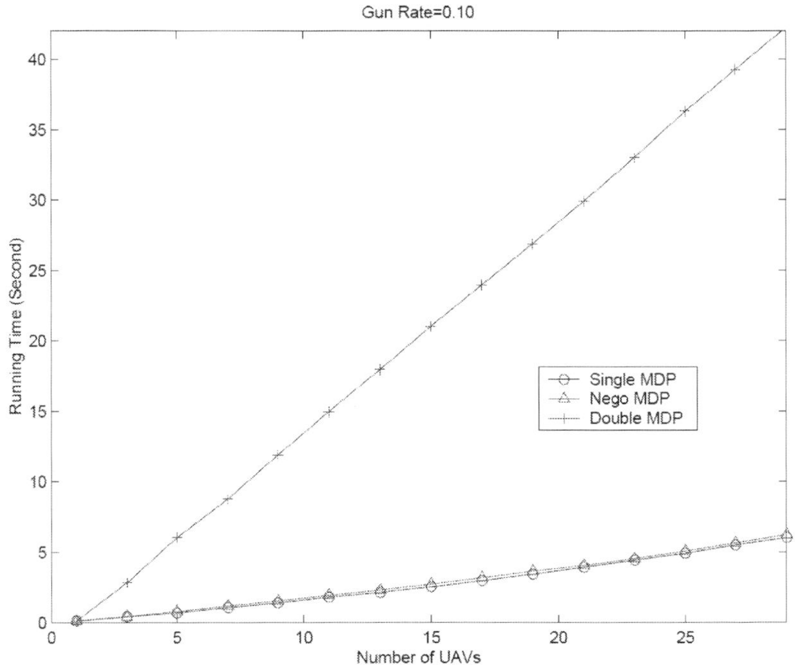

Figure 12. Running time comparison among Single-MDP, Nego-MDP and Double-MDP

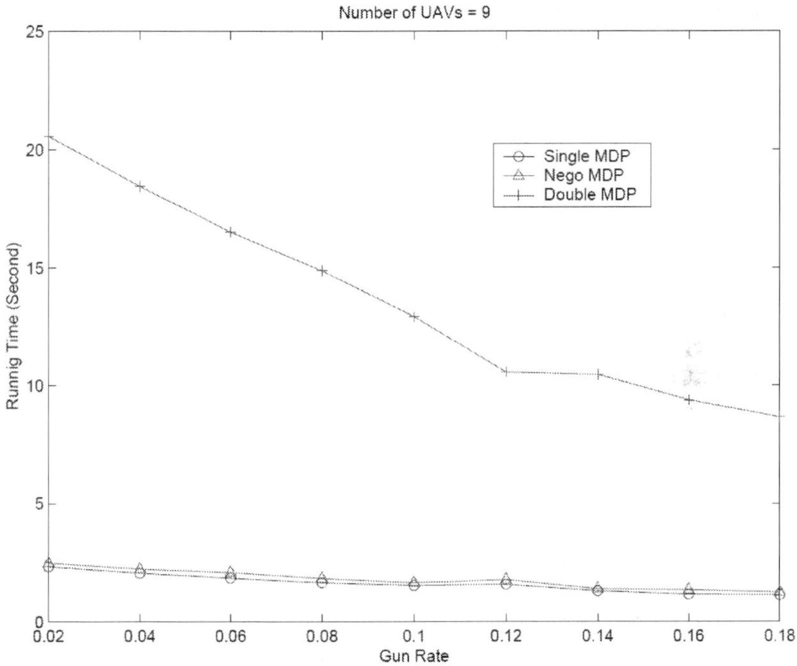

Figure 13. Running time comparison among Single-MDP, Nego-MDP and Double-MDP

4. Summary and Future Work

In this paper, we study a multi-agent based UAV system. Centralized MDP is complicated and unrealistic. In this decentralized model, UAVs have the same global objective which is to reach the target with maximum probability, but each UAV can make decision individually. Although the optimality problem is computational prohibitive, the heuristic results give us very important managerial insights. Based on the Nego-MDP and the Double-MDP algorithms, UAVs group themselves two by two dynamically, and find out the moving directions very effectively. Obviously, increasing the group size can improve the successful probability, but in the mean time, the algorithm will become more complicated.
The precise information on guns are very important for UAVs to reach the target effectively. So far, we only consider that the guns randomly walk on the lattice. That will be worthwhile if we can update the gun information dynamically. Further more, it will be interesting if UAVs can also attack guns. In that case, we need to introduce the game theory to figure out the best strategy for both the guns and the UAVs.

5. Acknowledgements

This work was funded in part by NSF Grants # 0075462, 0122173, 0325168, AFRL Contract # F30602-99-2-0525, and by UMAC Grant # RG016/02-03S.

6. References

Atkins, E.M.; Durfee, E.H. & Shin, K.G. (1996). Plan development using local probabilistic models, *Proceedings of the Conference on Uncertainty in Artificial Intelligence*, pp. 49-56

Barto, A.G.; Bradtke, S.J. & Singh, S.P. (1991). *Real-time learning and control using asynchronous dynamic programming*, Technical Report 91-57, University of Massachusetts, Amherst, Massachusetts

Dean, T.; Kaelbling, L.P.; Kirman, J. & Nicholson, A. (1993). Planning with deadlines in stochastic domains, *Proceedings of AAAI*, pp.574-579

Ginsberg, M.L. (1989). Universal planning: An (almost) universally bad idea. *AI Magazine*, Vol. 10, No. 4

Givan, R.; Leach, S.M. & Dean, T. (1997). *Bounded parameter Markov decision processes*, Vol. 1348, pp. 234-246, Springer

Howard, R.A. (1960). a. Dynamic programming and Markov processes, b. The Technology Press of The Massachusetts Institute of Technology and John Wiley & Sons, Inc.

Kaelbling, L.P.; Littman, M.L. & Cassandra, A.R. (1998). Planning and acting in partially observable stochastic domains, *Artificial Intelligence* 101

Kushmerick, N.; Hanks, S. & Weld, D. (1994). An algorithm for probabilistic least-commitment planning, *Proceeding of AAAI*, pp. 1073-1078

Littman, M.L.; Dean, T.L. & Kaelbling, L.P. (1995). On the complexity of solving Markov decision problems, *Proceedings of UAI-95*

Lovejoy, W.S. (1991). A survey of algorithmic methods for partially observed Markov decision processes, *Annals of Operations Research*, Vol. 28, pp. 47-65

McLain, T.W.; Chandler, P.R. & Pachter, M. (2000). A decomposition strategy for optimal coordination of unmanned air vehicles, *Proceedings of the American Control Conference*, pp. 369-373, Chicago

Meuleau, N.; Hauskrecht, M.; Kim, K.E.; Peshkin, L.; Kaelbling, L.P.; Dean, T. & Boutilier G. (1998). Solving very large weakly coupled Markov decision processes, *Proceedings of the Fifteenth National Conference on Artificial Intelligence*

Monahan, G.E. (1982). A survey of partially observable Markov decision processes: Theory, models, and algorithms, *Management Science*, Vol. 28, pp. 1-16

Musliner, D.J.; Durfee, E.H. & Shin, K.G. (1995). World modeling for the dynamic construction of real-time control plans, *Artificial Intelligence*, Vol. 74, pp. 83-127

Schoppers, M.J. (1987). Universal plans for reactive robots in unpredictable environments, *Proceeding of International Joint Conference on Artificial Intelligence*, pp. 1039-1046

Sutton, R.S. (1990). Integrated architectures for learning, planning and reacting based on approximating dynamic programming, *Proceedings of the seventh International Conference on Machine Learning (Austin, Texas)*, Morgan Kaufmann

Watkins, C.J. & Dayan, P. (1992). Q-learning, *Machine Learning*, Vol. 8, pp. 279-292

Xuan, P.; Lesser, V. & Zilberstein, S. (1999). Communication in multi-agent Markov decision processes, *Technical report*, University of Massachusetts at Amherst

21

Forced Landing Technologies for Unmanned Aerial Vehicles: Towards Safer Operations

Dr Luis Mejias[1], Dr Daniel Fitzgerald[2], Pillar Eng[1] and Xi Liu[1]
Australian Research Centre for Aerospace Automation[3]
Australia

1. Abstract

While using unmanned systems in combat is not new, what will be new in the foreseeable future is how such systems are used and integrated in the civilian space. The potential use of Unmanned Aerial Vehicles in civil and commercial applications is becoming a fact, and is receiving considerable attention by industry and the research community. The majority of Unmanned Aerial Vehicles performing civilian tasks are restricted to flying only in segregated space, and not within the National Airspace. The areas that UAVs are restricted to flying in are typically not above populated areas, which in turn are one of the areas most useful for civilian applications. The reasoning behind the current restrictions is mainly due to the fact that current UAV technologies are not able to demonstrate an Equivalent Level of Safety to manned aircraft, particularly in the case of an engine failure which would require an *emergency* or *forced landing.*

This chapter will preset and guide the reader through a number of developments that would facilitate the integration of UAVs into the National Airspace. Algorithms for UAV Sense-and-Avoid and Force Landings are recognized as two major enabling technologies that will allow the integration of UAVs in the civilian airspace.

The following sections will describe some of the techniques that are currently being tested at the Australian Research Centre for Aerospace Automation (ARCAA), which places emphasis on the detection of candidate landing sites using computer vision, the planning of the descent path/trajectory for the UAV, and the decision making process behind the selection of the final landing site.

2. Introduction

The team at the Australian Research Centre for Aerospace Automation (ARCAA) has been researching UAV systems that aims to overcome many of the current impediments facing the widespread integration of UAVs into civillian airspace. One of these impediments that the group identified in 2003 was how to allow a UAV to perform an emergency landing.

---

[1] Dr Luis Mejias, Pillar Eng and Xi Liu are with the Queensland University of Technology
[2] Dr Daniel Fitzgerald is with the ICT centre CSIRO
[3] ARCAA is a joint venture between Queensland University of Technology and CSIRO

An emergency or forced landing (in the case of unpowered flight), is where the aircraft is required to perform an unplanned landing due to the occurrence of some onboard emergency, such as an engine failure. This capability is an inherent component for benchmarking the performance of the manned aviation industry, therefore the group identified this as a key impediment to overcome in order to allow UAVs to operate over populated areas in civilian airspace (Fitzgerald, Walker et al. 2005; Fitzgerald 2007). Hence, it is believed that UAVs must therefore be provided with the ability to safely terminate the flight through a range of emergency scenarios. A UAV plummeting uncontrollably into the middle of a busy freeway or a school yard is a risk that the public will be unwilling to except. In this context, a UAV emergency landing system will be an important component towards enabling routine missions in civilian environments.

To date, no commercial system is available that allows a UAV to decide autonomously on the safest area to land in an unknown environment. Safety systems currently available to UAVs only allow the aircraft to fly towards a pre-defined safe landing area from a database of known safe landing locations. However, these systems must be preprogrammed with up-to-date information, thus requiring a continuous communications link between a human operator and the air vehicle to ensure that the latter will not attempt to land at an unsuitable location.

An alternative would be to have a system onboard the UAV that can process information in a similar way to a human pilot in emergency situations that require the aircraft to land. Therefore, the objective of this research is to develop an onboard capability that allows the UAV to select a suitable landing site, and then manoeuvre the UAV autonomously to land at this location. If this functionality is realised, it will bring UAVs one step closer to flying in civilian airspace above populated areas.

The research described in (Fitzgerald 2007) has reduced the technical risks for a vision-based emergency landing system and hence in the past year, a number of research programs have begun at ARCAA to complement this research. It has now been proposed that a complete prototype system suitable for flight trials be developed. A range of flight test scenarios will be evaluated on the prototype system (encompassing a range of altitudes and different terrain), and will be conducted with the relevant approvals from the Civil Aviation Safety Authority (CASA) of Australia.

This chapter will describe the different research programs and results to date, and how these will be integrated to form a complete prototype system ready for flight testing.

These research programs can be classified into the three broad areas of:

- Visual identification and classification of UAV forced landing sites;
- Trajectory planning and tracking for autonomous aircraft forced landings; and
- Multilevel decision-making for high-level reasoning during the descent

These form the basis of this chapter and their use in developing a real-time implementation and system for flight testing.

3. Machine Vision Landing Site Selection

Over a 3 year period a computer vision based approach has been developed and optimised. This approach, which mimics human processes, identifies emergency landing sites that are obstacle free. More specifically, the landing sites are chosen based on their size, shape and slope as well as their surface type classification. Subsequent algorithms have been

developed that allow automatic classification of the candidate landing site's surface (based on back propagation neural networks). At the time of writing we are implementing the approach described in this section in real time and are conducting flight trials of a small UAV to further demonstrate this approach.
The remainder of this section will describe this vision-based techniques to find potential UAV landings sites from aerial imagery. For a detailed description of this work please refer to (Fitzgerald 2007).

3.1 Candidate Landing Site Selection Framework

The aim for the selection of candidate landing sites for the UAV forced landing problem is to locate regions from aerial imagery that were of similar texture, free of obstacles and large enough to land in. Regions that met the criteria would be further classified according to their surface type to aid in the choice of the most suitable landing region.

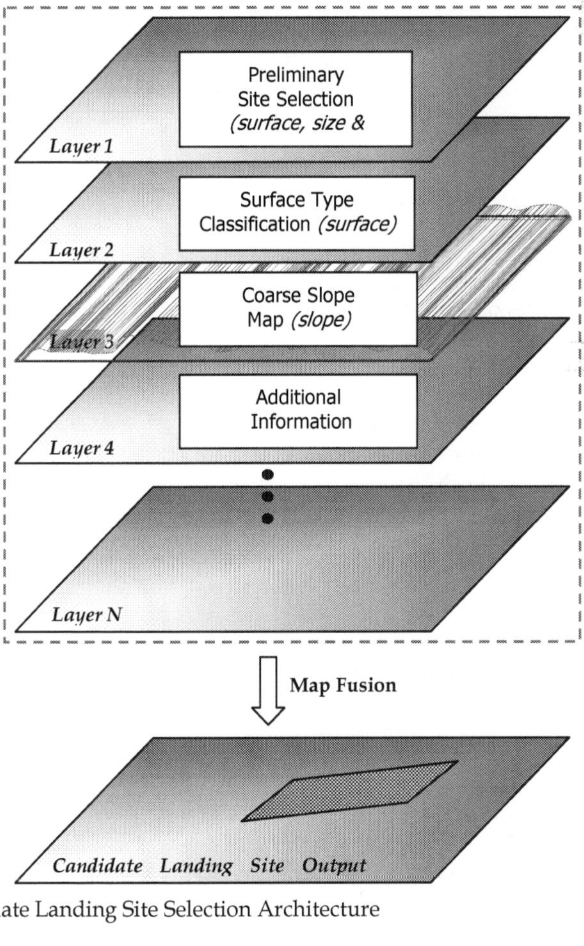

Figure 1. Candidate Landing Site Selection Architecture

This criterion for UAV landing site selection is based on the elements that a human pilot considers during a forced landing. Section 4 describes a number of factors that a human pilot considers when selecting the most appropriate landing site - Size; Shape; Slope; Surface; Surroundings; and S(c)ivilisation. These factors are more commonly known as the six S' and are cues used by pilots when selecting a landing site during a forced landing. The approach that has been developed for selecting a candidate landing site for a UAV forced landing has been approached with these factors.

With these human factors in mind, Figure 1 illustrates the architecture of the approach developed to locate candidate UAV forced landing sites.

This architecture incorporates many of the elements that human pilots use when deciding upon the most suitable forced landing site. It also leaves open the ability to add additional layers of information at a later stage if information is available. Information from each layer can be fused together by using a number of linguistic or fuzzy rules. For example areas free of objects (identified in the Preliminary Site Selection layer), with a *flat* slope (Coarse Slope Map layer) and classified as *grass* (Surface Type Classification layer) would be given a *very safe* output map value. This final output map has similarities to the T-Hazard map in (Howard & Seraji, 2001, 2004) where a number of data sources are fused together with linguistic rules.

The remaining indicators used by human pilots in a forced landing situation, that are not considered in the architecture of Figure 1 are civilisation, wind and surroundings.

Civilisation is not a valid indicator for choosing candidate landing sites for the UAV forced landing problem, as the civilisation criterion is an indication of proximity to life critical services for a human pilot. It is important for a human to be as close as is safely possible to populated areas, so that any medical attention can be administered as quickly as possible after a forced landing. A UAV does not have this problem as there are no humans on board. The concern for a UAV forced landing is to avoid people entirely. Here the best option is to head for large areas free of obstacles.

Wind is the final additional indictor included by (CASA, 2001) that is not covered by the traditional six S' used by pilots when selecting a suitable landing site during a forced landing. Knowledge of wind direction is important for a pilot's decision on the final approach direction (for a conventional landing) and also to determine the maximum range the aircraft is able to glide to. A tailwind can increase the distance an aircraft is able to glide, however a head wind will reduce this range. Wind will be discussed in Section 5.

Errors in the maximum glide range estimate of the UAV introduced by wind has been mitigated by introducing a buffer between the theoretical maximum glide distance (based on the particular UAV) and the range of landing sites included in the output of the coarse slope map generation discussed next. Any decrease in the ranges to potential landing sites considered can be used. A figure of 85 % of the theoretical maximum range distance has been used for this research. Additionally, a decreasing scale can also be adopted for the inclusion of landing sites as you move away from the UAV's current position. The result is that suitable areas close by can be given more weight that areas towards the extremities of the UAV's glide range.

Finally, surroundings refer to the identification of objects in the image such as trees, buildings and powerlines. For the human forced landing case, objects such as powerlines or fences often cannot be seen, but are inferred by other objects in the image. For example, the presence of a house or building is likely to indicate a power line nearby, just as a road is likely to indicate the presence of fences near by. Humans use this knowledge in the choice for an appropriate approach path to the final landing site.

The elements discussed above are part of the multilevel decision making and determining the most appropriate approach path to the chosen landing site. These areas of research are covered in Sections 4 and 5 of this chapter, respectively.

The information from the 3 (or more) layers can be fused together by a set of linguitic rules or by some other technique resulting in a weighted map of potential landing sites. A higher level decision making process (described in Section 4) will take these information to decide the best landing site.

3.1.1 Preliminary Site Selection

The preliminary site selection layer is responsible for extracting regions from the aerial image that are large enough for a UAV forced landing and that do not contain obstacles. The approach extracts these areas directly, without the need for image segmentation. This results in a process that is fast and suits the forced landing application specifically.

The approach was broken down into two steps - a region sectioning and a geometric acceptance. These techniques now will be described briefly.

Region Sectioning Phase

The region sectioning phase is responsible for finding areas in the image that are of similar texture and that are free of objects. The approach uses two measures that are augmented together to create a map of suitable areas.

The first measure uses a well known technique, the Canny edge detector [Canny, 1986] on the entire image, followed by a line expansion algorithm. It was observed then further assumed that regions in the image that contained no edges corresponded to areas that contained no obstacles. Additionally, since boundaries between different objects - for instance grass and bitumen - usually have a distinct border, areas with no edges should corresponded to areas of similar texture (ie: the same object, for example a grass field).

This assumption was made after studying a number of edge gradient maps similar to the one shown in Figure 2. Peaks in the plot correspond to edges in the image. The figure shows clear evidence that the areas free of obstacles in Test Image 1 (refer top Figure 4) correspond to areas with a low number of edges. Subsequent observations on different images were made to verify this assumption. Additionally, clear borders (corresponding the edges in the Canny edge detection image) could be observed between different regions in the image which is also desired for the forced landing problem.

A line expansion algorithm immediately follows the edge detection, and involves the examination of the pixels of all edges found. For each pixel found, the algorithm inspects the surrounding pixels within a certain search radius. If another edge pixel is found, the algorithm will set all pixels within this radius to a "1". This is shown below in Figure 3.

This calculation is performed by knowing how much distance each pixel equates to on the ground (pixel ground resolution). Based on a number of assumptions the pixel ground resolution for the image can be determined from:

- Height above ground (for example: 2500 ft; approx 762 meters);
- Image dimensions (for example: 720 x 576 pixels);
- Camera viewing angle (for example: 35.0 x 26.1 degrees).

The pixel resolution values are used at different stages of the algorithm to determine measures such as the landing site pixel dimensions and line expansion radius values. These can be altered according to the landing requirements of the UAV - dependant on the UAV class.

The edge detection measuring layer outputs a layer map that contains a 1 for every pixel location corresponding to an edge. The following figure shows a number of images with the edge detection measure shown (Canny edge detection plus the line expansion algorithm).

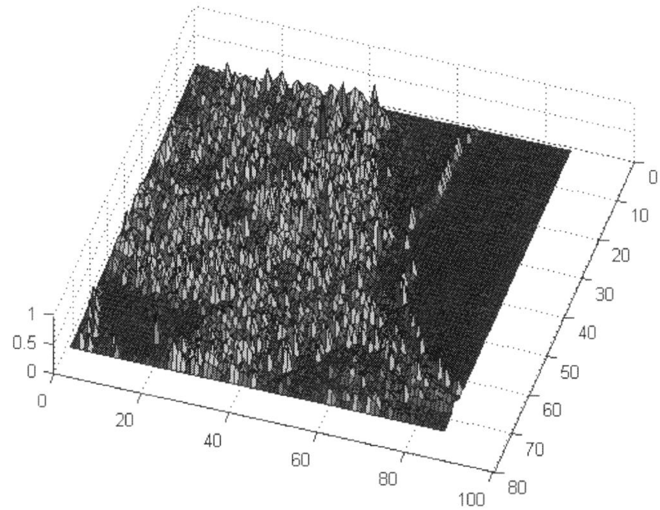

Figure 2. Edge Gradient Map of Test Image 1

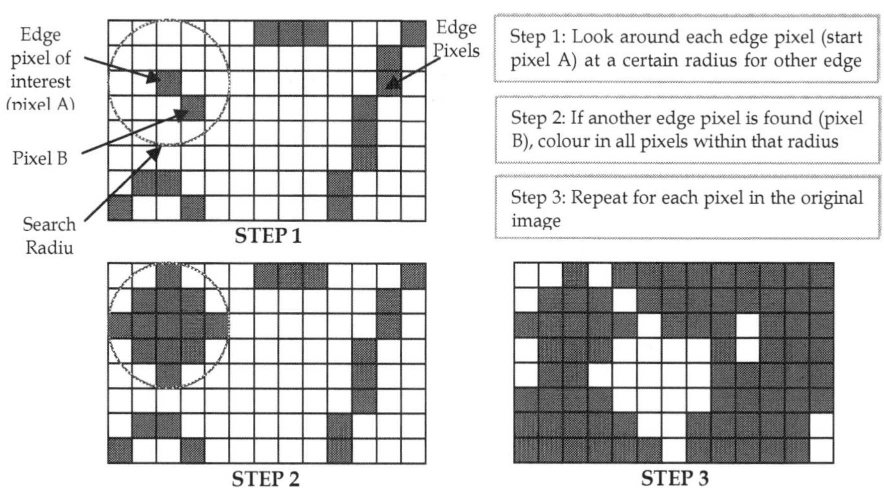

Figure 3. Line Expansion Algorithm

This final step of the algorithm ensures a suitable boundary is placed between obstacles detected and potentially safe areas to land in. The search radius size in this algorithm can be altered depending on the UAV's height above ground level, to maintain this suitable safety zone.

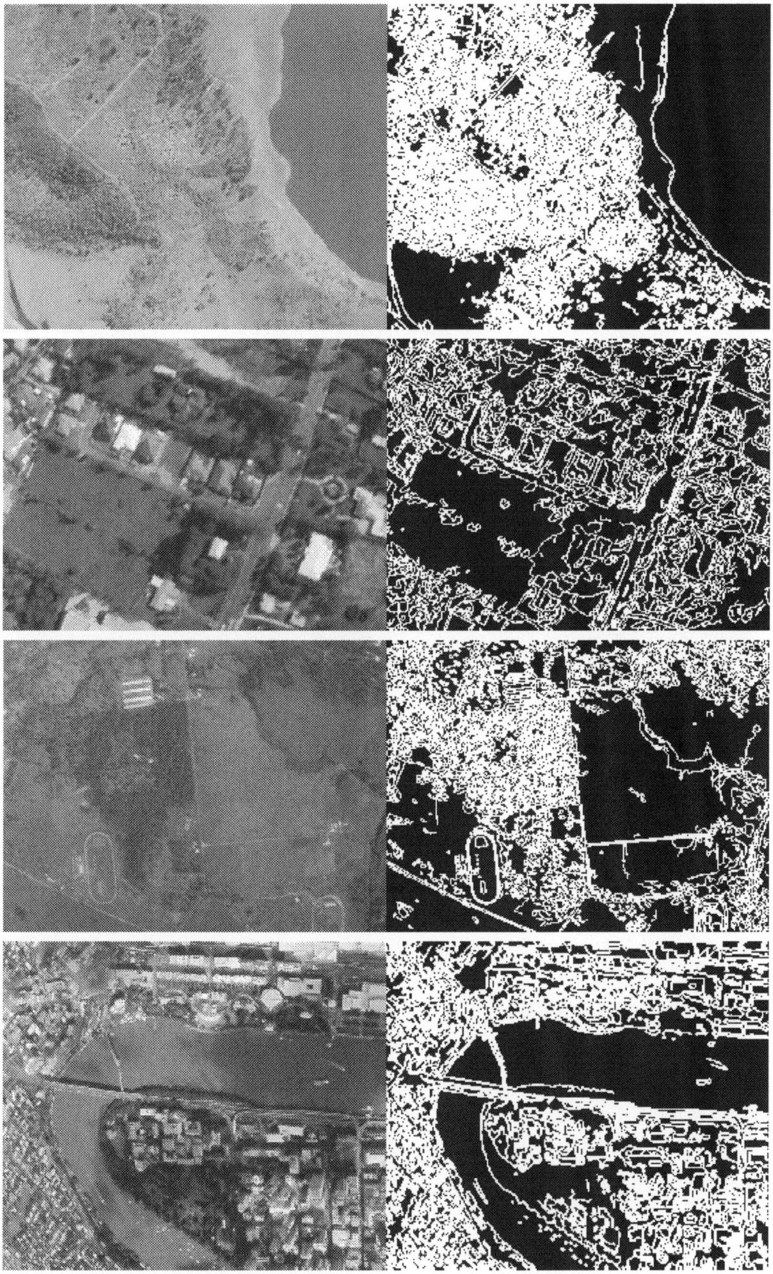

Figure 4. Edge Detection Measure Output for a Number of Test Images

The second measure was an intensity measure that was required to help distinguish between man-made and natural objects in the image. It was observed through experimentation that man-made regions in the images often exhibited the highest intensity values in the image. These objects included roof tops, cars and sheds - anything that had a high reflectance. Natural surfaces such as grass and trees generally were of lower intensity.
The intensity measure was formulated by dividing the intensity image space into 10 equal partitions and labelling any pixel in the highest valued intensity region a 1 and the other lower intensity valued pixels a value of 0. Any pixel with a value of 1 was considered to be a man-made region and assigned a *very unsafe* value and would not be considered as a potential UAV forced landing site. The pixels with intensity values in the top 10% of the intensity scale were removed from landing site consideration as they were assumed to be man made objects.
The following figure (Figure 5) shows some of the test image dataset with the made-man areas identified by the intensity measure shown overlaid on the original images. The intensity measure images are shown also for clarity.
In summary, the edge detection measure is used for the identification of objects in the image based on the assumption that edges occur between boundaries of objects. This assumption is valid under the condition that there is sufficient contrast between objects or regions in the image and that the spatial resolution is large enough. It follows that as the height above the ground decreases, the better the algorithm is at defining boundaries between objects and detecting smaller objects.
The intensity measure aims to eliminate some of the man made objects in the scene. These usually correspond to white building or roof tops, as these are most likely to reflect the sun the most to the airborne sensor. It is important to note that the algorithm does not perform any intensity stretching, that is, intensities are not scaled. This means that the algorithm remains valid over differing terrain, where a large number of man-made objects can be eliminated in dense urban environments or only a few objects eliminated in rural regions. Note also that the algorithm appears to work well accross the greatly differing altitudes used in our testing. One disadvatage, is that on more overcast days or at earlier or later times of the day, some objects may not be detected.
The two measures were combined by performing a bit OR across each individual pixel in the two measure maps. The resultant output map pixels will have a "1" in every location that had a "1" in either the edge detection measure or the intensity measure map - *unsafe* pixels will have a value of "1" and *safe* pixels will have a value of "0".

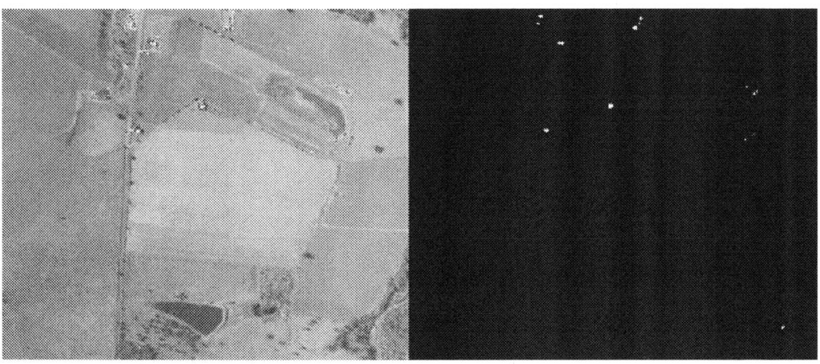

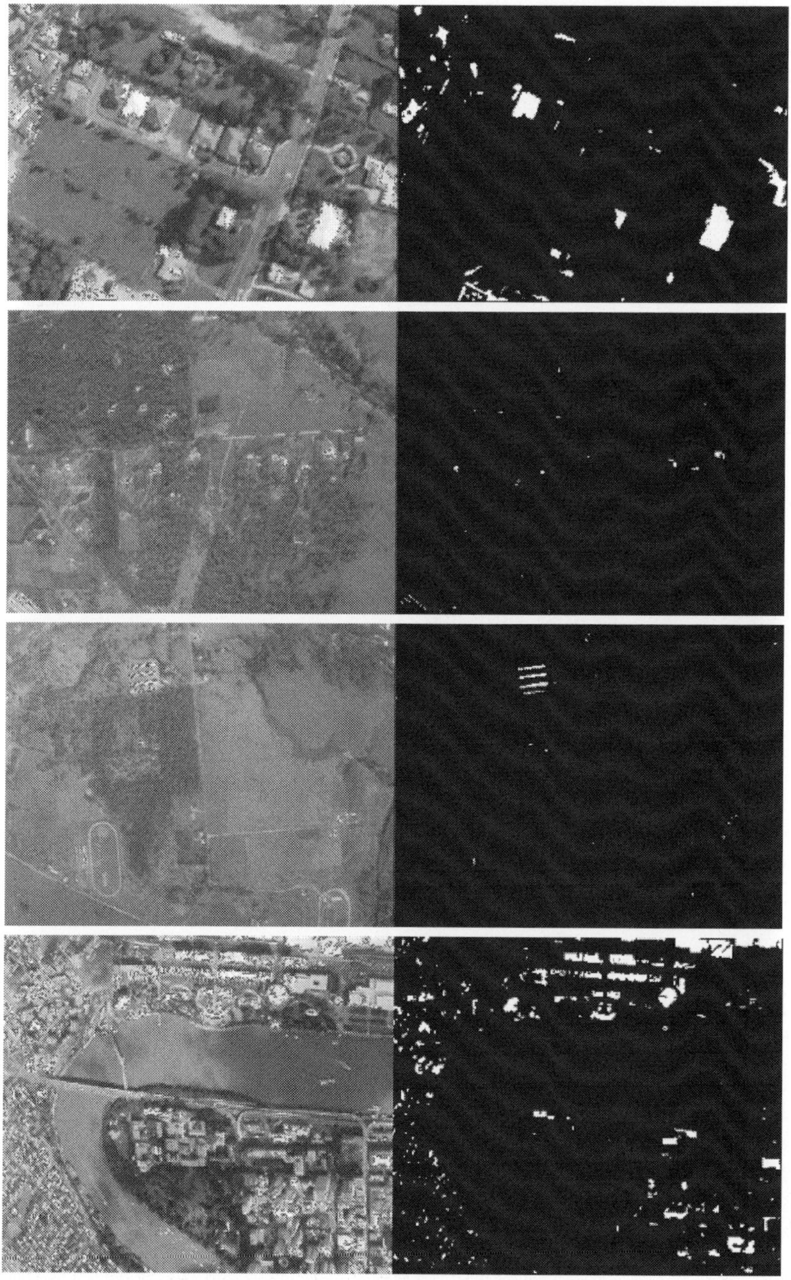

Figure 5. Intensity Measure Result for a Number of Test Images

The Geometric Acceptance phase then takes this map and looks for areas of appropriate size and shape for a UAV forced landing that contain pixels with values only equal to "0".

Geometric Acceptance Phase

The geometric acceptance phase locates landing areas of a given size and shape suitable for a UAV forced landing. All areas that have a suitable size and shape will be then labelled as candidate landing sites and will be considered in the next phase of surface type classification.

The algorithm to perform the geometric test in the geometric acceptance phase involves the use of four masks. The masks are rectangular in shape and scalable (size is determined by the pixel resolution calculation). They are also rotated in a number of orientations, catering for approaches to the candidate landing site from different directions. The four masks (labelled A-D) are shown in the figure below.

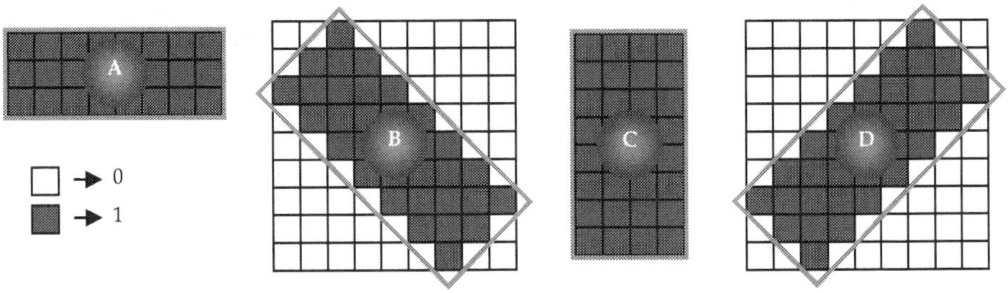

Figure 6. Landing Site Matrix Mask Definitions

Additional mask rotations could have been used, however four masks were chosen, as it was observed that the four masks gave adequate coverage. The example outputs shown in Figure 7 illustrates this point, with the output of the Preliminary Site Selection layer map shown for each test image. As can be seen the masks perform well in extracting every possible area clear of obstacles available. Additionally, the use of only four masks keeps the processing time to a minimum.

The masks are individually moved over the output map from the combination of the edge detection and intensity measure described in the previous section. The image area that the mask passes over is scanned to determine whether or not the area contains entirely *safe* pixels. If all pixels in the area are SAFE, then the area is marked as a candidate landing location.

To perform the scanning check, each mask is represented by a matrix with a "1", indicating that it is part of the mask and a "0" representing that it is not part of the mask (refer Figure 6). For instance, mask B, contains both "0" and "1" elements - the elements that are marked as "1" are members of the mask. For a particular region being tested, if all pixels under the pixels in the mask containing a "1" are *safe* pixels (value of "0") then the pixels in this region on the final Preliminary Site Selection map are set to a *safe*value. This process is repeated across the entire image.

Figure 7 shows some example outputs of the Preliminary Site Selection algorithm. *Safe* areas are shown in white, and the *unsafe* areas in black. Results based on analysis of hundreds of images are summarised at the end of this chapter.

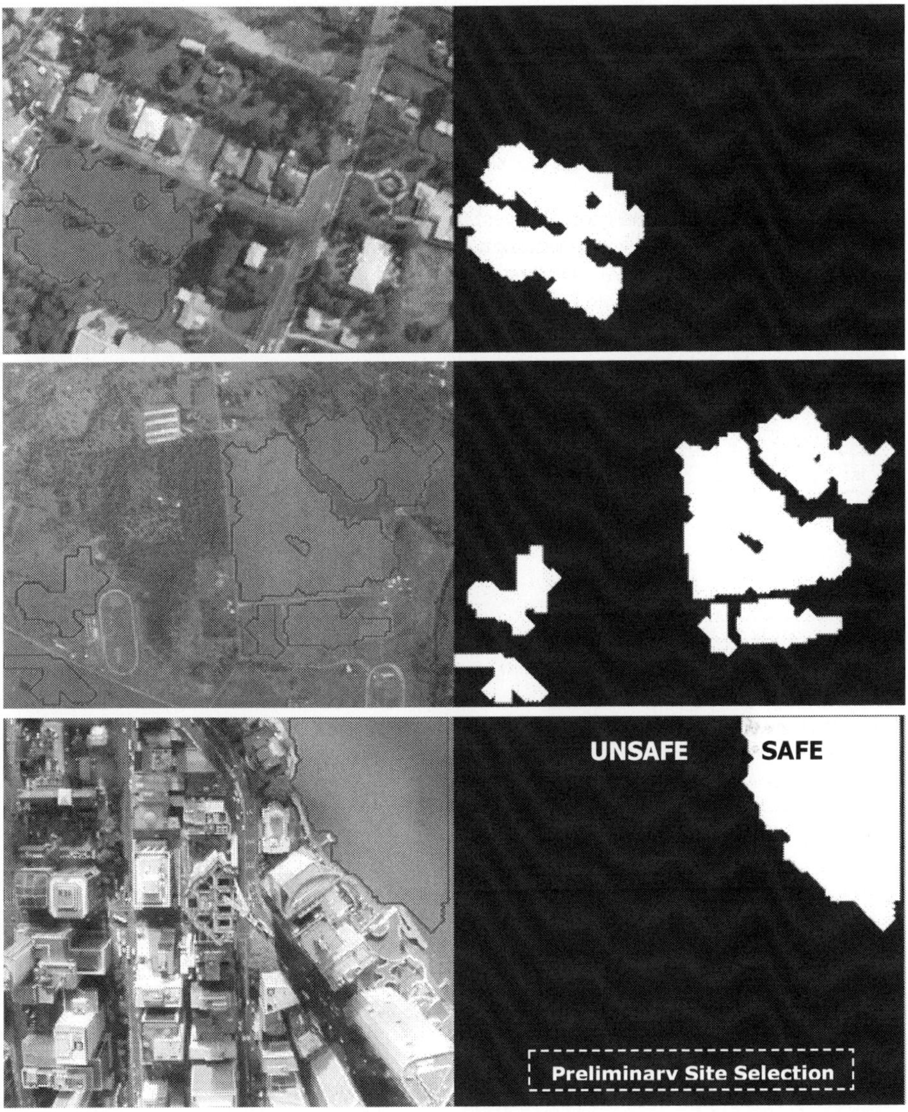

Figure 7. Preliminary Site Selection Output Map Examples

3.1.2 Surface Type Classification

The machine vision techniques presented in Section 3.1.1, locate candidate landing sites based on their size, shape and slope. The surface type classification layer is then concerned with labelling each candidate landing site with the surface type so the best landing site can

be chosen from those available. This problem falls into the well studied areas of texture classification, pattern classification and the field of automated image indexing.
Our novel controbutions in this area fall in the areas of developing generic semi-automated approaches for developing classification systems. In particular, we automate the process of splitting our dataset into subclasses from generic classes (for example, *grass, trees, water*) to improve classification accuracy of the system. This approach is good as a human operator can tell easily the difference between class samples such as *grass* or *water*, etc, given the contextual information of the whole seen, however can find it difficult to decide on a number of suitable sub-classes and in turn which samples belong to a particular subclass. Examples of subclasses that a human may find it hard to distinguish between for say *grass*, could be *gree grass* and *brown grass*.
Another important contribution lies in finding a suitable selection of input features for the classification system. We present a method for finding optimal features (the input feature optimisation algorithm - IFO), where the feature space is automatically studied for nearly 90 different features from the literature for each subclass, and the algorithm determines which features are good at distinguishing between subclasses.
Using these algorithms with a back-propogation neural network classifier yielded improvements over our original methods tested by around 10%. Full details of the approach can be found in (D.L. Fitzgerald, 2007) and detailed results to date are given at the end of this chapter.

3.1.3 Coarse Slope Estimation

This section addresses the gathering of a coarse slop estimate for the immediate area of interest for the UAV forced landing, and does fall outside the machine vision area, but is included in this section for completeness.
The incorporation of slope into the selection of a suitable forced landing area is a fundamental component of the forced landing process. Slope is one of the key indicators that a human pilot uses to select a forced landing site. A human pilot uses depth of perception and other visual cues to determine the slope of the terrain below. Inferring slope from vision based methods has been well studied in the literature however a robust solution to the problem is still yet to be found. Some of the problems include occlusions of features from shadows and other objects, effects of differing lighting conditions and also problems with reconstruction of 3D from higher altitudes. This particular research is tackling the problem of candidate landing site selection from relatively high altitudes, thus methods based solely on machine vision are seen to not offer the robustness in the design for the aforementioned reasons.
Coarse slope estimation in our sense, refers to finding the slope of the ground in the surrounding area. It is *coarse*, as we are talking about freely available height DEM (digital elevation map) data at a resolution of 3 arc-seconds (latitude and longitude). This corresponds to height elevation readings at approximately 90m spacings. The data that was sourced to demonstrate this technique was from the National Geospatial-Intelligence Agency (NGA) and NASA Shuttle Radar Topography Mission (SRTM). The heights are referenced to the WGS84 geoid which is useful as GPS altitude is referenced to this same geoid. The idea of estimating the slope is to discount areas in the surrounding region.
The augmentation of the vision payload with the GPS and DEM data will provide a number of useful pieces of information that will assist in the selection of a candidate landing site for

a UAV forced landing. The following is a summary of the information that has been identified as useful for this problem:

- Calculating pixel resolution;
- Calculating the coarse slope map for each image frame; and
- Defining a slope map for the global operational area.

The knowledge of the pixel resolution in the image is important, because this information is used to define and construct the landing site dimensions for finding sites large enough to land in. If a particular UAV requires 20x100m to land in for example, then this must be converted into the appropriate pixel dimensions for the Preliminary Site Selection algorithm as described above.

The information from the coarse slope map will contain measures of slope for the pixels in each image frame and will be able to be fused directly to the other layers of information to provide a final output map of candidate landing sites.

Finally, a slope map for the global operational area will allow higher order navigation processes to guide the UAV toward areas within the glide range where the slope of the terrain is suitable. These areas may be outside the field of view of the vision sensor and so this slope information plays a vital role in assisting mission planning type decisions for navigation. This information becomes increasingly important if no suitable candidate landing sites exist below the UAV's current position, with the knowledge that the UAV only has a finite time before it reaches the ground.

The full details of this research are presented in (D.L. Fitzgerald, 2007), however we will briefly run over the basical concepts.

Calculating Pixel Resolution

Pixel resolution is based on the field of view of the camera, the number of pixels in the camera's image plane and the height above the ground being viewed. It follows that for this problem some averaging of the terrain heights within the field of view of the camera is required to generate the best estimate of pixel resolution for the complete image frame.

Note the following figure.

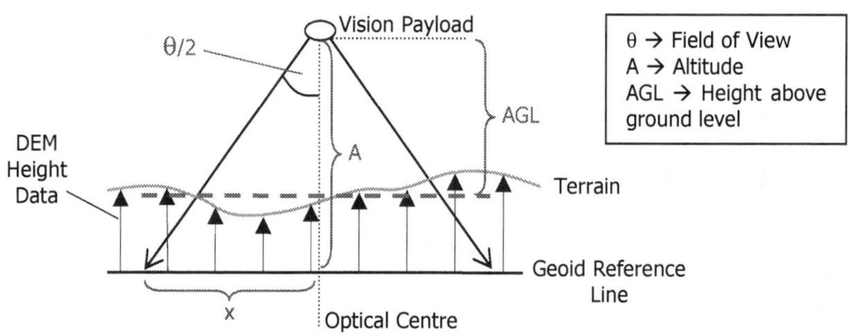

Figure 8. Finding the Average Height Above Ground

The first step is to determine which of the DEM data points are going to be included in the calculation for the average height above ground level (AGL). The points chosen will be those that fall within the view of the camera, extended to the geoid reference line. On the above figure, this would include the DEM Height Data arrows within the x range indicated on both sides of the optical centre of the vision payload.

The pixel resolution can then be determined from this estimate of the height above ground. This calculation uses the distance spanned across the image (based on the AGL) divided by the number of pixels across the image. This final calculation used to determine the pixel resolution is shown in (1).

$$\Delta Px = \frac{(A - H_{av})\tan\left(\frac{\theta}{2}\right)}{\frac{NumPixels}{2}} = \frac{2(A - H_{av})\tan\left(\frac{\theta}{2}\right)}{NumPixels} \tag{1}$$

Calculating the coarse slope map for each image frame

The purpose of the coarse slope map is to provide measures of slope for the pixels in each the image frame that can be fused directly with the other layers of information. The problem is to solve where each DEM data point is projected to on the image plane. When laid out appropriately as in Figure 7, it can be seen that the projection problem involves the solution to 2 similar triangles.

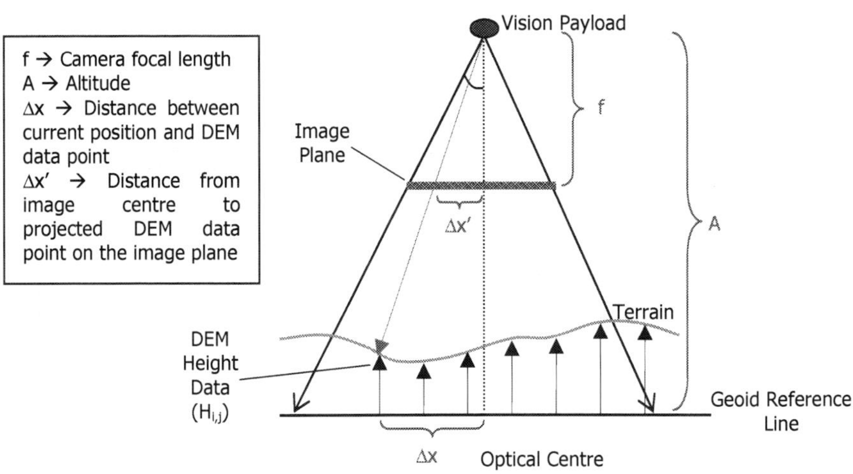

Figure 9. Solving the Terrain Projection Problem

The position of the DEM data point as projected onto the image plane is solved in the x and y directions by the following equations:

$$\Delta x' = f.\frac{\Delta x}{A - H_{i,j}} \tag{2}$$

$$\Delta y' = f.\frac{\Delta y}{A - H_{i,j}} \tag{3}$$

The Δx and Δy positions are added or subtracted from the position of the centre of the image (optical centre) to obtain the location for each height data point. The distribution of these points is totally dependent on the contour of the terrain below. DEM data points are continued to be projected onto the image plane and beyond the boundary, such that all pixels in the image are able to be assigned a slope value as discussed below.

The next step is to determine the slope measures that lie between the data points. This involves looking at the sets of 4 DEM data points and determining the maximum slope between them. Once this is done all pixels that lie between these 4 points are labelled with the appropriate slope measure. An example of this is shown in the following figure.

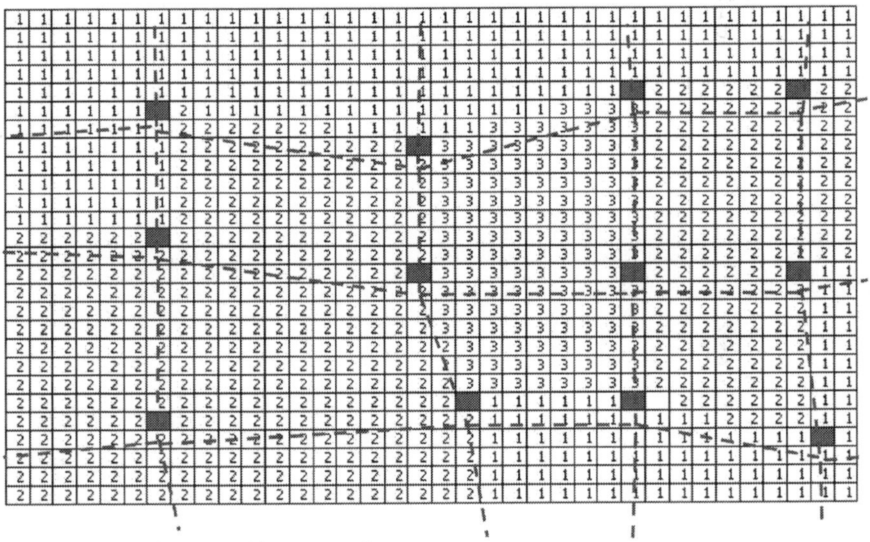

Figure 10. Example Image Slope Map Output

The highlighted pixels/squares labelled in the figure above indicate a DEM data point location. Pixels between these data points are assigned a value depending on the maximum slope in the region. A "1" indicates a Flat region, "2" indicates a Sloped region and a "3" indicates a "Steep" region. Dotted lines have been overlaid to highlight the connections between the DEM data points and the slope boundary locations.

Defining a slope map for the global operational area

A slope map for the global operational area will allow higher order navigation processes to guide the UAV toward areas within the glide range where the slope of the terrain is suitable. These areas may be outside the field of view of the vision sensor and so this slope information plays a vital role in assisting mission planning type decisions for navigation. This information becomes increasingly important if no suitable candidate landing sites exist below the UAV's current position, with the knowledge that the UAV only has a finite time before it reaches the ground.

This is an extension of the previous theory presented in this chapter. The only additional input required is the glide performance of the aircraft, and once this is defined, the maximum horizontal distance from the current position can be solved.
The image below shows the height map above the S 260 30″ 0′; E 1520 30″ 0′ area at an altitude of 400m using a function that we have written.

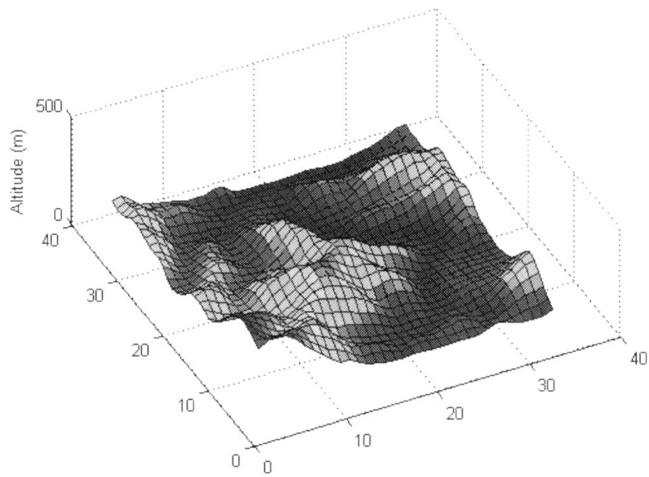

Figure 11. Example Height Map at 400m altitude

The corresponding slope map was then constructed for this entire area and is shown in Figure 11 & Figure 12. Note how the sloped and flat areas are related in the two figures.

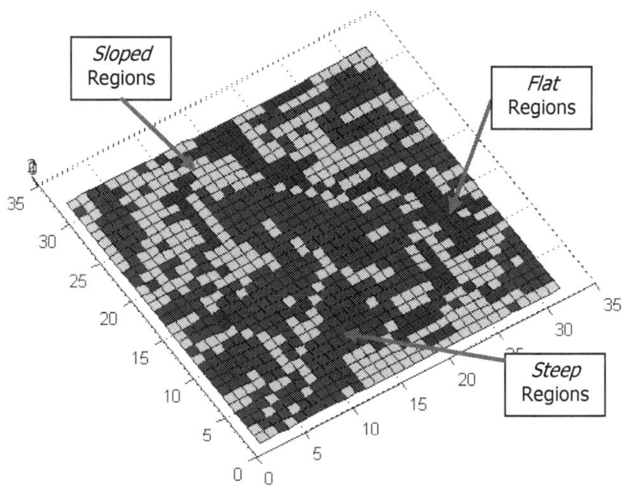

Figure 12. Example Slope Map at 400m Altitude

As discussed, the glide performance of the aircraft will dictate how far the aircraft can travel before reaching the ground, and this can then be used to define the slope map for the operational area.

3.2 Results

A series of flight trials were performed using a Cessna 172 aircraft with a 100% success rate for locating large open landing areas. 92% of these large open landing areas were considered to be completely free of obstacles, with only 8% having small obstacles such as trees. These obstacles were missed by the algorithm due to the resolution of the camera vs the current height above ground. As the UAV descends however, these obstacles would be detected and the descent planner could take the appropriate action.

The surfaces of these large open areas were also classified to accuracies of over 93% - for example grass, water, etc. This classification information will be used by the descent planner to select the most suitable landing site from the ones available.

The recommendation is that additional information be used to select the most appropriate landing area to compliment the research. Based on this recommendation, it is proposed that additional maps will be produced by the landing site subsystem to highlight keep-out areas such as roads and buildings.

Summary

This section has presented the techniques we have developed to find potential landing site areas for a fixed wing UAV. The testing to date has involved the use of manned aircraft to evaluate the Landing Site Selection algorithm and surface classification techniques and algorithms we have developed. The next phase is to implement these algorithms in real time and evaluate the performance on a small UAV in a forced landing scenario.

4. Decision making: Choosing the right landing site

One of the most important tasks in the initial stages of a forced landing is to decide on a feasible landing site, and then how to best approach this landing site. These two aspects are closely related to the multicriteria decision analysis and the trajectory planning and tracking component of the overall approach, respectively. This section will shed light on the main concepts behind the challenging decision-making process, which in reality is continuosly validated and updated throughout much of the descent should new information yield a more appropriate landing site.

4.1 Multiple Criteria

According to the Australian Civil Aviation Safety Authority's latest Visual Flight Rules flight guide (CASA 2001), there are seven criteria to selecting the optimum site for a manned aircraft forced landing. These include:

- Wind
- Surrounding
- Size and Shape
- Surface and Slope
- S(c)ivilisation

When applied in the context of UAVs, many of these factors still hold their significance, and a number of other variables also come into consideration which are not explicitly stated for piloted aircraft. These include the aircraft dynamics, the uncertainty of sensor data and the method of estimating wind.

Also to be considered is the geometrical relationship between the various candidate sites. As the aircraft descends, the number of available landing sites will rapidly decrease. Thus, it is generally better to glide towards several possible sites in close proximity than to one that is isolated, as this keeps multiple landing site options open for as long as possible. This is important so as to have several alternatives if obstacles are detected on the candidate landing sites at lower altitudes.

The number of structures and the population density that lies in the descent path to each site must also be accounted for if applicable, as it would be safer to fly over empty terrain than a populated area, in case further mishaps occur. These points, along with other factors which remain to be identified, will be evaluated to reach an optimal, verifiable decision on which candidate landing site the aircraft should aim for.

Further investigations will also be conducted in order to identify any other elements that affect this decision process, possibly including surveys and simulations involving experienced pilots and/or UAV controllers.

4.2 Multiple Objectives

The complexity of the forced landing decision process due to multiple criteria is further increased by multiple objectives that must be met. In many cases, these objectives may be conflicting, and thus compromises must be made such that the most critical objective/s could be achieved.

According to the Civil UAV Capability Assessment (Cox, Nagy et al. 2004), in the event of an emergency landing the UAV needs to be able to respond according to the following objectives and in the following order:

1. Minimize expectation of human casualty;
2. Minimize external property damage;
3. Maximize the chance of aircraft survival; and
4. Maximize the chance of payload survival.

In many scenarios, the best landing site for meeting Objectives 3 and 4 may compromise the more important objectives (1 and/or 2), or vice versa. This complex trade-off between the risks and uncertainties involved with each possible choice is but one example of a difficult problem that the multi-criteria decision-making system must face.

4.3 Decision Making

The Decision Making module will initially have predeveloped contingency plans from map data to give fast, reflex responses to emergencies. These contingency plans will guide the aircraft towards known landing sites initially, or large flat areas identified from slope map data. The Guidance and Navigation module (discussed in the next section) will constantly make estimates of the wind speed and direction, which will be taken as input for decision making. The aircraft dynamics will also be known and necessary restraints applied when judging the feasibility of a decision. As the aircraft descends, the vision-based Landing Site Selection module will continously analyse the terrain that the aircraft is flying over. Possible landing sites, buildings, and roads will be identified, including the associated uncertainties of objects in each map. With this information the Decision Making module will be able to continuously validate and update its decision in real-time.

It is expected that uncertainties will reduce as the aircraft descends, however the options available will also reduce. It may be very likely that an initially selected landing site will eventually be deemed unsuitable by the Landing Site Selection subsystem, and an alternative must be sought after. It is the responsibility of the Decision Making subsystem to be prepared for such situations by maximizing the number of alternative choices available.

The research in this area is focussing on the development of a multi-agent based architecture, where multiple events require layered decision schemes. Different software agents that handle different events during the landing process will be in constant interaction and communication throughout the descent in order to handle all the different events.

From the literature review, it was concluded that there are essentially two broad classes of multi-criteria decision analysis methods; one follows the outranking philosophy and builds a set of outranking relations between each pair of alternatives, then aggregate that according to some suitable technique. The other essentially involves determining utility/value functions for each criterion, and determining the 'utility' of each alternative based on each criterion, then aggregating those with a suitable technique to find the overall utility of the alternative.

Many of the existing techniques are not designed for 'decision making'; rather they are intended as 'decision aid' methods, and hence some only generate additional information for the human decision maker to make the final decision with. Decision making is in many ways a subjective matter, as discussed earlier, in most cases there is no 'best' decision, and it is subject on the preference information given by human decision makers. Due to the nature of the forced landing situation, where decisions made could potentially lead to damage to property or even harm life, it is critical then that the decision making system to be developed must be based on justifiable and generally accepted preference data. This means that the technique chosen should require preference data that is clear and understandable by people who don't understand the mathematics of method, and also that the technique should be as transparent as possible for purposes of accountability.

Additional requirements used to evaluate the various techniques include the ability to handle uncertainty in terms of input data, and the assumptions made regarding the decision problem. A number of the discussed techniques are currently under trial, such as PROMETHEE [Brans, 2005] and MAUT [Dyer, 2005]. Promethee is an outranking method that requires relatively simple preference data in terms of criterion weights and preference functions. Maut which is based in Expected Utility Theorem makes the assumption of independence, which essentially means that only the probability distribution of risks of individual criterion are considered, and they don't affect each other. This may be unrealistic for the forced landing scenario, yet it can be addressed by using fuzzy Choquet Integrals, which addresses synergy and redundancy between criteria.

The technique of most interest does not readily fit in to either of the main families of multicriteria decision analysis techniques, and that is the decision rules approach, and the one of specific interest is dominance-based rough set approach (DRSA). This method takes samples of decisions made by human experts, and analyses them to determine the minimum set of decision rules expressed in the form of "if…, then…" statements. These statements are then used to evaluate the alternatives in the multi-criteria decision problem, and aggregated with an appropriate aggregation technique such as the Fuzzy Net Flow Score. There is the capacity to deal with inconsistent preference information from the human decision makers by using the rough sets, and fuzzy sets can be implemented to address uncertainty in the input data. This method is the most transparent and understandable of all of those investigated so far, and is being treated as the most promising technique for use in this research.

5. Trajectory Planning and Tracking: Commanding the platform to land in the right place

The development of a UAV platform capable of precision flight, addressing safety and reliability as main concerns, is the logical progression for future UAVs in civilian airspace. Achieving this realization will not be limited to designing advanced control laws and/or flight control systems, since these UAVs will be mainly used to support reconnaissance and surveillance roles. For these applications, computer vision can offer its potential, providing a natural sensing modality for feature detection, tracking and visual guidance of UAVs.

An important part of the fixed-wing aircraft forced landing problem is how to navigate to land on a chosen site in unknown terrain, while taking into account the operational flight envelope of the UAV and dynamic environmental factors such as crosswinds and gusts, small flying objects and other obstacles in the UAV glide path. Static obstacles such as buildings, telegraph/light poles and trees on the perimeter of the chosen landing site will also be considered as they may interfere with the approach glide path of the UAV.

5.1 Vision-based Navigation in the Literature

In order to command the aircraft to the desired landing site, visual information plays a crucial role in the control of the platform. Using the visual information to control the displacement of an end effector is referred to in the literature as *visual servoing* (Hutchinson, Hager et al. 1996). It is envisaged that the location of the candidate landing sites in the image should be used to command the aircraft while is descending.

Previously (Mejias, Roberts et al. 2006) has demonstrated an approach to command the displacement of a hovering vehicle using an Air Vehicle Simulator, AVS (Usher, Winstanley et al. 2005). This task required the development of suitable path planning and control approaches to visually manoeuvre the aircraft during an emergency landing. In this approach the vehicle had to navigate through a scaled operating environment equipped with power lines and artificial obstacles on the ground and find a safe landing area.

5.2 Preliminary Results in Dynamic Path Planning using a Fixed-Wing UAV

Currently, work is underway to develop robust path/trajectory planning and tracking algorithms, and initial simulations using the MATLAB Simulink programming environment have provided valuable feedback on the designs trialled. In these simulations, an AeroSim model of an Aerosonde UAV was modified and expanded to include blocks for flight controls, path planning, GPS waypoint navigation, wind generation, wind correction and an interface to FlightGear. By running MATLAB and FlightGear concurrently, the user is able to visualize the UAV flying in a manner as dictated by the Simulink model.

At present, the primary focus of this simulation is to evaluate the dynamic path planning capability for a UAV performing a forced landing in changing wind conditions. This simulation is intended to serve as a tool in the design and testing of a visual servoing and path planning system for automating a fixed-wing UAV forced landing. It will be further enhanced to model complex, uncooperative environments with hazards such as buildings, trees, light poles and undulating terrain, as well as machine vision for use in the feedback control loop.

5.2.1. Wind Compensation

In the current forced landing simulation, the initial wind velocities are given by uniformly distributed random numbers that are updated every sixty seconds. These numbers generate the initial W_{North}, W_{East} and W_{Down} components, which are then multiplied by a continuous square wave giving the profile shown in Figure 13. The values of W_N, W_E and W_D were chosen based on the wind rose generated for Brisbane, Australia, and combined to give a maximum wind velocity of 60 kts, which can arise from any direction. A wind rose is a diagram that summarises the occurrence of winds at a location, showing their strength, direction and frequency. The wind rose used in the simulation represented wind measurements taken at 9 a.m. from 1950 to 2000, and are published by the Australian Government Bureau of Meteorology.

Note that gusts have not been modelled in the simulation, instead, the input wind is assumed to blow with a constant magnitude and direction for sixty seconds, before changing magnitude and direction for the next sixty seconds. Whilst this does not necessarily represent the wind conditions found in an actual descent, it does present a challenging wind shift scenario for the simulations to date. Future simulations will include wind gusts.

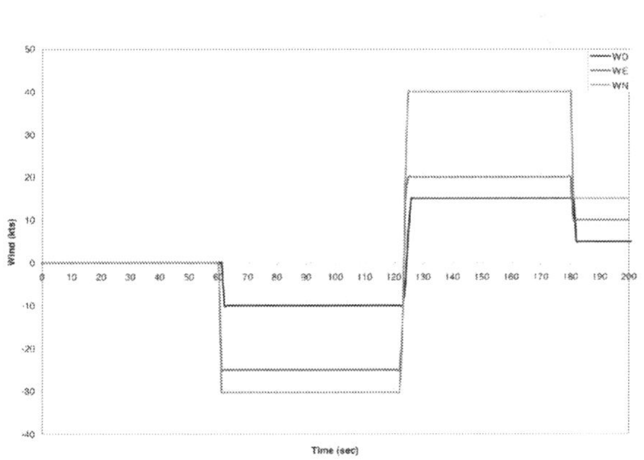

Figure 13. Wind components (W_N: Green, W_E: Pink, W_D: Blue). These components are used to compute the resultant wind vector incident on the UAV

Correction for wind is performed using the principles of vector algebra to compute the wind correction angle, which is compared with the current aircraft heading and passed as input to the UAV flight planning subsystem. From Figure 14, suppose that waypoint B is 600m (0.32 nmi) north-east (045° true) of waypoint A and the UAV glides from A to B, maintaining a heading of 045° true and a constant True Airspeed (TAS) of 37kts. A wind velocity of 340°/9.7kts coming from the south-east will cause the UAV to drift to the left.

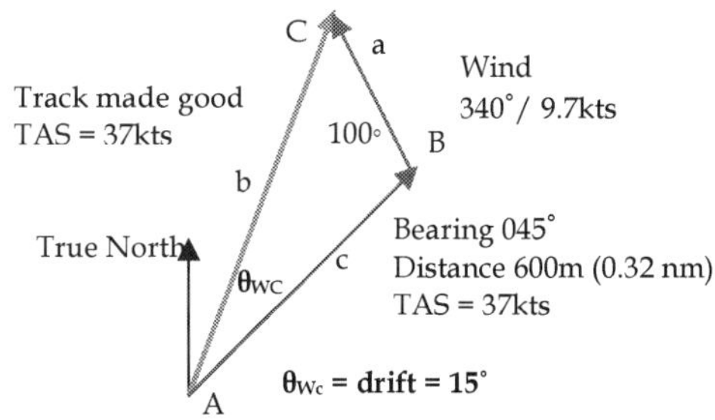

Figure 14. Wind Triangle Calculations

This implies that the wind correction angle supplied to the flight planning subsystem must be 15° in the opposite direction, such that the "track made good" will converge on the "required track" to target.

5.2.2 Path Planning

In this simulation, the path planning algorithm generated a series of waypoints, which formed a flight path along which the UAV was guided to land at the chosen landing site. The waypoints were extracted from the forced landing circuit pattern as outlined in (CASA 2001). Table 1 gives the coordinates of the idealised waypoints for a right-hand circuit pattern, and Figure 15 shows their relationship to the landing site. Note that a similar pattern for a left-hand circuit pattern can also be generated.

Waypoint	**Longitud (rads)**	**Latitude (rads)**	**Alt (ft)**
High Key	0.4782	2.6725	**2500**
Low Key	0.4783	2.6722	**1700**
End Base	0.4786	2.6721	**1200**
Decision Height	0.4786	2.6723	**670**
Overshoot1	0.4787	2.6724	**400**
Aimpoint	**0.4784**	**2.6725**	**13**

Table 1. Waypoints - Left-hand Approach Circuit Pattern

Based on the initial position of the UAV, the path planning algorithm then generated a modified table of waypoints which included the aim point, and all or a combination of the other waypoints listed in Table 1. The UAV then flew to these new waypoints using the great-circle navigation method defined in (Kayton and Fried 1997), using a set of Proportional-Integral-Derivative (PID) controllers to control the airspeed and bank angle Figure 15 depicts three possible flight paths generated using the planning algorithm described.Fixed-Wing Simulation Results

To test the performance of the path planning algorithm, a Monte Carlo simulation consisting of 500 automated landings was conducted. The simulations were run with randomised initial aircraft positions, attitudes and wind velocities. In this simulation we observed that the majority of landings had a radial miss distance between 0 and 400m from the aimpoint, which is located one-third along the length of the landing site from the direction of final approach. The value of the miss distances can be attributed to several factors; the relative spacing between the waypoints, how the path planning algorithm selects the waypoints for the UAV to navigate to and the fact that the UAV is constrained to fly with a positive 3 degree pitch attitude. However, from these tests it was observed that 151 landings lay within the site boundaries, corresponding to approximately 32% of the total population. While this figure was not exemplary, it did present a baseline for subsequent refinements to the navigation and path planning algorithms to improve upon.

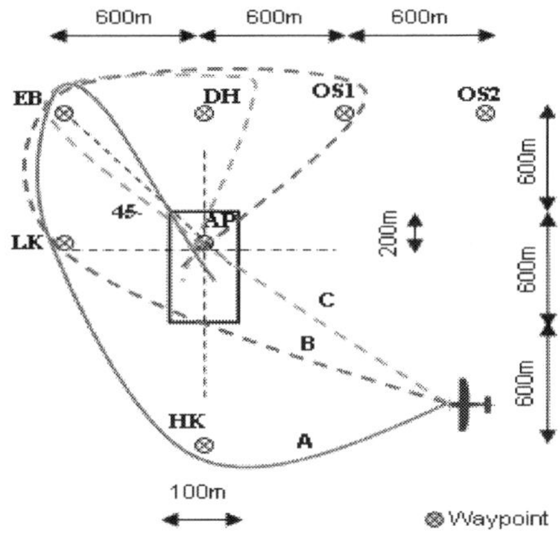

Figure 15. Forced Landing Circuit Patterns. HK=high key, LK=low key, EB=end base, DH=decision height, OS1=overshoot 1, AP=aimpoint.

Figure 16 shows a plan and isometric view of the aircraft trajectory during one simulated landing manoeuvre. The green arrows depict the direction of the changing wind affecting the aircraft during flight. The path described by the red line is the trajectory computed by the path planning algorithm, while the blue line is the actual path that the aircraftdescribes. The designated landing area is illustrated by a thick green line.

5.3 Remodelled Path Planning and Tracking Algorithm

To improve upon the algorithm described above, a remodelled path planning, tracking and control strategy has been implemented that does not restrict the robotic aircraft to following trajectories developed for human pilots. In addition, a model of a Boomerang 60 UAV was chosen for evaluation, as this represented the platform to be used in future flight tests and has manoeuvring capabilities similar to the Aerosonde. One other major difference between the remodelled algorithms and that based on the CASA forced landing circuit is that the latter will only attempt to guide the UAV from an initial point of engine failure to a point where the aircraft is aligned with the landing site and trimmed for final descent. Usually, this final approach point is 400-500 ft above ground level and it has been assumed that the Global Positioning System (GPS) signals will be available throughout the descent to this altitude. Below this altitude, GPS signals could be affected by multipath errors, dropouts, jamming and other adverse conditions, hence other sensors, such as machine vision, would need to be employed to enhance the accuracy and integrity of the existing navigation and guidance system. The descent from final approach to touchdown will be the topic of future research.

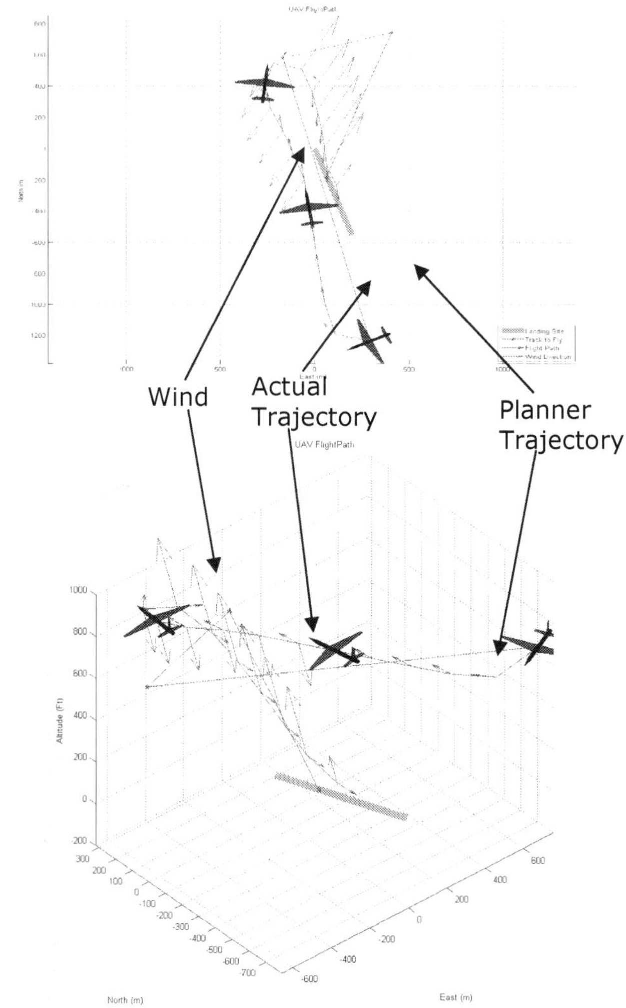

Figure 16. Detailed view of the forced landing simulations under changing wind. Green arrows indicate the direction of the wind

5.3.1 Improved Path Planning

The new path planning strategy is based on the concept of Dubins circles [Dubins, 1957] and the work presented in [Ambrosino, 2006]. Given a desired start and end position, the shortest path to the goal can be constructed geometrically in the xy-plane by the union of an arc of circumference, a segment and again an arc of circumference. There is no constrain on the radii of the arcs except that they should not exceed the minimum turn radius of the aircraft.

To translate the path into 3-D, the portions of path described by the arcs are then transformed into that traced segments of helix, and the two segments are then joined together by a straight line. This line has an angle of elevation γ_P that does not exceed the maximum dive angle, γ_{sc} of the aircraft. Should the aircraft be initially higher than the described path allows, the planning algorithm will extend the path such that the aircraft can circle to lose altitude (in reality flying along *n* spirals of a helix) before joining the path at the start of the first helix.
A representative example of the planned trajectory in 2-D is shown in Figure 17a. The desired initial and final positions are indicated by red arrows, and these positions are linked by the shortest (optimal) Dubins path highlighted in red. Note that the radii of the initial and final arcs of circumference are different, but are both greater than the minimum turn radius of the aircraft. The final 3-D flightpath is depicted in red in Figure 17b.

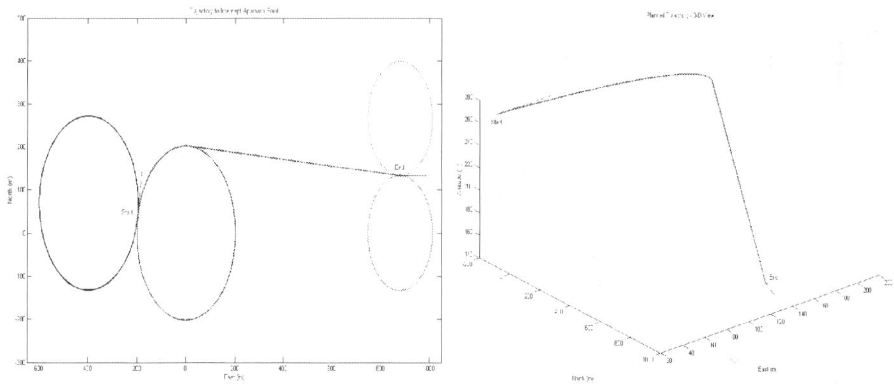

a) **b)**

Figure 17. (a) The planned descent trajectory in 2-D. The optimal path is shown in red, joining the desired start and end positions (red arrows). (b) The planned descent trajectory in 3-D

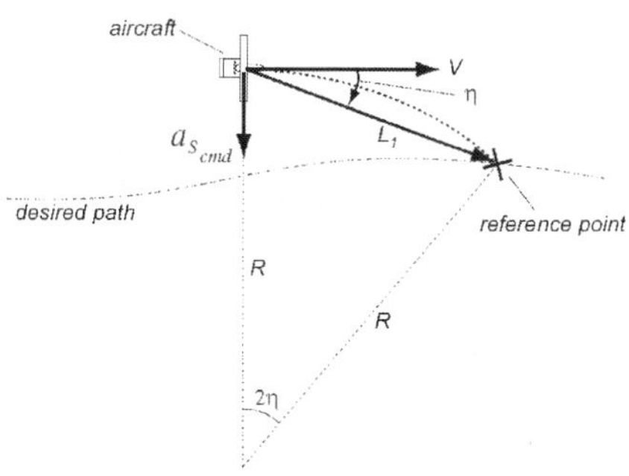

Figure 18. Diagram for guidance logic (Park, 2007)

5.3.2 Trajectory Tracking

The new trajectory tracking strategy is based primarily on the work of [Niculescu, 2001] and [Park, 2007]. Essentially, the guidance logic selects a reference point on the desired trajectory, and then generates a lateral acceleration command using the reference point. The lateral acceleration command is determined by

$$a_{S_{cmd}} = 2\frac{V^2}{L_1}\sin\eta \tag{4}$$

Where V is the airspeed of the aircraft, L_1 the distance to the reference point, forward of the vehicle, and η the angle between the V and L_1 vectors. The sign of η controls the direction of acceleration, which is equal to the centripetal acceleration required to follow an instantaneous circular segment with radius R. The relationships between V, L_1, η and R are depicted in Figure 18.

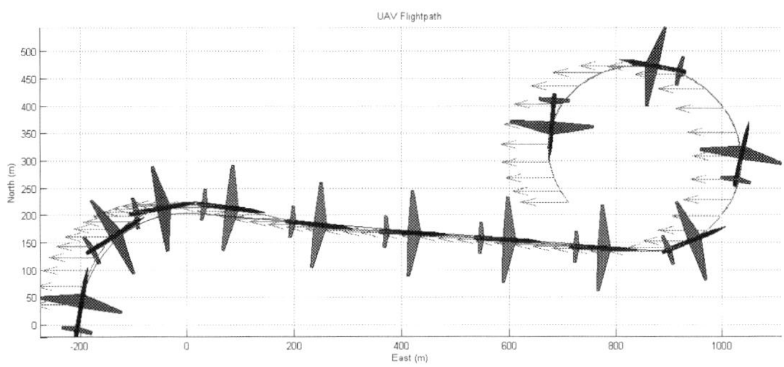

a)

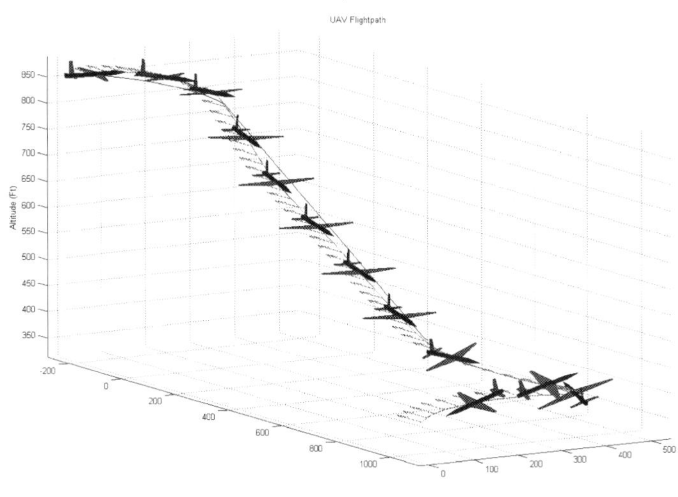

b)

Figure 19. (a) 2-D view of the flightpath using the new path planning and trajectory tracking algorithms. Wind is indicated by the green arrows. (b) 3-D view of the same flightpath

To track the path in the longitudinal plane, the algorithm compares the desired path angle, γ_d with the current aircraft path, γ_{sc} and feeds the error into a set of PID gain schedules. In calculating γ_d, the algorithm uses the speed polar curve specific to the aircraft to determine the desired airspeed and path angle, γ_w to achieve in ambient wind conditions, then compares this value with the original path angle, γ_p. The speed polar curve is used to determine the glide angle that would produce the minimum loss of altitude in wind. The error between γ_w and γ_p is then multiplied by a gain such that the aircrafty will still converge on the path to fly. Figures 19a and 19b show an example of the performances of the new trajectory planning and tracking algorithms. Wind is blowing from the east at 5 m/s (10 kts) and is indicated by the green arrows. The solid black line shows the path-to-track and the actual flightpath is depicted by the solid, blue line. The lateral cross-track error at the final approach point (which can be considered as the aimpoint when comparing with the initial algorithms) is approximately 1.1 m, while the longitudinal cross-track error (altitude error) is just over 4 ft. These results demonstrate that the new algorithms can be feasibly implemented on a real UAV for flight testing.

6. Conclusions

A number of research programs have been presented in this chapter and an overview of the approaches used in a Forced Landing system for UAVs. This overview aims to present the methodology for the development of a system capable of flight trials for a UAV forced landing.

Research in this area by the group over the past three years has seen the technical risk of an autonomous UAV forced landing system decrease, and the group is now confident that with the existing results and new research objectives, future flight tests will demonstrate this level of capability that is missing in UAVs today. The capability for a UAV to be able to land in an unknown environment with no human input is something that must be solved if UAVs are to fly above populated areas in civillian airspace.

Many complex decision making problems still remain to be investigated. These problems involve multiple conflicting objectives, and it is most often true that no dominant alternative will exist that is better than all other alternatives. Generally it is impossible to maximize several objectives simultaneously, and hence the problem becomes one of value tradeoffs. The tradeoff issue often requires subjective judgement of the decision maker, and there may be no right or wrong answers to these value questions.

It is believed that the approach presented will allow the progression of this novel UAV forced landing system from the development and simulation stages through to a prototype system that can demonstrate this important capability for UAVs to the research community.

7. Acknowledgement

The authors would like to thank Richard Glassock for piloting the test aircraft, as well as Dmytriy Stepchenkov and Lachlan Mutch for providing engineering support in the field.

8. References

Ambrosino, G., Ariola, M., Ciniglio, U., Corraro, F., Pironti, A., & Virgilio, M. (Eds.). (2006) *45th IEEE Conference on Decision & Control*, . San Diego, U.S.A

Azencott, R., Durbin, F., & Paumard, J. (1996). Robust recognition of buildings in compressed large aerial scenes. Paper presented at the *Image Processing*, 1996. Proceedings., International Conference on.

Brans, J. P., & Mareschal, B. (2005). Promethee Methods (Vol. 78): Springer New York

Canny, J. F. A computational approach to edge detection. *IEEE Trans Pattern Analysis and Machine Intelligence, 8*(6), 679-698.

CASA. (2001). *VFR Flight Guide*: Civil Aviation Safety Authority Australia (CASA) - Aviation Safety Promotion Division.

Chen, K., & Blong, R. (2002). Extracting building features from high resolution aerial imagery for natural hazards risk assessment. Paper presented at the *Geoscience and Remote Sensing Symposium*, 2002. IGARSS '02. 2002 IEEE International.

Cox, T. H., Nagy, C. J., Skoog, M. A., & Somers, I. A. (2004). *Civil UAV Capability Assessment.*

Dyer, J. S. (2005). MAUT. Multiattribute Utility Theory (Vol. 78): Springer New York.

Dubins, L. E. (1957). On Curves of Minimal Length with a Constraint on Average Curvature with prescribed Initial and Terminal Positions and Tangents. *American Journal of Mathematics*, 79, 471-477.

Fitzgerald, D. L. (2007). *Landing Site Selection for UAV Forced Landings Using Machine Vision.* Phd Thesis, Australian Research Centre for Aerospace Automation (ARCAA), Queensland University of Technology.

Fitzgerald, D. L., Walker, R. A., & Campbell, D. (2005). A Vision Based Forced Landing Site Selection System for an Autonomous UAV. Paper presented at the 2005 *IEEE International Conference on Intelligent Sensors*, Sensor Networks and Information Processing (ISSNIP).

Howard, A., & Seraji, H. (2001). Intelligent Terrain Analysis and Information Fusion for Safe Spacecraft Landing. *NASA - Jet Propulsion Laboratory, Technical Document.*

Howard, A., & Seraji, H. (2004). Multi-sensor terrain classification for safe spacecraft landing. *IEEE Transactions on Aerospace and Electronic Systems, 40(4)*, 1122-1131.

Hutchinson, S., Hager, G. D., & Corke, P. I. (1996). A tutorial on visual servo control. *Robotics and Automation, IEEE Transactions on , Volume: 12, Issue: 5 , Oct.*, 651 -670.

Kayton, M., & Fried, W. R. (1997). *Avionics Navigation Systems* (2 ed.): John Wiley & Sons, Inc., New York.

Mejias, L., Roberts, J., Corke, K. U., & Campoy, P. (2006). Two Seconds to TouchDown. Vision-Based Controlled Forced Landing. Paper presented at the *IEEE/RSJ International Conference on Intelligent Robots and Systems*, Beijing, China.

Niculescu, M. (2001). Lateral Track Control Law for Aerosonde UAV. Paper presented at the *39th AIAA Aerospace Sciences Meeting and Exhibi.*

Park, S. J. D., & How, J. P. (2007). Performance and Lyapunov Stability of a Nonlinear Path-Following Guidance Method. *Journal of Guidance, Control, and Dynamics,*, 30, 1718-1728..

Usher, K., Winstanley, G., Corke, P., Stauffacher, D., & Carnie, R. (2005). Air vehicle simulator: an application for a cable array robot. Paper presented at the *IEEE International Conference on Robotics and Automation*, Barcelona, Spain

22

Design Considerations for Long Endurance Unmanned Aerial Vehicles

Johan Meyer, Francois du Plessis and Willem Clarke
University Johannesburg
South Africa

1. Abstract

The arena of Unmanned Aerial Vehicles (UAVs) has for many years been dominated by the defence industries. The reason for this can be attributed to the complexity and cost of designing, constructing and operating of these vehicles. An additional contributing factor is the legislative issues around operating an unmanned aircraft in civilian airspace. However in resent years advances in micro-electronics especially Micro Electronic Mechanical Systems (MEMS) and advanced composite manufacturing techniques have placed the design and construction of UAVs in the domain of the commercial civilian users. A number of commercial UAV applications have emerged where the legislative requirements for operating of a UAV in segregated airspace can be met. UAVs are extremely well suited for the dull, dirty and dangerous tasks encountered in performing surveying and surveillance applications. For these tasks the primary design considerations in the design of the UAV would be the propulsion system, the guidance and control system and the payload system.

2. Introduction

The European Unmanned Vehicle Association identifies five main categories of UAVs (Sarris, 2001):

- **Close range** - fly in a range of less than 25 km. Usually extremely light;
- **Short range** - operate within a range of 25-100km.
- **Medium range** - Able to fly within a range of 100-200km. Need more advanced aerodynamic design and control systems due to their higher operational performance.
- **Long range** - Fly within a range of 200-500km. Require more advanced technology to carry out complex missions. Need satellite link in order to overcome the communication problem between the ground control systems and aircraft created by the curvature of the earth.
- **Endurance** - Operate in a range more than 500km, or can stay in the air for more than 20 hrs. This is considered the most sophisticated of the UAV family due to their high capabilities.

This chapter presents design considerations for UAVs which can be categorized as endurance UAVs. The design considerations discussed include the primary UAV systems namely, propulsion system, navigation and control system and sensor payload system.

Renewable energy sources is an attractive alternative to the conventional fossil fuel based propulsion systems. The renewable energy sources evaluated for long endurance UAV applications include solar energy, hydrogen fuel cells and energy storage sources such as batteries and super capacitors. The advantages and disadvantages of each source are presented. Results from an algorithm developed for the selection of the optimal energy source based upon the application requirements and constraints are presented. Implementation issues around a solar powered UAV for long endurance applications are discussed. The results of the feasibility study of using solar power for long endurance UAVs are presented.

Advances in the development of MEMS based inertial sensors have enabled the development of low cost inertial navigation systems for use in UAV applications. An overview of the field on inertial navigation is presented along the advances in inertial sensors. Some considerations that impact the selection of the inertial sensors and the final design of the navigation system are presented. A number of options for the improvement of the navigational performance of the low-cost inertial sensors by combining the sensor data with the measurements form other sensors such as GPS, cameras and altimeters are discussed.

Another aspect of UAV design is the control system. Autonomous control is of paramount importance for the safe and successful operation of unmanned aircraft that is operated out of visual range. The development of UAV autopilots is presented by looking at the classical and the modern approaches to control system development.

UAVs are extremely well suited for applications where the payload consists of optical image sensors such as cameras. Cameras offer powerful lightweight sensors suited for a variety of tasks. In keeping with commercial trends (in contrast to the military environment) some of the functionality associated with the operation of long range UAVs can be "out sourced" by using existing infrastructure. This effectively increases the risk of an unsuccessful UAV mission (in a commercial sense), but lowers the cost of development and operations.

Image processing solutions invariably implies large processing power, which leads to high power requirements. By using the pervasiveness and low cost of the mobile communications technology and industry, some of the processing can be "outsourced" to a powerful ground processing station. This opens up a Pandora's box of opportunities, i.e. real-time scene identification, automated mission control, visual cues, speed and orientation estimations, super resolution of low resolution sensory information, etc. Recent trends see high speed, parallel processing Graphics Processing Units (GPU) integrated into low powered mobile phones. This can provide high speed processing power on-board the UAV for complex image processing applications. A balance need to be maintained between local, on-board processing requirements (e.g. for reliable navigational purposes) and higher level functionality (associated with the mission).

The next section presents the design considerations for long endurance UAV applications with regard to renewable energy propulsion sources, navigation and control aspects and payload applications.

3. Electrical Power Sources for Long Endurance UAV Applications

Unmanned Aerial Vehicles are ideally suited for long endurance applications, but to be able to make full use of this feature, effective power sources need to be developed to ensure the long endurance functionality of the propulsion system and onboard equipment. For a UAV the flight endurance is in direct relationship to the total weight of the craft. In order to maximise flight endurance the need for high density energy sources are created. The objective of this section is to present design considerations for high energy density, cost effective, non-carbon emitting and renewable energy sources. The following energy sources and combinations thereof are considered:

- Lithium Polymer (Li-Po) Batteries;
- Super Capacitors (SC);
- Photo Voltaic (PV) Cells; and
- Hydrogen Fuel (FC) Cells.

Lithium Polymer batteries and super capacitors are in essence only energy storage mediums. However in the context of UAV power sources these energy stores can be considered as energy sources for supplying power to the UAV and associated onboard equipment.

3.1 Solar Energy

Solar energy refers to the solar power collected from solar irradiance by photovoltaic cells. The power output of photovoltaic cells depends primarily on the absolute value and spectral distribution of irradiance in the plane of the photovoltaic cell and the resulting operational temperature (Luque & Hegedus, 2003). Much research has been conducted with regards to the factors influencing solar power generation which is beyond the scope for this chapter. The total amount of energy produced by the photovoltaic cells is a function of the geographical position (latitude, longitude, and altitude), time of the year, atmospheric absorption and efficiency of the photovoltaic cells. The Linke turbidity factor (Muneer et al., 2004) is used to characterize the clearness of the sky. The lower this factor, the clearer the sky, the larger the beam irradiation and the lower the relative fraction of the diffuse irradiation. For higher altitudes, the absorption is lower because of less radiation scattering by the atmosphere which lowers the Linke turbidity factor. Typical values for the Linke turbidity factor are listed in Table 1.

T_{LK}	**Sky Condition**
1	Pure sky
2	Very clear sky
3	Clear sky
5	Summer with water vapour
7	Polluted urban industrial

Table 1. Typical values for the Linke turbidity factor

The amount of solar energy available per day for propulsion of a solar powered UAV is not only dependant on the clearness of the atmosphere but also highly dependant on the time of the year. UAVs operating during the summer months, when the available energy is at its highest, has approximately 2.2 times the energy available relative to operation during winter months when the available energy is at its lowest. Even when a worst case summer day (with Linke turbidity equal to 5) is compared with a clear winter's day (with Linke turbidity equal to 2), the ratio of available solar energy in summer is still on the order of 1.5 to the available solar energy in winter. This difference in available solar energy is mainly contributed by the distance between the earth and the sun increasing in winter and the smaller sun angle and the shortening in day light hours as a result of the inclination of the earth axis. When designing solar powered UAVs, consideration has to be given to the expected operating time of the year. Designing for minimum available solar energy conditions may result in an over design by a factor of 2 under maximum available solar energy conditions. An over design factor of 2 has significant negative impact on the UAV airframe design in terms of size and cost. A positive impact may be that the excess solar energy available may be used to overcome the increased aerodynamic drag when the UAV is flying at faster speeds. This may result in UAVs which can be operated at higher flying speeds during the summer months, when the available solar energy is at a maximum. Figure 1 shows the theoretically available energy which can be collected in the southern hemisphere at a latitude of 25 degrees by a photovoltaic array of 1 m^2 with an efficiency of 16% as a function of the time of the year. Another issue to consider is matching the output of the photovoltaic cells to the input of the energy storage medium, which can make a great difference in the efficiency of power utilization, thus some form of maximum power point tracker will be required (Luque & Hegedus, 2003).

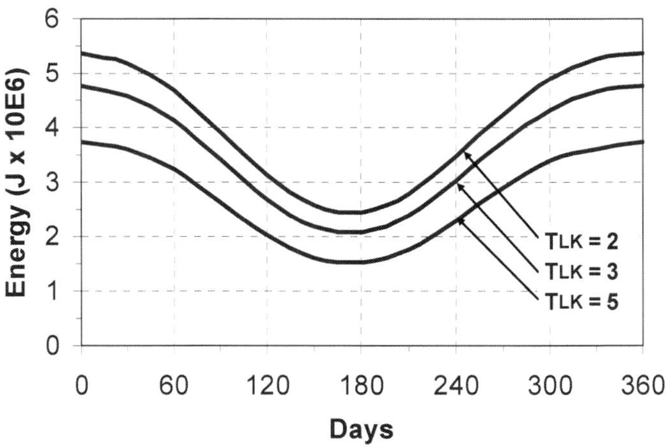

Figure 1. Available solar energy per day which can be collected by a 1 m^2 photovoltaic array with a 16% efficiency for various values of Linke turbidity factor where day 1 corresponds to the summer solstice in the southern hemisphere

3.1.1 Solar Energy Advantages

Advantages of using photovoltaic cells as energy sources can be summarized as follows:

- Very little maintenance required;
- Photovoltaic cells are non-polluting; and
- Essentially no operating cost. The greatest cost of photovoltaic cells is the initial acquisition costs.

3.1.2 Solar Energy Disadvantages

Disadvantages of photovoltaic cells are the high initial procurement cost and the fact that it can only generate electrical power during daylight hours which necessitates the use of energy storage mediums. Secondly photovoltaic cell efficiency is rather low in the rage of 14% to 18% for commercially available terrestrial grade cells. Space grade cells have efficiencies as high as 24% (Luque & Hegedus, 2003). The dependence on the atmospheric conditions makes solar energy a lot less predictable and seasonal variations will affect the design.

3.2 Lithium Polymer Cells

Lithium polymer cells are constructed using a flexible, foil-type case containing an organic solvent. In lithium-ion cells a rigid case presses the electrodes and the separator onto each other whereas in polymer cells external pressure is not required because the electrode sheets and the separator sheets are laminated onto each other Lithium polymer batteries, the next generation power source (van Schalkwijk et al., 2002) since no metal battery cell casing is needed, the weight of the battery is reduced and it can be formed to shape. The denser packaging without inter cell spacing and the lack of metal casing increases the energy density of Li-Po batteries to over 20% higher than that of a classical Li-ion batteries. Lithium polymer cells are considered fully charged when the cell terminal voltage reaches 4.2 V and are fully discharged when the cell terminal voltages decreases to a voltage of 3.0 V. (van Schalkwijk et al., 2002). A variety of Li-Po batteries are commercially available consisting of different series and parallel configuration of cells making up the required battery voltage and current characteristics.

3.2.1 Lithium Polymer Cell Advantages

In UAV applications the obvious benefit of using Li-Po batteries is the higher energy density offered. Advantages of Li-Po batteries can be summarized as follows:

- High energy density;
- Low self-discharge properties;
- The flexible casing of the polymer batteries allow for design freedom in terms of profile thicknesses; and
- Low maintenance.

3.2.2 Lithium Polymer Cell Disadvantages

Lithium polymer cells have special recharging procedures necessitating the use of specifically designed chargers. Charging Lithium cells is the most hazardous aspect of the batteries. Cell count and terminal voltage are of utmost importance when charging Li-Po batteries. It is important not to exceed both the high voltage limit of 4.2 V and the low voltage limit of 3.0 V. Exceeding these limits can permanently harm the battery and may result in a fire hazard. When series cells configurations are employed to obtain the required battery terminal voltage cell balancing circuits are required to ensure an even voltage distribution over the cell stack. Failure to balance the series cell stack may cause overcharging of individual cells with the associated fire hazards. Lithium polymer batteries are also subject to ageing effects, due to this the expected lifetime of such batteries is limited. Disadvantages of Lithium polymer batteries can be summarized as follows:

- Special charging circuits required to maintain cell voltage within safe limits;
- Subject to aging;
- Subject to cell balancing for series stack configurations;
- High procurement cost; and
- Requires special disposal processes.

3.3 Hydrogen Fuel Cells

Polymer electrolyte membrane fuel cells (PEMFC) are constructed using a solid polymer as electrolyte, absorbent electrodes combined with a platinum catalyst. Hydrogen gas is recombined with oxygen gas producing electricity with water vapour as emission. Onboard storage of the hydrogen would be required for UAV applications. Alternatively, hydrogen may be manufactured onboard the UAV from electrolysis of water using solar energy. A closed loop system could be operated whereby the water from of the PEMFC can be electrolyzed into oxygen and hydrogen for later re-use. Oxygen is generally obtained from the surrounding air. Operating temperatures are relatively low around 80 °C, enabling quick starting and reduced wear. Platinum catalysts are required for operation and to reduce corrosion. Polymer electrolyte membrane fuel cells are able to deliver high energy densities at low weight and volume, in comparison to other fuel cells (Barbir, 2005).

3.3.1 Fuel Cell Advantages

The advantages of using PEMFCs can be summarized as follows:

- Relative high efficiency;
- High energy density;
- Low noise;
- Non carbon producing only water emission; and
- Low maintenance.

3.3.2 Fuel Cell Disadvantages

The biggest disadvantages of using PEMFCs are the initial procurement cost and the safety issues regarding the storage of the onboard hydrogen gas. PEMFCs also suffer from a limited lifetime.

3.4 Super-Capacitors

Super capacitors, (SC) or Ultra-capacitors are also known as Electric Double Layer Capacitors (EDLC). Super capacitors have a double layer construction consisting of two carbon electrodes immersed in an organic electrolyte. During charging, ions in the electrolyte move towards electrodes of opposite polarity; this is caused by an electric field between the electrodes resulting from the applied voltage. Consequently, two separate charged layers are produced. Even though the capacitors have a similar construction to batteries, their functioning depends on electrostatic action. No chemical action is required; the effect of this is an easily reversible cycle with a lifetime of several hundreds of thousands of cycles (Conway, 1999).

3.4.1 Super-Capacitors Advantages

The advantages of Super capacitor energy sources can be summarized as follows:

- High cell voltages are possible, but there is a trade-off with storage capacity;
- High power density;
- No special charging or voltage detection circuits required;
- Very fast charge and discharge capability; and
- Life cycle of more than 500,000 cycles or 10-12 year life time.

3.4.2 Super-Capacitors Disadvantages

Disadvantages of Super capacitors can be summarized as follows:

- Low energy density;
- Low power to weight ratio when compared to current battery technology;
- Moderate initial procurement cost; and
- High self discharge rate.

3.5 Electrical Power Source Comparison

When considering an electrical energy source for powering of an UAV the before mentioned advantages and disadvantages must be compared. The key performance parameters for energy sources in UAV applications were identified as:

- Energy density (Wh/kg);
- Energy unit cost (Wh/$); and
- Lifespan (years).

Figure 2 shows a comparison of the key performance parameters for the energy sources considered.

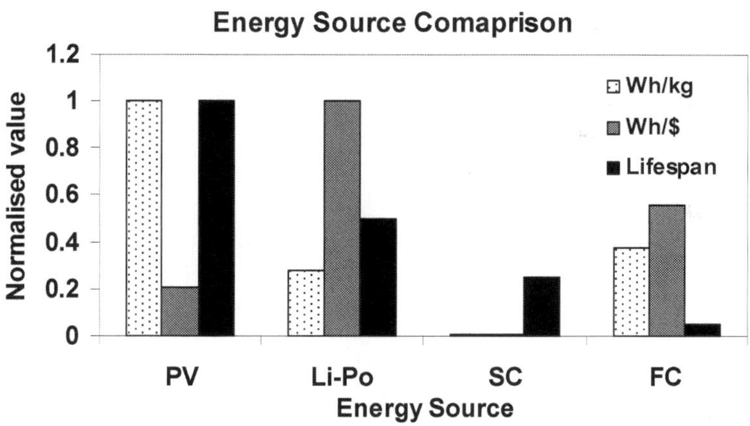

Figure 2. Normalised comparison of the key performance parameters for the energy sources considered

An algorithm was designed which is capable of determining the most applicable selection of the energy source, from the sources considered for a given UAV application. The UAV application is quantified by the user input of the following parameters:

- The electrical power required by the application;
- The maximum allowed weight of the power source;
- The required time duration for the supply of electrical power. This can also be considered as the flight duration; and
- The maximum cost limit for the power supply.

The results from the selection algorithm are presented in Table 2.

Flight Duration (h)	Power Source Weight (kg)	Required Power (W)	Solution 1	Solution 2
2	5	40	PV	Li-Po
10	4	20	Li-Po	FC
12	10	150	FC	PV-Li-Po
21	10	200	PV-Li-Po	None
24	20	200	FC	PV-Li-Po

Table 2. Applicable energy source for various UAV application requirements

Form Table 2 follows for longer flight durations and high power ratings the preferred solution is hydrogen fuel cells or a photovoltaic-lithium polymer battery hybrid power supply. For lower power supply weight budgets and relatively short flight time durations the demands are met by using photovoltaic cells or lithium polymer battery supplies. Hybrid solutions are preferred above the fuel cell solution at lower weight requirements due to the photovoltaic cells advantage in power density, but the initial costs exceed that of fuel cells. However considering the life expectancy of the hybrid solutions this option becomes more attractive. A fuel cell solution is and all round good option and the relative simplicity of such a system makes this option very attractive.

From the results presented in Table 2 one of the ideal power solutions for long endurance UAV flights is the hybrid solution between the photovoltaic cells and Li-Po batteries. Considerations for implementation of hybrid photovoltaic cells and Li-Po batteries are presented in the next section.

3.6 Considerations for the Implementation of a Hybrid PV-LiPo UAV Power Source

Charging of a Li-Po battery requires the application of a 4.2 V voltage source across each cell in the battery with the current limited to the C rating of the battery. This is referred to as constant voltage, constant current charging. In Li-Po batteries the individual cells are connected in a series configuration in order to achieve the desired battery terminal voltage. The series connecting of cells requires that the cells are all identical in capacity and state of charge. This may not always be true. High discharge rates can cause cell imbalances as does cell aging. Series connection of Li-Po cells in batteries may lead to cell drifting, which poses serious dangers when charging. The proposed solution is therefore to connect all the Li-Po cells in parallel forcing the cells to the same state of charge. There are additional benefits of having the cells in parallel such as:

- Each cell ages the same amount if equal current sharing is implemented significantly reducing the charging safety hazards;
- Increased redundancy when cell stacks are set up so that the remaining sources pick up the slack of a faulty cell; and
- All cells in the stack are forced to same state of charge.

There are however a number of disadvantages to parallel connection of cells in battery stacks which may include the following:

- Cell isolation circuitry may be required to prevent a faulty short circuited cell to discharge the remaining cells; and

- Lower battery terminal voltage which would necessitate the use of boost converters to obtain the required operational voltage.

Charging a Li-Po cell at 1C for 1 hour will ensure that at least 70 % of the rated power will be charged. Considering the graph shown in figure 3, it is observed that the power input into the cell starts to fall off once the voltage of the cell reaches 4.1 V. By paralleling Li-Po cells in an overly large battery stack a battery is created which is able to store all the available energy generated by the photovoltaic cells. Thus it would be best to use a battery stack, consisting of a number of cells connected in parallel, with output voltage regulation using a self adjustable booster circuit. The power rating of the battery should be such that 70 % of the battery power capacity coincides with the maximum total power obtainable from the PV cells for the flight duration. This arrangement not only simplifies the circuit design by eliminating the need for specialised current limiting charging circuits but also ensures that all the available power generated by the photovoltaic cells can be stored without the need for complex switching circuits to switch the excess energy to the remaining batteries.

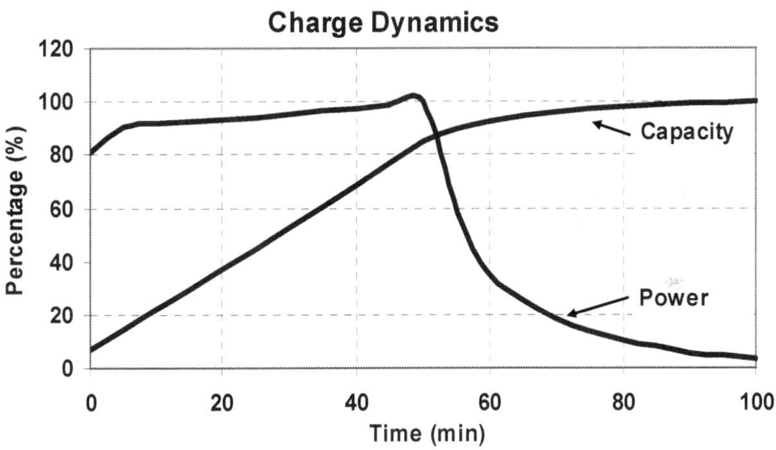

Figure 3. Power input into the battery per unit of time as function of time. The capacity of the battery is indicated as a percentage

In order to evaluate the feasibility of the proposed electrical power source a simulation was performed of the typical power flow during a long duration UAV flight

3.7 Long Endurance Solar Powered UAV Flight Simulation

For the evaluation of long endurance low altitude solar flight a simulation model was realized. The simulation model includes a radiation model which is parameterized by the geographical position and the time of the year (Bird & Halstrom, 1981). A facility is provided whereby the flight start time can be entered as an offset to the simulation time. The model of the solar panel takes into account the solar array size and efficiency. The power generated by the solar array is then made available to the battery model. A UAV aircraft model was implemented which computes the required power for level flight.

Electrical power required by the avionics is added to the power required for level flight to obtain the total power requirements for the UAV flight. Figure 4 shows the power progression graph for the power flow of the UAV during a long duration UAV flight of 48 hours. The flight was started at 18:00 in the evening on summer solstice with a fully charged battery.

Better insight into the power balance requirements can be obtained by plotting the energy flow graph of the UAV as in figure 5. It can be seen that the energy collected by the solar array always exceeds the energy used by the UAV.

The contribution of the fully charged battery can be seen as the flight starts at 18:00 in the evening. The initial battery charge is nearly depleted at 06:30 in the following morning from where charging resumes to full capacity at 17:00 in the afternoon. As the available light diminishes towards the evening the battery starts to discharge to supplement the dwindling available solar energy. At about 06:00 in the morning of the second day the battery is again nearly discharged when the following charge cycle begins.

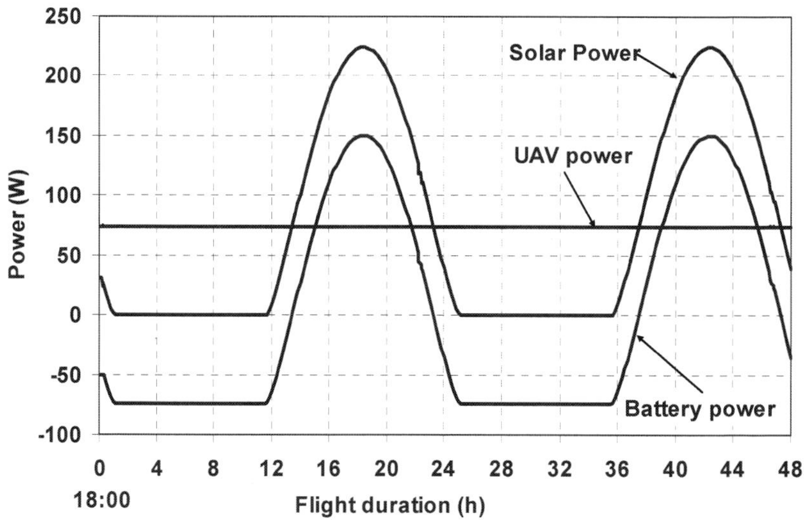

Figure 4. Power flow for a long endurance solar powered UAV requiring 74 W for maintaining level flight. A solar array of 1.2 m^2 with efficiency of 16 % is collecting the solar energy. Negative power means power being supplied by the battery

From the work presented it can be concluded that the primary consideration in the design of a PV-LiPo hybrid power supply for low altitude long endurance UAV is the available solar energy which can be collected by the photovoltaic array mounted on the wing surface of the UAV. The available solar energy is very strongly dependant on the time of year and the clearness of the atmosphere. During operation of long endurance flights the UAV must be trimmed for optimal speed ensuring maximum endurance. The capacity of the energy store can be determined by the power and energy requirements of the UAV. By taking these considerations into account it would be possible to a PV-LiPo hybrid power supply to power a low altitude long endurance UAV capable of demonstrating sustained flight.

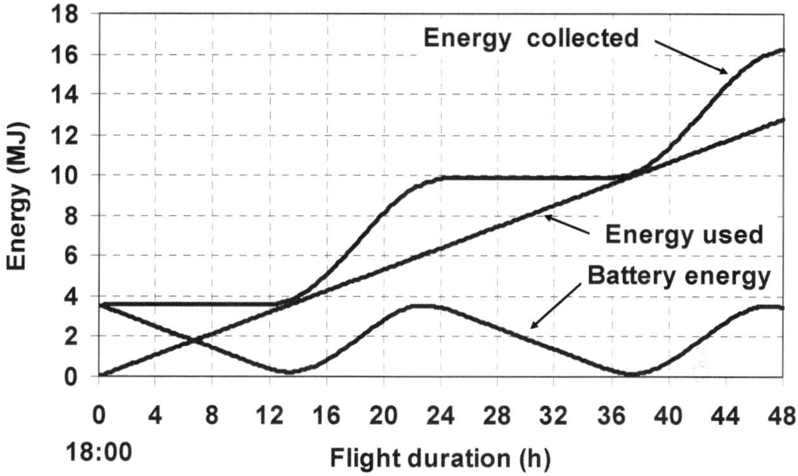

Figure 5. Energy flow for a long endurance solar powered UAV for a flight duration of 48 hours

4. Navigation and Control

Consider the following example of a situation that frequently arises. Let's say company ABC Imaging have expert knowledge in the development of some high-powered thermal imaging sensor and they consider it a profitable business opportunity to integrate the thermal imaging payload with a UAV and sell the complete product with the sensor as an integrated payload. How should they go about the integration of their payload with the UAV as their expertise is in the field of thermal imaging sensors and not autonomous aircraft? It is suggested that this company follow one of the following three options:

1. Recruit a whole team of engineers with UAV knowledge and build their own system, including the airframe and all the subsystems, from scratch - usually resulting in excessive cost and delays in getting the integrated product into the market;
2. Acquire some of the key components such as the airframe, the control system, communication links, etc. as building blocks and integrate these with their imager to obtain the UAV system. This could mean that some of the minor components needed for the cooperation and management of the various subsystems be developed; or
3. Obtain an off-the-shelve fully functional UAV as an aerial platform and integrate their imager as payload with this platform.

It should be clear that each one of the options have its advantages and disadvantages. Option 1 is a high risk alternative that should probably be taken if the company is eager to establish a long-term presence in the UAV field and expand their technical capability beyond their current field of expertise. Options 2 and 3 usually results in lower risk and lower product-into-market delays, with Option 3 probably resulting in the shortest development time with the least amount of flexibility.

The perspective that will be presented in this section does not necessarily present one of the options as superior to any of the others. The choice of the developmental strategy is a business decision that needs to be evaluated on the merit of the position of the business. It is, however, suggested that small companies with focused payload expertise should generally follow Option 3, somewhat larger companies or ones that have a broader base of technical knowledge could opt for Option 2 with the large companies with an established engineering presence usually deciding on Option 1 or a combination of Option 1 and 2.
The objective of this section is therefore to provide an overview of the autopilot development process and address key technologies that are represented in this process. It will be assumed that the reader does not necessarily have a background in navigation and control. A conceptual overview of some of the key technologies will therefore be presented as part of the content of this section. As the focus of this chapter is on the design considerations of a UAV and not necessarily on the research aspect of control and navigation, predominantly mature, proven techniques of control and navigation are presented. Emphasis will therefore be placed on the design constraints without going into too much of the supporting theory.
As the focus of this chapter is on long endurance UAVs, which is generally considered to be fixed-wing aircraft, the rest of this section will discuss building blocks that can be used in the development of such systems. Apart from complete UAVs and commercially available autopilots, most of the other building blocks are generic enough to be used on different UAV configurations such as airships or rotorcraft.
(Valavanis, 2007) presents a good overview of the development of autonomous helicopter and quadrotor UAV systems. Also included in this book are a variety of subjects on the supporting operations and subsystems needed to develop a successful UAV system. This reference is considered as a good reference for both the inexperienced as well as the established UAV systems engineer who need to gain insight into all the components and subsystems that are needed in modern UAVs.

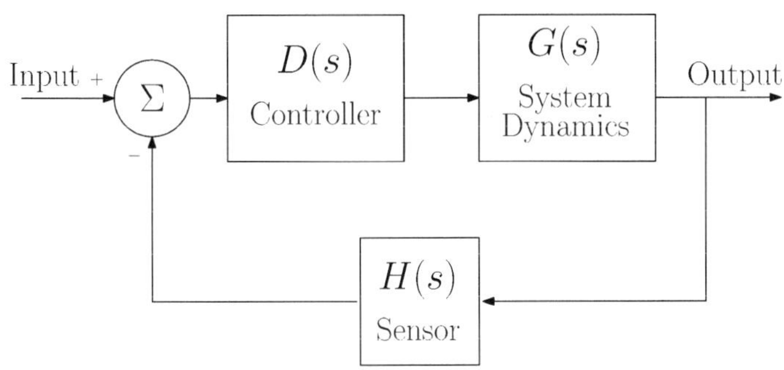

Figure 6. Generic control system block diagram

4.1 Navigation Systems

This section will consider the navigation and control (N&C) system components and subsystems by first looking at the algorithmic aspects involved in the development of the N&C system and then looking at the hardware on which the algorithms can be implemented and that is required to perform the various tasks. The actuator, although being modelled as part of the controller, will be considered as being part of the airframe and will therefore not be discussed any further. An overview of a generic control system is presented in figure 6. From this block diagram it can be seen that the core building blocks of any control system are the

- Dynamics model of the system - known as the plant;
- The sensor that measures the variables of interest; and
- The actual control algorithm as implemented on the control processor. The actuator is not indicated on the diagram, but is considered to be part of the controller.

4.1.1 Navigation

The sensor system of an aircraft is generally called the navigation system. This system constitutes one of the most critical components on any aircraft and it is of even greater importance on a UAV where there is no pilot onboard that can use his own senses to act as a backup "navigation system". Due to the criticality of the navigation system, it usually consists of a combination of sensors that are complementary in nature. From an intuitive perspective it should be apparent that the variables of interest for autonomous UAV control are the aircraft

- Altitude;
- Position;
- Velocity;
- Acceleration; and
- Orientation (attitude).

The selection of sensors is therefore supplementary in nature to ensure that a trustworthy set of the most critical parameters are available at all times. Apart from the availability of the critical system parameters, it is also necessary to perform some calculations on these parameters and combine the various measurements to provide the optimal set of system variables at all times.

The next two subsection present a discussion of the fundamental algorithms involved with navigation and then discuss the process of combining the measurements from various sensors into an optimal measurement.

4.1.1.1 Navigation Algorithm

The core element of aircraft navigation systems is the inertial navigation system (INS). The INS consists of a sensor block called the Inertial Measurement Unit (IMU) and the navigation processor used to implement the navigational equations. (Titterton and Weston, 2004) present a comprehensive discussion on the various inertial sensors as well as the navigational equations necessary for successful INS development.

The IMU consists of a triad of 3 orthogonal gyroscopes and a triad of 3 orthogonal accelerometers. Most modern navigation systems make use of strapdown navigation where the IMU is fixed to the body of the aircraft. The gyroscopes therefore measure the angular rotational rate of the aircraft body in the body (x,y,z) coordinate system (also known as the body frame) relative to inertial space. The accelerometers measure the aircraft acceleration in the body frame in terms of the body x,y and z components.

The advantage of classical inertial navigation is a self-contained system that does not require any external measurements to operate and the navigation sensors are therefore not susceptible to external sources of interference.

The navigational equations that are applied to the IMU output to obtain the complete set of navigation parameters is presented in block diagram form in Figure 7. It consists of the double integration of the measured acceleration to obtain the aircraft velocity and then the position. In a strapdown navigation implementation the acceleration is measured in the aircraft body axis and the position of the system is required in a local level coordinate system called the navigation frame. To determine the aircraft acceleration in the navigation frame, it is necessary to determine the orientation of the aircraft body frame relative to this navigation frame and use this description of the relative orientation between the two axis systems to convert the acceleration measurement from the body frame to the navigation frame. This is where the gyroscopes come into play. As presented in figure 7, the gyro cluster measurement of the aircraft angular rate relative to the inertial frame is integrated to determine the system's orientation (known as the attitude). The attitude can be presented in terms of Euler angles (roll, pitch and yaw), in terms of a direction cosine matrix or in terms of a quaternion representation (Titterton & Weston, 2004; Stevens & Lewis, 2003; Farrell,2008; Rogers 2007; Groves, 2008; Farrell & Bath, 1999). The attitude representation is used to convert the accelerations to the navigational frame. Some additional corrections for the effect of the Earth's rotational rate on the gyros, changes in gravity as a function of position and altitude and the Coriolis effect resulting from the relative rotation between the two frames is also needed as presented in Figure 7.

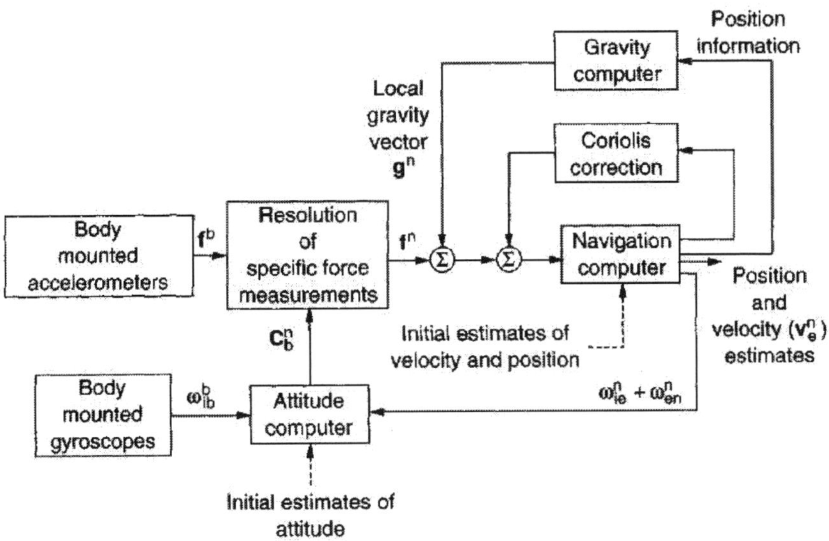

Figure 7. Inertial navigation system algorithm block diagram. Taken from Titterton & Weston (2004)

The navigation equations presented in Figure 7 is collectively known as the INS. The output of INS consists of the complete system position, velocity, acceleration, orientation and angular rates of rotation. These parameters are used by the control system as the measurements of the system motion on which the control action will be performed.

In some situations the complete navigational solution is not required and only the aircraft orientation and heading is required by the control system. A standalone GPS unit is then used for position information. Such a unit is called an attitude and heading reference unit (AHRS). The AHRS consist of a simplification of the INS equations presented in Figure 7 by not performing the computations to determine the aircraft position and velocity.

Pure inertial navigation solutions have the advantage of being immune to interference as it does not depend on data received from external sensors. To be usable in an independent configuration, it is necessary to use reasonably high quality sensors. The reason being both gyros and accelerometers are subject to significant sensor measurement noise and the noise directly impacts the navigation solution. Any errors in the accelerometers are integrated twice and any gyro errors are effectively integrated three times during the calculation of the aircraft position. The result is that even a small error in any of the sensors cause the navigational solution to rapidly deteriorate.

IMUs used to perform pure inertial navigation portray very low levels of sensors noise. These very high quality inertial sensors (known as inertial grade IMUs) are very expensive, typically costing in the range of millions of US-dollars and can generally be used to navigate autonomously for periods of time ranging from a couple of hours up to a number of days and even longer depending of the quality of the sensor. Lower quality inertial sensors used to navigate autonomously for about an hour are called sub-inertial grade sensors while the lowest quality sensors are called tactical grade sensors. Sub-inertial grade sensors are generally about an order cheaper than the inertial grade sensors, but are still about an order more expensive than the tactical grade sensors. The problem with tactical grade sensors are, although relatively cheap, they cannot be used as the primary sensors for navigation for any period longer than a couple of minutes before the accumulation of errors start to dominate the navigational solution.

4.1.1.2 Hybrid navigation

The accumulation of errors in pure inertial navigation system and the high cost associated with inertial grade sensors have resulted in the practise of aiding the navigational solution with other sensors. Probably the most widely used navigational aide is the Global Positioning System (GPS), which provides high accuracy position and velocity measurements with bounded errors which can be used to improve the pure inertial navigational solution when combined with the output of an INS (Steven & Lewis, 2003; Farrell, 2008; Rogers, 2007; Groves, 2008; Farrell & Bath, 1999). GPS and INS measurements are combined in an optimal way using an estimation algorithm known as the Kalman filter.

Various configurations for the combination of the data exist, but the actual functionality of the Kalman filter does not differ much. The Kalman filter is a statistical estimator using the noise properties of the various sensors to determine the weighting factors when combining the measurements. References on the applications of Kalman filters to inertial navigation are found in (Steven & Lewis, 2003; Farrell 2008; Rogers, 2007; Groves, 2008; Farrell & Bath, 1999; Maybeck, 1994). A more detailed discussion on Kalman filtering and optimal estimation is presented in (Maybeck, 1994; Simon, 2006; Zarchan & Musoff, 2005; Lewis et. al, 2008). A typical use of the Kalman filter in navigation aiding is the complementary filtering scheme (Bar-Shalom & Li; 2001) as presented in Figure 8.

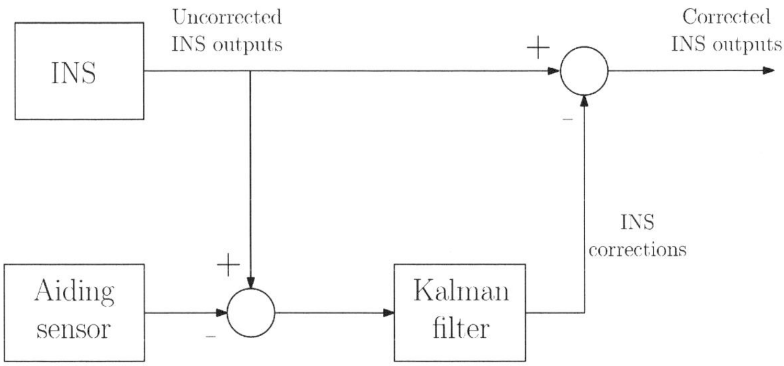

Figure 8. Complementary filtering as used for INS aiding

One of the disadvantages of GPS that became clear in recent years during the Afghanistan and Iraq wars is it can be jammed, either from specifically designed jamming equipment of by accident by transmitters operating in the GPS frequency band. This has motivated significant research into the use of alternative sensors for navigation aiding. The use of optical sensors as vision-based navigation aids is receiving much attention due to the sensor being independent of external transmitters like GPS. The advent of high-powered image processing hardware is contributing to the increased momentum of this method of aiding. Imaging sensors are used as aiding sensors in the following configurations:

- Optical altimeter, where two or more cameras are used to develop an altimeter through stereo vision;
- Feature detection in Simultaneous Localization and Mapping (SLAM), where the image is processed to determine the location of features in the scenes and these features are either correlated with known locations of features or used to develop a map used to update the navigational information;
- Range sensor through stereo vision, where the sensor is not purely used as an altimeter, but where distance to identified features in the image is determine from the combination of multiple images; and
- Angular rate sensor, where the system's angular rate can be determined using optical flow techniques.

Apart from GPS and optical sensors, other sensors frequently used in navigational aiding on UAVs are:

- Magnetometer, utilising the Earth's magnetic field to determine orientation and sometimes position information;
- Barometric sensors, using air pressure to determine the aircraft altitude;
- Laser altimeters, providing very high accuracy altitude measurements; and
- Radar, a ground-based sensor predominantly under military control but sometimes available for civilian use.

4.1.2 Calibration and alignment

Two critical tasks associated with inertial navigation that can severely influence the navigation accuracy are the calibration and alignment of the INS. Calibration is the process whereby the errors associated with the IMU are determined and stored in software for use in an online compensation routine. IMU calibration is an extremely specialized area that requires expensive test equipment. This is an area that should be ventured into with great care and it is therefore suggested that pre-calibrated inertial sensors be used whenever possible. By following this approach the inner technical workings of the IMU is hidden from the application engineer who simply want to use the sensor as a black-box measurement device.

Alignment is the process whereby the initial position and orientation of the IMU is determined to act as initial conditions for the INS integration equations. Traditionally, strapdown inertial systems depended on gyrocompass alignment (also known as static alignment or ground alignment) of the system as described by (Britting, 1971). In this method the accelerometers are used to determine the initial estimate of tilt (roll and pitch) angles of the aircraft through a process known as analytical alignment. Once the tilt angles are known, the rotational speed of the Earth is measured by the gyros to determine the heading of the IMU. The initial alignment results are refined using a Kalman filter while the heading converges to the actual value. The whole process usually takes about 10 to 15 minutes. Position is either determined by positioning the aircraft at a known position on the runway, or from a high accuracy GPS position measurement. Disadvantages of this approach to alignment is that the gyros need to be quite accurate (at least in the high end of the sub-inertial grade sensors) to detect the rotational rate of the Earth performing the heading alignment. In fact, the need for accurate heading alignment is the biggest driving factor when specifying accuracy of the gyros in the IMU. If alternative alignment techniques are used, gyro requirements can be significantly reduced, thereby reducing the total IMU cost. (Farrell, 2008) presents some alternatives for alignment with the use of a magnetometer to determine the heading angle being the most widely used method for low-grade IMUs which is not sensitive enough to detect that Earth's rate of rotation.

4.1.3 Design Considerations: Long Endurance Navigation

From the discussion on navigation presented in the previous sections, the following design considerations can be highlighted for long endurance missions:

- Proper calibration of the navigation sensors and high accuracy alignment before the start of the mission are of utmost importance to ensure good navigation data for long endurance missions.
- Low grade inertial sensors cannot be used to perform autonomous navigation without being aided by an additional sensor. GPS is the logical choice for aiding the low grade IMUs by means of Kalman filter mechanization.
- A high accuracy reliable GPS receiver is essential if lower accuracy inertial sensors are used.
- Although not discussed here, the navigation computations and the method and sequence in which these computations are performed on the navigation computer is very important as any delays in data resulting in skewing or stale data could be catastrophic to the system performance.

4.2 Aircraft Control Systems

The International Aerial Robotics Competition[1] (IARC) has been playing a significant role in accelerating the development of UAVs and specifically in the advancement of the control strategies. Initially when the competition started in 1991, the focus was strongly on the development of the control algorithms, but since then the control systems technology has matured to such a degree that the control system is usually regarded as a "hidden technology" with more focus being placed on the development of the payload sensors, sensor intelligence, autonomous mission planning and collaboration between multiple vehicle systems. The control system developments occurred as part of the IARC have had a significant impact on the control systems of commercial products as many of the algorithms developed have matured into usable products.

A vast array of control algorithms exist that have been implemented with varying degrees of success. In general these control strategies can be divided into the areas of classical, modern and state-space control as presented in the following three subsections.

4.2.1 Classical Control

The term classical control refers to the oldest and most established method of control. A number of techniques fall in this category which consists of a control loop being designed per controlled variable, resulting in a large number of independently designed control loops for large multi-variable systems.

The classical approach to fixed-wing UAV control can be divided into a number of actions, being:

- Attitude stabilization;
- Longitudinal control; and
- Lateral control.

Controllers are applied to perform each one of these actions by means of multiple control loops. The attitude stabilization loop is usually the innermost loop with the highest control bandwidth. The lateral and longitudinal control loops require lower bandwidth actions and these are closed around the attitude loop to control the aircraft altitude, velocity, position and heading. Figure 9 and figure 10 present the classical autopilot multi-loop configurations for the longitudinal (altitude hold) and lateral (heading hold) modes, respectively. In both these systems, the innermost loops stabilize the aircraft rates (q is the pitch rate and p is the roll rate), followed by the aircraft orientation (theta is the pitch angle and phi the roll angle). The outermost loops control the true parameters of interest, being the altitude in the longitudinal mode and the heading in the lateral mode.

---

[1] http://iarc.angel-strike.com/ ;
http://en.wikipedia.org/wiki/International_Aerial_Robotics_Competition

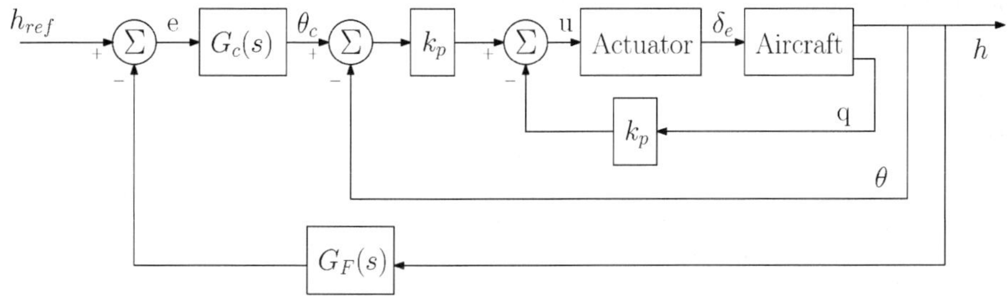

Figure 9. Classical altitude hold autopilot system

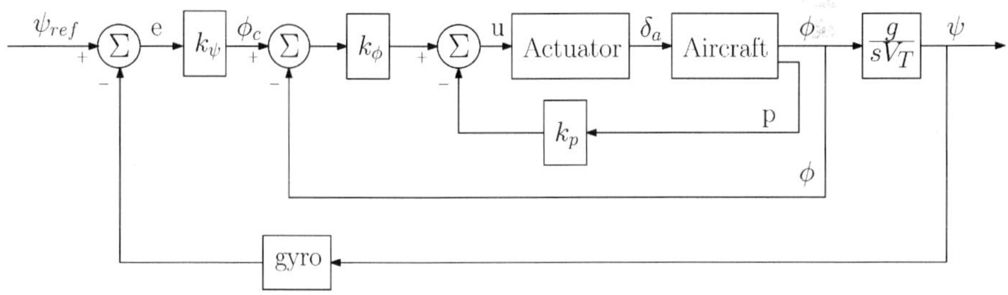

Figure 10. Classical heading hold autopilot system

Probably the most prevalent control strategy amongst all areas of closed-loop control is the proportional-plus-integral-plus-derivative (PID) type of controllers for which the block diagram is presented in Figure 11. The controller consists of three terms that can be combined or left out as required. The advantage of PID control and the reason that this technique is so widely used is the fact that the P (proportional), I (integral) and D (derivative) terms can be very effectively "tuned" in a very short time to obtain a controller that will perform an adequate control action. The "tuning" of the controller can be performed without a dynamics model of the system or without any knowledge of the dynamics of the system. More formal methods could also be used to determine the controller parameters, but the experience-based tuning of the controller terms based on some heuristic rules is most commonly followed.

These heuristic rules can be defined as follows[2]:

- The proportional term KP is used to reduce the system rise time and to reduce the steady state error of the system, but it cannot be used to completely eliminate the error. Used on its own, it is the simplest possible controller and often results in an adequate system response.
- The integral term KI is used to eliminate the steady-state error, but the expense of dramatically slowing down the system response and possibly even making the system unstable.

[2] http://www.engin.umich.edu/group/ctm/PID/PID.html

- The derivative term KD has a stabilizing effect on the system response by decreasing the settling time and the overshoot of the system. It does not have any significant impact on the steady-state error.

These guidelines are generally correct, but deviations from them can be experienced when any of the three parameters is changed, thereby influencing the balance between the three terms. After some trial and error, a general method for tuning the PID terms can be determined. A guideline often followed, is to tune the proportional term until satisfied, then add integral control and finally add the derivative control as necessary.

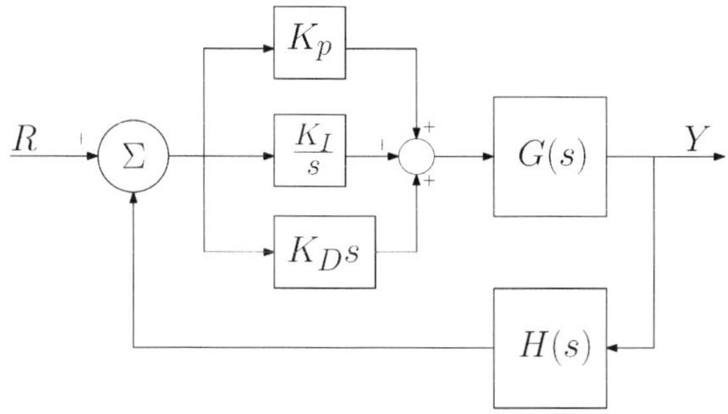

Figure 11. A PID controller

Another classical technique is the lead-lag type of controller which is more of a purist approach to PID control with more options during the control approach. Lead-lag controllers are usually designed using formal design methods such as the frequency domain techniques (using Bode plots) or the root locus method. These techniques are based on a proper dynamics model of the system.

The book of (Franklin et. al., 2006) is an excellent introductory source of information and tools on the design of classical and state space control systems. (Stevens & Lewis, 2003; Roskam, 1979; McRuer et. Al, 1973; Blakelock, 1991) present a detailed discussion of various aspects of aircraft control from the classical perspective.

4.2.1.1 Dynamics model

The use of the proportional-plus-integral-plus-derivative (PID) control strategy on low-cost UAV systems in recent years has resulted in less emphasis being placed on the development of a dynamics model for the aircraft as an adequate controller can be obtained through the tuning of the controller parameters. However, most methods of control system design depend heavily on the availability of an accurate dynamics model of the aircraft. A high fidelity deterministic flight controller cannot be obtained without using such a model. A detailed treatise on dynamic and aerodynamic modelling of aircraft is beyond the scope of this chapter. It is a fairly complex field combining knowledge from a number of disciplines to derive the model to be used in the development of the control system and in a computer simulation to test the control system. (Stevens and Lewis, 2003) presents a thorough discussion on the development of the complete dynamic and aerodynamic model needed for the design of a fixed wing autopilot. (Roskam, 1979) looks in more detail at the methods used to obtain the stability derivatives used to describe aerodynamic forces and moments on the aircraft. It must be emphasized that a proper grasp of the fields of aircraft modelling and simulation are needed to design trustworthy aircraft autopilot systems.

4.2.2 Modern Control

Since the 1960's the development of larger, more complex systems have necessitated that the methods of control design be updated to accommodate these systems. The state-space approach to control system design (generally known as modern control in comparison to the traditional methods that was known as classical control) was developed to address this issue. A number of design techniques based on the state-space approach is in operation on practical systems.

The non-linear nature of the full aircraft dynamics model means that either the high complexity non-linear design techniques will need to be followed, or that a linear approximation to the non-linear model be determined. The last option is usually taken where a number of linear system models are identified by linearizing the non-linear model at a couple of operating points that represent the actual operational conditions of the aircraft. A separate e set of controller gains is determined for each operating point with a different controller being selected based on the operational conditions - a process known as gain scheduling.

The LQR techniques (Sinha, 2007; Skogestad & Postlethwaite, 2005; Stevens & Lewis, 2003) is one technique that are used as part of such a gain scheduling strategy. It requires that a relative high accuracy system model be available for the control system design. Based on the noise properties of the system model and the measurements, optimal feedback control gains are determined that will result in the fastest possible system response while limiting the required control effort needed to achieve this performance. Stevens & Lewis, 2003 presents a detailed discussion of the application of this technique to the design of aircraft control systems where the complexity of the systems and the number of parameters that need to be tuned require modifications to the standard LQR techniques.

For some implementations uncertainty exists in the system model or in the operational environment in which the controller will be used. Under these conditions robust control design techniques are used. The most widely used robust control techniques are the linear quadratic regulator / loop-transfer recovery (LQR/LTR) and the H-Infinity (H-inf or H∞) techniques. Sinha, 2007; Skogestad & Postlethwaite, 2005; and Stevens & Lewis, 2003 discuss these techniques from both theoretical and practical perspectives. The H-Infinity ((H-inf or H∞)) design technique is a robust frequency-domain method of designing controllers for systems that exhibit uncertainty either in terms of the system model used to design the controller, or in terms of the operational environment. The LQR/LTR approach is based on the separation of the controller design into a Kalman filter observer and then a state feedback controller. The advantage of this approach, which is very popular amongst aircraft designers, is that the state feedback controller (the linear quadratic regulator) has some inherent properties that result in a robust controller which is aided with the LTR technique. Stevens & Lewis, 2003 discusses the application of this technique to the autopilot design of fixed-wing aircraft.

4.2.4 Control system hardware

On most UAVs the type of onboard systems and sensors are constrained in terms of:

- Size;

- Weight; and
- Electrical power requirements.

Given the total system cost an important driving factor on most UAVs, the selection of the actual avionics and sensor hardware are of critical importance. This section will attempt to address some of the available options in terms of computer hardware and sensors available for navigation use on a fixed-wing UAVs. An invaluable source of information is the UAS-Info website[3] containing the annual publication of the UVS-International Yearbook. Apart from articles of general interest, the yearbooks contain sorted lists of available IMUs, autopilots, complete UAV systems, payload sensors, power plants and much more.

4.2.4.1 Inertial Sensors

In general the accuracy of an aircraft inertial sensors is directly proportional to the cost of the sensors. The number of manufacturers of both inertial and other sensors are so vast that a discussion is beyond the scope of this chapter. References for available sensors and subsystems are the Reference Section of the UVS-International Yearbook[4] under the heading Autopilots & Inertial Measuring Units and the Association for Unmanned Vehicle Systems International (AUVSI) website[5] and their monthly magazine Unmanned Systems.

The inertial sensor pack is probably the most critical aspect governing the total UAV navigational and positioning performance. It will also be the most expensive component on the UAV, only to be surpassed by very high quality integrated electro-optical sensor systems. The normal design constraints of weight, power and size play slightly less of a role when it comes to the inertial sensors as some accommodation will be made to be able to host the sensor which guarantees mission success. Important factors to consider when buying the inertial sensors are listed below:

- Is the IMU pre-calibrated?;
- Is the sensor-pack accurate enough to allow for gyro-compass alignment? If not, an additional magnetometer or a sensor of equivalent performance including a magnetometer will be required.
- Does the sensor pack have sufficient dynamic range for the type of application?
- Is the environmental specification of the sensor pack compatible with the operational domain of the aircraft? Aspects to consider are shock and vibration, temperature, moisture and operational range in terms of height above sea level.
- Does the system have adequate interface connectors for the type of application? Some requirements could be moisture proof, dustproof or vibration proof connectors.
- Is the data-rate of the sensor output compatible with the requirements of the control system?
- Is the sensor data time-tagged or is the data-latency such that it can be ignored in the system design? Delays in inertial measurements which cannot be precisely characterised results in degraded system performance.

There are four types of inertial sensors available and listed in order of reducing cost are ring laser gyros, mechanical gyros, fibre-optic gyros and MEMS (Micro-Machined ElectroMechanical Systems). Due to the high accuracy inertial sensors being an enabling

---

[3] http://www.uvs-info.com
[4] http://www.uvs-info.com
[5] http://www.auvsi.org

technology for the development of UAVs and for guided weapons alike, high accuracy inertial sensors are governed by ITAR (International Trading in Arms Regulations). The result is a significant amount of time and effort usually have to be spent to obtain the necessary export permits and end-user certificates before the systems are received from the manufacturer. Fortunately the combination of GPS with a low-cost INS usually results in a system of sufficient accuracy. Once again, the combination of these two sensors may also be subject to ITAR, leaving the end-user no other option than buying a low-cost IMU or AHRS, convert this into an INS, and then removing the inaccuracies in the system through the combination with GPS or other sensors.

For the other sensors acting as aiding sensors in the system, the same type of design considerations as for the inertial sensors can be used, with the most emphasis probably being placed on:

- Interface connector integrity;
- Data interface protocol and data latency;
- Sensor accuracy;
- Weight, power and size; and
- Environmental robustness.

4.2.4.2 Control Processor

One of the design considerations for a UAV is whether the complete system will be designed or whether mature building blocks will be used to develop the system. When the hardware for the flight management and flight control system is considered, the "buy" or "develop" issue plays a very important role. Designing an in-house processor board for the control and navigation system does have the advantage of tight coupling between the processor and the various sensor systems, potential low power requirements, low weight and custom form factors, but such custom-built hardware could easily result in immature hardware. Combine the immature hardware with control and navigation algorithms that are itself still under development and the result could be the system responsible for the safe execution of a mission actually putting the mission in danger. It suggested, unless the weight, power (both computational and electrical) and physical size requirements absolutely necessitates it, the decision should rather be made to buy mature hardware and implement the (perhaps immature) algorithms on it.

A number of different hardware products are available for use as onboard processing systems on the UAV. The most widely used and most mature commercially available hardware platform is probably the PC/104 architecture single board computers (SBCs). An example of a flight computer based on the PC/104 architecture developed at the University of Stellenbosch in South Africa[6] is presented in figure 12. One of the primary advantages of the PC/104 architecture is it has a standard form factor. The PC/104 bus (or extensions thereof in the form of the PC/104+ or PCI/104 busses) forms the main communications backbone enabling various cards with the same form factor, but different functionality, to be stacked together building a more compact computer system with higher functional capability than what could have been fitted onto a single card. The stacking of the various PC/104 form factor cards can be seen in Figure 12. The three larger cards shown in the image are PC/104 cards.

[6] http://esl.ee.sun.ac.za/

Some of the PC/104 cards available include:

- The main processor board;
- Power supply boards;
- Analog to digital converter cards that usually include digital to analog converters with some digital input/output pins as well;
- Counter / timers cards;
- Direct servo controller cards used to directly power the control system actuators;
- Serial port expansion cards;
- CAN controller cards; and
- Wireless communication and GPS cards.

A huge number of possible manufacturers and distributors of PC/104 based hardware exist worldwide. It is suggested that a local agent with a good track record as a trustworthy supplier of a well-known international brand be used to supply the hardware.

Another topic of great importance when deciding on the hardware to be used, is the operating system used onboard the flight control or navigation computer. If dedicated hardware is designed, the algorithms are usually implemented using assembly language or the C programming language. This approach has the advantage of direct control over the algorithm sequence of execution, but, as was previously said, it could be at the cost of immature hardware.

Figure 12. PC/104 based flight computer developed at the University of Stellenbosch, South Africa[7]

Most hardware vendors provide a version of embedded real-time Linux customized to work on their respective hardware platforms. The use of a high-level operating system poses significant advantages, especially when interfacing to complex navigational sensor

[7] http://esl.ee.sun.ac.za/

systems. The availability of a realtime operating system is a serious consideration when the decision for the flight computer hardware is made. Independent realtime operating systems such as QNX, VxWorks and Winriver Linux do exist, but these are usually expensive and it is not guaranteed for the software to work with all embedded hardware.
(Valavanis, 2007) presents a discussion on the development of an embedded processing system for a UAV based on commercially available hardware. Although not addressed here, the packaging of such a processing system is very important to ensure that the system will continue to operate under all environmental conditions.

4.2.5 Commercial Solutions

4.2.5.1 Autopilots

For the company that is more interested in the mission and the gathering of the payload data than in the actual development of the control and navigation system, it is an option to consider one of the following commercial-off-the-shelve (COTS) autopilot systems. An almost exhaustive list of available UAV autopilots is published in the reference section of the UVS-International Yearbook[8]. The available systems vary in maturity and reliability of the algorithms and the hardware with fixed wing autopilots generally being more mature than the helicopter autopilots of the same companies.
Some autopilot manufacturers embed the inertial sensors within the autopilot system, thereby eliminating the need for an additional, external sensor system. Although the inertial sensors of the integrated autopilot solutions is not necessarily the highest grade of sensors, the tight integration scheme between the autopilot and the sensor system result in adequate performance. Most high quality commercial autopilots do provide the capability to integrate external inertial sensors with the autopilot. The ability of the system to handle external inertial sensors should be an important design consideration if the accuracy gained from this approach is crucial to the system performance. External inertial sensors often integrated with the autopilot include high accuracy differential GPS systems, laser altimeters or laser scanners, optical sensor information or various onboard or ground-based radar measurement sensors. The inertial sensor integration cost could be excessive and should be addressed during the negotiation and budgeting phases.
As would be expected, most commercial solutions are the products from companies in the USA. This fact has the added disadvantage of the product potentially being subject to ITAR which could make it very difficult for some companies and countries of gaining access to the technology. Not all USA based autopilots are ITAR regulated and fortunately a number of high quality autopilot systems are developed by non-USA companies which are generally available for commercial UAV applications.
Some of the more well-known and commonly used autopilot systems for fixed-wing UAV aircraft are:

- Micropilot - Canada
- weControl - Switzerland
- UAV Navigation - Spain
- Cloudcap - USA
- Microbotics - USA

[8] http://www.uvs-info.com

- Mavionics - Germany

These systems are relatively mature and present a significant variation in cost and performance. The following aspects should be considered during the selection of a commercial of the shelf (COTS) autopilot:

- Weight and size of the hardware;
- Power requirements;
- The system's ability to handle vibration, shock and the excessive temperatures. Very high temperatures are sometimes experienced inside the UAV avionics bay on the runway, followed by very low temperatures a few minutes later at high altitudes;
- Control algorithm used and the accuracy thereof both in terms of positioning and aircraft stabilization;
- The method of aircraft characterization and autopilot adaptation to the airframe;
- Autonomous "return to home" functionality in case of lost of communication with the ground station;
- The availability of a compatible ground control station (GCS) for the autopilot and whether the system includes the necessary communication data link to connect the autopilot with the GCS; and
- The range of altitudes under which the system can operate.

The system's reliability can be determined from available operational data, if available. It is important to look at the following aspects:

- Number of operational units;
- Total amount of flight hours gained by these units;
- Mean time between failures under normal operational conditions; and
- Number of crashes during operational use.

When the long endurance flight requirement is considered the following concepts should receive attention:

- Absolute sensor accuracy and the drift in the total navigation solution are important to consider.
- Temperature and cooling requirements could play a significant role during long flights as the exposure of a system to high temperatures for long periods of time usually leads to component failure.
- The aggressiveness of the autopilot control strategy. A very aggressive control strategy could result in reduced UAV fuel efficiency resulting in reduced duration of the mission.

It is important to note the purchase of a COTS autopilot system should always be combined with some kind of support contract. Contracts usually last for a period of one year from date of purchase, but some companies provide support for the lifetime of the product. Along with the contract it is important to negotiate the terms and conditions associated with the customization of the autopilot to the airframe. The autopilot setup is usually performed by the manufacturer or their agent and the cost associated with the setup alone could be up to 50% of the actual autopilot cost, excluding travel and accommodation fees. It is therefore advised that the complete product lifecycle cost be considered before a final system is selected. The aspects needed to be considered and budgeted for are:

- Actual system hardware cost;
- System maintenance (repair) fees;
- Annual licensing fees (if applicable);

- Cost associated with customization of the standard system, such as DGPS integration or specific data capturing requirements;
- Installation and setup fees; and
- Travel and accommodation cost for the engineers performing the installation and setup.

It is recommended that a team of technicians of engineers be allocated to take part in the setup process and a technology transfer arrangement be made with the autopilot supplier to train the team for any installations of future systems. The team must also be capable of performing all the maintenance and support on the system, thereby reducing the cost of such exercises. In addition to these comments, it should also be noted most companies are open to negotiations on the unit price if multiple units are purchases. Academic pricing structures are also usually available if the product is used for research applications.

4.2.5.2 Complete UAV solutions

Once the scope of the project and the level of technical depth the company is willing to pursue as part of the UAV development have been determined, it is sometimes found to be the most logical and cost effective option to buy the complete UAV system and integrate the locally developed payload with the airframe. The complete system would consist of the airframe, communication, control and navigation and the power supply and would provide the capability to simply integrate the scientific or commercial payload with the UAV.

An example of such an approach that proved to be very successful is the Maldives AUAV Campaign[9] performed by the Scripps Institute of Oceanography in the USA. In this experiment a scientific payload developed by Scripps was integrated with a Manta UAV[10] to determine the concentration of aerosols in the various atmospheric layers above the Maldives Islands. The objective was to perform the scientific expedition without having to perform a technical venture into the unknown field of UAVs, resulting in the buy and integrate philosophy that was followed.

The Aerosonde UAV[11] developed by the AAI Corporation, Australia, has also been known to be used for scientific operations. It was the first UAV to cross the Atlantic Ocean, thereby directly meeting the long endurance requirements laid out in this chapter. It has also been used to gather data on hurricanes, tropical cyclones, carry a NASA microspectrometer for the gathering of high resolution atmospheric data and to fly extensive scientific missions in the Arctic region.

As with the autopilots and the inertial sensors, the UVS-International Yearbook[12] presents an almost exhaustive list of available UAV solutions in the Reference Section under the heading Civil/Commercial, Research & Dual Purpose UAS. It is suggested that UAV manufacturers with systems of interest be contacted directly to determine how products comply with the guidelines set out in this chapter. It should be remembered the abundance of different UAV manufacturers and designs are predominantly due to different missions and operational requirements.

[9] http://www-abc-asia.ucsd.edu/MAC/secure/Index.htm
[10] http://www.acrtucson.com/UAV/manta/index.htm
[11] http://www.aerosonde.com
[12] http://www.uvs-info.com

4.3 Certification

Another aspect that is often overlooked when a COTS autopilot or a complete system is the issue of UAV certification for operation in civilian airspace. This topic is of key concern whenever UAV operations are being planned as autonomous aircraft are generally considered to be a safety risk to commercially operated aircraft. In all circumstances the local civil aviation authority should be contacted to determine the system operational requirements. Worldwide the operational requirements are still being formulated and the approval of a UAV operation is usually performed on a case-by-case basis. The normal process is to lay down a safety case for the operation of the planned UAV and to prove with the safety case the risk to the public and to civilian aviation have been adequately reduced through the addition of redundant sensor systems, certification of the autopilot system or through very strict operational procedures.

Some long-range UAVs operating in civilian airspace are the Global Hawk, Predator and the Aerosonde systems. Looking at these systems, it appears the whole certification process could be eased if the following building blocks are present in the UAV:

- RPV (Remotely Piloted Vehicle) mode, where a ground-based pilot controls the aircraft position and velocity and the autopilot acts in an assistive mode to stabilize the airframe while not being granted complete control over the system's motion. This is a great advantage for line-of-sight operations in particular high-risk areas;
- The availability of a transponder (both a radio for voice communication and an IFF (Identification Friend or Foe)) system to identify the aircraft within the civilian airspace; and
- The certification of the autopilot by the local civil aviation authority in the country where it was designed.

Apart from these guidelines, it must be remembered that the whole certification issue is still much debated and no final agreement has been reached as to when a UAV is considered to be safe for autonomous operation in civilian airspace.

5. Image Based Payload Systems

Long endurance UAVs are extremely well suited for applications where the payload consists of optical image sensors. However, in designing an image-based payload system for an endurance-based UAV application, certain additional requirements need to be considered. In this section we will consider these requirements in more detail. First we will provide a short overview of the typical image-based applications for UAVs in general, but more specifically those applicable to endurance UAVs. This will be followed by a discussion on the different aspects to be considered in the design of image-based payloads.

The US Department of Defense (DoD) categorizes payloads (Department of Defence, 2005) into the four general categories:

- Sensors (electro-optical, radar, signals, meteorological, chem-bio);
- Relay (communications, navigation signals);
- Weapons; and
- Cargo (leaflets, supplies), or combinations of these.

Almost all UAV applications reported sport an image-based payload in one form or another. The range of applications of these image-based payloads is varied, from surveillance and reconnaissance applications to remote sensing applications. However, image-based navigation and tracking is becoming more prevalent, with the imaging system assisting with the navigation of the UAV as an additional sensor input to the other navigational sensors (e.g. GPS and inertial system).

Research in the past has focused on low-level control capability with the goal of developing controllers which support the autonomous flight of a UAV from one way-point to another (Doherty, 2004). Focus on research has moved from low-level control towards a combination of low-level and decision-level control integrated in sophisticated software architectures.

Image-based payloads have increased in sophistication, functionality and algorithm complexity. The ever decreasing size and cost of sensors and increasing processing power have ensured that more functionality can be provided for the weight and power budgets.

For sensor technology, miniaturisation of a specific sensor technology is a forerunner to cost implosion. Once sensors get cheap, the consumer market reaps them up in consumer products and possibly the high-volume military market (infantry) (Wilson, 2002). The automotive industry (a high volume market of approximately 60 million cars produced every year) is usually the last adopter, because it represents the worst combination of low cost and high reliability.

Situational awareness in the automotive market involves all kinds of sensors. Sensors with integrated lasers for 3D imaging or ranging will become significant, as can be seen in the parking-distance sensors now available in most vehicles, and the infrared nigh-vision detectors used by some cars for collision avoidance. For COTS design involving sensors, one should therefore carefully consider the life-cycle of sensor technology in the automotive industry.

With the number of sensors per system (including UAVs) increasing, the finite failure probability per channel (uncertainties per channel) could concatenate in the end until you can't trust the output. Such a multi-sensor approach requires a new system architecture. One has to calibrate from a systems point of view, where any combination of sensors can fail and the system will still function accurately. The system must be taught the net error, rather than compartmentalising the uncertainty, and how to accommodate it with correction algorithms. It is smart sensors combined with smart systems design (Wilson, 2002).

In considering the design of a long-endurance UAV, the design of the image-based payload system should be carefully considered to fit into the overall design specification of the long-endurance UAV. As was previously stated, the major factors in the design of a long endurance UAV is power efficiency and weight.

Intelligent UAVs will play an equally important role in civil applications. There is a desire for more sophisticated UAV platforms where the emphasis is placed on development of intelligent capabilities and on abilities to interact with human operators and additional robotic platforms.

Civilian and commercial missions of UAVs will drive the synthesis of UAV technologies (especially image-based payloads) to ensure a high degree of safety, facilitating the integrating of UAVs within the air traffic management system (Okrent, 2004). This is an important point to consider if UAVs are to be adopted on a large scale for use in a commercial sense.

UAV maturity should be understood at a number of levels: improvement in mission reliability and readiness, in operational safety and increased autonomous operation while at the same time experiencing a significant reduction in cost.

5.1 Overview of Applications

Before considering the detailed design considerations for long endurance imaging payloads, it is necessary to evaluate the range of applications for the imaging system, where each application drives its own design requirements. The list of image-based applications for UAV payloads presented here is bound to increase in the future and become out of date before publication, but it is worthwhile to list some of the image-based applications currently reported in the literature. Some specific applications are elaborated on more completely to identify key design issues.

5.1.1 Scientific Missions

A number of researchers throughout the world have documented the use of UAVs to perform scientific investigation of remote areas and aerial surveys for specific missions - land mapping, tropospheric sampling etc. (Niranjan , 2007). One of the essential tasks for any aerial explorer is to be able to perform scientifically valuable imaging surveys. The high-level of mobility, and unique observational perspective, provided by aerial vehicles makes them potentially extremely valuable tools.

Scientific research of any nature (environmental, atmospheric, archaeological, pollution, etc.) can be carried out by UAVs equipped with the appropriate payloads. An inconclusive list of scientific applications reported are the following (Okrent, 2004; Sarris, 2001):

- Atmospheric research;
- Geological surveys;
- Hurricane evolution and research;
- Oceanographic observations;
- Volcanoes study and eruption alert;
- Weather forecasting;
- Archaeological surveys;
- High accuracy terrain mapping
- Atmospheric heating measurement programmes;
- Cloud study programmes;
- Glacier and ice sheet programmes;
- River flow and discharge missions;
- Ozone layer studies and monitoring;
- and many others.

The payload for scientific applications is often more than just a visible imaging system, including multispectral imaging, LIDAR, and other types of sensors, mostly in combination.

Satellites are the proverbial work house when it comes to scientific image-based applications, especially with regards to monitoring and analyzing the natural environment. However, satellites and High Altitude Long Endurance (HALE) UAVs have complementary applications: Satellites provide worldwide coverage while HALE UAVs provide regional and local applications. HALE UAVs also have lower altitude, and therefore provide the opportunity of higher resolutions, typically 10cm to 5cm.

Most scientific studies utilising UAVs require data collection over extended periods of time and are therefore well suited to long endurance UAVs (Awan, 2007).

Photogrammetric recording of archaeological sites

Archaeological site recordings often use photogrammetric methods to obtain accurate 3D models. Photogrammetry is a research area concerned with obtaining reliable and accurate measurements from noncontact imaging (Fisher, 2005). These applications require a combination of GPS/INS stabilisers for the imaging platform for highly reliable image acquisition (Eisenbeiss, 2006). Standard helicopters or airplanes with photogrammetric cameras are too expensive and do not allow flying close to objects or have the capacity for complicated manoeuvring. While the GPS/INS unit enables semi-automatic navigation along a predefined flight path, the stabilisers ensure a stable flight attitude and thus a highly reliable image acquisition (Eisenbeiss, 2005).

In a recent report (Bendeaa, 2007) an archaeological site survey was documented. The survey was conducted at an altitude of 100m. Each photogrammetric flight plan was designed setting the UAV speed at 15 m/s, which meant a shooting time interval of about 1,5 s (60 m) and 2,5 s (100 m) to get adequate frame overlaps.

For any photogrammetry application camera calibration is of great importance for the accuracy of the final product. This has an implication in the design of a UAV, especially when off-the-shelf cameras are used (e.g. Digital SLR cameras with zoom capabilities), as vibration becomes a factor, as well as lens movement during zooming operations.

The internal orientation parameters (focal length, coordinate of principal point and radial lens distortion) of the camera was estimated using a self-calibration approach. An a priori estimation was performed using a self-developed calibration software, but it was found a non stability of the radial distortion parameters. This was probably due to the fact that the lens is retractable, that means it move in and out during switching on/off operations.

A comparison between high altitude and low altitude UAVs found that a higher height allows the pilot to better control the UAV speed and to follow a straight line. On the other hand low altitude images are more detailed, allowing an improvement of the accuracy during the 3D feature extraction. The obtained 3D accuracy was found to be suitable for archaeological evidences mapping at a very large map scale.

5.1.2 Emergency Missions

UAVs are well suited to emergency missions. Reports on UAVs used for emergency missions almost always include imaging payloads, often with real-time video streaming. Some typical applications reported are the following (from (Okrent, 2004)):

- Disaster operations management;
- Fire fighting;
- Oil slick observations;
- Flood watch;
- Volcano monitoring;
- Catastrophic situation assessment;
- Search and rescue (looking for survivors from shipwrecks, aircraft accidents, etc.);
- Hurricane watch;
- Earthquake monitoring; and
- Nuclear radiation monitoring.

Most of these missions require extended time in the air, and also real-time video to the ground control station. This implies a good communications link with relatively large bandwidth, which impacts on the power budget of the UAV.

Forest fire detection and monitoring

Forest fire fighting often happens in rough terrains which need immediate line-of-sight communications to fire fighters. UAVs are starting to be employed in supporting the fire fighting efforts (Awan, 2007; Merino, 2002; Merino, 2006).

Frequent updates concerning the progress of a forest fire are essential for effective and safe fire fighting. Fire fighters need frequent and high quality information of a fire's development to conduct an effective and safe fire fighting mission (Casbeer, 2005). Low altitude UAVs can capture high resolution imagery and broadcast frequent updates to fire crews. In fire monitoring missions, the objective is to image the perimeter of the fire and upload the location of the fire perimeter (with associated imagery) to the base station as frequently as possible.

Since fire is growing and changing directions, UAVs need intelligent path planning ability using limited real-time information. They also need the intelligence to return to base (Casbeer, 2005).

Fire fighting monitoring requires that the UAV have long endurance. Depending on the area to be covered, a high speed communications link will be required over reasonable distances. This has a direct impact on the power budget of the endurance UAV.

Fire detection

Some of the image processing issues to address in a detection application (Merino, 2002) are techniques for fire segmentation and techniques for fire geo-location. Feature matching methods are also used to track points employed by geo-referencing. To geo-locate fire features, a very accurate camera position estimation is required. Image processing techniques can be used besides GPS to help with the geo-referencing of the features.

In (Merino, 2006), multiple heterogeneous UAVs are used in detecting and localising fires. The heterogeneity here refers to different sensors employed on the various UAVs, all working in unison towards the same mission goals. Although it increases the complexity, the authors reported that it also provided several advantages, i.e. different sensors in the air which may be impossible to carry on one UAV.

The UAVs are reported to have motorised pan and tilt units to allow orientating the cameras independently from the body of the vehicle. Encoders are used to measure the pan and tilt angles. Various combinations of low cost infrared and visual video cameras are employed. The UAVs were further equipped with DPGS, gyroscopes, and IMUs. Some of the software functions reported in the software architecture are:

- Feature matching (to differentiate fire pixels from background pixels);
- Fire segmentations (employing various processing functions, i.e. histograms, thresholding, look up tables);
- Geo-referencing;
- Motion compensation (image stabilisation).

For geo-location, the sensors onboard the UAVs in (Merino, 2006) are used to compute, in a global and common coordinate frame, the position and orientation of each UAV itself and also of the sensors that are carried on board. For the latter, the UAV attitude angles measured by the IMU units have to be combined with those of the pan and tilt devices.

A lot of attention is placed on the mission aspects, especially with regards to the multitude of UAVs. The mission is decomposed in the following stages: fire search, fire confirmation, fire observation. The recovered position of the fire could be determined within 1 meter of the actual position.

One issue of concern is that of image compression for bandwidth conservation, as it may influence the image processing algorithms. Experimental work was needed to find the right compression factor. All communications were carried out using 802.11 wireless systems. Experiments demonstrated that if the UAV motion is smooth, speed moderate and the distance to the wireless base shorter than 100 meters a robust communication channel can be established. For other conditions a more robust data link would be required. As image processing is carried on board, the required bandwidth is considerably reduced and can be sustained by a radio-modem data link.

It was also found that sensor coordinate system can generate serious problems, especially in light of multiple UAVs cooperating.

Search and rescue support

The UAV support for search and rescue missions is a complicated task requiring thousands of hours of search over large and complex terrains, again signalling the requirement for long endurance. The discussion to follow is mostly extracted from (Adams, 2007), and clearly shows the various issues involved in the application (which was wilderness search and rescue). However, also see (Westall, 2007) for a maritime search and rescue missions.

When studying classical search theory (in the sense of "search and rescue"), a critical factor in designing an optimal search is determining the instantaneous probability of detection by one glimpse of the target. An observer that must make a target classification in real-time must progress slowly enough to ensure enough glimpses of the potential target to obtain a satisfactory probability of detection. The goal of 100% target detection is in conflict with the goal of searching the largest area possible. This implies that long UAV flight hours will be logged.

The basic steps for a successful UAV searching for evidence include:

- Aiming the camera to make it likely that visual evidence appears in the video; and
- Identifying the evidence's location in order to guide the rescue team to the missing person.

Imagery is acquired by planning a path, flying the UAV and controlling the camera viewpoint to ensure that imagery is obtained of the complete search area. During this phase, the control variables are the speed and path of the camera's footprint over the ground.

Finding items of interest in the provided imagery is a challenging task for an autonomous algorithm. Design variables at this stage are pixel density, field of view, image stability, contrast between the sign (evidence) and background.

Locating a sign of the target with a UAV to constrain the search requires three activities:

- Analysing imagery;
- Localising the sign; and
- Refining the imagery.

In analysing imagery, the goal is to find the target's location. The key variable is the probability that a human can detect a target in an image given a set of image features, which is influenced by how the information is obtained and presented.

Localising the target is called geo-referencing. Localisation can be performed autonomously by using a GPS, the UAVs pose, triangulation, terrain information, and image features.

Some additional design considerations and technologies reported are the following (Adams, 2007):

- The ability to enhance raw video through stabilisation, mosaicking, and image enhancement;

- The ability to autonomously maintain height above ground; and
- The ability for the UAV and the support team to coordinate effectively.

5.1.3 Surveillance Missions

Airborne surveillance is applicable to a wide range of applications. The main objective of these kind of missions is to enable UAVs to acquire and interpret data in real-time, followed by decision-making in terms of signalling an alarm, while flying over a targeted area (Kontitsis, 2004).

The first applications for U.S. military unmanned aerial vehicles was surveillance, to be the "eyes in the sky" in operations where it was too dangerous to send manned aircraft or just too expensive.

"Eye-in-the-sky" surveillance could assist law-enforcement agencies in the protection of citizens and the integrity of borders. Road traffic, pipelines, power cables, forests, volcanoes, etc. could be continuously or selectively monitored.

Some typical applications reported are the following (from (Okrent, 2004)):

- International border patrol;
- Environmental monitoring;
- Law enforcement;
- Road traffic monitoring and control;
- Coastline monitoring;
- Maritime patrol;
- Drug traffic monitoring; and
- Crop and harvest monitoring.

Surveillance missions often have a endurance requirement, high altitude ability, video transmission, etc.

Road traffic surveillance

A good example of surveillance missions is that of traffic surveillance. A good research survey is presented in (Puri, 2004) providing an overview of the types of research being conducted. Traffic surveillance is a prime example of long endurance flights.

The mission of roadway transportation agencies is to focus on the needs of the travelling public. This requires collection of precise and accurate information about the state of the traffic and road conditions. It also requires timely information on road emergencies.

Aerial views provide better perspective with the ability to cover a large area and focus resources on the current problems. It is mobile and able to be present in both space and time.

Satellites are not ideal in road traffic monitoring applications for the following reasons:

- The transitory nature of their orbits make it difficult to obtain the right images for continuous problems (e.g. traffic tracking); and
- Cloud cover on days with bad weather results in bad image quality.

UAVs can move at higher speeds than ground vehicles, fly in potentially dangerous conditions, view a whole set of network of roads at a time and inform the base station of emergency or accidental sites.

Typical applications of UAVs in traffic monitoring are the following (Puri, 2004; Nordberg, 2002; Rathinam, 2005):

- Incident response;
- Monitoring of freeway conditions;
- Coordination between a network of traffic signals;

- Traveller information;
- Emergency vehicle guidance;
- Track vehicle movements in an intersection;
- Measurement of typical roadway usage;
- Detection of specific road events (overtaking, U-turns);
- Estimation of various features of a vehicle such as velocity and type;
- Monitor parking lot utilisation; and
- Estimate origin-destination flows.

UAVs for traffic monitoring may be equipped with a range of interchangeable imaging devices and sensors:

- Day and night real-time video cameras;
- Infrared cameras;
- Multi-spectral and hyper-spectral sensors;
- Thermal sensors;
- Synthetic aperture radar (SAR);
- Moving target indicator radar;
- Laser scanners;
- Chemical, biological and radiological sensors;
- Road weather information systems to record the necessary information such as weather, fire and floods; and
- Communications hardware to relay data to the ground station.

From a software architecture point of view, a three-layer agent architecture seems to be prevalent (Coradeschi, 1999):

- A process layer for image processing and flight control;
- A reactive layer that performs situation-driven task execution; and
- A deliberative layer mainly concerned with planning and monitoring.

Barriers to the wide scale adoption of UAVs in unrestricted air space are the lack of standards and regulations international in the various civil aviation authorities. The FAA requires UAVs to have onboard "detect, see and avoid" capabilities to prevent in-air collisions. A fail-safe option for the mission must automatically apply if the ground to UAV communications link fails. Further, the communications regulatory environments also regulate the licensing of spectrum, which may inhibit the real-time transmission of high quality images due to a lack of bandwidth.

Based on the above discussion, some of the following observations can be made (from (Nordberg, 2002)):

- Tracking requires a certain amount of planning to be made by higher levels of system to predict where and how a vehicle is going to move on the ground, manage occlusions on the ground, etc.
- Another scenario is to make the UAV assist a ground vehicle to either intercept a tracked vehicle or to get a specific location in an urban area with traffic jams. This requires the UAV to be able to detect and estimate the size of the traffic jams, and to find free paths around it.
- A UAV should be able to autonomously navigate between points, track ground vehicles, and estimate various features and detect events related to individual vehicles or pairs or even large sets of vehicles.

- Advanced image processing tasks (road detection (Kim, 2005), landmark tracking, and motion detection) require spatial or spatio-temporal filtering of the camera images. This includes filtering for detection of lines/edges, estimation of their orientation, detection of corners and other local symmetries.

5.1.4 Communications Missions

The long endurance nature of UAVs with high altitude capabilities make UAVs suitable to act as communication relay stations (Okrent, 2004). Some applications in this regard are:

- Broadband communications;
- GPS augmentation system;
- Telecommunication relay service; and
- Pseudo satellite.

5.1.5 Industrial Applications

With the commercialisation of UAV airframes, the lower cost of COTS-based vision payloads and increasing experience, it is to be expected that industrial applications should follow. Some examples of industrial applications are the following (Okrent, 2004; Sarris, 2001):

- Crop spraying;
- Nuclear factory surveillance;
- Mining and exploration;
- Power line monitoring;
- Pipe line monitoring; and
- Agricultural applications.

Some discussion on some of these industrial applications will now be provided:

High voltage power line monitoring

Power lines need to be monitored for several reasons, e.g. detecting cracks in isolators, monitoring undergrowth to prevent flashovers due to fires, etc.

Most often power line monitoring is done by foot patrolling or by using helicopters. Both of these methods require manpower and time, both costly. Helicopters are used for this purpose; however this is a dangerous and costly option. For these missions, pilots have to fly the helicopter at low altitude and close to the power lines. This is a tedious job with a high level of fatigue as the pilot has to keep the helicopter on a fixed altitude. UAVs are very well suited to this (and similar) kind of problem (Awan, 2007).

For power line monitoring missions, images are usually recorded by the UAVs video camera and transmitted to the ground station, where the operator can interrupt the flight plan at any time to change the zoom of the power line monitoring camera (Sarris, 2001).

The image-based payload plays an important part in these missions. Some of the issue to consider are the following for these kinds of missions:

- Endurance. The power lines to be monitored may be tens to hundreds of kilometres in length, and constant refuelling or charging may delay the process.
- Automated flight. As a result of the distances and times involved, the requirement is that navigation need to be automated (through GPS way-points), but will also include visual tracking of the power lines.
- Zoom levels. Tracking of the power lines require vision at some distance. Isolators and possible cracks need to be investigated much closer.

- Multispectral imaging. The imaging payload for this application may require multispectral imaging for various purposes, i.e. visual imaging for navigation and general inspection, ultra-violet to detect corona effects, etc. IR can also be used for tracking purposes.
- Automation of visual monitoring and detection. The level of repetitiveness for this kind of mission is high, with the result that artefacts of importance may be missed due to fatigue and boredom of an operator.
- Large on-board processing power and memory. The image processing algorithms involved are complex, in addition to the high resolution required. This will impact on the processing power and the memory on-board the aerial platform.

Surveillance of pipelines

Pipeline conditions are causing damage and environmental issues, with an increasing demand for some form of inspection technology. Pipelines are also vulnerable targets for terrorists and therefore need to be monitored from a security point of view (Awan, 2007).

Similar to traffic monitoring missions, satellite monitoring of pipelines presents some problems. An additional problem is that the plumes (methane) leaked from pipe lines disperse into the atmosphere rapidly without being properly detected. Therefore UAVs may be the best option to perform this task. Presently, most of the pipelines are monitored with helicopters or ground inspections. Many gas companies in the USA, Russia, France and Germany are using UAVs for pipeline monitoring on an experimental basis.

The design considerations for monitoring pipelines are similar to that of power line monitoring, although there may be some differences in the multispectral imaging and image processing algorithms.

There are various environmental and human activities that need to be monitored on a routine basis to ensure the safety of the huge investment in gas pipelines. Some of the monitoring activities are:

- Construction Work;
- Earth Movement and Excavation;
- Lying of pipes, cables etc.;
- Erection of buildings;
- Soil upheaval and erosion;
- Water logged surfaces;
- Plantation of shrubs and trees; and
- Discolouring of vegetation.

The purpose of an aerial imaging platform in such a scenario is to perform object detection and location, and moving vehicles. The spatial requirement for aerial images translates to the following:

- Monitoring strip of 200m
- Metallic machinery 2m X 1m
- Objects e.g. pipes 0.2 m X 5 m
- Excavations 0.5m X 5m 0.5m
- Tree tops diameter : more than 2 m
- Location accuracy : 5m

The UAV platform should also ensure weather independency and surveillance frequency of more than once in a week.

Mineral exploration and exploitation

The remote sensing payload ability of some UAVs has enabled it to collect data for exploration of minerals deposits (Awan, 2007).

Agricultural applications

UAVs with IR and visible-light optical sensor payloads take images of crops. These images are combined into colour graphics to show how the crop is growing and indicate areas of blight. In addition to the above, moisture levels in the soil and the amount of plant life can also be measured.

In Japan, UAVs are used extensively for the purpose of crop spraying.

5.1.6 Navigation

The autopilots found in most of the reported autonomous or semi-autonomous UAVs require real time information of the UAV to accomplish their tasks, e.g. the attitude, velocity, position and acceleration. This data is then passed on to the flight control system which, based on a PID control law, actuate the servos to deflect the control surfaces. It is important that if onboard processing of payload is required that there is redundant processor along with flight control processor (Niranjan , 2007).

In certain environments (urban or in forests), satellite visibility may be limited. In these cases, the GPS accuracy may be negatively impacted. In such cases, it is necessary to determine position and attitude through other means, such as through visual means (Caballero, 2005). An example of such an application will be that of a Mars rover.

Image processing methods are also used for safe landing, on stationary bases or on moving targets such as ships.

5.2 Design Considerations for Image-Based Payload Systems

5.2.1 General Considerations

In designing image-based payload systems for UAV applications, one should consider the following generic issues:

- Flexibility - the ability of an UAV system to perform not only its original mission, but additional missions which were not originally envisioned. This should be done without change to the system;
- Adaptability - the ability of a system to perform not only its original mission;
- Upgradeability - the ability of a system to be changed (or reconfigured) enabling it to perform additional missions;
- Reliability - the ability of a system to be flexible, adaptable and/or upgradeable while still being able to operate for many years; and
- Scalability - the ability of a system to perform its original mission to a much greater or smaller extent.

In general, the major design decision for long endurance UAVs will be based on a trade-off between power requirements, weight, functionality, reliability and cost.

5.2.2 Integration Considerations

A UAV can be defined as consisting of several subsystems, where each subsystem provides some necessary functionality. The concept of "loose coupling" whereby subsystems are not tightly integrated, but rather coupled in a manner allowing for at least semi-independent evolution, helps engineers design highly complex systems (Dahlgren, 2006).

The concept of a "performance range" for subsystem specifications will likely lessen the number of hard technical requirements, enhance the opportunity for system evolution, decrease program risk, and improve the opportunity for major systems to be delivered within the overall performance, schedule and budgetary constraints.

5.2.3 Image Processing Algorithms

Image processing is a general term covering all forms of processing of captured image data. Often, the term machine vision is also used, which is a general term for processing image data by a computer. There is a tendency to use the term "machine vision" for practical vision systems, such as industrial vision systems (Fisher, 2005) used in manufacturing.

The image processing field is a mature field, and an extensive knowledge and application based is available. There is an extensive set of models, algorithms and computational structures that are ready for implementation.

There is no shortage of models and algorithms for image processing; however there is a shortage of effective, well engineered implementations. Implementing an algorithm into a hardware system is a non-trivial task (Barrows, 2002) that will require skill, patience and experimental work.

A good start is to read any one of several machine vision books available. A development tool commonly used is Matlab, together with an image processing toolbox which provides a set of basic image processing algorithms. General image processing frameworks are available (e.g. Intel's OpenCV) that can shorten custom development time; however this depends on the hardware and operating software architecture.

For long endurance UAV applications, the choice of image processing algorithms will play an important role in the design consideration. As the complexity of the algorithm (and hopefully the functionality as well) increases, the processing requirements will also increase. Another design consideration is also the resolution of the images to be processed by the algorithms. As the image resolution increases, the number of pixels to process increases too, adding to the processing burden.

Another approach, depending on the application, is to do just enough image processing on-board the UAV (e.g. for navigational purposes) and then do the complex processing at the ground station. In this case, one should consider the trade-off between memory requirements and communications requirements. However, long endurance UAVs by definition may capture data (images) for a long time during its mission, which implies that it may require access to large on-board memory.

Some common image processing tasks will be now discussed in the context of long endurance UAVs:

Image enhancements

Image enhancement is a general term covering a number of image processing operations that alter an image in order to make it easier for humans (or another algorithm) to perceive (Fisher, 2005). There are several techniques to enhance the resulting image, with increasing degrees of complexity. Obviously, the specific application will drive the development and complexity of the image enhancement algorithms. The first rule is to capture the best possible image in the first place, by considering the optical path design (including the sensor), the resolution, shutter speed, aperture, etc.

Some of the simpler image enhancements that can be done algorithmically are the following:

- Image denoising;
- Contrast adjustments;
- Histogram equalisation;
- Colour enhancements;
- Blurring;
- Sharpening;
- Cropping;
- Scaling; and
- Rotating.

These operations can usually be implemented effectively on normal processors (CPUs) or digital signal processors (DSPs). Translations (e.g. rotation) can become computing intensive.

However, in challenging applications, more advanced image enhancement algorithms will be required. As an extreme example, atmospheric turbulence (or heat shimmer) can become a severe problem on a relatively hot day on images taken over a large horizontal distance. This example may be quite relevant to long endurance-based UAV missions. The resulting image may blur and lose too much detail to be useful. However, compensating for this is still an open research problem with complex algorithms, requiring high levels of processing power.

Super-resolution of images is also possible using several low resolution images of the same object viewed from different angles. The implication of this is that in video sequences, frame rate can possibly be exchanged with resolution - again depending on the application.

Object and event recognition

An interesting consideration for long endurance surveillance-type missions is the resulting long video sequences that may result. Most of these footage will be eventless and therefore of no interest. Automated scanning of this video footage (in addition to enhancements) for significant events or objects will greatly increase the usability of the data (Li, 2005).

However, for navigational purposes, object recognition may be important.

Image Mosaicking

Image mosaicking is the composition of several images, to provide a single, larger image with covering a wider field of view (Fisher, 2005). Depending on the application, images collected may be required to be stitched before further processing is possible (Horcher, 2004). For this purpose, good feature points are necessary. Good feature points invariable require good edges or corners, as often found in roads, forest edges, and a stream course. However, with more uniform textures in pictures, determining good feature points can become a problem. Better feature identification algorithms (e.g. SIFT) may be an option. However, state information from the on-board payload sensors may be an additional input.

Design considerations here will be the following:

- Large resolution images may be difficult to process on-board the UAV; and
- Use state information from the other sensors as additional input.

5.2.4 Image and Video Compression

In the design of long endurance UAVs, one should also consider the use of image and video compression as part of the design trade-off. When considering compression, redundancy is removed, substantially reducing the size of the images. Typical UAV video sequences may be well suited to good compression ratios, as the scenery doesn't change too often.

In the selection process of an image compression algorithm, one should consider the lossy nature of the algorithms. Certain compression algorithms permanently remove information without too much loss of visual quality (called lossy compression). This may be adequate for visual perception, and if that is the only application of the video, one should seriously consider this. An example of such a lossy compression algorithm for images is the jpeg compression algorithm. However, if high quality images are required, especially for image processing algorithms, one should rather consider loss-less compression algorithms.

Some benefits of using compression algorithms are the following:

- Reduced onboard memory requirements for the same sequence;
- Longer flight times as more images may be stored on the same onboard memory footprint;
- Reduced bandwidth requirements from the onboard communications link; and
- Higher quality pictures for the same memory footprint;

There are certain issues to consider:

- Increased processing requirements, depending on the algorithm employed;
- The trade-off between compression ratios and the quality of the images;
- The delays introduced by the compression algorithms on video transmission; and
- The sensitivity of the compressed images to bit errors. Bit errors may be introduced by several noise sources, e.g. EMI onboard and normal transmission errors. Error correction coding should be considered when compression is introduced. This may add to the processing complexity.

In general, taking the operating environment into consideration, one should experiment with different algorithms, compression ratios and error correction coding schemes.

5.2.5 Communications Link

The communications link is an important design consideration for any UAV design, especially as it relates to the power budget. However, special consideration should be given to the imaging payload requirements for the communications link. As the distance between the UAV and the ground control station increases, the signal-to-noise ratio of the transmission link will deteriorate, with the introduction of additive errors. Further, if there are reflections in the signal due to mountains and buildings, fading of the transmission link can be expected.

Local signal interference may also reduce the throughput rate. In cases where the IEEE 802.11 series of protocols are used (e.g. WiFi) for the transmission links, one should consider the contention nature of the communications link and its effect on throughput.

Images and video streams invariantly translate into high bandwidth requirements for the communications link. As the resolution increases, the bandwidth requirements increase. In video sequences, an additional consideration is the frame-rate. The higher frame-rate may require higher bandwidth, but may be required to detect fast-moving artefacts.

Some design considerations for the communications link with regard to the image payload are the following:

- Consider using only the necessary frame rate for video necessary. This will be determined by the requirements of the application, e.g. sampling rate of object movement, averaging of frames, etc.

- Consider using the necessary resolution. Again, this directly impacts the bandwidth requirements of the data link;
- Perform some image processing onboard to lower the transmission requirements. As an example, event detection can be used to only transmit sequences of interest; a region of interest can be determined on-board to transmit only those areas to the ground station;
- Consider the use of image compression, together with the trade-off required;
- If real-time imaging is not required, consider increasing the on-board memory for storage of large sequence;
- If real-time imaging is required, determine the delay, image quality, etc. required and incorporate it into the communication link design;
- Consider the various modulation schemes, communications protocols, as well as Medium Access Layer (MAC) protocols for optimal usage of the available frequencies;
- Evaluate the different frequency bands available for use in the region, i.e. the various ISM-band frequencies, the bandwidth afforded and the transmission properties of the band.

In most cases, a combined trade-off will need to be made, taking a multi objective optimisation approach.

5.2.6 Processing Considerations

Image processing algorithms by nature are processing intensive. Image processing algorithms are also dependent on image resolution. As the complexity of algorithms increase, the requirement for processing power can also be expected to increase.

The typical selection of the processing unit for image processing applications are usually made from various families of Digital Signal Processors (DSPs), Field Programmable Gate Arrays (FPGAs), and Central Processing Units (CPUs).

The expectations are that Moore's law will continue for at least the next 5 to 10 years, and provide increasingly capable processing hardware (Barrows, 2002).

Parallel processing

It is well known that most image processing algorithms are well suited to parallel processing architectures. Massively parallel digital processing is no longer exotic but mainstream, and will become even more important in the future. The lower cost of parallel architectures is driven by higher volumes. Note that by using parallel processing, Moore's Law growth rate is doubled. However, this increase is only achieved if the applications are implemented to utilize parallel architectures (Barrows, 2002). A simple increase in processing cores (as done in the Intel and AMD processors) increases the inter-process communications burden. The thread processing model introduces a level of complexity, and run into scaling problems.

Since much of image processing consists of repeating the same instruction over and over again on different data, it makes sense to use parallel architectures in future machine vision systems. There remains the challenge of knowing what algorithms to implement in such a processor for a given application.

In the past few years, the use of Graphics Processing Units (GPUs) for general purpose processing has steadily increased. The recent (2008) population of GPU architectures are all based on stream processing architectures and very well suited to image processing. The literature on image processing algorithms on GPUs is exploding with most papers reporting at least an order of a magnitude (or two) jump in speed. GPU technology is driven by the gaming market, which has exploded in that past two years due to the proliferation of gaming consoles. The implication of these high volumes is the low cost of these GPU cards. The interesting consequence of this is the real-time execution of complex algorithms.

However, there are a few caveats to consider before selecting GPU technology, especially as it relates to long endurance UAVs (Wosylus, 2008):

- GPUs are approaching the 2 TFLOPs (1012 floating point operations per second) on a single card at a low cost, providing in most cases more processing power than required. However, the power requirements of these cards are extremely large (150 watts or more), and definitely not suitable for most long endurance UAV applications;
- Programming for GPU architectures is not simple and requires a fair amount of retraining. Porting algorithms to the GPU can be time consuming and costly;
- Standard GPU cards for consumer gaming computers are often discontinued after just a few months, this being the typical lifecycle for standard computer boards intended for the consumer market. This has serious implications on long term support and modularity;
- The image processing bottleneck will probably not be based on the processing power anymore, but on the data flow between sensors, CPU and the GPU;
- Reliance on products from the consumer market (such as GPU cards), will be rewarded by significant expenditure during the product's lifecycle: frequent driver updates, limited reliability due to fan failures;
- The physical dimensions of consumer boards and their cooling designs often conflict with the embedded principles of compact dimensions, simplified cooling and standardised form factors; and
- The option of designing proprietary graphics capabilities is even more problematic, since the components used could be discontinued before the finished design goes to market.

To utilise the power of GPUs the following suggestions are made:

- Consider the selection of newer embedded boards that are released with PCI-e ports, allowing for the integration of GPU-based boards into the design;
- Lower power versions of GPU cards are available, but with a subsequent decrease in processing power. Through experimentation, determine the minimum level of processing required and select that GPU card - this will limit the power requirements;
- Develop algorithms on an open platform (we suggest OpenGL) to gain as much hardware independence as possible. This will also provide some protection from hardware changes and newer versions of the standard, as OpenGL is backwards compatible;
- Some embedded processing boards are released with an integrated, low power GPU onboard, which are OpenGL compliant; and
- Distribute the processing of the algorithms in such a way that the compute intensive algorithms run on the ground station where a GPU can be introduced, while lighter processing tasks takes place on-board the UAV. However, this translates the problem from a processing problem into a communications problem, i.e. the high bandwidth required to transmit the video stream to the ground station. This will obviously be dictated by the specific application. Algorithms intended for navigational purposes, as

will be safety and avoidance tasks will most likely stay on-board, while more mission-related algorithms may be transmitted to the ground station for processing.

For long endurance UAV payloads, the processing power has a direct affect on the power consumption.

5.2.7 Mounting

In the development and selection of an image-based payload system, special attention must be given to the mounting of the UAV payload, as the importance of mounting cannot be stressed enough (Kahn, 2001).

Some reasons to regard the mounting of the payload system are the following:

- A good mount will save the system from failure many times over;
- A good mount will make the electronics run more reliable; and
- A good mount design must be tested by extensive experimentation.

As part of a long endurance mission, the payload will be submitted to prolonged and high levels of vibration. For a vision-based payload, the vibration may adversely affect the basic functioning, but also the accuracy of the devices. For this reason, one need to conduct a wide range of vibration tests of components and modules to get an idea of the level of vibration each device can withstand. The payload vibration mount must then be designed to meet the needs of the most sensitive device.

A rule of thumb is to try to make the device that is to be damped as heavy as possible to increase its inertia (Kahn, 2001). The more the inertia, the more force is needed to accelerate the box. This large inertia works with the damping material to isolate the hardware. Therefore, from a vibration damping point of view it is advisable to place devices together in a single box (to increase weight) and then to isolate the box as a whole. However, this may introduce additional EMI problems.

The materials to be used in the mount should also be selected depending on the level of damping needed. The damping material under consideration should have two key properties:

- Strong and tear resistant; and
- Soft and flexible.

Additional considerations for the mounting of the long endurance UAV payload are the following:

- It should provide good stability, especially if accurate measurement sensors are to be mounted on it;
- If required, it should provide weather protection;
- It should allow for easy access to the payload to affect changes and replace modules;
- It should provide general protection for the electronics; and
- It should be light weight.

5.2.8 COTS Consideration

One of the biggest cost-raising factors that manufacturers have to deal with in the UAV industry is the purchasing of parts. The relative small (but growing) number of suppliers in relation to the manufacturers gives them the opportunity to suppress the market by providing their products at high prices and low variety.

A lot of the sub-assembly parts come from the area of high technology, which makes them expensive to acquire by nature (Sarris, 2001).

"Plug-and-play" design allows for any new device, so long as it satisfies the standards of the system, to be inserted into the system without needing to do any software configuration. Advantages of using COTS components are the following:

- Less expensive; and
- Utilise industry standards in power, size, and communications data buses.

Some COTS industry standards to consider when designing a long endurance UAV are the following:

- Form factor (physical dimensions), e.g. PC104 for processing boards;
- Processor type;
- Communications; and
- Power requirements.

One issue to take cognisance of is that many modules of a supplier are of the "closed-modular" design, modular within a specific vendor's range (Kahn, 2001). By restricting the mating characteristics of modules it becomes difficult and expensive to take advantage of new technology. This will be loosely coupled.

Maintaining the modularity while using industry standards allows for future upgrades and low-cost parts replacements.

5.2.9 Optical Components

In the design and selection of optical imaging components, there are several issues to consider. Again, the weight factor comes into play, but certainly also quality factors (lens quality, camera quality, electrical interfaces and connectors). One should consider the whole optical path when designing the image-based payload. A suggestion is to consider a good machine vision handbook on the design of the different aspects involved (See Handbook of Machine Vision (Hornberg, 2006)).

Image and video standards are evolving all the time and the standards are driving the costs down. HDTV formats are being introduced in consumer cameras, allowing for high definition video. This aspect should be taken into consideration when selecting imaging components as following standards generally drives down the cost and increase modularity.

Image stabilization is critical to obtaining usable information (Department of Defence , 2005). Technology improvements in stabilization technology (electromechanical and electromagnetic) permit nominal sensor mounting systems to achieve stabilization accuracies in the tens of micro radians.

5.2.10 Human-Machine Interface

The design of an operator interface (ground control station) and the UAV autonomy is essentially a problem of human-robot or human-machine interaction, and a "catch-all" solution for all aircraft and all applications is unlikely to emerge (Adams, 2007).

UAV-human interaction design is fundamentally interrelated with UAV autonomy design, and a multi-dimensional trade-off between precision, response time, neglect tolerance, portability, and team size. It is desirable to develop autonomy and operator interfaces that span multiple application domains as much as possible.

The capabilities of a particular combination of airframe, autopilot, and control algorithm delineate the set of affordances that frame the human-UAV interaction space, and the set of constraints on the kinds of tasks that can be performed.

Increased AUV autonomy may produce:

- Higher neglect tolerance;
- Decreased operator workload; and
- Better fan-out ratios.

Autonomy can also result in negative consequences (Adams, 2007):

- Reduced situational awareness;
- Difficulty in supervising autonomy;
- Increased interaction time; and
- Increased demands on humans and autonomy, the "stretched system" effect.

5.2.11 EMI/RFI

EMI/RFI integration is the hardest part to understand (Kahn, 2001). A computer with an oscillator may not work together with other components, e.g. it interferes with the radio control subsystem, and can cause complete loss of control (Kahn, 2001). For long endurance UAV missions, this can become a problem as long distances come into play with subsequent lowering of signal to noise rations in the communications/data ling. Shielding should be carefully consider. However, as remarked in (Kahn, 2001), it is mostly a trial and error process to find part causing problems.

5.2.12 Geo-referencing

Photogrammetry in general deals with the three dimensional object reconstruction from two dimensional imagery (Cramer, 2001). To fully utilise UAV-acquired, remotely sense imagery for applications such as change detection or situational awareness, the imagery must be geometrically corrected and registered to a map projection. This process is referred to as geo-referencing which requires the scaling, rotating and translating from the image coordinate system to the map coordinate system (Hruska, 2005).

Traditionally, ground control points were used to register the imagery. However there are a number of issues with using ground control points with UAV missions:

- There is a high cost associated with the collection of ground control points;
- With many missions performed by UAVs, the ground control points are either not available or impossible to obtain;
- The small footprint of low-altitude UAV acquired imagery complicates the geo-referencing process because such imagery lack distinguishable ground control points.

The introduction of low-cost inertial and optical sensors has advanced the field of vision-based navigation. In these systems, information extracted from digital images is combined with inertial measurements to estimate position, velocity, and attitude.

Direct referencing high-resolution still imagery from small UAVs requires tight integration between the Global Positioning System (GPS), an Inertial Measurement Unit (IMU) and an imaging sensor to acquire exterior orientation parameters at the time of image exposure (Veth, 2008).

Sensor placement should be driven by two objectives (Hruska, 2005):

- Isolating the inertial measurement unit (IMU) from the airframe vibrations, avoiding differential movements; and
- Minimizing boresight misalignment (any offsets between the GPS antenna, IMU and the perspective centre if the imaging sensor.

Synchronisation refers to referencing the individual sensor components to a common clock and input trigger (Hruska, 2005). Even small errors associated with improper synchronisation can have a serious impact on the direct referencing process and mission success.

Prior to the use of the system for mapping purposes, calibration must be performed. This concerns both the calibration of individual sensors (aerial camera or an IMU) and the system calibration due to the aircraft mount. Thereby the differences in orientation between the IMU and the image sensor(s) - known as boresight orientation(s), must be determined (Legat, 2006, Skaloud, 2003).

The relative orientation between the inertial and optical sensors is a critical quantity which must be determined prior to operation of the system. The accuracy with which the relative orientation is known effectively sets the lower bound for the navigation accuracy of the system.

The ultimate goal of the alignment process is to determine the relative orientation between an optical and inertial sensor, and more specifically, the sensitive axes of both sensors. The methods used to estimate this orientation fall into two categories: mechanical and estimation-based techniques.

Mechanical techniques use mechanical measurements (such as laser theodolites) to determine the relative orientation between known fiducials on each sensor. This method requires knowledge of the relationship between the sensitive axes of the sensors and the external reference fiducials, which is subject to unknown manufacturing errors. In addition, depending on the required accuracy, this method requires external equipment which is not really suitable for use in the field.

Estimation-based alignment techniques utilise actual sensor measurements while subjecting the system to known conditions. In one method, the sensors are mounted on a calibrated pendulum while imaging a reference pattern. The orientation of the scene detected by the optical sensor, combined with the current local gravity vector, is used to estimate the relative orientation and inertial sensor biases.

These current approaches require dedicated equipment which would increase the difficulty of field calibrations. In addition, these "captive" techniques separate the calibration and navigation functions of the system and cannot compensate for time-varying errors due to temperature changes, flight profile, flexure modes, etc.

In the most ideal case, all sensors are attached to a common rigid mounting structure, preventing variations of their relative positions and orientations. However, this is not always achievable and the effect of this must be considered in the design.

Several accuracy problems can be experienced related to reliability (instrument and/or data quality) or incorrect use (unstable sensor mount, uncompensated effects due to platform stabilisation, or inadequate datum/projection transformation procedures). A common error is the placement of the GPS antenna at another, remote position on the UAV. With a stabilised IMU/GPS platform, rotations of the stabilised platform introduce position offset change (lever arm) of the GPS antenna from the IMU (Legat, 2006).

Potential error sources in direct geo-referencing that may diminish the quality of data (Legat, 2006) are the following:

- Lack of rigidity of the sensor assembly, including lever-arm variations caused by the platform stabilisation;
- Synchronisation errors or non-compensated sensor delays, e.g. time shifts between a camera trigger command and the actual shutter time of the camera;

- Calibration errors of sensor lever arms and boresight angles or failure to apply these parameters correctly; and
- Erroneous settings or assumptions of the GPS/INS processing.

5.2.13 Power Requirements

The goal of the payload system is to gather and provide accurate date. Any disruption in power or disturbances may affect the integrity of the information.

The power system helps to transmit interference between components of the UAV. As an example, components start and stop at different times, causing power spikes on the power bus. This can reset other devices, causing another spike, eventually leading to periodic interference. In (Kahn, 2001), the author experimented with different configurations and components and suggested the use of DC-DC converters, as it will save valuable time in system integration. These converters are isolated, separating the interference and power spikes from one device to the rest of the power bus. However, one should determine through experimentation which devices work together.

Sensors (including optical) are getting smaller, and more energy efficient. However, as sensors get smaller, power dissipation densities increases substantially (Wilson, 2002). The direction all sensors will be moving is toward un-cooled. The main reason for this is cost. Un-cooled IR sensors would also reduce the weight, size and power requirements. The final limitation on the size of IR sensors is dependent on how far you want to see and the optics you put in front of it. The electronics will continue to shrink and get cheaper, but at some point you reach the limit on physical size because of the optics you're looking through.

Every component of the aircraft, sensor, and data link strives for small size, weight, and power consumption. For endurance UAVs, batteries with high power/weight ratios are important to maximize sensor capability and endurance.

5.2.14 Data Storage

One of the consequences of long endurance UAV mission is that large amounts of data will be collected and stored, mostly onboard. Onboard storage of sensor data in the terabyte class is a goal that the USA Department of Defence is pursuing (Department of Defence , 2005). Storage of complex imagery or phase history of radar data onboard can substitute for the extremely wideband data links required for near-real-time relay. Similarly, storage of the full output of a hyper spectral sensor will allow transmission of selected bands during a mission and full exploitation of data post-mission.

A 1.4 Terabyte storage capability coupled with an imagery index system and IP-enabled interface has been demonstrated on Global Hawk (Department of Defence , 2005). Known as the Advanced Information Architecture (AIA), this system permitted the capture of over 3 days of full resolution Global Hawk imagery and enabled users to access the imagery using internet search tools. The storage system and IP server were constructed using COTS components and integrated into the existing space allocated to the recorder suite.

5.2.15 Software Design Considerations

System software development and integration is a major challenge for UAVs as payload functionality and complexity increases. UAVs are getting more commercialised and long term support of complete systems will be required. A consequence of this is that careful consideration will have to be given to software engineering aspects. As can be expected, different tools, modelling abstractions and engineering skills are utilized at various stages of a typical UAV development project.

In (Joshi, 2002) the author suggests a software framework as a domain-specific software architecture that provides the means to interconnect software subsystems. A software framework needs to provide the "plumbing" necessary to integrate software at different levels of abstraction. Frameworks provide partially-implemented patters of behaviour that are finalized for a particular application.

The use of a software framework can drastically reduce development time while creating robust and flexible application software. A software framework will also provide well-defined integration mechanisms that help reduce development time.

A suggested UAV software framework that combines a hierarchical software component architecture is suggested in (Joshi, 2002) but similarly in (Wilson, 2002):

- Software is developed at various levels of functionalities in a UAV;
- At the lowest level, the drivers that interface with the hardware are found. This layer abstracts the hardware in a meaningful way for use at higher levels of functionality;
- The middle level is the low-level servo controls, necessary to achieve primitive control behaviours. It needs periodic sample loops running at fixed rates; and
- The last level adds higher levels of intelligence and controls. Typically event-driven behaviour, conditional logic and sensor-based decision making.

Modern UAVs are inherently distributed systems involving multiple processor nodes (Heintz, 2004). Therefore, a software framework for UAVs should also provide a communications infrastructure to enable deployment of distributed systems. Adaptation of distributed software to maintain the best possible application performance in the face of changes in available resources is an increasingly important and complex problem (Karr, 2001). The use of middleware for real-time embedded applications is found in several UAV applications, with most referring to the Corba middleware architecture.

Key characteristics of software frameworks and their impact on UAVs are listed in the following:

- It should be component-based. Component-based design facilitate reusability, clarity, reconfigurability and interoperability;
- The separation of interfaces from implementations. Benefits derived from this is replaceability (components can be replaced with others), extensibility (software can be extended), and the ability to leverage domain expertise;
- Hierarchical architecture. This provides complexity management (software elements can be packaged into coherent components), high-level programming, and scalability and layering (higher levels of abstraction can be realized by combining low level components); and
- Enhance integration capabilities.

6. Conclusion

Traditionally the field of inertial navigation systems was limited to high cost applications. However, recent advances in low cost MEMS based inertial sensors have resulted in the development of inertial navigations systems for the commercial market. The availability of these low cost navigation systems enabled a number of applications ideally suited to commercial UAV operators. In this chapter a number of applications for long endurance

UAV were discussed. Long endurance UAV flights require a number of aspects to be taken into consideration during the design phase. In section two an overview of potential renewable power sources for long endurance UAVs were presented. It was shown how a hybrid combination of photovoltaic cells and Li-Po batteries can fulfil the requirements of a long endurance UAV power source. Fuel cell power sources are attractive power sources for shorter duration UAV flights.
Section three showed by coupling the low cost inertial navigation system to a suitable control system how a complete navigation solution can be provided for long endurance UAV flights. A number of control techniques are discussed enabling the construction of autopilots for autonomous flight applications.
The field of image processing is a rapidly developing field. Since imaging sensors are ideal UAV payload sensors, advances in image processing directly benefits many UAV applications. A number of sensor payload design consideration are discussed with regard to long endurance UAV missions.
In an overview paper in 2007, Kenzo Nonami (Nonami, 2007) indicated the following aspects as important future research areas for UAV civilian use:

- Formation flight control (for data relay, in-air refuelling, observation work) with a possible flight control accuracy in the cm-order;
- Integrated hierarchical control in order to fly different classes of UAVs simultaneously. An example sited is that of coordinating various sizes of UAVs and MAVs with a larger supervisory UAV;
- High altitude flight, e.g. flights in the stratosphere for scientific observation missions;
- High precision trajectory following flight;
- All weather flight;
- Collision-avoidance systems;
- Intelligent flight control and management systems; and
- Advanced reliability.

Long endurance UAVs are an empowering technology for a number of previously unexploited applications which are now within reach of the civilian commercial market.

7. References

Adams, J.A.; Cooper, J.L.; Goodrich, M.A.; Humphrey, C.; Quigley, M.; Buss, B.G. & Morse, B.S. (2007); Camera-equipped mini UAVs for wilderness search support: *Task analysis and lessons from field trials*, Byuhcmi Technical Report 2007-1

Awan, S.H. (2007); UAV applications in atmospheric research, *MSc Aerodynamics*, Cranefield University School of Engineering

Barbir, F. (2005); PEM Fuel Cells: Theory and Practice (Sustainable World Series); *Academic Press*; ISBN-10: 0120781425, 978-0120781423

Barrows, G.L. (2002); Future Visual Microsensors for Mini/Micro-UAV applications, *Proceedings of the Workshop on Cellular Neural Networks and their Applications*

Bar-Shalom, Y. & Li, X. (2001); Estimation with Applications to Tracking and Navigation; John Wiley & Sons; ISBN - 047141655X; Hoboken, NJ, USA

Bendeaa, H.; Chiabrandoa, F; Tonolob, F.G. & Marenchino, D. (2007); Mapping of archaeological areas using a low-cost UAV - The Augusta Bagiennorum test site,

Proceedings of the 21st International CIPA Symposium, 01-06 October 2007, Athens, Greece

Bird, R. E. & Halstrom, R. L. (1981); A Simplified Clear Sky Model for Direct and Diffuse Insolation on Horizontal Surfaces; Solar Energy Research Institute; SERI/TR-642-761

Blakelock, J.H. (1991); Automatic Control of Aircraft and Missiles - Second Edition; Wiley InterScience; ISBN - 0471506516; Hoboken, NJ, USA

Britting, K.R.(1971); Inertial Navigation Systems Analysis; Wiley Interscience; ISBN - 047110485X; Hoboken, NJ, USA

Caballero, F.; Merino, L.; Ferruz, L. & Ollero, A. (2005); A visual odometer without 3D reconstruction for aerial vehicles. Applications to building inspection, *Proceedings of the 2005 IEEE International Conference on Robotics and Automation* Barcelona, Spain

Casbeer, D.W. & Li,S.M.(2005); Forest fire monitoring with multiple small UAVs, *Proceedings of the 2005 American Control Conference*, June 8-10, 2005, Portland, USA, pp.3530-3535

Conway, B. E. (1999); Electrochemical Supercapacitors: Scientific Fundamentals and Technological Applications; Springer; ISBN-10: 0306457369, 978-0306457364

Coradeschi, S.; Karlsson, L. & Nordberg, K. (1999); Integration of vision and decision-making in an autonomous airborne vehicle for traffic surveillance, *Computer Vision systems*, Springer, Berlin/Heidelberg

Cramer, M. (2001); On the use of Direct Georeferencing in Airborne Photogrammetry, *Proceedings of the 3rd. International Symposium on Mobile Mapping Technology*, January 2001, Cairo

Dahlgren, J.W. (2006); Real Options and Value Driven in Spiral Development, *Proceedings of the CCRTS 2006*, March 31, 2006

Department of Defence (2005); Unmanned Aircraft Systems Roadmap: 2005 - 2030,

Doherty, P. (2004); Advanced Research with Autonomous Unmanned Aerial Vehicles, *Proceedings on the 9th International Conference on Principles of Knowledge Representation and Reasoning*

Eisenbeiss, H.& Zhang, L. (2006); Comparison of DSMs generated from mini UAV imagery and terrestrial laser scanner in a cultural heritage application, *Proceedings of IAPRS* vol. 36, part 5, Dresden 25-27 September 2006

Eisenbeiss, H.; Lambers, K. & Sauerbier, M. (2005); Photogrammetric recording of the archaeological site of Pinchango Alto (Palpa, Peru) using a mini helicopter (UAV), *Proceedings of the 33rd CAA Conference*, Tomar, Portugal, 21-24 March 2005

Farrell, J.A. & Bath, M. (1999); The Global Positioning System & Inertial Navigation; McGraw Hill; ISBN - 0-07-022045-X; New York, USA

Farrell, J.A. (2008); *Aided Navigation - GPS with High Rate Sensors*; McGraw Hill; ISBN - 978-0-07-149329-1; New York, USA

Fisher, R.B.; Dawson-Howe, K.; Fitzgibbon, A.; Robertson, C.& Trucco, E. (2005); *Dictionary of Computer Vision and Image Processing*, John Wiley & Sons, Ltd, England, 2005

Franklin, G.F. & Powell, J.D. & Emami-Naeini, A. (2006): *Feedback Control of Dynamic Systems* (Fifth edition); Addison-Wesley, ISBN - 978-0131499300, Reading, MA.

Groves, P.D. (2008); Principles of GNSS, *Inertial and Multisensor Integrated Navigation Systems*; Artech House; ISBN - 1-580-53255-1; Boston, USA

Heintz, F. & Doherty, P. (2004); A distributed architecture for UAV experimentation, In the *Proceedings of the AAAI Workshop on Anchoring Symbols to Sensor Data*, 2004.

Horcher, A. & Visser, R.J.M. (2004); Unmanned aerial vehicles: Applications for natural resource management and monitoring, *Proceedings of the Council Of Forest Engineering*, 2004.

Hornberg, A.(2006); *Handbook of Machine Vision*, Wiley-VCH, 2006, ISBN 3-527-40584-4

Hruska, R.C.; Lancaster, G.D.; Jerry. H.; Harbour, L. & Cherry, S.J. (2005); Small UAV-Acquired, High-resolution, Georeferenced Still Imagery, *Proceedings of the Wildlife Society 12th Annual Conference*, Baltimore, Maryland,28-30 May 2005

Joshi, R.; Bose, A. & Breneman, S. (2002); Efficient development of UAV electronics using software frameworks, *Proceedings of the AIAA 1st Conference and Workshop on Unmanned Aerospace Vehicles*, Portsmouth, Virginia, May 20-23, 2002

Kahn, A.D. (2001); The Design and Development of a Modular Avionics System, *Master Thesis*, Georgia Institute of Technology, April 2001.

Karr, D.A.; Rodrigues, C.; Krishnamurthy, Y.; Pyarali, I.; Loyall, J.P. & Schantz, R.E.(2001); Application of the QuO Quality-of-Service Framework to a Distributed Video Application, *Proceedings of the Third International Symposium on Distributed Objects and Applications* (DOA'01), pp. 299, 2001.

Kim, Z.W. (2005); Realtime road detection by learning from one example, *Proceedings of the Seventh IEEE Workshops on Application of Computer Vision*, 2005. vol.1, pp.455-460, 5-7 Jan. 2005

Kontitsis, M.; Valavanis, K.P. & Tsourveloudis, N. (2004); A UAV Vision System for Airborne Surveillance, *Proceedings of IEEE International Conference on ICRA '04*, pp. 77-83, 26 April-1 May 2004

Legat, K. & Skaloud, J. (2006); Reliability of direct georeferencing - A case study on practical problems and solutions, *Report to EuroSDR Commission*, 2006

Lewis, F.L. & Xie, L. & Popa, D. (2008); Optimal and Robust Estimation with an Introduction to Stochastic Control Theory; Taylor & Francis Group LLC; ISBN - 0-849-39008-7; Boca Raton, FL, USA

Li, Y.; Atmosukarto, U.; Kobashi, M.; Yuen, J. & Shapiro, L.G. (2005); Object and event recognition for aerial surveillance, *Proceedings- SPIE the international society for optical engineering*, 2005, vol. 5781, pp. 139-149

Luque, A. & Hegedus, S. (2003); *Handbook of Photovoltaic Science and Engineering*; John Wiley & Sons; ISBN 0471491969, 9780471491965; Hoboken, NJ, USA

Maybeck, P.S. (1994 - reprint); *Stochastics Models, Estimation, and Control* - Volume 1; Navtech Books & Software Store;

McRuer, D. & Ashkenas, I. & Graham, D. (1973); *Aircraft Dynamics and Automatic Control*; Princeton University Press; ISBN - 0691024405; Princeton, NJ, USA

Merino, L. & Ollero, A. (2002); Computer vision techniques for fire monitoring using aerial images, *Proceedings of IECON 02 (28th Annual Conference of the Industrial Electronics Society)*, vol.3, pp.2214-2218, 5-8 Nov. 2002

Merino, L.; Caballero, F.; Martinez-de Dios, J.R.; Ferruz, J. & Ollero, A. (2006); A cooperative perception system for multiple UAVs: application to automatic detection of forest fires, *Journal of Field Robotics*, vol.23, issue 3-4, pp.165-184, 2006

Muneer, T.; Gueymard, C. & Kambezidis, H. (2004); *Solar Radiation and Daylight Models*; Butterworth-Heinemann; ISBN 0750659742, 9780750659741

Niranjan, S.; Gupta, G.; Sharma, N.; Mangal, M. & Singh, V. (2007); Initial Efforts toward Mission-specific Imaging Surveys from Aerial Exploring Platforms, *Proceedings of the Map World Forum*, January 22 - 25, 2007, Hyderabad, India.

Nonami, K. (2007); Prospect and Recent Research & Development for Civil Use Autonomous Unmanned Aircraft as UAV and MAV, *Journal of System Design and Dynamics*, Vol.1, No.2 Special Issue on New Trends of Motion and Vibration Control pp.120-128, 2007

Nordberg, K.; Doherty, P.; Farneback, G.; Forssen, P.; Granlund, G.; Moe, A. & Wiklund, J.(2002); Vision for a UAV helicopter, *Proceedings of IROS'02, Workshop on aerial robotics*, October, 2002

Okrent, M.(2004); Civil UAV Activity within the framework of European Commission Research, *Proceedings of the AIAA 3rd "Unmanned Unlimited" Technical Conference*, Workshop and Exhibit, Chicago, Illinois, Sep. 20-23, 2004

Puri, A. (2004); A survey of Unmanned Aerial Vehicles (UAV) for Traffic Surveillance, *Technical Report* 2004

Rathinam, S.; Kim, Z.; Soghikian, A. & Sengupta, R. (2005); Vision based following of locally linear structures using an Unmanned Aerial Vehicle, *44th IEEE Conference on Decision and Control*, 2005 and 2005 European Control Conference, pp.6085-6090, 12-15 Dec. 2005

Rogers, R.M. (2007); *Applied Mathematics in Integrated Navigation Systems* - Third Edition; American Institute of Aeronautics and Astronautics; ISBN - 1-56347-927-3; Reston, VA, USA

Roskam, J. (1979) ; *Airplane Flight Dynamics and Automatic Flight Controls*; Roskam Aviation and Engineering Corp.; ISBN - 1-884-88517-9; Lawrence, Kans.:

Sarris, Z.(2001); Survey of UAV applications in Civil Markets, *Proceedings of the 9th Mediterranean Conference on Control and Automation*, Dubrovnik, Croatia, June 27-29, 2001

Simon, D. (2006); *Optimal State Estimation*; Wiley InterScience; ISBN - 0-471-70858-5; Hoboken, NJ, USA

Sinha, A. (2007); Linear Systems - *Optimal and Robust Control*; Taylor & Francis Group LLC; ISBN - 0-8493-9217-9; Boca Raton, FL, USA

Skaloud, J.& Schaer, P.(2003); Towards a More Rigorous Boresight Calibration, *Proceedings of the ISPRS International Workshop on Theory, Technology and Realities of Inertial/GPS/Sensor Orientation*, Commission 1, WG I/5, Castelldefels, Spain, Semptember 9-12

Skogestad, S. & Postlethwaite, I. (2005); *Multivariable Feedback Control Analysis and Design*; John Wiley & Sons; ISBN - 0-470-01168-8; Chichester, England

Stevens, B.L. & Lewis, F.L. (2003); *Aircraft Control and Simulation* - Second Edition; John Wiley & Sons Inc.; ISBN - 0-471-37145-9; Hoboken, NJ, USA

Titterton, D.H.; & Weston, J.L.(2004). *Strapdown Inertial Navigation Technology* - Second Edition, American Institute of Aeronautics and Astronautics; ISBN - 1-56347-693-2; Reston, VA, USA

Valavanis, K.P., (2007) *Advances in Unmanned Aerial Vehicles - State of the Art and the Road to Autonomy*; Springer, ISBN - 978-1-4020-6113-4, Dordrecht, The Netherlands

Van Schalkwijk, W. A. & Scrosati, B. (2002); *Advances in lithium-ion batteries*; Springer; ISBN 0306473569, 9780306473562

Veth, M.; Anderson, R.; Webber, F. & Nielsen, M. (2008); Tightly-Coupled INS, GPS, and Imaging Sensors for Precision Geolocation, *Proceedings of the 2008 National Technical Meeting of the Institute of Navigation*, January 28 - 30, 2008, San Diego, California

Westall, P.; Carnie, R.J.; O'Shea, P.; Hrabar, S. & Walker, R.A. (2007); Vision-based UAV maritime search and rescue using point target detection, *Proceedings of the 12th Australian International Aerospace Congress*, 19-22 March 2007

Wilson, J.R.(2002); The Incredible Shrinking Sensor, *Military & Aerospace Electronics Magazine*, March 2002

Wosylus, A. (2008); High-end graphics conquers embedded computing applications, In *Boards & Solutions - The European Embedded Computing Magazine*, Vol.2, pp.6-10, April 2008

Zarchan, P. & Musoff, H. (2005); *Fundamentals of Kalman Filtering: A Practical Approach* – Second Edition; American Institute of Aeronautics and Astronautics; ISBN - 1-563-47694-0; Reston, VA, USA

23

Tracking a Moving Target from a Moving Camera with Rotation-Compensated Imagery

Luiz G. B. Mirisola and Jorge Dias
Institute of Systems and Robotics - University of Coimbra
Portugal

1. Introduction

In our previous work [Mirisola and Dias, 2007b, Mirisola and Dias, 2008], orientation measurements from an Attitude Heading Reference System (AHRS) compensated the rotational degrees of freedom of the motion of the remotely controlled airship of Fig. 1. Firstly, the images were reprojected in a geo-referenced virtual horizontal plane. Pure translation models were then used to recover the camera trajectory from images of a horizontal planar area, and they were found to be especially suitable for the estimation of the height component of the trajectory. In this paper, the pure translation model with best performance is used to recover the camera trajectory while it images a target independently moving in the ground plane. The target trajectory is then recovered and tracked using only the observations made from a moving camera and the AHRS estimated orientation, including the camera and AHRS onboard the airship, as it is shown in Fig. 2(b), and results in a urban people surveillance context with known ground truth. To compare our pure translation method with an image-only method, the camera trajectory is also recovered by the usual homography estimation and decomposition method, and the target is also tracked from the corresponding camera poses.

GPS also can be utilized to recover the airship trajectory, but GPS position fixes are notoriously less accurate in the altitude than in the latitude and longitude axes, and this uncertainty is very significant for the very low altitude dataset used in this paper. Uncertainty in the camera orientation estimate is the most important source of error in tracking of ground objects imaged by an airborne camera [Redding et al., 2006], and its projection in the 2D ground plane is usually anisotropic even if the original distribution is isotropic. The Unscented Transform [Julier and Uhlmann, 1997], which has been used to localize static targets on the ground [Merino et al., 2005], is thus used to project the uncertainty on the camera orientation estimate to the 2D ground plane, taking into account its anisotropic projection.

Kalman Filters are utilized to filter the recovered trajectories of both camera and the tracked target. In the airship scenario, the visual odometry and GPS position fixes can be fused together by the Kalman Filter to recover the airship trajectory. The target trajectory is represented, tracked, and filtered in 2D coordinates. In this way the full geometry of the camera and target motion is considered and the filters involved may utilize covariances and constants set to the physical limits of the camera and target motion in actual metric units and coordinate systems. This should allow for more accurate tracking than when only pixel coordinates in the images are utilized.

Figure 1. An unmanned airship and detailed images of the vision-AHRS system and the GPS receiver mounted onto the gondola

1.1 Experimental Platforms

The hardware used is shown in fig. 1. The AHRS used are Xsens MTi [XSens Tech., 2007] for the airship experiment and a Xsens MTB-9 for the people tracking experiment. Both AHRS models use a combination of 3-axes accelerometers, gyroscopes and magnetic sensors to output estimates of their own orientation in geo-referenced coordinates. They output a rotation matrix ${}^{W}\mathbf{R}_{AHRS}|_i$ which registers the AHRS sensor frame with the north-east-up axes. The camera is a Point Gray Flea [Point Grey Inc., 2007], which captures images with resolution of 1024 × 768 pixels, at 5 fps. The camera is calibrated and its images are corrected for lens distortion [Bouguet, 2006], its intrinsic parameter matrix $\mathbf{K}$ is known, and f is its focal length. To establish pixel correspondences in the images the SURF interest point library is used [Bay et al., 2006].

1.2 Definitions of Reference Frames

The camera provide intensity images $\mathbf{I}(x, y)|_i$ where x and y are pixel coordinates and i is a time index. Besides the projective camera frame associated with the real camera (CAM) and the coordinate system defined by the measurement axes of the AHRS, the following other reference frames are defined:

- World Frame $\{\mathcal{W}\}$: A LLA (Latitude Longitude Altitude) frame, where the plane $z=0$ is the ground plane. It is origin is an arbitrary point.
- Virtual Downwards Camera $\{\mathcal{D}\}|_i$: This is a projective camera frame, which has its origin in the centre of projection of the real camera, but its optical axis points down, in the direction of gravity, and its other axes (i.e., the image plane) are aligned with the north and east directions.

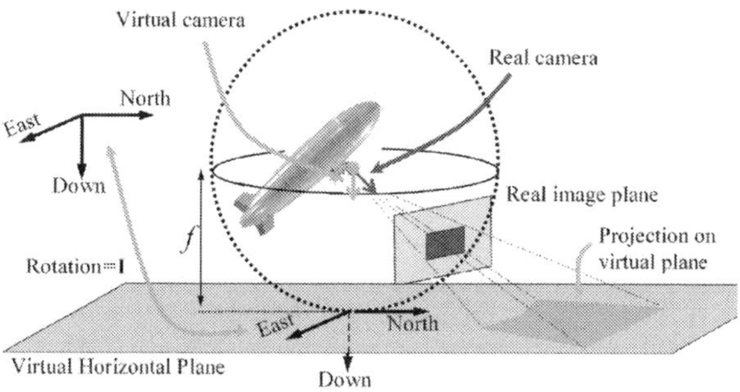

(a) The virtual horizontal plane concept

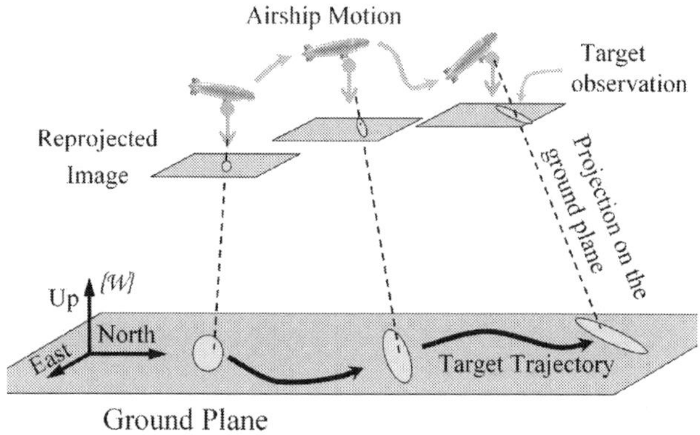

(b) Target observations projected in the ground

Figure 2. Tracking an independently moving target with observations from a moving camera

1.3. Camera-AHRS Calibration and a Virtual Horizontal Plane

The camera and AHRS are fixed rigidly together and the constant rotation between both sensor frames $^{AHRS}\mathbf{R}_{CAM}$. is found by the Camera-Inertial Calibration Toolkit [Lobo and Dias, 2007].

The translation between both sensors frames is negligible and considered as zero. The AHRS estimates of its own orientation are then used to estimate the camera orientation as $^{\mathcal{W}}\mathbf{R}_{CAM}|_i = {}^{\mathcal{W}}\mathbf{R}_{AHRS}|_i \cdot {}^{AHRS}\mathbf{R}_{CAM}$. The knowledge of the camera orientation allows the images to be projected on entities defined on an absolute NED (North East Down) frame, such as a virtual horizontal plane (with normal parallel to gravity), at a distance f below the camera center, as shown in Fig. 2(a). Projection rays from 3D points to the camera centre intersect this plane, projecting the 3D point into the plane. This projection corresponds to the image of a virtual camera such as defined in Sect. 1.2. It is performed by the infinite homography [Hartley and Zisserman, 2000], which depends on the calculation of the rotation between the real and virtual camera frames: $^{\mathcal{D}}\mathbf{R}_{CAM}|_i = {}^{\mathcal{D}}\mathbf{R}_{\mathcal{W}} \cdot {}^{\mathcal{W}}\mathbf{R}_{CAM}|_i$ where the rotation between the $\{\mathcal{D}\}|_i$ and $\{\mathcal{W}\}$ frames is, by definition, given by:

$$^{\mathcal{D}}\mathbf{R}_{\mathcal{W}} = \begin{bmatrix} 1 & 0 & 0 \\ 0 & -1 & 0 \\ 0 & 0 & -1 \end{bmatrix} \tag{1}$$

1.4. Recovering the Camera Trajectory with a Pure Translation Model

Suppose a sequence of aerial images of a horizontal ground patch, and that these images are reprojected on the virtual horizontal plane as presented in section 1.3. Corresponding pixels are detected between each image and the next one in the temporal sequence. The virtual cameras have horizontal image planes parallel to the ground plane. Then, each corresponding pixel is projected into the ground plane, generating a 3D point, as shown in figure 3(a). Two sets of 3D points are generated for two successive views, and these sets are directly registered in scene coordinates. Indeed, as all points belong to the same ground plane, the registration is solved in 2D coordinates. Figure 3(b) shows a diagram of this process.

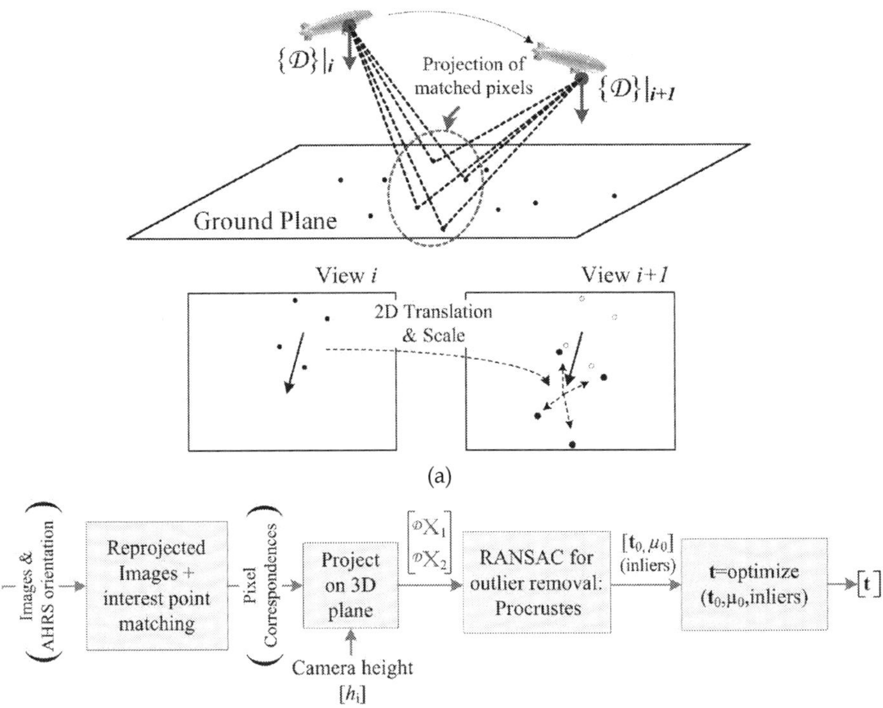

(b)

Figure 3. Finding the translation between successive camera poses by 3D scene registration

Each corresponding pixel pair $(\mathbf{x}, \mathbf{x}_0)$ is projected by equation (2) yielding a pair of 3D points $(\mathbf{X},\mathbf{X}')$, defined in the $\{\mathcal{D}\}|_i$ frame:

$$\boldsymbol{X} = \begin{bmatrix} \frac{(x_x - n_x)\cdot h_i}{f} \\ \frac{(x_y - n_y)\cdot h_i}{f} \\ h_i \end{bmatrix}, \boldsymbol{X}'(\mathbf{t}) = \begin{bmatrix} \frac{(x'_x - n_x)\cdot(h_i - t_z/t_w)}{f} + \frac{t_x}{t_w} \\ \frac{(x'_y - n_y)\cdot(h_i - t_z/t_w)}{f} + \frac{t_y}{t_w} \\ h_i - t_z \end{bmatrix} \quad (2)$$

where $\mathbf{x} = [x_x, x_y, 1]^T$, $\mathbf{x}' = [x'_x, x'_y, 1]^T$, again in inhomogeneous form, h is the camera height above the ground plane, $\mathbf{t}$ is defined as a four element homogenous vector $\mathbf{t} = [t_x, t_y, t_z, t_w]^T$. The $\mathbf{t}$ value which turns $\mathbf{X}'(\mathbf{t}) = \mathbf{X}$ is the translation which registers the $\{\mathcal{D}\}|_i$ and $\{\mathcal{D}\}|_{i+1}$ frames, and which must be determined. If there are n corresponding pixel pairs, this projection yields two sets of 3D points, $\mathcal{X} = \{\mathbf{X}_k \,|\, k = 1 \ldots n\}$ and $\mathcal{X}' = \{\mathbf{X}'_k \,|\, k = 1 \ldots n\}$.

An initial, inhomogeneous, value for $\mathbf{t}_0$ is calculated by the Procrustes registration routine [Gower and Dijksterhuis, 2004]. It finds the 2D translation and scale factor which register the two point sets taken as 2D points, yielding estimates the x and y components of $\mathbf{t}_0$ and of the scale factor μ_0. The inputs for the Procrustes routine are the configurations $\mathcal{X}$ and $\mathcal{X}'(0)$.

From μ_0 and the current estimate of the camera height an initial estimate the vertical component of $\mathbf{t}_0$ can be calculated, as $\mu_0 = (h_i - t_z)/h_i$. Outliers in the pixel correspondences are removed by embedding the Procrustes routine in a RANSAC procedure. Then $\mathbf{t}_0$ is used as an initial estimate for an optimization routine which minimizes the registration error between $\mathcal{X}$ and $\mathcal{X}'(\mathbf{t})$, estimating an updated and final value for $\mathbf{t}$.

The optimization variables are the four elements of $\mathbf{t}$, with equation (2) used to update $\mathcal{X}'(\mathbf{t})$. The function to minimize is:

$$\min_{(t_x, t_y, t_z, t_w)} \sum_{k=1\ldots n} dist\left(\boldsymbol{X}'_k(\mathbf{t}), \boldsymbol{X}_k\right) \quad (3)$$

The same process could be performed with an inhomogeneous, three element $\mathbf{t}$. But, as it is the case with homography estimation, the over-parameterization improves the accuracy of the final estimate and sometimes even the speed of convergence. In this case the extra dimension allows the length of the translation to change without changing its direction.

For datasets where the actual camera orientation is almost constant or the error in the orientation estimate is less significant, the algebraic Procrustes procedure obtains good results alone, with no optimization at all. Indeed, if the assumptions of having both image and ground planes parallel and horizontal are really true, with outliers removed, and considering isotropic error in the corresponding pixel coordinates, then it can be proved that the Procrustes solution is the best solution in a least squares sense. But the optimization step should improve robustness and resilience to errors, outliers and deviations from the model, and still exploit the available orientation estimate to recover the relative pose more accurately than an image-only method.

More details and other pure translation models are shown in [Mirisola and Dias, 2007b, Mirisola and Dias, 2008]

1.5. Filtering the Camera Pose

The camera trajectory is recovered as a sequence of translation vectors t, considered as velocity measurements which are filtered by a Kalman Filter with a Wiener process acceleration model [Bar-Shalom et al., 2001]. The filter state contains the camera position, velocity and acceleration. The filter should reduce the influence of spurious measurements and generate a smoother trajectory. The process error considers a maximum acceleration increment of 0.35 m/s^2, and the sequence of translation vectors is considered as a measurement of the airship velocity, adjusted by the sampling period of 0.2 s. The measurement error is considered as a zero mean Gaussian variable with standard deviation of 4 m/s in the horizontal axes and 1 m/s in the vertical axis. The camera pose ${}^{w}\mathbf{X}_C(i)$ is taken from the filter state after the filtering.

2. Tracking of Moving Targets

Once the camera pose is known, a moving target is selected on each reprojected image. Problems such as image segmentation or object detection are out of the scope of this paper. Nevertheless, to track its position on the plane, the target coordinates on the virtual image must be projected on the reference $\{\mathcal{W}\}$ frame, considering the error in the camera position and orientation. Figure 4 summarizes this process which is detailed in this section.

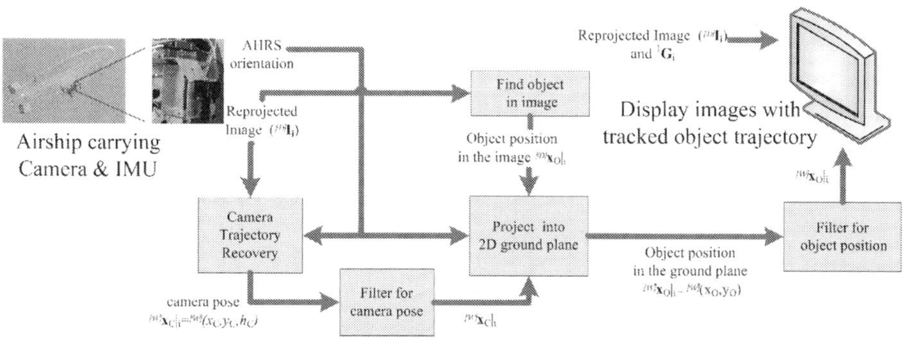

Figure 4. A block diagram of the tracking process

2.1. Target Pose Measurement: Projecting from Image to World Frame

The target coordinates in the image are projected into the ground plane by equation (2), and then these coordinates are transformed into the $\{\mathcal{W}\}$ frame by the appropriate rotation - equation (1) - and translation (the origin of the $\{\mathcal{D}\}$ frame is ${}^{w}\mathbf{x}_C$ in the $\{\mathcal{W}\}$ frame).

The actual generation of reprojected images in the virtual horizontal plane does not by itself improve the measurement of the target position on the ground. Interest point matching could be performed with the original images, and only the coordinates of the matched interest points need to be actually reprojected on the virtual horizontal plane in order to apply the pure translation method. Nevertheless, interest point matching is faster or more robust if the rotation-compensated images are used [Mirisola and Dias, 2007a, Mikolajczyk et al., 2005], and the reprojected images can be drawn on the ground plane forming a map as in figures 5 and 6.

The measurement of the position of a target imaged on the ground is known to be very sensitive to errors in the camera orientation [Redding et al., 2006]. Therefore the uncertainty of the camera 6D pose is propagated with the Unscented Transform, which has already been used to estimate the position of static targets observed from a low altitude UAV [Merino et al., 2005, Redding et al., 2006]. The actual errors in the camera position and orientation are unknown. The covariance found by the KF of section 1.5 are used for the camera pose, and the camera orientation estimate is supposed to have zero mean Gaussian error with standard variation of 5°.

Therefore, given a sequence of camera poses with the respective images and an object detected on these images, this projection generates a sequence of 2D coordinates with anisotropic covariance ellipses for the target pose on the ground plane.

2.2. Filtering of Target Pose

The target pose is tracked in the 2D reference frame, and filtered by a Kalman Filter similar to the filter described in section 1.5, although the state position, velocity and acceleration are now 2D. The process error considers a maximum acceleration increment of 1 m/s^2, and the Unscented Transform supplies measurements of the target position with covariance matrices which are considered as the measurement error.

The target observations projected in the ground plane have high frequency noise, due to errors in the camera position and orientation estimate, and in the target detection in each image. This is clearly seen in the trajectories of Fig. 11 where the ground truth trajectory is a sequence of straight lines. These errors are accounted for by the Unscented Transform to estimate a covariance for the target observation, but nevertheless, the original target trajectory is filtered by a low pass filter before the input of the Kalman Filter. Analyzing the spectrum of the trajectory of the walking person, most of the energy is concentrated below 1 Hz. As the frequencies involved are too small, a low pass filter with too large attenuation or too small cut frequency would filter out true signal features such as going from zero velocity to motion in the beginning of the movement, and introduce delays in the filtered signal after curves. Therefore after empirical testing, the low pass filter parameters were set to a cut frequency of 2 Hz and attenuation of −10 dB. Thus the input of the Kalman Filter is a better conditioned signal, and the final trajectory is smoother.

3. Results

3.1 Tracking of a Moving Target from Airship Observations

Firstly, an object of known dimensions in the ground was observed, and the height of the camera estimated from its image dimensions, eliminating the scale ambiguity inherent to relative pose recovery from images alone. This was done a few seconds in the image sequence before the images shown. Then the airship trajectory was recovered by the model of Sect. 1.4. Only the Procrustes procedure was necessary as the optimization did not improve the results.

Figure 5(a) shows the recovered airship trajectories using the method of section 1.4 (red circles) and by the standard homography estimation and decomposition method (green crosses). The blue squares show the GPS measured trajectory. In the ground the target (a moving car) trajectory derived from the airship trajectory recovered by our method is shown as blue stars.

The trajectories recovered by the Procrustes method are shown again in figure 5(b). The images projected in the ground plane by using equation (2) to find the coordinates of their corners in the ground plane and drawing the image in the canvas accordingly. One every three images is drawn.

Figure 6 shows a 2D view of the target trajectory on the ground over the corresponding images for the pure translation (a) and image-only (b) methods. The error in height estimation for the image only method is apparent in figure 6(b) as an exaggeration in the size of the last images. The same low pass and Kalman filters were used with both methods.

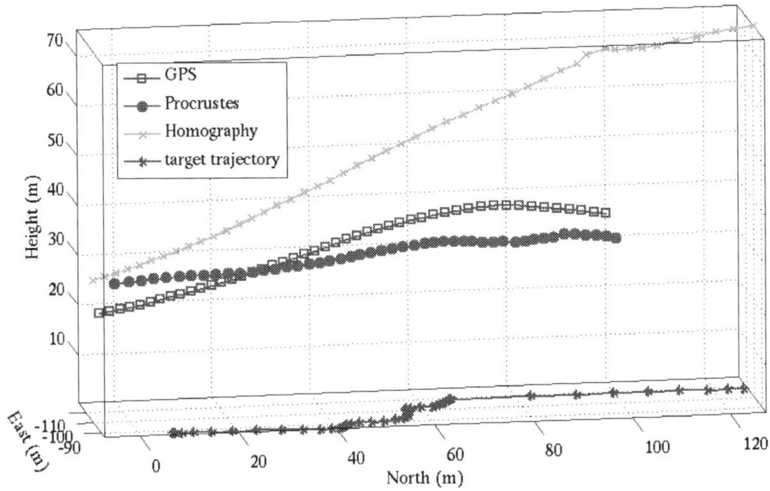

(a)

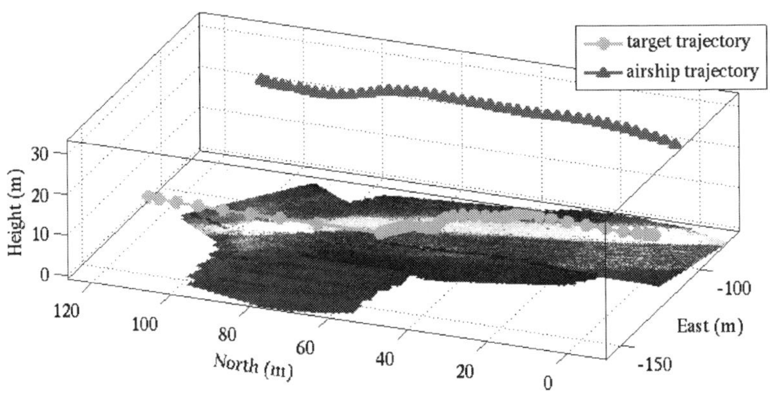

(b)

Figure 5. A 3D view of the recovered trajectories: (a) Airship trajectories from GPS, pure translation and image-only method. Target trajectory derived from pure translation airship trajectory. (b) Trajectories recovered by the pure translation method, with registered images drawn on the ground plane

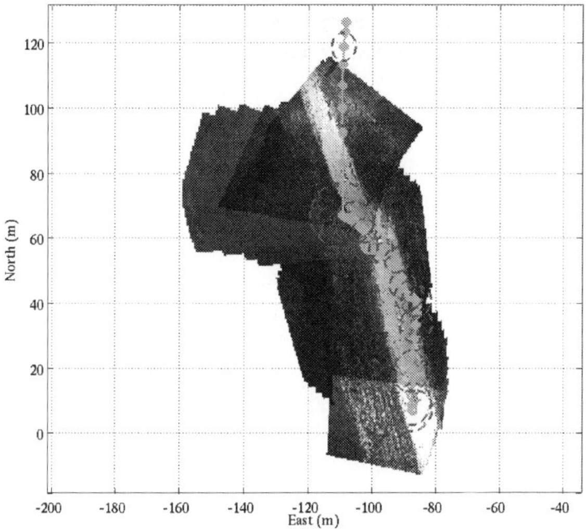

(a) Pure translation method

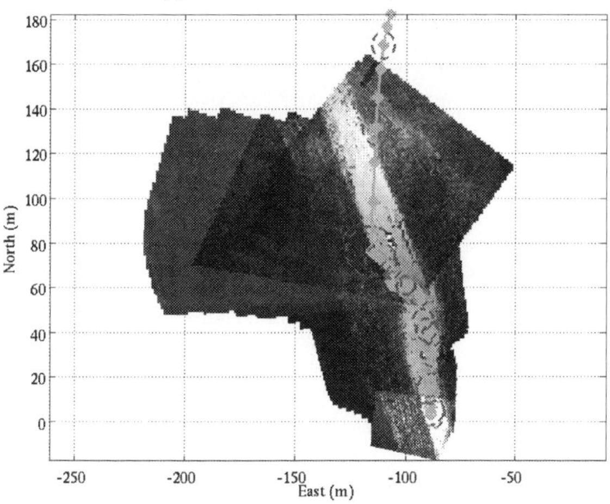

(b) Image-only method

Figure 6. Tracking a car from the airship with the pure translation and the image only methods. The green circles are the target trajectory with one standard deviation ellipses drawn in red

3.2. Tracking after fusing GPS and Visual Odometry

In this experiment, the car has been driven in a closed loop in the ground while the airship was flying above it. To recover the airship trajectory, the translation recovered by the visual odometry was fused with GPS position fixes in a Kalman Filter with a constant acceleration model [Bar-Shalom et al., 2001]. The usage of this model does not imply that the actual acceleration of the vehicle or target is constant; it is just the approximation used by the filter. The GPS outputs standard deviation values for its position fixes (shown as the red ellipses and red points in Fig. 7), and the translation vectors from the visual odometry are interpreted as a velocity measurement between each pair of successive camera poses, with a manually set covariance smaller in the vertical axis than in the horizontal ones. The GPS absolute position fixes keep the estimated airship position from diverging, while the visual odometry measurements improve the trajectory locally. The fused airship trajectory is shown as green crosses in figure 7, while the target observations are shown as blue points in the ground plane. The target could not be continuously observed, therefore the straight lines (for example the straight lines crossing the path) indicate where observations were missing and resumed at some other point of the path.

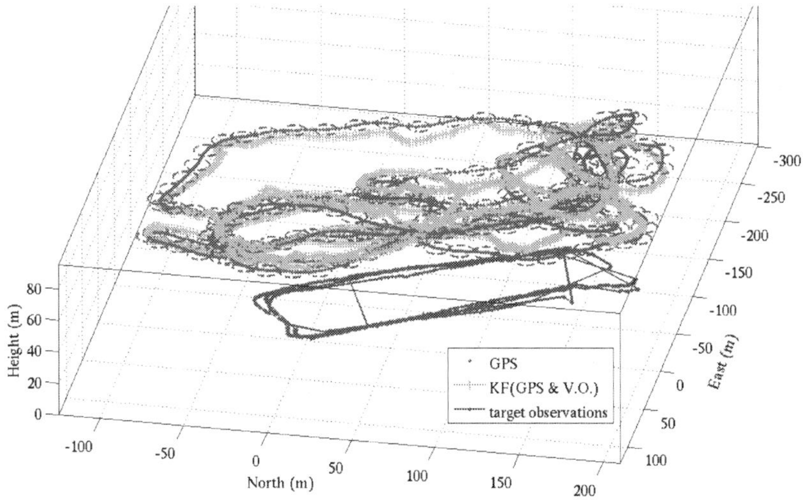

Figure 7. The airship trajectory from GPS and from the fusion of GPS and visual odometry, with the target observations shown in the ground plane

Figure 8(a) shows the target trajectory drawn over a satellite image of the flight area. The airship trajectory was taken directly from GPS. Figure 8(b) shows the same target trajectory obtained when the airship trajectory is recovered by a Kalman Filter fusing both visual odometry and GPS. In both figures, the squares show the coordinates of the target observations in the ground plane, the circles show the target trajectory filtered by its own Kalman Filter, and the crosses indicate that the target is "lost". The airship can not keep observing the target continuously, thus when there are not observations for an extended period of time the tracked trajectory diverges. If the target position standard deviation becomes larger than 30 m than the target is declared "lost" and the filter is reinitialized at the next valid observation. Fusing the visual odometry with GPS resulted in a smoother trajectory for the tracked target.

3.3. Tracking People with a Moving Surveillance Camera

The method described in Sect. 2 was applied to track a person moving on a planar yard, imaged by a highly placed camera which is moved by hand. The camera trajectory was recovered by the Procrustes method with the optimization described by equation (3) which improved the results (the AHRS was the less accurate MTB-9 model). The large squares in the floor provide a ground truth measure, as the person was asked to walk only on the lines between squares. The ground truth trajectories of the camera and the target person are highlighted in Fig. 9(a), and Fig. 9(b) shows the recovered trajectories with the registered images in the top. The camera height above the ground was around 8.6 m, and each floor square measures 1.2 m.

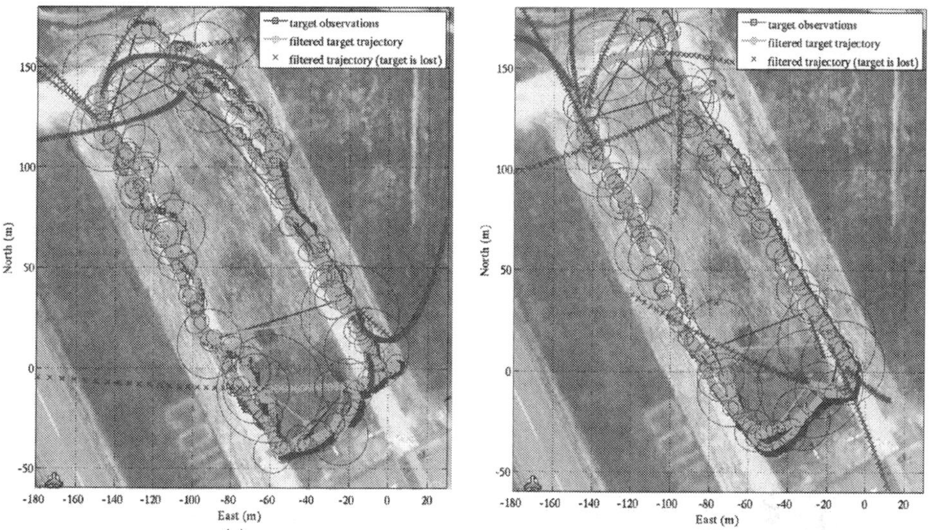

(a) (b)

Figure 8. The target trajectory over a satellite image of the flight area. The car followed the dirty roads. In (a), the airship trajectory was taken from GPS, in (b) a Kalman Filter estimated the airship trajectory by fusing GPS and visual odometry

Figure 10 shows the target observations projected in the ground plane before (squares) and after (circles) applying a low pass filter to the data. Figure 11(a) shows a closer view of the target trajectory to be compared with Fig. 11(b). In the latter case, the camera trajectory was recovered by the homography model. The red ellipses are 1 standard deviation ellipses for the covariance of the target position as estimated by the Kalman Filter. In both figures, the large covariances in the bottom right of the image appear because the target was out of the camera field of view in a few frames, and therefore its estimated position covariance grew with the Kalman filter prediction stage. When the target comes back in the camera field of view the tracking resumed. The solid yellow lines are the known ground truth, marked directly over the floor square tiles in the image. Comparing the shape of the tracked trajectories is more significant than just the absolute difference to the ground truth, as the image registration itself has errors. The tracked trajectory after recovering the camera trajectory with the pure translation model appears more accurate than when the homography model is used. The same low pass filter and Kalman Filter were used to filter the target observations in both cases generating the target trajectories shown.

(a)

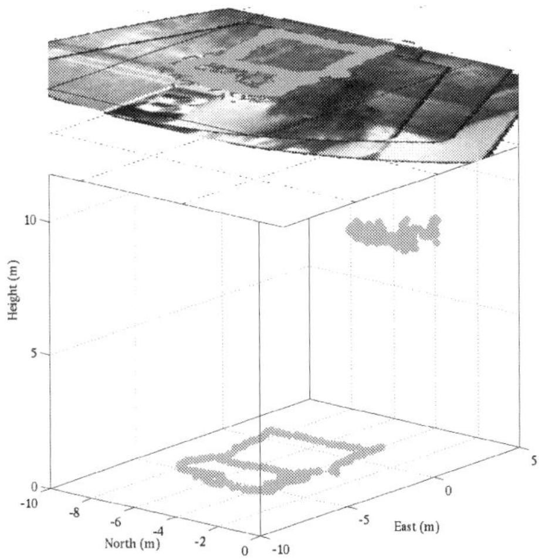

(b)

Figure 9. A photo with highlighted trajectories of camera and target person (a). A 3D view of the recovered trajectories, using the pure translation method to recover the camera trajectory (b)

4. Conclusions and Future Work

Our previous work on camera trajectory recovery with pure translation models was extended, with the same images being used to recover the moving camera trajectory and to track an independently moving target in the ground plane. The better accuracy of the camera trajectory recovery, or of its height component, resulted in better tracking accuracy. The filtering steps were performed in the actual metric coordinate frame instead of in pixel space, and the filter parameters could be related to the camera and target motion characteristics.

With a low altitude UAV, GPS uncertainty is very significant, particularly as uncertainty in its altitude estimate is projected as uncertainty in the position of the tracked object, therefore recovering the trajectory from visual odometry can reduce the uncertainty of the camera pose, especially in the height component, and thus improve the tracking performance.

Visual Odometry can also be fused with GPS position fixes in the airship scenario, and the improvements in the recovered airship trajectory translate in a smoother recovered trajectory for the moving target in the ground. As the GPS position fixes keep the system from diverging, the tracking can be performed over extended periods of time.

In the urban surveillance context these methods could be applied to perform surveillance with a camera carried by a mobile robot, extending the coverage area of a network of static cameras. The visual odometry could also be fused with other sensors such as wheel odometry or beacon-based localization systems.

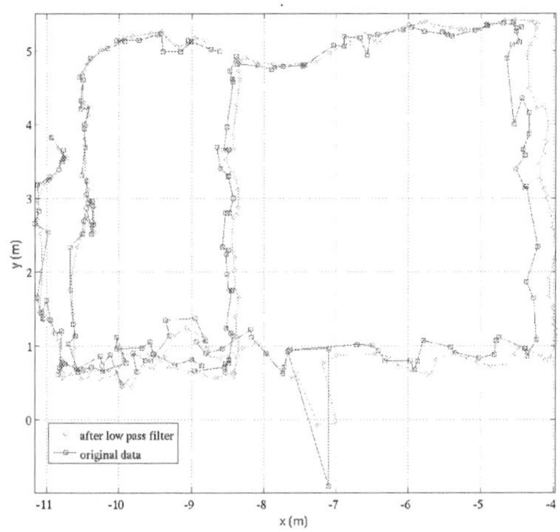

Figure 10. A low pass filter is applied to the observed target trajectory before the input of the Kalman Filter

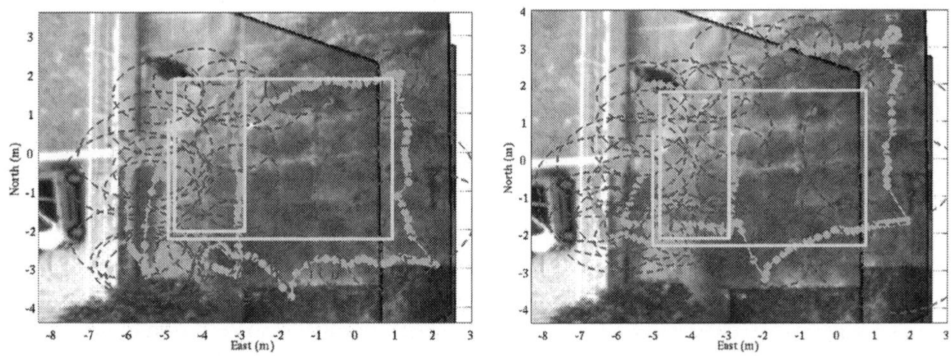

(a) (b)

Figure 11. The tracked trajectory of the target person. In (a) the camera trajectory was recovered by the pure translation method, in (b) by the image-only method

5. References

Bar-Shalom, Y., Li, X. R., & Kirubarajan, T. (2001). *Estimation with Applications to Tracking and Navigation*. John Willey & Sons, Inc.

Bay, H., Tuytelaars, T., & van Gool, L. (2006). SURF: Speeded Up Robust Features. *In the Ninth European Conference on Computer Vision*, Graz, Austria.

Bouguet, J. (2006). *Camera Calibration Toolbox for Matlab*. http://www.vision.caltech.edu/bouguetj/calib_doc/index.html.

Gower, J. C. & Dijksterhuis, G. B. (2004). Procrustes Problems. *Oxford Statistical Science Series*. Oxford University Press.

Hartley, R. & Zisserman, A. (2000). *Multiple View Geometry in Computer Vision.* Cambridge University Press, Cambridge, UK.
Julier, S. J. & Uhlmann, J. K. (1997). A new extension of the kalman filter to nonlinear systems. *In International Symposium in Aerospace/Defense Sensing, Simul. and Controls,* Orlando, FL, USA.
Lobo, J. & Dias, J. (2007). Relative pose calibration between visual and inertial sensors. *International Journal of Robotics Research,* 26(6): 561-575.
Merino, L., Caballero, F., de Dios, J., & Ollero, A. (2005). Cooperative fire detection using unmanned aerial vehicles. In *IEEE International Conference on Robotics and Automation,* pages 1896-1901, Barcelona, Spain.
Mikolajczyk, K., Tuytelaars, T., Schmid, C., Zisserman, A., Matas, J., Schafialitzky, F., Kadir, T., & van Gool, L. (2005). A comparison of affine region detectors. *International Journal of Computer Vision,* 65(7):43 - 72.
Mirisola, L. G. B. & Dias, J. (2007a). Exploiting inertial sensing in mosaicing and visual navigation. *In 6th IFAC Symposium on Intelligent Autonomous Vehicles,* Toulouse, France
Mirisola, L. G. B. & Dias, J. (2007b). Trajectory recovery and 3d mapping from rotation compensated imagery for an airship. In *IEEE Int. Conf. on Robots and Systems (IROS07),* San Diego, CA, USA.
Mirisola, L. G. B. & Dias, J. (2008). Exploting inertial sensing in vision based navigation with an airship. *Journal of Field Robotics.* (submitted for publication).
Point Grey Inc. (2007). www.ptgrey.com.
Redding, J., McLain, T., Beard, R., & Taylor, C. (2006). Vision-based target localization from a fixed-wing miniature air vehicle. *American Control Conference, Minneapolis,* MN, USA.
XSens Tech. (2007). www.xsens.com.

24

An Open Architecture for the Integration of UAV Civil Applications

E. Pastor, C. Barrado, P. Royo, J. Lopez and E. Santamaria
Dept. Computer Architecture, Technical University of Catalonia (UPC)
Spain

1. Introduction

The current Unmanned Aerial Vehicles (UAVs) technology offers feasible technical solutions for airframes, flight control, communications, and base stations. In addition, the evolution of technology is miniaturizing most sensors used in airborne applications. Hence, sensors like weather radars, SAR, multi spectrum line-scan devices, etc. in addition to visual and thermal cameras are being used as payload on board UAVs. As a result UAVs are slowly becoming efficient platforms that can be applied in scientific/commercial remote sensing applications (see Fig. 1 for the most common subsystems in an UAV).

UAVs may offer interesting benefits in terms of cost, flexibility, endurance, etc. Even remote sensing in dangerous situations due to extreme climatic conditions (wind, cold, heat) are now seen as possible because the human factor on board the airborne platform is no longer present. On the other side, the complexity of developing a full UAV-system tailored for remote sensing is limiting its practical application. Currently, only large organizations like NASA or NOAA have enough budget and infrastructure to develop such applications, and eventually may lease flight time to other organizations to conduct their experiments.

Even though the rapid evolution of UAV technology on airframes, autopilots, communications and payload, the generalized development of remote sensing applications is still limited by the absence of systems that support the development of the actual UAV sensing mission. Remote sensing engineers face the development of specific systems to control their desired flight-profile, sensor activation/configuration along the flight, data storage and eventually its transmission to the ground control. All these elements may delay and increase the risk and cost of the project. If realistic remote sensing applications should be developed, effective system support must be created to offer flexible and adaptable platforms for any application that is susceptible to use them.

In order to successfully accomplish this challenge, developers need to pay special attention to three different concepts: the flight-plan, the payload and the mission itself. The actual flight-plan of the UAV should be easy to define and flexible enough to adapt to the necessities of the mission. The payload of the UAV should be selected and controlled adequately. And finally, the mission should manage the different parts of the UAV application with little human interaction but with large information feedback. Many research projects are focused on only one particular application, with specific flight patterns and fixed payload. These systems present several limitations for their extension to new missions.

This research introduces a flexible and reusable hardware/software architecture designed to facilitate the development of UAV-based remote sensing applications. Over a set of embedded microprocessors (both in the UAV and the ground control station) we build a distributed embedded system connected by a local area network. Applications are developed following a service/subscription based software architecture. Each computation module may support multiple applications. Each application can create and subscribe to available services. Services can be discovered and consumed in a dynamic way like web services in the Internet domain. Applications can interchange information transparently from network topology, application implementation and actual data payload.

This flexibility is organized into a user-parameterizable *UAV Service Abstraction Layer* (*USAL*). The USAL defines a collection of standard services and their interrelations as a basic starting point for further development by users. Functionalities like enhanced flight-plans, a mission control engine, data storage, communications management, etc. are offered. Additional services can be included according to requirements, but all existing services and inter-service communication infrastructure can be exploited and tailored to specific needs. This approach reduces development times and risks, but at the same time gives the user higher levels of flexibility and permits the development of more ambitious applications.

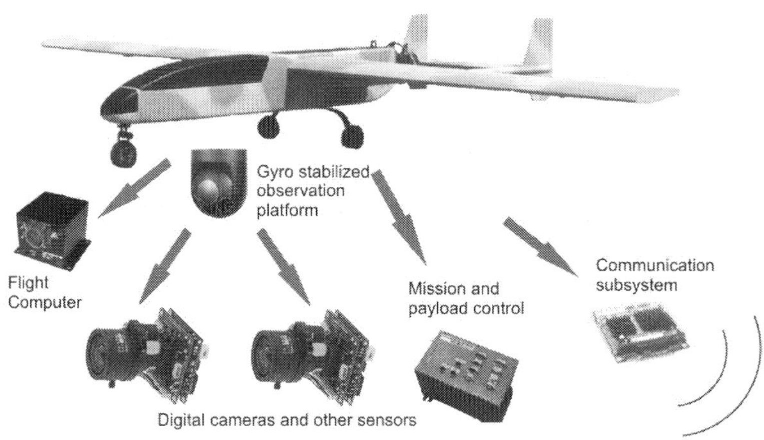

Figure 1. Common elements in a civil UAV system

This chapter is organized as follows. Section 2 generally describes the underlying service oriented technologies that will be applied to create the USAL. Section 3 overviews the architecture of the USAL and describes the most relevant services that are included in the USAL to facilitate the development of UAV applications. Section 4 details the *Virtual Autopilot System* (VAS) that permits the USAL architecture to abstract from autopilot details. Section 5 describes the Flight Plan Manager service that together with its RNAV based dynamic flight-plans constitute the core of the navigation capabilities inside the USAL. Finally, Section 6 concludes the chapter and outlines future research and development directions.

2. System Overview

This section describes the architecture we propose for executing UAV civil missions: a distributed embedded system that will be on board the aircraft and that will operate as a payload/mission controller. Over the different distributed elements of the system we will deploy software components, called services, which will implement the required functionalities. These services cooperate for the accomplishment of the UAV mission. They rely on a middleware (Lopez et al. 2007) that manages and communicates the services. The communication primitives provided by the middleware promote a publish/subscribe model for sending and receiving data, announcing events and executing commands among services.

2.1 Distributed Embedded Architecture

The proposed system is built as a set of embedded microprocessors, connected by a Local Area Network (LAN), in a purely distributed and scalable architecture. This approach is a simple scheme which offers a number of benefits in our application domain that motivates its selection.

Development simplicity is the main advantage of this architecture. Inspired in the Internet applications and protocols, the computational requirements can be organized as services that are offered to all possible clients connected to the network.

Extreme flexibility is given by the high level of modularity of a LAN architecture. We are free to select the actual type of processor to be used in each LAN module. Different processors can be used according to functional requirements, and they can be scaled according to computational needs of the application. We denominate node to a LAN module with processing capabilities.

Node interconnection is an additional extra benefit in contrast with the complex interconnection schemes needed by end-to-end parallel buses. While buses have to be carefully allocated to fit with the space and weight limitations in a mini/micro UAV, the addition of new nodes can be hot plugged to the LAN with little effort. The system can use wake-on-LAN capabilities to switch on/off a node when required, at specific points of the mission development.

2.2 Service Oriented Approach

Service Oriented Architecture (SOA) is getting common in several domains. For example, Web Services in the Internet world (W3C, 2004), and Universal Plug and Play (UPnP, 2008) in the home automation area. The main goal of SOA is to achieve loose coupling among interacting components. We name the distributed components services. A service is a unit of work, implemented and offered by a service provider, to achieve desired final results for a service consumer. Both provider and consumer are roles played by software agents on behalf of their owners.

The benefits of this architecture are the increment of interoperability, flexibility and extensibility of the designed system and of their individual services. In the implementation of a system we want to be able to reuse existing services. SOA facilitates the services reuse, while trying to minimize their dependencies by using loosely coupled services.

Loose coupling among interacting services is achieved by employing two architectural constraints. First, services shall define a small set of simple and ubiquitous interfaces, available to all other participant services, with generic semantics encoded in them. Second, each interface shall send, on request, descriptive messages explaining its operation and its capabilities. These messages define the structure and semantics provided by the services. The SOA constraints are inspired significantly by object oriented programming, which strongly suggests that you should bind data and its processing together.

In a network centric architecture like SOA, when a service needs some external functionality, it asks the network for the required service. If the system knows of another service which has this capability, its reference is provided to the requester. Thus the former service can act as a client and consume that functionality using the common interface of the provider service. The result of a service interface invocation is usually the change of state for the consumer but it can also result on the change of state of the provider or of both services. The interface of a SOA service must be simple and clear enough to be easy to implement in different platforms, both hardware and software. The development of services and specially their communication requires a running base software layer known as middleware.

2.3 Middleware

Middleware-based software systems consist of a network of cooperating components, in our case the services, which implement the business logic of the application. The middleware is an integrating software layer that provides an execution environment and implements common functionalities and communication channels. On top of the middleware, services are executing. Any service can be a publisher, subscriber, or both simultaneously. This publish-subscribe model eliminates complex network programming for distributed applications. The middleware offers the localization of the other services and handles their discovery. The middleware also handles all the transfer chores: message addressing, data marshalling and demarshalling (needed for services executing on different hardware platforms), data delivery, flow control, message retransmissions, etc. The main functionalities of the middleware are:

- *Service management*: The middleware is responsible of starting and stopping all the services. It also monitors their correct operation and notifies the rest of system about changes in service behaviour.
- *Resource management*: The middleware also centralizes the management of the shared resources of each computational node such as memory, processors, input/output devices, etc.
- *Name management*: The services are addressed by name, and the middleware discovers the real location in the network of the named service. This feature abstracts the service programmer from knowing where the service resides.
- *Communication management*: The services do not access the network directly. All their communication is carried by the middleware. The middleware abstracts the network access, allowing the services to be deployed in different nodes.

Fig. 2 shows the proposed UAV distributed architecture. Services, like the Video Camera or the Storage Module, are independent components executing on a same node located on the aircraft. Also on board, there is another node where the Mission Control service executes. Both nodes are boards plugged to the LAN of the aircraft. The mission has also some services executing on ground. The figure also shows two of them: the Ground Station service and a redundant Storage Module.

Each node of the UAV distributed architecture executes also a copy of the Service Container software (see Fig. 2). The set of all Service Containers compose the middleware which provides the four functionalities described above to the application services. This includes acting as the communication bridge between the aircraft and the ground. The middleware monitors the different communication links and chooses the best link to send information to ground or to air. From the service point of view the middleware builds a global LAN network that connects the LAN on ground and the LAN on board.

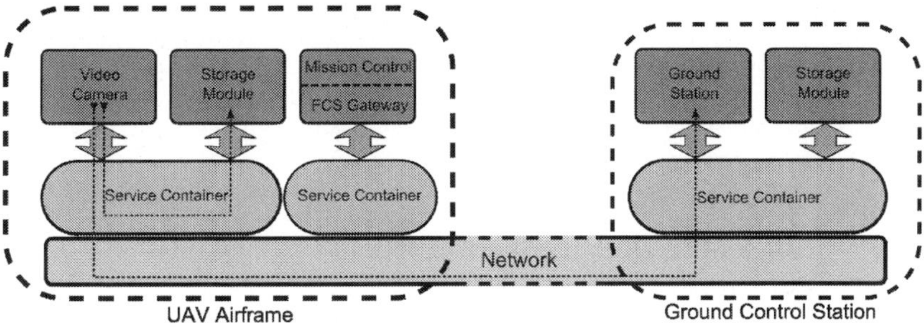

Figure 2. Overview of the architecture implementing the underlying middleware

2.4 Communication Primitives

For the specific characteristics of a UAV mission, which may have lots of services interacting many-to-many, we propose four communication primitives based in the Data Distribution Services paradigm. It promotes a publish/subscribe model for sending and receiving data, events and commands among the services. Services that are producing valuable data publish that information while other services may subscribe them. The middleware takes care of delivering the information to all subscribers that declare an interest in that topic. Many frameworks have been already developed using this paradigm, each one contributing with new primitives for such open communication scenario. In our proposal we implement only a minimalistic distributed communication system in order to keep it simple and soft real-time compliant. Next, we describe the proposed communication primitives, which have been named as *Variables, Events, Remote Invocations* and *File Transmissions*.

Variables are the transmission of structured, and generally short, information from a service to one or more services of the distributed system. A Variable may be sent at regular intervals or each time a substantial change in its value occurs. The relative expiry of the Variable information allows to send it in a best-effort way. The system should be able to tolerate the lost of one or more of these data transmissions. The Variable communication primitive follows the publication subscription paradigm.

Events also follow the publication-subscription paradigm. The main difference in front of Variables is that Events guarantee the reception of the sent information to all the subscribed services. The utility of Events is to inform of punctual and important facts to all the services that care about. Some examples can be error alarms or warnings, indication of arrival at specific points of the mission, etc.

Remote Invocation is an intuitive way to model one-to-one interactions between services. Some examples can be the activation and deactivation of actuators, or calling a service for

some form of calculation. Thus, in addition to Variables and Events, services can expose a set of functions that other services can invoke or call remotely.
In some cases, there also exists the need to transfer continuous media with safety. This continuous media includes generated photography images, configuration files or services program code to be uploaded to the middleware. The *File Transmission* primitive is used basically to transfer long file-structured information from a node to another.

3. USAL: UAV Service Abstraction Layer

Providing a common infrastructure for communicating isolated avionics services is not enough for keeping the development and maintenance costs for UAV systems low. The existence of an open-architecture avionics package specifically designed for UAV may alleviate the development costs by reducing them to a simple parameterization. The design of this open-architecture avionics system starts with the definition of its requirements. These requirements are defined by the type of UAV (mini or tactical UAV in our case) and the mission objectives. From the study and definition of several UAV missions, one can identify the most common requirements and functionalities that are present among them (UAVNET, 2005; RTCA, 2007; SC-203, 2007; Cox et. al., 2006).
These common requirements have been identified and organized leading to the definition of an abstraction layer called UAV Service Abstraction Layer (USAL). The goal of the USAL is twofold (Pastor et. al., 2007; Royo et. al., 2008):

- Reduce "time to marked" when creating a new UAV system. The USAL together with middleware will simplify the integration of all basic subsystems (autopilot, communications, sensors, etc) because it will already provide all required glue logic between them.
- Simplify the development of all systems required to develop the actual mission assigned to the UAV. In most cases reducing this complexity to specifying the desired flight plan and sensor operation to some of the services available in the USAL and parameterizing the specific services to be used every time.

Using the USAL allows to abstract the UAV integrator from time consuming and complex underlying implementation details. The USAL and middleware offer a light weight service-based architecture with a built-in inter-service communication infrastructure. A large number of available services can be selected to create specific configurations, while new services can be easily created by inheriting exiting ones.

3.1 USAL Service Types

Even though the USAL is composed of a large set of available services, not all of them have to be present in every UAV or in any mission. Only those services required for a given configuration/mission should be present and/or activated in the UAV. Available services have been classified in four categories according to the requirements that have been identified (see Fig. 3).
The principal element is the computing system that should orchestrate the overall mission. This system may be extended with additional systems specific to the mission like image processing hardware accelerators, etc. Storage and communication management should be also included by default. This set of standard plus user-defined services that control the mission intelligence are named *Mission Services*.

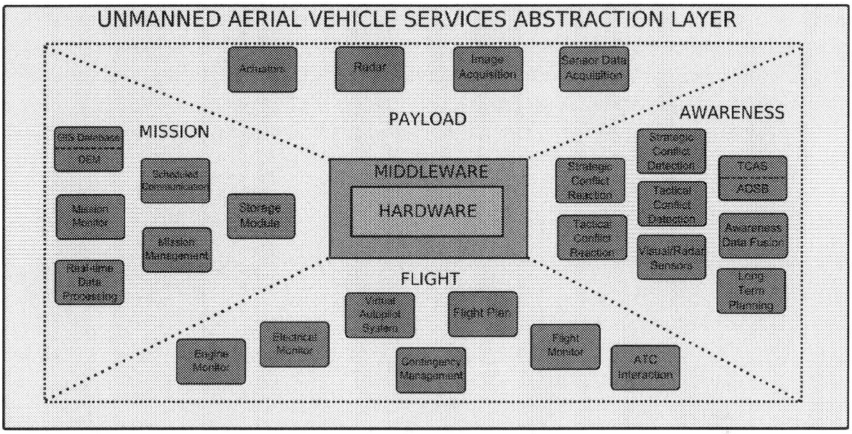

Figure 3. Overview of the USAL service-based architecture

The next relevant system is the UAV autopilot. USAL considers the autopilot as a co-processor; it provides the system with a specific set of primitives that control the flight in the short term. The autopilot operation is supervised by a Flight Plan Manager that abstracts users from autopilot peculiarities and offers flight plan specifications beyond classical way point navigation, thus improving operational capabilities. Additional services help improving the safety and reliability of the operation. The services in charge of the flying capabilities of the UAV are named *Flight Services*.

Successful integration of UAV in non-segregated aerospace will require a number of features to be included in the UAV architecture. Interaction with cooperative aircrafts through transponders, TCAS or ADS systems; and detection of non-cooperative aircrafts through visual sensors, should be implemented and the UAV must inform the pilot in command or automatically react following the operational flight rules for UAV that are currently being developed (EUROCONTROL, 2003; FAA, 2008). However, for certain cases, e.g. flying in segregated airspace, such services may not be necessary. Services that manage the interaction of the UAV with the surrounding airspace users, air traffic management or conditions are named *Awareness Services*.

Payload includes all those other systems carried on board the UAV. We divide them in data acquisition systems (or input devices) and actuators (or output devices). Input devices can be flight sensors (GPS, IMU, Anemometers) and earth/atmosphere observation sensors (visual, infra-red and radiometric cameras, chemical and temperature sensors, radars, etc.) Output devices are few or even do not exist in UAV civil missions because of the weight limitations: flares, parachutes or loom shuttles are examples of UAV actuators. Services controlling these devices are named *Payload Services*.

3.2 Mission Services

From the end-users point of view, earth observation is the main target of a UAV flight. For this reason the user selects a geographic area that constitutes the initial objective of the observation. Also, the user must define the input sensors to activate and the conditions for mission updates.

The DO-304 RTCA (RTCA, 2007) describes 12 scenarios, selected from a list of 70 as representative of UAV missions. Scenarios are representative of different types of airframes and navigation conditions. Regarding their navigation procedures, we have classified them in three types, with increasing complexity:

- Static Area Surveillance. The navigation pattern over the surveillance area is given before take-off. This navigation is the base of the following RTCA scenarios: Communicator Repeater (Scenario 7), Courier (Scenario 8), Border Surveillance (Scenario 16), Coastal Border Control (Scenario 26), High Altitude Communication Relay (Scenario 40).
- Target Discovering. The UAV System has a recognition service that can detect on-the-fly some objects or behaviours in the static area of surveillance. This navigation is the base of the following RTCA scenarios: Perimeter Defence (Scenario 3), OAV Police Operation (Scenario 10), and Mass Casualty Analysis (Scenario 29). Although in some scenarios the mission continues after target discovering, the mission intelligence is relegated to ground operators decision and control.
- Dynamic Target Tracking. The most advanced missions are those that assume intelligence Mission Services, with the capacity to redefine the surveillance area with no human intervention. Hurricane Chase (Scenario 2), Coast Guard Reconnaissance and Surveillance (Scenario 37) are scenarios where this situation is given at some level.

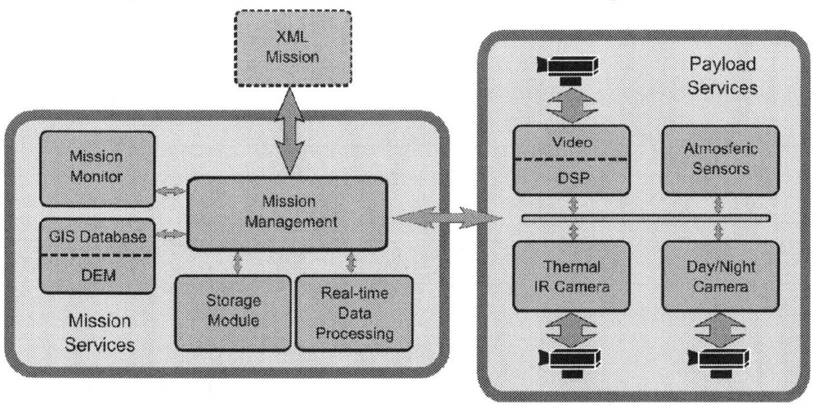

Figure 4. Overview of the available mission and payload service category

In general, the most complex missions include the simplest missions as part of its objective. For example, in order to track a moving object, its previous discovering is needed, which is based on an area surveillance mission. For the Static Area Surveillance the only condition needed is the end-of-mission condition, but for the other two the end-users have to give the condition for Target Recognition.

The USAL offers a number of predefined services to implement a wide range of missions, namely the *Mission Manager*, the *Real-Time Data Processing*, the *Storage*, the *Scheduled Communications*, the *GIS/DEM Database* and *Mission Monitor* (see Fig. 4). These services can be adapted to requirements by means of parameters (e.g. specific flight plan, sensors to be activated, information flows, etc), by adding specific software code to be executed or by adding specific user defined services. Next we describe in more detail some of them.

The *Mission Manager* (MMA) is the orchestra director of the USAL services. This service supervises the flight services and the payload services; as well as the coordination of the overall operation. The MMA executes a user defined automata with attached actions (i.e. service activations) at each state or transition. Actions can be predefined built-in operations or specific pieces of user code. In particular the MMA is capable of modifying the actual flight plan by redefining its parameters or by defining new stages or legs.

The *Real-Time Data Processing* (RDP) gives the intelligence for complex missions. The RDP offers predefined image processing operations (accelerated by FPGA hardware if available) that should allow the MMA to take dynamic decisions according to the actual acquired information.

The *Mission Monitor* (MMO) shows to end-users human friendly useful information about the mission. For example, during a wildland fire monitoring mission, it may present the current state of the fire front over a map. The MMO is executed on the ground and should be highly parameterized to fit the specific requirements of each mission.

3.3 Flight Services

Many autopilot manufacturers are available in the commercial market for tactical UAVs with a wide variety of selected sensors, sizes, control algorithms and operational capabilities. However, selecting the right autopilot to be integrated in a given UAV is a complex task because none of them is mutually compatible. Moving from one autopilot to another may imply redesigning from scratch all the remaining avionics in the UAV.

Current commercial UAV autopilots also have two clearly identified drawbacks that limit their effective integration with the mission and payload control inside the UAV:

- Exploiting the on-board autopilot telemetry by other applications is complex and autopilot dependent. Autopilot's telemetry is typically designed just to keep the UAV state and position under control and not to be used by third party applications.
- The flight plan definition available in most autopilots is just a collection of waypoints statically defined or hand-manipulated by the UAV's operator. However, no possible interaction exists between the flight-plan and the actual mission and payload operated by the UAV.

Flight services are a set of USAL applications designed to properly link the selected UAV autopilot with the rest of the UAV avionics (Santamaria et al., 2008; Santamaria et al., 2007), namely the *Virtual Autopilot Service*, the *Flight Plan Manager Service*, the *Contingency Service*, the *Flight Monitor Service*, etc. (see Fig. 5).

The *Virtual Autopilot Service* (VAS) is a system that on one side interacts with the selected autopilot and is adapted to its peculiarities. VAS abstracts the implementation details from actual autopilot users. From the mission/payload subsystems point of view, VAS is a service provider that offers a number of standardized information flows independent of the actual autopilot being used.

The *Flight Plan Manager* (FPM) is a service designed to implement much richer flight-plan capabilities on top of the available autopilot capabilities. The FPM offers an almost unlimited number of waypoints, waypoint grouping, structured flight-plan phases with built-in emergency alternatives, mission oriented legs with a high semantic level like repetitions, parameterized scans, etc. These legs can be modified by other services in the USAL by changing the configuration parameters without having to redesign the actual flight-plan; thus enabling the easy cooperation between the autopilot and the UAV mission.

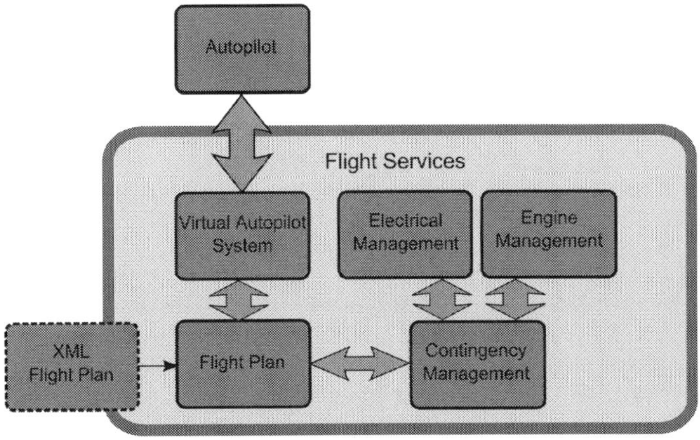

Figure 5. Overview of the available flight service category

The *Contingency Management* services are a set of services designed to monitor critical parameters of the operation (like battery live, fuel, flight time, system status, etc.). In case contingencies are detected, actions will be taken in order to preserve the safety and integrity of the UAV: from flight termination, mission abort or system re-cycle.

3.4 Awareness Services

A UAV System is a highly instrumented aircraft and has no pilot on board. With these conditionings the more suitable flight rules for a UAV are IFR, however for remote sensing missions the advantages of UAV systems is precisely its capacity for flying at any altitude, where VFR aircrafts are found.

UAVs must rely on its instrumentation equipment to properly inform the pilot in command on the ground or substitute the pilot capacities in VFR conditions. The awareness services are responsible for such functionalities. Flight Services are in charge of the aircraft management in normal conditions while the Awareness Services are in charge of monitoring surrounding situations and overtake aircraft management in critical conditions. In this case mission services come to a second priority, until flight conditions become again normal. The list of awareness services is seen in Fig. 6.

The *Awareness Data Fusion* (ADF) is a service designed to collect all available data about air vehicles surrounding our UAV, terrain and meteorological conditions. All this information can be obtained either by on board sensors or even through an external provider.

The *Tactical/Strategic Conflict Detection* services will analyze the fused information offered by the ADF in order to detect potential collision conflicts with objects/terrain/bad climate. Depending on the type of conflict, different types of reaction procedures will be activated. While reaction is executed the ADF will keep monitoring that the conflict is really being avoided.

The *Tactical/Strategic Reaction* services, will implement avoidance procedures according to the severity of the conflict. Tactical reaction is designed in such a way it can overtake the FPM in order to execute a radical avoidance manoeuvre. Once completed, the FPM will retake the control. A strategic reaction will command the FPM to slightly modify its selected flight plan trying to avoid the conflict but at the same time retaining the original mission requested by the Mission Manager.

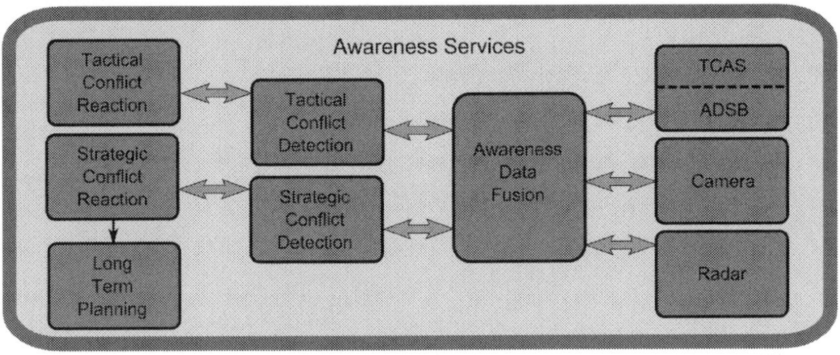

Figure 6. Overview of the available flight service category

3.5 Payload Services

Payload services are defined for sensor devices, mainly raw data acquisition sensors that need to be processed before being used in real-time or stored for post-mission analysis. The complete list of services is directly related to available sensors, and except for most classical cameras they need to be created or adapted by the end user. However, USAL offers pre-build skeletons that should be easily adapted for most common devices.

4. The Virtual Autopilot System (VAS)

The Virtual Autopilot Service is the component that will interact between the autopilot and the rest of the components of the USAL (Royo et. al., 2008). After studying several UAV autopilots we have seen that their functioning and capabilities are very similar, however their implementation details greatly differ (Haiyang et al., 2007). To improve the flexibility of a UAV it is needed to abstract the concrete autopilot used. The main objective of the VAS is to implement this abstraction layer, to isolate the system of autopilot hardware changes. As every UAV mission needs to control autonomously the aircraft following a list of waypoints, the VAS clearly belongs to the minimum services required in the USAL.

4.1 VAS Functional Overview

The VAS is specially suited to work in conjunction with the Flight Plan Manager service (FPM). The latter service stores the flight plan of the mission which is composed of a complex hierarchy of waypoints that the UAV must fly. The VAS will command the hardware autopilot to guide the UAV flight and it will ask the FPM for new waypoints as the autopilot is consuming them.

VAS operates similarly as drivers work on operating systems, abstracting away the implementation details from actual autopilot users. This service may also offer all required flow of data to any other application on board the UAV.

The flight planning capabilities of all autopilots are dissimilar, but generally limited to simple waypoint navigation and in some cases they include automatic take-off and landing modes. From the point of view of the current missions or applications being developed by the UAV using a simple waypoint-based flight plan may be too restrictive. In addition to the VAS the Flight Plan Manager service (FPM) is designed to implement much richer flight-plan capabilities than offered by the actual autopilot. The FPM offers structured flight plan phases with built-in emergency alternatives, leg based path description, mission oriented legs with a high semantic level like repetitions, parameterized scans, etc. Fig. 5 provides a schematic view of the relation between VAS and FPM inside the USAL.

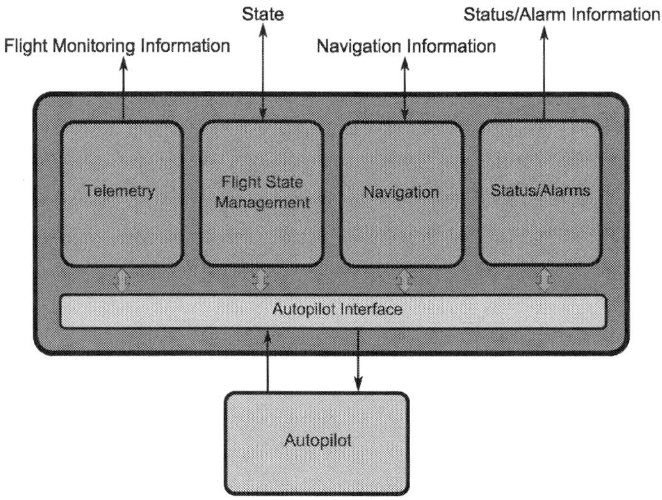

Figure 7. Information flow inside the Virtual Autopilot System

Given that the real autopilot capabilities will be much simpler than those available in the FPM, additional waypoints will be generated according to requirements. Internal waypoints will be dynamically fed into the autopilot through the VAS during the mission time, therefore transforming the FPM in a virtual machine capable of *executing* flight plans. As a result, combining both the abstraction mechanism provided by the VAS and the increased flight plan capabilities of the FPM, we obtain a highly capable platform that can be easily integrated to perform much more efficiently a number of valuable missions.

4.2 VAS Architecture

VAS is a service that provides a standardized interface to the particular autopilot on board. Since it directly interacts with the selected autopilot it needs to be adapted to its peculiarities. For other services in the UAV, the VAS is a service provider that offers a number of information flows to be exploited. Given that not all autopilots are equal, VAS follows a contract between it, as a service provider, and its potential clients. This means that all the information provided by this service is standardized independently of the autopilot being used. VAS belongs to the set of services defined in the USAL and will provide flight monitoring and control capabilities to other services.

The inclusion of the VAS greatly improves the flexibility of the system. The autopilot unit can be replaced by a new version or a different product, but this change will have no impact on the system except for the VAS. Another important motivation is to provide an increased level of functionality. VAS should permit operation with a virtually infinite number of waypoints, thus overcoming a limitation present in all studied UAV autopilots. It will also be able to check the plausibility of these waypoints. This increased level of functionality includes the capability to take control of the flight and generate new waypoints in response to contingencies when services in charge of navigation control fail.

UAV autopilots available today are similar in their operation and capabilities, though their implementation details greatly differ. The key to carry out a correct abstraction is to offer in the VAS interface the common functionality and data that can be found in any autopilot.

This information will be organized in the following four groups. The first group relates to the need of the autopilot to acquire and process attitude and position data. The second one is needed to determine the path that the aircraft will follow. The third, gives information about its current status and possible alarms. Finally, the last one is added to the VAS design to provide the aforementioned increased level of functionality. This last group will change the autopilot states when necessary. Fig. 7 shows the different parts of the VAS service. As displayed in the figure, monitoring and status/alarms information are outgoing flows, while navigation and state management have input/output direction. These four groups of interfaces are described more in detail in the following subsections.

4.3 Flight Monitoring Services

There are several ways to expose the information generated by the sensors of the autopilot. Usually, the autopilot manufacturer groups all this information in large packets of data, which are sent via a radio modem at a certain frequency. In our service-based architecture, the VAS service will offer this information over a LAN to all the services that need this information. The information will be semantically grouped in a way that this information relates to parameters, situations or attitudes of the aircraft, independently of the real autopilot hardware and sensors. The remainder of the section shows all the Flight Management Information which the VAS will publish the rest of the services on the network.

- UAV Angles: Indicates the instantaneous angular position of the system. Roll, pitch and yaw are offered in radians.
- UAV Acceleration: Indicates the instantaneous acceleration vector of the system. (X, Y, Z) vector is offered in meters per square second.
- UAV Rate of Turn: Indicates the instantaneous rate of turn of the system. (X, Y, Z) turn vector is offered in radians per second.
- UAV Position: Indicates the instantaneous position of the system. Latitude, longitude, altitude (MSL) and barometric pressure altitude are offered in radians and meters respectively.
- UAV Speed: Indicates the instantaneous speed vector of the system. North, east and down speed vector is offered in meters per second.
- UAV Air Speed: Indicates the instantaneous speed vector of the system relative to the air. North, east and down speed vector is offered in meters per second.
- Wind Estimate: Indicates wind direction estimation around the system. North, east and downwind vector is offered in meters per second.
- Mission Time: Indicates the current mission time in seconds since the VAS start-up or the last mission time reset.

In general, UAV autopilots provide lots of information from all its sensors; however only a little part of them is really useful to implement a system to develop missions. Flight Monitoring Information has been chosen thinking in what information could be useful for the mission development.

4.4 Navigation Services

In the USAL architecture, the Flight Plan Manager service is in charge of generating the navigation commands to the VAS. In most cases these commands will take the form of waypoints or requests for changing the autopilot state. The VAS feeds the autopilot with its internal waypoints as it consumes system waypoints and commands.

The next group of information is the Output Service Navigation group. This information basically states where to the UAV is going at any moment, in which direction is moving and which waypoint is flying:

- Current Waypoint: Indicates the current waypoint where the system is flying to. This information is offered as an event every time the autopilot switches from one waypoint to the next. Also offered as a function. Waypoints are indicated using the USAL waypoint specification format.
- Previous Waypoint: Indicates the previous waypoint where the system was flying to. This information only offered as a function and it is intended to compute the previous leg that the autopilot was flying. Waypoints are indicated using the USAL waypoint specification format.
- Runway Situation: Indicates the position of the selected runway for normal operations. Runways are defined as runway entry points, runway direction and runway lengths. This information only offered as a function and it is intended to confirm the data stored in the VAS.
- Alternative Runway Situation: Indicates the position of the selected runway for alternative operations. This runway is intended to be used in case of emergency if the UAV cannot reach the preferred runway (not enough fuel, battery, structural damage, etc.).
- UAV Direction: Indicates selected direction of the UAV flight. This direction will depend on the selected autopilot mode: waypoint navigation, directed mode, hold mode, etc. It identifies its target bearing, airspeed and altitude.
- VAS State: Indicates the actual state of the VAS system. The supported states and a brief description are included latter in the section. State is reported each time the VAS switches from one state to another; and each time a state change is requested but cannot be fulfilled.

All of this service information will be mainly used for the FPM in order to interact dynamically with VAS and the mission services. In case the FPM or some other service related to the UAV flight fails, the VAS can take control of the UAV. This emergency state implies a change in the flight state management to Safe mode until the flight plan is recovered or the UAV lands safely. To support the Safe mode the VAS incorporates extra functionalities. One of them is the ability to change the main runway to a closer one. Therefore, the VAS stores alternative runways just in case a contingency situation happens and an emergency landing is needed.

The next group of information is the Input Navigation Service group. This information basically tells the VAS configuration parameters for the autopilot operation, as well as parameters to configure the operative parameters of the states in which the VAS may operate:

- QNH Ground Pressure: Sets the QNH pressure value at the main runway for the pressure altimeter.
- Ground Level Altitude: Set the ground level altitude to the autopilot. The VAS does not have a digital elevation model, so this is the only way for the system to know the ground level.
- Max Mission Time: Once the VAS system starts up a Mission Time timer starts. The Max Mission Time sets up the limit operation time. If this limit is surpassed an alarm will be raised.
- New Waypoint: The system uses this packet to feed up the autopilot with the mission waypoints. With this packet we will compose the flight plan of the mission, that is, the points over which the UAV will fly.
- UAV Speed: Set target indicated airspeed. This packet only has effect once in directed state. However, each waypoint defines the speed which the UAV will have to arrive over that waypoint.
- UAV Altitude: Set target altitude. As UAV speed, this packet only have effect once in directed state and each waypoint defines the altitude which the UAV will have to fly over that waypoint.
- New Main Runway: Set coordinates of runway. The main runway is where the UAV will land. In principle the main runway is where the UAV will take off. However, the UAV will sometimes need a closer runway for landing, especially if it has happened any contingency and the UAV needs to land in order to solve the problem.
- New Alternative Runway: Set coordinates of alternative runway. In the alternative runway, the UAV will save alternative places where the UAV can land. The purpose of this packet is to store a runway closer to where the UAV is flying each moment.
- Change VAS Mode: Sets the current VAS state. With this packet some services as FPM, ground station service or awareness service can switch the VAS states.
- Clear Waypoints: This packet is used to clear all the flight plan waypoints. Sometimes, we may want to change the mission flight plan for another one. First of all, we have to cancel the actual flight plan with this packet and then we can upload a new one.
- Skip Leg: Legs specify the path that the plane must follow in order to reach a destination waypoint from the previous one. In some cases we may need to skip a whole leg instead of all the waypoints. This packet permits skip one leg and pass to another.
- Discard Leg Waypoint: The FPM controls the entire mission flight plan. In some cases, the FPM may be interested in discard a range of waypoints of the flight plan. These waypoints can take several legs or can be part of a leg.

4.5 VAS and Autopilot Status/Alarms

The VAS is the interface between the autopilot and the rest of the system. So it is in charge of informing the status of the autopilot and its own status. An autopilot is a complex hardware that needs to be monitoring every time. With this group of packets we can monitor the autopilot and the VAS status; when any part of these devices has a failure the VAS will send an alarm to the network. All of these alarms are sent as events for two reasons. First, because the alarms are very important for the system and it is needed that these notifications safely arrive to all the services that process them. Second, we will only need to know the status when something is wrong. They are not periodical information like the telemetry flows.

The most important alarms are the ones that monitor the status of the link between the UAV and the ground control station (Ground Communication Loss Alarm) and the links between the GPS receiver and the satellites (GPS Autopilot Alarm). Another important situation that should be controlled is the voltage of the different hardware components of the UAV. Low voltages clearly indicate problems in power system (batteries, generators or the signal conditioning system) and usually will end in a global malfunctioning of the system.

The VAS also provides detailed alarms for notifying problems in the different sensors inside the autopilot (Accelerometer, Gyroscope, Magnetometer, Anemometer and Pressure altimeter). UAV operation can usually continue however on a degraded mode. In this case, the UAV will usually try to return to base for maintenance.

Finally, some alarms manage the specific operation of the VAS service. The Waypoint Range alarm will be raised when the FPM (or the service on charge of providing the waypoint list to the autopilot) tries to feed a waypoint that it is not coherent with the rest of the flight plan (too far away from the previous waypoint). This usually indicates a problem with the flight plan service or the flight plan itself.

Another specific situation that is notified by the UAV, it is the case when a main runway has not been programmed. This runway is needed for some calculations and it is also the place where the UAV will try to carry an emergency landing. Finally, in case of an internal error the VAS will raise the Process Error alarm, indicating that it has detected some abnormal behaviour in its internal calculations.

4.6 VAS Operational States

During its operation an autopilot has different forms of guidance: take-off, waypoint navigation, directed flight, landing, etc. Depending on the implementation and the capabilities of the selected autopilot, these behaviours or states will differ. Some sort of standardization and adaptation is needed if the VAS has to make interoperable different autopilots.

In this section we are going to describe a general overview of the VAS operational states. Commercial autopilots are much focused on flight states, however a mission can be composed of many different states, for example, and we can have different behaviours for the contingencies, which need different types of response. Many autopilots solve this sort of problems just coming back to base station. However, we want the UAV to be able to enter in safe states where it can try to recover the situation.

Obviously, not all the commercial autopilots will implement the full range of states provided by the VAS and it is responsibility of each VAS implementation to fulfil the whole functionalities needed. For example, in case that a concrete autopilot does not have take-off mode, its companion VAS will have to implement a take-off algorithm.

Fig. 8 shows all the VAS states. This is a graph diagram in which each node corresponds to one state. In each state, the UAV develops several tasks in order to achieve the goals of the mission. All the related states are grouped together. The initial state inside each group is showed with an arrow on the top right box state corner. The rest of the arrows show the transitions between the different states. The diagram is descendant from the beginning of the mission to the end, although, in some case, there are horizontal transitions.

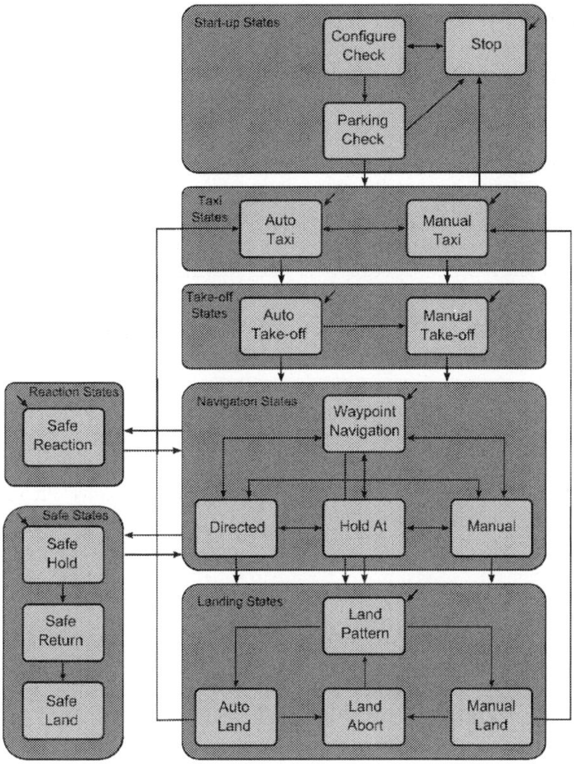

Figure 8. VAS Operational States and Transitions

Start-Up States. The initialization of the system is arranged in the Start-Up set of states. The initial state when the system is switched on is the Stop state. In this state, the UAV is stopped with all the devices of the system on. The next state is the Configure & Check state. During this state the UAV begins to configure itself; this means that all the services needed to accomplish the mission are deployed and configured over the hardware in the UAV. Finally, services and their associated hardware are checked for correct functioning. All of these operations will be done with the UAV on ground still inside the hangar.

When all services are properly configured, the UAV can change to the following state: Parking & Check. It is well known that the behaviour of some subsystems can change when the engine is on, and in the parking state a new check is done after the engine is started. Some sensors, servo mechanisms and communication modules will be checked again when the engine is working. Also during this state we will monitor the engine parameters in order to start the mission in optimal engine conditions. If along this state we detect any anomaly in any service, the UAV comes back to the Stop state. If the UAV pass all check procedures, the UAV can move to taxi states.

Taxi States. At this moment, the UAV is ready and may proceed to the runway to start the flight. The VAS can execute this operation in two ways, as auto taxi or as manual taxi. With the auto taxi the UAV look up an adequate runway and goes to it by itself. This is a complex operation as it means that it automatically takes into account air traffic controller interaction, wind estimation, etc. In other cases, the operator has to drive the UAV to the correct runway, in this situation the VAS is working in manual taxi mode.

While the UAV is on ground the UAV speed will be limited. Also, if the UAV is in auto taxi and anything is wrong the VAS can give the control to the operator changing to manual taxi. When the UAV is in the runway heading the VAS can pass to the take-off states. For airborne operation a flight plan it is needed into the VAS. If the VAS does not detect a loaded flight plan, it only permits manual manipulation.

Take-off States. After the taxi states the UAV is prepared to start the mission. The mission starts with the take off state; this operation is one of the most dangerous because the aircraft begins to fly and the UAV usually does not have enough altitude to respond to any contingency.

This operation can also be developed manually or automatically. If the UAV takes off automatically and any contingency happens the operator can take the UAV control, to solve the problem. However, if we have decided to take-off manually we will remain in manual state, until the VAS can change to Waypoint Navigation state.

Navigation States. At this point of the diagram, the UAV is flying to a secure altitude. When the UAV has arrived to this altitude, it will change automatically to Waypoint Navigation which is the first state of the navigation group. The navigation states are composed by waypoint navigation state, directed state, hold at state and manual state.

In waypoint navigation the UAV follows the waypoints that have been uploaded previously in the VAS. While the UAV is in directed state it will maintain a specific altitude, airspeed and bearing. When the UAV is in Hold At state, it executes a hold pattern around the indicated waypoint. Finally, in manual state the operator has direct control over the airframe's control surfaces. The operator can switch from one Navigation state to each other as he commands from the ground station.

Landing States. When the mission has been successfully finished, the UAV has to go back home and prepare for landing. In order to carry out this task the UAV switches to landing states. This group is composed of Land Pattern state, Land Abort state, Auto Land and Manual Land states. During the the Land Pattern state, the UAV will fly an approximation pattern in order to prepare for the landing. After this state we can choose between auto landing and manual landing.

In both cases, if some problem occurs during the landing, the UAV can switch to the Land Abort state. In this state the UAV climbs up to a secure altitude and goes back to the land pattern state to try to land again. If the landing has been achieved normally, the UAV will be braking on the runway. When the UAV speed is low enough, the UAV will switch to the Taxi state again and will continue to the hangar to finish the mission.

Safe States. If any failure occurs during the navigation states, the VAS can switch to the safe states. These states are composed by Safe Hold state, Safe Return state and Safe Land state. When we have a failure in the system, the UAV will change the state to the Safe Hold. In this state the UAV will remain trying to recover from the failure.

After a timeout the UAV will switch to Safe Return state. In this state the UAV will return to the closest emergency runway in order to start the landing pattern. After the landing pattern the UAV will change to the Safe Land state. In this state a landing flight plan is generated. This flight plan is generated according to the runway situation. These three states compose the safe states; however the system has another safe state: Safe Reaction.

Safe Reaction State. The Safe Reaction is on charge of analyzing the environment around the aircraft and generating reactions in case of a quick evasive response is needed. When the Awareness services detect a dangerous situation, they generate an alarm and the VAS switches to Safe Reaction. In this state the UAV will be commanded by the Tactical Reaction service to solve the problem.

5. Mission Aware Flight Planning

Previous sections have introduced the proposed UAV architecture. In this section we detail the specification mechanism that will be used to describe the UAV flight plans. Most current UAV autopilot systems rely on lists of waypoints as the mechanism for flight plan specification and execution. This approach has several important limitations: (1) It is difficult to specify complex trajectories and it does not support constructs such as forks or iterations. (2) It is not flexible because small changes may imply having to deal with a considerable amount of waypoints and (3) it is unable to adapt to mission circumstances. Besides (4) it lacks constructs for grouping and reusing flight plan fragments. In short, current autopilots specialize in low level flight control and navigation is limited to very basic go to waypoint commands.
We believe that it is necessary to improve current UAV operation with higher level constructs, with richer semantics, and which enable flight progress to be aware of mission variables must be introduced. For that reason a new flight plan specification mechanism is introduced.

5.1 Flight Plan Overview

This section describes the major characteristics of the specification mechanism that will be used to program the FPM service. Some ideas are based on current practices in commercial aviation industry for the specification of RNAV procedures (EUROCONTROL, 2003; FAA, 2008) which is briefly described in the next paragraphs.
Radio Area Navigation (RNAV) is a method of navigation that takes advantage of the increasing amount of navigation aids (including satellite navigation) and permits aircraft operation on any desired flight path. RNAV procedures are composed of a series of smaller parts called legs. To translate RNAV procedures into a code suitable for navigation systems the industry has developed the "Path and Termination" concept. Path Terminator codes should be used to define each leg of an RNAV procedure. Leg types are identified by a two letter code that describes the path (e.g., heading, course, track, etc.) and the termination point (e.g., the path terminates at an altitude, distance, fix, etc.).
Our specification mechanism makes use of the Path Terminator concept to describe basic legs. A subset of RNAV legs applicable to GPS navigation is also of interest. These elements are brought to the UAV field and extended with additional constructs. New control constructs such as iterative legs and intersection legs are added. Reverse traversal of legs belonging to an iterative construct is supported. And adaptivity is increased by means of parametric legs. Further details are given in the next subsections, which describe the flight plan structure and its elements.
The flight plan represents the instructions that will be given to the FPM. A flight plan follows a hierarchical structure and is composed of stages, legs and waypoints (see Fig. 9).
Stages are the largest building blocks within a flight plan. They organize legs into different phases that will be performed in sequence. Legs specify the path that the plane must follow in order to reach a destination waypoint from the previous one. Several primitives for leg specification are available.

A waypoint is a geographical position defined in terms of latitude/longitude coordinates. Waypoints can be named or unnamed depending on whether they are associated to a fix. A fix corresponds to a geographical position of interest with a name and description. Another distinction is made between fly-by and fly-over waypoints. Fly-by waypoints will be used when the aircraft should start turning before reaching the waypoint. Fly-over waypoints require the aircraft to fly over them before initiating a turn. A waypoint may also be accompanied by altitude and speed change indications.

Optionally, a partial flight plan to be carried out if an emergency occurs can be associated to a flight plan. This emergency plan will be superseded by emergency plans specified at stage or leg level. In this way the user will be able to choose the appropriate level of granularity for alternate plans specification. A partial flight plan follows the same structure as a normal flight plan but contains only those stages necessary to fly from the current position to the landing runway of choice.

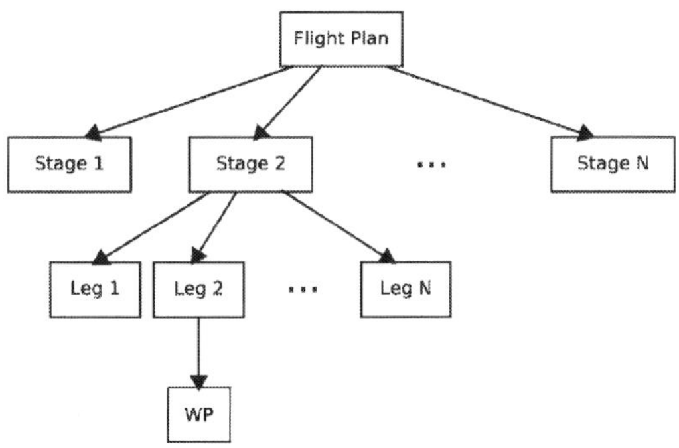

Figure 9. Overview of the Flight Plan Hierarchy

Stages constitute high-level building blocks for flight plan specification and are used to group together legs that seek a common purpose. They correspond to flight phases that will be sequentially executed. Fig. 10 shows a complete flight plan with all stages in the context of a fire fighting mission. The flight plan starts with a Take-Off. Once the aircraft is airborne it will perform a Departure Procedure that will connect with an En Route leading to the forest fire site. Upon arrival the Mission stage will start. When this stage concludes the UAV will enter another En Route stage leading to the landing area. There an Arrival Procedure will be executed to connect with the final Approach and Land stages. Each one of these stages will be composed of a set of legs as described in the following paragraphs.

5.2 Detailed Leg description

A *leg* specifies the flight path to get to a given waypoint. In general, legs contain a destination waypoint and a reference to their next. Most times legs will be flown in a single direction, but within iterative legs reverse traversal is also supported. There are four different kinds of legs:

- *Basic legs*: Specify leg primitives such as "Direct to a Fix", "Track to a Fix", etc.
- *Iterative legs*: Allow for specifying repetitive sequences.
- *Intersection legs*: Provide a junction point for legs which end at the same waypoint, or a forking point where a decision on what leg to fly next can be made.

- *Parametric legs*: Specify legs whose trajectory can be computed given the parameters of a generating algorithm, e.g. a scan pattern.

Basic Legs. A number of basic legs are available to the flight plan designer. They are referred to as basic legs to differentiate them from control structures like iterative or intersection legs and parametric legs. All of them are based on already existing ones in RNAV. Its original name is preserved.

- *Initial Fix* (IFLeg): Determines an initial point. It is used in conjunction with another leg type (e.g. TF) to define a desired track.
- *Track to a Fix* (TFLeg): Corresponds to a straight trajectory from waypoint to waypoint. Initial waypoint is the destination waypoint of the previous leg.
- *Direct to a Fix* (DFLeg): Is a path described by an aircraft's track from an initial area direct to the next waypoint, i.e. fly directly to the destination waypoint whatever the current position is.
- *Radius to a Fix* (RFLeg): Is defined as a constant radius circular path around a defined turn center that terminates at a waypoint. It is characterized by its turn center and turn direction.
- *Holding Pattern*: Specifies a holding pattern path. There are three kinds of holding patterns: Hold to an Altitude (HALeg), Hold to a Fix (HFLeg) and Hold to a Condition (HCLeg). In all cases the initial waypoint, the course (azimuth) of the holding pattern and the turn direction must be specified. The distance between both turn centers and the diameter of the turn segments is also needed.

The three available holding types differ in how they are terminated. Hold to an Altitude terminates when a given altitude is reached, therefore the target altitude and the climb rate must be indicated. A Hold to a Fix is used to define a holding pattern path, which terminates at the first crossing of the hold waypoint after the holding entry procedure has been performed. The final possible type is the Hold to a Condition. In this case the holding pattern will be terminated after a given number of iterations or when a given condition no longer holds (regardless of the number of iterations).

Iterative Legs. A complex trajectory may involve iteration, thus the inclusion of iterative legs. An iterative leg has a single entry (i.e. its body can be entered from a single leg), a single exit and includes a list with the legs that form its body. Every time the final leg is executed an iteration counter will be incremented. When a given count is reached or a specified condition no longer holds the leg will be abandoned proceeding to the next one.

Intersection Legs. Intersection legs indicate points where two or more different paths meet and where decisions on what to do next can be made. All joins and forks will end and start at an intersection leg.

Parametric Legs. In many occasions the dynamic characteristics of the mission environment will make previous leg types insufficient. Parametric legs provide an increased level of adaption to changes that occur during mission time. With parametric legs the flight path is dynamically generated according to input values. Eventually a library of different parametric legs will be available, complete enough so that a wide range of missions can be performed. With the use of parametric legs two goals are achieved. First, complex trajectories can be generated with no need to specify a possibly quite long list of legs. Second, the UAV path can dynamically adapt according to the input values.

Conditions. There are several points in the flight plan where conditions can be found, namely in holding patterns, iterative and intersection legs. For intersection legs, they are necessary in order to determine what path to follow next. For the rest of legs they will let the

FPM know when to leave the current leg and proceed to the next one. Conditions will not be directly specified in the flight plan. Instead, each leg that depends on a condition will contain a reference which will be used to identify the condition. Conditions will be stored and processed separately. When the outcome of a condition is set the FPM will be notified so that this change is taken into account for waypoint generation.
Conditions can be based on elapsed flight time, whether some task has been completed, on counters and on operator input. Operator input will always override any automatically generated outcome. In case of conflict or if the condition cannot be resolved we will resort to the default value or await user input.

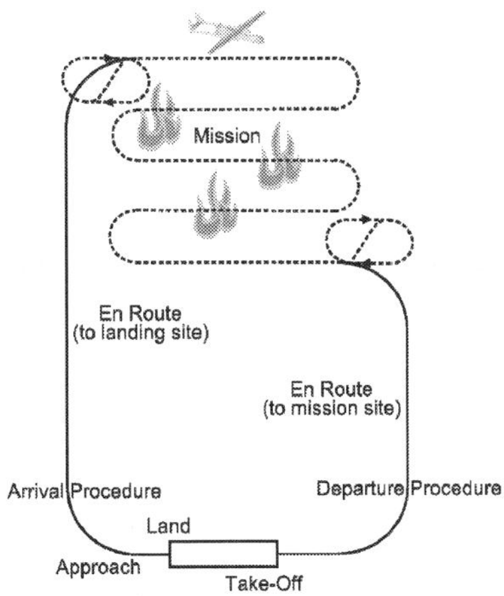

Figure 10. Flight Plan Application Example

5.3 Flight Plan Manager

This section presents the USAL service responsible for processing and executing the proposed flight plans: the Flight Plan Manager service (FPM). The FPM forms part of a wider collection of services that provide the UAV with all its capabilities. The FPM belongs to the flight services category, it will collaborate with other on-board and remote services in order to execute the given flight plan.
The FPM is responsible for executing flight plans, but it doesn't operate in a stand-alone fashion. To achieve its goals it collaborates with other services. The main service the FPM collaborates with is the Virtual Autopilot Service (VAS), which is the only service with direct access to the installed autopilot. Apart from the VAS, the most relevant services the FPM interacts with are the Ground Control Station (GCS) and the Mission Management service (MMA). The GCS provides monitoring and control capabilities to a human operator. The Mission Management Service is in charge of evaluating conditions and updating parametric legs.
In order to execute the flight plan the FPM will send navigation commands to the VAS. These commands mainly consist in waypoints the aircraft has to fly to. Since the flight plan is specified in terms of legs some translation process is needed for converting them into the waypoint sequences expected by the VAS.

Computing waypoint sequences that approximate the legs specified in the flight plan is the main task of the FPM execution engine. This flow of waypoint commands is the main form of interaction between the FPM and the VAS. Sometimes, these waypoints will be accompanied by additional fields indicating speed and altitude change requests.

Other commands related to waypoint management include clearing all sent but pending waypoints. This will be necessary, for instance, when an emergency which forces to execute an alternative plan occurs. The FPM will also issue partial cancellation commands as would be the case when ignoring a number of waypoints in order to directly jump to a given leg. Another type of command will allow the FPMS to change the VAS operation mode to request special operations such as taking-off or landing.

The Ground Control Station is the interface to the system that human operators will interact with. As such it must be able to manage and control several aspects of the flight plan and its execution. To this end, the FPM provides the following operations:

- Load flight plan: Load the flight plan. Before starting any mission a flight plan must be submitted to the UAV.
- Set initial leg: Since each stage of the flight plan can store multiple paths, when there is more than one possibility for the first flight plan stage, the GCS operator will indicate which one to start with.
- Start: Initiate aircraft flight. An automatic or manual take-off, depending on UAV capabilities and configuration will be performed.
- Pause: The aircraft will fly a holding pattern until commanded to resume flight plan execution.
- Manual: Inform the FPM that we are going into manual mode.
- Resume: Switch from paused or manual mode to normal automatic operation.
- Stop: Stops FPM operation.
- Goto leg: Fly directly to the given leg skipping intermediate ones without abandoning automatic mode of operation.
- Update flight plan: Update command provided for modifying the flight plan. A waypoint can be moved, the values of a parametric leg can be updated and other leg parameters such as speed and altitude can be changed.
- Set condition result: Set the result of any of the conditions the flight plan depends on.
- Trigger emergency return: The FPM will switch to the emergency flight plan defined for the current leg, stage or flight plan.

Apart from the listed operations, the FPM also provides a number of information flows that enable monitoring. This data consists in the position of the aircraft in flight plan terms, i.e. what are the current stage, leg and other leg-related information such as current iteration of an iterative leg, etc. It also provides information about the current operating state of the service.

One of the main features of our flight plan specification mechanism is that it enables the UAV to adapt to mission circumstances. This can be done in two manners: first, with the possibility of using conditions in different leg types. Second, by using parametric legs, whose final form depends on the values of the input parameters.

Conditions mark a point where a decision can be made about what path to follow next. The decision-making process is not directly done by the FPM. The flight plan contains references to condition identifiers. The outcome of these conditions will be set by the Mission Management Service. The MMA will also decide what the actual parameter values for parametric legs are. These functions are currently done from the GCS but the inclusion of the MMA will provide the UAV with a high degree of automation.

The main task of the FPM consists in generating the sequence of waypoints that will direct the UAV flight. At the same time the FPM has to respond to commands sent from the GCS and be aware of situations where VAS control is taken over by other services. This results in the FPM operating in a number of states as seen in Fig. 11. A description of each one of these states follows:

- *On Command*: The FPM can be either on command or on standby. If on command it has control over the VAS and determines the path followed by the aircraft. When a contingency occurs the main flight plan is replaced by the corresponding emergency plan and waypoint generation starts over. If the flight plan is updated all non-flown waypoints affected by the change will be cancelled and replaced by new ones. *On Command* is a super-state that encompasses two sub states, namely *Auto* and *Paused.*
- *Auto*: When in this state the FPM is generating and forwarding waypoints to the VAS.
- *Paused*: The FPM has received a pause command. This directly translates to issuing a change of state command to the VAS requiring it to perform a holding pattern. When the resume command is received it will notify the VAS that waypoint navigation can continue.
- *On Standby*: This state is entered when the FPM is notified that some other service has taken control over the VAS. It can be entered because another service is trying to avoid a collision or because the VAS is now under manual control. It differs from the *Paused* state in that the FPM is not going to send any message to the VAS. It just waits and tries to recover and continue flight execution once it regains control.

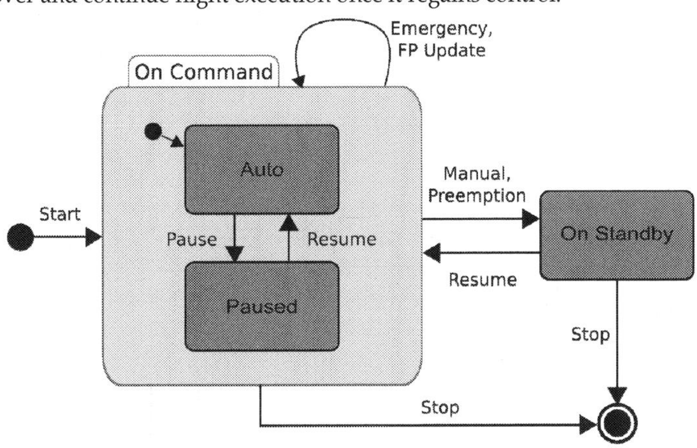

Figure 11. Flight Plan Execution Automata

Once the submitted flight plan has been loaded into the FPM, an internal representation is generated and the service is ready to start waypoint generation. The flight plan is represented by a tree whose root node corresponds to the whole flight plan (see Fig. 9). Stages are located at the next level of the tree, legs follow. At this point some degree of recursion can be found due to iterative legs, whose children legs form the body of the iterative structure. Finally each leg has an associated waypoint.

When the start command is received a traversal of the tree begins. The execution engine goes through each one of the flight plan stages, processing the legs they contain and generating waypoints as appropriate. Legs which present curved paths are approximated by sequences of waypoints.

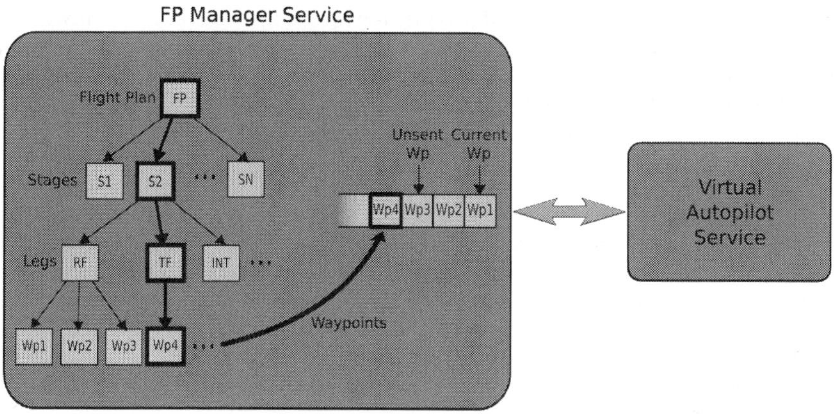

Figure 12. Overview of the Flight Plan execution procedure by the FPM

Waypoint generation works in a decoupled manner from the FPM operation depicted in Fig. 12. The FPM is implemented following a producer-consumer model. There is one worker thread in charge of waypoint generation. Each new waypoint is stored in a queue. The consumer will pass the waypoints from the queue on to the VAS. The head of the queue contains the waypoint the VAS is heading to. The queue is also used to keep track of unsent waypoints. Using this scheme, only the consumer needs to be aware of petitions or changes happening in the system (a waypoint has been reached, the flight plan updated, manual mode entered, etc.) and command the generation engine to take appropriate action if necessary (e.g. restart at a given leg). When a condition is encountered the producer will block and no waypoints will be generated until the condition outcome is available.

At this point the consumer will monitor the state of the queue and ensure that a decision is made in time, otherwise a default path, if present, will be taken. If the queue becomes empty (all waypoints have been flown), the VAS will be commanded to perform a holding pattern.

6. Conclusions

The class of mini/micro UAVs will become available to many research institutions, universities, and private companies to develop their research or commercial applications. However, current technology offers solutions for most components in the UAV system except systems to support and automate the actual sensing mission.

This chapter has introduced USAL, a service-oriented architecture designed to support the development of remote sensing applications. USAL offers a number of pre-defined services that can be easily parameterized to the specific needs of the application. Additional services can be introduced to incorporate new functionalities and at the same time reusing available services. A middleware designed to support the type of inter-service communications required by the USAL is also introduced.

A novel flight plan specification to be employed within the USAL has been also introduced. The proposed specification mechanism improves on the common list of waypoints approach by providing RNAV-like legs as the main unit for flight plan construction. Besides the leg concept is extended to include higher level control structures for specifying iterative behaviour and branching. The decision making for this control structures can be based on mission variables, therefore enabling the UAV to respond depending on mission time information. The adaptability of the system to mission time circumstances is also greatly improved by the inclusion of parametric legs. With parametric legs, the flight path of the aircraft is dynamically generated depending on its input parameters.

7. Acknowledgments

This work has been partially funded by Ministry of Science and Education of Spain under contract CICYT TIN 2007-63927; and by the Innovative Studies Programme of the EUROCONTROL Experimental Centre, EUROCONTROL Innovative Studies: www.eurocontrol.int/eec/public/standard page/WP Innovative Studies.html.

8. References

W3C Working Group (2004), W3C Note 11: Web Services Architecture, *http://www.w3c.org/TR/ws-arch*, February 2004

UPnP Forum (2008), UPnP Device Architecture 1.0, *http://www.upnp.org/specs/arch*, April 2008

UAVNET Thematic Network (2005), European Civil Unmanned Air Vehicle Roadmap *http://www.uavnet.com*, March 2005

RTCA (2007), DO-304: Guidance Material and Considerations for Unmanned Aircraft Systems, March 2007

Santamaria, E.; Royo, P.; Lopez, J.; Barrado, C.; Pastor, E. & Prats, X. (2007). Increasing UAV capabilities through autopilot and flight plan abstraction, *Proceedings of the 26th Digital Avionics Systems Conference*, Dallas (TX), October 2007, AIAA/IEEE

Santamaria, E; Royo, P.; Barrado, C.; Pastor, E.; Lopez, J. & Prats, X. (2008). Mission Aware Flight Planning for Unmanned Aerial Systems, *Proceedings of AIAA Guidance, Navigation, and Control Conference and Exhibit*, Honolulu (HI), August 2008, AIAA

Pastor, E.; Lopez, J. & Royo, P. (2007). UAV Payload and Mission Control Hardware/Software Architecture. *Aerospace and Electronic Systems Magazine*, Vol.22, No.6, June 2007 3-8

Lopez, J.; Royo, P.; Pastor, E.; Barrado, C. & Santamaria, E. (2007). A Middleware Architecture for Unmanned Aircraft Avionics, *Proceedings of 8th Int. Middleware Conference*, NewPort (CA), November 2007, ACM/IFIP/USEUNIX

Royo, P.; Lopez, J.; Pastor, E.; & Barrado, C. (2008). Service Abstraction Layer for UAV Flexible Application Development, *46th AIAA Aerospace Sciences Meeting and Exhibit*, Reno, Nevada, January 2008, AIAA

Cox, T.H.; Somers, I. & Fratello D.J. (2006). Earth Observations and the Role of UAVs: A Capabilities Assessment, 2006, NASA

Haiyang, C.; Yongcan, C. & YangQuan, C. (2007). Autopilots for Small Fixed-Wing Unmanned Air Vehicles: A Survey, *Proceedings of International Conference on Mechatronics and Automation (ICMA)*, Harbin, China, 2007, IEEE

FAA (2008). Aeronautical Information Manual, Official Guide to Basic Flight Information and ATC Procedures, *U.S. Federal Aviation Administration*, 2007.

EUROCONTROL (2003). Guidance Material for the Design of Terminal Procedures for Area Navigation, *European Organisation for the Safety of Air Navigation*

Schmidt, D.C. (2002). Middleware for Real-Time and Embedded Systems, *Communications of the ACM*, June 2002

25

Design, Implement and Testing of a Rotorcraft UAV System

Juntong Qi, Dalei Song, Lei Dai and Jianda Han
State Key Laboratory of Robotics, Shenyang Institute of Automation, CAS
Graduate School, Chinese Academy of Sciences
People's Republic of China

1. Introduction

Unmanned aerial vehicles (UAV) are useful for many applications where human intervention is considered difficult or dangerous. Traditionally, the fixed-wing UAV has been served as the unit for these dangerous tasks because the control is easy. Rotorcraft UAV (RUAV), on the other hand, can operate in many different flight modes which the fixed-wing one is unable to achieve, such as vertical take-off/landing, hovering, lateral flight, pirouette, and bank-to-turn. Due to the versatility in maneuverability, helicopters are capable to fly in and out of restricted areas and hover efficiently for long periods of time. These characteristics make RUAV applicable for many military and civil applications.

However, the control of RUAV is difficult. Although some control algorithms have been proposed (Sanders et al., 1998, Garratt et al., 2003, Enns et al., 2000, Bijnens et al., 2005, Koo et al., 1998, Jiang et al., 2006), most of them were verified by simulation instead of real experiments. One reason for this is due to the complicate, nonlinear and inherently unstable dynamics, which has cross coupling between main and tail rotor, and lots of time-varying aerodynamic parameters. Another reason is that the flight test is in high risk. If a RUAV lost its control, it would never be stabilized.

Shenyang Institute of Automation, Chinese Academy of Sciences (SIA, CAS) as a national research institute focus its future research on RUAV 5 years ago. Until now, we have 3 types of experimental platforms for advanced control algorithm research demonstration. ServoHeli-20 (Fig.1) is a model class platform which has 20 kilograms takeoff weight. ServoHeli-40 (Fig.2) and ServoHeli-110 (Fig.3) are engineering class platforms for highway patrol, electrical line patrol and photography. They have 40 and 110 kilograms takeoff weight and have finished full autonomous flight control experimental demonstration. In SIA, the control algorithm research on RUAV involves in navigation, advanced flight control, 3D path planning and fault tolerant control.

This paper details the development of an unmanned helicopter testbed - ServoHeli-20 (Qi et al., 2006) (Fig.3), and the experiments performed toward achieving full autonomous flight. The brief of this paper is as follow: the ServoHeli-20 platform is introduced in Section II. The introduction of sensor package is in Section III. The modeling of the RUAV system is presented in the Section IV. In Section V, we introduce an independent-channel control scheme as a baseline control of the platform. In Section VI, an overview of fault tolerant control research is presented. In the end, we conclude our work and discuss some future research issues.

Figure 1. ServoHeli-20 airframe

Figure 2. ServoHeli-40 airframe

Figure 3. ServoHeli-110 airframe

2. ServoHeli-20 Platform Description

As the basic airframe of the RUAV system, we chose the small-scaled model helicopter which is available in the market. Such a choice is easy for us to exchange the accessories and cost low price.

ServoHeli-20 aerial vehicle (Fig.4) is a high quality helicopter which is changed by us using a RC model helicopter operating with a remote controller. The modified system allows the payload of more than 5 kilograms, which is sufficient to take the whole airborne avionics box and the communication units. The fuselage of the helicopter is constructed with sturdy ABS composite body and the main rotor blades are replaced with heavy-duty carbon fiber reinforced ones to accommodate extra payloads. The vehicle is powered by a 90-class glow plug engine which generates 3.0hp at about 15000 rpm, a displacement of 14.95cc and practical angular rate ranging 2,000 to 16,000 rpm. The full length of the fuselage is 1260mm as well as the full width of it is 160mm. The total height of the helicopter is 410mm, the main rotor is 1600mm and the tail rotor is 260mm.

Designing the avionics box and packing the box appropriately under the fuselage of the helicopter are two main tasks to implement of the UAV helicopter system. In the actual flight environment, the weight and the size of the avionics box are strict limited. Our airborne control box, which is shown in Fig.5, is a compact aluminum alloy package mounted on the landing gear. The center of gravity of the box lies on the IMU device where is not the geometry center of the system that ensure the navigation data form IMU accurate. The digital compass and the IMU which are taken as the horizontal center of the gravity of the avionics system to locate and the other components are installed on the same line.

Figure 4. Implemented rotorcraft UAV

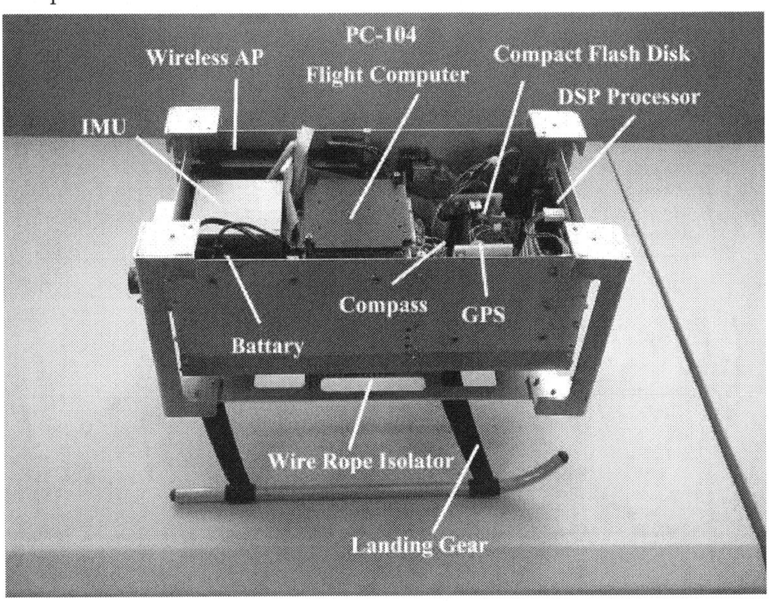

Figure 5. The avionics control system

The original landing gear of the model helicopter is plastic, which is no enough room to install the designed avionics system in the fuselage of the helicopter. While, we re-design a landing gear with aluminum alloy and make a larger room under the fuselage of the model helicopter for the control box. To avoid the disciplinary vibration about 20Hz caused by characteristic of the helicopter, ENIDINE® aviation wire rope isolators which are mounted between the avionics box and the changed landing gear are chosen. They are comprised of stainless steel stranded cable, threaded through aluminum alloy retaining bars, crimped and mounted for effective vibration isolation.

3. Flight Control System of the ServoHeli-20

3.1 Flight Control Computer

The onboard avionics system is responsible for the overall RUAV's main managements including navigation, autonomous control, communication and so on. The PC-104 flight computer, communication components and sensor units including inertial measure unit, global position system (GPS), digital compass, air-press altimeter, ultrasonic sensor are installed onboard.

In our research in the project, PC-104 computer system is used as the onboard computer which is shown in Fig.6.

Figure 6. PC-104 computer system

The flight computer installed in avionics box is a typical industrial embedded computer system, so-called PC-104 which the whole system is kept as compact and light-weight as possible. The PC-104 has the ISA or PCI bus which features a 108.2cm×115.06cm footprint circuit board. Our flight computer system consists of a main CPU board and some other peripheral boards such as DC-DC power supply board, 8-channel serial communication device and PWM generation board.

The main CPU board has a Celeron processor at 400MHz with 256MB SDRAM, fully compatible with the real-time operation system such as QNX. Hard drive or other equivalent mass-storage device for booting and running an operation system and storing useful sensor data is needed to the flight computer. The Compact Flash (CF) card by KingStone® is a 1GB flash RAM device and is suitable for air environment. The Eurotech®

PC-104 processor board has only two serial ports, which are not enough for collection data from more than two sensors that communicate with serial port. As a result, a serial port expender is packed with the main CPU board providing RS-232 / 485 communications. In order to control the Futaba® model helicopter servos, a PWM generation board is needed. Take price and reliable as consideration, DSP board is chosen to generate 5-channel PWM signal as well as capture the PWM signal encoded by remote controller when the system run at manual mode. Such a design is to ensure that the system can run independently at manual mode, which the close loop of the servo control is not through PC-104 processor except for receiving the commands of servos from serial port. Li-Ion battery serves as a power supplement to the overall onboard system such as flight computer system, sensors, communication units and servos through DC-DC converter. Eurotech® ACS DC-DC converter is mounted to meet the system design requirement of converting a 9-40V DC input voltage to multi-voltage power outputs including +5V, +3.3V and +12V with overload protection. An optional onboard microprocessor monitors the temperature of the module and protects it by turning the module off when temperatures exceed 85 centigrade-degree. The power supplement of the overall avionics system as well as five servos that control the helicopter is powered by a Li-Ion battery pack which has the capacity of 78WH at the output of 19V.

3.2 QNX Real-Time Operation System

To our flight control system, a real-time operation system (RTOS) is required for the onboard computer system. After carefully consideration and comparison, QNX Neutrino RTOS is selected as the operation system, which is ideal for embedded real-time applications. It can be scaled to very small size and provides multitasking, threads, priority-driven preemptive scheduling, and fast context-switching - all essential ingredients of an embedded real-time system. The applied program can be coded and debugged in the remote windows-host computers and can be executed in the airborne computer system independently, which provides great convenience during the flight experiments without modify the program in onboard computer.

3.3 Ground Control Station

The ground station mainly includes the ground control computer, the ground development computer, model helicopter remote controller, wireless-LAN access point, video signal receiver, the antennas of the communication devices and ground power source. The role of the ground station is to issue control commands to the onboard avionics system and monitor its real-time status. The pre-scheduled trajectory and commands as well as the synchronized sensor data are transmitted and received by wireless APs.

The ground control computer is a laptop, which sends the pre-scheduled commands and trajectory to the airborne flight computer. The program of the ground control computer is developed by Visual C++.The interface of it is presented in the Fig.7. The whole picture of ground station is shown in the Fig.8.

The development computer is used for the onboard software development of QNX Neutrino RTOS as well as the DSP processor. QNX Momentics IDE which is an integrated development environment of the QNX system is installed in the computer as a windows-host to modify the remote flight computer programme. As the same to the QNX system, Code Composer Studio IDE (CCS) is also setup in the development computer to change the programme in the DSP which is the PWM signal generator in avionics box. The 9-channel RC controller which is at 72MHz radio communication signal is used in manual mode in system modelling.

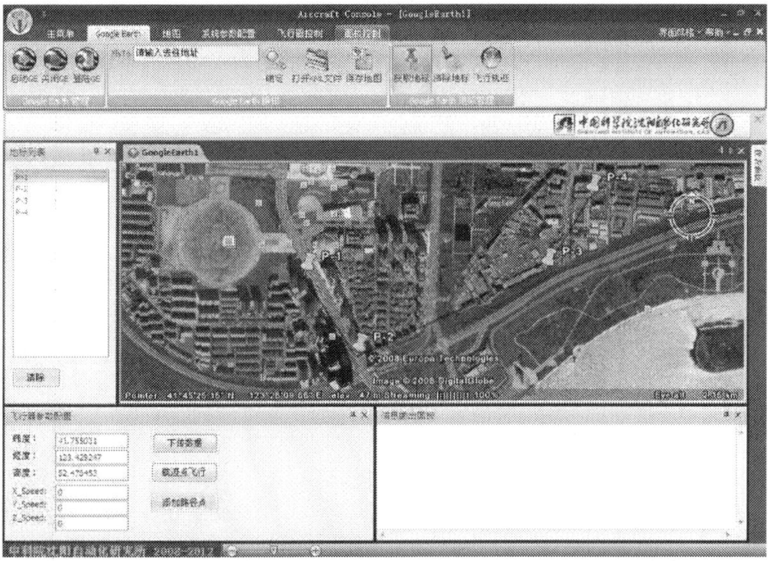

Figure 7. Interface of the ground control computer

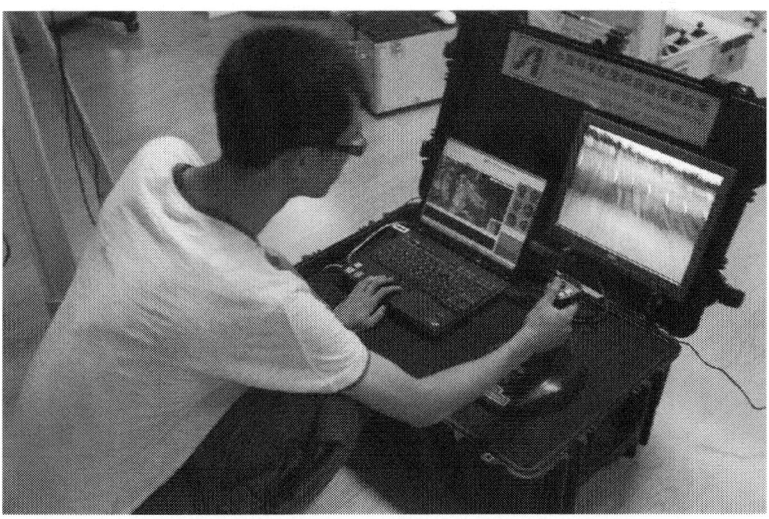

Figure 8. Ground station interface

3.4 Sensors for Attitude and Position Estimation

In order to navigate following a desired trajectory while stabilizing the vehicle, the information about helicopter position, velocity, acceleration, attitude, and the angular rates should be known to the guidance and control system. The rotorcraft UAV system is equipped with sensors including inertial sensor unit, GPS, digital compass, rotor speed sensor, air-press altimeter and ultrasonic sensor to obtain above accurate information about the motion of the helicopter in association with environmental information.

The Crossbow IMU300, which is shown in Fig.9, is a six-axis measurement system designed to measure the linear acceleration along three orthogonal axes and rotation rate around three orthogonal axes. It employs on board digital processing to provide application-specific outputs and to compensate for deterministic error sources with in the unit. Solid-state MEMS sensors make the IMU300 product responsive and reliable.

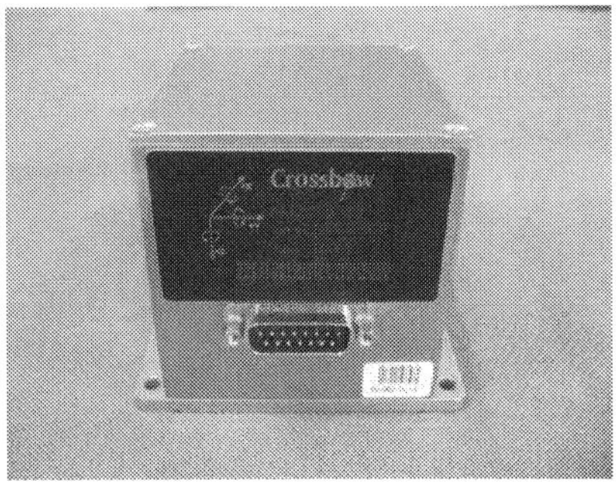

Figure 9. Crossbow IMU

Hemisphere GPS, which is shown in Fig.10, is a space-based satellite radio navigation system developed by a Canada company. GPS provides three-dimensional position and time with the deduced estimates of velocity and heading. The GPS provides position estimates at up to 10 Hz. For operation, the GPS and the antenna are installed on the host aerial vehicle.

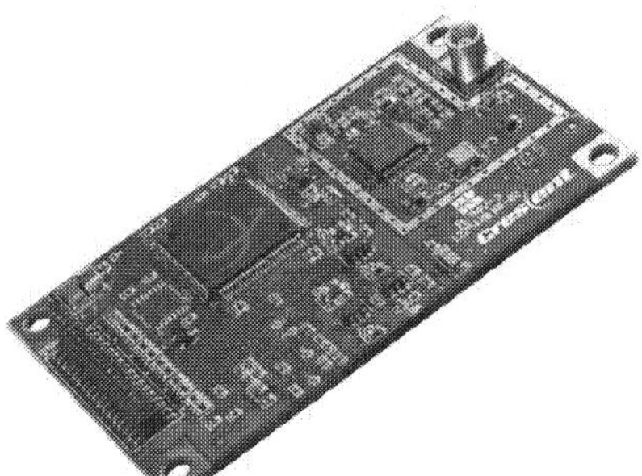

Figure 10. Hemisphere OEM GPS

HMR3000 digital compass, which is presented in Fig.11, is an electronic compass module that provides heading, pitch and roll output for navigation and guidance systems. This compass provides fast response time up to 20 Hz and high heading accuracy of 0.5 degree with 0.1 degree resolution.

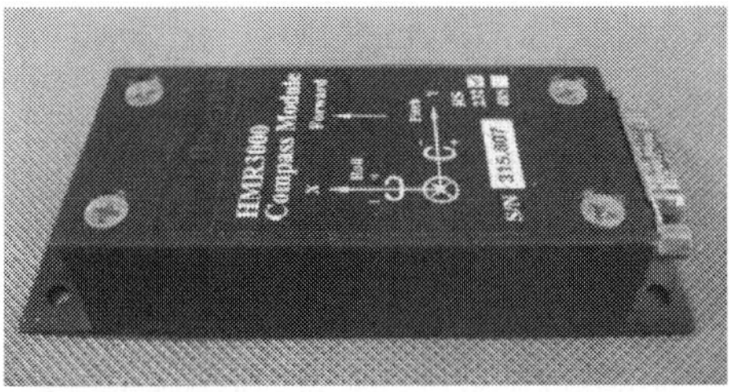

Figure 11. HMR3000 digital compass

In order to get the accurate altitude information of the vehicle, an air-pressure altimeter that collecting data higher than 5 meters as well as an ultrasonic sensor that getting the information on other situations is equipped under the avionics box.

The update rate of all sensors is ranging from 10-100Hz, which is enough for implementation for advanced control algorithms.

3.5 Passive and Active Vibrations Isolation

In our avionics box, we use rate gyros and accelerometers to measure rates about three axes, and accelerations along 3 axes; processor is used to extract absolute roll and pitch. However, in the real flight environment, the sensors will subjected to rotor frequency vibrations; both the rate and acceleration readings are grossly inaccurate; consequential, so to is the attitude estimation.

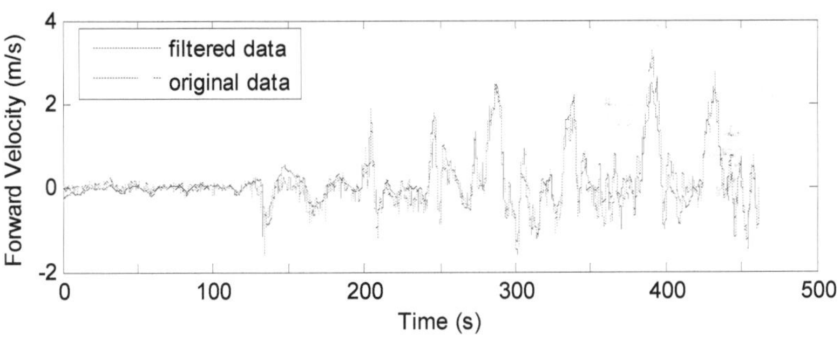

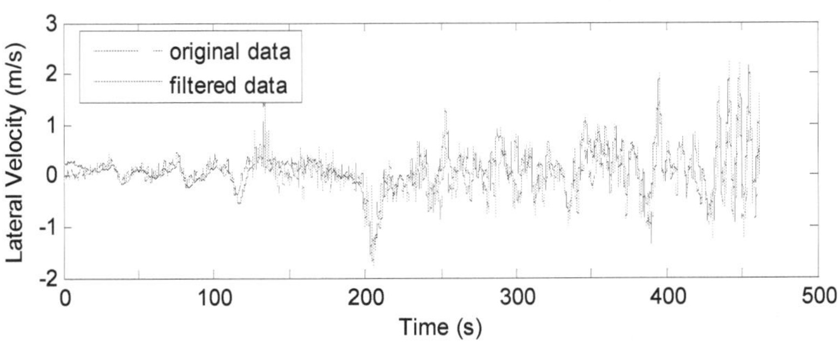

Figure 12. Velocities before and after isolation

In order to isolate the unit from these frequencies, we use the passive and active isolation method. The passive method is that the sensors are spring mounted inside the main avionics box. With the foam damping included, the isolation can act as a passive effect. However, the active method is the Kalman filter way to isolate the vibrations and biases. A typical plot of the forward and lateral velocities before and after isolation is given in Fig.12.

4. Rotorcraft UAV Modelling

For effective hovering identification, the original model in reference (Mettler et al., 2004) is decomposed into three groups (longitudinal, lateral, and yaw-heave coupling), and a semi-decoupled model is obtained. Each group has a decoupled system matrix, and the coupling characteristics is presented only in the control matrix. Thus, the number of unknown parameters and control inputs are reduced and the control loops are semi-decoupled. Then, to identify the unknown parameters in the MIMO semi-decoupled model, a new cost function is proposed to make the traditional method of SISO system frequency estimation (Bendat et al., 1993) applicable to the MIMO state-space models. The proposed cost function is presented in the addition form of the frequency error of every input-output pair for transfer matrix, and the parameters are identified by minimizing the cost function. The simplified model and proposed identification method free the selection of initial estimation and constraint is not required.

Take the yaw - heave model for example. We have got the numerical model, and then other serial of input data was verified using the proposed model. The blue line is the measurement of the yaw rate from the real flight as well as the red line is calculated by the numerical model and the real flight input. As is shown in the Fig13, the estimation output is similar to the real flight data and we can conclude that the proposed modeling method is useful to the rotorcraft UAV.

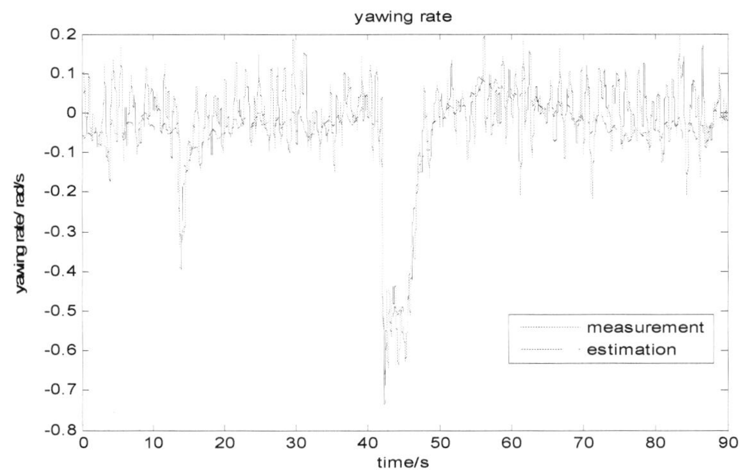

Figure 13. Rotorcraft UAV modeling verification

5. Independent-Channel Control Scheme

Simulation studies have shown that a better strategy for the control of a small-scaled helicopter is to use the flight controller as consisting of two cascaded controllers: an inner loop and an outer loop. The inner loop, that has the faster dynamics, is designed as the attitude controller which takes desired attitude angles as inputs and generates the actuator commands that will result in the desired attitude. The outer-loop controller, which controls the slower translational rate variables, takes desired velocity or position as input and generates desired angels to the inner loop. The overall flight control scheme is shown in Fig.14, while five linear controllers are designed to control the engine speed, height, yaw, lateral and longitudinal motion.

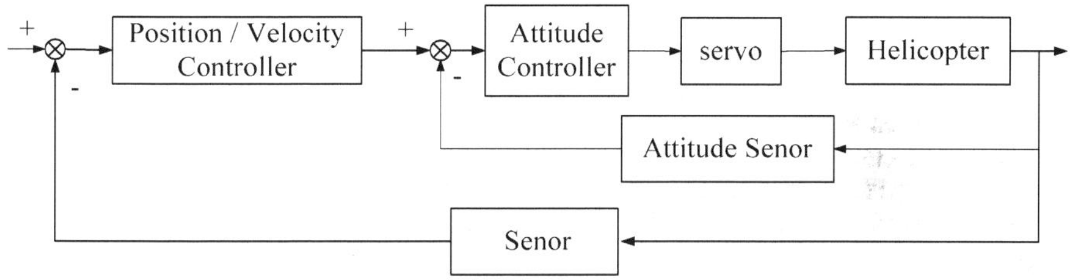

Figure 14. Overall fight control scheme

5.1 Engine Speed Control

The engine speed during the hover-envelope flights is maintained at 1200 rpm as a result of the experiments in manual mode. To get a steady rotary speed of the engine, a PID controller is used in feedback control which from the speed sensor to the throttle demands. When collective pitch changing, the power of the engine will change as a result of it. A feed forward term from the collective pitch is introduced to compensate for the extra loading experienced. The engine control scheme is shown in Fig.15.

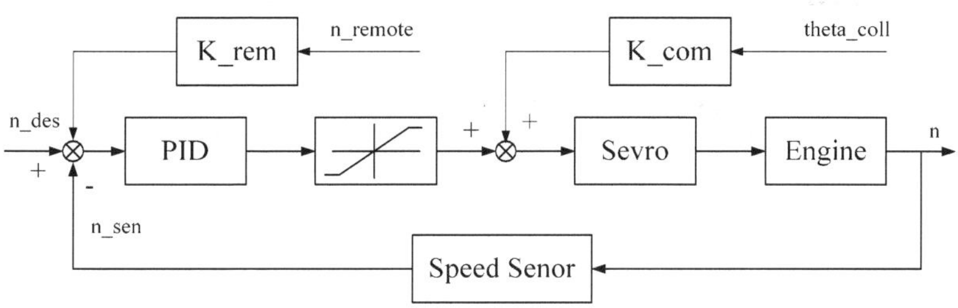

Figure 15. Engine speed control scheme

5.2 Height Control

The height control is a one loop scheme which a PI controller using feedback from the height sensor generates collective pitch demands in Fig.16.

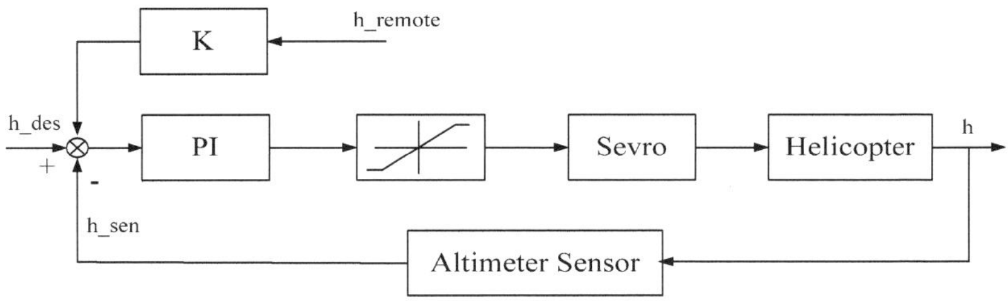

Figure 16. Height control scheme

5.3 Yaw Control

The yaw control two loop structure is presented in Fig.17. As is shown in this figure, the inner loop is a yaw rate stabilization loop which proportional control using yaw rate feedback from the IMU output demands to the rudder servo and the outer loop uses the scheme from the digital compass output to the yaw rate input.

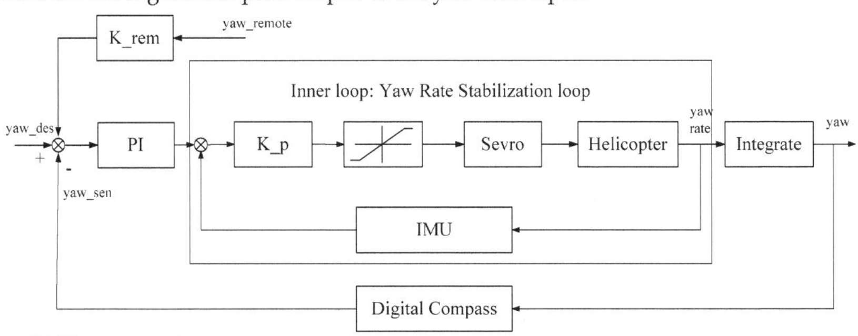

Figure 17. Yaw control scheme

5.4 Lateral and Longitudinal Motion Control

Similar to the yaw control scheme, IMU, digital compass and GPS are used as the feedback sensors to maintain the lateral and longitudinal position with simple proportional and PI controllers. The inner loop is pitch / roll rate stabilization component as well as the outer loop serves as the position feedback unit which is shown in Fig.18.

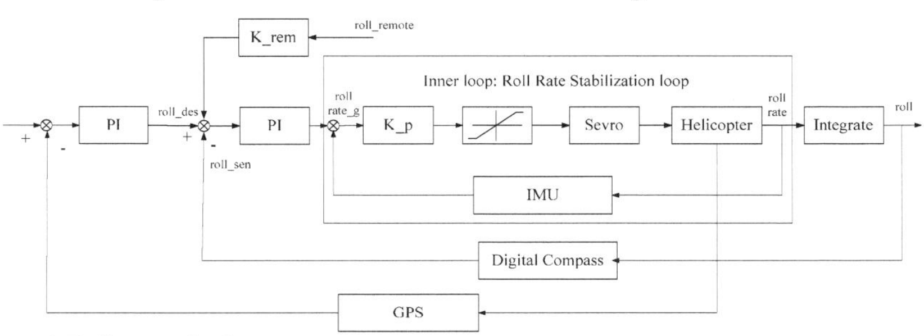

Figure 18. Roll control scheme

5.5 Autonomous Flight Result

A two-loop control scheme for the rotorcraft UAV system was design and tested using the ServoHeli-20 platform. We design some specified trajectories to be flown. These trajectories were selected in order to evaluate the inner loop and outer loop response over several different sequences of inputs. We selected a tunnel way to be followed, as is shown in the Fig.21. The proposed controller handled this flight trajectory with minimal error, Fig.22. Fig.19 and Fig.20 show that the angles and velocities which controlled by inner loops are also get a stable response.

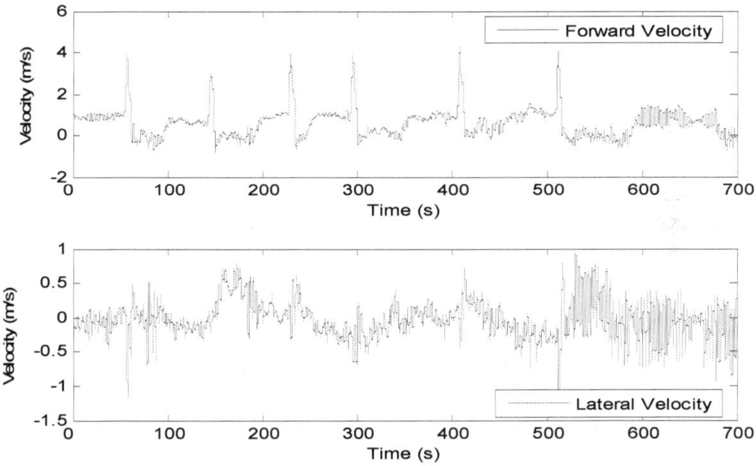

Figure 19. Forward and lateral velocity during the flight

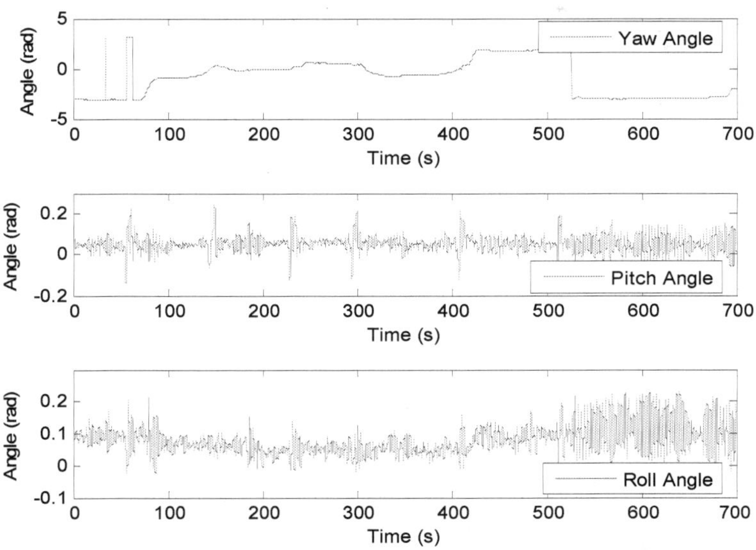

Figure 20. 3-axes angles during the flight

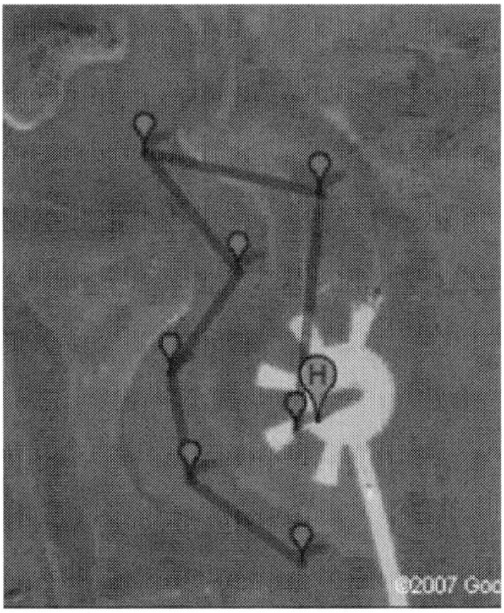

Figure 21. Trajectory in the Google Map

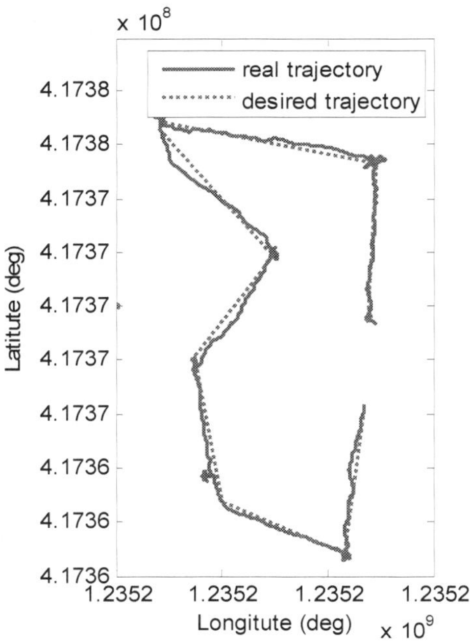

Figure 22. Desired and real trajectory

6. Fault Tolerant Control Research on Rotorcraft UAV

6.1 Fault Tolerant Control Architecture for RUAV

Fig.23 is the overall architecture for RUAV fault tolerant control. The white part is the conventional 3-layer UAV control architecture which including mission planning, path planning and robust flight control. Based on this part, we introduce the mission re-planning and path re-planning to the upper and middle level as while as a reconfigurable flight control to the lower level. In this section, we propose the sensors and actuators failure detection algorithms and demonstrate their effectiveness with simulations.

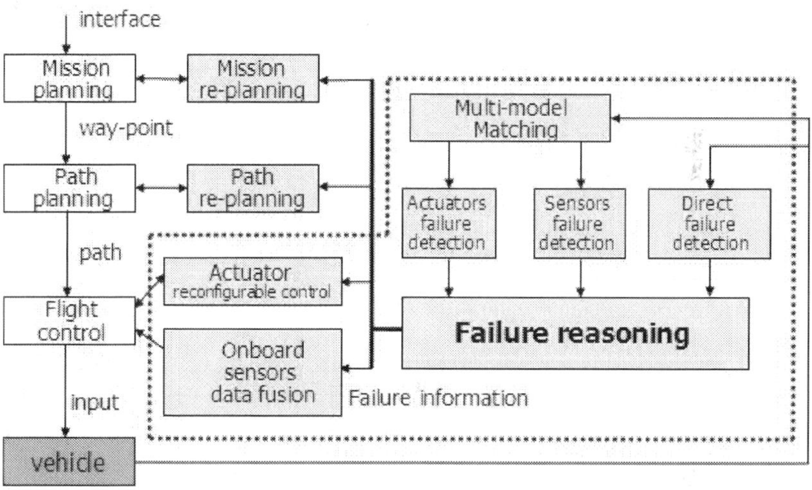

Figure 23. Overall fight control scheme

6.2 Wavelet Transform Based Sensors Failure Detection Algorithms

By use of wavelet transforms that accurately localize the characteristics of a signal both in the time and frequency domains, the occurring instants of abnormal status of a sensor in the output signal can be identified by the multiscale representation of the signal. Once the instants are detected, the distribution differences of the signal energy on all decomposed wavelet scales of the signal before and after the instants are used to claim and classify the sensor faults. Synthetic data simulated by means of a computer using real flight data from ServoHeli-20 RUAV, which is designed and implemented by ourselves, have verified the effectiveness of the proposed method. (Qi et al., 2006, 2007)

The sensors of the navigation system with different mechanism also have different performance. We can not get the ideal fault detection results using the traditional fault detection techniques. In order to accompany the short control period and the highly update rate, we use the parallel wavelet analyzer, which is shown as Fig.24.

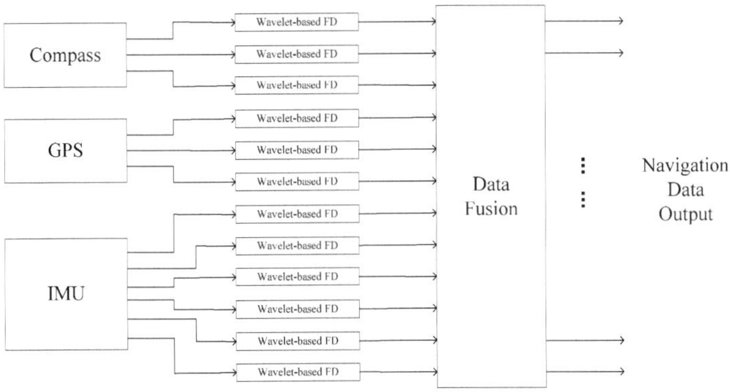

Figure 24. Wavelet-transform based sensors failure detection

6.3 Adaptive Filter Based Actuators Failure Tolerant Control Algorithms

Due to the inherently unstable dynamics, either flight test or real application of a RUAV is in high risk while a minimal failure may lead to the whole system collapse. In our recent research (Qi et al., 2007, 2008), a novel adaptive unscented Kalman filter (AUKF) is proposed for onboard failure coefficient estimation and a new fault tolerant control method is designed against the actuator failure of RUAV. The filter method with adaptability to statistical characteristic of noise is presented to improve the estimation accuracy of traditional UKF. The algorithm with the adaptability to statistical characteristic of noise, named Kalman Filter (KF)-based adaptive UKF (Fig.25), is proposed to improve the UKF performance. Such an adaptive mechanism is intended to compensate the lack of a prior knowledge. By introducing the actuator health coefficients (AHCs) into the dynamics equation of a RUAV, the proposed AUKF is utilized to online estimate both the flight states and the AHCs (Fig.26). A fault adaptive control is further designed based on the estimated states and AHCs. The comparisons between the adaptive-UKF-based fault tolerant control and the normal-UKF-based one show the effectiveness and improvements of the proposed method.

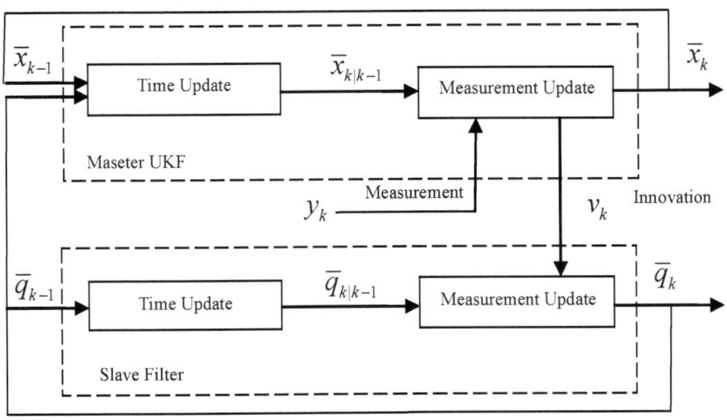

Figure 25. KF-based adaptive UKF

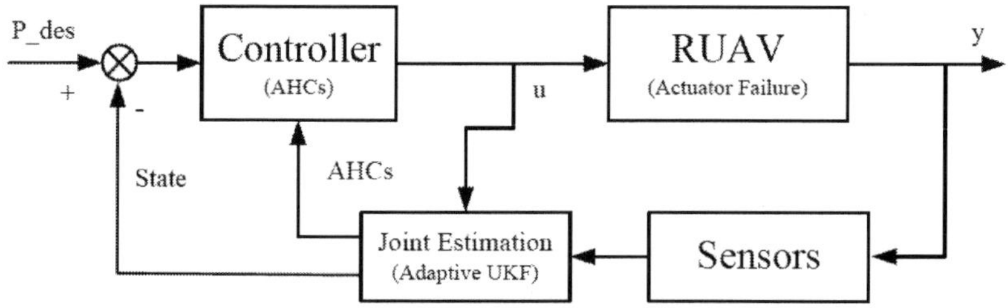

Figure 26. Actuator fault tolerant control scheme

7. Conclusions

This paper describes the current status of the ServoHeli-20 autonomous helicopter. We have introduced the system implementation of the rotorcraft UAV and control scheme for model scaled helicopter. A remote-controlled model helicopter is selected as the basic helicopter, which is changed to adapt to the heavy load. We also introduce the sensors and algorithm for attitude and position estimation. The two loop linear control scheme is presented in this paper for RUAV system and is a simple but useful control law in unmanned aerial vehicle experiments. Then we introduce our recent research in RUAV fault tolerant control algorithms.

The rotorcraft UAV system has been tested successfully for full autonomous flight including autonomous take off and landing. The next step is to integrate the visual and IMU estimation into a unified sensor suite and to develop advantage autonomous flight control algorithm for maneuverable fight.

8. References

Sanders, P. DeBietto, A. Eric F. (1998). Hierarchical Control of Small Autonomous Helicopters, *Proceedings of the 37th IEEE Conference on Decision & Control*, pp. 3629-3634, Tampa, Florida USA, December 1998.

Garratt, M.A. MAIAA, Chahl, J.S. (2003) Visual Control of an Autonomous Helicopter, *Proceedings of 41st Aerospace Sciences Meeting and Exhibit* pp. 6-9, Reno, Nevada, January 2003.

Enns, R. and Si, J. (2000) Helicopter ight control design using a Learning control approach, *Proceedings of the 39th IEEE Conference on Decision and Control*, pp. 1754-1759, Sydney, Australia, 2000.

Bijnens, B. Chu, Q. Voorsluijs, M. (2005) Adaptive feedback linearization flight control for a helicopter UAV, *Proceedings of the AIAA Guidance, Navigation, and Control Conference and Exhibit*, San Francisco, California, August 2005.

Koo, T. J. Shim, D. H. Shakernia, O. (1998) Hierarchical Hybird System Design on Berkeley UAV, *International Aerial Robotics Competition*, August 1998.

Jiang, Z. Han, J. (2006) Enhanced LQR Control for Unmanned Helicopter in Hover, *Proceedings of the 1st International Symposium on Systems and Control in Aerospace and Astronautics*, pp. 1438-1344, Harbin, China, January 2006.

Qi, J. Zhao, X. Jiang, Z. Han, J. (2006) Design and implement of a rotorcraft UAV testbed, *Proceedings of the 2006 IEEE international conference robotics and biomimetics*, Kunming,, China, Decemeber, pp. 109 – 114, 2006

Mettler, B. Dever, C. & Feron, E. (2004) Scaling Effects and Dynamic Characteristics of Miniature Rotorcraft, *Journal of Guidance Control and Dynamics*, Vol. 27, (3), pp.466-478, 2004.

Bendat J.S. and Piersol A.G., Engineering Application of Correlation and Spectral Analysis, *John Wiley & Sons*, USA, 1993.

Qi, J. Han, J. (2007) Application of Wavelets Transform to Fault Detection in Rotorcraft UAV Sensor Failure, *Journal of Bionic Engineering*, Vol.4 No.4 pp 265-270, 2007.

Qi, J. Yu, Q. Wang, H. Han, J. (2007) A Wavelet-Based Approach to Rotorcraft UAV Sensor Failure Detection, *International Conference on Intelligent Unmanned Systems*, Bali, Indonesia, 2007

Qi, J. Zhao, X. Jiang, Z. Han, J. (2007) Adaptive Neural-Network Scheme for Rotorcraft UAV Sensor Failure Diagnosis, *Lecture Note in Computer Sciences*, vol. 4493, pp. 589-596, 2007.

Qi, J. Jiang, Z. Zhao, X. Han, J. (2007) Fault Detection Design for RUAV with an Adaptive Threshold Neural-Network Scheme, *Proceedings of 2007 IEEE International Conference on Control and Automation*, Guangzhou, CHINA - May, pp.554-559, 2007

Qi, J. Wang, H, Jiang, Z. Zhao, X. Zhao, X. Han, J. (2007) Adaptive UKF-based Rotorcraft UAV Fault Adaptive Control for Actuator Failure, *AIAA Guidance, Navigation, and Control Conference and Exhibit*, AIAA-2007-6315, 2007

Qi, J. Jiang, Z. Zhao, X. Han, J. (2007) UKF-based Rotorcraft UAV Fault Adaptive Control for Actuator Failure. *IEEE International Conference on Robotics and Biomimetics 2007*, Sanya, China, pp 1545-1550, 2007

Qi, J. Zhao, X. Jiang, Z. Han, J. (2006) Design and Implement of a Rotorcraft UAV Testbed, *Proceedings of the 2006 IEEE International Conference on Robotics and Biomimetic*, Kunming, China, pp. 109-114, 2006

Qi, J. Wu, Z, Han, J. (2008) Rotorcraft UAV Actuator Failure Estimation with KF-based Adaptive UKF Algorithm, *2008 American Control Conference*, Seattle, U.S, 2008 [Accept]

Qi, J. Han, J. (2008) Fault Adaptive Control for RUAV Actuator Failure with Unscented Kalman Filter, *Third International Conference on Innovative Computing, Information and Control*, Dalian, China, 2008 [Accept]

Attitude and Position Control of a Flapping Micro Aerial Vehicle

Hala Rifaï[1], Nicolas Marchand[1] and Guylaine Poulin[2]
[1]*GIPSA-Lab, Control Systems dept. (Université de Grenoble),*
[2]*Unité Imagerie et Cerveau (Université de Tours)*
France

1. Introduction

Inspired from the natural flight, the flapping Micro Aerial Vehicles (MAV) combine the advantages of the rotary and fixed airfoils. They are able to achieve vertical taking off and landing, stationary flight and are characterized by their high maneuverability, soft noise and use the unsteady aerodynamics in order to develop higher lift force and theoretically reduce their energy consumption. They also get benefit from their biomimetic shape in order to execute discrete missions. The main disadvantages of such airfoils remain the complexity of analyzing the mechanisms adopted by insects during flight and maneuvers (Dudley, 2002) besides the technological reproduction of these techniques on flying robots (Hedrick & Daniel, 2006). Their development is constrained by the necessity of using low computational embedded systems, tiny sensors and actuators to ensure the free autonomous flight. Moreover, the conventional aerodynamic theory, well known for fixed aircrafts, fails for flapping wings airfoils due to the low Reynolds numbers and the influence of the unsteady airflows on the wings besides the high degrees of under actuation.

Micro aerial vehicles may be used for numerous indoor and outdoor civil applications (monitoring buildings, forests, cities, seism or high voltage lines, preventing forests fires, inspecting high monuments, intervening in narrow and dangerous environments for rescuing, gaming), military applications where its discretion thanks to its biomimetic behaviour is an advantage (spying and investigating) or even for exploring other planets like Mars (Thakoor et al., 2003).

Researches in flapping flight domain attract biology, aeronautic, robotic and avionic communities. The progress in microelectronic technologies, materials, sensors, actuators, embedded computational systems, communication tools, etc. is helping the feasibility and development of these aircrafts.

Therefore, flapping micro aerial vehicles are in a full rise nowadays; different projects are held all over the world. The present work lies within the scope of the French project OVMI[1] (Objet Volant Mimant l'Insecte) financed by the national agency for scientific research. It

[1] The OVMI project involves the IEMN (Valenciennes, Lille - France) for microelectronic study and prototype design, the ONERA (Palaiseau - France) for fluid mechanics modeling, the SATIE (Cachan - France) for energy aspects, the GIPSA-lab (Grenoble - France) for modeling and control.

aims to design, develop and control a silicon-based flapping robot, of scale one, mimicking the insect in flight and size, taking into consideration fluid mechanics and energetic aspects.

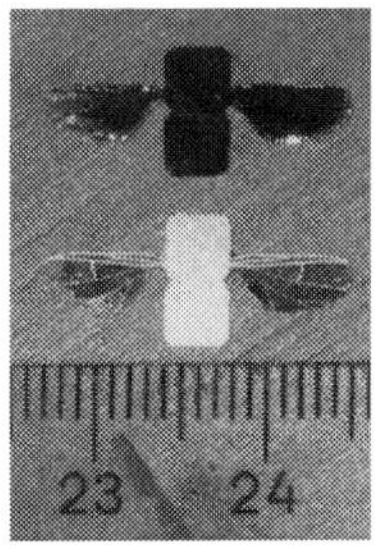

Figure 1. Centimetre scale prototype of the OVMI project

The goal of the present chapter is to develop control laws able to ensure the control of a flapping micro aerial vehicle in a three dimensional space, i.e. the attitude and position stabilization should be established.

Few of previous works have treated the control problem. Attitude stabilization of flapping airfoils has been treated using the linearized dynamics of the system to compute a proportional derivative controller (Deng et al., 2002). A linear quadratic optimal control law is tested in (Deng et al., 2003). State feedback controllers are proposed in (Schenato et al., 2002b; Schenato et al., 2004). Note that some of these control laws are computed using sensors measurements like halteres, ocelli, magnetometer and optic flow sensors (Deng et al., 2006a). A sensor's measurements based control law is computed in (Reiser et al., 2004) within upwind flow. The position control of flapping MAV is treated in (Schenato et al., 2002a) through the control of the vertical force and the torques: the control law is bounded and computed using a pole placement based on the linearized dynamics of the system. A state feedback control acting directly on the position is computed in (Schenato et al., 2001). A linear quadratic Gaussian control is proposed in (Deng et al., 2006b) based on some sensors measurements. (Dickson et al., 2006) have proposed a control law using a feedback of angular and linear vertical velocities using optic flow sensor's measurement in order to avoid obstacles in a tunnel. Time and distance optimal controls are proposed in (Sriram et al., 2005) aiming to control the MAV movement in a horizontal plane. An optimal control is also proposed in (Tanaka et al., 2006) in order to control the movement of the body in a vertical plane. Backstepping control laws are developed in (Rakotomamonjy, 2006) in order to stabilize separately the forward, vertical and pitch movements of a flapping aerofoil.

Nevertheless, the control laws developed in the literature present some limitations. Linear control laws are not robust with respect to external disturbances (wind, rain drop, shock, etc.). Therefore, nonlinear control should be used. Two techniques are widely used: the input-output linearization and the backstepping. While the first one brings the problem back to the linear case, the second depends on the system's inertia. Moreover, the proposed control laws are not bounded, except in (Schenato et al., 2002a). However, in the latter, the control is computed using the linearized dynamics and, consequently, does not ensure the global stability of the system.

In the present work, bounded state feedback nonlinear control laws of the flapping body's position and orientation are proposed. They are bounded in order to respect the maximum limit of the actuators driving the flapping wings. Moreover, they are very simple and have a

low computational cost, which makes them suitable for embedded implementations. Besides, they are independent from the model's parameters like the inertia matrix, for example.
The control laws are designed using the averaged model over a wing beat period and applied to the time varying system. This strategy is efficient for high-frequency systems like flapping micro aerial vehicles: the aerodynamic forces and torques affect the aircraft's behavior only by their mean values since the body's dynamics are much slower than the flapping wings' ones.
The chapter is organized as follows. In section 2, some mathematical background is recalled. In section 3, a simplified model of a flapping MAV is proposed. The average model is computed in order to determine and test the control laws. The problem is stated in section 4. In section 5, a bounded control law is presented in order to stabilize the body's attitude. The control of the position is developed in section 6. The dimensions of the flapping MAV are given in section 7. The results of simulations and some robustness tests are presented in section 8. Finally, section 9 presents some conclusions and introduces future works.

2. Mathematical Background

In the present paragraph, some definitions and properties used in this work are recalled.
The body's attitude is represented by quaternion (Shuster, 1993). It is a vector of four elements defining a rotation about an axis $\vec{e}$ of an angle v. It is given by

$$q = \begin{pmatrix} \cos\frac{v}{2} \\ \sin\frac{v}{2}\vec{e} \end{pmatrix} = \begin{pmatrix} q_0 \\ \vec{q} \end{pmatrix} \tag{1}$$

The quaternion respects a unit norm defined by the Hamilton space

$$\mathbb{H} = \{q \mid q_0^2 + \vec{q}^T\vec{q} = 1\} \tag{2}$$

The inverse of a unit quaternion q is determined by

$$q^{-1} = \left[q_0, -\vec{q}^T\right]^T \tag{3}$$

The product of two quaternions q and Q, represented by $\otimes$, is defined by

$$q \otimes Q = \left[(q_0Q_0 - \vec{q}^T\vec{Q}), (q_0\vec{Q} + Q_0\vec{q} + \vec{q} \wedge \vec{Q})^T\right]^T \tag{4}$$

The quaternion error defining the error between a current orientation given by a quaternion q and a desired one given by q_d is determined by

$$q_e = q \otimes q_d^{-1} \tag{5}$$

The rotation matrix representing the rotation of angle v about axis $\vec{e}$ is expressed, function of the quaternion, by

$$R(q) = (q_0^2 - \vec{q}^T\vec{q})I_3 + 2(\vec{q}\vec{q}^T + q_0\hat{q}) \tag{6}$$

Where

$$R(q) \in SO(3) = \{R(q) \in \mathbb{R}^{3x3} : R^T(q)R(q) = I, \det R(q) = 1\} \tag{7}$$

A quaternion q and its opposite $-q$ define the same attitude: they have the same rotation matrix. Hence, they represent two rotations about axis $\vec{e}$, one of angle v and the other of angle $(2\pi - v)$.
A sign function is defined as

$$\text{sign}(x) = \begin{cases} 1 & \text{if} \quad x \geq 0 \\ -1 & \text{if} \quad x < 0 \end{cases} \tag{8}$$

A classical saturation function is defined by (M is the saturation bound)

$$\text{sat}_M(x) = \begin{cases} x & \text{if} \quad |x| \leq M \\ M\,\text{sign}(x) & \text{if} \quad |x| > M \end{cases} \tag{9}$$

A differentiable function, bounded between ± 1 is defined by

$$\sigma(x) = \begin{cases} \text{sign}(x) & |x| > 1 + \mu \\ e_1 x^2 + e_2 x + e_3 & x \in [-1 - \mu, -1 + \mu] \\ x & x \in [-1 + \mu, 1 + \mu] \\ -e_1 x^2 + e_2 x - e_3 & x \in [1 - \mu, 1 + \mu] \end{cases} \tag{10}$$

Where $e_1 = \frac{1}{4\mu}, e_2 = \frac{1}{2} + \frac{1}{2\mu}, e_3 = \frac{\mu^2 - 2\mu + 1}{4\mu}$. If the function is bounded between $\pm M$, then $\sigma_M(\cdot) = M\sigma(\cdot)$.
The derivative of σ is given by

$$\dot{\sigma}(x) = \begin{cases} 0 & |x| > 1 + \mu \\ 2e_1 x + e_2 & x \in [-1 - \mu, -1 + \mu] \\ 1 & x \in [-1 + \mu, 1 + \mu] \\ -2e_1 x + e_2 & x \in [1 - \mu, 1 + \mu] \end{cases} \tag{11}$$

with $\dot{\sigma}_M(\cdot) = M\dot{\sigma}(\cdot)$.
A level function is defined by

$$\gamma(x, L, M) = \begin{cases} M & \text{if} \quad |x| > L \\ M + L - |x| & \text{if} \quad |x| \leq L \end{cases} \tag{12}$$

An integrator chain is a system of the form

$$\begin{cases} \dot{x}_i = x_{i+1} & i \in \{1, \ldots, n-1\} \\ \dot{x}_n = u \end{cases} \tag{13}$$

Where u is the control input and n the system's order.

System (13) can be controlled using a bounded control law $-\tilde{u} \le u \le \tilde{u}$ using Teel's approach (Teel, 1992). The control law is then given by

$$u = -sat_{M_n}(y_n + sat_{M_{n-1}}(y_{n-1} + \ldots + sat_{M_2}(y_2 + sat_{M_1}(y_1))\ldots)) \quad (14)$$

Where y_k is defined by

$$y_{n-j} = \sum_{i=0}^{j} \frac{j!}{i!(j-i)!} x_{n-i} \quad (15)$$

The control (14), bounded between $\pm M_n$ globally stabilize the system (13) for $M_j < \frac{1}{2} M_{j+1}$ with $j \in \{1,\ldots,n-1\}$, the poles of the system are $(-1,\ldots,-1)$.

This control has been generalized by (Marchand, 2003) using variable saturation bounds.

$$u = -sat_{M_n}(y_n + sat_{\gamma_{n-1}(y_n, L_n, M_{n-1})}(y_{n-1} + \ldots + sat_{\gamma_2(y_3, L_3, M_2)}(y_2 + sat_{\gamma_1(y_2, L_2, M_1)}(y_1))\ldots)) \quad (16)$$

Where γ_j, $j \in \{1,\ldots,n-1\}$, is the level function defined by (12), with $M_n := \tilde{u}$, $L_j := M_j$ for $j \in \{2,\ldots,n\}$ and $M_j = \frac{1}{1.00001} L_{j+1}$ for $j \in \{1,\ldots,n-1\}$, n is the integrator chain order.

The control law (13) has also been generalized by (Johnson and Kannan, 2003) for a pole placement at $(-a_1, -a_2, \ldots, -a_n)$. The coordinate transformation can be written as

$$y_{n-j} = a_{j+1} \sum_{i=0}^{j} C(j,i) x_{n-i} \qquad j \in \{0,\ldots,n-1\} \quad (17)$$

Where $C(j,i)$ is a function representing the sum of product combination of the system's poles (Johnson and Kannan, 2003).

3. Micro Aerial Vehicle Model

The goal of this work is to develop low computational cost control laws suitable for embedded implementation. The complete model of a flapping wing MAV will be presented at first. Then, this model is simplified such that it is based only on the steady aerodynamics and on a simple wing movement parameterization. The simplified model is averaged, thereafter, and used to compute the control laws.

3.1 Wings movement parameterization

The flapping wing is considered as a rigid body associated to a frame $\Re^W(\vec{r}, \vec{t}, \vec{n}, \psi, \phi, \theta)$ (see Fig. 1). The axis $\vec{r}$ is oriented from the wing base to its tip; the axis $\vec{t}$ is parallel to the wing chord, oriented from trailing to leading edge and the axis $\vec{n}$ is perpendicular to the wing plane oriented so that the three-sided frame $(\vec{r}, \vec{t}, \vec{n})$ is direct. The angles (ψ, ϕ, θ) are used to specify the position of the wing through three rotations about the wings axes

$(\vec{r},\vec{t},\vec{n})$ respectively. The flapping angle ϕ defines an up and down movement of the wing. The rotation angle ψ defines a rotation of the wing about its longitudinal axis and the deviation angle θ defines the orientation of the stroke plane. Furthermore, the wing is characterized by other complex phenomena like the flexion and the torsion. Flexibility allows the wing to be more resistant to turbulence and provides a gentler flight than a same size rigid wing. Torsion allows the wing to twist and provides aerodynamic stability without the need of a tail.

The wings frames should be indexed left $\Re_l^w(\vec{r}_l,\vec{t}_l,\vec{n}_l,\psi_l,\phi_l,\theta_l)$ and right $\Re_r^w(\vec{r}_r,\vec{t}_r,\vec{n}_r,\psi_r,\phi_r,\theta_r)$ for the left and right wings respectively.

Angles ϕ and ψ are assumed to vary according to saw tooth and pulse functions respectively, so that the wing changes its orientation at the end of each half stroke (see Fig. 2). In order to use actuators for 2 degrees of freedom only, the wings are supposed to beat in the mean stroke plane; therefore angle θ is taken to zero. The temporal variation of the wings angles is given by (18).

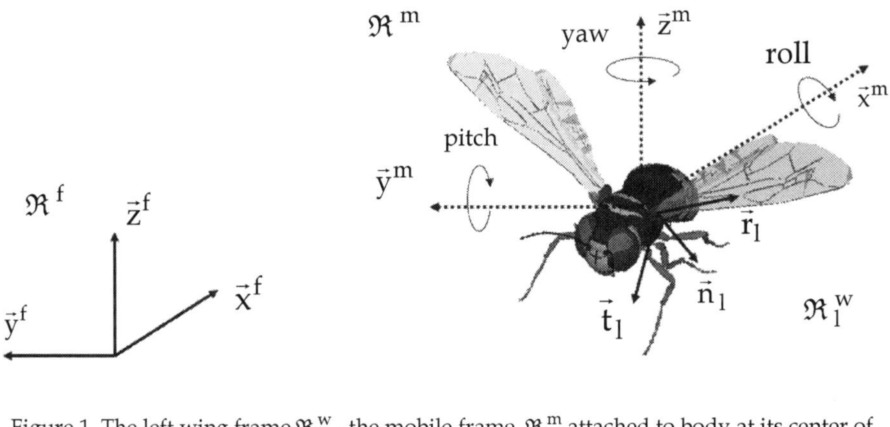

Figure 1. The left wing frame $\Re_l^w$, the mobile frame $\Re^m$ attached to body at its center of gravity, and the fixed frame $\Re^f$

$$
\begin{aligned}
\phi(t) &= \begin{cases} \phi_0\left(1-\dfrac{2t}{\kappa T}\right) & 0 \le t \le \kappa T \\ \phi_0\left(2\dfrac{t-\kappa T}{(1-\kappa)T}-1\right) & \kappa T < t \le T \end{cases} \\
\psi(t) &= \psi_0 \,\mathrm{sign}(\kappa T - t) \qquad 0 \le t \le T \\
\theta(t) &= 0 \qquad 0 \le t \le T
\end{aligned} \tag{18}
$$

where sign is defined by (8), T is the wingbeat period, κ is the ratio of downstroke duration to the wingbeat period, ϕ_0 and ψ_0 are respectively the amplitudes of flapping and rotation angles. The last two parameters considered for both left and right wings will be taken as control variables, as explained in the following.

Note that this should not be understood as the real movement of the wing, but as a reference input of the actuators. The actuators that exist on the market, in particular the piezoelectric ones, have a very fast dynamics; their influence on the MAV's movement is consequently almost not detectable. They operate in a resonant mode, thus ensuring the movement of the wings at a predefined frequency. The alternative voltage applied to the actuator is delivered by an electronic converter. This one should be conceived especially for piezoelectric actuators, which are reactive loads (Janocha and Stiebel, 1998; Campolo et al., 2003) and present a non linear behavior (hysteresis, creep) that can be compensated using an adapted control strategy (Kuhnen et al., 2006). Therefore, it is necessary to use a low-level controller in order to control the flapping wings. The input of this controller is the reference signal (amplitudes of the flapping and rotation angles) computed via the control law of the system. Thus, the local controller and the actuator behave as a first order filter that has a low response time so the steady regime is established very quickly (19).

$$\ddot{A} = \ddot{A}_r - \lambda_1(\dot{A} - \dot{A}_r) - \lambda_2(A - A_r) \quad (19)$$

With A is the amplitude of the flapping or rotation angle (actuator output), A_r is the reference amplitudes (actuator input). λ_1 and λ_2 are fixed so that the time constant of the local actuation loop is verified.

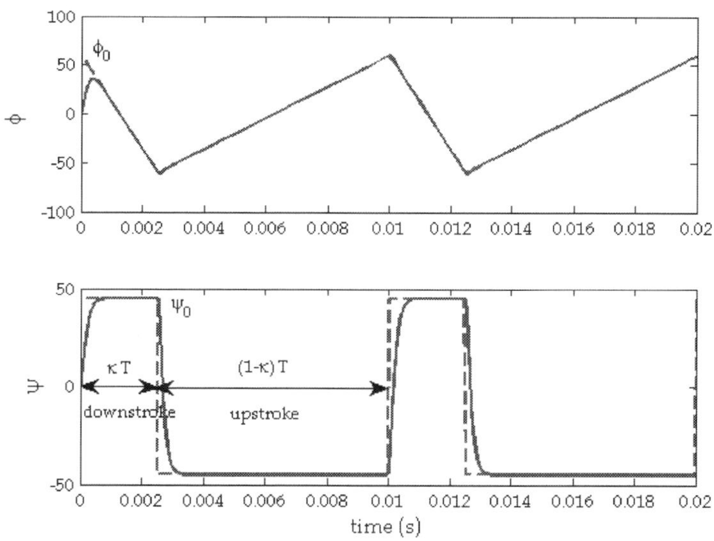

Figure 2. The flapping angle ϕ and the rotation angle ψ : the theoretical angles are plotted with red dashed line (actuators input) and the real angles (actuators output) with blue continuous line

3.2 Aerodynamic Forces and Torques

Different mechanisms act synergistically to produce the aerodynamic force in flapping flight (Dickinson et al., 1999; Sane, 2003): delayed stall, rotational circulation, added mass, wake capture, etc. The first one is developed during the translational movement of the wing (the flapping movement), while the others are generated due to the rotation of the wing about the radial axis $\vec{r}$. These forces are perpendicular to the wing surface, which means they are collinear to the wing normal vector $\vec{n}$. They are applied at the aerodynamic center of the wing, located at $x_1 = 0.65L$ and at $x_0 = 0.25l$, respectively from the wing base and leading edge; L is the wing length and l is the wing chord. In the present work, only the steady aerodynamic force, the added mass force and the rotational lift will be considered. The others are neglected since they have a minor contribution, on the one hand, and are difficult to model, on the other hand.

- *Steady aerodynamic force*: This force is due to the air pressure on the flapping wing surface and has the opposite direction of the wing velocity. It is given by

$$f_s^w = -\frac{1}{2}\rho C_w S_w v^w \left|v^w\right| \tag{20}$$

ρ is the air density, S_w is the wing's surface, v^w is the wing's velocity, C_w is a coefficient of the aerodynamic force applied on a wing.

$$C_w = \begin{cases} C(1+C_f) & 0 < t < \kappa T \\ C(1-C_f) & \kappa T < t < T \end{cases} \tag{21}$$

Where $C \approx 3.5$ is the aerodynamic force coefficient, derived empirically in (Dickinson et al., 1999; Schenato et al., 2003) and C_f is a coefficient chosen so that the aerodynamic force is 20% greater during downstroke than during upstroke. This dissymmetry between the two half strokes can be justified based on (Dudley, 2002). During downstroke, the dorsal side of the wing is opposite to the air flow. The supination opposes the ventral side of the wing to the flow. Consequently, the effective area of the wing is reduced and the orientation of the air circulation about the wing reverses, leading to a wing camber alteration. Therefore, downstroke lift is likely to be higher than that of upstroke, so that the averaged force over a single wingbeat period should at least balance the body's weight. The position of the aerodynamic center of the wing is given by

$$\vec{p}^w = [x_1, 0, 0]^T \tag{22}$$

such that it is belongs to the radial axis $\vec{r}$. $\vec{p}^w$ is expressed in the mobile frame $\Re^m$ by applying the rotation matrix R_w^m (the left and right wings are indexed l and r respectively, R_l^m is the rotation matrix from $\Re_l^w$ to $\Re^m$ and R_r^m from $\Re_r^w$ to $\Re^m$)

$$\begin{aligned} \vec{p}_l^m &= R_l^m \vec{p}^w \\ \vec{p}_r^m &= R_r^m \vec{p}^w \end{aligned} \tag{23}$$

The derivative of the left and right aerodynamic forces expressed in the mobile frame $\Re^m$ is given by $\vec{v}^m_{l,r} = \dot{\vec{p}}^m_{l,r}$. The projection of the velocity in the wing frame is then given by $\vec{v}^w = R^w_m \vec{v}^m$.

Note the relative velocity due to vortices is not considered in this work. A recent work on fish modeling seems to show that the effect, on the overall motion, of this phenomenon as well as the nonlinear dynamic phenomenon, characteristic of small Reynolds numbers, can be shrewdly taken into account with a modification of the masses and parameters of the system (Boyer et al., 2006).

- *Added mass force*: The added mass phenomenon is due to the additional fluid mass acceleration developed around the wing when it accelerates and rotates. It can be modeled by (Rakotomamonjy et al., 2004) ($\ddot{\phi}$ is the second-order derivative of the flapping angle ϕ)

$$f^w_{ma} = \frac{\pi}{4}\rho l^2 x_1 L\ddot{\phi} \tag{24}$$

- *Rotational force*: The wing rotating about its span-wise axis, during pronation or supination, causes the deviation of the ambient fluid. As a reaction to this phenomenon, the wing generates additional rotational circulation (Sane, 2003). This force can be modeled based on (Rakotomamonjy et al., 2004) and using the simplification considered in the present work, as ($\dot{\psi}$ is the first-order derivative of the rotation angle ψ)

$$f^w_r = \frac{1}{2}\pi\rho l^2 L v^w \dot{\psi} \tag{25}$$

The total aerodynamic force generated by a wing during the flapping flight is the sum of these three forces

$$f^w = f^w_s + f^w_{ma} + f^w_r \tag{26}$$

As mentioned before, the aerodynamic force is perpendicular to the wing surface, thus collinear to the vector $\vec{n}$: $\vec{f}^w = f^w \vec{n}$.

Projecting the aerodynamic force generated by the left and right wings into frame $\Re^m$ (R^m_w is the rotation matrix from $\Re^w$ to $\Re^m$)

$$\vec{f}^m_{l,r} = R^m_w \vec{f}^w_{l,r} \tag{27}$$

and summing up, the global aerodynamic force can be obtained

$$\vec{f}^m = \vec{f}^m_l + \vec{f}^m_r \tag{28}$$

The aerodynamic force has two components, the thrust that ensures a forward movement of the MAV, and the lift that ensures a vertical one.

Angular viscous torque is negligible with respect to the aerodynamic torque (Schenato et al., 2003). The aerodynamic torque is the cross product of the force and its center of application. It is given by

$$\vec{\tau}^m = \vec{p}_l^m \wedge \vec{f}_l^m + \vec{p}_r^m \wedge \vec{f}_r^m \tag{29}$$

3.3 Body's dynamics

The movement of the wings generates the aerodynamic force and torque (20, 24, 25, 29). The body is thus subject to the aerodynamic force and torque, the viscous and gravitational forces. These forces generate consequently the displacement and the maneuvers of the MAV. The translational and rotational movement of the body is computed through the dynamic equations:

$$\begin{aligned}
\dot{\vec{P}}^f &= \dot{\vec{V}}^f \\
\dot{\vec{V}}^f &= \frac{1}{m} R^T(q)\vec{f}^m - c\vec{V}^f - \vec{g} \\
\begin{pmatrix} \dot{q}_0 \\ \dot{\vec{q}} \end{pmatrix} &= \frac{1}{2}\begin{pmatrix} -\vec{q}^T \\ I_3 q_0 - \hat{q} \end{pmatrix} \vec{\omega}^m \\
\dot{\vec{\omega}}^m &= J^{-1}(\vec{\tau}^m - \vec{\omega}^m \wedge J\vec{\omega}^m)
\end{aligned} \tag{30}$$

Where $\vec{P}^f \in \mathbb{R}^3$ and $\vec{V}^f \in \mathbb{R}^3$ are respectively the linear position and velocity of the body's center of gravity relative to the fixed frame $\mathfrak{R}^f$. $\vec{\omega}^m$ is the angular velocity with respect to the mobile frame $\mathfrak{R}^m$ attached to the insect's body on its center of gravity. c is the viscous coefficient and $\vec{g}$ the gravity vector. $\vec{f}^m \in \mathbb{R}^3$ and $\vec{\tau}^m \in \mathbb{R}^3$ are respectively the aerodynamic force and torque vectors. $J \in \mathbb{R}^{3\times 3}$ is the inertia matrix of the body relative to $\mathfrak{R}^m$ and I_3 is the identity matrix. q is the quaternion defining the attitude of the body relative to $\mathfrak{R}^f$ (3). $R^T(q)$ is the rotation matrix defined by (6,7).

3.4 Average model

Generally, insects have a high wingbeat frequency. The averaging theory (Khalil, 1996, Bullo, 2002, Vela, 2003) shows that the averaged dynamics of high frequency oscillating systems are a good approximation of the system. Consequently, the mean model is computed using the averaged dynamics (aerodynamic force and torque) over a wingbeat period.

In this work, the amplitudes of the wings angles are chosen to be the control variables. Denoting by $u = \left(\phi^l(t), \phi^r(t), \psi^l(t), \psi^r(t), \theta^l(t), \theta^r(t)\right)$ the flapping, rotation and deviation angles for left and right wings, $v = \left(\phi_0^l, \phi_0^r, \psi_0^l, \psi_0^r, \theta_0^l, \theta_0^r\right)$ the amplitudes of the angles (18), $f(x, u, \dot{u})$ the system defined by (30) and T the wingbeat period, the following theorem can be established (Bullo, 2002).

Theorem 1. Consider a time varying system and its average over a period T

$$\begin{cases} \dot{x} = f(x,u,\dot{u}) \\ u = g(v,t) \\ v = h(x) \\ g(v,t) = g(v,t+T) \end{cases} \tag{31}$$

$$\begin{cases} \dot{x} = \bar{f}(\bar{x},\bar{v}) \\ \bar{f}(x,v) = \frac{1}{T}\int_0^T f(x,g(v,t),\dot{g}(v,t))dt \\ \bar{v} = h(\bar{x}) \end{cases} \tag{32}$$

where $\{x,\bar{x}\} \in \mathbb{R}^n, u \in \mathbb{R}^m, v \in \mathbb{R}^p$ and all functions and their derivatives are continuous up to the second order. If $\bar{x} = 0$ is an exponentially stable equilibrium point for the averaged system (32), then there exists k>0 such that $|x(t) - \bar{x}(t)| < kT$ for all $t \in [0,\infty)$. Moreover the original system (31) has a unique, exponentially stable, T-periodic orbit $x_T(t)$ with the property $|x_T(t)| < kT$.

Theorem 1 shows that an exponentially stable equilibrium state for the averaged dynamics of a high frequency oscillating system is also an equilibrium state for the oscillating (time variant) system. Thus, a stabilizing control of the averaged system (32), which is equivalent to a rigid body, will stabilize the time varying system (31).

As mentioned before, the amplitudes of the wings angles are chosen to be the control variables. Only the steady aerodynamic force is used to compute the average model. The relation between the angles defining the wings kinematics and the mean force and torque, averaged over a wingbeat period, can be written as follows.

$$\begin{aligned} -\alpha\left[\phi_0^r \sin\phi_0^r \sin\psi_0^r + \phi_0^l \sin\phi_0^l \sin\psi_0^l\right] &= \bar{f}_x \\ -\beta\left[\phi_0^r \sin\phi_0^r \cos\psi_0^r + \phi_0^l \sin\phi_0^l \cos\psi_0^l\right] &= \bar{f}_h \\ \beta x_1\left[\phi_0^{r^2} \cos\psi_0^r - \phi_0^{l^2} \cos\psi_0^l\right] &= \bar{\tau}_1 \\ \alpha x_1\left[\phi_0^{r^2} \sin\psi_0^r - \phi_0^{l^2} \sin\psi_0^l\right] &= \bar{\tau}_3 \end{aligned} \tag{33}$$

System (33) can be written in a compact form as:

$$\Lambda(\phi_0^l, \phi_0^r, \psi_0^l, \psi_0^r) = (\bar{f}_x, \bar{f}_h, \bar{\tau}_1, \bar{\tau}_3) \tag{34}$$

For any averaged state feedback control of the force and torque,

$$(\bar{f}_x, \bar{f}_h, \bar{\tau}_1, \bar{\tau}_3) = U(\bar{x}) \tag{35}$$

the wings angles amplitudes can be computed

$$(\phi_0^l, \phi_0^r, \psi_0^l, \psi_0^r) = \Lambda^{-1}(U(\bar{x})) \tag{36}$$

Finally, $h(\cdot) = \Lambda^{-1}(U(\cdot))$ in (32).

3.5 Saturation set

Physically, the flapping and rotation angles are bounded. Considering that

$$\begin{aligned} 0 \le \phi_0 \le \tilde{\phi}_0 \\ -\tilde{\psi}_0 \le \psi_0 \le \tilde{\psi}_0 \end{aligned} \tag{37}$$

for both left and right wings, system (33) defines a convex set Ω in the mean control variables $(\bar{f}_x, \bar{f}_h, \bar{\tau}_1, \bar{\tau}_3)$ (see Fig. 5). $\Omega_{\bar{\tau}_1,\bar{\tau}_3}$ and $\Omega_{\bar{f}_x,\bar{f}_h}$ are the projection of Ω on the planes $(\bar{\tau}_1, \bar{\tau}_3)$ and $(\bar{f}_x, \bar{f}_h)$ respectively. Therefore, anywhere in the set Ω there exists a wing configuration $(\phi_0^l, \phi_0^r, \psi_0^l, \psi_0^r)$ producing the mean desired forces and torques $(\bar{f}_x, \bar{f}_h, \bar{\tau}_1, \bar{\tau}_3)$.

4. Problem Statement

Considering the mean behavior over a wingbeat period of system (30), the MAV is approximated by a rigid body subject to external forces and torques. Therefore, the averaged state of the time varying model $\bar{x}$ is equivalent to a rigid body state x^{rb}. Therefore, the following equivalence can be established

$$\underbrace{\Lambda(\phi_0^l, \phi_0^r, \psi_0^l, \psi_0^r)}_{\text{flapping wings}} = \underbrace{(\bar{f}_x, \bar{f}_h, \bar{\tau}_1, \bar{\tau}_3)}_{\text{rigid body}} \tag{38}$$

The problem is thus transformed to a classic rigid body control, and is applied to the flapping wings body by computing the flapping and rotation angles given the control forces and torques. The strategy proposed in the present work consists of controlling the orientation of the flapping MAV as a first step, then to stabilize its position, in hovering mode, based on the attitude control. This strategy is adopted since the translational dynamics of the system (30) depends on the rotational ones, but the rotational dynamics are independent of the translational ones. (30) can be written in the form of a cascade system

$$\begin{cases} \dot{x} = f(x, y) \\ \dot{y} = g(y, u) \end{cases} \tag{39}$$

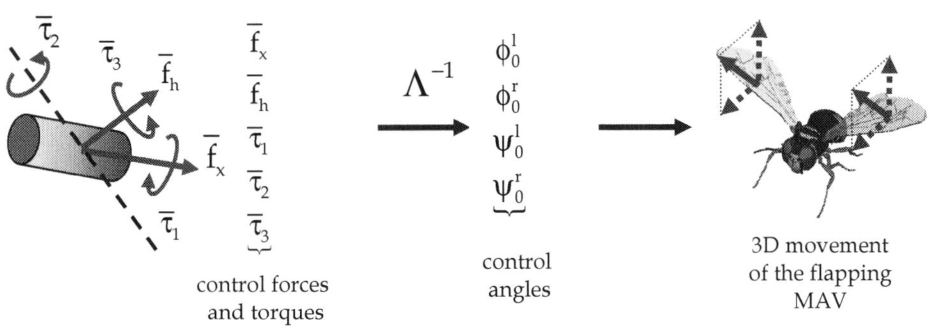

Figure 3. The transition from a rigid body control to a flapping MAV control

The control of the thrust $\bar{f}_x$ and lift $\bar{f}_h$ forces, besides the roll $\bar{\tau}_1$ and yaw $\bar{\tau}_3$ torques is achieved by controlling the wings angles amplitudes. The control of the pitch movement is performed by a small mass moving longitudinally inside the body and changing its center of gravity. This can be realized using the Electrowetting on Dielectric technique (Renaudin et al., 2004). The control of the lateral movement is accomplished by tilting the MAV sideway by acting on the roll angle. The lift $\bar{f}_h$ will have then a lateral component besides its vertical one. This movement is adopted by the majority of insects to ensure the lateral displacement. In the following, a coupling between the lift and roll angle is considered for this purpose.

At convergence, the position should converge to the desired value while the linear and angular velocities, the roll, pitch and yaw angles (or quaternion) to zero.

$$\begin{cases} \vec{P} & \rightarrow & P_d \\ \vec{V} & \rightarrow & 0 \\ q & \rightarrow & q_I \\ \vec{\omega} & \rightarrow & 0 \end{cases} \quad \text{as} \quad t \rightarrow \infty \tag{40}$$

The bloc diagram of the MAV is represented on Fig. 4.

The blocs *Control forces* and *Control torques* will be detailed thereafter. The bloc *Wings angles amplitudes* determines the angles amplitudes using (36). Given the amplitudes, the *Wings parameterization* is given by (18). The *Simulator* computes the linear and rotational position and velocity based on (20-30). The bloc *Averaging* computes the average state over a wingbeat period; the average state will be used to compute the control laws.

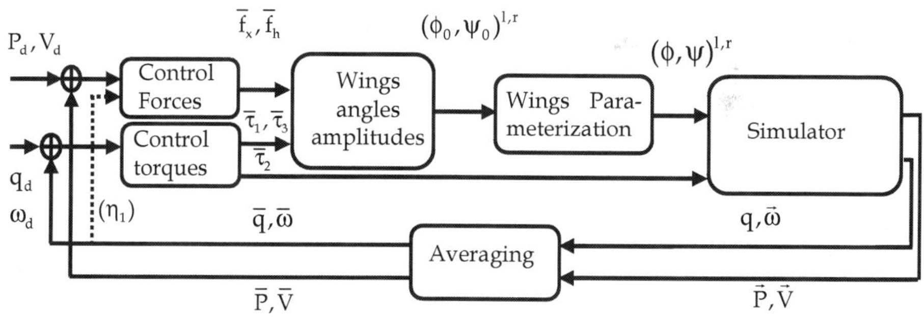

Figure 4. The bloc diagram of the flapping MAV

5. Attitude control

The control law applied in this paragraph is supposed to drive the body to a desired orientation q_d, while the angular velocity should vanish (40). The proposed attitude stabilizing control torque is a bounded state feedback based in its formulation on the model of a rigid body (Guerrero-Castellanos et al., 2007) (equivalent to the averaged model of the flapping body) and applied to the time variant model (flapping MAV).

This control law is extremely simple and therefore suitable for an embedded implementation. Moreover, the control law is robust with respect to aerodynamic coefficient errors and does not require the knowledge of the body's inertia. Let $\bar{\tau} = [\bar{\tau}_1, \bar{\tau}_2, \bar{\tau}_3]^T$ be the roll, pitch and yaw control torques.

$$\bar{\tau}_j = -\mathrm{sat}_{\tilde{\tau}_j}\left(\lambda_j\left[\delta_j\bar{\omega}_j + \mathrm{sign}(q_0)\mathrm{sat}_{M_{1,j}}(\bar{q}_j)\right]\right) \tag{41}$$

where $j \in \{1,2,3\}$, $\mathrm{sign}(q_0)$ takes into account the possibility of 2 rotations to drive the body to its equilibrium orientation; the one of smaller angle is chosen. $\bar{\omega}_j$ and $\bar{q}_j$ are the averaged angular velocities and quaternion over a single wingbeat period, representing the time varying angular velocities and quaternion of a rigid body. λ_j and δ_j are positive parameters. Differently from (Guerrero-Castellanos et al., 2007), δ_j has been added in order to slow down the convergence of the torque relative to the angular velocity. $\mathrm{sat}_{M_{1,j}}$ and $\mathrm{sat}_{\tilde{\tau}_j}$ are saturation functions with $M_{1,j}$ and $\tilde{\tau}_j$ the saturation bounds: $M_{1,j} \geq 1$, $\tilde{\tau}_j \geq \delta_j(2M_{1,j} + \varepsilon_j)$ and $\varepsilon_j > 1$. The $\tilde{\tau}_j$'s are chosen in order to respect input saturations: wings Euler angles and body's length. Based on (33, 37), the maximum flapping and rotation amplitudes, $\tilde{\phi}_0$ and $\tilde{\psi}_0$, define a set $\Omega_{\bar{\tau}_1,\bar{\tau}_3}$ of admissible torques (see Fig. 5). The saturation bounds $\tilde{\tau}_1$ and $\tilde{\tau}_3$ are adjusted in (41) so that $\bar{\tau}_1$ and $\bar{\tau}_3$ remain in the limits of $\Omega_{\bar{\tau}_1,\bar{\tau}_3}$, which guarantees not to exceed the maximum angles. $\tilde{\tau}_2$ should respect the saturation induced by the length of the body, since the pitch torque is generated by a small mass moving inside it.

The asymptotic stability of the closed loop system has been shown in (Guerrero-Castellanos et al., 2007) for rigid bodies using the following Lyapunov function (the added parameter δ_j does not change the proof).

$$V = \frac{1}{2}\bar{\omega}^T J\bar{\omega} + \kappa((1-\bar{q}_0)^2 + \vec{\bar{q}}^T\vec{\bar{q}}) \tag{42}$$

Therefore, $\bar{\omega} \to 0$ and $\bar{q} \to q_I$ (based on the rigid body case). By means of the averaging theory, $|\vec{\omega} - \bar{\omega}| < k_1 T$ and $|q - \bar{q}| < k_2 T$ for $k_{\{1,2\}} > 0$ and T the wingbeat period.

6. Position control

Neglecting the viscous force $c\vec{V}^f$ acting on the MAV's body by supposing that it is moving at low speeds, the translational subsystem (30) can be transformed into a chain of integrators. $c\vec{V}^f$ will be considered as a disturbance term in simulations. Supposing that after a sufficiently long time, the MAV is stabilized over the pitch and yaw axes $(\eta_2 = \eta_3 = 0)$ thanks to the control law (41), thereby the rotation matrix defines solely a rotation about the roll axis $\vec{x}^m$. The normalized translational subsystem, augmented of a state representing the integral of the position, can be written ($\vec{P}^f = [P_x, P_y, P_z]^T$ is the current position)

$$\begin{cases} \dot{p}_1 & = & p_2 \\ \dot{p}_2 & = & p_3 \\ \dot{p}_3 & = & v_x \end{cases} \tag{43}$$

$$\begin{cases} \dot{p}_4 & = & p_5 \\ \dot{p}_5 & = & p_6 \\ \dot{p}_6 & = & -v_h \sin(\eta_1) \\ \dot{p}_7 & = & p_8 \\ \dot{p}_8 & = & p_9 \\ \dot{p}_9 & = & v_h \cos(\eta_1) - 1 \end{cases} \tag{44}$$

$p = \frac{1}{g}\left(\overline{\int P_x^f}, \bar{P}_x^f, \bar{V}_x^f, \overline{\int P_y^z}, \bar{P}_y^f, \bar{V}_y^f, \overline{\int P_z^f}, \bar{P}_z^f, \bar{V}_z^f \right) = (p_1, \ldots, p_9)$ is the averaged state of the translational subsystem, $v_x = \frac{\bar{f}_x}{mg}, v_h = \frac{\bar{f}_h}{mg}$ where $\bar{f}_x$ and $\bar{f}_h$ are respectively the control thrust and lift, η_1 is the roll angle and 1 is the normalized gravity (Hably et al., 2006).

The averaged normalized system (43, 44) will be used to compute the normalized control thrust v_x and lift v_h. As for (41), the proposed controls are bounded and have a low computational cost.

6.1 Stabilization of the forward movement

System (43) defines a triple integrator chain. It can be stabilized using the control based on nested saturations (14) with the variable saturation bound (16). The variable change given in (17) developed to the third order for poles placement in $(-a_1, -a_2, -a_3)$

$$\begin{bmatrix} y_1 \\ y_2 \\ y_3 \end{bmatrix} = \begin{bmatrix} a_1 a_2 a_3 & a_3(a_1 + a_2) & a_3 \\ 0 & a_1 a_2 & a_2 \\ 0 & 0 & a_1 \end{bmatrix} \begin{bmatrix} x_1 \\ x_2 \\ x_3 \end{bmatrix} \tag{45}$$

Which can be written in a compact form as $y = \Pi x$, with $\Pi_{i,j}$ the element at the i'th row and j'th column. v_x can then be written as

$$\begin{aligned} v_x = & -\sigma_{\tilde{v}_x}\Big(\Pi_{x_{3,3}} p_3 + \sigma_{\gamma_2(\Pi_{x_{3,3}} p_3, L_{x3}, M_{x2})}(\Pi_{x_{2,3}} p_3 + \Pi_{x_{2,2}} p_2 + \ldots \\ & \sigma_{\gamma_1(\Pi_{x_{2,3}} p_3 + \Pi_{x_{2,2}} p_2, L_{x2}, M_{x1})}(\Pi_{x_{1,3}} p_3 + \Pi_{x_{1,2}} p_2 + \Pi_{x_{1,1}} p_1))\Big) \end{aligned} \tag{46}$$

where $\tilde{v}_x$ is the saturation bound of v_x and respects the saturation $\Omega_{\bar{f}_x, \bar{f}_h}$ (see Fig. 5) in order to guarantee wings angles lower than the maximum values. $\sigma(\cdot)$ is the saturation function defined in (10) and Π_x is the transformation matrix relative to the forward movement. The asymptotic stability of (p_1, p_2, p_3) is proven based on (Johnson and Kannan, 2003; Marchand, 2003).

6.2 Stabilization of the lateral and vertical movements

The coupling between the roll angle η_1 and normalized lift force v_h is explicitly shown in (44). η_1 behaves like an intermediate input to (44) transforming the problem to a VTOL (Vertical Taking Off and Landing) one. η_1 should converge to a desired value η_{1d} given by

$$\eta_{1_d} = \arctan\left(\frac{-v_y}{v_z + 1}\right) \tag{47}$$

Where v_y and v_z will be determined thereafter. The normalized lift is determined by

$$v_h = \sqrt{v_y^2 + (v_z + 1)^2} \tag{48}$$

The desired quaternion is computed by

$$q_d = \left[\cos\frac{\eta_{1_d}}{2}, \sin\frac{\eta_{1_d}}{2}, 0, 0\right]^T \tag{49}$$

The flapping MAV should track a desired angular velocity given by

$$\left[0, \vec{\omega}_d^T\right]^T = 2\dot{q}_d \otimes q_d^{-1} \tag{50}$$

$$\vec{\omega}_d = \left[\dot{\eta}_{1_d}, 0, 0\right]^T \tag{51}$$

Where q_d^{-1} is the quaternion inverse (3) and $\otimes$ the quaternion product (4). The derivative of the roll angle is given by ($\dot{v}_y$ and $\dot{v}_z$ will be determined thereafter)

$$\dot{\eta}_{1_d} = \frac{-\dot{v}_y (v_z + 1) + v_y \dot{v}_z}{v_y^2 + (v_z + 1)^2} \tag{52}$$

The quaternion error is computed using (5) and the error of angular velocity by $\vec{\omega}_e = \vec{\omega} - \vec{\omega}_d$. Applying control torque (41) on the error dynamics, the convergence of η_1 to η_{1_d} is ensured. System (44) is then transformed into two independent triple integrators.

$$\begin{cases} \dot{p}_4 = p_5 \\ \dot{p}_5 = p_6 \\ \dot{p}_6 = v_y \end{cases} \qquad \begin{cases} \dot{p}_7 = p_8 \\ \dot{p}_8 = p_9 \\ \dot{p}_9 = v_z \end{cases} \tag{53}$$

Applying the same control law as for the forward movement, v_y and v_z are computed

$$\begin{aligned} v_y = & -\sigma_{\tilde{v}_y}\Big(\Pi_{y_{3,3}} p_6 + \sigma_{\gamma_2(\Pi_{y_{3,3}} p_6, L_{y3}, M_{y2})}(\Pi_{y_{2,3}} p_6 + \Pi_{y_{2,2}} p_5 + \dots \\ & \sigma_{\gamma_1(\Pi_{y_{2,3}} p_6 + \Pi_{y_{2,2}} p_5, L_{y2}, M_{y1})}(\Pi_{y_{1,3}} p_6 + \Pi_{y_{1,2}} p_5 + \Pi_{y_{1,1}} p_4))\Big) \end{aligned} \tag{54}$$

$$\begin{aligned} v_z = & -\sigma_{\tilde{v}_z}\Big(\Pi_{z_{3,3}} p_9 + \sigma_{\gamma_2(\Pi_{Z_{3,3}} p_9, L_{z3}, M_{z2})}(\Pi_{z_{2,3}} p_9 + \Pi_{z_{2,2}} p_8 + \dots \\ & \sigma_{\gamma_1(\Pi_{z_{2,3}} p_9 + \Pi_{z_{2,2}} p_8, L_{z2}, M_{z1})}(\Pi_{z_{1,3}} p_9 + \Pi_{z_{1,2}} p_8 + \Pi_{z_{1,1}} p_7))\Big) \end{aligned} \tag{55}$$

Where $\sigma(\cdot)$ is defined by (10). $\tilde{v}_y$ and $\tilde{v}_z$ should verify

$$\tilde{v}_h = \sqrt{\tilde{v}_y^2 + (\tilde{v}_z + 1)^2} \tag{56}$$

$\tilde{v}_h$ should respect the saturation set $\Omega_{\bar{f}_x, \bar{f}_h}$ in order to guarantee admissible wings angles amplitudes ϕ_0 and ψ_0. Finally, in order to evaluate the desired angular velocity, the derivative of v_y and v_z are computed

$$\begin{aligned}\dot{v}_y = -\dot{\sigma}_{\tilde{v}_y}\Big(&\Pi_{y_{3,3}}p_6 + \sigma_{\gamma_2(\Pi_{y_{3,3}}p_6, L_{y3}, M_{y2})}(\Pi_{y_{2,2}}p_5 + \Pi_{y_{2,3}}p_6 + \ldots \\ &\sigma_{\gamma_1(\Pi_{y_{2,2}}p_5+\Pi_{y_{2,3}}p_6, L_{y2}, M_{y1})}(\Pi_{y_{1,1}}p_4 + \Pi_{y_{1,2}}p_5 + \Pi_{y_{1,3}}p_6))\Big).\{\Pi_{y_{3,3}}r_y + \ldots \\ &\dot{\sigma}_{\gamma_2(\Pi_{y_{3,3}}p_6, L_{y3}, M_{y2})}(\Pi_{y_{2,2}}p_5 + \Pi_{y_{2,3}}p_6 + \ldots \\ &\sigma_{\gamma_1(\Pi_{y_{2,2}}p_5+\Pi_{y_{2,3}}p_6, L_{y2}, M_{y1})}(\Pi_{y_{1,1}}p_4 + \Pi_{y_{1,2}}p_5 + \Pi_{y_{1,3}}p_6))\Big).\Big[\Pi_{y_{2,2}}p_6 + \ldots \\ &\Pi_{y_{2,3}}r_y + \dot{\sigma}_{\gamma_1(\Pi_{y_{2,2}}p_5+\Pi_{y_{3,3}}p_6, L_{y2}, M_{y1})}(\Pi_{y_{1,1}}p_4 + \Pi_{y_{1,2}}p_5 + \Pi_{y_{1,3}}p_6).(\Pi_{y_{1,1}}p_5 + \Pi_{y_{1,2}}p_6 + \Pi_{y_{1,3}}r_y)\Big]\Big\}\end{aligned} \tag{57}$$

$$\begin{aligned}\dot{v}_z = -\dot{\sigma}_{\tilde{v}_z}\Big(&\Pi_{z_{3,3}}p_9 + \sigma_{\gamma_2(\Pi_{z_{3,3}}p_9, L_{z3}, M_{z2})}(\Pi_{z_{2,2}}p_8 + \Pi_{z_{2,3}}p_9 + \ldots \\ &\sigma_{\gamma_1(\Pi_{z_{2,2}}p_8+\Pi_{z_{2,3}}p_9, L_{z2}, M_{z1})}(\Pi_{z_{1,1}}p_7 + \Pi_{z_{1,2}}p_8 + \Pi_{z_{1,3}}p_9))\Big).\{\Pi_{z_{3,3}}r_z + \ldots \\ &\dot{\sigma}_{\gamma_2(\Pi_{z_{3,3}}p_9, L_{z3}, M_{z2})}(\Pi_{z_{2,2}}p_8 + \Pi_{z_{2,3}}p_9 + \ldots \\ &\sigma_{\gamma_1(\Pi_{z_{2,2}}p_8+\Pi_{z_{2,3}}p_9, L_{z2}, M_{z1})}(\Pi_{z_{1,1}}p_7 + \Pi_{z_{1,2}}p_8 + \Pi_{z_{1,3}}p_9))\Big).\Big[\Pi_{z_{2,2}}p_9 + \ldots \\ &\Pi_{z_{2,3}}r_z + \dot{\sigma}_{\gamma_1(\Pi_{z_{2,2}}p_8+\Pi_{z_{3,3}}p_9, L_{z2}, M_{z1})}(\Pi_{z_{1,1}}p_7 + \Pi_{z_{1,2}}p_8 + \Pi_{z_{1,3}}p_9).(\Pi_{z_{1,1}}p_8 + \Pi_{z_{1,2}}p_9 + \Pi_{z_{1,3}}r_z)\Big]\Big\}\end{aligned} \tag{58}$$

$\dot{\sigma}(\cdot)$ is defined in (11), r_y and r_z are computed by

$$\begin{aligned} r_y &= -v_h \sin(\eta_1) \\ r_z &= v_h \cos(\eta_1) - 1 \end{aligned} \tag{59}$$

The asymptotic stability of $(p_4, \ldots, p_9)$ is proved based on (Johnson and Kannan, 2003; Marchand, 2003).

Applying the proposed control law, $\bar{P} - P_d \to 0$ and $\bar{V} - V_d \to 0$. By means of Theorem 1, $|\vec{P} - \bar{P}| < k_3 T$ and $|\vec{V} - \bar{V}| < k_4 T$ for $k_{\{3,4\}} > 0$ and T the wingbeat period.

7. MAV dimensions

Diptera insect (Dudley, 2002) is the model adopted for simulations. It has a mass of 200mg and a wingbeat frequency of 100Hz. Its maximum flapping angle amplitude is 60°. The wing is supposed to rotate up to 90° about its span-wise axis. The wingspan and wings surface are assumed respectively to 2L=3cm and $2S_w = 1.14\text{cm}^2$, so that a vertical ascendant movement can be achieved using flapping angles amplitudes lower than the maximum values. Using these numerical values, admissible sets for control forces $\Omega_{\bar{f}_x, \bar{f}_h}$ and torques $\Omega_{\bar{\tau}_1, \bar{\tau}_3}$ can be defined (33).

$\Omega_{\bar{\tau}_1,\bar{\tau}_3}$ has been approximated to the largest ellipse E_r that fits inside $\Omega_{\bar{\tau}_1,\bar{\tau}_3}$ (see Fig. 5) for computation simplification reasons. Therefore, the control torques should respect an ellipsoidal admissible E_r set defined by

$$[\bar{\tau}_1\ \bar{\tau}_3]P_r[\bar{\tau}_1\ \bar{\tau}_3]^T \leq 1 \tag{60}$$

where P_r is positive definite matrix representing the ellipse's semi-axes. Practically, if $\bar{\tau}_1 \geq \tilde{\tau}_1$ (41), $\bar{\tau}_1$ could be saturated to $\tilde{\tau}_1$ and $\bar{\tau}_3 = 0$. To avoid a null yaw control torque in this case, 70% of $\tilde{\tau}_1$ will be attributed to $\bar{\tau}_1$, $\bar{\tau}_3$ will be computed by (60) defining then a set Ω_r (see Fig. 5). This choice is justified by the necessity to bring the MAV on the flat (horizontal plane) first. The admissible set of thrust and lift forces $\Omega_{\bar{f}_x,\bar{f}_h}$ is drawn in Fig.5. It can be approximated to the largest semi-ellipse E_t that fits inside (E_t almost coincides with $\Omega_{\bar{f}_x,\bar{f}_h}$), ($P_t$ is positive definite matrix representing the ellipse's semi-axes)

$$\begin{array}{rcl} \left[\bar{f}_x\ \bar{f}_h\right]P_t\left[\bar{f}_x\ \bar{f}_h\right]^T & \leq & 1 \\ \bar{f}_h & \geq & 0 \end{array} \tag{61}$$

A fixed saturation level inside E_t is attributed to $\bar{f}_h$ since it will be decomposed in $mg\bar{v}_y$ and $mg\bar{v}_z$ (for computation simplification reasons). The saturation bound is computed such that more power is attributed to the lift since it is associated to the roll movement (99% of E_t's vertical semi-axis is attributed to $mg\bar{v}_h$): the MAV is brought to the horizontal plane rapidly. The saturation bound $mg\bar{v}_x$ of the thrust $\bar{f}_x$ satisfies the semi-ellipse's equation. The saturation set of the control forces is Ω_t.

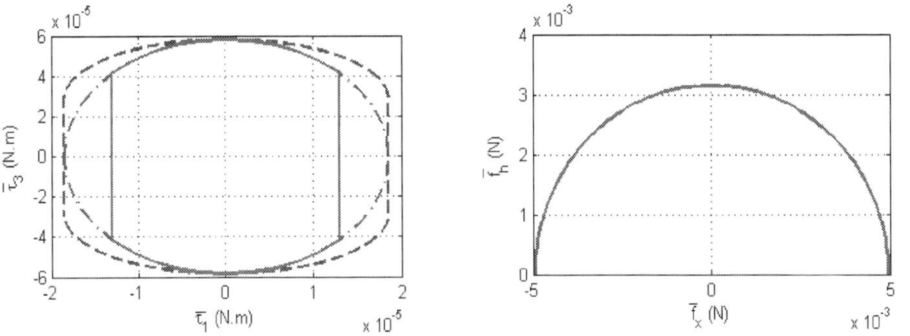

Figure 5. Yaw torque versus roll torque (left), defining the saturation set $\Omega_{\bar{\tau}_1,\bar{\tau}_3}$ (dashed blue line) approximated to an ellipse E_r (red dot-dashed line) then to a set Ω_r (green continuous line). Lift versus thrust (right), defining the saturation set $\Omega_{\bar{f}_x,\bar{f}_h}$ approximated to a semi-ellipse E_t, that almost coincides with $\Omega_{\bar{f}_x,\bar{f}_h}$ (red dot-dashed line), then to the set Ω_t (green continuous line)

8. Simulations and robustness tests

The control laws are tested in simulations using the complete model. The initial position is (1m, 1m, -1m) and the initial orientation (-40°, -25°, 50°). The choice for the poles placement is the following: $(-a_{1_x},-a_{2_x},-a_{3_x})=(-3,-3,-3)$ for the forward dimension, $(-a_{1_y},-a_{2_y},-a_{3_y})=(-3.5,-3.5,-3.5)$ for the lateral one, and $(-a_{1_z},-a_{2_z},-a_{3_z})=(-2.5,-2.5,-2.5)$ for the vertical one.

The evolution of the linear position and velocity and the control forces are shown in Fig. 6. Due to the poles placement, the system's dynamics are accelerated which causes an overshoot. The saturation bound of the control thrust $\overline{f}_x$ depends on the value of the control lift $\overline{f}_h$. Besides, $\overline{f}_h$ does not converge to 0 but to mg in order to balance the MAV's weight (in hovering mode). The roll, pitch and yaw angles, angular velocities and the control torques are plotted on Fig. 7. The dependence of the roll, lateral and vertical movements can be clearly noticed. The wings angles amplitudes are presented on Fig. 8.

8.1 Robustness with respect to external disturbances

The external disturbances are assimilated to external forces and torques applied to the body. The MAV is perturbed at t=7s during 10 wingbeat periods. The magnitude of the disturbances, over the three axes, is $(5.10^{-3},5.10^{-3},3.10^{-3})\mathrm{N}$ for the forces and $(3.10^{-5},3.10^{-5},3.10^{-5})\mathrm{N}\cdot\mathrm{m}$ for the torques, values considered in $\Re^m$. Note that a rain drop weighs about $5.10^{-6}\mathrm{N}$ (almost 1000 times lighter than the disturbance). Such high values of the disturbances are simulated to show the importance of the saturations in the divergence avoidance. Even though the disturbance is the lowest along the vertical axis, its influence on the vertical movement is the highest (Fig. 9). The control torques and forces cooperate in order to overcome the disturbances and ensure stability. They reach the saturation bounds in order to use their maximum power (Fig. 9, 10 and 11). The evolution of the angles and angular velocity (Fig. 10) zoomed around the saturation (Fig. 11) shows that the MAV executes many turns around its axes, the angular velocity reaches very high values. The relation between the roll and the yaw saturation bounds is shown clearly on Fig. 11. The disturbance is also detectable on the wings angles amplitudes which saturate (Fig. 12).

8.2 Robustness with respect to aerodynamic errors

The robustness of the control law is also tested for a bad estimation of the aerodynamic coefficient C, known to be difficult to identify. This property is essential for real time implementation where the flapping MAV can execute missions in different areas having different aerodynamic characteristics. An additive error is introduced to C

$$C_{dist}=C-\Delta C \tag{62}$$

where ΔC is a stochastic parameter subject to a uniform distribution, such that C_{dist} varies within the interval [2, 3.5]. A different value of ΔC is applied at each wingbeat period. Such a quick variation of the aerodynamic coefficient is not realistic; it is simulated only to emphasize the control law robustness. The influence of the aerodynamic coefficient variation is shown in Fig. 13 and 14. The stochastic aspect can be seen on the vertical position P_z and velocity V_z besides the lift force $\bar{f}_h$. The control lift has then a greater value $(\bar{f}_h > mg)$ at convergence in order to balance the reduction of the aerodynamic coefficient.

9. Conclusions and future works

The present chapter has presented a new strategy of controlling flapping MAVs. It is based on a bounded state feedback control of the forces and the torques. The control takes into consideration the saturation of the actuators driving the flapping wings. They are based on the theory of cascade, aiming to stabilize the attitude of the MAV, while driving the body to a desired position, associating the translational movement to the rotational one. The controls are extremely low cost, therefore suitable for an embedded implementation. They are computed using a simplified model of a flapping MAV (averaged over a wingbeat period). This model is based only on the steady aerodynamics. The control laws are applied at each wingbeat period. Different robustness tests are performed, especially with respect to the simplifications adopted in the proposed model, besides external disturbances, modeling errors, parameters uncertainties, etc.

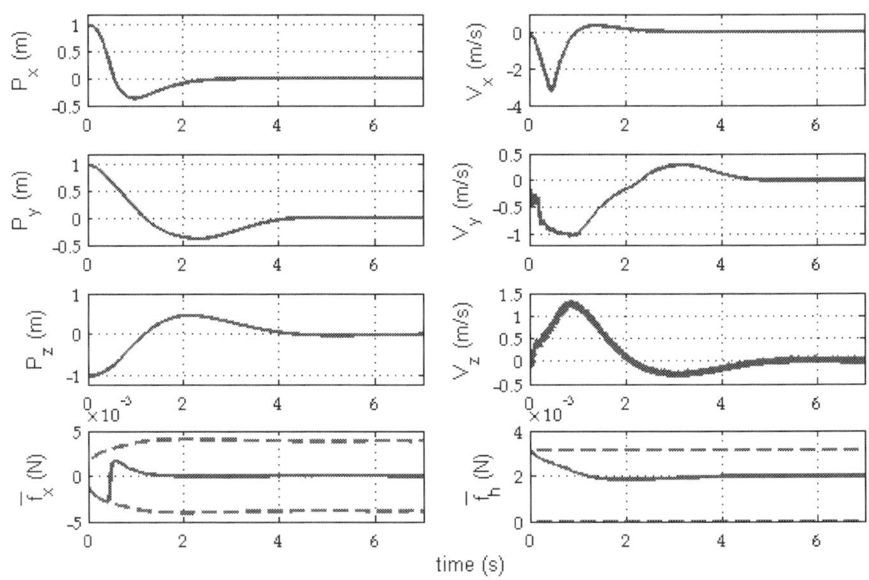

Figure 6. The linear position (left) and velocity (right) of the MAV and the control forces (bottom), the saturation bounds are plotted with the red dashed line

Future works consist of developing bounded control laws based directly on a minimum number of sensors measurements in order to ensure the stability in three dimensions.

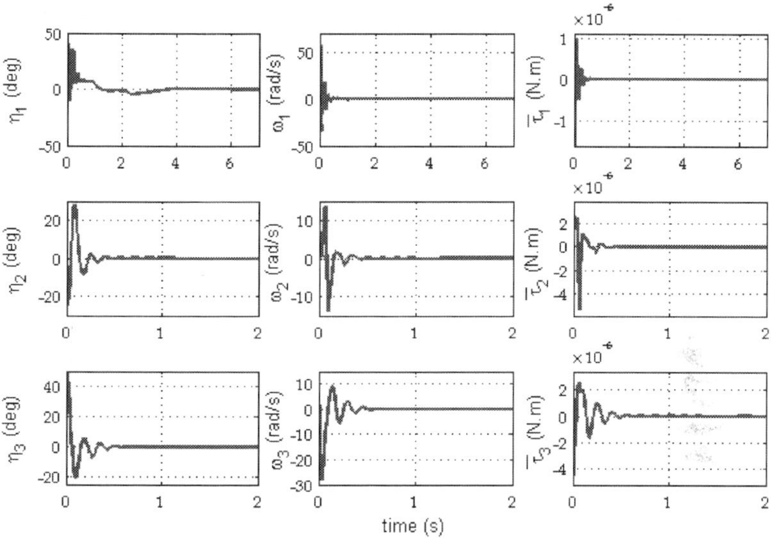

Figure 7. The roll, pitch and yaw angles (left), angular velocity (middle) and the control torques (right) of the MAV. The pitch and yaw components are zoomed to the first 2s

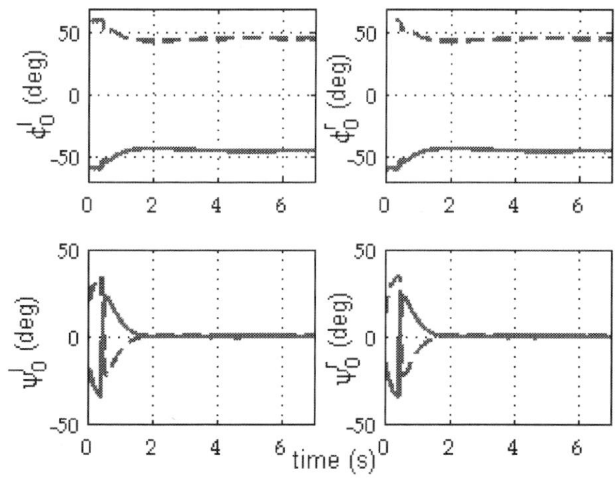

Figure 8. The envelops of the flapping and rotation angles amplitudes

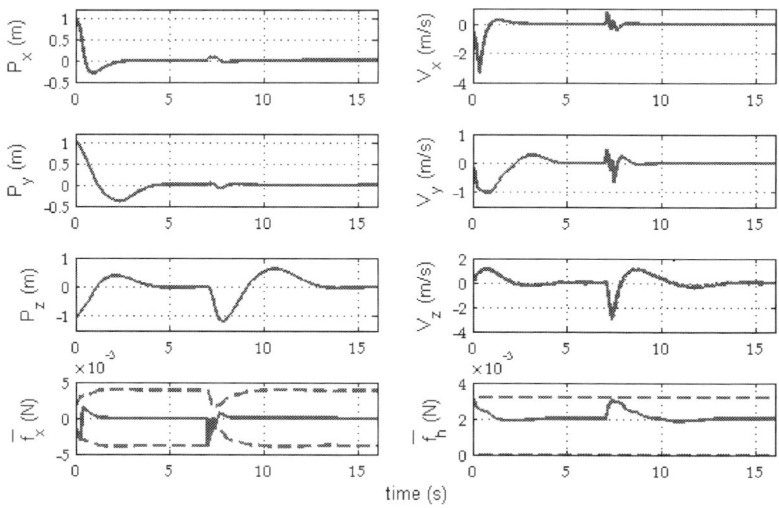

Figure 9. Robustness with respect to external disturbances: Evolution of the linear position (left) and velocity (right) and the control forces (bottom)

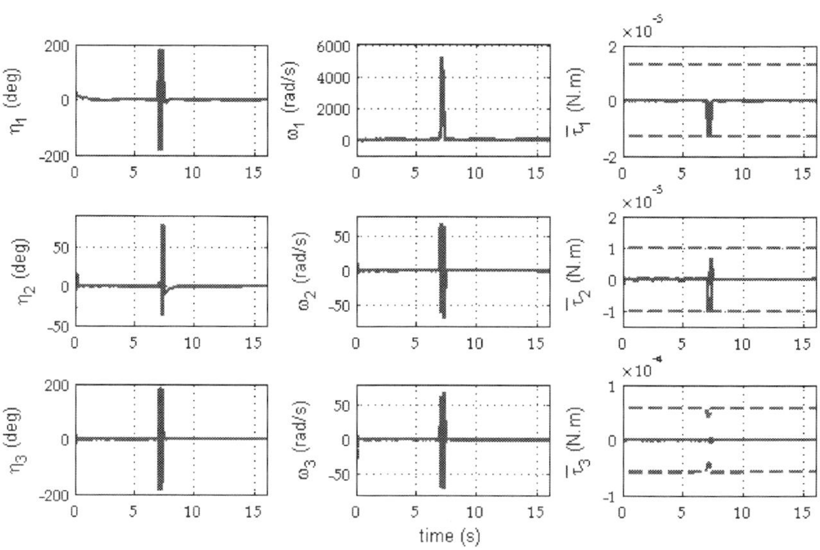

Figure 10. Robustness with respect to external disturbances: Evolution of the roll, pitch and yaw angles (left), the angular velocities (middle) and the control torques (right)

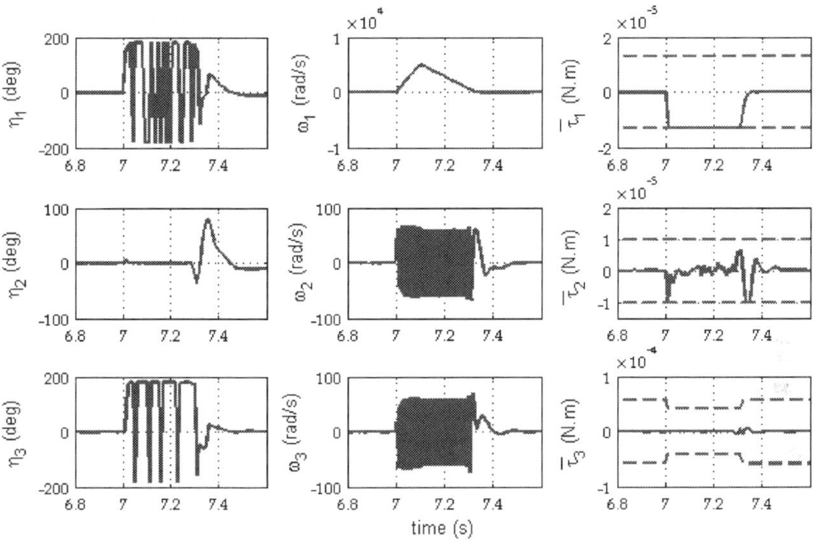

Figure 11. Robustness with respect to external disturbances: Evolution of the roll, pitch and yaw angles (left), the angular velocities (middle) and the control torques (right) zoomed around the disturbance

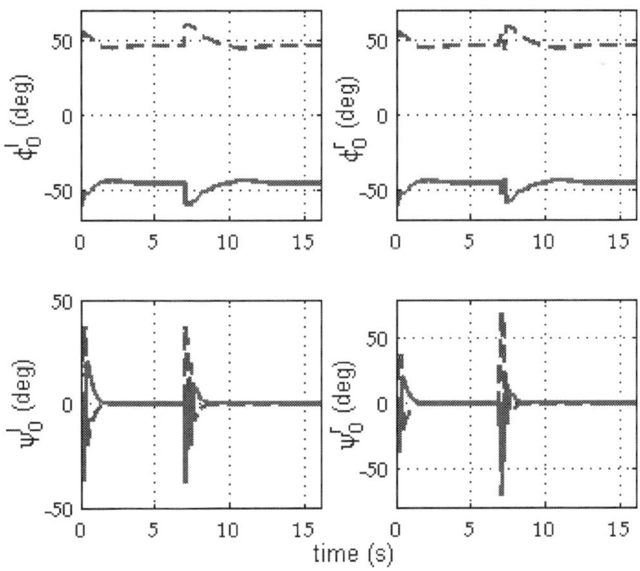

Figure 12. Robustness with respect to external disturbances: The envelops of the wings angles amplitudes

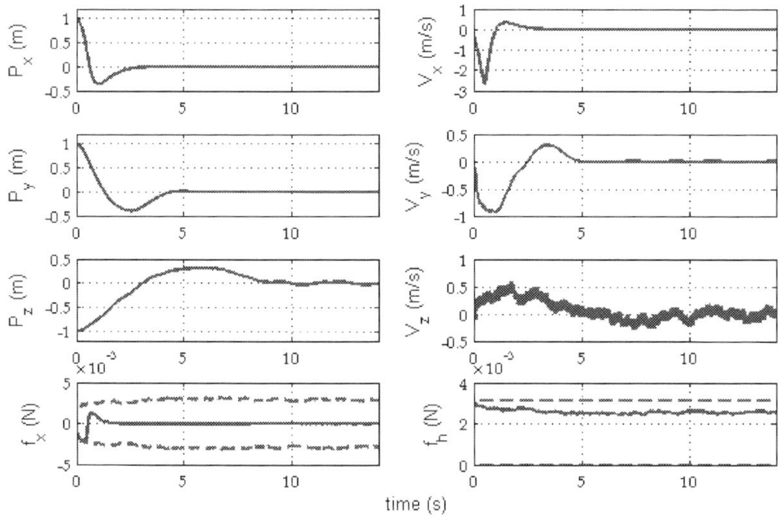

Figure 13. Robustness with respect to the aerodynamic coefficient: Evolution of the linear position (left) and velocity (right) and the control forces (bottom)

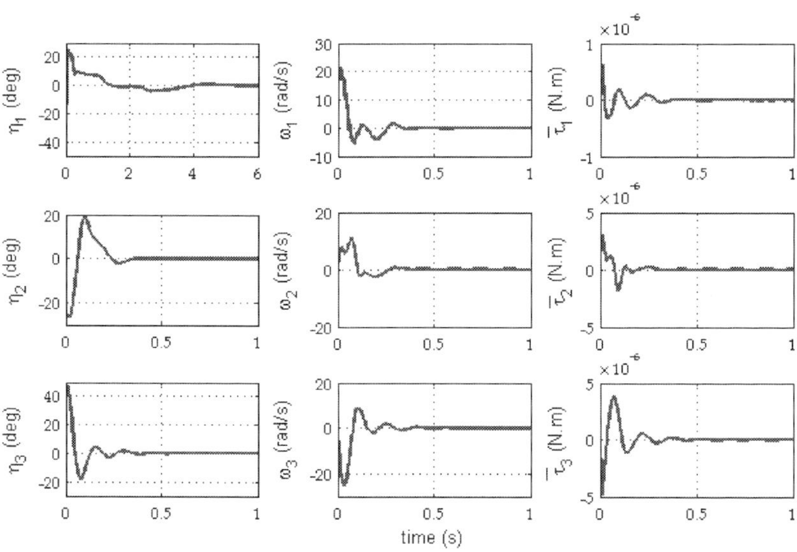

Figure 14. Robustness with respect to the aerodynamic coefficient: Evolution of the roll, pitch and yaw angles (left), the angular velocities (middle) and the control torques (right) zoomed to the first second for the pitch and yaw

10. References

Boyer, F.; Porez, M. & W. Khalil, Macro-continuous computed torque algorithm for a three-dimensional eel-like robot, *IEEE Transaction on Robotics and Automation*, Vol. 22, No. 4, 2006, page numbers (763–775).

Bullo, F. (2002). Averaging and vibrational control of mechanical systems. *SIAM Journal on Control and Optimization*, Vol. 41, No. 2, page numbers (452-562)

Campolo, D.; Sitti, M. & Fearing, R. S. (2003). Efficient charge recovery method for driving piezoelectric actuators with quasi-square waves. *IEEE Trans. on Ultrasonics, Ferroelectrics and Frequency Control*, Vol. 50, No. 3, page numbers (237-244)

Deng, X. ; Schenato, L. & Sastry, S. (2002). Model identification and attitude control scheme for a micromechanical flying insect. *Proceedings of the 7th International Conference on Control, Automation, Robotic and Vision*, pp. 1007-1012, Singapore.

Deng, X.; Schenato, L. & Sastry, S. (2003). Model identification and attitude control for a micromechanical flying insect including thorax and sensor models. *Proceedings of the IEEE Int. Conference on Robotics and Automation*, pp. 1152-1157, Taipei, Taiwan.

Deng, X.; Schenato, L.; Wu, W.-C. & Sastry, S. (2006a). Flapping flight for biomimetic robotic insects : Part I- system modeling. *IEEE Transactions on Robotics*, Vol. 22, No. 4.

Deng, X.; Schenato, L.; Wu, W.-C. & Sastry, S. (2006b). Flapping flight for biomimetic robotic insects : Part II- flight control design. *IEEE Transactions on Robotics*, Vol. 22, No. 4, page numbers (789-803)

Dickinson, M.; Lehmann, F.-O. & Sane, S. (1999). Wing rotation and the aerodynamic basis of insect flight. *Science*, Vol. 284, No. 5422, page numbers (1954-1960)

Dickson, W.; Straw, A.; Poelma, C. & Dickinson, M. (2006). An integrative model of insect flight control. *Proceedings of the 44th AIAA Aerospace Sciences Meeting and Exhibit*, Reno, USA

Dudley, R. (2002). *The biomechanics of insect flight: form, function, evolution*. Princeton University Press

Guerrero-Castellanos, J.; Hably, A.; Marchand, N. & Lesecq, S. (2007). Bounded attitude stabilization : Application on four rotor helicopter. *Proceedings of the 2007 IEEE Int. Conf. on Robotics and Automation*, pp. 730-735, Roma, Italy.

Hably, A.; Kendoul, F.; Marchand, N. & Castillo, P. (2006). Further results on global stabilization of the PVTOL aircraft. *Proceedings of the Second Multidisciplinary International Symposium on Positive Systems : Theory and Applications*, pp. 303-310, Grenoble, France.

Hedrick, T. & Daniel, T. (2006). Flight control in the hawkmoth Manduca sexta : the inverse problem of hovering. *The journal of experimental Biology*, Vol. 209, No. 16, page numbers (3114-3130).

Janocha, H. & Stiebel, C. (1998). New approach to a switching amplifier for piezoelectric actuators. *Proceedings of ACTUATOR 98, 6th International Conference on New Actuators*, pp. 189-192, Bremen, Germany.

Johnson, E. & Kannan, S. (2003). Nested saturation with guaranteed real poles. *Proceedings of the American Control Conference*, volume 1, pp. 497-502.

Khalil, H. (1996). *Nonlinear Systems*. Prentice-Hall.

Kuhnen, K.; Janocha, H.; Thull, D. & Kugi, A. (2006). A new drive concept for high-speed positioning of piezoelectric actuators. *Proceedings of the 10th International Conference on New Actuators*, pp. 82-85, Bremen, Germany.

Marchand, N. (2003). Further results on global stabilization for multiple integrators with bounded controls. *Proceedings of the IEEE Conference on Decision and Control, CDC'2003*, volume 5, pp. 4440-4444, Hawai, USA.

Rakotomamonjy, T. (2006). *Modelization and flight control of a flapping-wing Micro Air Vehicle.* PhD thesis, University Paul Cezanne Aix Marseille.

Rakotomamonjy, T.; Le Moing, T. & Ouladsine, M. (2004). Simulation model of a flapping-wing micro air vehicle. *Proceedings of the European Micro Aerial Vehicle Conf., EMAV* 2004, Braunschweig, Germany.

Reiser, M. ; Humbert, J.; Dunlop, M.; Del Vecchio, D.; Murray, R. & Dickinson, M. (2004). Vision as a compensatory mechanism for disturbance rejection in upwind flight. *Proceedings of the American Control Conference*, volume 1, pp. 311-316, Boston, Massachusetts, USA.

Renaudin, A.; Zhang, V.; Tabourier, P.; Camart, J. & Druon, C. (2004). Droplet manipulation using SAW actuation for integrated microfluidics. *Proceedings of the μTAS*, pp. 551-553, Malmö, Sweden.

Sane, S. (2003). Review The aerodynamics of insect flight. *The journal of experimental Biology*, 206,23, page numbers (4191-4208).

Schenato, L. ; Campolo, D. & Sastry, S. (2003). Controllability issues in flapping flight for biomimetic micro aerial vehicles (MAVs). *Proceedings of the IEEE International Conference on Decision and Control*, Las Vegas, USA.

Schenato, L.; Deng, X. & Sastry, S. (2001). Flight control system for a micromechanical flying insect : Architecture and implementation. *Proceedings of the IEEE International Conference on Robotics and Automation*, pp. 1641-1646, Seoul, Korea.

Schenato, L. ; Deng, X. & Sastry, S. (2002a). Hovering flight for a micromechanical flying insect : Modeling and robust control synthesis. *Proceedings of the 15th IFAC World Congress on Automatic Control*, Barcelona, Spain.

Schenato, L., Wu, W.-C., and Sastry, S. (2002b). Attitude control for a micromechanical flying insect via sensor output feedback. *Proceedings of the 7th International Conference on Control, Automation, Robotic and Vision*, pp. 1031-1036, Singapore.

Schenato, L.; Wu, W.-C. & Sastry, S. (2004). Attitude control for a micromechanical flying insect via sensor output feedback. *IEEE Journal of Robotics and Automation*, Vol. 20, No. 1, page numbers (93-106).

Shuster, M. (1993). A survey of attitude representations. *Journal of astronautical sciences*, Vol. 41, No. 4, page numbers (439-517).

Sriram; Gopinath, A.; Van Der Weide, E.; Kim, S.; Tomlin, C. & Jameson, A. (2005). Aerodynamics and flight control of flapping wing flight vehicles : A preliminary computational study. *Proceedings of the 43rd Aerospace Sciences Meeting*, Reno, USA.

Tanaka, F.; Ohmi, T.; Kuroda, S. & Hirasawa, K. (2006). Flight control study of an virtual insect by a simulation. *JSME Int. Journal*, Vol. 49, No. 2, page numbers (556-561).

Teel, A. (1992). Global stabilization and restricted tracking for multiple integrators with bounded controls. *Systems & Control Letters*, Vol. 18, No. 3, page numbers (165-171).

Thakoor, S.; Cabrol, N.; Lay, N.; Chahl, J.; Soccol, D.; Hine, B. & Zornetzer, S. (2003). Review: The benefits and applications of bioinspired flight capabilities. *Journal of Robotic Systems*, Vol. 20, No. 12, page numbers (687-706).

Vela, P. A. (2003). *Averaging and Control of Nonlinear Systems.* PhD thesis, California Institute of Technology.

UAV Trajectory Planning for Static and Dynamic Environments

José J. Ruz[1], Orlando Arévalo[1], Gonzalo Pajares[2] and Jesús M. de la Cruz[1]
[1]*Dept. of Computer Architecture and Automatic Control, Complutense University of Madrid*
[2]*Dept. of Software Engineering and Artificial Intelligence, Complutense University of Madrid*
Spain

1. Introduction

An unmanned aerial vehicle (UAV) is a robotic aircraft that can fly with either a remote input from a ground-based operator, or autonomously without human intervention based on pre-programmed flight plans (How et al., 2004). UAVs offer advantages over conventional manned vehicles in many applications because they can be used in situations otherwise too dangerous for manned vehicles and without being weighed down by the systems required by a pilot. UAVs are currently receiving much attention in research because they can be used in a wide variety of fields, both civil and military, such as reconnaissance, geophysical survey, environmental and meteorological monitoring, aerial photography, and search-and-rescue tasks. Most of these missions are usually carried out in threatened environments, and then it is very important to fly along a route which keeps the UAV away from known threats. Detection radars are one of the main threats for an UAV, but there are others that should also be avoided, such as fires, electric storms, radio shadowing zones, no flight zones, and so on. One of the main goals in many UAV's projects has been to establish the route that maximizes the likelihood of successful mission completion taking into account all known information about technological constraints, obstacles and threat zones on a static environment (Richards & How, 2002). Some papers that investigate path planning for UAVs presume that the location of the threats and their presence are deterministically known at planning-time, and interpret a path which avoids possible threat regions as an optimal path (Borto, 2000). However more recent projects are examining the possibilities of UAVs as realistic autonomous agents working on dynamic environments where threat zones called pop-up are present (Zengin & Dogan, 2004). The true presence of these types of zones is only known at flying-time, but the location and knowledge about the probability of appearance can be known at planning-time.

In this chapter we will present an approach to trajectory optimization for UAV in presence of obstacles, waypoints, and risk zones. The approach has been implemented on SPASAS (*System for Planning And Simulation of Aerial Strategy*), an integrated system for definition of flight scenarios, flight planning, simulation and graphic representation of the results developed at Complutense University of Madrid. The system uses two alternative methods for trajectory generation: mixed integer linear programming (MILP) and a modification of the A* algorithm, depending on the characteristics of the scenario between two waypoints.

The risk zones constitute the dynamic elements of the scenario, and they consist of pop-ups with a known future probability of appearance. For each pop-up the system generates at planning-time several feasible evasion manoeuvres, qualified by a set of route parameters (fuel consumption, assumed risk and spent time) used at flight-time as decision variables to optimize the route. The work extends our preliminary results on trajectory generation over static environment (Ruz et al., 2006), taking into account the knowledge about pop-ups in the trajectory design. The route planner that is proposed in this work allows the UAV to make a decision among several alternative routes considering both, current state of the UAV and probabilities of pop-up threats appearance in future time. The planning is carried out in three steps. First, an initial and optimal path planning is designed taking into account only the static elements of the scenario. Second, alternatives routes are calculated to bypass each pop-up zone having an appearance probability greater than zero. Finally, these alternatives are attached to the original flying plan, and given to an upper layer module in charge of making decisions according to the imposed limitations of fuel, time, and risk.
The chapter is organized as follows. Section 2 presents the MILP techniques as a method for trajectory generation of UAVs. This section encompasses the formulation of constraints for UAV's dynamic, obstacle's avoidance, target reaching, radar's avoidance and target within radar zones. For the last set of constraints a linear approximation with indicator 0-1 variables is introduced so that the non-linear and non-separable terms of the radar detection function could be linearized. Section 3 describes the path planning approach based on a modification of the A* algorithm. Section 4 explains how the planner presented in this work modifies the optimal trajectory to avoid pop-up threats. This section examines the three parameters utilized in the election of an alternative route that bypasses the threats: mean risk, flying time and fuel, and poses the corresponding decision making as an Integer Linear Programming (ILP) problem. In section 5, we present the implementation and some results of the planner. Finally, section 6 presents the conclusions and future works.

2. MILP as a method for trajectory planning of UAVs

Mixed Integer Linear Programming (MILP) is a powerful mathematical programming framework that extends continuous linear programming to include binary or integer decision variables to encode logical constraints and discrete decisions together with the continuous vehicle dynamics. The approach to optimal path planning based on MILP was introduced in (Schouwenaars et al., 2001). The UAV's trajectory generation is faced as a 3D optimization problem under certain conditions (see Fig. 1) in the Euclidean space, characterized by a set of *decision variables*, a set of *constraints* and the *objective function*. The *decision variables* are the UAV's state variables, i.e. position and speed. The *constraints* are derived from a simplified model of the UAV and the environment where it has to fly on. These constraints include:

- Dynamics constraints, such as a maximum turning force which causes a minimum turning radius(C_R), as well as a maximum climbing rate (C_C).
- Obstacles avoidance constraints like no-flight zones (C_O)
- Target reaching constraints of a specific waypoint or target (C_T).
- Radar avoidance constraints (C_R)

The *objective function* includes different measures of the quality in the solution of this problem, although the most important criterion is the minimization of the total flying time to reach the target. However, when there are threats that put in risk the UAV mission, for example missiles guided by radars, it becomes necessary to incorporate to the objective function some term that minimizes the risk of UAV detection (R_D). In this work the MILP approach for path planning is expanded to include linear constraints derived from a non-linear risk detection model of the UAV by radars. The MILP alternative to plan a static route will be mainly used when the environment contains threats that put under certain risk the UAV's mission, for instance the tracking and the shooting of missiles guided by radars.

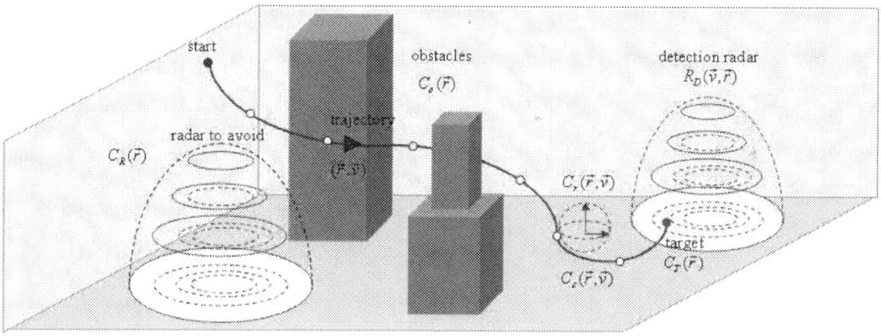

Figure 1. Trajectory generation as a MILP constrained optimization problem

2.1 Dynamic constraints

The trajectory optimization is constrained by a time-discrete dynamics to obtain the state of the system on every time t + 1, from its previous state at time t. This model represents the UAV with its limits in speed, climbing rate and turning rate, caused by a maximum magnitude in the input force u(t) that the aircraft might be put under (Bellingham & How, 2002).

Therefore the dynamics is expressed in (1).

$$\begin{bmatrix} \vec{v}(t+1) \\ \vec{r}(t+1) \end{bmatrix} = A \cdot \begin{bmatrix} \vec{v}(t) \\ \vec{r}(t) \end{bmatrix} + B \cdot \vec{u}(t) \tag{1}$$

The matrices A and B are the state matrices (kinematics matrices) of the UAV, and t the discrete time. Speed and force magnitudes are constrained by the following equations which are convenient linearizations of the original quadratic form:

$$\begin{aligned} & v_x \cos(\varphi_n)\sin(\theta_m) + v_y \sin(\varphi_n)\sin(\theta_m) + v_z \cos(\theta_m) \le |\vec{v}_{\max}| \\ & u_x \cos(\varphi_n)\sin(\theta_m) + u_y \sin(\varphi_n)\sin(\theta_m) + u_z \cos(\theta_m) \le |\vec{u}_{\max}| \\ & \varphi_n = 2\pi n / N; \qquad \theta_m = \pi m / M \\ & n = 1,2,3,\dots N; \qquad m = 1,2,3,\dots M \end{aligned} \tag{2}$$

where the maximum speed and input force are in the right side of (2). N and M are the number of points used to approximate the spherical space closed by all the possible directions of a vector in 3D (Schouwenaars et al., 2004). The angle θ is the zenith and the angle φ is the azimuth of vector $\vec{u}$ represented in spherical coordinates.

Additionally, in order to model the climbing rate of a UAV, the components of speed and acceleration in the Z axis must be constrained according to its manoeuvring capabilities (3).

$$\begin{aligned} &\forall t \\ &v_{z\_\min} \le v_z(t) \le v_{z\_\max} \\ &u_{z\_\min} \le u_z(t) \le u_{z\_\max} \end{aligned} \tag{3}$$

At the same time the third component of the UAV's position must be limited to one side of the Z axis, which in our case is the positive one, because the implemented referential frame has been the ENU (East-North-Up → {x,y,z}).

2.2 Obstacles avoidance constraints

As mentioned in the beginning of this section, the optimization considers box obstacles avoidance, which is modelled by knowing the lower south west and the upper north east vertexes of the no-flight zones (Kuwata & How, 2004), as expressed by the constraints in (4):

$$\begin{aligned} x_i(t) &\le x_{i\min}^k + M_o \cdot \delta_j^k(t) \\ x_i(t) &\ge x_{i\max}^k - M_o \cdot \delta_j^k(t) \end{aligned} \tag{4}$$

where k identifies the obstacle, (x_1min,x_2min,x_3min) is its lower south west corner and (x_1max,x_2max,x_3max) its upper north east corner, M_o is a bound for $x_i(t)$ to enable/disable some of the constraints, and δ_j^k are the indicator variables. The jth condition is then relaxed when $\delta_j^k = 1$, and enabled when $\delta_j^k = 0$. By (5) it is imposed that at least one of these constraints is active.

$$\begin{aligned} &\sum_{j=1}^{6} \delta_j^k(t) \le 5 \\ &\forall k \ / \ k = 1,2,\dots,K_{obst} \ \wedge \ \forall t \ / \ t = 0,1,\dots,T_{\max} \end{aligned} \tag{5}$$

2.3 Target reaching constraints

The constraints in (6) must be satisfied for all instant during the flight, to impose the UAV to get to its final destination, such as a waypoint or a target.

$$\begin{aligned} &x_i(t) - x_{i_f} \le M(1-\lambda_t) \\ &x_i(t) - x_{i_f} \ge -M(1-\lambda_t) \\ &\sum_0^T t \cdot \lambda_t \le t_{arrival}, \quad \forall t \end{aligned} \tag{6}$$

In the arrival constraints, (x_f,y_f,z_f) is the target location, λ_t is a binary indicator variable to enable/disable each constraint, and $t_{arrival}$ is the time to be included as a term of the objective function to be minimized (Ruz et al., 2006).

2.4 Radars avoidance constraints

The treatment developed to avoid threat zones where there are enemy units such as tracking radars and missile launchers is based on a distribution which corresponds to the probability for the UAV to be detected, tracked and shot. This probability distribution is characteristic for every single type of tracking radar (7), where c_1 and c_2 are constants obtained by a curve fitting routine. This fit is previously made using the information provided by the radar's specifications.

$$P_t = \frac{1}{1 + c_2 \left(R^4 / \sigma \right)^{c_1}} \tag{7}$$

In the tracking probability distribution the value σ represents the RCS (radars cross section) in squared meters, while the variable R is the distance in meters between the UAV and the radars. In Fig. 2 it is observed that a radius equal to 50Km could be considered as the maximum radar's range, from which P_t totally diminishes.

Thus, for any maximum accepted value of detection, tracking, or killing probability its corresponding minimum radius R_{min} to avoid the radar's zones is computed. This avoidance condition is then expressed in (8), where a linearization of the quadratic form of all the possible directions for a vector in the Euclidean space is used once again to define the constraints in the MILP model.

$$\begin{aligned} & x_R \cos(\varphi_n)\sin(\theta_m) + y_R \sin(\varphi_n)\sin(\theta_m) + z_R \cos(\theta_m) \geq \left| \vec{R}_{\min} \right| \\ & \varphi_n = 2\pi n / N; \;\; \theta_m = \pi m / M; \;\; n = 1,2,...N; \;\; m = 1,2,...M \end{aligned} \tag{8}$$

The coordinates x, y and z are the relative coordinates of the UAV to each radar, making the aircraft able to approximate then differently, depending on the types of radars and on the interests in any mission in particular.

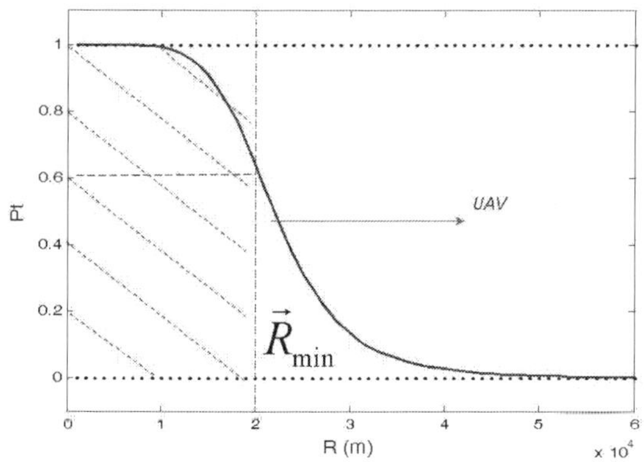

Figure 2. Profile of a tracking probability distribution with a RCS = 1.0m²

2.5 Constraints for targets within radar zone

When the UAV must reach a target within a radar zone, the detection risk must be minimized. To address such a situation we have used a model based on idealizing geometrical and physical UAV properties (Murphey et al., 2003). The model assumes that in the radar zone the UAV approaches to the target holding a fixed height (h). The detection risk D is proportional to the UAV's RCS, σ, reciprocal to the fourth power of the distance between the UAV and the radar, r_R and independent of the UAV speed (Figure 3).

$$D = \sigma / r_R^4 \tag{9}$$

The UAV is considered to be an ellipsoid with the axis of symmetry determining the direction of the UAV trajectory (Murphey et al., 2003).

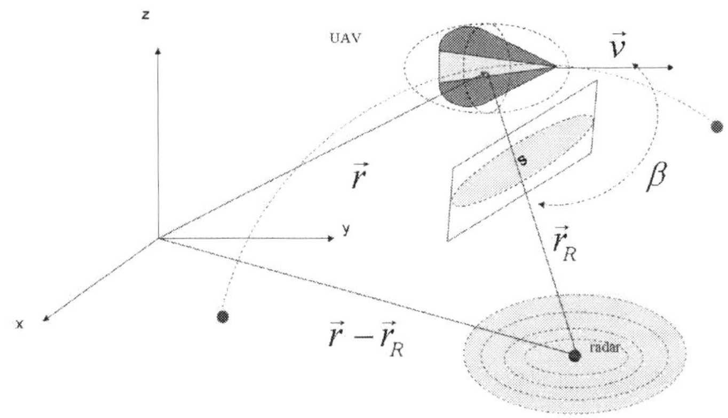

Figure 3. Layout of an UAV under radar detection risk

The UAV's *RCS* exposed to the radar is proportional to the area of the ellipsoid's projection onto the plane orthogonal to vector $\vec{r}_R$, as shown in Figure 3. Therefore, the *RCS* is given by:

$$\sigma = kS \tag{10}$$

Where k depends on the radar technical characteristics, and S is the ellipsoid's projection given by Eqn.11.

$$S = \pi\sqrt{a^2 \sin^2 \beta + b^2 \cos^2 \beta} \tag{11}$$

β is the angle between vectors $\vec{r}_R$ and $\vec{v}$ (velocity), and the ellipsoid is defined by semi-axis a and b. As the altitude is fixed (h), β will be a function of x, y, v_x, and v_y, then the detection risk might be expressed as in Eqn.12, such that:

$$D(x,y,v_x,v_y) = k\sqrt{\frac{(hv_y)^2 + (hv_x)^2 + (v_x y - v_y x)^2 \left(\frac{b^2}{a^2}\right)(v_x x + v_y y)^2}{(v_x^2 + v_y^2)(x^2 + y^2 + h^2)^9}} \tag{12}$$

In order to include the detection risk function (Eqn.12) in a MILP framework, it has been approximated by a piecewise first order linearization D_L showed in Eqn.13.

$$D_L = \sum_i (a_i x(t) + b_i y(t) + c_i v_x(t) + d_i v_y(t) + e_i)\delta_i(t) \tag{13}$$

where a_i, b_i, c_i, d_i, and e_i are the coefficients of the ith hyper-plane inside the ith domain, $\delta_i(t)$ is a binary indicator variable identifying the domain $[lx_i, ly_i, v_x, v_y]$ -$[ux_i, ux_i, v_x, v_y]$ that must take value 1 only in this domain, hence it must be constrained to:

$$\begin{aligned} &\sum_{i=1}^{p} lx_i\delta_i \le x(t) \le \sum_{i=1}^{p} ux_i\delta_i \\ &\sum_{i=1}^{q} ly_i\delta_i \le y(t) \le \sum_{i=1}^{q} uy_i\delta_i \\ &\sum_{i=1}^{p\cdot q} \delta_i(t) = 1 \end{aligned} \tag{14}$$

To change the positive and negative range of $x(t)$, $y(t)$, $v_x(t)$, $v_y(t)$ to the [0, 1] interval, we apply the following change of variables:

$$\begin{aligned} &x(t) = x^b(t)[u_x - l_x] + l_x \\ &y(t) = y^b(t)[u_y - l_y] + l_y \\ &v_x(t) = v_x^b(t)[u_{vx} - l_{vx}] + l_{vx} \\ &v_y(t) = v_y^b(t)[u_{vy} - l_{vy}] + l_{vy} \\ &0 \le x^b(t), x^b(t), v_x^b(t), v_y^b(t) \le 1 \end{aligned} \tag{15}$$

where u and l are upper bounds and lower bound of x, y, v_x and v_y , respectively. To linearize the products $V^b(t)\delta_i(t)$ (where $V^b(t)= x^b(t), y^b(t), v^b(t), v^b(t)$) of a binary variable $\delta_i(t)$ and a continuous variable $V^b(t)$, we replace $V^b(t)\delta_i(t)$ for

$$W_i^b(t) = x_i^b(t), y_i^b(t), v_{x,i}^b(t), v_{y,i}^b(t) \tag{16}$$

and impose the following constraints.

$$\begin{aligned} &W_i^b(t) \le \delta_i(t) \\ &W_i^b(t) \le V_i^b(t) \\ &W_i^b(t) \ge V_i^b(t) + \delta_i(t) + 1 \end{aligned} \tag{17}$$

So, piecewise detection risk function will be:

$$\begin{aligned} &D_L = \sum_k A_k x_k^b + B_k y_k^b + C_k v_{x,k}^b + D_{ij} v_{y,k}^b + E_k\delta_k \\ &where \quad A_k = a_{ij}(u_x - l_x), \ldots \end{aligned} \tag{18}$$

The coefficients are calculated by selecting five points over the hyper-surface of D, and solving the corresponding system of equations. Fig. 4 shows the original function D and the piecewise approximation D_L. The relative error of D achieved with this approximation was, for the implemented number of linearization, under 5%.

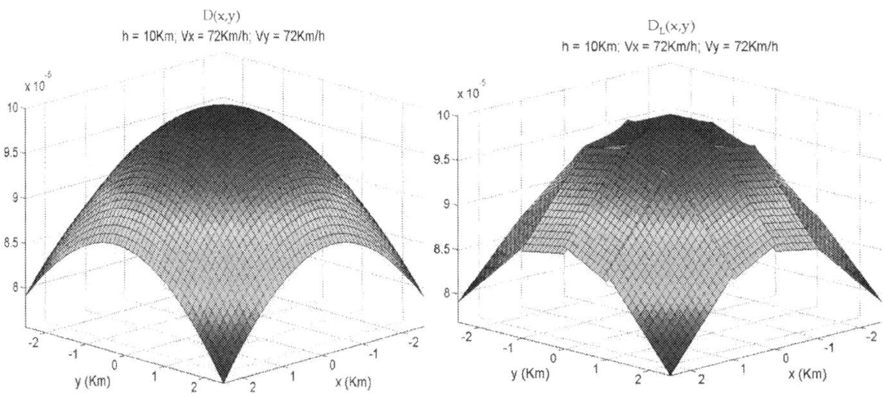

Figure 4. Radar detection function D(x,y) and its linearized radar function D_L(x,y)

3. Trajectory generation using A*

It is well known that the A* is a general AI search algorithm which is highly competitive with other path finding algorithms, and yet very easy to implement (Melchior et al., 2003; Szczerba et al., 2000; Trovato, 1996) . For UAV's trajectory generation it has the advantage to present a good performance when the terrain is quite irregular and the manoeuvres must be done close to the ground's surface. The main drawback is its high computational cost in 3D path planning problems, but it is reasonable to deal with this fact when an appropriate scaling is done. In the case under study the A* algorithm has been modified to adapt it and improve its capabilities to optimize trajectories for UAVs under conditions like no-flight zones, scarped terrains, and radars, also called ADUs (air defence units) which might appear dynamically during the flight. The modifications done to the original A* algorithm are mainly logical restrictions which evaluate whether the successors of every possible node of the UAV's trajectory build a minimum turning radius curve, while avoiding physical obstacles and threats.

3.1 Flying direction constraint

The constraint imposed on the flying direction is the computation of the angle between the arriving direction and every possible departing direction, for all the nodes of the trajectory. The angle ϕ is calculated supported by the definition of the dot product (19).

$$\phi = arc\cos\left(\frac{\vec{V}_f \cdot \vec{V}_i}{\left|\vec{V}_f\right| \cdot \left|\vec{V}_i\right|}\right) \tag{19}$$

Only successors with angle values less than a threshold (set to $\pi/2$ in this work) are allowed during the expansion. Fig. 5 shows a 2D view of the space explored by the A* algorithm when put under this condition.

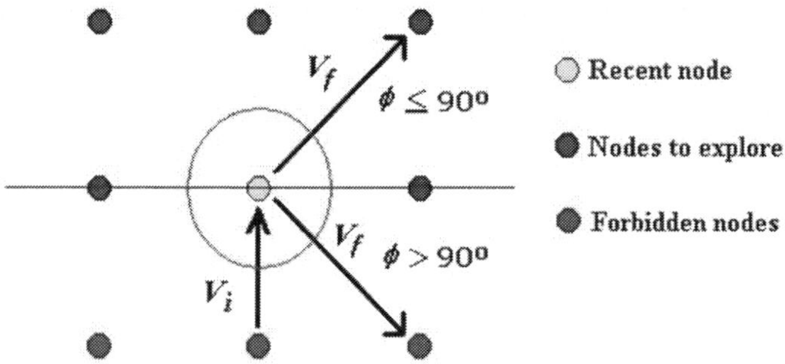

Figure 5. Nodes explored by the 3D A* algorithm

This modification not only intends to emulate the fixed wings aircraft limitation to turn immediately around when flying at high speeds, but also reduces the computing time because the algorithm expands less nodes than original version.

3.2 UAV's inertia constraint

The existence of waypoints in the designed missions leads us to impose a set of conditions of continuity on them, for the kinematical expressions of the UAV (20), and over all the passages from one section to the next one.

$$\begin{aligned} &r(t+\Delta t) \rightarrow r(t); \\ &v(t+\Delta t) \rightarrow v(t); \\ &a(t+\Delta t) \rightarrow a(t); \\ &\forall t \wedge \Delta t \rightarrow 0 \end{aligned} \tag{20}$$

This is modelled within the A* algorithm by constraining the first node to expand from each waypoint according to the initial conditions. Thus, the initial flying direction is given to the algorithm so it can project it onto all the possible directions towards the expanded nodes.
The projection with the highest value is selected, and determines the next node to be incorporated to the trajectory. This modification on the algorithm tries to model the UAV's inertia, and the behavior of the resulting trajectory depends on the A* resolution, which should be properly calibrated with the size of the scenario and the UAV's speed.

3.3 Refinement of the A* space

The resolution of the A* space is related to the ratio between the total number of nodes and the size of the flying space. For example, a defined longitudinal density ρ_i along the i-th axis will be the total number of nodes N_i chosen to represent that real dimension into the A* space, over the total length L_i of the scenario on this axis (21).

$$\rho_i = \frac{N_i}{L_i} \tag{21}$$

From this definition it is possible to infer that the more the number of nodes in the A* space the higher the resolution of the algorithm. However, there is an important limitation for the selected resolution, because this one carries a combinatory of the order $\sim N_i^3$ when looking for the optimum path. For spaces with approximately 100 nodes per dimension this results in a high computational cost. But a lower resolution might be not enough when used to represent a space with surface irregularities. To save computing time without loosing information of the terrain we have carried out a fine sampling between each pair of nodes expanded by the algorithm, as seen in Fig. 6.

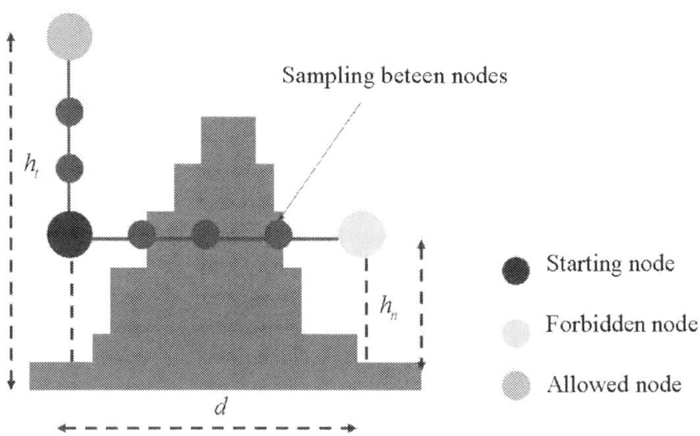

Figure 6. Improving the resolution with a finer sampling between nodes of the A* space

The sampling introduces a combinatory of the order $\sim N_i^2$ to the searching task of the algorithm that, thanks to this modification only looks for the coordinates of the space in the *X-Y* plane, and compares the sampling point's height with the altitude of the surface, to discriminate nodes that could hide abrupt changes of the soil's profile, like cliffs or mountain peaks. In this way the algorithm remains computationally bearable while the surface is included in the path planning with an acceptable resolution.

4. Modification of trajectories in dynamic environments

The scenario under study is made up of one UAV, no-flight zones, obstacles (buildings), terrain, fixed ADUs, and pop-up ADUs as dynamic elements. The uncertain appearance of these last elements encourages the search of a routine to optimize decisions made during planning time and, if possible, on line, depending on how the mission is evolving in time. Fig. 7 shows the way the UAV updates information about previously known dynamic elements and re-launch the decision making ILP routine if needed, or even discovers new pop-ups and, if it has enough computing resources, re-executes the planning from the bypassing up to the decision making process, trying to follow again the main trajectory.

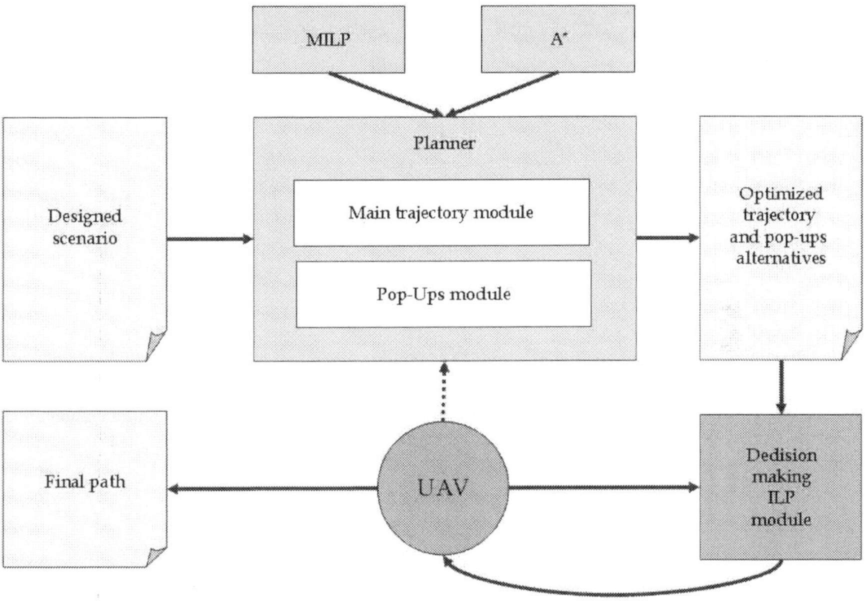

Figure 7. Layout of the path generator

The original trajectory is modified according to a decision which is made based on the following three elements: mean risk, flying time, and fuel consumption. The ILP formulation computes the optimum alternatives to be chosen (Ruz et al., 2007).

4.1 Cumulative mean risk parameter

We start from the computation of the kill probability (P_K) on each point of a trajectory, from which a measure of the cumulative mean risk for that particular trajectory is computed. This computation of the probability of kill can be made by any available analytical, statistical, or even empirical model of the interactions between the aircraft and the enemy defence units, as seen before in (7). It is also possible to count with tabulated values of P_K, for several ranges of the variables that affect this probability such as UAV's position and RCS.

From the kill probability P_{Ki} at any trajectory point, the survival probability P_{Si} associated to the enemy i-th ADU is computed as follows in (22).

$$P_{Si}(\vec{r},\vec{v}) = 1 - P_{Ki}(\vec{r},\vec{v}) \tag{22}$$

The total survival probability P_{ST} is computed by multiplying the following three factors (23):

1. The P_{Si} product from 1 to the number N of known ADUs.
2. The Pop-Up ADU survival probability affecting the point, P_{SPU}.

3. A weighting factor α embedding the Pop-Up appearance probability when flying towards its location.

We have chosen the product combination based in the assumption that all the ADUs work in cooperation.

$$P_{ST}(\vec{r},\vec{v}) = \alpha P_{SPU}(\vec{r},\vec{v})\prod_{i=1}^{N} P_{Si}(\vec{r},\vec{v}) \tag{23}$$

The weighting factor α, ranging in (0, 1), is defined as the complementary probability of the pop-up appearance probability P_{APU}, at any state $(\vec{r},\vec{v})$ of the aircraft (24). It should be strictly greater than zero and less than one, because the fact to exactly assign any of these values to a P_{APU} probability makes the ADU turn into a nonexistent or fixed one respectively, which is already considered in the last term of the product.

$$\alpha = 1 - P_{APU} \tag{24}$$

Then, it is clear that the more probable a pop-up arises in front of the UAV during the flight the less probable the UAV will survive to it, bringing a greater cumulative mean risk to the chosen trajectory.

Once the total survival probability to a set of N cooperating ADUs is computed, including the unexpected activation of more threats, the total probability of kill P_{KTm} in the state $m = (\vec{r},\vec{v})$ is given by the complement to one (25).

$$P_{KTm}(\vec{r},\vec{v}) = 1 - P_{STm}(\vec{r},\vec{v}) \tag{25}$$

Hence, we define the cumulative mean risk of a trajectory R_K as the average of the total kill probabilities of all the M points which form this trajectory (26). This concept will be used as a parameter to characterize the group of alternatives to build a final trajectory under a decision making formulation.

$$R_K = \frac{1}{M}\sum_{m=1}^{M} P_{KTm}(\vec{r},\vec{v}) \tag{26}$$

The risk is calculated as a mean value, based on the discrete time system assumption. The points of the trajectory are approximately equally spaced since the flying speed is constant. If the time were continuous the integral form to calculate a mean value would be used instead of the sum showed in (26).

4.2 Cumulative flying time parameter

The flying time parameter is simply a way to characterize alternative trajectories in terms of the cumulative time needed by the aircraft to achieve them, assuming a constant flying speed. Thus, for an M-points time discrete path, with all points equally spaced in Δt time, the total flying time T is given by (27).

$$T = \sum_{m=1}^{M} \Delta t_m = M\Delta t \tag{27}$$

There is also a way to normalize the cumulative time parameter, with the aim to compare different alternatives of a trajectory or even different trajectories. If we define the amount τ as the time to go along the minimum length/time trajectory between any pair of points (straight line), it is possible to define a normalized cumulative flying time factor f_t (28), where the zero value represents a characterization for the mentioned minimum path.

$$f_t = 1 - \frac{\tau}{T} \tag{28}$$

4.3 Cumulative fuel factor parameter

Since the UAV's trajectory generation is represented as a 3D optimization problem, it might be formulated with an objective function and a set of constraints in a Cartesian referential frame where (x,y,z) is the UAV's position. Among the constraints there are kinematical and dynamical limitations of the system, which is an air vehicle unable to make stationary flights. Furthermore, the linear approach and the time discrete character of the solution led us to the matrix representation already shown in (1), with the limits that produce a minimum turning radius possible to achieve.

A more convenient expression for the limits of speed and acceleration as a function of their components in $\boldsymbol{R}^3$ is in (29), where again the maximum limits are in the right side of the inequalities.

$$\begin{aligned} v_x^{\,2} + v_y^{\,2} + v_z^{\,2} &\le \left|\vec{v}_{\max}\right|^2 \\ u_x^{\,2} + u_y^{\,2} + u_z^{\,2} &\le \left|\vec{u}_{\max}\right|^2 \end{aligned} \tag{29}$$

It is possible to reorder the constraint for the maximum turning rate making normalization for each of the acceleration's components (30), where the angle θ is the zenith and the angle φ is the azimuth of vector $\vec{u}$ represented in spherical coordinates.

$$\frac{u_x \cos(\varphi)\sin(\theta)}{\left|\vec{u}_{\max}\right|} + \frac{u_y \sin(\varphi)\sin(\theta)}{\left|\vec{u}_{\max}\right|} + \frac{u_z \cos(\theta)}{\left|\vec{u}_{\max}\right|} = C_t \tag{30}$$

Thus, in (31) it is shown the constraint for the signal C_t, which is a normalized input signal for each discrete time t. It represents a way to measure the acceleration applied to the system, needed to change the flying direction at any time step. This signal can be considered directly bounded to the aircraft fuel consumption because it might be the control signal, or the actuator signal used to change the UAV's course.

$$C_t \le 1 \qquad \forall t \tag{31}$$

Finally, in (32) we define the fuel consumption factor as the average of the fuel consumed along the t trajectory points.

$$F_c = \frac{1}{T}\sum_{t=1}^{T} C_t \tag{32}$$

4.4 Decision making module

In every mission the path designers might count with very accurate information about most of the elements involved in the flying environment, which can be provided and confirmed by several sources during the planning time. However, it is also possible to possess a minimum knowledge about uncertain or dynamic elements characterized by a probability of appearance, and that might represent a threat for the UAV's path.

The strategy proposed in this work implements an initial path planning taking into account only the well known and fixed components of the scenario, to obtain the main optimum trajectory which will be followed by the UAV. After having a main route, the knowledge of non-static elements, such as pop-up radars, is included in the scenario for considering only those pop-ups that actually may be a serious threat for the UAV. Once the actual threats have been discriminated from all the originally counted, a local avoidance strategy is computed, using MILP or A* algorithm, to bypass the pop-ups. These alternatives are all attached to the original flying plan, and given to an upper layer module in charge of making decisions according to the imposed limitations; let's say fuel consumption, time, and risk. It is right here where an optimum decision making process will increase the chances of a successful mission.

Suppose there is a mission to go from a starting point to an objective, as seen in figure 8, and that the originally planned trajectory might be affected by three independent unforeseen threats, characterized by their corresponding appearance probability P_{PU}. Therefore, each of them has an associated probability factor α of nonappearance, which assigns certain weight to the survival probability of the aircraft against those pop-ups.

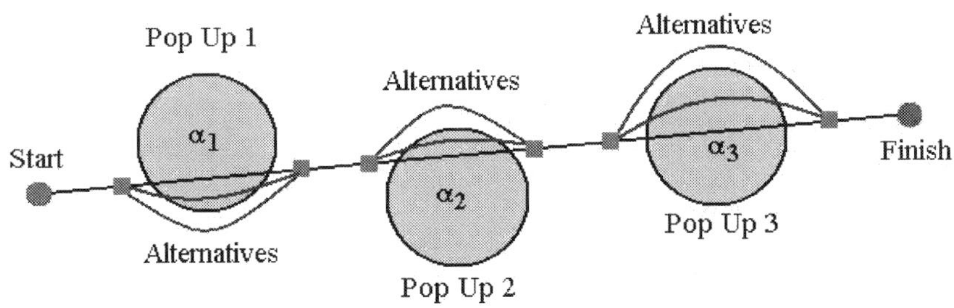

Figure 8. Trajectory decision map with three possible pop-ups and the corresponding three alternatives to avoid each of them

If the number n_a of alternatives is the same for each pop-up in particular, it's easy to compute the total amount of combined alternatives. In this case, the combinatory leads us to a total amount of alternatives ($n_a^{N_{pu}}$), which is the number of alternatives by pop-up powered to the total number of pop-ups N_{pu}.

All the alternatives have their characteristic parameters to be processed in a decision making algorithm that seeks and find the optimum final trajectory, based not only on the recent and past information at the moment of the decision, but also on the probability of future events.

The choice of the optimal sequence of alternatives that will compose the final planned route can be posed as an ILP problem. The cumulative time and fuel consumption parameters will be the constraints, and the cumulative mean risk the objective (minimum) function. The mentioned objective function is given by (33), where L is the total number of pop-ups affecting the original trajectory (pre-planned trajectory without pop-ups), and the indexes $\{i,j,\ldots,w\}$ range over all the alternatives for each one of the affecting pop-ups.

$$J=\left(\sum_{i} R_{K1i}\delta_{1i}+\sum_{j} R_{K2j}\delta_{2j}+\ldots+\sum_{w} R_{KLw}\delta_{Lw}\right) \tag{33}$$

In this objective function the coefficients R_{Klm} are the cumulated mean risk of each alternative, and the variables δ_{lm} binary variables associated to the chosen alternative among all the possible ones for each pop-up. Therefore, the variables must be constrained (34), to guarantee that only one of the alternatives is selected at the time of making a specific decision.

$$\sum_{i=1}^{I}\delta_{1i}=1;\ \sum_{j=1}^{J}\delta_{2j}=1;\ \ldots;\ \sum_{w=1}^{W}\delta_{Lw}=1 \tag{34}$$

The rest of the constraints refer to the upper limit assigned to the accepted cumulative time factor (35), and to the maximum cumulative fuel consumption factor (36). Both limits can be set based on the UAV's dynamics, and on its fuel consumption model.

$$\frac{\left(\sum_{i} T_{1i}\delta_{1i}+\sum_{j} T_{2j}\delta_{2j}+\ldots+\sum_{w} T_{Lw}\delta_{Lw}\right)}{L}\leq T_{\max} \tag{35}$$

The T_{lm} coefficients are the cumulative time factor of every computed alternative, and $T_{\max}$ is the limit accepted for the time factor of the mission.

$$\frac{\left(\sum_{i} F_{c1i}\delta_{1i}+\sum_{j} F_{c2j}\delta_{2j}+\ldots+\sum_{w} F_{cLw}\delta_{Lw}\right)}{L}\leq F_{c\max} \tag{36}$$

The coefficients F_{clm} are the cumulated fuel consumption factor of every computed alternative, and $F_{c\max}$ is the upper limit for the mean fuel consumption along the global trajectory.

5. Implementations and results

A path planning software platform was developed implementing both, MILP and A* algorithm trajectory optimizers. The MILP model takes advantage of the powerful CPLEX 9.0 solver through the ILOG CPLEX package (ILOG, 2003), to find the solution for optimum trajectories in the space of the discrete UAV's variables of state. The A* algorithm was coded in JAVA language, using the JRE system library jre1.5.0_06. The metric used as the heuristic was the Euclidean distance.

Figure 9 shows the resulting trajectory computed in a scenario where there are mountains, waypoints, and pop-up radars only. The black solid line would be the optimum path whenever the pop-ups don't get enabled during the UAV's approximation. The yellow dashed line is an alternative calculated during the planning time to safely escape from the threat that possibly causes a mission fail.

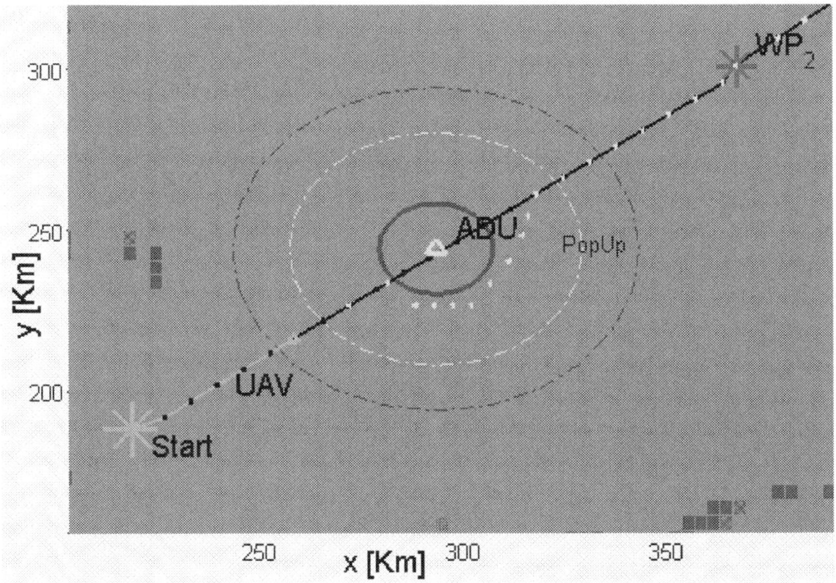

Figure 9. Computed trajectory (black) with the alternative (yellow) to avoid one pop-up

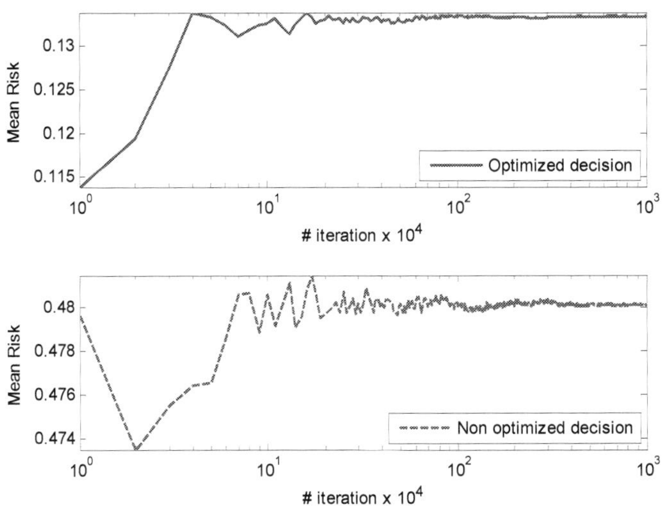

Figure 10. Cumulative mean risk after 10^7 Monte Carlo simulations

A Monte Carlo simulation was done to evaluate the decision making strategy proposed in this work (Berg & Chain, 2004), where a probability of future appearance assigned to every threat pop-up is taken into account to activate them, while the parameters of risk, time and fuel are constrained in an ILP model. This strategy was compared with the simple decision made on the basis of the consumed fuel and the spent time, which are only past and present sources of information. Figure 10 shows the cumulative mean risk of both strategies, after 10^7 iterations, where there were three pop-ups, with three alternative trajectories each. The probabilities of appearance were 0.5, 0.2, and 0.8, for the pop-ups affecting the original trajectory in that chronological order. As mentioned in section 1 these probabilities are provided by expert knowledge prior to the mission design. Depending on the selected alternative the mean risk accumulated different values. The more direct is the route, the more risky it is, while the less time it spends. The greater the turning radius of the route is, the less the fuel is consumed. It might be possible to find trajectories with the maximum time spent along it without having the higher fuel consumption. Constraints over the time factor (0.35) and the fuel consumption (0.40) were imposed into the ILP decision making model, to obtain the optimum final global path.
The histograms in figure 11 of the two simulated strategies show the advantages of choosing the optimum decision plan, because the constraints over spent time and fuel consumption are never violated, while the cumulated risk in minimized. The strategy that only considers past and present information doesn't violate the time and fuel criteria either, but its response to the cumulated risk is very poor because the most probable pop-ups is not the necessary the first one to appear.

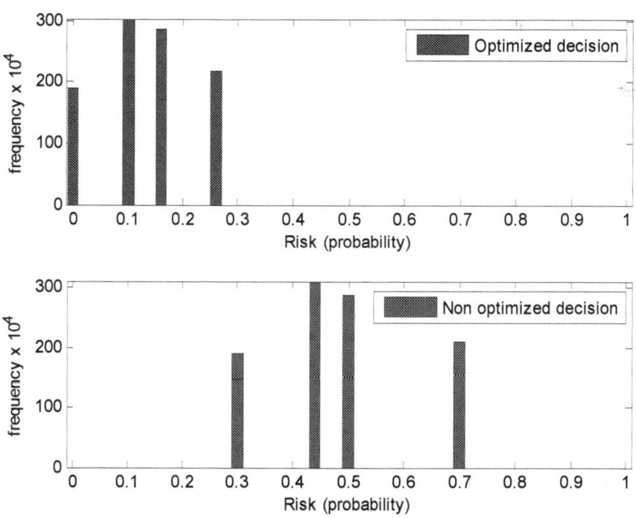

Figure 11. Histogram of the mean risk after 10^7 Monte Carlo simulations

Finally, figure 12 shows the results when the UAV's trajectory must reach its target within a radar zone. The detection risk is minimized respect to objective function (37)

$$J = \mu_1 t_{arrival} + \mu_2 D(x, y, v_x, v_y) \tag{37}$$

where D is the nonlinear radar detection function, μ_1 and μ_2 are weights which consider the importance of flight time and acceptable threat concerning to a particular mission.

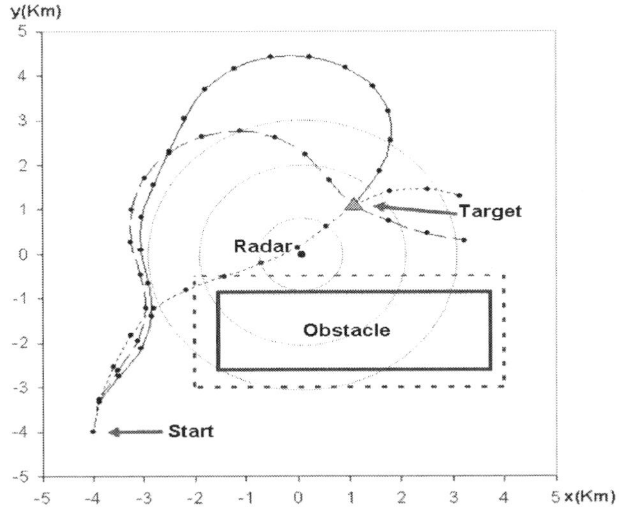

Figure 12. Comparison of three trajectories with target in radar zone

The UAV tries to avoid the radar detection by maintaining the biggest possible distance, compatible with the values μ_1 and μ_2, and controlling the RCS it presents to the radar. The trajectories plotted in Figure 12 shows that the UAV does not fly directly to the target, and when a higher risk of detection is even accepted, the UAV will use a more direct and risky trajectory ($\mu_1 = 1$, $\mu_2 = 2.7e4$). It can be observed that when the UAV is next to the target and the admitted risk is low ($\mu_1 = 1$, $\mu_2 = 2.8e4$), its trajectory tries to approach radially to the radar, minimizing its RCS. Over a no radar zone ($\mu_2 = 0$) the flying trajectory goes directly to the target.

6. Conclusions and future work

We have presented the trajectory generation module of SPASAS, an integrated system for definition of flight scenarios, flight planning, simulation and graphic representation of the results developed at Complutense University of Madrid. The module uses two alternative methods, MILP and a modification of the A* algorithm, and considers static and dynamic environmental elements, particularly pop-ups. Both methods have been implemented and a Monte Carlo simulation was done to evaluate the decision making strategy proposed.

The results showed the advantages of choosing the optimum decision plan that considers the known values of the probability of appearance of pop-up threats in the future. The possibility to update the information concerning the pop-up's appearance probabilities, available fuel, time, and even the assumed risk, and then re-launch a decision making routine to optimize the chosen alternatives has been proven, since the ILP model provides a solution affordable in real time (~1s).

When the UAV must reach a target within a radar zone, the detection risk is minimized using an efficient MILP formulation that approximates the continuous risk function with hyper-planes implemented with integer 0-1 variables.

For the future, we are already working in three main objectives: (a) the use of rotary-wing UAVs such as quad-rotors, (b) the introduction of video cameras onboard UAVs, and (c) the design of coordination algorithms for a fleet of UAVs. Rotary-wing UAVs will incorporate more maneuverability than conventional fixed-wing UAV, since they can take-off and land in limited space and can easily hover above a target. Cameras onboard will allow the use of vision-based techniques for locate and track dynamic perimeters as is needed in tasks such as oil spill identification or forest fires tracking. Finally, a team of UAVs will get an objective more efficiently and more effectively than a single UAV.

7. Acknowledgements

This research was funded by the Community of Madrid, project "COSICOLOGI" S-0505/DPI-0391, by the Spanish Ministry of Education and Science, project "Planning, simulation and control for cooperation of multiple UAVs and MAVs" DPI2006-15661-C02-01, and by EADS(CASA), project 353/2005.

The authors would like to thank Tomas Puche, Ricardo Salgado, Daniel Pinilla and Gemma Blasco from EADS(CASA), and Bonifacio Andrés, Segundo Esteban and José L. Risco from UCM, for their contribution to SPASAS project.

8. References

Bellingham, J. & How, J. (2002) Receding Horizon Control of Autonomous Aerial Vehicles. *Proc. of American Control Conference.*

Berg, B. & Chain, M. (2004) Monte Carlo Simulations and their Statistical Analysis. *World Scientific,* ISBN 981- 238-935-0.

Borto, S. (2000) Path planning for UAVs. *Proceedings of the American Control Conference.* pp. 364-368

How, J.; King E. & Kuwata, Y. (2004) Flight Demonstrations of Cooperative Control for UAV Teams. AIAA *3rd Unmanned Unlimited Technical Conference, Workshop and Exhibit.*

ILOG, Inc. ILOG CPLEX 9.1 (2003) User's guide, http://www.ilog.com/products/cplex.

Kuwata, Y. & How, J. (2004) Three Dimensional Receding Horizon Control for UAVs. *AIAA Guidance, Navigation, and Control Conference and Exhibit.*

Melchior, P.; Orsoni, B.; Lavialle O.; Poty A. & Oustaloup, A. (2003) Consideration of obstacle danger level in path planning using A* and fast-marching optimisation: comparative study. *Signal Process.* Vol. 83,11, pp. 2387-2396.

Murphey, R.; Uryasev, S. & Zabarankin, M. (2003) Trajectory Optimization in a Threat Environment. *Research Report 2003-9, Department of Industrial & Systems Engineering.* University of Florida.

Richards, A. & How, J. (2002) Aircraft Trajectory Planning with Collision Avoidance Using MILP. *Proceedings of the IEEE American Control Conference.* pp. 1936-1941.

Ruz, J.; Arévalo, O.; Cruz J. & Pajares, G. (2006) Using MILP for UAVs Trajectory Optimization under Radar Detection Risk. *Proc. of the 11th IEEE Conference on Emerging Technologies and Factory Automation*. ETFA'06, pp. 957-960.

Ruz, J.; Arevalo, O.; Pajares, G. & Cruz, J. (2007) Decision Making among Alternative Routes for UAVs in Dynamic Environments. *12th IEEE Conference on Emerging Technologies & Factory Automation*. ETFA'07,pp. 997-1004.

Schouwenaars, T.; Moor, B.; Feron, E. & How, J. (2001) Mixed Integer Programming for Multi-Vehicle Path Planning. *Proceedings of the European Control Conference*. pp. 2603-2608

Schouwenaars, T.; How, J. & Feron, E. (2004) Receding Horizon Path Planning with Implicit Safety Guarantees. *Proceedings of American Control Conference*. pp. 5576-5581.

Szczerba R.; Galkowski, P.; Glicktein, I. & Ternullo, N. (2000) Robust algorithm for real-time route planning. *IEEE Trans. Aerosp. Electron. Syst*. Vol. 36, 3, pp. 869-878.

Trovato, K. (1996) A* Planning in Discrete Configuration Spaces of Autonomous Systems. *PhD dissertation*. Amsterdam University.

Zengin, U. & Dogan, A. (2004) Probabilistic Trajectory Planning for UAVs in Dynamic Environments. *Proc. of AIAA 3rd Unmanned Unlimited Technical Conference, Workshop and Exhibit*. pp. 1-12.

28

Modelling and Identification of Flight Dynamics in Mini-Helicopters Using Neural Networks

Rodrigo San Martin Muñoz, Claudio Rossi and Antonio Barrientos Cruz
Universidad Politécnica de Madrid, Robotics and Cybernetics Research Group
Spain

1. Introduction

Unmanned Aerial Vehicles have widely demonstrated their utility in military applications. Different vehicle types - airplanes in particular - have been used for surveillance and reconnaissance missions. Civil use of UAVs, as applied to early alert, inspection and aerial-imagery systems, among others, is more recent (OSD, 2005). For many of these applications, the most suitable vehicle is the helicopter because it offers a good balance between manoeuvrability and speed, as well as for its hovering capability.

A mathematical model of a helicopter's flight dynamics is critical for the development of controllers that enable autonomous flight. Control strategies are first tested within simulators where an accurate identification process guarantees good performance under real conditions. The model, used as a simulator, may also be an excellent output predictor for cases in which data cannot be collected by the embedded system due to malfunction (e.g. transmission delay or lack of signal). With this technology, more robust fail-safe modes are possible.

The state of a helicopter is described by its attitude and position and the characteristics of its dynamics system correspond to those of a non-linear, multivariable, highly coupled and unstable system (Lopez, 1993). The identification process can be performed in different ways, on analytical, empirical or hybrid models, each with its advantages and disadvantages.

This Chapter describes how to model the dynamic of a mini-helicopter using different kinds of supervised neural networks, an empirical model. Specifically, the networks are used for the identification of both attitude and position of a radio controlled mini helicopter. Different hybrid supervised neural network architectures, as well as different training strategies, will be discussed and compared on different flight stages. The final aim of the identification process is to build a realistic flight model to be incorporated in a flight simulator.

Although several neural network-based controllers for UAVs can be found in the literature, there is little work on flight simulator models. Simulators are valuable tools for in-lab testing and experimenting of different control algorithms and techniques for autonomous flight. A model of a helicopter's flight dynamics is critical for the development of good a simulator. Moreover, a model may also be used during flight as predictor for anticipating the behaviour of the helicopter in response to control inputs.

The Chapter first focuses on two neural-network architectures that are well suited for the particular case of mini-helicopters, and describes two algorithms for the training of such neural-network models. These architectures can be used for both multi-layer and radial-based hybrid networks. The advantages and disadvantages of using neural networks will also be discussed. Then, a methodology for acquiring the training patterns and training the networks for different flight stages is presented, and an algorithm for using the networks during simulations is described. The methodology is result of several years of experience in UAVs. Finally, the two architectures and training methods are tested on real flight data and simulation data, and the results are compared and analysed.

2. Network Architectures for Modelling Dynamics Systems

Modelling a dynamic system like a mini-helicopter, requires estimating the effect of both the inputs and the system's internal state on the outputs (Norgaard et al., 2001). Considering the system's identification by means of state variables, a dynamic system can be described in a discrete space as shown in (1), where v is the state variable, x the input and y the output.

$$\begin{aligned} v(k+1) &= \Phi[v(k), x(k)] \\ y(k) &= \Psi[v(k), x(k)] \end{aligned} \tag{1}$$

The first equation shows the dependence of the state at a certain time with regard to the state and inputs in a previous instant, similar to recurrent neural networks, for example Hopfield Neural Network (Hopfield, 1982). The second equation shows the dependence of the outputs in instant k with regard to the state and inputs in the same instant, as in non-recurrent neural networks, for example Radial Basis (RB) or multi-layer Perceptron (MLP) Neural Network (Freeman & Skapura, 1991). These equations suggest the use of a mixed neural network with recurrent and non-recurrent properties (Narendra & Parthasarathy, 1990). The recurrent component is used to describe the system's state, while the non-recurrent component defines the system's outputs. There are two mains types of mixed networks, Jordan's (Jordan, 1986) and Elman's (Elman, 1990). In this case, theses networks are not applicable. For this reason, will be used a new proposed hybrid networks which is suitable for modelling of flight dynamics in mini-helicopters.

2.1 Jordan's Network

Jordan's network (Jordan, 1986) consists of a multi-layer network with external inputs **X=[x_1 x_2 ... x_n]** and contextual neurons **C=[c_1 c_2 ... c_m]**, that represent the internal states (see Figure 1). These contextual neurons are recurrent because they use both their previous output **C(t-1)** and the previous system's output as inputs. This means that they store the systems' past states adjusted by μ (store weight), as shown in (2).

This architecture works, basically, as a multi-layer network with the particularity that the network's external input and contextual neurons create a new input vector **U=[x_1 x_2 ... x_n c_1 c_2 ... c_m]**. In (2) it is possible to observe that the i-th output corresponds to the composition of the outputs of every layer, as in a non-recurrent network.

$$\begin{aligned} C_i(k) &= \mu \cdot C_i(k-1) + y_i(k-1) \forall i = 1,2,\ldots,m \\ C_i(k) &= \sum_{j}^{k-1} (\mu^{j-1} y_i(k-1)) \forall i = 1,2,\ldots,m \\ y_i(k) &= S_i(\sum_{j=1}^{j=g} N_j(\sum_{h=1}^{n+m} u_h \cdot w_{jh}) \cdot w_{ij}) \forall i = 1,2,\ldots,m \end{aligned} \tag{2}$$

Where u_i corresponds to the elements of vector **U**, $\mathbf{w_{jh}}$ are the weights in the hidden layers (**N**), $\mathbf{w_{ij}}$ are the weights in the output layer (**S**) and μ is the weight of time constant.

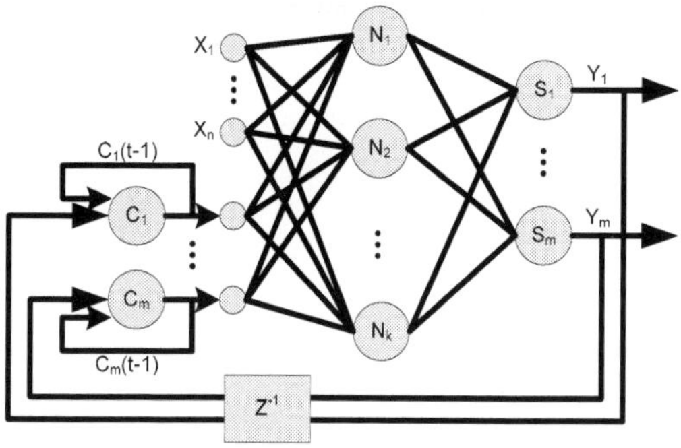

Figure 1. Jordan's network, where Z^{-1} is a sample delayed in the time

Rewriting (2) results in (3), where **Y** it is defined by a function whose inputs are the elements u, which are defined by vectors **C** and **X**, and **C** is defined by a weighted sum of **Y,** which is nothing more than **C** and **X**.

$$\begin{aligned} C(k+1) &= F[C(k), X(k)] \\ Y(k) &= G[C(k), X(k)] \end{aligned} \tag{3}$$

2.2 Elman's Network

Elman's network (Elman, 1990) (Fig. 2), unlike Jordan's, does not feed back the contextual neuron output or the system output, as shown in (4). Instead, it feeds back the output of an intermediate layer.

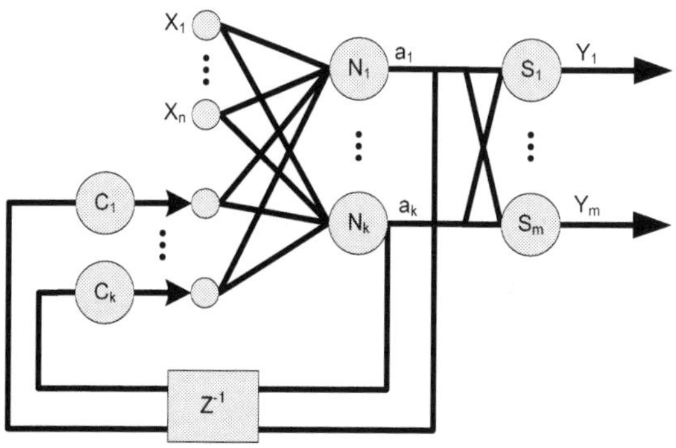

Figure 2. Elman's network

The input vector is exactly the same as in Jordan's network, the only difference being the value of the contextual neurons, as defined in (4). In this case, the contextual neurons do not preserve past states, but save only the last state value.

$$C_i(k) = a_i(k-1) \forall i = 1,2,\ldots,k \tag{4}$$

2.3 Proposed Hybrid Network

Other neural network architectures are possible, in which the context neurons appear only in the first layer of the network. The architecture of (Narendra & Parthasarathy, 1990) tries to generate the contextual neurons at all levels, that is, both for the outputs and for the inputs. This idea is the basis of the architecture that will be developed in this work.

The proposed hybrid network consists of two blocks: the first is the recurrent component with context neurons for the inputs and outputs. These neurons use past states as inputs and the number of states is defined automatically by the training algorithm or by means of some stochastic model. The second block is a non-recurrent network that may be either an MLP or an RB. Fig. 2.3 shows a hybrid network in which Block A consists of a non-recurrent network and Block B corresponds to context neurons (recurrent network). Block's B neurons do not follow Elman's or Jordan's architecture, but rather a mix: the feedback is **d** past system outputs and **h** past system inputs. The number of past states to use is adjustable, as is the number of neurons in the hidden layer of Block A. This flexibility is necessary because the system is unknown and the network must adapt during the training process to attain the minimum error.

The external inputs are represented by vector $\mathbf{X=[x_1\ x_2\ ...\ x_n]}$ and these are stored in the contextual neurons $\mathbf{C_{xi1}}$ to $\mathbf{C_{xid}}$. The outputs are represented by vector $\mathbf{Y=[y_1\ y_2\ ...\ y_m]}$ which is also stored in contextual neurons $\mathbf{C_{yi1}}$ to $\mathbf{C_{yih}}$. As opposed to Jordan's network (2) these contextual neurons keep **d** past inputs and **h** past outputs, which yields the vectors $\mathbf{C_{xi}=[x_i(k\text{-}1)\ ...\ x_i(k\text{-}d)]}$ and $\mathbf{C_{yi}=[y_i(k\text{-}1)\ ...\ y_i(k\text{-}h)]}$ for the i-th iteration. Thus, the input vector for the non-recurrent network in Block A is $\mathbf{U=[X\ C_{x1}\ ...\ C_{xn}\ C_{y1}\ C_{ym}]}$. This means that **U** contains **n** elements from the input vector **X**, **n** elements for every previous state **d**, **m** elements from the output vector **Y** and **m** elements for every previous state **h**. To sum up, the amount of elements for vector **U** is: $\mathbf{n+(n \bullet d)+(m \bullet h)}$.

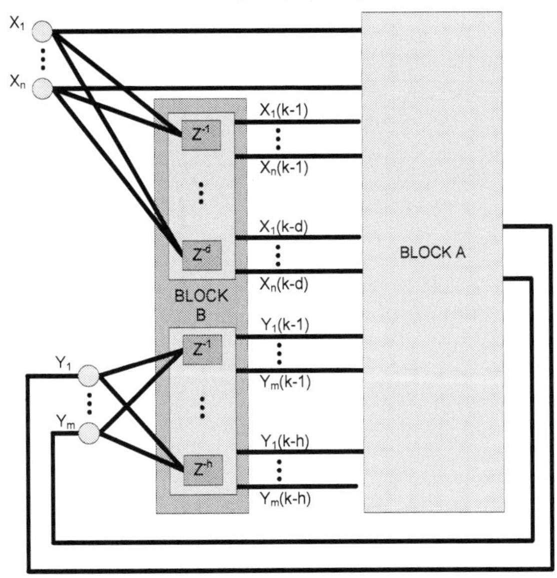

Figure 3. Hybrid network

A slightly modified Elman's network is very close to this idea, the main difference being that Elman's feeds back from the hidden layer. The modification consists of a feedback loop that involves not only a delay $\mathbf{Z^{-1}}$, but also a delay block similar to **Block B** in the hybrid network (Fig. 3). On the other hand, the Jordan's network does feed back the output, but does not consider all previous states with the same weight. Equation (2) shows how each previous value is stored in the memory of the contextual neuron itself, which causes the value of the contextual neuron to be the result of adding all the weighted previous states. In other words, the hybrid network has a finite number of previous states that generate the same amount of contextual neurons; in contrast, Jordan's network generates a contextual neuron with an infinite amount of previous states by performing a weighted sum of them.

When creating a hybrid network, principles from both networks are taken into account, but the possibility of additional modifications for future improvement is left open. For example, consider a contextual neuron for storing the rest of the previous states, as done by Jordan. Also, it is possible to feed back the output of the first layer, as suggested by Elman. The possibilities are manifold, but it is necessary to choose one architecture in order to be able to start performing any tests.

As mentioned before, **Block B** corresponds to the recurrent stage, where the previous states are stored, and it is an integral part of the network architecture. In the case of a system with substantial inertia, the order of the delays (**d** and **h**) will increase. **Block A** corresponds to the non-recurrent stage, which performs the system output signal tracking and is able to operate internally both with Multi-layer Perceptron and Radial Basis networks.

Similar to the mixed networks mentioned previously, the hybrid network can be converted to the form (3), where functions **F** and **G** will depend on the architecture selected for **Block A**, either an RB or MLP network, as well as the internal transfer functions of the recurrent networks. The following section describes the hybrid network used for the UAV system and justifies its architecture.

3. Proposed Hybrid Network Architecture for the Identification of a Mini-Helicopters (UAV)

For the identification of a system like the mini-helicopter, some adjustments need to be made to the hybrid network and the simulation strategy must be planned. A helicopter flight is based on the angle-of-attack and angular velocity of its blades. These values define the attitude and lift that, in turn, change the position of the aircraft (Lopez, 1993). Due to these circumstances, two stages are considered in the identification process: computing the attitude using the control commands as inputs, and then using the attitude to obtain, in this case, the vehicle position. Hybrid networks can be used to model both stages (Fig. 2.3).

In summary, the dynamic system to be modelled corresponds to a system which, after receiving some control commands, modifies its attitude (θ, ϕ, ψ) and consequently, its position (**X, Y, Z**) (as shown in Fig. 4).

The neural network's outputs are the helicopter's attitude and position, and its inputs are the roll, pitch and yaw cyclic steps and the collective, labelled Croll, Cpitch, Cyaw and Ccole, respectively. Croll and Cpitch control the cyclic angle-of-attack of the main rotor blades, Cyaw commands the tail rotor and Ccole is a combination of the main rotor blade's angle-of-attack and the engine throttle. These control signals are the same ones a human pilot uses to command a mini-helicopter and represent the inputs of the radio-controller.

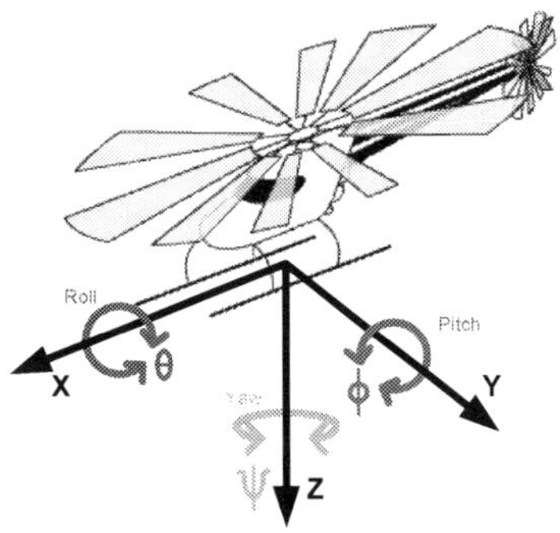

Figure 4. Attitude angle and position

With this architecture, based on two hybrid networks, two training methods are possible. The first connects the system as a daisy chain: the output of the attitude system's training is used as the input for the position system's training. The second training method places the systems in parallel and uses real flight data to train both networks. It is important to note that both training methods require carefully selected parameters: number of inputs n and outputs m, order of contextual neurons for inputs d and outputs h and type of network in Block A (MLP or RB). The method used to set these parameters will be described in the following sections.

3.1 Training Architectures

The data obtained from the avionics (attitude: θ roll, ϕ pitch, ψ yaw, and position: X, Y, Z) and the radio transmitter (control commands: Croll, Cpitch, Cyaw, Cole), are the patterns used for the training. The position and attitude degrees of a helicopter's flight system are depicted in Fig.4, where its position being defined by its attitude.

As mentioned above, there are two training methods:

Daisy chain Architecture: the attitude system is trained as a single and isolated system, which is possible thanks to the previous knowledge of the attitude data (roll θ, pitch ϕ, yaw ψ). The values obtained from the attitude network, estimated data, (roll', pitch' and yaw') are used as inputs for training the position network.

For the attitude system, the training pattern for **Block $\mathbf{A_a}$** is vector $\mathbf{P_a}$ (5), with an external input vector $\mathbf{X_a}$ consisting of the different radio control commands (Croll, Cpitch, Cyaw and Ccole), another vector name $\mathbf{C_{in\_a}}$, that corresponds to the input contextual neurons, and finally $\mathbf{C_{out\_a}}$, which represents the attitude system's feedback output contextual neurons. The training pattern is represented by vector $\mathbf{T_a}$ (6), which is the real attitude data provided by the avionics.

$$
\begin{aligned}
P_a &= [X_a, C_{in\_a}, C_{out\_a}] \\
X_a &= [Croll(t), Cpitch(t), Cyaw(t), Cole(t)] \\
C_{in\_a} &= [C_{Croll}, C_{Cpitch}, C_{Cyaw}, C_{Cole}] \\
C_{out\_a} &= [C_{roll}, C_{pitch}, C_{yaw}] \\
C_i &= [i(t-1),..., i(t-d)] \forall i = Croll, Cpitch, Cyaw, Cole \\
C_j &= [j(t-1),..., j(t-h)] \forall i = roll, pitch, yaw
\end{aligned} \tag{5}
$$

$$T_a = [roll\,(t),\, pitch\,(t),\, yaw\,(t)] \tag{6}$$

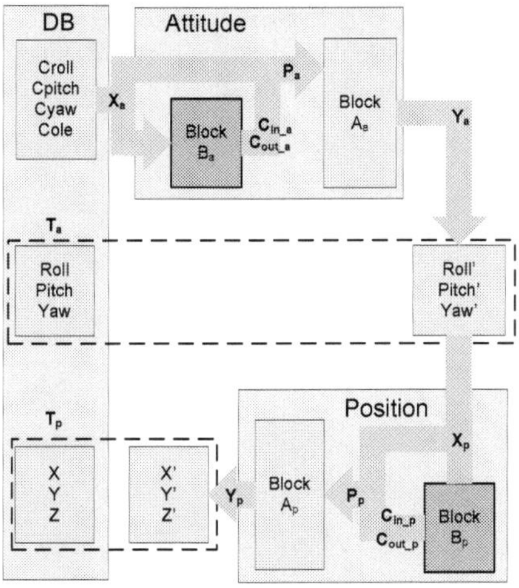

Figure 5. Daisy chain Architecture

Once the patterns are obtained, the training of **Block $\mathbf{A_a}$** starts by comparing the desired system output $\mathbf{T_a}$ (6) with the current network output $\mathbf{Y_a}$ (7).

$$Y_a = [roll',\, pitch',\, yaw'] \tag{7}$$

All necessary adjustments are performed with this *error* according to the training rules of a recurrent network. Basically, as long as the network is correctly trained, a minimum error is expected in the comparison between the desired output $\mathbf{T_a}$ and the real output $\mathbf{Y_a}$.

After adequate training of the attitude, the network is used as a simulator and its $\mathbf{Y_a}$ vector (7) is used as the input pattern for the position system. This pattern will be very similar to the one used in the previous network, the main difference being the input vector $\mathbf{X_p}$ (8), which does not have avionics-acquired values but simulated data from the previous network. The output pattern for the training is vector $\mathbf{T_p}$ (9), which contains the avionics-acquired (GPS) position.

$$\begin{aligned} P_p &= [X_p, C_{in\_p}, C_{out\_p}] \\ X_p &= [roll'(t),\, pitch'(t),\, yaw'(t)] \\ C_{in\_p} &= [C_{roll'}, C_{pitch'}, C_{yaw'}] \\ C_{out\_p} &= [C_X, C_Y, C_Z] \\ C_i &= [i(t-1),\ldots,\, i(t-d)]\forall i = roll', pitch', yaw' \\ C_j &= [j(t-1),\ldots,\, j(t-h)]\forall i = x, y, z \end{aligned} \tag{8}$$

$$T_p = [x(t),\, y(t),\, z(t)] \tag{9}$$

The value obtained at the output of the network is $\mathbf{Y_p}$ (10)

$$Y_p = [x'(t),\, y'(t),\, z'(t)] \tag{10}$$

Decoupled Training Architecture: the only difference between the training architectures (Fig. 6.) is the input data used for the training process of the position network: a decoupled trainer uses the external input $\mathbf{X_p}$ (11), which contains avionics-acquired attitude values instead of simulated data. The attitude network training is identical.

$$\begin{aligned} P_p &= [X_p, C_{in\_p}, C_{out\_p}] \\ X_p &= [roll(t), pitch(t), yaw(t)] \\ C_{in\_p} &= [C_{roll}, C_{pitch}, C_{yaw}] \\ C_{out\_p} &= [C_X, C_Y, C_Z] \\ C_i &= [i(t-1),..., i(t-d)] \forall i = roll, pitch, yaw \\ C_j &= [j(t-1),..., j(t-h)] \forall i = x, y, z \end{aligned} \tag{11}$$

The process of training both networks (attitude and position) is, in this case, independent, and is done in parallel, because the attitude network outputs are not used for the position network training.

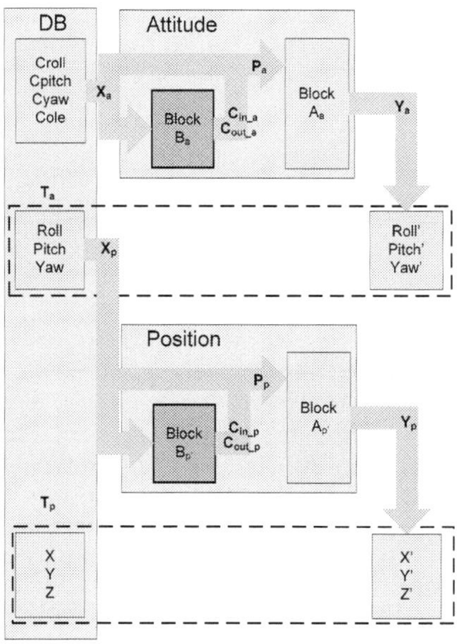

Figure 6. Decoupled architecture

The **training** and **simulation** errors of the attitude network are the same for both architectures, as the training process is the same. The **training** error for the position network is lower when the decoupled architecture is used, and one could assume that this would lead to a lower **simulation** error (when real-time data from the avionics is used to simulate the UAV's position). However, this is not the case. Real-time simulation of the UAV works in daisy chain: flight-data is fed to the attitude network, which in turn feeds the position network. Position networks trained with the daisy-chain architecture have a lower **simulation** error because they have been trained with simulated data and have *learned* to compensate for the simulation error.

3.2 Proposed Hybrid Network Architecture

The modelled system has significant inertia and dynamics. Therefore, the order of the contextual neurons depends on the correlation between the command signals and attitude, and between attitude and position. This correlation is expressed as the delay between a significant change in the inputs and a significant change in the outputs (i.e., inertia). Careful analysis of flight data shows that the delay fluctuates between 500 and 900 ms (the sampling period is 100 ms). Considering worst-case scenarios, a delay of 10 samples was used for the output contextual neurons $\mathbf{C}_{out}$ and a delay of 5 samples was used for the input contextual neurons $\mathbf{C}_{in}$, for both the attitude and the position system. This decision is based on extensive tests that have proven that these values provide the best performance.
The contextual neuron order affects the training patterns and their conformation as well as the number of neurons in the hidden layer of the MLP, which is based on the amount of inputs (Li et al., 1988). The input vector for the attitude network of **Block** $\mathbf{A}_a$ is formed as follows: 4 direct inputs from the radio transmitter $\mathbf{X}_a$ (5), which also generates vector $\mathbf{C}_{in\_a}$ (5) with an order $\mathbf{d}$ = 5, for a total of 20 contextual neurons; the output vector $\mathbf{Y}_a$ (7) is used to create vector $\mathbf{C}_{out\_a}$ (5) with an order $\mathbf{h}$ = 10, which results in 30 neurons. All of them yield the input vector $\mathbf{P}_a$ (5), with 54 neurons. This number is kept fixed both for the Radial Basis networks and the MLP. The input vector $\mathbf{P}_p$ (11) for the position system of **Block** $\mathbf{A}_p$ has 48 neurons: 3 direct inputs, corresponding to the attitude outputs $\mathbf{X}_p$ (11) that in turn generate the 15 contextual neurons of vector $\mathbf{C}_{in\_p}$, with order $\mathbf{d}$ = 5; the contextual neurons for the output is vector $\mathbf{C}_{out\_p}$ (11), with order $\mathbf{h}$ = 10 and 30 elements.
For both the MLP and the RB the training error threshold is 10^{-5}. The number of epochs for the RB networks is subject to the number of neurons in the hidden layer and is, therefore, variable. For the MLP the number of epochs is fixed at 40,000.

4. Training Pattern Generation

The success of systems identification depends on a good experimental method, even more so with a model based on neural networks. Moreover, it is well known that a wide range of flight scenarios is needed to successfully train the network. This is why it is important to choose the flight data carefully and assess its quality. The data-acquisition equipment is described below, and its capabilities are known. However, there are other important factors to be considered, for example: sampling intervals, wind speed, GPS precision variations, hardware malfunction, vibration, air temperature, etc. For all these reasons many data-gathering flights are needed to guarantee representative and high-quality samples for different kind of conditions and actions (take-off, hover, etc.).
The modelled UAV is an in-house prototype with a 5 kg payload capacity and embedded avionics and control systems, built with a 26cc Benzin Trainer radio-controlled helicopter by Vario. This UAV was developed within Project DPI 2003-01767 of the *Ministerio de Educación y Ciencia of Spain.*
The dynamic system to be modelled corresponds to a system which, after receiving some control commands, modifies its attitude (θ, ϕ, ψ) and consequently, its position (**X, Y, Z**) (as shown in Fig. 3.1). Both the attitude and position are acquired and pre-processed by the avionics with three sensors: a Mircroinfinity A3350M IMU (referential unit), a Honeywell HMR3000 compass and a Novatel OEM-4 DPGS (capable of using RT-2 corrections for 2 cm accuracy). The avionics system is built around a PC-104 board, the OctagonPC/770 with a 1GHz Pentium III processor, running Redhat Linux 8.0 with a Linux/RT kernel. Power is

supplied by an HPWR104+HR DC/DC converter and two 4LP055080+pc 14,8V-2000mAh battery packs, offering two hours of autonomy.

The system's output (attitude, position and related linear/angular velocities and accelerations) and input (control commands) signals are stored synchronously in a data file. Additional information is appended (e.g. GPS signal quality, servo PWM input, etc.) such that different flight phases and actions (take-off, landing, etc.) can be identified and used to build different training patterns for the neural network (Nguyen & Prasad, 1999).

4.1 Acquisition Procedure

The experiments are performed with a helicopter controlled by means of a radio controller in the hands of an expert pilot who commands the vehicle through a set of predefined manoeuvres.

A validation procedure has been established, repeated for each flight, that is, essentially, the first quality filter for the flight data. This procedure, shown in Fig. 7, consists of a permanent *health* evaluation of the helicopter's hardware and software. Usually the hardware (Fig. 7.a) is verified in the lab with routine tests and benchmarks; batteries are fully charged for each test and periodically tested. After a successful boot of the avionics computer, the communication links between the helicopter and the ground station are verified and the DGPS quality is asserted (Fig. 7.b). Then the pilot does an extensive pre-flight verification (e.g. radio-controller range, servo condition, etc.). If all the requirements are met, the system is ready for take-off. The main objective of these tests is to guarantee flawless operation of the helicopter.

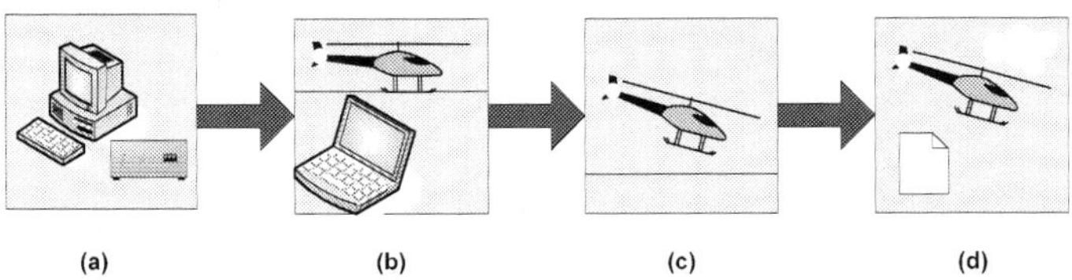

Figure 7. Data acquisition procedure (a) Hardware checking in the lab (b) Ground check of the communications and position systems (c) low height flight (d) data acquisition

A low-height flight ensures that all subsystems are operating correctly and that atmospheric conditions are within pre-established limits (Fig. 7.c). As part of routine checks or when hardware/software malfunction is suspected, the flight data is stored for detailed examination (Fig. 7.c). These tests may reveal subtle problems, such as the intermittent loss of the radio link with the ground station.

Once the system and environmental conditions are considered satisfactory, the system is set up to obtain the experimental data (Fig. 7.d) based on the flight plan. This plan includes five flight stages: start, take-off, manoeuvre, landing and end (see Fig. 8).

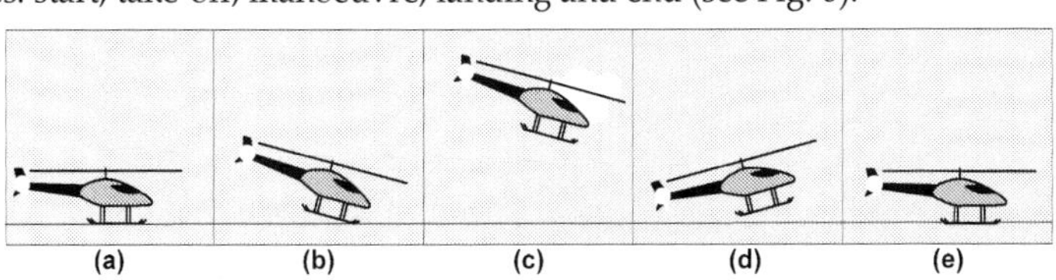

Figure 8. Flight stages stored in the text file. (a) start (b) take-off (c) flight (d) landing (e) end

Start stage: data from the helicopter standing on the ground, initial conditions. (Motor state: off)
Take-off: data from the moment that the helicopter is standing on the ground until it reaches the cruise height, before performing any manoeuvre. These values are affected by the ground effect.
Flight: data from the manoeuvres chosen for the current session (the manoeuvre plan).
Landing: data from the moment the landing procedure starts until the helicopter stands on the ground and stops.
End stage: data after landing; this data and the start data are necessary to check the correct operation of the equipment. (Motor state: off)

4.2 Data Selection Criteria

Flight data is analysed after the data acquisition process. The purpose of these inspections is to validate data quality before the data is used to build the training patterns. Two sets of criteria are established: the first is signal quality, altered by environmental conditions and equipment state. The second set is form: the type of flight performed and the similarity between the desired flight-manoeuvre plan and the actual flight.
Quality criteria: the objective here is to separate samples into suitable and unsuitable. It is important to note that suitability is defined by the requirements of different tasks. For example, samples may be suitable for simulation or training or observation, depending on their quality and significance. The quality criteria are:

- **Atmospheric:** represents the reliability of the data depending on the weather conditions present during the acquisition process, e.g., wind speed.
- **Position data quality:** in general the quality of the GPS solution for position must be better than narrow float (Novatel, 2002).
- **Attitude data quality:** in order to ensure the reliability of this data set, the attitude data obtained from the IMU at the start and end stages is compared (Fig. 8.a and Fig. 4.2.e). Considering a flat surface for the take-off and the landing manoeuvres, roll and pitch must be similar and close to zero.
- **Timing quality:** this criterion verifies the periodicity of the samples (i.e. timestamps). The sampling period is 100 ms and various malfunction conditions may lead to significant deviations. Data with a sampling period of more than 200 ms is considered to be of low quality.

Any sample that does not satisfy all quality criteria is marked unsuitable for training and discarded immediately.
Form criteria: these criteria define the experiment or type of flight. Not all flights are aimed at data acquisition since there are test and training flights (see Fig. 7 b and c). There are three flight types:

- **Standardisation:** corresponds to tests that bring the equipment to its ranges limits, e.g. signal limits for sensors or radio controller. In most cases, these tests are carried out while on the ground.
- **Test Flight:** used for system analysis. The data is not stored for further training, simulations or standardisations; it is only used to correct and measure acquisition errors like, for example, atmospheric conditions.

- **Displacements and Hovering Flight:** corresponds to lateral, longitudinal and vertical displacements and hovering. Different manoeuvres make it possible to train the network under different conditions and scenarios.

4.3 Pattern Transformation

System identification with neural networks requires pre-processing the flight data: normalization, periodicity and yaw adjustments.

Normalisation: in general transfer functions of neural networks operate in the ranges [-1 1] or [0 1]. Therefore, the pattern data must be normalised. Equations (12) and (13), respectively, normalise each entry x_k of vector $\mathbf{X=[x_1...x_k...x_n]}$ using the maximum and minimum values of X.

$$x_{k\,norm} = \frac{2 \cdot (x_k - \min(X))}{(\max(X) - \min(X))} - 1 \tag{12}$$

$$x_{k\,norm} = \frac{(x_k - \min(X))}{(\max(X) - \min(X))} \tag{13}$$

Periodicity: training a network is considered a process in discrete time. Therefore, it is necessary that samples be obtained periodically. Due to malfunction, the sampling period may not be constant, and depending on the severity of the deviation, samples may be discarded (timing quality factor) or interpolation/extrapolation algorithms may be applied.

Yaw reference: roll and pitch are easily validated since the helicopter's attitude while standing on a horizontal surface must be very close to zero. The yaw reference is the magnetic north and does not necessarily begin or end at zero. To simplify the network training, the initial yaw is considered as an offset so the initial values are close to zero.

5. Hybrid Network Algorithm Description

The training and simulation algorithms were developed with MATLAB, using the Neural Network Toolbox (Demuth & Beale, 2004) and many custom tools that had to be developed since standard toolboxes are not suited to our particular architecture and dynamic characteristics of the modelled system.

5.1 Training Algorithm

Fig. 9 shows the training block for a hybrid network used MLP architecture. Although the training block of a RB network is similar (Freeman & Skapura, 1991), the propagation is different. Pattern **P** is obtained from the data adaptation algorithm and is used for training the hybrid network. The first samples of the state variables **P** are considered as the system's initial values and then the input values are propagated - or their equivalent in the case of the RB network. The system output **Y**, which is calculated for that input sample, enables the calculation of the corresponding errors between **Y** and the train pattern **T**, which are stored in a delta error (**Ed**), however the network's weights are left unchanged. Then the next sample is obtained and the state variables are replaced with the output **Y** calculated previously and result in a new propagation. The cycle continues until the end of the epoch (**i=nsample**). Once the epoch reaches its end, the training error **E** is calculated, the parameters adjusted (weights and bias) and the training parameters are updated (momentum and learn rate).

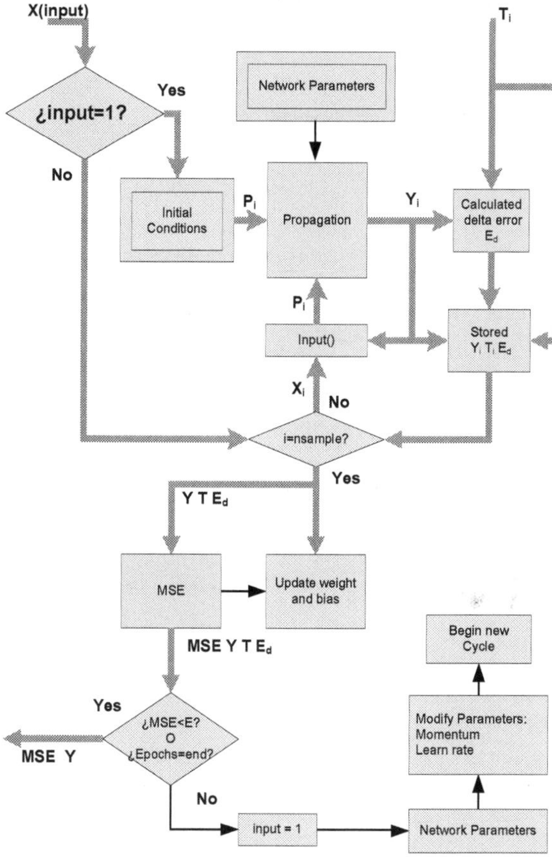

Figure 9. Training Algorithm for Hybrid Network

The process is interrupted when the target error ***E*** is attained or the maximum number of epochs (***end***) is reached. The training process is considered successful only if ***Error*** is lower than E, and trainings that reach ***end*** epochs are re-run with adjusted parameters (momentum, learning factor, number of neurons for the hidden layer, etc.).

The specific training algorithms for the decoupled and daisy chain architectures, both for MLP and RB networks, are adaptations of this generic description, defined in Section 3.1.

5.2 Simulation Algorithm

Figure 10 shows the decoupled simulation algorithm for a generic network, which can be applied to attitude or position networks (the only difference is the input vector **X**). The algorithm will run until the final condition is reached (***while(input exists)***), which is the same as saying that it will run until it iterates through all the input data. The algorithm is capable of running with real-time input data.

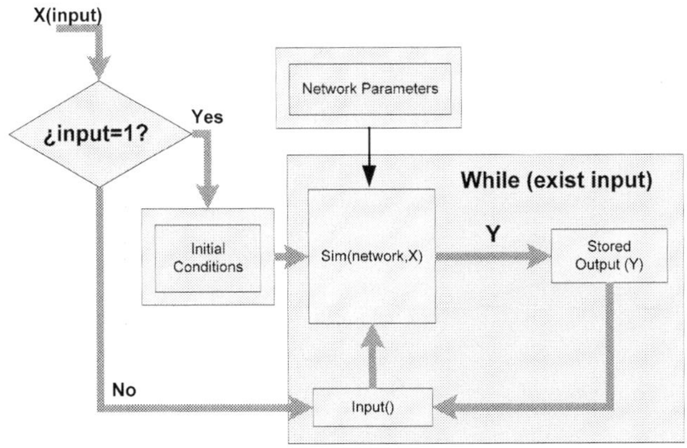

Figure 10. Simulation Algorithm for hybrid network

The first sample (***input=1***) sets the initial conditions, which are determined empirically when running in real-time with real-flight data or obtained from the input file when running with stored flight data. After this initialisation step, the system iterates through the simulation process (***sim(network,input)***) and the output vector **Y** is stored in a matrix, to be used as input in successive iterations.

6. Test and Results

The training patterns were built with data from 9 flight sessions that adhere to the quality and form criteria. Each session has approximately 3 minutes of high-quality flight data. The system identification process will be used to model complete flights and the 5 flight stages (see Fig. 8). The objective is to have different models for each type of manoeuvre and compare the performance of the training architectures and network types (MLP vs. RB).

6.1 Complete Flight vs. Flight stages

The dynamic behaviour of the helicopter is different for each one of the flight stages described in Section 4. For example, the ground effect is present during take-off and landing procedures but is nonexistent above a certain altitude. This is why it is necessary to differentiate between flight stages and to analyse the performance of *universal* models vs. groups of models specialised in different stages.

Fig. 11 shows the results for attitude simulation with MLP (Fig. 11.a) and RB (Fig. 11.b) based networks, both for complete-flight and flight-stage models. Table 1 compares the performance of complete-flight models with the average performance of flight-stage models for three stages (take-off, manoeuvres, landing). Finally, Table 2 shows the mean square error (MSE) for the flight-stage models whose average appears in Table 1.

Table 1 shows that the differences in performance observed between complete-flight and flight-stage models for the attitude simulation are significant. Thus, this experiment has not been repeated for the position simulation since the attitude errors will be propagate.

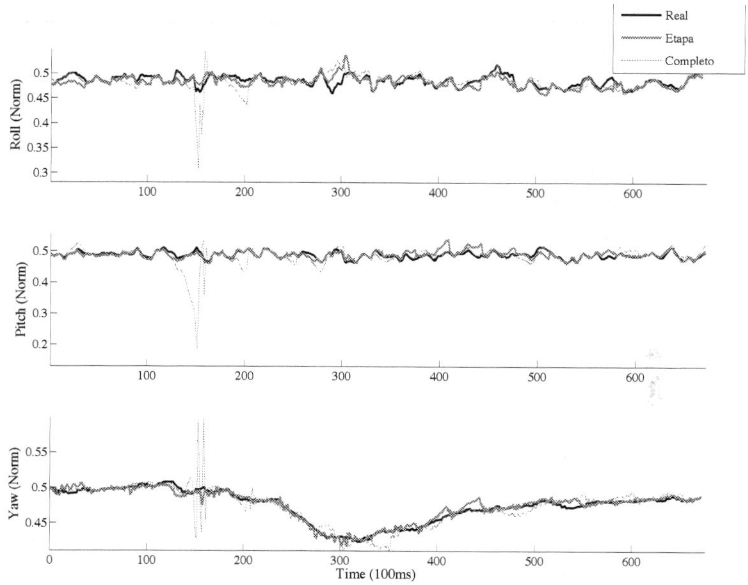

Figure11.a Attitude simulation with MLP

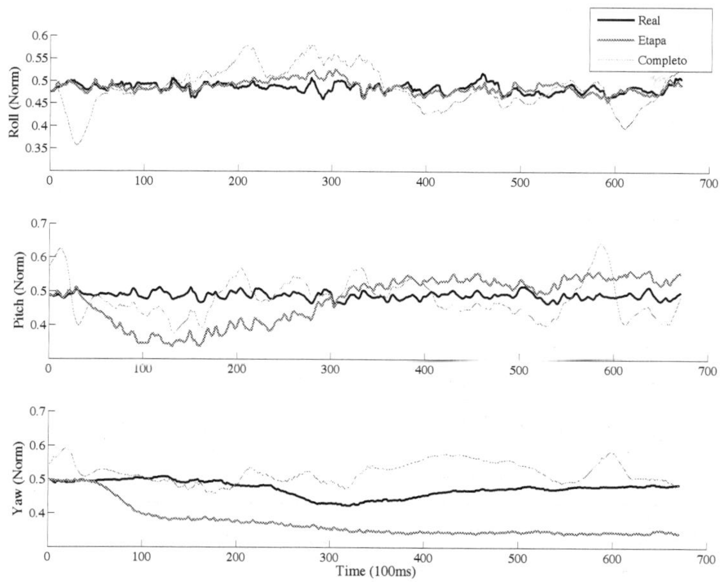

Figure 11.b Attitude simulation with RB

	Complete-flight	Average flight-stages
Attitude	MLP	MLP
Roll	2,27E-04	4,17E-06
Pitch	7,48E-04	2,82E-06
Yaw	1,08E-04	3,51E-06
	RB	RB
Roll	4,56E-03	3,47E-07
Pitch	2,44E-03	3,00E-07
Yaw	5,91E-03	4,23E-07

Table 1. Comparison MSE for complete-flight models with flight-stage models

	Take-off	Manoeuvres	Landing
Attitude	MLP	MLP	MLP
Roll	4,17E-6	8,53E-5	3,13E-6
Pitch	2,82E-6	9,57E-5	4,76E-6
Yaw	3,51E-6	2,63E-5	2,81E-6
	RB	RB	RB
Roll	3,47E-7	1,72E-4	2,34E-6
Pitch	3,00E-7	5,01E-3	2,27E-6
Yaw	4,23E-7	1,18E-2	6,14E-7

Table 2. MSE flight-stage models (take-off, manoeuvres and landing)

6.2 Radial Basis vs. Multi-layer Perceptron

Radial Basis and MLP networks were compared for the same flight stage. Fig. 12.a shows the attitude simulation for both networks versus the real attitude. Fig. 12.b shows the position simulation for each network during take-off versus the real position.

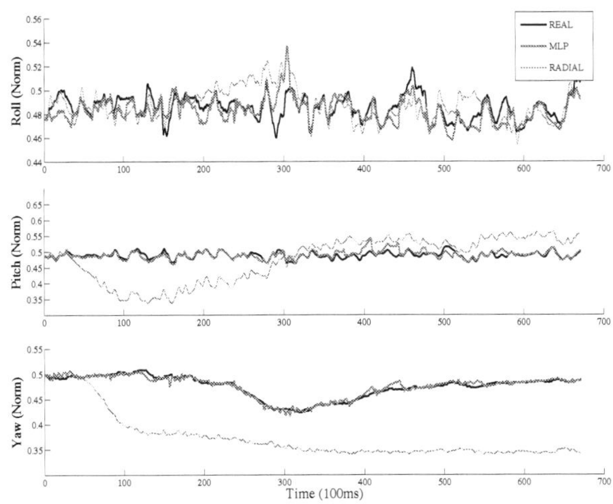

Figure 12.a. Real flight attitude vs. simulated flight attitude with MLP and RB

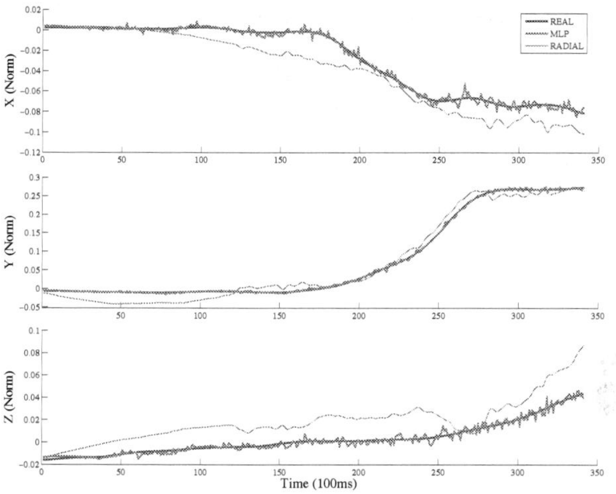

Figure 12.b. Real takeoff position vs. simulated takeoff position with MLP and RB

Table 3 summarises the errors for the degrees of freedom of attitude and position, respectively.

	Takeoff		Flight		Landing	
Attitude	MLP	Radial	MLP	Radial	MLP	Radial
Roll	4.17E-06	3.47E-07	8.53E-05	1.72E-04	3.13E-05	2.34E-05
Pitch	2.82E-06	3,.00E-07	9.57E-05	5.01E-03	4.76E-05	2.27E-05
Yaw	3.51E-06	4.23E-07	2.63E-05	1.18E-02	2.81E-05	6.14E-06
Position	MLP	Radial	MLP	Radial	MLP	Radial
X	8.13E-06	2.08E-04	2.06E-04	2.53E-01	8.66E-05	6.76E-05
Y	6.69E-06	3.39E-04	1.13E-04	6.37E-01	6.25E-05	9.25E-05
Z	7.33E-06	3.20E-04	1.09E-04	7.16E-03	8.43E-05	4.17E-05

Table 3.MSE for RB and MLP

In general the MLP shows a better performance, even though the RB signal tracking is remarkable. However the MLP requires 18 hours of training while the RB is trained in minutes. It is important to note that each simulation step is 'instantaneous' because it only requires few matrix multiplications.

It must be noted that RB networks have a better performance for local approximations (Chen et al., 1991) (Craddock & Warwick, 1996), which is confirmed by the data shown in Table 3 for the position simulation for take-off and landing.

6.3 Decoupled Training vs. Daisy Chain Training

Both scenarios require a trained attitude network which is the result of the previous sections analysis. The best attitude models are used with MLP or RB networks to simulate the position.

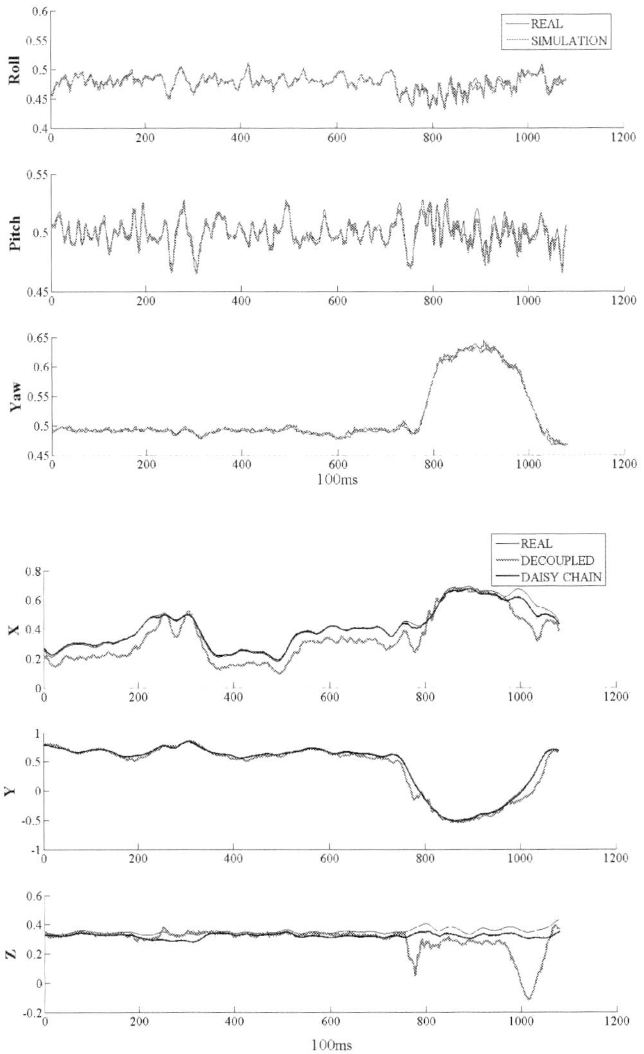

Figure 13. Attitude simulation for flight manoeuvres (Up) and position simulation for flight manoeuvres (down), with Daisy chain and Decoupled architectures

Fig. 13 shows the values of simulated versus real data for both systems (control-signals-to-attitude and attitude-to-position) for the two training architectures. Note that for the attitude simulation no distinction is made between the two architectures because the results are equivalent. Table 4 shows the quantitative results in terms of the MSE for six different flights sessions. It is important to note that these sessions are *not* the ones used for training. Both the graphic and quantitative results show that the daisy chain architecture has a better performance.

	Flight A	Flight B	Flight C	Flight D	Flight E	Flight F
Daisy chain						
X	1.50E-04	1.13E-04	2.13E-04	2.06E-04	1.25E-04	9.13E-05
Y	2.36E-04	1.44E-04	4.55E-04	1.13E-04	8.34E-05	1.49E-04
Z	8.58E-05	9.38E-05	8.23E-05	1.09E-04	1.09E-04	1.06E-04
Decoupled						
X	5.85E-03	2.00E-04	6.08E-03	5.99E-03	1.97E-04	3.87E-03
Y	1.28E-04	3.28E-03	6.35E-04	1.19E-03	9.20E-04	1.09E-03
Z	9.84E-02	3.77E-04	1.73E-02	6.85E-04	1.12E-04	2.55E-03

Table 4. Comparison of MSE for real vs. simulated position, for different flight sessions (only the flight manoeuvre stage is considered)

7. Conclusion

The main objective of this study was to show a new method for identifying dynamic and complex systems as in UAVs, by means of supervised neural networks. The methods described and the results obtained confirm the viability and performance of this alternative method.

It was observed that even though a decoupled architecture has a lower training error, the daisy chain architecture has better simulation performance. This is due to the fact that a network trained with a decoupled architecture does not consider the error propagated from the attitude simulation to the position model.

Empirical results show that the helicopter is modelled better if different flight stages are treated independently. One of the most significant variations in the vehicle dynamics is ground effect, a variation that cannot be accommodated by a single neural network. This means that different models are needed not only for take-off, landing and free flight, but also for particular manoeuvres in different physical and weather conditions.

In general, a better performance was observed with MLP, although the signal tracking achieved by the Radial Base network is remarkable. The main difference lies in the error rate, which in this case is better for MLP. But while an RB network needs 30 minutes training and yields the results mentioned above, MLP needs 18 hours.

It must be taken into account that Radial Basis networks have a better performance for local approximations, as shown in the fig. 6.2.b and table 3 corresponding to take-off and landing. On the other hand, global approximations are better with MLP. This is also shown in the flight versus simulated data figures, where MLP exceeds the error rates of the Radial Basis networks.

The importance of this work lies in the confirmation of neural networks as a valid tool for dynamic system identification. Although this is only a first approach, the results obtained with this identification are very encouraging and an excellent basis for future work, where each flight stages have its hybrid networks like an fuzzy systems models each flight stages in Sugeno works (Nguyen & Prasad, 1999).

8. References

Chen, S.; Cowan, C.F.N. & Grant, P.M. Orthogonal Least Squares Learning Algorithm, for Radial Basis Function Networks: *IEEE Transactions on neural networks*, Vol.2, n°2, March 1991.

Craddock, R.J. & Warwick, K. Multi-Layer Radial Basis Function Networks An Extension to the Radial Basis Function: *IEEE International Conference on Neural Networks*, 1996.

Demuth, H. & Beale, M. Neural Networks TOOLBOX@ Version 4.0, The Math Works, USA, 2004.

Elman, J. Finding structure in time: *Cognitive Science*, Vol. 14, pp. 179-211, 1990.

Freeman, J.A & Skapura, D.M. Neural Networks. Algorithms Applications, and Programming Techniques, Addison-Wesley, Massachusetts, USA, 1991.

Hopfield, J. Neural networks and phisycal systems with emergent collective computational abilities: *Proceedings of the National Academy of Science, National Academy of Sciences*, Vol.81, pp. 3088-3092, 1982.

Jordan, M. Serial order: A parallel distributed processing approach: *Technical report, Institute for Cognitive Science*. University of California, USA, 1986.

Li, J.H.; Michel, A.N. & Porod, W. Qualitative Analysis and Synthesis of a Class of Neural Netwroks: *IEEE Trans. Circuits Syst.*, Vol.35. N° 8, Aug.1988.

Lopez, J.L. Helicópteros: Teoría y diseño conceptual, ETSI Aeronáuticos, Madrid, Spain, 1993.

Narendra, K.S. & Parthasarathy, K. Identification and Control of Dynamical Systems Using Neural Networks: *IEEE Transactions on neural networks*, Vol.1 N° 1, March 1990.

Nguyen, H.T. & Prasad, N.R. Fuzzy modelling and control, Selected works of M.Sugeno, CRC Press, 1999.

Norgaard, M.; Ravn, O.; Poulsen, N.K. & Hansen, L.K. Neural networks for Modelling and Control of Dynamic Systems, Springer-Verlag, London, UK, 2001.

Novatel, OEM4 Family of Receivers: User Manual- Volume 1 & 2. Installation and Operation, NovAtel Inc., Canada, 2002.

OSD (Office of the Secretary of Defense), *Unmanned aircraft systems (UAS) roadmap, 2005-2030*, Secretary of Defensa, USA, 2005.

29

An Evasive Maneuvering Algorithm for UAVs in Sense-and-Avoid Situations

David Hyunchul Shim
KAIST
South Korea

1. Introduction

Highly autonomous unmanned aerial vehicles (UAVs) will need advanced flight management systems that will actively sense the surrounding environment and make a series of intelligent decisions to accomplish the given mission with minimum intervention from remotely located human operators. In near future, it is expected that UAVs will be found as a ubiquitous surrogate for manned vehicles in such fields as airborne sensing, payload delivery, and ultimately aerial combat. In that process, UAVs must be integrated into civilian or military airspaces along with other manned and unmanned aerial vehicles. However, such level of autonomy is yet to be fully developed. Reportedly, a German tactical UAV named LUNA had a close encounter with an Afghan Airline A300B4 in the sky over Kabul, Afghanistan on August 30, 2004. Attributed to a failure of the nearby air traffic control tower to follow standard procedures, two vehicles occupied the same airspace at the same time, no farther than 50 meters when closest. The UAV operator managed to command an evasive maneuver just a split second before impact. The strong wake of A300B4 blew the UAV into an unrecovered dive as seen by the onboard video system in Fig. 1. As exemplified in this rare but alarming event, the collision avoidance has to be incorporated into the flight management system especially when the vehicle is flying in a crowded airspace or at low altitudes where many obstacles such as terrain and buildings pose threat to safe flight.

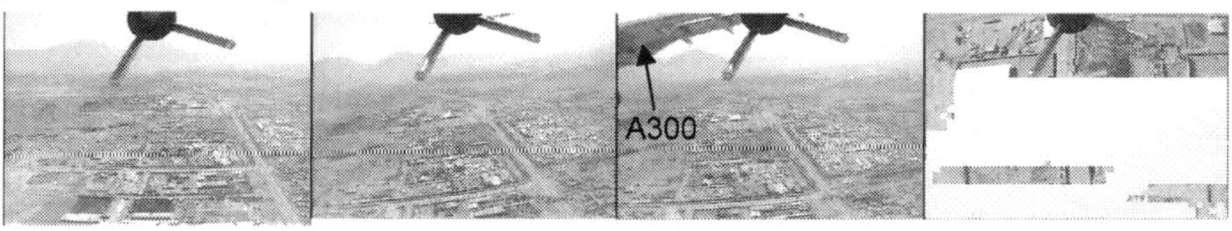

Figure 1. A near-miss incident of a UAV and A300 airplane (August 2004)

There are also increasingly many occasions that UAVs have to fly at a lower altitude where they are not free from collisions from obstacles such as terrain, buildings or power lines. In order for a UAV to avoid any imminent collision with other vehicles or such obstacles, it should be capable of sensing and tracking of objects, collision prediction, dynamic path planning and tracking. When the trajectories of objects on potential collision courses are predicted, a collision-free trajectory should be computed in real-time. There are a number of research results for real-time path planning (Bellingham et al, 2003; Dunbar et al, 2002; Milam et al, 2002). In the context of emergency evasive maneuver, however, one would expect that the vehicle may need to maneuver at its full dynamic capability, i.e., maximum turn rate, acceleration/deceleration, or climb/descent. In such cases, the inputs to control surfaces may saturate or the vehicle states, such as roll angle or cruise velocity, may reach the acceptable limits. In order to compute a plausible trajectory that the vehicle can actually fly along without exceeding its dynamic range, a proposed method should be capable of taking such limits into account when computing an evasion trajectory. In this article, we introduce a nonlinear model predictive control (NMPC) based approach, which can be applied to nonlinear dynamic systems with state constraints and input saturation, unlike most control theories available as now. One drawback of MPC is, as often pointed out, the heavy numerical load, which is now considered well within the reach of the latest CPU technology as demonstrated in (Shim & Sastry, 2006).

In this article, we present an MPC-based collision avoidance algorithm for safe trajectory generation and control of constrained nonlinear dynamic system with input saturation in real-time. We also introduce an active sensing method using a laser scanner. We consider a number of scenarios with moving vehicles or obstacles in the surroundings. The proposed approach is validated by a series of realistic simulations and experiments including a head-on collision and a flight in an urban canyon.

2. Real-time evasive Maneuvering using Model Predictive Control

In this section, we present the formulation of an NMPC-based approach for real-time safe trajectory generation during an evasive maneuver for avoiding collision. We consider scenarios that, when a UAV flies to a given destination, a collision with nearby flying or stationary obstacles are anticipated. The position information of obstacles is assumed to be directly measured or available from other sources including active communication with cooperating agents or an eye-in- the-sky.

2.1 NMPC Formulation

Suppose we are given a nonlinear time-invariant dynamic system such that

$$x(k+1) = f(x(k), u(k)) \tag{1}$$

$$y(k) = g(x(k)) \tag{2}$$

where $x \in X \subset \mathbb{R}^{n_x}, u \in U \subset \mathbb{R}^{n_u}$. The optimal control input sequence over the finite receding horizon N is obtained by solving the following nonlinear programming problem:

$$\text{Find } u^*(k),\ k = i, \ldots, i+N-1 \text{ such that} \tag{3}$$

$$u^*(k) = \arg\min V(x,k,u)$$

where

$$V(x,k,u) = \sum_{i=k}^{k+N-1} L(x(i),u(i)) + F(x(k+N)) \tag{4}$$

where $L(x,u)$, is a positive definite cost function and F is the terminal cost. Herein, $u^*(k),\ k=i,...,i+N-1$ is the optimal control sequence that minimizes $V(x,k,u)$ such that $V^*(x,k)=V(x,k,u^*(x,k))\le V(x,k,u),\ \forall u(k)\in U$. The cost function term L is chosen as

$$L(x,u)=\frac{1}{2}\left(x^r-x\right)^T Q\left(x^r-x\right)+\frac{1}{2}u^T Ru+S(x)+\sum_{l=1}^{n_o}P(x,\eta_l) \tag{5}$$

The first term penalizes the deviation from the original course. The second term penalizes the control input. $S(x)$ is the term that penalizes any states not in X as suggested in (Shim et al, 2003). Finally, $P(x_v,\eta_l)$ is to implement the collision avoidance capability in this NMPC framework: $P(x_v,\eta_l)$ is a function that monotonically increases as $\| x_v-\eta_l \|_2\rightarrow 0$, where $x_v\in\mathbb{R}^3$ is the position of the vehicle and η_l is the coordinates or l-th out of total n_o obstacles being simultaneously tracked.
The control input saturation can be facilitated by enforcing

$$u_i(k)=\begin{cases} u_i^{\max} & if\ u_i>u_i^{\max} \\ u_i^{\min} & if\ u_i<u_i^{\min} \end{cases} \tag{6}$$

during optimization. In this manner, one can find the control input sequence that will be always within the physical limit of the given dynamic system. We employ the optimization method based on indirect method of Lagrangian multiplier suggested in (Sutton & Bitmead, 2000).

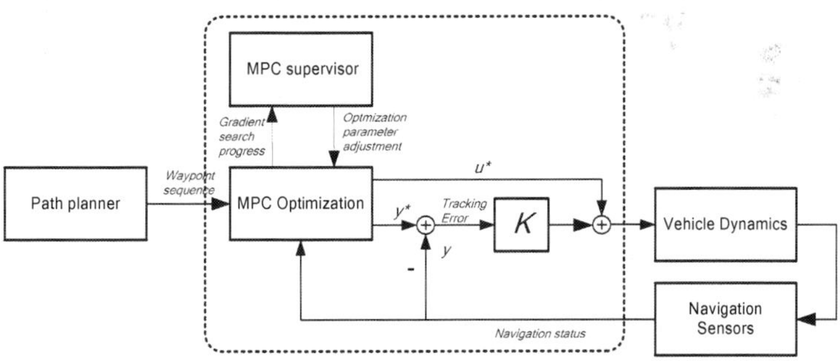

Figure 2. Flight control system architecture with MPC and explicit feedback loop

When an optimal control sequence is found at each epoch k, the control law is given as

$$u(k)=u^*(k)+K\left(y^*(k)-y(k)\right) \tag{7}$$

where K is a explicit feedback control gain, which can be found by approaches such as in (Shim, 2000). With $u^*(k),\ k=i,...,i+N-1$, one can find $y^*(k)$ by solving recursively the given nonlinear dynamics with $x(i)=x_0(i)$ as the initial condition. Ideally, if the dynamic model used in the prediction in the optimization problem is identical to the actual dynamics and there is no disturbance, the plant would behave as predicted. In the real world, such assumptions cannot be justified due to the inevitable model mismatch, disturbance, and many other reality factors satisfied. Therefore, with a tracking feedback controller in the feedback loop, the system can track the given trajectory reliably in the presence of disturbance or modeling error. The architecture of the proposed flight control system is given in Fig. 2.

2.2 Obstacle Sensing

For effective collision avoidance, it is very important to detect the location of obstacles of concern in an accurate and timely manner. Such information can be supplied by a preloaded map or information sent by other cooperative neighboring vehicles. However, such information may not be accurate, up-to-date, or always available if communication is lost. A local sensing is favored since it can provide up-to-second information. Obstacle detection can be done using active or passive sensors such as radar, laser scanners, or mono or stereo cameras and the choice depends on many factors such as operating condition, accuracy, and detection range. The laser scanning computes the distance to an object by measuring the time of flight (TOF) of the laser beam to make a round trip from the source to the reflected point on an object. The operation is straightforward and the measurement is very accurate, so it is suitable for short-range detection. However, as the detection range depends on the intensity of the light that radiates from the laser source, the range is limited by the maximum allowable intensity of the beam. Active radar has similar attributes since it operates in a similar principle. The resolution of radar sensing depends on the wavelength of radio wave used. Recently gigahertz-range radars are often used for its more accurate imaging capability at the expense of shorter detection range. Active sensing may not be desired when a covert operation is required.

Camera-based detection is attractive as it is a passive detection and the imaging device is usually much cheaper and smaller than comparable radar or laser sensors. However, cameras do not directly give the ranging information. A stereo camera system may be used to measure the distance by the parallax, but it is useful only when the objects are close enough. Optic flow can be also used for short-range detection (Rydergard, 2004; Hrabar, 2005). For long-range detection, the pixel area occupied by the obstacle can be the only visual cue to sense the existence and range. The resulted accuracy is usually much lower than the active sensing methods mentioned above. In this article, we choose to use a laser scanner for obstacle sensing.

A laser ranging sensor consists of a laser source, a photo-receptor and a rotating mirror for planar scanning. An accurate timing device measures the time lapse from the moment the laser beam is emitted to the moment the laser beam reflected on an object returns to the receptor. A rotating mirror reflects the laser beam in a circular plane, allowing for two-dimensional scanning. At each scan, the sensor reports a set of measurements that supplies the following measurement set:

$$Y_L = \{(d_n, \beta_n), n = 1, \ldots, N_{meas}\} \tag{8}$$

where d_n, β_n and N_{meas} represent the distance from an object, the angle in the scanning plane, and the total number of measurements per scan, respectively. Each measurement, i.e., the relative distance from the laser scanner to a scanned point in the laser-scanner coordinate system, can be written into a vector form such that

$$\mathbf{X}^L_{D/L}|_n = d_n\left(\cos\beta_n \mathbf{i}^L + \sin\beta_n \mathbf{j}^L\right) \tag{9}$$

where $\mathbf{i}^L$ and $\mathbf{j}^L$ are orthonormal unit vectors in X^L and Y^L directions on the scanning plane, respectively. *D*, *L*, *B*, and *S* represent scanned data, laser scanner, vehicle body coordinate system, and spatial coordinate system, respectively as shown in Fig. 3.

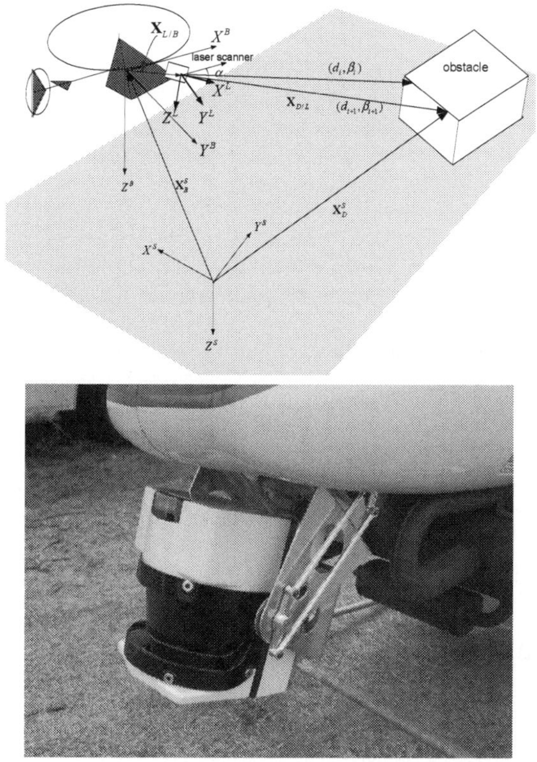

Figure 3. Coordinate transformation for laser scan data (left), a scanning laser mounted on a helicopter UAV (right)

The calculation of the spatial coordinates of detected points involves a series of coordinate transformations among three coordinate systems: body coordinate systems attached to the laser scanner and to the host vehicle, and the spatial coordinate system, to which the vehicle location and attitude are referred.

Each measurement vector in the laser scanner-attached coordinates is first transformed into the vehicle body coordinates and then the spatial coordinate system as following:

$$\begin{aligned}\mathbf{X}^S_{D/L} &= \mathbf{R}^{S/L}\mathbf{X}^L_{D/L} \\ &= \mathbf{R}^{S/B}\mathbf{R}^{B/L}(\alpha)\mathbf{X}^L_{D/L}\end{aligned} \tag{10}$$

$\mathbf{R}^{B/L}(\alpha)$ is the transformation matrix from the laser body coordinate L to vehicle body coordinate B where α is the tilt angle with respect to the vehicle body coordinate system.

$\mathbf{R}^{S/B}$ denotes the transformation matrix from vehicle body coordinates to spatial coordinates. Finally, the spatial coordinate of the obstacle is found by:

$$\begin{aligned}\mathbf{X}_D^S &= \mathbf{X}_{D/L}^S + \mathbf{X}_{L/B}^S + \mathbf{X}_B^S \\ &= \mathbf{R}^{S/B}\mathbf{R}^{B/L}(\alpha)\mathbf{X}_{D/L}^L + \mathbf{R}^{S/B}\mathbf{X}_{L/B}^B + \mathbf{X}_B^S\end{aligned} \quad (11)$$

Using (11), one can find the spatial coordinate of sampled points on obstacles by combining the raw measurement vector with the position, heading, and attitude of the vehicle, which are available from the onboard navigation system of the UAV. It should be noted that the detection accuracy in the spatial coordinate system not only depends on the laser scanner's accuracy itself, but also on the accuracy of the vehicle states.

To ensure conflict-free navigation in an airspace filled with obstacles, the laser scanner should scan the surroundings wide enough to find conflict-free trajectory. For example, if the laser scanner is installed to scan the area horizontally, an actuation in the pitch axis is necessary so that the scanner can cover the frontal area sufficiently higher than the rotor disc plane and lower than the landing gear. Fig. 3 shows an actuated laser scanner mounted on a helicopter UAV (Shim et al, 2006). The scanner is mounted on a tilt actuator with an encoder, which provides the tilt angle α in $\mathbf{R}^{B/L}(\alpha)$. Fig. 4 shows visualizations of laser scan data, which is obtained by (11). As can be seen, the shapes of objects can be accurately detected and reconstructed.

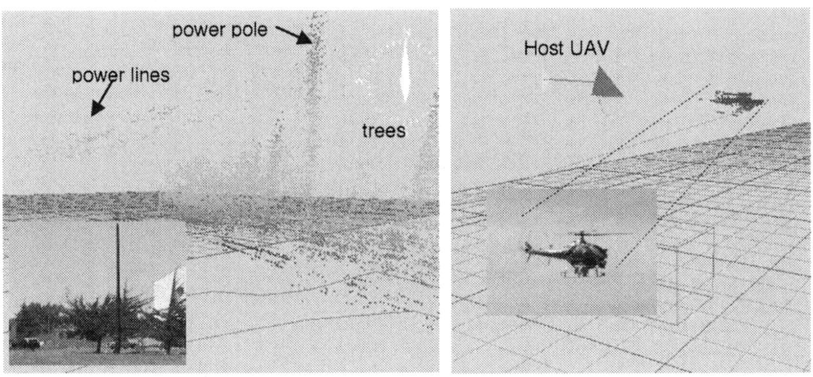

Figure 4. Point cloud of obstacles sensed by a laser scanner shown in Fig. 3 (left: scanning of ground-based objects of area shown in the inset image; right: scanning of a UAV airborne)

2.3 Trajectory Generation

For collision avoidance, we choose $P(x_v, \eta_l)$ in (5) such that

$$P(x_v, \eta_l) = \frac{1}{(x_v - \eta_l)^T G(x_v - \eta_l) + \varepsilon} \quad (11)$$

where G is positive definite and $\varepsilon > 0$ is to prevent ill conditioning when $\|x_v - \eta\|_2 \to 0$. One can choose $G = diag\{g_x, g_y, g_z\}$, $g_i > 0$ for an orthogonal penalty function. The penalty function (11) serves as a repelling field and has nonzero value for entire state space even when the vehicle is far enough from obstacles. The crucial difference from the potential field approach here is that we optimize over a finite receding horizon, not only for the current time as in the potential field approach. For obstacle avoidance, we consider two types of scenarios: 1) a situation when the vehicle needs to stay as far as possible from the obstacles even if no direct collision is anticipated and 2) a situation when the vehicle can be arbitrarily close to the obstacle as long as no direct conflict is caused. For the second situation, (11) can be enacted only when $\|x_v - \eta\|_2 \to \sigma_{\min}$, where $\sigma_{\min}$ is the minimum safety distance from other vehicles.

Since MPC algorithms optimize over the receding finite horizon into future, the predicted obstacles' trajectory over $k = i,...,i+N-1$ is needed in (11). It is anticipated that the inclusion of predicted obstacle locations in the optimization will produce more efficient evasion trajectory if the prediction is reasonably accurate. If the obstacle detection system is capable of estimating the current velocity in addition to the position of an obstacle, one can predict $\eta_l(k)$ by extrapolating it over N_p steps, namely *prediction Horizon*, using an equation such that

$$\eta_l(k+i) = \eta_l(k) + \Delta t v_l(k)(i-1) \tag{12}$$

It is noted that the prediction can be done in more elaborated way using a Kalman filter (Watanabe et al, 2005) if the dynamic characteristics is known at least partially in advance.

In this research, we propose a dual-mode strategy for the MPC-based collision avoidance system. In normal condition, we choose a parameter set that achieves good tracking performance. When the obstacle prediction algorithm using (12) predicts that a bogey may approach the host vehicle's future position within a cautionary margin σ_c such that $\|\eta_l(k+N_p) - y(k+N_p)\| < \sigma_c$, the MPC-based controller is switched to the evasion mode. The parameter set in (5) is then tuned for effective evasive maneuver to generate a conflict-free trajectory by lowering penalties on the large deviation (=tracking error) from the original course or an aggressive maneuver with large control effort if necessary. The control effort is also less penalized to allow for more aggressive maneuver. This approach is illustrated in Fig. 5. Optionally, if the predicted future trajectories of the host vehicle and bogeys get closer within the absolute safety margin $\sigma_a < \sigma_c$, the proximity penalty gain can be increased to allow for more clearance margins.

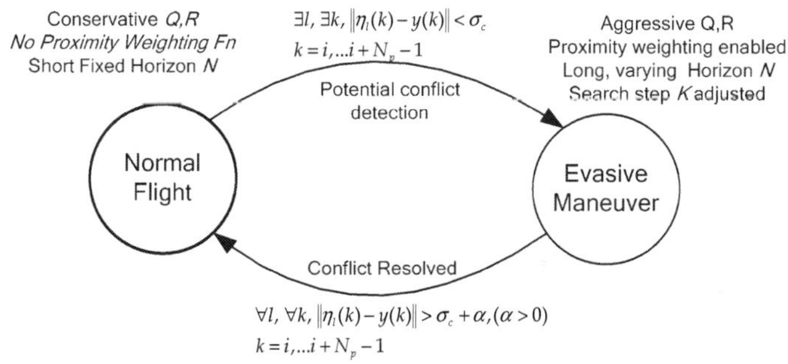

Figure 5. State transition diagram for flight mode switching algorithm

In the following section, we apply the proposed MPC algorithm for (a) one vehicle versus a non-cooperating vehicle and (b) one vehicle in an environment with obstacles.

3. Simulation and experiment results

3.1 One on One Situation

In this scenario, we consider a UAV cruising at 3m/s at 10 meters above the ground. Without loss of generality, we use a dynamic model for a rotorcraft UAV based on Yamaha R-50 industrial helicopter (Shim, 2000), whose specification is given in Table 1. The bogeys are staged to moves along a straight line at a constant altitude and speed at various incident angles. The detection range is simulated to be 50 meters based on a typical laser scanner and 100 meters for a hypothetic vision-based system. We investigate a fraction of these combinations of the factors mentioned above, which would highlight the performance of the proposed approach so that we may have the insight to the behavioral patterns and characteristics of the algorithm with a realistic detection.

In this scenario, we consider the case when a UAV encounters a bogey at various speed and incident angle. The horizon N is set to 100 with 40 ms of sampling time, so the prediction horizon spans over 4 seconds. For fixed obstacles, stationary obstacles 12 meters away can be considered in the optimization when cruising at 3 m/s. As expected, the moving obstacles will impose more challenges in detection and finding a safe evasion trajectory in a short time.

First, we consider the following cases: a bogey cruising towards the UAV at 2 m/s, 5 m/s, 15 m/s and 30m/s. The cautionary margin $\sigma_c = 50\,\text{m}$ and the absolute safety margin $\sigma_a = 10$ m. We judge the vehicles collide when the distance from each other is less than 5 m.

In Fig. 6, an example when a bogey closes in at 10 m/s, with 0° incident angle (head-on collision). As can be seen in the figures, the host UAV maintains sufficient margin, which decreases as low as 8 m/s, well above the minimal distance. For comparison, we consider when the horizon N is much shorter to demonstrate the advantage of the receding horizon approach. The simulation result when N is shortened to 20 (=0.8sec) and all other parameters are fixed as before is given in Fig. 5 as well. The result shows that the UAV manages to escape the collision, but the vehicle goes into a violent transient motion during the close fly-by interval. It is attributed that the short horizon length does not allow a sufficient time to predict the collision and then steer the vehicle away from the collision course. We also note the heading of the vehicle is implicitly determined by the optimization. In the following examples, we consider a set of different approach velocities and incident angles.

In Fig. 7, a number of approach velocities are tested. In Fig. 7-(d), the vehicle passes the bogey with 7m distance, which is considered as a bare minimum. It is expected that a longer horizon length will help to avoid the obstacle with a more sufficient margin. In Fig. 8, the trajectory planner shows a reliable performance in computing safe trajectories when the bogey flies in at various incident angles. In overall, the MPC-based collision avoidance algorithm demonstrates a satisfactory performance in various scenarios.

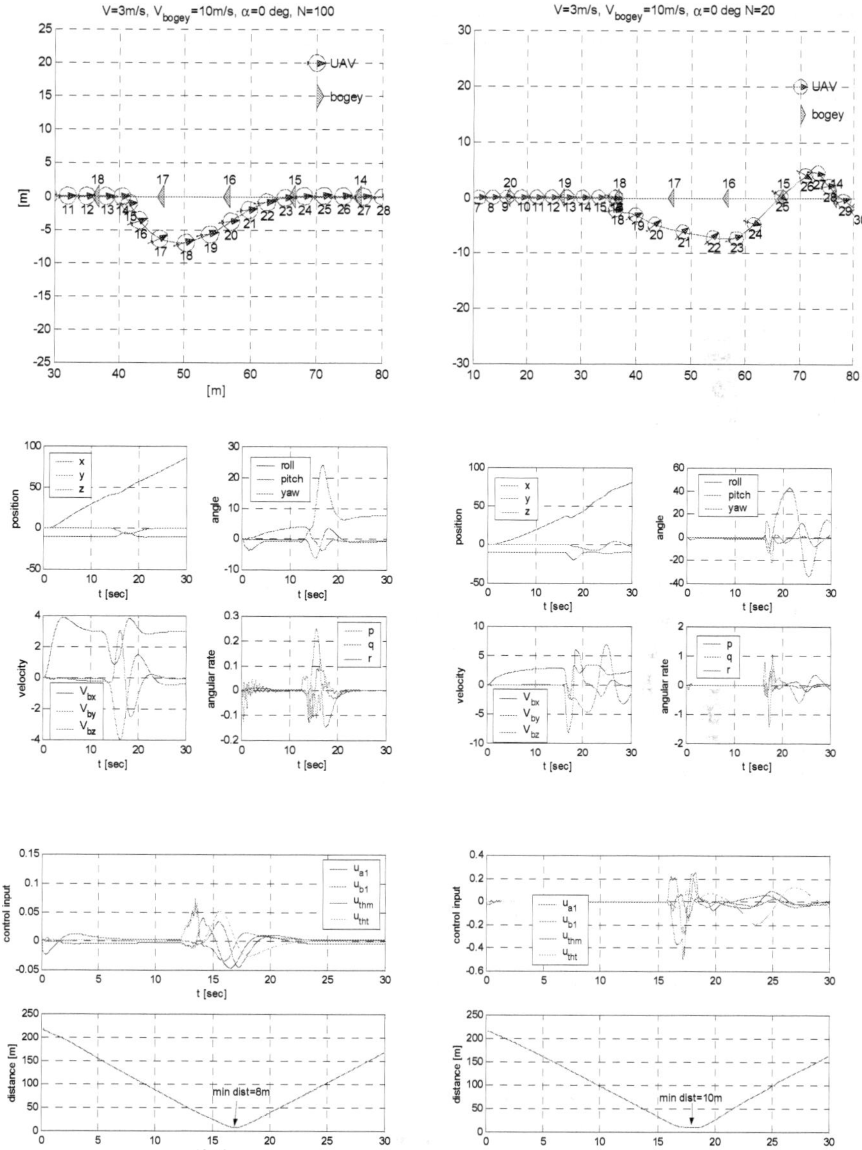

(a) N=100 (b) N=20

Figure 6. A head-on collision scenario with different horizon lengths, N=100 and 20. ($V_{cruise} = 3\text{m/s}$, $V_{bogey} = 10\text{m/s}$, α=0°)

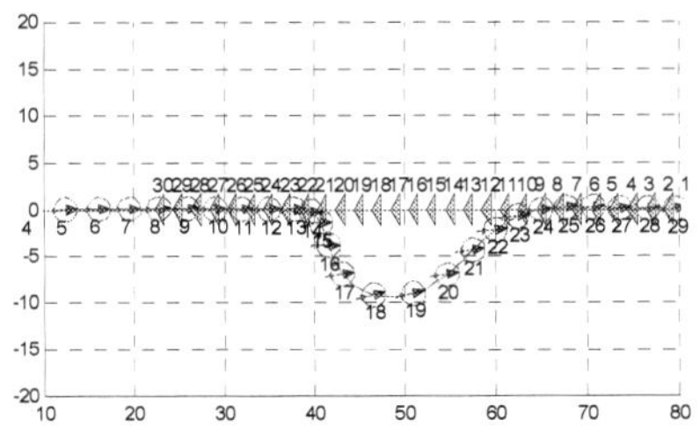

(a) $V_{cruise} = 2\,\mathrm{m/s}$

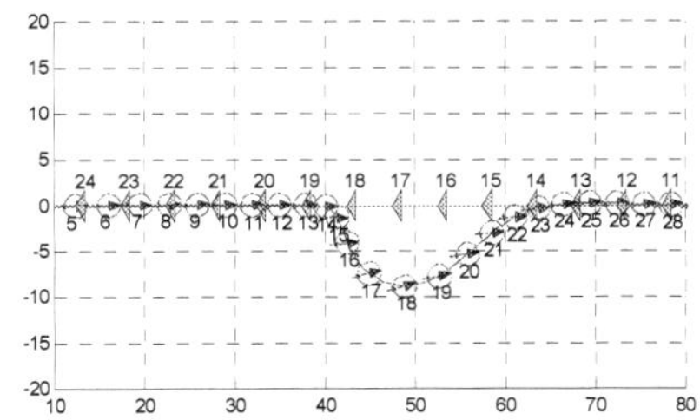

(b) $V_{cruise} = 5\,\mathrm{m/s}$

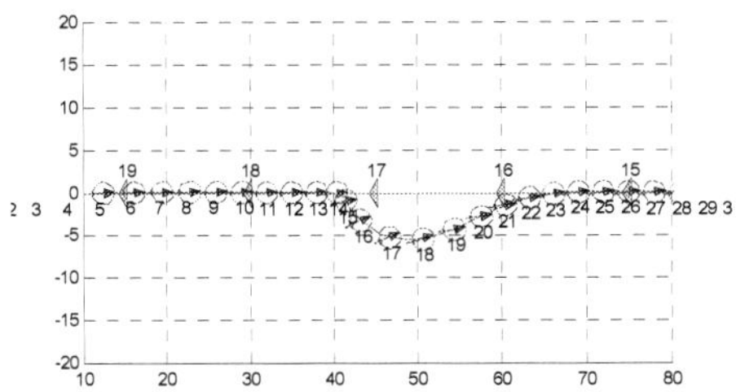

(c) $V_{cruise} = 15\,\mathrm{m/s}$

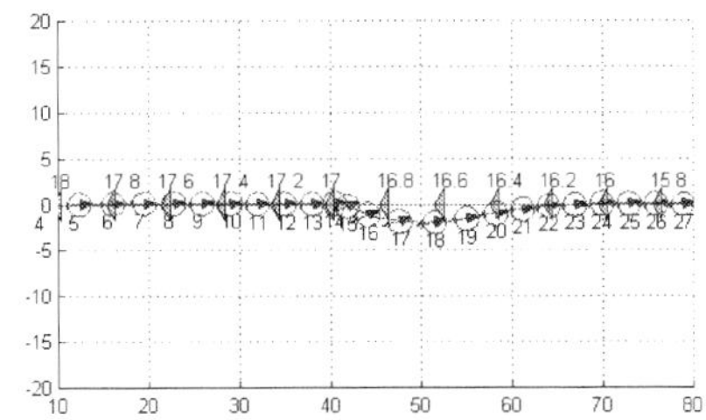

(d) $V_{cruise} = 30\,\mathrm{m/s}$

Figure 7. Various cruise velocity $V_{cruise} = 2, 5, 15, 30\,\mathrm{m/s}$ of host vehicle with $V_{bogey} = 10\,\mathrm{m/s}$

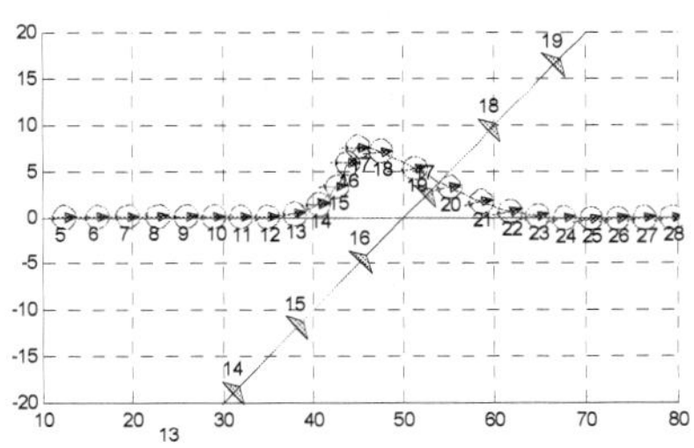

(a) α=45°

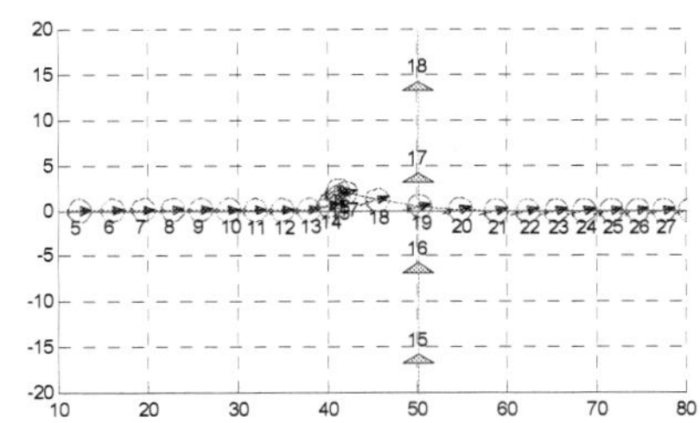

(b) α=90°

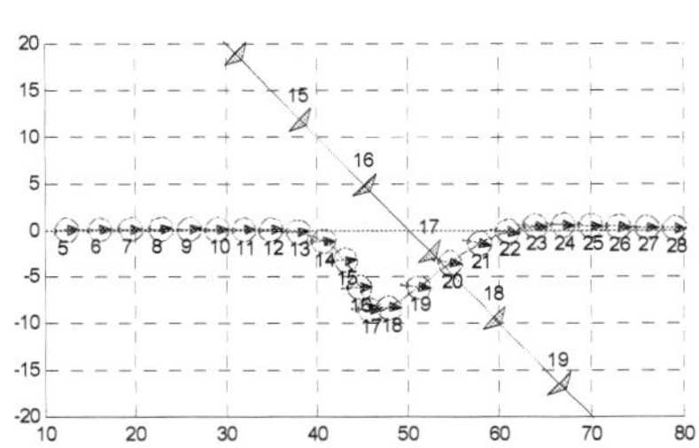

(c) α=135°

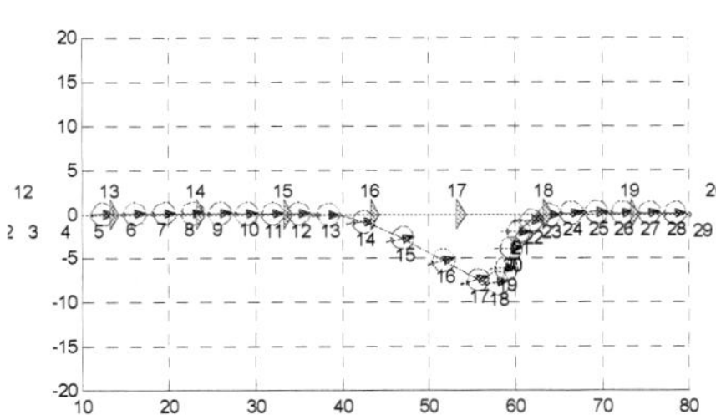

(d) α=180°

Figure 8. Various incident angles α=45°, 90°, 135°, 180°, $V_{cruise} = 3\,\mathrm{m/s}$ with $V_{bogey} = 10\,\mathrm{m/s}$

In order to validate the proposed algorithm experimentally, two helicopter UAVs (Table 1) are deployed in a collision course (Fig. 9 and 10). Two vehicles are initially flown manually 30 meters apart and commanded to trade their position while flying at 1.2 m/s. Then the MPC algorithm running on a notebook computer with Pentium 1.8 GHz CPU computes safe trajectories for each vehicle in MATLAB/Simulink environment. At each sampling time of 100 ms, each vehicle communicates with the centralized trajectory planner but not directly each other over a wireless channel to report the current position and receive a new waypoint. The experiment was performed in four separate occasions and the vehicles could fly to their own destination while avoiding collision. An experiment result set is shown in Fig. 10. It can be seen that the separation in the middle was about 12 meters from center to center of the vehicles and less than 9 meters from tip-to-tip.

Figure 9. Mid-air collision avoidance between two rotorcraft UAVs using real-time MPC

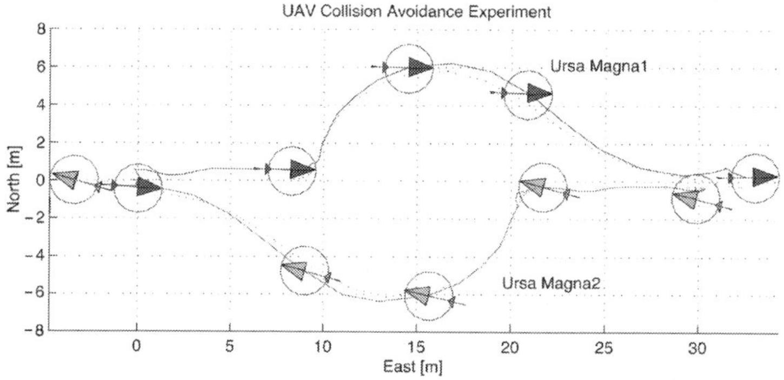

Figure 10. Trajectories of two UAVs Experiment result of Trajectories of of dynamic path planning for collision avoidance

Base platform	Yamaha R-50 Industrial Helicopter
Dimension	0.7 m(W) × 3.5 m (L) × 1.08 m (H)
Rotor Diameter	3.070 m
Weight	44 kg (dry weight) 20 kg (payload including avionics)
Engine	12 hp, 2 cycle air-cooled gasoline engine
Operation Time	Fuel: 40 minutes Avionics: 200 minutes
Onboard Systems	CPU: AMD K6 400MHz PC104 Boeing DQI-NP INS NovAtel GPS MillenRT-2 IEEE 802.11b Wireless Ethernet Ultrasonic altimeters SICK laser range finder (LMS-200)
Capabilities	Preloaded waypoint navigation Interactive waypoint navigation Trajectory tracking mode

Table 1. Specification of a testbed UAV

3.2 Obstacle Avoidance

In this section, we apply the proposed MPC-based algorithm to the navigation problem in a cluttered environment such as urban canyons. An obstacle sensing system is assumed to be combined with the avoidance algorithm. In this scenario, in place of using the current and linearly extrapolated position information of moving obstacles, the position of the nearest obstacle is used in (11). In other words, we need to find $\mathbf{X}_O^{\min}$, the vector from the *reference position* to the nearest point on an obstacle such that

$$\mathbf{X}_O^{\min}(\mathbf{X}_{ref}) = \arg\min_{\mathbf{X}_O \in S_{obs}} \left\| \mathbf{X}_O - \mathbf{X}_{ref} \right\|_2 . \tag{13}$$

Whilst the position of other vehicles can be treated as a point, obstacles cannot be effectively described as a point. Rather, they have complex shapes. Also, as the MPC algorithm solves over the finite horizon, the nearest obstacle from a future position of the vehicle in the prediction session changes. Theoretically, (13) demands a perfect knowledge on all obstacles in the surrounding environment, which assumes an ideal sensor capable of omni-directional scanning with infinite detection range through any other obstacles. Also, during the optimization, a hypothetic sensor should be moving along the trajectory of the state propagation over a finite horizon at each iteration step. Obviously, any realistic sensors would not provide such information. Finally, if the MPC algorithm is used as a reference trajectory generator, due to the inevitable tracking error, the range data is measured at the physical location of the vehicle, not on the reference trajectory. Therefore, in order to provide $\mathbf{X}_O^{\min}$ to the MPC-based trajectory generator during the optimization, it is important to maintain a local obstacle map caching recent measurements from onboard sensors.

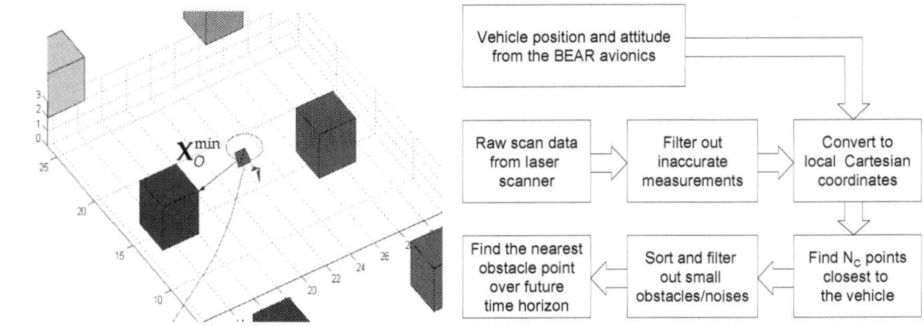

Figure 11. Finding nearest points at each state propagation during prediction (left) and local map building method for the nearest-point approach avoidance (right)

At each scan, the sensor provides N_{meas} measurements of the scan points from the nearby obstacles. Due to the imperfect coverage of the surroundings with possible measurement errors, each measurement set $\mathbf{X}_O^i$ is first filtered, transformed into local Cartesian coordinates, and cached in the local obstacle map repository. A first-in, first-out (FIFO) buffer is chosen as the data structure for the local map, whose buffer size is determined by the types of obstacles nearby. If the surrounding is known to be static, the buffer size can be as large as the memory and processing overheads permit. On the other hand, a more dynamic environment would require smaller buffer to reduce the chance to detect obstacles that may not exist anymore.

In order to solve (13), the measurement set in the FIFO is sorted in ascending order of $\left\|\mathbf{X}_O^i - \mathbf{X}_{ref}\right\|_2$ for all $\mathbf{X}_O^i$ in the local obstacle map, where $1 \le i \le N_C$. Prior to be registered in the database, any anomalies such as *salt-and-pepper noise* should be discarded. Also, the measurements are examined for any small debris, such as grass blades or leaves blown by the downwash of the rotor. Such small-size objects, not being serious threats for safety, are ignored. In order to eliminate these anomalies, we first discard measurements out of minimum and maximum detection range. Then we apply an algorithm that computes a bounding box that contains a series of subsequent points in the FIFO where the distance between the adjacent points in the sorted sequence is less than a predefined length. Then, if the volume of the bounding box is larger than a threshold of becoming a threat, the coordinates of the nearest point in the bounding box is found and used for computing (9). The procedure of the local obstacle map building method proposed above is illustrated in Fig. 11.

The proposed obstacle avoidance algorithm was validated in an actual flight test using the same helicopter UAV used in Section 3.1. The experiment design is carefully scrutinized for the safety: it is performed in a field with portable canopies simulating buildings, not with real ones. The canopies, measuring 3 × 3 × 3 meters each, are arranged as shown in Fig. 11. The distance between one side to the next adjacent side of canopies is set to 10 meters in the north-south direction and 12 meters in the east-west direction so that the UAV with 3.5 meter long fuselage can pass between the canopies with minimal safe clearance, about 3 meters from the rotor tip to the nearby canopy when staying on course.

For validation, an MPC engine originally used for the collision avoidance experiment introduced above is modified for urban navigation problem. The MPC engine is augmented with the local map builder using the laser range finder's data. The MPC with the local map building algorithm is implemented in C language for speed and portability. As shown in Fig. 12, the MPC path planner demonstrated its capability to generate a collision-free trajectory based on the original trajectory with intentional overlapping with buildings.

Figure 12. Aerial view of urban navigation experiment (dashed: given straight path, solid: actual flight path of UAV during experiment)

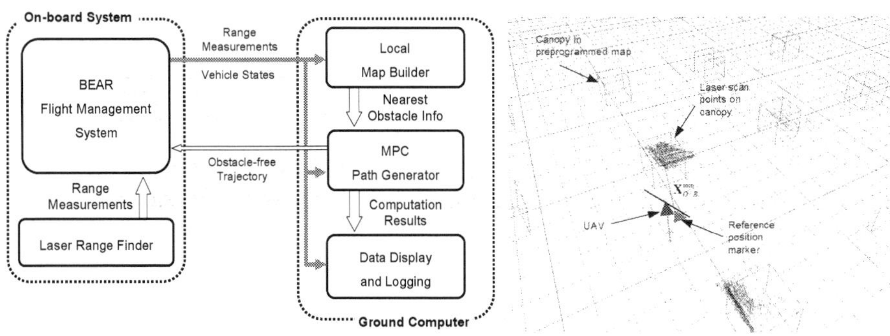

Figure 13. Overall system structure used in the experiments (left) and a snapshot of three-dimensional rendering during an urban exploration experiment (right)

A number of experiments for urban flight were performed. For obstacle detection, the vehicle is equipped with an LMS-200 from Sick AG, a two-dimensional laser range finder. It has 80 meters of detection range with 10 mm resolution. The scanner's measurement is sent to the flight computer via RS-232 and then relayed to the ground station running the MPC-based trajectory generator in Simulink. The trajectory generation module on MATLAB/ Simulink and the ground monitoring/commanding software were executed simultaneously on a computer with Pentium 4, 2.4 GHz with 512 MB RAM running Microsoft Windows XP. The laser scanner data is processed following the procedure described above. In Fig. 13, a three-dimensional rendering from the ground station software is presented. The display shows the location of the UAV, the reference point marker, $\mathbf{X}_O^{\min}$ to a point in the local obstacle map at that moment, and laser-scanned points as dots. During the experiments, the laser scanner was able to detect the canopies in the line of sight with outstanding accuracy, as well as other natural and artificial objects including buildings, trees and power lines. The processed laser scanned data in a form of local obstacle map is used in the optimization (5). The trajectory is then sent via IEEE 802.11b to the onboard flight management system at 10Hz. The overall system structure used in the experiments is shown in Fig. 13. The tracking layer controls the host vehicle to follow the revised trajectory. In the repeated experiments, the vehicle was able to fly around the obstacles with sufficient accuracy for tracking the obstacle-free trajectory, as shown in Fig. 12(solid line).

4. Conclusion

In this article, we presented a collision avoidance algorithm for UAVs using nonlinear model predictive control. The preview mechanism of receding horizon control is found ideal for such cases when the obstacles are moving. The proposed algorithm was also applied to obstacle avoidance problems, where onboard sensors combined with updated local map was combined with the MPC solver to compute conflict-free trajectories. Both cases are validated first in simulation and then in realistic experiments using helicopter UAVs. In each set of experiments, the proposed NMPC-based algorithms were able to run in real-time for computing conflict-free trajectories. The proposed algorithm will be further extended to vision-based sensing as well as the avoidance problems of fixed-wing UAVs.

5. References

Bellingham, J.; Kuwata, Y., & How, J. (2003). Stable Receding Horizon Trajectory Control for Complex Environments, *AIAA Conference on Guidance, Navigation, and Control,* Austin, Texas, August 2003.

Dunbar, W. B.; Milam, M. B.; Franz, R. & Murray, R. M. (2002). Model Predictive Control of a Thrust-Vectored Flight Control Experiment, *15th IFAC World Congress on Automatic Control*, Barcelona, Spain, July 2002.

Hrabar, S. E.; Corke, P. I.; Sukhatme, G. S.; Usher, K. & Roberts, J. M. (2005) Combined Optic-Flow and Stereo-Based Navigation of Urban Canyons for a UAV, *IEEE/RSJ International Conference on Intelligent Robots and Systems*, pp. 302-309, Edmonton, Alberta, Canada, August 2005.

Milam, M. B.; Franz, R. & Murray, R. M. (2002). Real-time Constrained Trajectory Generation Applied to a Flight Control Experiment, *IFAC World Congress on Automatic Control*, Barcelona, Spain, July 2002.

Polak, E. (1997). *Optimization: Algorithms and Consistent Approximations,* Springer-Verlag, ISBN 0-387-94971-2, New York, USA.

Rydergard, S. (2004), *Obstacle Detection in a See-and-Avoid System for Unmanned Aerial Vehicles,* Master's Thesis, Royal Institute of Technology, Stockholm, Sweden.

Shim, D. H. (2000). *Hierarchical Control System Synthesis for Rotorcraft- based Unmanned Aerial Vehicles,* Ph. D. thesis, University of California, Berkeley, 2000.

Shim, D. H.; Kim, H. J. & Sastry, S. (2003). Decentralized Nonlinear Model Predictive Control of Multiple Flying Robots, *IEEE Conference on Decision and Control*, Maui, HI, December 2003.

Shim, D. H.; Chung, H. & Sastry, S. (2006). Autonomous Exploration in Unknown Urban Environments for Unmanned Aerial Vehicles, *IEEE Robotics and Automation Magazine*, vol. 13, September 2006, pp. 27-33, ISSN1070-9932.

Sutton, G. J. & Bitmead, R. R. (2000). Computational Implementation of NMPC to Nonlinear Submarine, In: *Nonlinear Model Predictive Control,* volume 26, pp. 461-471, Institution Electrical Engineers, ISBN 0852969848, London, UK.

Watanabe, Y.; Calise, A. J.; Johnson, E. N. & J. H. Evers (2005). Minimum-Effort Guidance for Vision-based Collision Avoidance, *AIAA Guidance, Navigation, and Control Conference,* San Francisco, California, 2005.

30

UAS Safety in Non-segregated Airspace

Alan Simpson, Vicky Brennan and Joanne Stoker
Ebeni Limited
United Kingdom

1. Introduction

Unmanned Aerial Systems (UAS) are set to become part of everyday air traffic operations perhaps within the next few years; however there are significant challenges that need to be addressed in order to seamlessly introduce UAS into non segregated airspace. This chapter discusses some of the identified safety challenges in achieving this objective in the context of the current regulatory framework. It also takes a look at how one might rigorously argue the safety of UAS operations in non-segregated airspace from an Air Traffic Management (ATM) perspective. The chapter draws upon the experience of the authors' in the UAS domain, more specifically the lessons learnt from a number of safety assessments for flying UAS as Operational or General Air Traffic (OAT or GAT) inside and outside segregated airspace.

Most UAS operations are currently constrained to designated danger areas or within temporary restricted areas of airspace, commonly known as segregated airspace, or are flown under special arrangements over the sea. On some occasions, UAS operations are permitted in an extremely limited environment outside segregated airspace. To exploit fully the unique operational capabilities of current and future UAS and thus realise the potential commercial benefits of UAS, there is a desire to be able to access all classes of airspace and operate across national borders and airspace boundaries. Such operations must be acceptably safe but regulation should not become so inflexible or burdensome that the commercial benefits are lost.

The viability of the commercial market for UAS especially in the civil market is heavily dependent on unfettered access to the same airspace as manned civilian operations. Whilst it is essential that UAS demonstrate an equivalent level of safety compared to manned operations the current regulatory framework has evolved around the concept of the pilot-in-the-cockpit. There is a need to develop UAS solutions that assure an equivalent level of safety for UAS operations, which in turn will require adaptation of the current regulatory framework to allow for the concept of the pilot-not-in-the-cockpit without compromising the safety of other airspace users.

One of the major issues facing UAS operations is the demonstration of equivalence (in particular for See and Avoid) in the context of an evolving ATM environment. It is very important to understand that the current ATM environment is not static. Achieving equivalence with manned operations is not a fixed target as there are many significant changes proposed that aim to improve operational efficiency and performance or enhance safety. On the whole proposed changes to the ATM environment could be seen as

advantageous to UAS operations as more and more functions within the environment are automated thus there is a significant opportunity for the UAS industry to influence the shape of the future ATM environment to support wider UAS operations.
Assuring the safety of UAS operations in non-segregated airspace will therefore require a significant update to key elements of extant regulations and standards. UAS have yet to establish a good safety record[1] and there are many challenges both regulatory and technological to be resolved before such operations can become common place. Without a coherent regulatory framework (i.e. regulations, standards, etc.) for certifying UAS such systems must be argued as acceptably safe within the context of the whole operational Air Traffic environment[2]. However, regulators are taking steps towards providing this infrastructure. Wider and more detailed regulations and standards will likely form around the technologies that become available to resolve the operational and safety issues that UAS operations must address.

2. Current Regulatory Framework

2.1 Overview

The regulatory context for UAS operations in non-segregated airspace can be split into three main considerations; certification of air vehicle airworthiness, safe provision of air traffic services and licensing of operators, pilots, etc. Most of the regulatory work carried out to date has focussed on airworthiness certification as well as the licensing of UAS Pilots. There is also a need for regulation to include the Air Traffic Management environment and for the co-ordination of all regulatory activities, a point noted by EASA in a recent regulatory consultation paper (EASA A-NPA, 2005) to ensure the safety of air travel as a whole.
However, the regulatory framework is changing rapidly especially with respect to integration of UAS into the air traffic management environment, (Degamo, 2004) suggests that whilst the United States leads in UAS technology developments, they apparently lag behind the UK and Europe in legislative activities relating to Unmanned Aerial Vehicles (UAVs), more specifically they currently have no regulation or guidance for UAS operations; (Degamo, 2004) also provides a good background on UAS issues in general.

2.2 UK Regulations

(CAP722, 2008) provides guidance for UAS operations in UK airspace and has been compiled by the Civil Aviation Authority (CAA) Directorate of Airspace Policy (DAP) to cater for the growing capabilities and anticipated increase of pilotless aerial vehicles. (CAP722, 2008) details the overarching military and civil regulations applicable to UK UAS operation.
In addition, civil certification aspects of UAS are the responsibility of the Design and Production Standards Division of the CAA Safety Regulation Group (SRG). Their position set out within (Hatton & Whittaker, 2002) is that UAS should be granted permission to fly

[1] The current UAV accident rate is 100 times that of manned aircraft. According to US Air Force studies (Degammo, 2004) the accident rate is 50 times greater than for F-16, US commercial aircraft accident rate is 0.06 per million flight hours compared to Global Hawk at 1,600 per million flight hours; very few of these accidents resulted in third party losses.

[2] When a UAS Sense and Avoid specification is defined within the regulatory framework then it would be sufficient to certify such systems against this specification.

by qualifying for Certificates of Airworthiness by demonstrating compliance with defined airworthiness standards comparable to, and derived from, those applied to manned aircraft. The principles of the UK Policy for UAS are as follows:

- To operate within existing arrangements and regulations for air traffic
- No automatic right of airspace use
- An equivalent level of certification compliance
- No increased risk to existing users
- Operations must be "transparent"[3] to the controller
- Currently flights outside Danger Areas will be in temporary segregated airspace

In addition to the later point rules apply for operations in non-segregated airspace, including carriage of mandated equipment as per manned aircraft and provision of an acceptable sense and avoid system. Other important considerations include UAS control link reliability and security.

2.3 European Regulations

Several European agencies are engaged in activities aimed at the development of rules that can accommodate the Joint Aviation Requirements (JARs) and satisfy member nations' concerns over UAS safety and usage.

Prior to the formation of the European Aviation Safety Agency (EASA), the Joint Aviation Authorities (JAA) was an associated body of the European Civil Aviation Conference (ECAC) representing the civil aviation regulatory authorities of a number of European States who agreed to co-operate in developing and implementing common safety regulatory standards and procedures. One of the JAAs key functions was to develop and adopt Joint Aviation Requirements (JARs) in the fields of aircraft design and manufacture, aircraft operations and maintenance, and the licensing of aviation personnel. JARs applicable to UAS operations include (JAR 23, 1994 et al).

A Joint JAA/EUROCONTROL initiative on UAS (known as the UAV Task-Force) was established in September 2002 on the basis of a joint decision of the JAA and EUROCONTROL governing bodies. This decision was taken in reaction to the growing European UAS Industry and the recognised need for the authorities to commence work leading to European regulations for civil UAS. The non-existence of such regulations was seen as a major obstacle for further development of European UAS applications. The findings of the Final Report of the UAV-Task Force are documented in (JAA/EUROCONTROL, 2004).

Due to the significant interest in UAS operations shown by EASA, EUROCONTROL, the European Union and other National Organisations, the European Organisation for Civil Aviation Equipment (EUROCAE) set up a Working Group (WG-73 Unmanned Aerial Vehicles) to establish required recommendations and technical standards for UAS. The main objective of WG-73, as set out in (EUROCAE, 2006), is to ensure that UAS can operate safely within the airspace and are compatible with existing infrastructure and related systems and equipments.

---

[3] As defined by (CAP722, 2008) "A controller must not be expected to do anything different using Radio Telephony or landlines than he would for other aircraft under his control. Nor should he have to apply different rules or work to different criteria. UAS must be able to comply with instructions and with equipment requirements applicable to the class of airspace within which they intend to operate."

2.4 Future Air Traffic Management Environment

It is very important to understand that the air traffic regulatory framework is not static and achieving equivalence with manned operations is not a fixed target. The air traffic environment is constantly subject to changes that aim to improve operational efficiency, interoperability, performance, functionality or enhance safety. The technical changes proposed on the whole could be seen as advantageous to UAS operations as more and more functions within the environment are automated. There is a significant opportunity for the UAS industry to influence the shape of the future air traffic environment to support wider UAS operations. There is also clear evidence that UAS operations are being considered within the scope of some of these changes.

The most comprehensive package of changes proposed over the next 12 years or so is captured by the Single European Sky ATM Research or SESAR programme. The conceptual changes proposed by SESAR for the ATM environment are summarised in the ATM Target Concept (SESAR D3, 2007). There are many proposed changes, too many to describe here, but some may even affect the fundamental concepts in air traffic service provision. The most significant of these are:

- Simplification of the classification of airspace into managed and unmanaged, although managed airspace will also include high-density areas (e.g. around airports), dynamic and variable airspace reservations (known as "Moving ARES") and delegated separation provision arrangements.
- Introduction of co-operative separation provision and self separation provision modes in controlled airspace based on Airborne Separation Assistance Systems (ASAS) applications.
- Greater deconfliction capabilities, with less reliance on Controllers, allowing the use of Precision Trajectory Clearances, combined with automated trajectory control by speed adjustment.
- Development of Collision Avoidance systems to take into account the changes to separation provision modes highlighted above. Enhancements to Airborne Collision Avoidance System (ACAS) and Short Term Conflict Alert (STCA) are expected to take advantage of the availability of more detailed aircraft information and the sharing of information between ground and airborne surveillance systems.

The deployment timeline for the SESAR changes is documented in the ATM Deployment Sequence (SESAR D4, 2008). It should be noted that all of these ideas are still in research and the case for safety has yet to be made.

3. Operations in Non-Segregated Airspace

UAS operations need to be acceptably safe irrespective of the type of airspace where unmanned air vehicles are flown. UAS operations in designated areas of airspace, from which other air users are excluded, however, can significantly simplify[4] the problem of justifying that an acceptable level of safety is achieved. The justification becomes more complex when the air vehicle ventures into airspace shared with others, known as non-segregated airspace. The fundamental difference is that in segregated airspace the "system" in which the UAS operates can be bounded, controlled and is often unique to the UAS. In

[4] Notwithstanding that levels of airworthiness and pilot competence still need to be assured for the UAS, i.e. the air vehicle, its control system and its pilot in command.

non-segregated airspace the UAS must integrate with an air traffic environment developed over decades to support manned aircraft.

To add to the complexity, non-segregated airspace is further classified into seven types; from Class A to Class G, as currently defined by (ICAO Annex 2, 1990) although this may change in the future as part of the Single European Sky initiative. This means that in addition to baseline airworthiness certification, UAS must demonstrate that they:

- Will not undermine the safety of the provision of those services to the UAS or other air users.
- Meet the rules of the air applicable to the meteorological conditions and class of airspace.

Specifically this would include for example:

- Meeting the mandatory equipment requirements for the class of airspace to be flown in (known as Minimum Aviation Specification Performance)[5]
- Interfacing with the existing air traffic services provided in that airspace (eg full air traffic control, flight information service, radar advisory service, etc.)

(ICAO ATM Operational Concept Document, 2003) identifies three main components of Air Traffic Management; Strategic Conflict Management, Separation Provision and Collision Avoidance. Strategic Conflict Management encapsulates all pre-flight planning activities that take place to ensure demand, capacity and conflicts are managed prior to the real time situation. The Strategic Conflict Management component includes pre-flight processes such as airspace / procedure design and flight plan management. It is anticipated that this component will be implemented the same for manned and unmanned operations. A functional representation of the 3 components is shown in ig 1 below. The concept of See and Avoid covers both separation provision and collision avoidance and is discussed below as it represents one of the most challenging aspects of UAS operations.

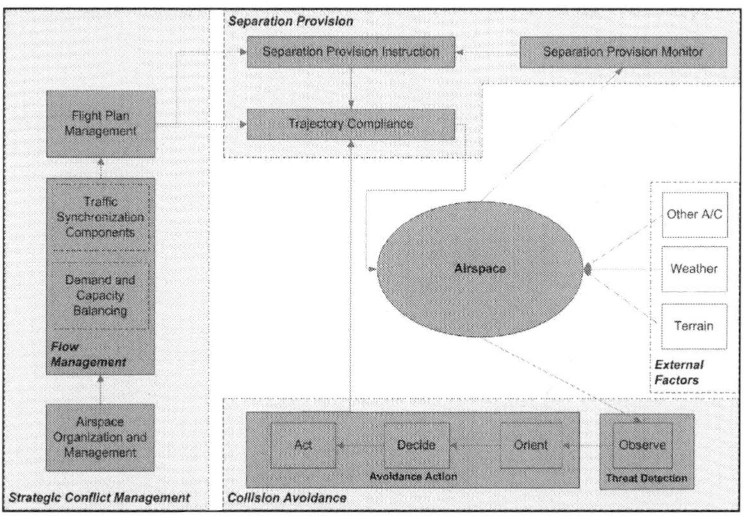

Figure 1. High-level Air Traffic Service Functional Model

[5] The amount of equipment implied could be well in excess of the weight limits for the air vehicle and this may be one of the major limitations on the type of air vehicle than can use non-segregated airspace.

3.1 Separation Provision

Separation provision is the tactical process of keeping aircraft away from other airspace users and obstacles by at least the appropriate separation minimum. Depending upon the type of airspace and, where applicable, the air traffic service being provided, separation provision can be performed by air traffic controllers or by the pilot-in-command. Where a controller is responsible for providing separation provision, the Separation Provision Monitoring and Demand for an aircraft are provided by the controller and the pilot is responsible for Trajectory Compliance. Where the pilot is responsible for Separation Provision, all these functions are performed by the pilot in accordance with the Rules of the Air.

Under Visual Flight Rules (VFR) in certain types of airspace, there is currently no specified minimum separation distance and the pilot of, for example, a manned aircraft arranges his trajectory using airborne radar and/or visual means to separate his flight path from other air users. In these scenarios for UAS operations (EUROCONTROL UAV-TF, 2007) defines a minimum separation distance of 0.5nm horizontally and 500ft vertically. The term Separation Provision should therefore be taken to include the actions necessary to provide physical separation between a UAS air vehicle and other air users of at least 0.5nm or 500ft, even though no separation minima is currently defined for manned operations.

3.2 Collision Avoidance

The Collision Avoidance component can be separated into pilot and collision avoidance system functions. Manned aircraft may be fitted with collision avoidance systems such as Traffic Alert and Collision Avoidance System (TCAS) II or elements thereof such as Secondary Surveillance Radar (SSR) Transponders[6]. Collision avoidance systems are designed to activate when separation provision has been compromised; although air traffic controllers can instigate collision avoidance action from a pilot, this mechanism would not be available to an autonomous UAS.

(SRC Policy Document 2, 2003) states that collision avoidance systems (referred to as Safety Nets) are not part of separation provision so must not be included in determining the acceptable level of safety required for separation provision. However, the collision avoidance performed by a pilot of a manned aircraft must be performed to an equivalent level of safety by the UAS whether piloted or autonomous.

The Safety Regulation Commission (SRC) Policy statement implies that UAS must provide an equivalent level of interaction with the Separation Provision component as provided by pilots. Furthermore the UAS separation provision system must maintain the level of safety (with respect to the scope of (ESARR 4, 2001)) without the need for a Safety Net. This implies that UAS need to provide independence between separation provision and collision avoidance systems.

3.3 See and Avoid

Current manned operations include provisions for pilot "See and Avoid" to implement (or augment depending on the class of airspace) the separation provision and collision avoidance functions. UAS operations need to provide an equivalent level of safety with a

[6] Mode A/C and S Transponders can be used by other aircraft fitted with TCAS.

"Sense and Avoid" capability to overcome the loss of manned "See and Avoid" capability[7]. However, it is important that separation provision and collision avoidance are addressed independently.

Firstly, if the Separation Provision component is working normally then the Collision Avoidance component is not under demand. Therefore in this environment the Collision Avoidance component should only provide situational awareness information[8]. Secondly, if Separation Provision fails in some way then the normal operation for the Collision Avoidance component is to act to avoid any imminent potential collisions. For this to work successfully a number of conditions must be satisfied; as a minimum the components should do as shown in Table 1.

Function	Collision Avoidance	Separation Provision
Be aware of all traffic in the vicinity	✓	✓
Implement and maintain appropriate separation minima with all other traffic		✓
Have criteria for when to implement traffic warnings (separation provision is potentially about to fail) and resolution warnings (separation has failed and immediate collision avoidance action is required)	✓	
Be able to identify traffic that is a collision (or near miss) threat, establish an appropriate avoidance response, taking into account other potential targets, and implement the response if the UAS pilot is unable to do so in time	✓	

Table 1. Sense and Avoid conditions

In addition since the UAS must integrate with the existing manned aircraft environment it must operate with extant co-operative and non-co-operative systems for surveillance and collision avoidance, inter alia:

- Other traffic must be able to 'see' the UAS air vehicle under all the conditions that other manned aircraft would be detected by another manned aircraft.
- Non-co-operative surveillance systems (e.g. Primary Surveillance Radar) must be able to 'see' the air vehicle.
- To cater for all potential air traffic scenarios a UAS Sense and Avoid system must be able to detect co-operative traffic (aircraft fitted with data link devices, e.g. Mode S transponders) and non-co-operative traffic (unfitted).

3.4 UAS Characteristics

The UAS encapsulates not only the air vehicle itself, but the entirety of equipment, people and procedures involved in the launch, control and recovery of the air vehicles. To establish

[7] Sense and Avoid capability should address many of the issues associated with pilot-not-in-the-cockpit, however, consideration also needs to be given to *inter alia*, emergency responses and off-tether operations, etc.

[8] This should not be taken to imply that the Collision Avoidance component must not be active, only that whilst the Separation Provision component is working correctly then the Collision Avoidance component should not interfere.

the potential differences in manned and unmanned operations, it is important to understand the specific characteristics of UAS that are potentially applicable to UAS operations.

A principle characteristic is physical separation of control of the air vehicle from the air vehicle itself. The UAS pilot will be remote from the UAV either on the ground or in another aircraft. The UAS pilot maintains control of the air vehicle through a UAS Control System via a UAS Control Link. The operation of the control link cannot be guaranteed under all conditions so the UAS must be able to work safely with or without the control link; this is referred to as flying on or off-tether.

The key characteristics that can affect UAS operations are as follows:

- **Conspicuity** - the visibility of the air vehicle to other airspace users is an important component in the Collision Avoidance component as well as when Separation Provision is the responsibility of the UAS pilot. This could be an issue for air vehicles that are smaller than manned aircraft, or those that present a poor signature for Primary Surveillance Radar.
- **Autonomous Operations** - One of the key characteristics of UAS's is the ability to operate under various conditions without human interaction. The necessity for human interaction, along with other factors such as safety, mission complexity and environmental difficulty determine the level of autonomy that the UAS can achieve. There are various taxonomies for classifying UAS autonomy for example Autonomy Levels For Unmanned Systems (Hui-Min Huangi, et al, 2005). However, it is not possible to define UAS operation in non-segregated airspace under any one classification as UAS may be expected to operate with varying degrees of autonomy depending on the circumstances.
- **Airworthiness** - UAS air vehicles (and as applicable control stations) must be fitted with certified equipment equivalent to that required for manned operation in the intended airspace; this may pose problems for smaller or lighter air vehicles due to space or weight constraints.
- **Flight Performance** - the manoeuvrability of a UAS air vehicle is important to understand. Currently, Air Traffic Controllers are required to understand flight performance characteristics of the types of aircraft that come under their control and provide separation provision instructions based on this understanding. This requirement for understanding will also need to apply to unmanned operations to ensure ATC instructions can be implemented.

4. A Definition of Acceptably Safe

(JAA/EUROCONTROL, 2004) defines acceptably safe in terms of achieving an equivalent level of risk with that for manned aircraft.

- *UAV Operations shall not increase the risk to other airspace users or third parties*

This definition rightly focuses on the equivalence in risk and not safety levels, regulation or certification[9] and cuts across the current debate on the certification versus safety target approach to assuring UAS safety, as discussed in (Haddon & Whittaker, 2002) and (EASA A-NPA, 2005).

---

[9] Achieving equivalence with manned aircraft through regulation/certification alone may be inadequate or overly prescriptive unless the impact on the risk is fully assessed.

However, it does rely on a fundamental assumption that current manned operations are acceptably safe and does lack the level of detail required to appreciate some of the issues facing UAS operations in non-segregated airspace, which include:

a. The level of acceptable risk for manned aircraft operations varies depending on the operational context.
b. The public perception of the UAS risk may demand a harsher consideration of risk than for manned operations.
c. In accordance with European ATM legislation (ESARR 3, 2000), the risk should also be reduced as far as reasonably practicable (AFARP).

In addition, levels of air traffic are predicted and expected to increase over the next few decades, which will also require an increasing level of safety. There is a significant demand[10] to make improvements to the existing Air Traffic environment to achieve this, and the opportunity that UAS technology[11] may provide to support this should not be ignored or overlooked.

4.1 Public Perception of UAV Risk

One of the key influences that will determine the direction and strength of the UAS market is acceptance by the general public. It is well understood that any public trust and support for UAS operations that exists today will evaporate as soon as a UAS air vehicle is involved in an accident, regardless of fault.

A public opinion survey undertaken in the United States in 2003, the findings of which are documented in (MacSween-George & Lynn, 2003), found that up to 68% of the public support cargo and commercial UAS applications and most were not concerned by UAS flying overhead. However, the survey also found that the majority of respondents would not support the use of UAS to fly passengers.

The CAA Directorate of Airspace Policy recently invited members of the UK aviation industry to attend a one day workshop (CAA, 2005) to discuss UAS matters and the effect that emerging systems may have on existing and future manned aviation activity. One of the syndicate sessions at this workshop was tasked with discussing public and aviation industry perceptions, the following issues were identified:

- Potential negative public perception due to lack of knowledge or concerns over UAS historical safety records.
- Perception from the current manned community in terms of lack of trust in shared airspace.
- Public concern on the safety and security implications of UAS.
- Lack of trust in the regulation of industry.

It is vitally important therefore to secure public acceptance via positive promotion of the capabilities, limitations and safety of UAS by active communication with all affected stakeholders.

---

[10] Under European Airspace Regulation, ANSPs are required to reduce risk As Far As Reasonably Practicable (ESARR 3, 2000).

[11] Or combining UAS and next generation manned technology.

4.2 A Practical Safety Criteria

From a safety perspective it is clear that the aim of the UAS industry, regulators and operators must be to ensure that the safety risk from UAS operations in non-segregated airspace shall be:

- No greater than for manned operations in the same operational context[12]
- Further reduced As Far As Reasonably Practicable

This supports the view proposed by Air Commodore Taylor as documented in (Taylor, 2005) but also takes into account the counter view expressed by (DeGarmo, 2004) and alluded to within (CAP72, 2008) by encouraging rather than mandating enhanced safety over and above manned operations. This is the basis on which (EUROCONTROL UAV-TF, 2007) was assessed in order to determine the safety requirements for such operations.

5. Safety Argument for UAS Operation in Non-segregated Airspace

The purpose of the safety argument presented below is to outline how the overall objective of "equivalent risk" can be broken down in relation to UAS operations in non-segregated airspace to a level where regulations can be defined that ensure that the resultant risk is acceptable in principle. In describing the safety argument some of the key issues and challenges facing the domain are described. Note that the safety argument is not specific to a type or class of air vehicle but rather to the concept of UAS operations in non-segregated airspace. This approach facilitates identification of specifications that are rigorous but avoid being implementation specific.

5.1 Top-level Safety Argument (Claim 0)

The overall objective for assuring that UAS operations will be safe is to show that they are and will continue to be acceptably safe (as defined in the previous section) within a clearly defined context. The context must include:

- Justification for the intended operational use.
- A definition of the operational scenarios (both mission and air traffic service related) that a UAS may face.
- Necessary assumptions (e.g. that current equivalent manned operations are acceptably safe).

Of necessity the argument must also consider all potential operational phases. An example scenario model for the latter phase of flight is shown in the Fig 3 below.

[12] This is a relative approach to assess risk. Within the air traffic management domain absolute safety targets are set for Air Traffic Service Providers in (ESARR 4, 2001) but the relative approach is still applied to manned operations although this will likely change over time. At some point in the future (sooner for some applications such as Area Navigation) UAS operations will also need to be compliant.

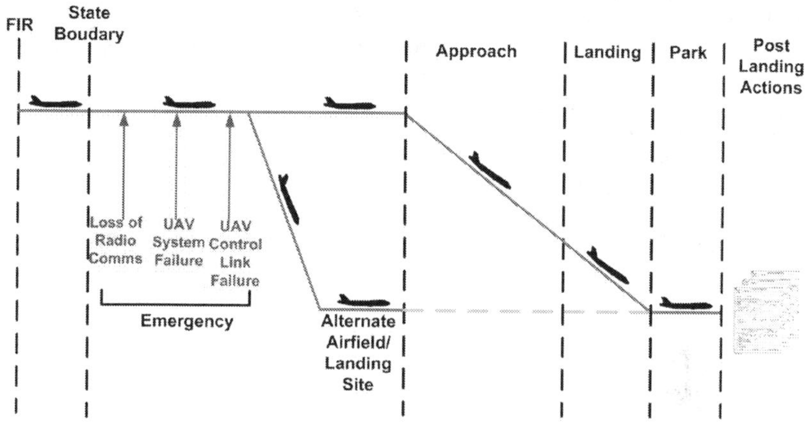

Figure 3. Example Scenario Model (1)

This top level goal can be shown to be met by demonstrating four principle safety goals:

1. Safety requirements are specified such that the safety criteria as discussed in section 4.2 is satisfied in principle.
2. Safety requirements are fully addressed in the relevant regulations and standards.
3. Safety requirements are developed at a level commensurate with the level of detail in regulations or standards.
4. UAS operations in non-segregated airspace fully satisfy the safety requirements within the regulations and standards in practice.
5. UAS operations in non-segregated airspace are monitored to ensure that the safety criteria continue to be satisfied in operation.

These principle goals are discussed in the following sections.

5.2 Safety Requirements for UAS Operations (Claim 1)

Safety requirements can be developed at almost any level of abstraction. For the purpose of setting regulation or standard specifications safety requirements need to reflect the level of detail determined by the scope and purpose of the regulation or standard. In turn the safety requirements need to be:

- Developed at a high-level but form a necessarily and sufficiently complete set to show the safety criteria are met.
- Based on validated models of UAS operation.
- Derived using an appropriate safety assessment methodology to include functional safety properties as well as integrity requirements.
- Realisable in implementation, although consideration as to whether the requirements are capable of implementation should not be limited by the capabilities of current UAS technology.

The safety requirements derived for (EUROCONTROL UAV-TF, 2007) were based on slightly more detailed models of UAS operations than those described in section 2. The principle conclusions of this work were:

- Despite the variety of airspace classifications, available ATM services, the multitude of possible scenarios and the different phases of flight etc. only three modes of operation needed to be considered, as follows:
  - Where ATC is responsible for separation provision.
  - Where the pilot in command is responsible for separation provision.
  - Where the air vehicle is not in contact with the pilot in command and so provides separation provision for itself.
- At the air traffic management functional safety level no distinction was drawn between manned and UAS operations, i.e. UAS operations do not introduce new hazards to the domain.
- Given the need for further research it was considered necessary to mandate that UAS pilots in command will require equivalent piloting skills to those of manned aircraft when flying in non-segregated airspace. However, this would be inadequate where pilots in command are responsible for more than one UAS at a time.
- Issues with requirements achievability were identified and are discussed further within section 5.4 below.

(SRC Policy Document 1, 2001) specifies a Target Level of Safety (TLS) for civil aircraft which is further apportioned within European ATM Regulation (ESARR 4, 2001) to ATM specific risks. These safety targets should be further apportioned by airspace users, Air Navigation Service Providers, etc. in order to set targets for specific operations. As this is often seen as too complex a task, many safety cases for European air traffic management concepts and systems rely on a relative argument, although not all, e.g. (EUROCONTROL RNAV, 2004), etc. In these cases UAS operations should demonstrate compliance with the specific defined absolute targets and safety requirements.

5.3 UAV Regulations and Standards (Claim 2)

Whilst there is a need for specific UAS regulations and standards for particular UAS technologies much of the regulations and specifications for non-segregated airspace operations already exist within the manned aircraft and air traffic regulations as outlined in section 2.2.

These provide a vital basis as advocated by (Haddon & Whittaker, 2002) for the creation of UAS specific regulations and standards. But regulations and standards need to be developed in accordance with derived safety requirements and not just based on the concept of "equivalence". The safety assessment work carried out for (EUROCONTROL UAV-TF, 2006) can be seen as a model for the development of other UAS regulations and standards to ensure that the overall objective of "equivalent risk" is achieved.

There is a need for regulations and standards to be developed in the context of commonly agreed safety requirements based on a whole "system of systems" model of UAS operations to ensure that each perspective is fully considered including pilots, industry, Air Traffic Controllers, Operators, Maintainers and regulators. It is of particular concern that at the moment UAS security is not clearly covered by any regulatory authority, yet ensuring and maintaining the security of control centres, data links, etc. is fundamental to the substantiation that operations are acceptably safety.

5.4 UAS Safety Requirements Implementation (Claim 3)

The principle conclusions with regards the implementation of safety requirements for UAS operations in non-segregated airspace are as follows:

- There needs to be independence between the implementation of the separation provision (strategic) and the implementation of collision avoidance (tactical separation provision).
- This is easier to achieve when an air traffic controller is responsible for separation provision and the pilot in command can control the air vehicle and is aware of other air users (see next bullet), since the air vehicle can be fitted with a collision avoidance system similar to TCAS II. However, there are unresolved concerns regarding the efficacy of TCAS II logic and UAS operations[13] and the level of risk reduction achieved by TCAS II, approximately 30% (EUROCONTROL ACAS, 2005) may be insufficient to achieve an equivalent level of risk.
- There are still some uncertainties with implementation of automated strategic and tactical separation provision systems to replace that currently performed by the pilot[14], i.e. the "Sense and Avoid" issue.
- The safety assessment conducted on the (EUROCONTROL UAV-TF, 2007) concluded that UAS sense and avoid technology offers the potential to improve threat detection and avoidance capability, especially given concerns with the effectiveness of human see and avoid capabilities. Achieving equivalence or even equivalent risk seems inadequate in this case. A comprehensive discussion of the issues is provided in (DeGarmo, 2004).
- UAV and Data Link reliability are key to minimising the workload impact on air traffic controllers arising from excessive instigation of emergency or contingency procedures.
- UAS operations must consider the scenario when the communication between the pilot in command and the air vehicle is unavailable. In this scenario the air vehicle must conform to a predefined flight plan so that UAS behaviour remains as deterministic as possible[15].
- Emergency Procedures may necessarily be different for UAS operations and as such UAS will need to be able to, for example, indicate when UAS are operating in isolation from the pilot in command (e.g. a unique transponder code), when to be provided increased separation provision, etc.
- For most of the risks identified no additional risk mitigation was identified within the air traffic domain that could further reduce the risk over and above manned operation. This leaves the challenge of achieving the AFARP safety criteria to the standards bodies and UAS implementers.
- There are other challenges that will arise during the implementation of the safety requirements, inter alia:

---

[13] Due for example to the significant reliance on the timeliness of pilot response to Resolution Advisories (RA), but such concerns need to be resolved in order to ensure that TCAS II is still working as effectively in single and multiple manned vs. UAS air vehicle encounters.

[14] JAR 91.113 Rights of Way Rules state that "regardless of whether an operation is conducted under instrumented flight rules (IFR) or visual flight rules (VFR), vigilance shall be maintained by each person operating an aircraft so as to see and avoid other aircraft".

[15] ATCO's consulted during the safety assessment suggested that errant UAS behaviour is probably no worse than manned military errant behaviour.

- The inadequacy of the current integrity of aeronautical data for terrain maps, obstacle heights GPS based navigation systems, etc. although this is being addressed through the European Commission Interoperability Mandates.
- Operating characteristics of current and future UAS air vehicles that may undermine principle safety assumptions in current safety cases for air traffic operations or concepts, e.g. the timeliness of pilot/UAS implementation of controller instructions.

5.5 Monitoring UAS Operations (Claim 4)

A programme of safety monitoring and improvement will need to be implemented by state regulators and other international bodies to ensure that UAS operations in non-segregated airspace remain acceptably safe. The safety assessment for (EUROCONTROL UAV-TF, 2007) did not identify any monitoring requirements in addition to those already recommended for manned OAT operations.

6. Conclusions

There is clearly a desire in industry to produce commercially viable UAS and concern that UAS regulations do not become over burdensome or inflexible. Whilst there is a wealth of existing regulation and standards for manned operations, there is still a need to ensure that the transition to UAS operations in non-segregated airspace does not jeopardise the safety of other airspace users, and perhaps even contributes to an improved level of safety in aviation, directly addressing issues with the public perception of the risk from UAS.

The work for EUROCONTROL DG/MIL has shown that the development and specification of regulations and standards can be subject to safety assessment, which can assure the completeness and correctness of the specifications whilst providing the rigorous evidence that the regulations and standards capture the safety requirements relevant to the their scope and purpose.

By applying this process at all levels of UAS regulation and standard setting it would be possible to ensure not only a cohesive approach to UAS regulation but also that UAS operations will not increase the risk to other airspace users and third parties. There is an accepted and recognised need for regulatory bodies to work together to ensure that all aspects of UAS regulation including Air Traffic Management, Vehicle Certification, Operation, Maintenance and Licensing, etc. interface correctly, taking into account the impact of issues within, and assumptions made by, each of the aspects as well as the practicalities and commercial viability of the final UAS solutions.

Notwithstanding that regulations and standards can be developed or updated to incorporate UAS operations such that they are acceptably safe, there remain many issues with the practical implementation of technology to achieve the essential safety requirements. The most relevant of these is the development of Sense and Avoid specifications that address the overarching safety requirements **and** can still be achieved in practice.

Consideration should also be given to pursuing the regulatory aspects of Sense and Avoid systems and ensuring the UAS operations are considered in all current and future ATM research, particularly SESAR, which may alter the concepts or technologies deployed. However, regulators and industry (e.g. through Work Group 73 of EUROCAE) steps towards providing the necessary UAS regulatory and standards infrastructure and specifications such as (EUROCONTROL UAV-TF, 2007) provide an important foundation. Wider and more detailed regulations and standards will likely form around the technologies that become available to resolve the operational and safety issues that UAS operations must address. There is still much scope for further research in the area of UAS regulation and implementation and programmes such as the UK DTI funded ASTRAEA project will help significantly to move the process forward.

7. References

CAA UAV Flights in UK Airspace Workshop - 13 October 2005 *Civil Aviation Authority Directorate of Airspace Policy* 8AP/15/19/02

CAP 722 UK Civil Aviation Authority Directorate of Airspace Policy Unmanned Aerial Vehicle Operations in UK Airspace - Guidance 28 April 2008

D. R. Haddon C. J. Whittaker. Aircraft Airworthiness Certification Standards for Civil UAVs, *UK Civil Aviation Authority* August 2002

EASA A-NPA Policy for Unmanned Aerial Vehicle (UAV) Certification Advance - Notice of Proposed Amendment (NPA) No 16/2005 A-NPA No 16/2005

ESARR3 Use of Safety Management Systems by ATM Service Providers Edition: 1.0 Edition Date: 17-07-2000

ESARR4 Risk Assessment and Mitigation in ATM Edition: 1.0 05 April 2001

EUROCAE Presentation Paper on the creation of WG- 73 "Unmanned Aerial Vehicles" WG-73 UAV *Presentation Paper* 30 Jan 06

EUROCONTROL Airborne Collision Avoidance System (ACAS) II *Post Implementation Safety Case* Edition: 1.1 12 December 2005

EUROCONTROL RNAV Preliminary Safety Case for Area Navigation in Final Approach Edition: 0.5 20 October 2004

EUROCONTROL UAV TF Specifications for the Use of Military Unmanned Aerial Vehicles as Operational Air Traffic Outside Segregated Airspace Version 0.6 2007

Hui-Min Huangi et al A Framework For Autonomy Levels For Unmanned Systems (*Proceedings of Unmanned Systems North America 2005*) June 2005

ICAO Annex 2 to the Convention on International Civil Aviation: Aerodromes. *Rules of the Air ICAO* Ninth Edition July 1990

ICAO ATM Operational Concept Document Appendix A AN-Conf/11-WP/4 September 2003

JAA/EUROCONTROL A Concept for European Regulations for Civil Unmanned Aerial Vehicles (UAVs) *UAV Task Force – Final Report* 11 May 2004

JAR 23 Joint Aviation Requirements for Normal Utility Aerobatic and Commuter Category Aeroplanes 11 March 1994

M. T. DeGarmo. Issues Concerning Integration of Unmanned Aerial Vehicles in Civil Airspace, MP 04W0000323 November 2004

MacSween-George Sandra Lynn. A Public opinion Survey – Unmanned Aerial Vehicles for Cargo Commercial and Passenger Transportation, The Boeing Company paper presented at the *2nd AIAA "Unmanned Unlimited" Systems Technologies and Operations Conference* 2003-6519

SESAR D3 DLM-0612-001-02-00a The ATM Target Concept - SESAR Deliverable D3 (available at www.sesar-consortium.aero) September 2007

SESAR D4 DLM-0706-001-02-00 The ATM Deployment Sequence - SESAR Deliverable D4 (available at www.sesar-consortium.aero) January 2008

SRC Policy Document 1: ECAC Safety Minima for ATM Edition 1.0 14 February 2001

SRC Policy Document 2: Use of Safety Nets in Risk Assessment and Mitigation in ATM Edition: 1.0 28 April 2003

Taylor Air Commodore Neil, The Challenge of Integrating UAVs into Mixed User Airspace Royal United Services Institute Defence Systems Summer 2005

31

A vision-based steering control system for aerial vehicles [1]

Stéphane Viollet, Lubin Kerhuel and Nicolas Franceschini
Biorobotics Dpt., Institute of Movement Sciences, CNRS and University of the Mediterranean
France

1. Introduction

Ever since animals endowed with visual systems made their first appearance during the Cambrian era, selection pressure led many of these creatures to stabilize their gaze. Navigating in 3-D environments (Collett & Land 1975), hovering (Kern & Varju 1998), tracking mates (N. Boeddeker, Kern & Egelhaaf 2003) and intercepting prey (Olberg et coll. 2007) are some of the many behavioural feats achieved by flying insects under visual guidance. Recent studies on free-flying flies have shown that these animals are able to keep their gaze fixed in space for at least 200ms at a time, thanks to the extremely fast oculomotor reflexes they have acquired (Schilstra & Hateren 1998). In vertebrates too, eye movements are also the fastest and most accurate of all the movements.
Gaze stabilization is a difficult task to perform for all animals because the eye actuators must be both :

- fast, to compensate for any sudden, untoward disturbances.
- and accurate, because stable visual fixation is required.

In the free-flying fly, an active gaze stabilization mechanism prevents the incoming visual information from being affected by disturbances such as vibrations or body jerks (Hengstenberg 1988) (Sandeman 1980)(Schilstra & Hateren 1998). This fine mechanism is way beyond what can be achieved in the field of present-day robotics.
The authors of several studies have addressed the problem of incorporating an active gaze stabilization system into mobile robots. A gaze control system in which retinal position measurements are combined with inertial measurements has been developed (Yamaguchi & Yamasaki 1994), and its performances were assessed qualitatively while slow perturbations were being applied by hand. Shibata and Schaal (Shibata et coll. 2001) designed a gaze control system based on an inverse model of the mammalian oculomotor plant. This system equipped with a learning network was able to decrease the retinal slip 4-fold when sinusoidal perturbations were applied at moderate frequencies (of up to 0.8Hz). Another adaptive image stabilizer designed to improve the performances of robotic agents was built and its ability to cope with moderate-frequency perturbations (of up to 0.6Hz) was tested (Panerai, Metta & Sandini 2002). Three other gaze stabilization systems inspired by the

---

[1] Part of this paper reprinted from L. Kerhuel, S. Viollet and N. Franceschini, IROS Conference, © 2007 with permission from IEEE.

human vestibulo-ocular reflex (VOR) have also been presented (two systems for mobile robots (Lewis 1997)(Viola 1989) and one for an artificial rat (Meyer et coll. 2005)), but the performances of these systems have not yet been assessed quantitatively on a test-bed. Miyauchi et al have shown the benefits of mounting a compact mechanical image stabilizer onboard a mobile robot moving over rough terrain (Miyauchi, Shiroma & Matsuno 2008). Twombly et al. has carried out simulations on a neuro-vestibular control system designed to endow a walking robot with active image stabilization abilities (Twombly, Boyle & Colombano 2006). In the humanoid research field, some robotic developments have addressed the need to stabilize the gaze by providing robots with visuo-inertial oculomotor reflexes (e.g.: (Panerai, Metta & Sandini 2000)). Wagner et al. built a fast responding oculomotor system (Wagner, Hunter & Galiana 1992), using air bearings and bulky galvanometers. An adaptive gaze stabilization controller was recently described, but the performances of this device were measured only in the 0.5-2Hz frequency range (Lenz et al. 2008). Recently, Maini et al. succeeded in implementing fast gaze shifts on an anthropomorphic head but without using any inertial-based oculomotor reflexes (Maini et al. 2008). None of the technological solutions ever proposed so far are compatible, however, with the stringent constraints actually imposed on miniature aerial robots.

The gaze stabilization mechanisms of flying insects such as flies, are based on fine oculomotor reflexes that provide the key to heading stabilization. These high performance reflexes are of particular relevance to designing tomorrow's fast autonomous terrestrial, aerial, underwater and space vehicles. As we will see, visually mediated heading stabilization systems require:

- mechanical decoupling between the eye and the body (either via a neck, as in flies, or via the orbit, as in vertebrates' "camera eye")
- active coupling between the robot's heading and its gaze, via oculomotor reflexes
- a fast and accurate actuator. Flies control their gaze using no less than 23 pairs of micro-muscles (Strausfeld 1976)
- a visual fixation reflex (VFR) that holds the gaze steadily on the target.
- a vestibulo-ocular reflex (VOR), i.e., an active inertial reflex that rotates the eye in counter phase with the head. Flies typically use an inertial reflex of this kind which is based on the halteres gyroscopic organ, especially when performing roll movements (Hengstenberg 1988). A similar system was also developed in mammals - including humans - some hundred million years later. Rhesus monkeys' VORs are triggered in the 0.5-5Hz (Keller 1978) and even 5-25Hz (Huterer & Cullen 2002) frequency range, and are therefore capable of higher performances than humans.
- a proprioceptive sensor which is able to measure the angular position of the eye in the head or in the body. Although the question as to whether this sensor exists in the primate oculomotor system is still giving rise to some controversy (Clifford, Know & Dutton 2000)(Dancause et al. 2007), it certainly exists in flies in the form of a pair of mechanosensitive hair fields located in the neck region (Preuss & Hengstenberg 1992), which serve to measure and compensate for any head-body angular deviations in terms of pitch (Schilstra & Hateren 1998), roll (Hengstenberg 1988) and yaw (Liske 1977).

In section 2, we will describe our latest aerial robot, which has been called OSCAR II. OSCAR II differs from the original (OSCAR I) robot (Viollet & Franceschini 2001) in that its eye is no longer *mechanically coupled* to the body: this configuration makes it possible for the gaze to be actively locked onto the target, whatever disturbances may be applied to the robot's body. In Section 3, we will describe the scheme underlying the fast, accurate control of the "eye-in-head" angle. In section 4, we will explain how we merged a gaze control system (GCS) with a heading control system (HCS). In sections 5 and 6, we will present the robot's yaw control strategy and describe the outstanding performances attained by the overall gaze and heading control systems, which are both able to counteract nasty thumps delivered to the robot's body. Finally, in section 7, we will discuss about a novel biomimetic control strategy which combines both gaze orientation and locomotion.

2. Eye-in-head or head-in-body movements : a key to forward visuomotor control

Many studies have been published on how the gaze is held still in vertebrates and invertebrates, despite the disturbances to which the head (or body) is subjected. For example, in humans, the Rotational Vestibulo Ocular Reflex (RVOR, (Miles 1998)) triggers a compensatory eye rotation of equal and opposite magnitude to the head rotation, so that the line of sight (the gaze) is stabilized. Studies on the human RVOR have shown that this inertial system responds efficiently with a latency of only about 10ms to sinusoidal head rotations with frequencies of up to 4 Hz (Tabak & Collewijn 1994) or even 6Hz (Gauthier et al. 1984), as well as to step rotations (Maas et al. 1989). Rhesus monkeys show very high VOR performances in the 0.5-5Hz (Keller 1978b) and even 5-25Hz (Huterer & Cullen 2002) frequency ranges, which means that monkeys are able to reject both slow and fast disturbances throughout this wide range of frequencies. The fly itself possesses an exquisite VOR-like reflex controlling the orientation of its head (Hengstenberg 1988). Figure 1 illustrates the outstanding performances achieved by the gaze stabilization systems of two different birds and a sandwasp. In the latter case, the authors nicely showed how the roll compensation reflex functioned in a wasp in free flight by maintaining the head fixed in space in spite of dramatic body rolls (amplitude up to 120° peak to peak) made to counter any lateral displacements (Zeil, Norbert Boeddeker & Hemmi 2008). Cancelling head roll prevents the wasp's visual system from being stimulated and therefore disturbed by rotational movements.

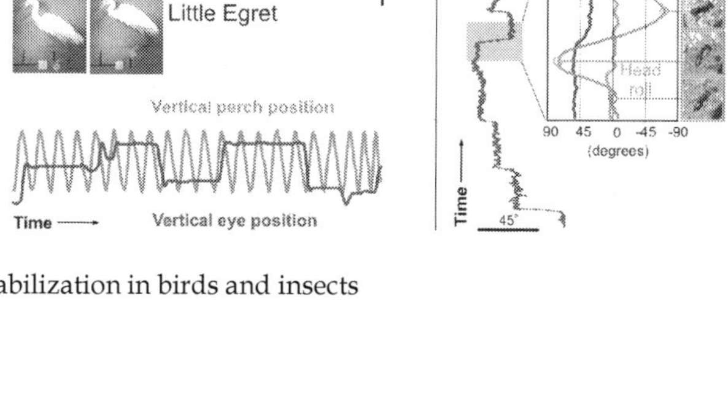

Figure 1. Gaze stabilization in birds and insects

Left: A night heron, Nycticorax nycticorax (top) and a little egret, Egretta garzetta (bottom) standing on a vertically oscillating perch. Note the long periods of perfectly stable eye position, interrupted by brief re-positioning head movements (From (Katzir et al. 2001)). Right: Horizontal gaze direction and head roll stabilization in a sandwasp (Bembix sp). Inset on the right shows thorax and head roll movements during a fast sideways translation to the left (see pictures) and a concurrent saccadic gaze shift to the right (From (Zeil, Boeddeker & Hemmi 2008)). Figure and legend reproduced from Zeil et al. with permission from Elsevier

In short, gaze stabilization seems to be a crucial ability for every animal capable of visually guided behavior. Even primitive animals such as the box jellyfish seem to be endowed with an exquisite mechanical stabilization system that holds the eyes oriented along the field of gravity (Garm et al. 2007).

3. Description of the OSCAR II robot

OSCAR II is a miniature (100-gram) cordless twin-engine aerial robot equipped with a single-axis (yaw) oculomotor mechanism (Fig. 2).

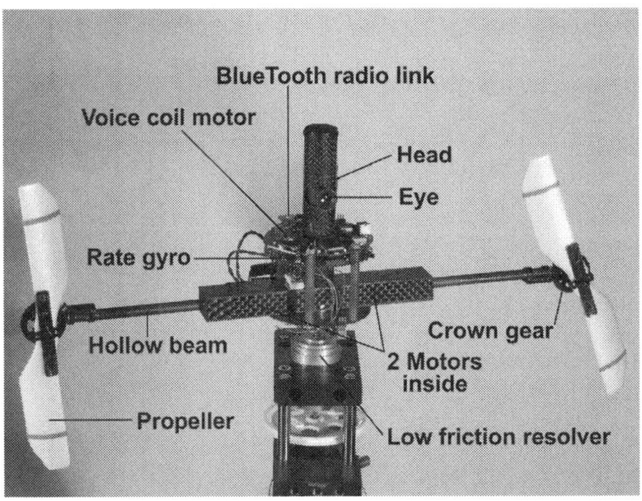

Figure 2. OSCAR II is a 100-gram aerial robot that is able to control its heading about one axis (the vertical, yaw axis) by driving its two propellers differentially on the basis of what it sees. The eye of OSCAR II is mechanically uncoupled from the head, which is itself fixed to the "body" A gaze control system (GCS in Fig. 6) enables the robot to fixate a target (a vertical white-dark edge placed 1 meter ahead) and to stabilize its gaze despite any severe disturbances (gusts of wind, slaps) that may affect its body. A heading control system (HCS in Fig. 6), combined with the GCS, makes the robot's heading catch up with the gaze, which stabilizes the heading in the gaze direction. OSCAR II is mounted on a low-friction, low-inertia resolver, so that its heading can be monitored

The robot is able to adjust its heading accurately about the yaw axis by driving its two propellers differentially via a custom-made dual sensorless speed governor (Viollet, Kerhuel

& Franceschini 2008). The robot's "body" consists of a carbon casing supporting the two motors. This casing is prolonged on each side by a hollow carbon beam within which the propeller drive shaft can rotate on miniature ball bearings. The robot's "head" is a large (diameter 15mm) carbon tube mounted vertically on the motor casing. Within the head, an inner carbon "eye tube" mounted on pivot bearings can turn freely about the yaw axis.
The robot's eye consists of a miniature lens (diameter 5mm, focal length 8.5mm), behind which an elementary "retina" composed of a single pair of matched PIN photodiodes scans the surroundings at a frequency of 10Hz by means of a fast piezo actuator (Physik Instrumente) driven by an onboard waveform generator circuit (for details, see (Viollet & Franceschini 2005)). The retinal microscanning movement adopted here was inspired by our findings on the fly's compound eye (Franceschini & Chagneux 1997). The microscanning of the two photoreceptors occurs perpendicularly to the lens' axis, making their line-of-sights deviate periodically in concert. For details on the whys and wherefores of the particular microscanning law adopted, readers can consult our original analyses and simulations of the OSCAR sensor principle (Viollet & Franceschini 1999). Basically, we showed that by associating an exponential scan with an Elementary Motion Detector (EMD), one can obtain a genuine *Angular Position Sensor* that is able to sense the position of an edge or a bar with great accuracy within the relatively small field-of-view available (FOV = ±1.4°, which is roughly equal to that of the human fovea). Interestingly, this sensor boasts a 40-fold better angular resolution than the inter-receptor angle in the task of locating an edge, and can therefore be said to be endowed with hyperacuity (Westheimer 1981). Further details about the performances (accuracy, calibration) of this microscanning visual sensor are available in (Viollet & Franceschini 2005).

4. Implementation of the robot's oculomotor system

In the human oculomotor system, the extra-ocular muscles (EOM) are often deemed to serve contradictory functions. On the one hand, they are required to keep the gaze accurately fixated onto a steady target (Steinman 1967), and on the other hand, they are required to rotate the eye with a very small response time: a saccade of moderate amplitude is triggered within only about 100 ms (Becker 1991). Figure 3 shows a top view scheme of the novel miniature oculomotor system we have built and installed in OSCAR II (figure 2).
The high performance human oculomotor system was mimicked by controlling the orientation of the eye-tube with an unconventional extra-ocular actuator: a Voice Coil Motor (VCM), which was initially part of a hard disk microdrive (Hitachi). A VCM is normally used to displace the read-write head in disk drive control systems (Chen et al. 2006) and it works without making any trade-off between high positional accuracy and fast displacement.
As VCM control requires an efficient position feedback loop. Whereas a simple PID controller was used in the original version (Kerhuel, Viollet & Franceschini 2007), we now used a state space approach by integrating a controller composed of an estimator cascaded with a state-augmented control gain K_{e0} (cf. figure 4) computed with a classical LQG method. This structure was used to servo the angular "eye in robot" position θ_{er} to the reference input $\theta_{er\_set\text{-}point}$ (see figure 4). θ_{er} was measured by placing a tiny Hall effect sensor in front of a micro magnet ($1mm^3$) glued to the eye-tube's rotation axis.

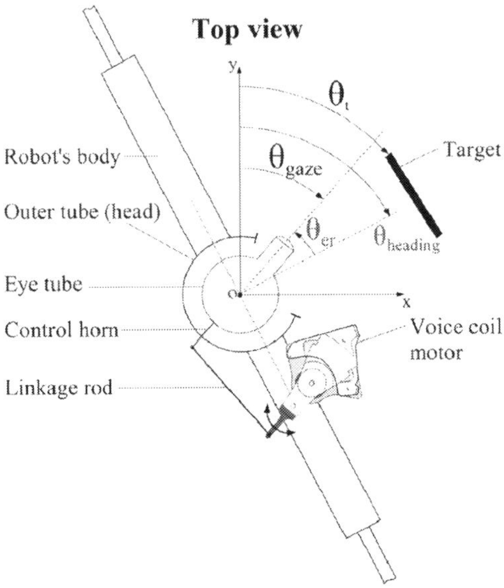

Figure 3. The OSCAR II oculomotor mechanism (top view). The central eye tube (equipped with its two-pixel piezo-scanning retina, not shown here) is inserted into a larger carbon tube (the "head"), which is mounted onto the robot's body. The eye tube is mechanically uncoupled from the head with one degree of freedom about the yaw axis. The angle θ_{er} between the robot's heading and the direction of the gaze is finely controlled (via the linkage rod and the control horn) by a micro Voice Coil Motor (VCM) that was milled out from a hard disk microdrive. The visual sensor's output is a linear, even function of $\theta_t - \theta_{gaze}$; it delivers 0 Volts when the gaze is aligned with the target (i.e., $\theta_{gaze} = \theta_t$). Adapted from (Kerhuel, Viollet & Franceschini 2007)

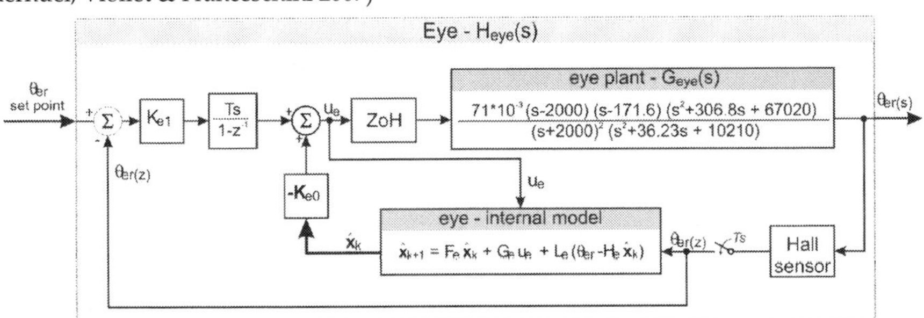

Figure 4. Block diagram of the Voice Coil Motor (VCM) servo system, which servoes the "eye in robot" angle θ_{er} (see figure 3) to the reference input $\theta_{er\_setpoint}$. In the internal state space model of the eye, both the command $U_e(z)$ and the measured angle $\theta_{er}(z)$ serve to estimate the 4 internal states of the eye's model, including its VCM actuator. The fifth external state is the integral of the eye's position error. A zero steady state error is classically obtained by augmenting the state vector and integrating the resulting angular position error

The step response shown in Figure 5 shows the very fast dynamics obtained with the closed-loop control of the eye-in-robot orientation, θ_{er}. We determined a rise time T_{rise} as small as 19ms and a settling time T_{settle} as small as 29ms (as compared to 44ms in the original version). With a 45-deg step (not shown here), a velocity peak of 2300°/s was reached, which is much higher than the 660°/s reached by our former PID controller (Kerhuel, Viollet & Franceschini 2007) and much higher than the saturation velocity (800°/s) of the human eye measured during a saccade (Maini et al. 2008). Unlike our robot's oculomotor control system (which is essentially linear), the human oculomotor control system is nonlinear, since the rise time increases typically with the saccade amplitude (Becker 1991).

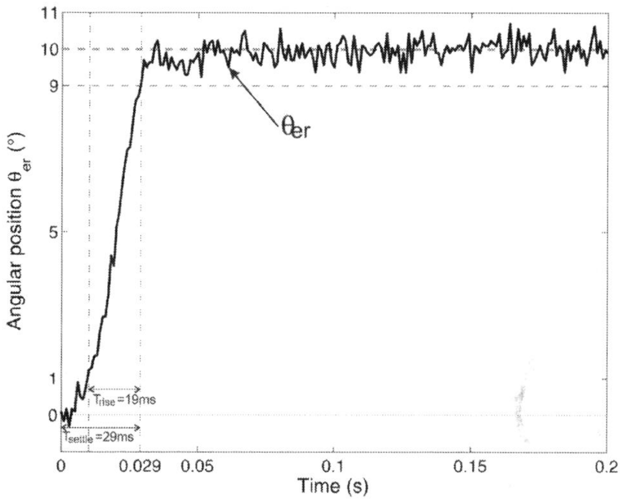

Figure 5. Closed-loop step response of the "Eye in Robot" angular position θ_{er} to a large (10 degrees) step input applied to the reference input $\theta_{er\ set\ point}$ (cf. figure 4). The voice coil motor actuator is controlled via a full state feedback controller that makes the settling time (T_{settle}) as small as 29ms. The angular position θ_{er} is measured with a miniature Hall sensor placed in front of a tiny magnet glued onto the eye's axis

5. A gaze control system that commands a heading control system

5.1 The gaze control system (GCS)

A VOR feedforward control pathway was implemented, which, like its natural counterpart, aims at counteracting any involuntary changes in heading direction. Like the semi circular canals of the inner ear, which give an estimate of the head's angular speed (Carpenter 1988), a MEMS rate gyro (analog device ADIS16100) measures the robot's body angular velocity. The VOR reflex makes θ_{er} follow any change in $\theta_{heading}$ faithfully but with opposite sign. In the frequency domain, this will occur only if the gain and phase of the transfer function relating θ_{er} to $\theta_{heading}$ are held at 0dB and 0deg, respectively, over the largest possible frequency range. This leads to the following theoretical expression for C_{VOR}:

$$C_{VOR_{th}}(s) = H_{gyro}^{-1}(s) H_{eye}^{-1}(s) \tag{1}$$

Stability problems caused by the high static gain introduced by the pseudo integrator $H_{gyro}^{-1}(s)$ led us to adopt an approximation noted $\hat{H}_{gyro}^{-1}(s)$. The expression of C_{VOR} therefore becomes:

$$C_{VOR}(s) = \hat{H}_{gyro}^{-1}(s) H_{eye}^{-1}(s) \qquad (2)$$

Figure 6 shows that the control signal U_e of the eye results from the difference of two control signals:

- U_v , an angular position signal arising from the *visual* (feedback) controller.
- U_{VOR}, an angular position signal arising from the *inertial* (feedforward) controller

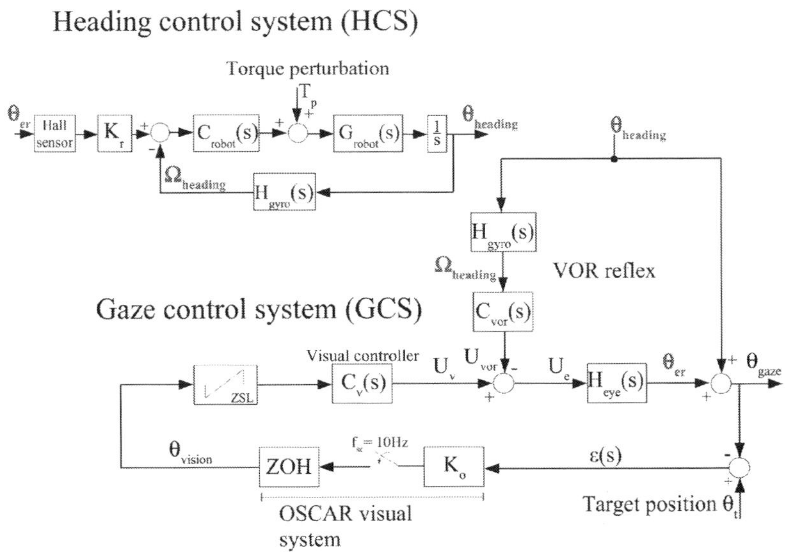

Figure 6. Block diagrams of the two interdependent control systems (an HCS and a GCS) implemented onboard the OSCAR II robot. The GCS keeps the gaze (θ_{gaze}) locked onto a stationary target (bearing θ_t), despite any heading disturbances (T_p). This system is composed of a visual feedback loop based on the OSCAR visual sensor (which acts as an "angular position sensing device") and a feedforward control system emulating the Vestibulo-Ocular-Reflex (VOR). The HCS servoes $\theta_{heading}$ to θ_{er} by adjusting the rotational speeds of the two propellers differentially. Since $\theta_{heading}$ is also an input disturbance to the GCS, any changes in heading (due to torque disturbances applied to the robot) is compensated for by a counter-rotation of the eye (θ_{er} angle). A null value of θ_{er} will mean that $\theta_{heading} = \theta_{gaze}$. Note that the two proprioceptive signals θ_{er} and $\Omega_{heading}$ given by the Hall sensor and the rate gyro (cf. Fig. 1), respectively, are used in both the GCS and the HCS. Adapted from (Kerhuel, Viollet & Franceschini 2007)

Therefore, if the robot's heading is subjected to a brisk rotational disturbance, the change in $\theta_{heading}$ will immediately be measured and compensated for by the VOR feedforward control system. The latter will impose a counter rotation of the eye of the similar amplitude but opposite sign. In Figure 6, it can be seen that $\theta_{heading}$ also acts as an *input disturbance* to the gaze control system (GCS). The control signal U_v derived from the visual controller $C_v(s)$ adjusts the orientation θ_{er} of the eye so as to compensate for this disturbance, thus holding the gaze θ_{gaze} effectively in the direction θ_t of the visual target (that is, making $\varepsilon(s) = 0$ in Figure 6, bottom right).

We established that θ_{er} is able to follow $\theta_{heading}$ faithfully over a very large frequency range (between 1Hz and 11Hz, data not shown here). The only limitations are due to the change we made in C_{VOR} (for the sake of stability) and the approximations made during the identification of the transfer functions $H_{gyro}(s)$ and $H_{eye}(s)$.

As described in section 9 (appendix), the visual controller $C_v(s)$ (see figure 6) is an integrator. This means that the visual controller copes with any target displacement without introducing any steady state error ($\varepsilon = \theta_t - \theta_{gaze}$ in figure 6). In other words, there is no "retinal slip error" in the steady state. To prevent runaway of the eye when it loses a target, we developed a special limiter (Viollet & Franceschini 2001), which we have called a Zero-Setting Limiter (ZSL), and introduced it upstream from the visual controller (figure 6). The purpose of this nonlinear block is to clamp the error signal back to zero whenever the latter becomes higher (or lower) than a specified positive (or negative) level. At a scanning frequency of 10Hz, the OSCAR II visual sensor inevitably introduces a latency of 100ms into the visual feedback loop. This latency is the main limiting factor in the process of rejecting any fast visual disturbances to which the robot is exposed. The VOR reflex acts in a complementary manner, dramatically improving the dynamics of gaze stabilization, and thus preventing the fixated target from straying outside the (narrow) field-of-view of the eye.

5.2 The heading control system (HCS)

One of the most novel features of the present study is the fact that the visuo-inertial reflex described above was combined with the heading control system of the OSCAR II robot. The HCS was designed to take the robot's yaw dynamics, given by the transfer function $G_{robot}(s)$, into account. The HCS involves (i) a measurement of the robot's yaw angular speed $\Omega_{heading}$ (given by the rate gyro), and (ii) a proportional-integral controller (included in $C_{robot}(s)$). In the steady state, the angle θ_{er} is null, which means that the HCS makes $\theta_{heading}$ equal to θ_{gaze} (zero steady-state error). In other words, the robot's heading catches up with the gaze direction: the robot orients itself where its eye is looking.

The use of the HCS (top part of figure 6) means that the robot's orientation ($\theta_{heading}$) is servoed to the eye-in-robot orientation (θ_{er}). These two angles are therefore actively coupled. The fact that the robot "carries the eye" means that $\theta_{heading}$ constitutes both an input disturbance to the GCS based on the OSCAR visual system and an input signal to the rate gyro. It is also worth noting that the rate gyro is involved in both the VOR reflex and the speed feedback loop of the HCS (see figure 6).

To summarize, both the GCS and the HCS act in concert and share the same two proprioceptive sensors: (i) the Hall sensor that delivers θ_{er} and the rate gyro that delivers $\Omega_{heading}$. Although the GCS and HCS loops are strongly interdependent, only the HCS involves the robot's dynamics. This means that the controllers present in the GCS can be tuned by taking only the dynamics of the disturbance $\theta_{heading}$, that needs to be rejected, into account. This greatly simplifies the design of the overall control system.

6. High performance gaze stabilisation system

The overall gaze control system does not require large computational resources. The two digital controllers (one dealing with the VCM based feedback control system, and the other with the propellers speed control system (Viollet, Kerhuel & Franceschini 2008)) were built using a custom-made rapid prototyping tool designed for use with Microchip dsPIC. All the controllers involved in the VOR and the visual feedback-loop were digitized using Tustin's method and implemented in the dSPACE environment.

To test our miniature gaze and heading control system, we applied drastic torque perturbations to the robot's body. For this purpose, we built a "slapping machine" consisting of a DC motor and a light wooden arm. The arm is attached to the shaft of an electromagnetic clutch. On powering the clutch, the DC motor suddenly delivers a high acceleration thump on one side of the robot's body. The slapping machine was placed so that the arm would hit the robot and brisk thumps were thus applied to the robot repetitively was fixating a contrasting edge placed 1m from the eye.

As can be seen from the HCS block diagram (figure 6, top), any torque perturbation T_p will be compensated for by the controller C_{robot}. Meanwhile, however, the torque perturbation will have led inevitably to a change of heading. Since $\theta_{heading}$ acts as an input disturbance to the GCS (see figure 6, top of GCS), any torque perturbation is also compensated for by a counter rotation of the eye-in-robot θ_{er}. This means that the robot re-orients its heading until θ_{er} becomes null again, thus automatically bringing the heading in line with the gaze.

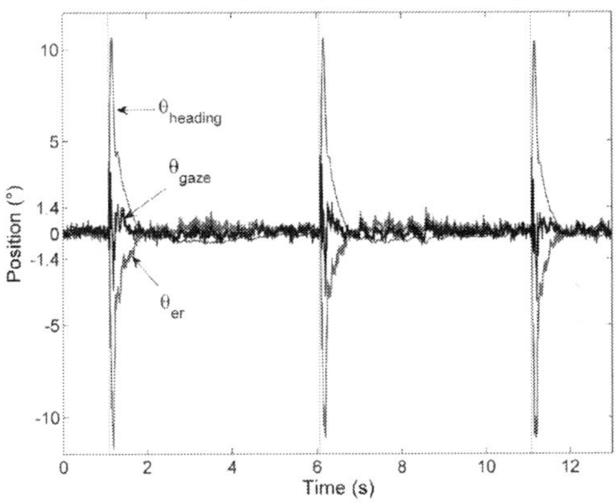

Figure 7. Reaction of the robot's orientation ($\theta_{heading}$), the "eye-in-robot" angle (θ_{er}) and the gaze (θ_{gaze}) to a sequence of 3 thumps delivered every 5 seconds (the thin vertical lines give the timing of each thump). The sudden yaw perturbation can be seen to have been counteracted swiftly, within 20ms by the VOR reflex, which succeeded in maintaining the robot's gaze (θ_{gaze}) close to the target position. The robot then reoriented itself more slowly (taking about 0.6 seconds) due to its slower body dynamics. Adapted from (Kerhuel, Viollet & Franceschini 2007)

The robot was mounted onto the shaft of a low friction, low inertia resolver which made it possible to accurately monitor the azimuthal orientation $\theta_{heading}$ (angular resolution of the resolver: 0.09°) - it should be stressed that the resolver is not involved in any control system whatsoever. As shown in Figure 7, the $\theta_{heading}$ was violently (and reproducibly) perturbed by three sudden slaps. The eye can be seen to have swiftly counter rotated in the robot's body (curve θ_{er}), keeping the gaze (curve θ_{gaze}) virtually locked onto the target, despite this untoward perturbation.

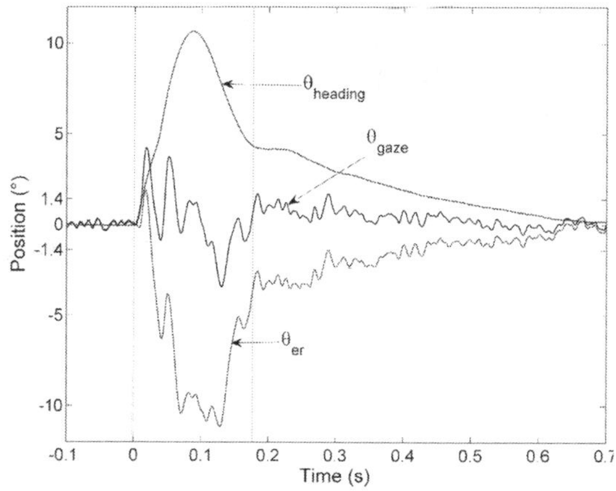

Figure 8. Magnified version of the second thump applied to the robot in figure 7. The time at which the thump was delivered is given by the left vertical line. The "eye-in-robot" profile (θ_{er} red curve) shows that the eye rotation immediately counteracts the robot's rotation ($\theta_{heading}$ blue curve), so that the gaze (θ_{gaze} black curve) remains quasi-steady. The robot's fast return phase (lasting between 0ms and 177ms) is mainly generated by the yaw rate inner loop combined with the action of the VOR. The $\theta_{heading}$ slow return phase (lasting between 177ms and 650ms) results from the control input signal θ_{er}. The VOR reflex operates quasi-instantaneously, whereas the robot's visual system has a relatively slow (10Hz) refresh rate. Adapted from (Kerhuel, Viollet & Franceschini 2007)

Figure 8 shows a close-up of the robot's eye and gaze responses to the second thump delivered as shown in figure 7. Time 0s corresponds here to the exact time when the thump was applied, as determined with a micro-accelerometer mounted at the tip of the inter-propeller beam. The robot's response can be decomposed into two phases:

- A fast phase (between 0ms and 177ms), when the perturbation was rejected, mostly by the yaw rate inner loop and the VOR via the reference input signal θ_{er} (cf. figure 6).
- A slow phase (lasting between 177ms and 650ms), when the perturbation was entirely rejected by both the VOR and the visual feedback-loop.

The eye position θ_{er} can be seen to counteract the robot's position $\theta_{heading}$ quasi perfectly (figure 8) thanks to the high speed dynamics of the eye's orientation feedback control system based on the VCM actuator. The eye's rotation is fast enough to keep the gaze θ_{gaze} locked onto the target. It is not possible to measure the robot's gaze (θ_{gaze}) directly (this would require an eye tracker or a magnetic search coil). The gaze was therefore calculated on the basis of the of the two measurable signals, $\theta_{heading}$ and θ_{er} (see figure 3):

$$\theta_{gaze} = \theta_{heading} + \theta_{er} \tag{4}$$

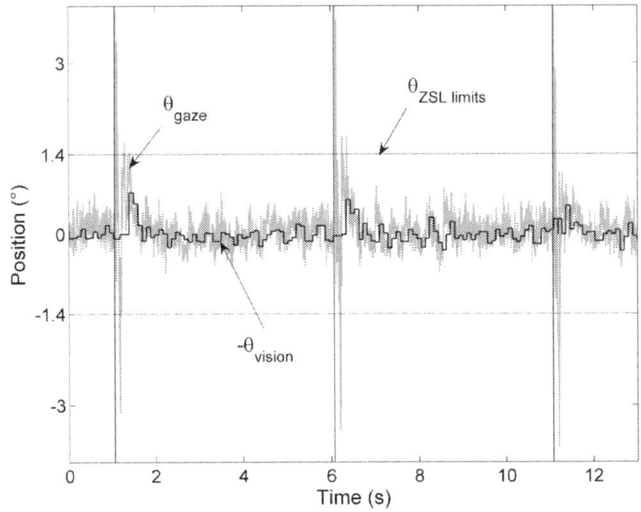

Figure 9. Gaze orientation (θ_{gaze}) compared with θ_{vision}, the gaze orientation to the target's orientation, as measured by the OSCAR sensor (see bottom left of figure 6), during the sequence of 3 thumps presented in figure 7. The two horizontal red lines delimit the field of view (±1.4°) of the eye. A gaze value greater than |1.4|° means that the target has wandered out of the field of view. The time during which the target strayed out of the visual field is so short (50ms, i.e. twice as short as the visual refresh period) that it does not impair the gaze stabilization performances. Adapted from (Kerhuel, Viollet & Franceschini 2007)

Figure 9 shows that the contrasting target (a white-dark edge) may actually wander out of the small, ±1.4° field of view of the eye for a very short time (50ms). The contrasting target keeps being "seen" by the eye, however, as shown by the θ_{vision} signal. The reason is that the time during which the target strays out of the visual field is so short (50ms, i.e. twice as short as the visual refresh period) that it does not impair the gaze stabilization performances.

7. Steering by gazing : an efficient biomimetic control strategy.

In addition to describing the use of suitably designed oculomotor reflexes for stabilizing a robot's gaze, the aim of this study was to present a novel concept that we call "steering by gazing". Many studies have addressed the question as to how vertebrates and invertebrates use their gaze during locomotion. These studies have shown that the locomotor processes at work in many species such as humans (Wann & Swapp 2000)(Schubert et al. 2003), flying insects (Collett & Land 1975)(Schilstra & Hateren 1998)(Zeil, Norbert Boeddeker & Hemmi 2008), crabs (Paul, Barnes & Varju 1998) and even bats (Ghose & Moss 2006) involve a gaze orientation component.

Figure 10 summarizes the various feedforward and feedback control systems involved in the control of a robotic platform such as OSCAR II. The control system depicted in figure 10 is a one input (θ_{target}) and two outputs (θ_{gaze} and $\theta_{heading}$) system. The "steering by gazing" control strategy aims at making θ_{gaze} and $\theta_{heading}$ (i.e., the complete robot) to follow any variation in

θ_{target}. The mechanical decoupling between the eye and the body is here modeled by the robot block where the unique control input signal is split into one input reference for controlling the eye's orientation and one error signal for controlling the robot's heading (cf. figure 10). For a stationary target, the control system will compensate for any disturbances applied to the body by holding the gaze locked onto the target. For a moving target, the control system will change both the eye's orientation and the robot's heading to track smoothly the target.

Let us look at the path involving the "vestibulo-ocular reflex" (VOR) and the eye blocks in figure 10. On this path, the VOR feedforward control can be identified between $\theta_{heading}$ and θ_{er}. The minus sign in Σ_2 means that any rotation of the head will be compensated for by a counter rotation of the eye.

The block diagram in figure 10 also shows two feedback loops controlling both a fast plant (the eye) and a slow plant (the robot's body):

- the eye's orientation is controlled by the *visual feedback-loop* (upper closed-loop in figure 10) and the *feedfoward control* based on the VOR block.
- the robot's heading is controlled by an *inertial feedback-loop* (lower closed-loop in figure 10) based on an estimate of the heading deduced from the robot's rotational speed measured by a rate gyro.

As shown in figure 10, these two control feedback-loops are merged by using the summer Σ_2 where the estimated robot's heading ($\hat{\theta}_{heading}$) becomes an *input disturbance* for the visual feedback-loop, whereas the retinal error becomes a *reference input* ($\theta_{heading\_ref}$) for the inertial feedback-loop.

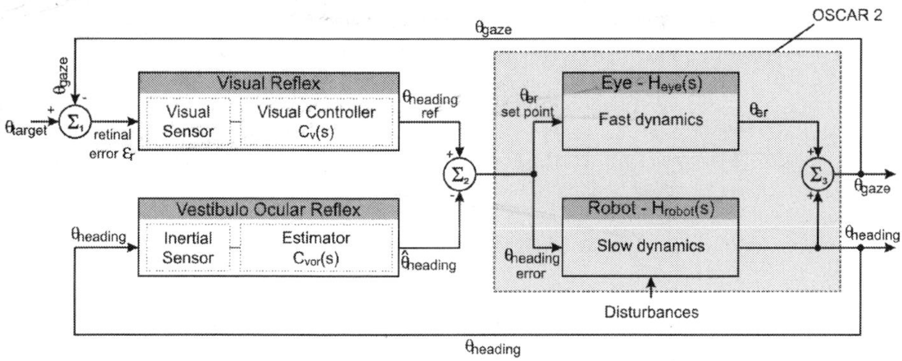

Figure 10. Generic block diagram of the "steering by gazing" control strategy. This is a control system where the input is the angular position of the target θ_{target} and its two outputs are the gaze orientation θ_{gaze} and the robot's heading $\theta_{heading}$. This system can be described in terms of Main-Vernier loops (Lurie & Enright 2000) where the reference input received by the slow heading feedback-loop is the $\theta_{heading\_ref}$ provided by the fast visual feedback loop (θ_{gaze}). This novel control system meets the following two objectives:

- keeping the gaze locked onto the visual target in spite of the aerodynamical disturbances (gusts of wind, ground effects, etc.) to which the robot is subjected
- automatically realigning the robot's heading $\theta_{heading}$ in line with the orientation of its gaze.

To summarize, the general control scheme presented in figure 10 enables any sighted vehicle:

- to hold its gaze locked onto a contrasting target such as an edge
- to stabilize its gaze despite any disturbances generated during its locomotion ("gaze stabilization")
- to track a moving target smoothly ("smooth pursuit")
- to orient or reorient its heading automatically in the line of sight, and hence toward the target ("steering by gazing")

8. Conclusion

Here we have described how a miniature tethered aerial platform equipped with a one-axis, ultrafast accurate gaze control system inspired by highly proficient, long existing natural biological systems was designed and implemented. The seemingly complex gaze control system (figure 6) was designed to hold the robot's gaze fixated onto a contrasting object in spite of any major disturbances undergone by the body. It was established that after being destabilized by a nasty thump applied to its body (cf. figures. 7,8 and 9), the robot:

- keeps fixating the target (despite the small visual field of its eye, which is no larger than that of the human fovea)
- reorients its heading actively until it is aligned with the gaze direction

Reorientation is achieved rapidly, within about 0.6 seconds (figure 8). The important point here is that the gaze itself is the fundamental (Eulerian) reference parameter, on which all the relevant motor actions (orienting the "eye in robot" and the "robot in space") are based.

This study considerably extends the scope of a former study, in which we developed a gaze control system but did not implement it onboard a robotic platform (Viollet & Franceschini 2005). Besides, the oculomotor mechanism we are now using is a novel version based on a voice coil motor (VCM) taken from a hard disk microdrive. This actuator, which is able to orient the gaze with a settling time as short as 19ms (figure 5) (i.e., faster than a human ocular saccade), was the key to the development of our ultrafast gaze stabilization system.

The two control systems (HCS and GCS) presented in figure 6 are strongly interdependent. The HCS is *actively* coupled to the GCS via the inputs θ_{er} (as measured by the Hall sensor) and $\Omega_{heading}$ (as measured by the rate gyro) it receives. Although the eye is mechanically uncoupled from the robot's body, the GCS is *passively* coupled to the HCS due to the fact that the robot carries the whole oculomotor system (and may therefore disturb the gaze orientation). By coupling the two control systems both actively and passively, we have established that the high performances of the robot's heading control system (HCS) result directly from the high performances of the gaze control system (GCS). In other words, a fast gaze stabilization is a key to fast navigation.

Our lightweight, robust gaze control system could provide a useful basis for the guidance of manned and unmanned air vehicles (UAVs) and benthic underwater (UUVs) vehicles, and especially for micro-air vehicles (MAVs) and micro-underwater vehicles (MUVs), which are particularly prone to disturbances due to fast pitch variations, wing-beats (or body undulations or fin-beats), wind gusts (or streams), ground effects, vortices, and many other kinds of unpredictable aerodynamic (or hydrodynamic) disturbances. Biological systems teach us that these disturbances can be quickly compensated for by providing robots with a visually mediated gaze stabilization system. In biological systems, locomotion is often based on visually-guided behavior where the gaze orientation plays the role of the pilot. The generic control scheme (figure 10) presented here is in an attempt to model this biomimetic "steering by gazing" strategy.

9. Appendix

$H_{gyro}(s) = K_g \frac{(\tau_2 s + 1)}{(\tau_1 s + 1)}$	With $\tau_1 = 4.3 * 10^{-3}$s , $\tau_2 = 1897.5$s and $K_g = 2.27 * 10^{-3}$
$\hat{H}^{-1}_{gyro}(s) = K_{ginv} \frac{(\tau_5 s + 1)}{(\tau_6 s + 1)}$	With $\tau_5 = 3.68 * 10^{-3}$s , $\tau_6 = 2.31$s and $K_{ginv} = 606.5$
$H_{eye}(s) = K_e \frac{(\tau_4 s + 1)}{(\tau_3 s + 1)}$	With $\tau_3 = 18.7 * 10^{-3}$s , $\tau_4 = 0.5 * 10^{-3}$s and $Ke = 226.3$
$H_{robot}(s) = \frac{K_r}{\frac{1}{W_r^2} s^2 + \frac{2\zeta_r}{W_r} s + 1}$	With $W_r = 39.9\,rad\,s^{-1}$, $\zeta_r = 0.27$ and $K_r = 0.7889$
$K_r = 6$	Pure gain
$C_{robot}(s) = K_c \frac{\tau_c s + 1}{\tau_c s}$	With $\tau_c = 7.4$s and $K_c = 3.7 * 10^{-3}$
$C_{vor}(s) = \hat{H}^{-1}_{gyro}(s)\ H^{-1}_{eye}(s)$	
$C_v(s) = \frac{K_0}{s}$	With visual sampling rate $T_{sc} = 0.1s$ and $K_0 = 0.0574$

10. Acknowledgements

The authors acknowledge the assistance of M. Boyron for the design of the miniature electronic boards including the piezo driver, the EMD and the control systems. We thank F. Paganucci and Y. Luparini for their help with the mechanical construction of the eye, L. Goffart, F. Ruffier and J. Serres for fruitful discussions and comments on the manuscript. This work was supported by CNRS, University of the Mediterranean, the French National Agency for Research (ANR, RETINAE project) and the French Defence Agency (DGA).

11. References

Becker, W., 1991. Vision and visual dysfunction (Vol 8). In *GR.H.S. Carpenter* (Ed) Macmillan Press, Ltd, pp. 95-137.

Boeddeker, N., Kern, R. & Egelhaaf, M., 2003. Chasing a dummy target: smooth pursuit and velocity control in male blowflies. In *Proc. R. Soc. Lond.B 270*. pp. 393-399.

Carpenter, R.H.S., 1988. Movements of the eyes, 2nd edt. In *PION*, London.

Chen, B.M. et al., 2006. *Hard Disk Drive Servo Systems 2nd Edition*, Springer, Berlin.

Clifford, R.W., Know, P.C. & Dutton, G.N., 2000. Does extraocular muscle proprioception influence oculomotor control ? *Br. J. Ophtalmol*, 84, 1071-1074.

Collett, T.S. & Land, M.F., 1975. Visual control of flight behaviour in the hoverflySyritta pipiens L. *Journal of Comparative Physiology A: Neuroethology, Sensory, Neural, and Behavioral Physiology*, 99(1), 1—66.

Dancause N., M. D. Taylor, E. J. Plautz, J. D. Radel, T. Whittaker, R. J. Nudo and A. G. Feldman, 2007. A stretch reflex in extraocular muscles of species purportedly lacking muscle spindles. *Exp Brain Res*, 180(1), 15—21.

Franceschini, N. & Chagneux, R., 1997. Repetitive scanning in the fly compound eye. In *Gottingen Neurobiol. Conf.* Gottingen, p. 279.

Garm A., M. O'Connor, L. Parkefelt and D-E. Nilsson, 2007. Visually guided obstacle avoidance in the box jellyfish Tripedalia cystophora and Chiropsella bronzie. *J Exp Biol*, 210(Pt 20), 3616—3623.

Gauthier, G.B. et al., 1984. High-frequency vestibulo-ocular reflex activation through forced head rotation in man. *Aviat Space Environ Med*, 55(1), 1—7.

Ghose, K. & Moss, C.F., 2006. Steering by hearing: a bat's acoustic gaze is linked to its flight motor output by a delayed, adaptive linear law. *J Neurosci*, 26(6), 1704—1710.

Hengstenberg, R., 1988. Mechanosensory control of compensatory head roll during flight in the blowfly Calliphora erythrocephala Meig. *J. Comp Physiol A*, 163, 151-165.

Huterer, M. & Cullen, K.E., 2002a. Vestibuloocular reflex dynamics during high-frequency and high acceleration rotations of the head on body in rhesus monkey. *J Neurophysiol*, 88, 13-28.

Katzir, G. et al., 2001. Head stabilization in herons. *J Comp Physiol [A]*, 187(6), 423—432.

Keller, E.L., 1978. Gain of the vestibulo-ocular reflex in monkey at high rotational frequencies. *Vis. Res.*, 18, 311-115.

Kerhuel, L., Viollet, S. & Franceschini, N., 2007. A sighted aerial robot with fast gaze and heading stabilization. In *Intelligent Robots and Systems, 2007. IROS 2007. IEEE/RSJ International Conference on*. p. 2634—2641.

Kern, R. & Varju, D., 1998. Visual position stabilization in the hummingbird hawk moth, Macroglossum stellatarum L. *J Comp Physiol A 182*, 225-237.

Lenz, A. et al., 2008. An adaptive gaze stabilization controller inspired by the vestibulo-ocular reflex. *Bioinspir Biomim*, 3(3), 35001.

Lewis, A., 1997. Visual navigation in a robot using Zig-Zag behavior. In *Proceedings of Neural Informations Processing Systems (NIPS)*. pp. 822-828.

Liske, E., 1977. The influence of head position on the flight behaviour of the fly, Calliphora Erythrocephala. *J. Insect Physiol*, 23, 375-179.

Lurie, B. & Enright, P., 2000. *Classical feedback control with Matlab* M. Dekker, éd., Marcel Dekker.

Maas E. F., W. P. Huebner, S. H. Seidman and R. J. Leigh, 1989. Behavior of human horizontal vestibulo-ocular reflex in response to high-acceleration stimuli. *Brain Res*, 499(1), 153—156.

Maini E., Manfredi L., Laschi C., and Dario P., 2008. Bioinspired velocity control of fast gaze shifts on a robotic anthropomorphic head. *Autonomous Robots*, 25(1), 37—58.

Meyer, J.-A., Guillot A., Girard B., Khamassi M., Pirim P., and Berthoz A., 2005. The Psikharpax project: towards building an artificial rat. *Robotics and Autonomous Systems*, 50(4), 211-223.

Miles, F.A., 1998. The neural processing of 3-D visual information: evidence from eye movements. *Eur J Neurosci*, 10(3), 811—822.

Miyauchi, R., Shiroma, N. & Matsuno, F., 2008. Compact Image Stabilization System Using Camera Posture Information. *J of field robotics*, 25, 268-283.

Olberg, R.M. et al., 2007. Eye movements and target fixation during dragonfly prey-interception flights. *J Comp Physiol A Neuroethol Sens Neural Behav Physiol*, 193(7), 685—693.

Panerai, F., Metta, G. & Sandini, G., 2002. Learning visual stabilization reflexes in robots with moving eyes. *Neurocomputing*, 48, 323-337.

Panerai, F., Metta, G. & Sandini, G., 2000. Learning VOR-like stabilization reflexes in robots. In *ESANN'2000 Proceedings*. Bruges,Belgium, pp. 95-102.

Paul, Barnes & Varju, 1998. Roles of eyes, leg proprioceptors and statocysts in the compensatory eye movements of freely walking land crabs (Cardisoma guanhumi) . *J Exp Biol*, 201 (Pt 24), 3395—3409.

Preuss, T. & Hengstenberg, R., 1992. Structure and kinematics of the prosternal organs and their influence on head position in the blowfly Calliphora erythrocephala Meig. *J Comp Physiol A 171*, 483-493.

Sandeman, D., 1980. Head Movements in Flies (Calliphora) Produced by Deflexion of the Halteres. *J Exp Biol*, 85(1), 43—60.

Schilstra, C. & Hateren, J.H.V., 1998. Stabilizing gaze in flying blowflies. *Nature*, 395(6703), 654.

Schubert M., C. Bohner, W. Berger, M. Sprundel and J. E J Duysens, 2003. The role of vision in maintaining heading direction: effects of changing gaze and optic flow on human gait. *Exp Brain Res*, 150(2), 163—173.

Shibata T., Tabata H. , Schaal S. and Kawato M., 2001. Biomimetic gaze stabilization based on feedback-error-learning with nonparametric regression networks. *Neural Networks*, 14, 201-216.

Steinman, R.M., 1967. Voluntary Control of Microsaccades during Maintained Monocular Fixation. *Sciences*, 155, 1577-1579.

Strausfeld, N., 1976. *Atlas of an Insect Brain*, Springer-Verlag, Berlin, Heidelberg.

Tabak, S. & Collewijn, H., 1994. Human vestibulo-ocular responses to rapid, helmet-driven head movements. *Exp Brain Res*, 102(2), 367—378.

Twombly, X., Boyle, R. & Colombano, S., 2006. Active Stabilization of Images Acquired on a Walking. In *Advances in visual computing. Part I-II : Second international symposium, ISVC 2006*. pp. 851-860.

Viola, P., 1989. Neurally inspired plasticity in oculomotor processes. In *Proceedings of Neural Informations Processing Systems (NIPS)*. pp. 290-297.

Viollet, S. & Franceschini, N., 2005. A high speed gaze control system based on the Vestibulo-Ocular Reflex. *Robotics and Autonomous systems*, 50, 147-161.

Viollet, S. & Franceschini, N., 1999. Biologically-inspired visual scanning sensor for stabilization and tracking. In *Intelligent Robots and Systems, 1999. IROS '99. Proceedings. 1999 IEEE/RSJ International Conference on*. p. 204—209vol.1.

Viollet, S. & Franceschini, N., 2001. Super-accurate visual control of an aerial minirobot. In *Autonomous minirobots for research and edutainment, AMIRE.*

Viollet, S., Kerhuel, L. & Franceschini, N., 2008. A 1-gram dual sensorless speed governor for micro-air vehicles. In *Control and Automation, 2008 16th Mediterranean Conference on.* p. 1270—1275.

Wagner, R., Hunter, I.W. & Galiana, H.L., 1992. A Fast Robotic Eye/head System: Eye Design and Performance. In *Proc. of IEEE Engineering in Medicine and Biology Society.* pp. 1584-1585.

Wann, J.P. & Swapp, D.K., 2000. Why you should look where you are going. *Nat Neurosci,* 3(7), 647—648.

Westheimer, G., 1981. *Visual hyperacuity,* Berlin: Ottoson, Sensory Physiology 1, Springer.

Yamaguchi, T. & Yamasaki, H., 1994. Velocity Based Vestibular-Visual Integration in Active Sensing System. In *Proc. IEEE Intern. Conf. on Multisensor Fusion and Integration for Intelligent System.* Las Vegas, USA, pp. 639-646.

Zeil, J., Boeddeker, N. & Hemmi, J.M., 2008. Vision and the organization of behaviour. *Curr Biol,* 18(8), R320—R323.

32

Robust Path-Following for UAV Using Pure Pursuit Guidance

Takeshi Yamasaki, Hiroyuki Takano and Yoriaki Baba
National Defense Academy
Japan

1. Introduction

This chapter presents a description of a guidance and control system on robust path-following flight for Unmanned Aerial Vehicle (UAV). Guidance and control systems on path-following flight for UAV, most based on a tracking-error-correction approach, have been studied for several years (Blajer, 1990; Baba et al., 1996, 2002; Kaminer et al., 1998; Boyle & Chamitoff, 1999; Ochi et al., 2002; Park et al., 2004; Rysdyk, 2006). However, these systems present the difficulty of following in large tracking-error situations, e.g. steep curved path-following or tracking under wind turbulence, which might cause control saturation or divergence because the control command tends to increase as the tracking-error becomes large. One coauthor proposed a variable gain method using fuzzy logic (Baba & Takano, 1998), which gathers and weighs on control laws according to tracking-error quantities. It performed well, but it was necessary to set some design points and to conduct several gain tunings before applying fuzzy logic.

This chapter presents a novel, simple, yet robust guidance method for path-following UAV (Sato et al., 2006; Yamasaki et al., 2007). The methodology uses pure pursuit guidance instead of traditional tracking-error correction-based methods. Pure pursuit guidance (e.g. Machol et al., 1965) demands only one gain-tuning. It produces guidance commands that are not large, with no relation to tracking error quantities. It can avoid control divergence because the pure pursuit guidance system generates guidance commands in relation to the line-of-sight angle (the angle formed by the UAV velocity vector and the line-of-sight to the target point), which is, at most, π radian. That angle might be much less than the tracking error, e.g. 10 m. For that reason, a robust path-following UAV can be realized assisted by pure pursuit guidance (Park et al., 2007). A target point for the UAV to orient is necessary for applying pure pursuit guidance for path-following. We introduced a receding virtual waypoint, which is placed at a proper point along a desired trajectory. This is the point of difference from the error-based reference point described in the literature (Park et al., 2007). The desired path can be provided easily in a form of a function of the trajectory arc length using cubic-spline interpolation based on some waypoints through which the UAV is presumed to fly. Once the path is generated in the form of the function of the arc length, the receding virtual waypoint, which is the target point for pure pursuit guidance, is calculable using the cubic-spline function based on the UAV flight arc length added a few seconds ahead of the future flight path length, as inferred from the UAV dynamics. This added term provides a future desired maneuver to the UAV, which compensates the UAV's response delay because of the UAV dynamics. Guidance commands for the UAV orientation are generated using the pure pursuit guidance with a receding virtual waypoint. The UAV can change its heading by altering the attitude. The system employs a dynamic inversion technique for the UAV's attitude control. Dynamic inversion techniques (Brockett, 1978; Lane & Stengel, 1998; Snell et al., 1992; Baba et al., 1996), which enable the UAV to perform a nonlinear path-following ability and to maintain high maneuverability, are well-known control schemes of nonlinear dynamic systems. A dynamic inversion technique with two-time scale separation (Menon et al., 1987; Snell et al., 1992; Azam & Singh, 1994) is applicable for UAV control, which allows the order of the dynamic inversion controller to be small. The pure pursuit guidance lets the UAV head to the target at any time. Because the pure pursuit guidance can, by its nature, head to a target point only and never works for the velocity control along the velocity vector directly, a velocity control system is necessary to maintain a desired distance from the given trajectory. The velocity controller is designed based also on the dynamic inversion approach, considering the aircraft dynamics.

First, the body axes are explained along with full, rigid body, six-degree-of-freedom, nonlinear equations of motion. Then we will discuss the guidance and control system for UAV, which is based on the pure pursuit guidance law with a receding virtual waypoint, and the two-time scaled dynamic inversion technique, along with velocity control. Several simulations are conducted to demonstrate its path-following robustness under some random wind turbulence and steeply curved path-following situations. We also show a comparison to the tracking-error-correction-based method.

2. Symbols and Notations

x, y, z = Components of position vector w.r.t. body axes, [m]

x_I, y_I, z_I = Components of position vector w.r.t. inertial axes, [m]

u, v, w = Components of airplane velocity along the x, y, z body axes, respectively, [m/s]

p, q, r = Components of airplane angular velocity about the x, y, z body axes, respectively, [rad/s]

ϕ, θ, ψ = roll, pitch, and yaw angle, respectively, [rad] or [deg]

$\delta_e, \delta_a, \delta_r$ = Deflection of elevator, aileron and rudder, respectively, [rad] or [deg]

L = Lift, [N]

D = Drag, [N]

F_X, F_Y, F_Z = Guidance forces w.r.t. body axes, [N]

$F_{\tilde{X}}, F_{\tilde{Y}}, F_{\tilde{Z}}$ = Guidance forces w.r.t. wind axes, [N]

F_T = Thrust, [N]

$\boldsymbol{F}_d$ = Desired guidance force vector

q_D = Dynamic pressure, [N/m^2]

$g, \boldsymbol{g}$ = Acceleration of gravity, [m/s^2] and gravity vector

$V, \boldsymbol{V}$ = Resultant velocity, [m/s] and velocity vector

ρ = Air density, [kg/m^3]

m = Mass, [kg]

S = Wing area, [m²]
$b, \bar{c}$ = Wing span and wing mean aerodynamic chord, [m]
h = Altitude, [m]
α, α_0 = Angle of attack (AoA) and zero lift angle of attack, [rad] or [deg]
β = Side slip angle, [rad] or [deg]
N = Navigation constant
$K_\alpha, K_\beta, K_\phi$ = Slow mode controller gains
K_p, K_q, K_r = Fast mode controller gains
$\boldsymbol{a}_d$ = Desired acceleration vector
C_L, C_D = Lift and drag coefficients
$C_{L\alpha}$ = Lift-curve slope, [/rad]
C_{D0} = Zero lift drag coefficient
$R, \boldsymbol{R}$ = Relative distance, [m] and line of sight (LOS) vector
$E(\beta, -\alpha)$ = Rotation matrix from the body axes to the wind axes
I_X, I_Y, I_Z = Moments of inertia, [kg m²]
C_X, C_Y, C_Z = Aerodynamic force coefficients
C_l, C_m, C_n = Aerodynamic moment coefficients
$[\ ]_C$ = Concerned with a UAV
$[\ ]_d$ = Indicates a desired value or command
$[\ ]_W$ = Concerned with wind axes
$\dot{[\ ]}$ = Time derivative

3. Equations of motion

Figure 1 portrays the coordinate system used for this study. The complete, six-degree-of-freedom (6DOF), nonlinear, rigid-body dynamics for fixed-wing UAV with respect to the body-fixed axis system depicted in Fig. 1 are modeled by the following 12 first-order differential equations.

$$\dot{u} = rv - qw - g\sin\theta + \frac{q_D S C_X}{m} + \frac{F_T}{m} \tag{1}$$

$$\dot{v} = pw - ru + g\cos\theta\sin\phi + \frac{q_D S C_Y}{m} \tag{2}$$

$$\dot{w} = qu - pv + g\cos\theta\cos\phi + \frac{q_D S C_Z}{m} \tag{3}$$

$$\dot{p} = \frac{I_Y - I_Z}{I_X} qr + \frac{I_{XZ}}{I_X}(\dot{r} + pq) + \frac{q_D S b}{I_X} C_l \tag{4}$$

$$\dot{q} = \frac{I_Z - I_X}{I_Y} pr + \frac{I_{XZ}}{I_Y}(r^2 - p^2) + \frac{q_D S \bar{c}}{I_Y} C_m \tag{5}$$

$$\dot{r} = \frac{I_X - I_Y}{I_Z} pq + \frac{I_{XZ}}{I_Z}(\dot{p} - qr) + \frac{q_D Sb}{I_Z} C_n \tag{6}$$

$$\dot{\phi} = p + q\sin\phi\tan\theta + r\cos\phi\tan\theta \tag{7}$$

$$\dot{\theta} = q\cos\phi - r\sin\phi \tag{8}$$

$$\dot{\psi} = q\sin\phi\sec\theta + r\cos\phi\sec\theta \tag{9}$$

$$\dot{x}_I = u\cos\theta\cos\psi + v(\sin\phi\sin\theta\cos\psi - \cos\phi\sin\psi) + w(\cos\phi\sin\theta\cos\psi + \sin\phi\sin\psi) \tag{10}$$

$$\dot{y}_I = u\cos\theta\sin\psi + v(\sin\phi\sin\theta\sin\psi + \cos\phi\cos\psi) + w(\cos\phi\sin\theta\sin\psi - \sin\phi\cos\psi) \tag{11}$$

$$\dot{z}_I = -u\sin\theta + v\sin\phi\cos\theta + w\cos\phi\cos\theta \tag{12}$$

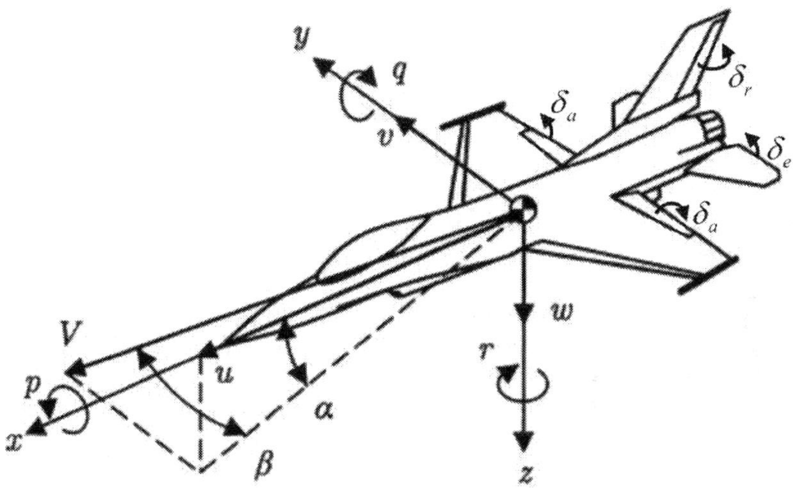

Figure 1. Coordinate system

For them, we assumed the aerodynamic coefficients appeared in the above equations as follows.

$$C_X = C_X\left(\alpha, \beta, \frac{\bar{c}q}{2V}, \delta_e\right) \tag{13}$$

$$C_Y = C_Y\left(\alpha, \beta, \frac{bp}{2V}, \frac{br}{2V}, \delta_a, \delta_r\right) \tag{14}$$

$$C_Z = C_Z\left(\alpha, \beta, \frac{\bar{c}q}{2V}, \delta_e\right) \tag{15}$$

$$C_l = C_l\left(\alpha, \beta, \frac{bp}{2V}, \frac{br}{2V}, \delta_a, \delta_r\right) \tag{16}$$

$$C_m = C_m\left(\alpha, \beta, \frac{\bar{c}q}{2V}, \delta_e\right) \tag{17}$$

$$C_n = C_n\left(\alpha, \beta, \frac{bp}{2V}, \frac{br}{2V}, \delta_a, \delta_r\right) \tag{18}$$

Equations (1-3), Eqs. (4-6), Eqs. (7-9), and Eqs. (10-12) respectively represent force equations, moment equations, rotational kinematic equations, and navigation equations. The first nine equations govern the dynamics. The navigation equations are used only for simulations to calculate the UAV position with respect to the earth frame.

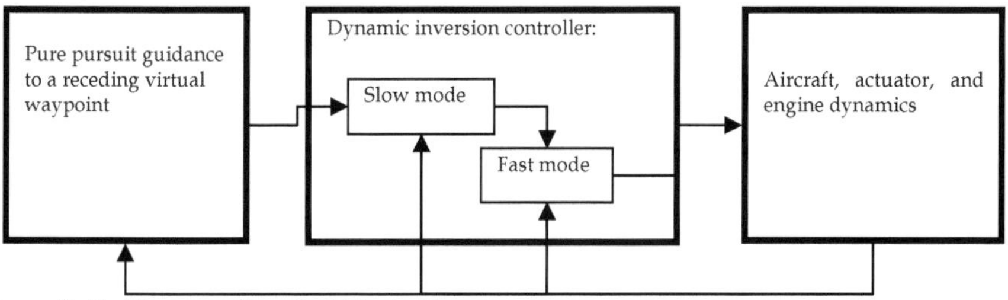

Figure 2. Guidance and control block diagram

4. Guidance and Control System

Figure 2 portrays a block diagram of the guidance and control system for a path-following UAV, which employs the pure pursuit guidance law for calculating the guidance command, in conjunction with a two-time-scaled dynamic inversion method (Baba et al., 1996, 2002; Sato et al., 2006; Yamasaki et al., 2007) consisting of a slow-state and a fast-state dynamic mode to generate guidance forces for nonlinear dynamic motions of the UAV. This diagram does not include the velocity control, which is discussed later.

4.1 Receding virtual waypoints

A desired trajectory might be used to guide a UAV along a predetermined path such as a runway approach and landing. It is useful to generate the desired trajectory from several waypoints. For this study, the cubic spline function is used because it gives the minimum curvature (Blajer, 1990; Jackson & Crouch, 1991). Figure 3 portrays an interpolated trajectory by the cubic spline function. The circles labeled "WAYPOINT" in Fig. 3 are waypoints that are only used to generate a spline interpolated-trajectory for convenience. They differ from the receding virtual waypoint described later.

A trajectory calculated from cubic spline function can be represented using continuous functions of time or the arc length of the trajectory. The arc length is chosen as the independent variable (Baba et al., 1996, 2002; Ochi et al., 2002; Yamasaki et al., 2007), which enables the UAV to fly at various speeds. A virtual waypoint should be set as the target for the pure pursuit guidance on the generated trajectory along which the trajectory-tracking UAV are to be guided. The virtual waypoint, which is moved in each time step, is obtainable using the cubic spline function of a distance instead of using the arc length of the desired trajectory. The distance is calculated based on the UAV's flight distance added using a few-seconds more future flight length. Therefore, this virtual waypoint determination approach uses only the flight distance information and obviates extra information such as the tracking error components described in the literature (Park et al., 2007). If the UAV deviates from the desired trajectory, the virtual waypoint is set at a receded point in each time step according to the flight distance, including a flight distance deviated from the desired trajectory, which can avoid rapid correction control. Consequently, this virtual waypoint is designated as "a receding virtual waypoint." Pure pursuit guidance with the receding virtual waypoint can alleviate control saturation, different from the case of tracking-error-correction-based control.

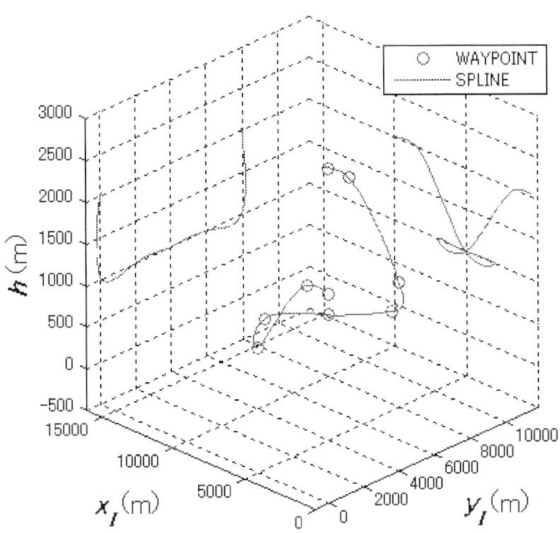

Figure 3. Interpolated trajectory using spline function of the arc length (c.f. WAYPOINT differs from the receding virtual waypoint)

4.2 Pure pursuit guidance law

Generally, pure pursuit guidance (PPG) is not used to guide a UAV for rendezvous or intercept purposes if the UAV velocity becomes 1.5 times greater than the target velocity because the guidance forces tend to be diverted when the relative distance between the UAV and the target becomes zero (Machol et al., 1965). However, those problems will not occur for a path-following purpose in keeping the relative distance because target points recede as the UAV moves. The relative distance can be maintained at a few degrees, not zero. Although the proportional navigation (PN) guidance might also be considered, the error distance from the desired trajectory is prone to remain at some few degrees because PN guidance tends to keep a line of sight (LOS) angle constant (Zarchan, 2002). In contrast, pure pursuit guidance always guides the UAV to orient in the waypoint direction in spite of both velocities. Therefore, we use the pure pursuit guidance for this study. If the target point velocity is set as zero, the characteristics of PN guidance are almost equal to that of PPG with respect to the UAV orientation direction. Assuming that the UAV pursues a receding virtual waypoint with velocity of V_C as portrayed in Fig. 4, the following equation is satisfied if the line of sight (LOS) vector $\boldsymbol{R}$ and the UAV's velocity vector $\boldsymbol{V}_C$ are heading in the same direction.

$$\boldsymbol{V}_C \times \boldsymbol{R} = \boldsymbol{0} \tag{19}$$

To guide the UAV into the pure pursuit course, the following acceleration feedback is needed for Eq. (19).

$$\boldsymbol{a}_d = \frac{N(\boldsymbol{V}_C \times \boldsymbol{R}) \times \boldsymbol{V}_C}{V_C R} \tag{20}$$

Therein, N is the navigation constant. Its value is set as 2 for this study, as in a previous study (Park et al., 2007). The following equation might be preferred to using Eq. (20) for implementation purposes in the real system, especially for a large LOS angle, in applying the real LOS angle feedback.

$$\boldsymbol{a}_{d,real} = \frac{N(\boldsymbol{V}_C \times \boldsymbol{R}) \times \boldsymbol{V}_C}{V_C R} \frac{\sigma}{\sin \sigma} \tag{21}$$

In that equation, σ is the LOS angle. The equation means that the desired acceleration is exactly proportional to the LOS angle, as measured from, for example, a seeker's angle. However, we use Eq. (20) in this study for simplicity. From Newton's second law, the required guidance force, along with the gravity compensation, can be expressed as

$$\boldsymbol{F}_d = m(\boldsymbol{a}_d - \boldsymbol{g}) \,, \tag{22}$$

where vector $\boldsymbol{g}$ is a gravity vector with respect to the body axes. Thereby, the desired guidance commands are obtained.

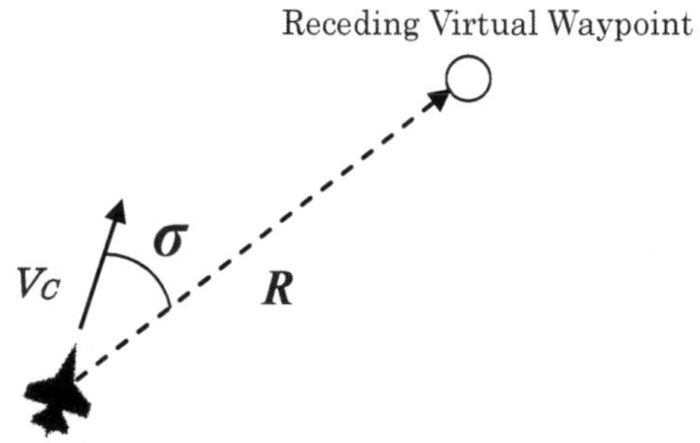

Figure 4. Position view of the UAV and the receding virtual waypoint

4.3 Derivation of command values

The desired guidance forces are obtained by changing the direction and magnitude of the force vector acting on the UAV. The direction and magnitude of the force vector can be changed generally by altering the UAV's slow dynamic states, i.e., angle of attack (AoA), sideslip angle, and roll (or bank) angle. The desired AoA, sideslip angle, and roll (or bank) angle is calculable from the desired forces in Eq. (22). The guidance forces shown in Eq. (22) related to the body axes must be transformed to wind-axis forms.

$$\begin{pmatrix} F_{\tilde{X}} \\ F_{\tilde{Y}} \\ F_{\tilde{Z}} \end{pmatrix} = E(\beta,-\alpha)\begin{pmatrix} F_X \\ F_Y \\ F_Z \end{pmatrix} \tag{23}$$

In that equation, $E(\beta,-\alpha)$ represents the rotation matrix from the body axes to the wind axes. Before deriving the desired lift, L_d and the desired roll angle, ϕ_d, we assume coordinated flight, which means zero-sideslip angle. The desired AoA, sideslip angle and bank angle can be expressed by the following simple equations using guidance forces in Eq. (22) (Baba et al., 2002, Yamasaki et al., 2006, 2007).

$$\alpha_d = \frac{1}{C_{L\alpha}}\left(\frac{L_d - F_T \sin\alpha}{q_D S}\right) + \alpha_0 \tag{24}$$

$$\beta_d = 0 \tag{25}$$

$$\phi_d = \phi + \Delta\phi_d \tag{26}$$

In those equations, the following pertain.

$$L_d = \sqrt{F_{\tilde{Y}}^{\ 2} + F_{\tilde{Z}}^{\ 2}} \tag{27}$$

$$\Delta\phi_d = \tan^{-1}\frac{F_Y}{-F_Z} \tag{28}$$

The desired additional roll angle appeared in Eq. (28) is determined geometrically with respect to the body axes.

4.4 Dynamic inversion

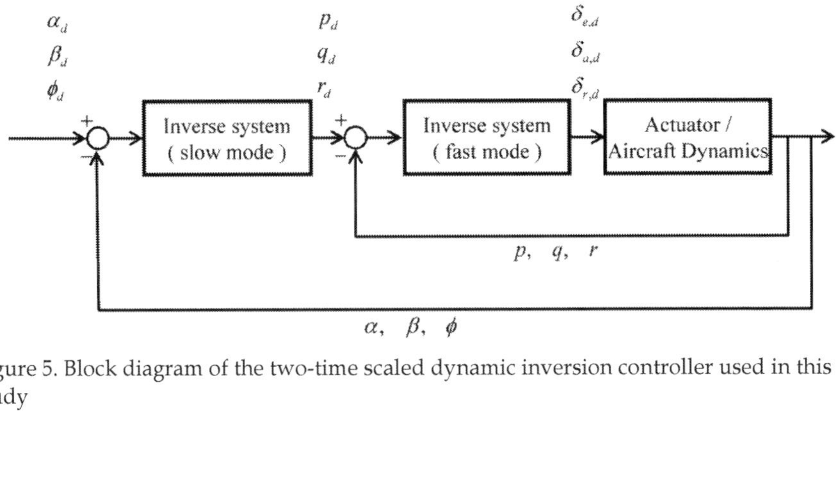

Figure 5. Block diagram of the two-time scaled dynamic inversion controller used in this study

To maintain high maneuverability, a nonlinear dynamic controller is designed. The dynamic inversion (DI) approach is a well-known scheme for nonlinear dynamical control design. We apply for the DI approach for a conventional fixed-wing UAV. In considering control for such a UAV, attitude control is achieved by rotating the UAV. The rotation is achieved by actuating the control surfaces such as elevators, ailerons, and a rudder. The attitude control is generally conducted using two loops: rotation control loop for fast-time scale state dynamics, and attitude control loop for slow-time scale state dynamics. We used the two-time scaled DI approach (Menon et al., 1987; Snell et al., 1992; Azam & Singh, 1994) to achieve the desired AoA, sideslip angle, and roll angle determined from the previous section. The two-time scaled DI method can reduce the controller order, which simplifies the design process. It can avoid tiresome control design problems arising in non-minimum phase systems. Figure 5 shows the schematic two-time scaled DI controller block diagrams used for this study (Baba et al., 1996; Yamasaki et al., 2006,2007). The inner loop in Fig. 5 corresponds to the fast-state dynamics: the states p, q, and r are controlled by the rudder, elevators, and ailerons. The outer loop corresponds to the slow-state dynamics: states α, β, and ϕ, which are controlled by the body rate p, q, and r. Regarding the outer loop, it is assumed that the transient dynamics of the fast states occur so quickly that they have a negligible effect on the slow states. In the inner loop, slow-time scale variables such as V, α, and β are assumed to remain constant. The fast-time scale controller attempts to maintain the body rates p_d, q_d, and r_d close to their values in the outer solution.

To obtain the slow state dynamics related to V, α, and β, the following relations are used.

$$u = V\cos\alpha\cos\beta \tag{29}$$

$$v = V\sin\beta \tag{30}$$

$$w = V\sin\alpha\cos\beta \tag{31}$$

The following are obtained after some algebraic manipulations of the above relations with Eqs. (1-3).

$$\dot{V} = -\frac{q_D S}{m}C_{D_W} + \frac{F_T}{m}\cos\alpha\cos\beta + g(\cos\phi\cos\theta\sin\alpha\cos\beta + \sin\phi\cos\theta\sin\beta - \sin\theta\cos\alpha\cos\beta) \tag{32}$$

$$\dot{\alpha} = -\frac{q_D S}{mV\cos\beta}C_L + q - \tan\beta(p\cos\alpha + r\sin\alpha) + \frac{g}{V\cos\beta}(\cos\phi\cos\theta\cos\alpha + \sin\theta\sin\alpha) - \frac{F_T\sin\alpha}{mV\cos\beta} \tag{33}$$

$$\dot{\beta} = \frac{q_D S}{mV}C_{Y_W} + p\sin\alpha - r\cos\alpha + \frac{g}{V}\cos\beta\sin\phi\cos\theta + \frac{\sin\beta}{V}(g\cos\alpha\sin\theta - g\sin\alpha\cos\phi\cos\theta - \frac{F_T\cos\alpha}{m}) \tag{34}$$

The last two equations and Eq. (7) form the slow-state dynamics related with α, β, and ϕ. The desired outer loop slow-state dynamics used for this study are specified by Eq. (35).

$$\begin{pmatrix} \dot{\alpha}_d \\ \dot{\beta}_d \\ \dot{\phi}_d \end{pmatrix} = \begin{pmatrix} K_\alpha(\alpha_d - \alpha) \\ K_\beta(\beta_d - \beta) \\ K_\phi(\phi_d - \phi) \end{pmatrix} \equiv \begin{pmatrix} U_\alpha \\ U_\beta \\ U_\phi \end{pmatrix} \tag{35}$$

Therein, subscript d represents the desired value. The terms α_d, β_d, and ϕ_d are the commanded AoA, sideslip angle, and roll angle given by the guidance part described in Eqs. (24-26). The bandwidths K_α, K_β, and K_ϕ should be slightly lower than or equal to the bandwidths of the UAV dynamics. Solving Eqs. (33-34) and (7) for $(p, q, r)^T$, and replacing the respective $(p, q, r)^T$ and $(\dot{\alpha}, \dot{\beta}, \dot{\phi})^T$ terms to the desired $(p_d, q_d, r_d)^T$ and $(U_\alpha, U_\beta, U_\phi)^T$ terms, one obtains the following.

$$\begin{pmatrix} p_d \\ q_d \\ r_d \end{pmatrix} = \begin{pmatrix} -\cos\alpha\tan\beta & 1 & -\sin\alpha\tan\beta \\ \sin\alpha & 0 & -\cos\alpha \\ 1 & \sin\phi\tan\theta & \cos\phi\tan\theta \end{pmatrix}^{-1} \begin{pmatrix} U_\alpha - A_\alpha \\ U_\beta - A_\beta \\ U_\phi \end{pmatrix} \tag{36}$$

In that equation, the following pertain.

$$A_\alpha = \{-q_D S C_L - F_T \sin\alpha + mg(\cos\phi\cos\theta\cos\alpha + \sin\theta\sin\alpha)\}/(mV\cos\beta) \tag{37}$$

$$A_\beta = \{q_D S C_{Y_W} - F_T \cos\alpha\sin\beta + (\sin\theta\cos\alpha\sin\beta + \sin\phi\cos\theta\cos\beta - \cos\phi\cos\theta\sin\alpha\sin\beta)\}/(mV) \tag{38}$$

Equation (36) represents the inverse system of the slow-state dynamics, which yields the fast states' desired values.
The following matrix-form equation, transformed from the moment Eqs. (4-6), models the fast-state dynamics.

$$\begin{pmatrix} \dot{p} \\ \dot{q} \\ \dot{r} \end{pmatrix} = \begin{pmatrix} A_p \\ A_q \\ A_r \end{pmatrix} + \begin{pmatrix} B_{pa} & 0 & B_{pr} \\ 0 & B_{qe} & 0 \\ B_{ra} & 0 & B_{rr} \end{pmatrix} \begin{pmatrix} \delta_a \\ \delta_e \\ \delta_r \end{pmatrix} \tag{39}$$

The following are used in that equation.

$$A_p = \{I_{XZ}(I_X - I_Y + I_Z)pq + (-I_Z^2 + I_Y I_Z - I_{XZ}^2)qr + I_Z q_D Sb[C_{l_\beta}\beta + (b/2V)(C_{l_p}p + C_{l_r}r)] + I_{XZ} q_D Sb[C_{n_\beta}\beta + (b/2V)(C_{n_p}p + C_{n_r}r)]\}/(I_X I_Z - I_{XZ}^2) \tag{40}$$

$$A_q = \{(I_Z - I_X)pr + I_{XZ}(r^2 - p^2) + q_D S\bar{c}[C_{m0} + C_{m_\alpha}\alpha + (\bar{c}/2V)C_{m_q}q]\}/I_Y \tag{41}$$

$$A_r = \{(I_{XZ}^2 - I_X I_Y + I_X^2)pq - I_{XZ}(I_X - I_Y + I_Z)qr + I_{XZ} q_D Sb[C_{l_\beta}\beta + (b/2V)(C_{l_p}p + C_{l_r}r)] + I_X q_D Sb[C_{n_\beta}\beta + (b/2V)(C_{n_p}p + C_{n_r}r)]\}/(I_X I_Z - I_{XZ}^2) \tag{42}$$

$$B_{pa} = q_D Sb(I_Z C_{l_{\delta_a}} + I_{XZ} C_{n_{\delta_a}})/(I_X I_Z - I_{XZ}^2) \tag{43}$$

$$B_{pr} = q_D Sb(I_Z C_{l_{\delta_r}} + I_{XZ} C_{n_{\delta_r}})/(I_X I_Z - I_{XZ}^2) \tag{44}$$

$$B_{qe} = q_D S\bar{c} C_{m_{\delta_e}}/I_Y \tag{45}$$

$$B_{ra} = q_D Sb(I_{XZ}C_{l_{\delta_a}} + I_X C_{n_{\delta_a}}) \big/ (I_X I_Z - I_{XZ}{}^2) \tag{46}$$

$$B_{rr} = q_D Sb(I_{XZ}C_{l_{\delta_r}} + I_X C_{n_{\delta_r}}) \big/ (I_X I_Z - I_{XZ}{}^2) \tag{47}$$

The desired closed-loop fast-state dynamics used for this study are specified as shown below.

$$\begin{pmatrix} \dot{p}_d \\ \dot{q}_d \\ \dot{r}_d \end{pmatrix} = \begin{pmatrix} K_p(p_d - p) \\ K_q(q_d - q) \\ K_r(r_d - r) \end{pmatrix} \equiv \begin{pmatrix} U_p \\ U_q \\ U_r \end{pmatrix} \tag{48}$$

The terms p_d, q_d, and r_d are commanded, roll, pitch, and yaw rate given by the slow-state control calculated from Eq. (36). The bandwidths K_p, K_q, and K_r should sufficiently exceed the bandwidths of the outer loop: α, β, and ϕ loop, to avoid coupling between the inner and outer loop dynamics. The bandwidth corresponds to the body rate cut-off frequencies. Solving Eq. (39) for $(\delta_e, \delta_a, \delta_r)^T$ and replacing the respective $(\delta_e, \delta_a, \delta_r)^T$ and $(\dot{p}, \dot{q}, \dot{r})^T$ to desired $(\delta_{e,d}, \delta_{a,d}, \delta_{r,d})^T$ and $(U_p, U_q, U_r)^T$, the desired control surface values are obtained as follows.

$$\begin{pmatrix} \delta_{a,d} \\ \delta_{e,d} \\ \delta_{r,d} \end{pmatrix} = \begin{pmatrix} B_{pa} & 0 & B_{pr} \\ 0 & B_{qe} & 0 \\ B_{ra} & 0 & B_{rr} \end{pmatrix}^{-1} \begin{pmatrix} U_p - A_p \\ U_q - A_q \\ U_r - A_r \end{pmatrix} \tag{49}$$

Equation (49) represents the inverse system of the fast states dynamics, yielding the desired control surface commands.

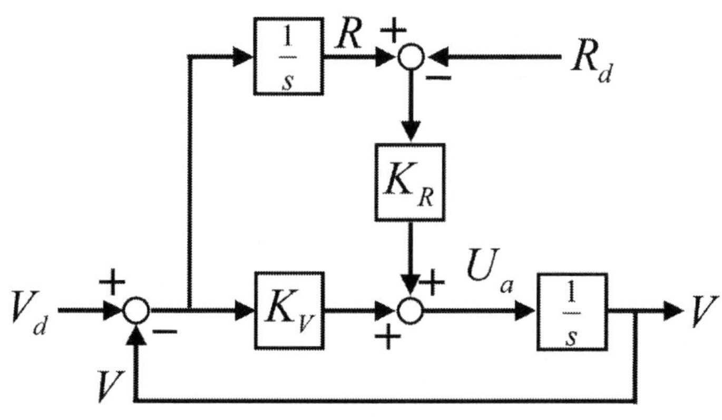

Figure 6. Block diagram of the velocity control

4.5 Velocity control

The dynamic inversions described above are used only for attitude control and never work directly for the velocity control. In considering the path-following capabilities of a UAV, one must take into account its velocity control, which is necessary to maintain a desired position along the given path. Figure 6 shows a block diagram illustrating the desired velocity dynamics used for this study. The desired velocity-loop dynamics depicted in Fig. 6 are specified as

$$\dot{V}_d = K_R(R - R_d) + K_V(V_d - V) \equiv U_a \,, \tag{50}$$

where K_R and K_V respectively signify the relative distance feedback gain and the velocity feedback gain. The velocity inverse dynamics are derived from Eq. (32) by solving for the thrust input; then, replacing $\dot{V}$ by a pseudo-input U_a as

$$F_{T,d} = \{mU_a - mg(-\sin\theta\cos\alpha\cos\beta + \cos\theta\sin\phi\sin\beta + \cos\theta\cos\phi\sin\alpha\cos\beta) + q_D S C_{D_W}\}/(\cos\alpha\cos\beta) \,. \tag{51}$$

Equation (51) yields the desired thrust force commands so that the desired velocity Eq. (50) holds in the velocity dynamics of the UAV.

The UAV, as well as aircraft, has no active brake in the air. Therefore, velocity is generally controlled gradually relative to an attitude control. The gains K_R and K_V shown in Eq. (50) must be determined taking account of this fact. The desired velocity control system depicted in Fig. 6 is represented as shown below.

$$\frac{V}{V_d} = \frac{K_V s + K_R}{s^2 + K_V s + K_R} \tag{52}$$

The characteristic equation of a general quadratic system is defined with damping ratio ζ and natural angular frequency ω_n.

$$s^2 + 2\zeta\omega_n s + \omega_n^2 = 0 \tag{53}$$

Comparing the characteristic equation in the above equation with the desired velocity control system, the characteristics of the desired velocity dynamic system are represented as follows.

$$K_R = \omega_n^2 \tag{54}$$

$$K_V = 2\zeta\omega_n \tag{55}$$

Because vibration of the velocity response should be alleviated for smooth path-following purpose, ζ is set around 1; ω_n is determined using the following approximated equation, which is based on the approximated long-period mode for the UAV as well as general aircraft so that the velocity response converges gradually.

$$\omega_n \cong \frac{\sqrt{2}g}{V} \tag{56}$$

Thereby, the UAV can follow the given path with PPG with DI attitude controller and DI velocity controller. Figure 7 portrays a schematic diagram of the guidance and control system described thus far.

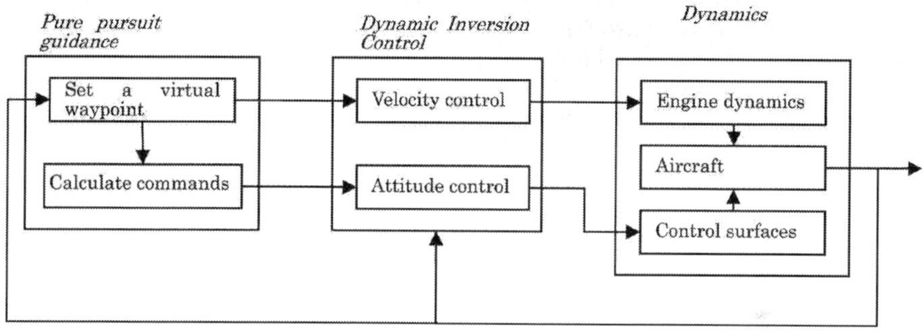

Figure 7. Schematic diagram of the guidance and control system

5. Simulations

These simulations use the YF-16 aircraft model (Gilbert et al., 1976) as a UAV, which includes detailed aerodynamics coefficients, the engine model, and the actuator time lags to demonstrate the guidance and control system in nonlinear 6 degree-of-freedom simulations. The following assumptions simplify the problem and demonstrate the total system.

1. The ambient atmosphere is stationary.
2. The aircraft is symmetric about the x-z plane.
3. The aircraft mass is constant and the engine momentum is negligible.
4. The aircraft model is available and aerodynamic uncertainties are negligible.
5. Information of the inertial position and velocity, angular velocity, and Euler angles is available.

Assumptions 1 and 4 might be violated in some situations, but robustness for model uncertainties, external disturbances, and measurement noise might be compensated with the inner loop design of the dynamic inversion controller (Snell, 1992; Brinker & Wise, 1996; Boyle & Chamitoff, 1999) instead of using the proportional control appearing in Eq. (48), along with, for example, an extended Kalman filter, as appearing in the literature (Tanaka et al., 2006).

Parameters	Values
Mach number	0.6 [Mach]
Initial position	(0.0, 0.0, 2000.0) [m]
Initial heading	0.0[deg]
R_d	500 [m]
N	2
$K_\alpha, K_\beta, K_\phi$	4 [rad/s]
K_p, K_q, K_r	8 [rad/s]
Actuator time constant	0.1 [s]

Table 1. Simulation Settings

5.1 Simulations in stationary atmosphere

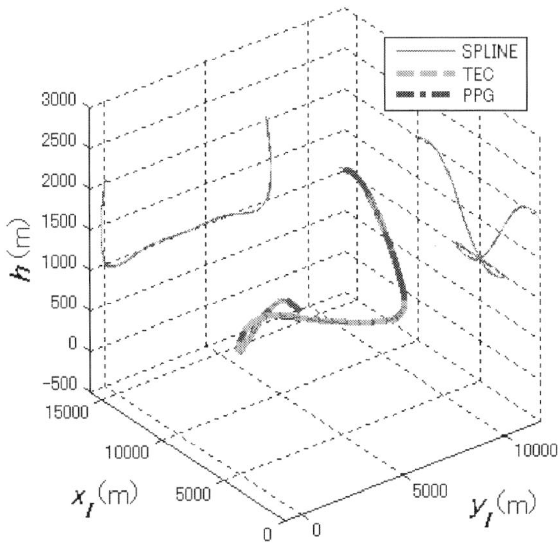

Figure 8. Tracking flight along a gentle curved path

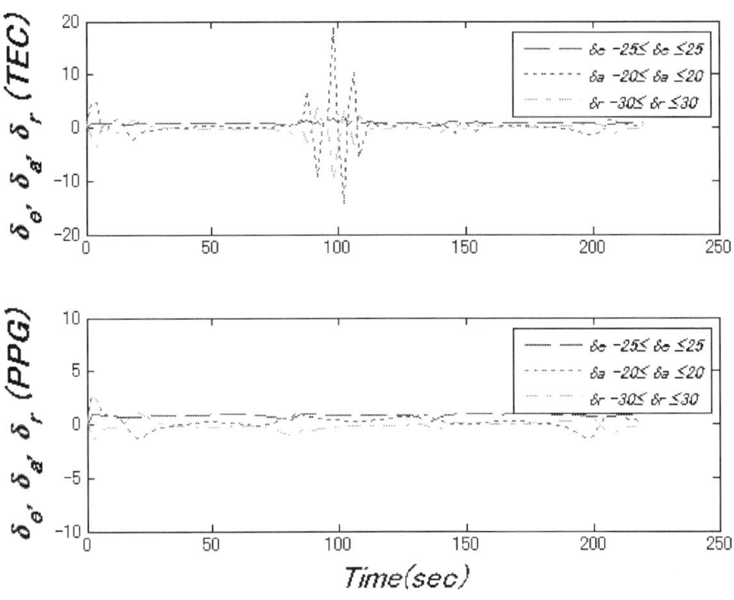

Figure 9. Control commands during path-following along the gentle curved path (deg)

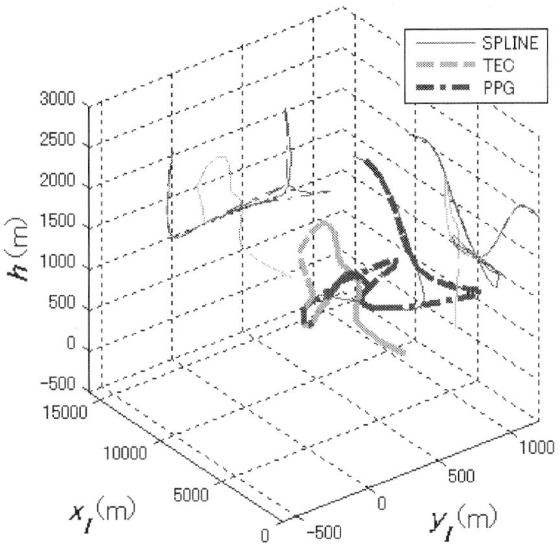

Figure 10. Tracking flight along a sharply curved path

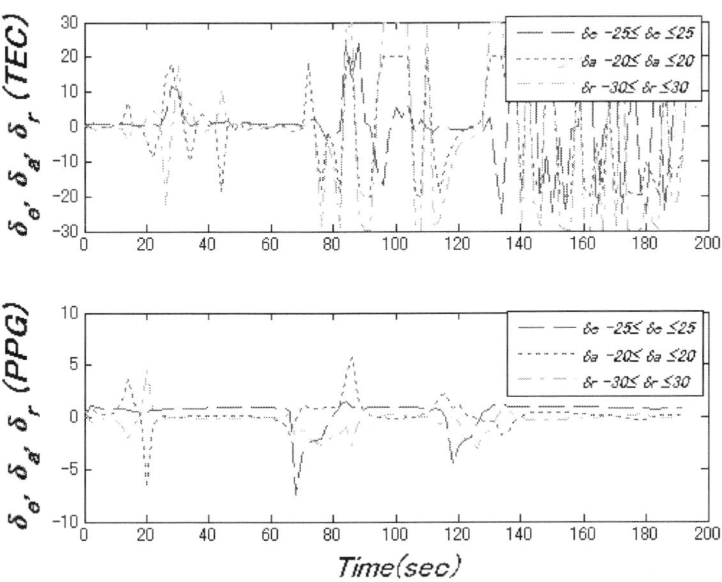

Figure 11. Control commands during path-following along the sharply curved path (deg)

Figures 8–11 portray simulation results. In Fig. 8, the solid lines labeled "SPLINE" mark the desired flight trajectories; the dashed lines labeled "TEC" and the broken dotted lines "PPG" respectively portray the conventional tracking-error-correction-based method (Baba, 1996) (TEC) and the pure pursuit guidance-based method with the receding virtual waypoint (PPG), along with their (X, h)-plane and (Y, h)-plane projections. Figure 8 shows that both TEC and the proposed PPG methods have good trajectory-tracking capability in the gentle curved trajectory. Figure 9 shows that no control saturation exists throughout the tracking flight. However, in the case of a sharply curved trajectory tracking flight, the proposed PPG method performs the high tracking ability, although the UAV deviates greatly from the desired trajectory twice whereas the conventional TEC-based method can not follow the trajectory when the UAV is deviated from the trajectory (Fig. 10). Figure 11 shows that the TEC-based method causes control saturation, although the proposed PPG method does not.

5.2 Simulations in wind turbulence circumstance

The Dryden wind turbulence model (Parris, 1975) is used for wind perturbation generation. A wind shear model is added to the wind turbulence for setting some averaged wind speed. Turbulence scales and its intensities are determined from prior specifications (MIL-F-8785C, 1980) assisted by linear interpolation for the lack of information range. The wind shear model presented in the specifications (MIL-F-8785C, 1980) is used in the simulations for averaged wind speed generation. As in the previous section, two simulations are conducted: a gentle curved path following and a sharply curved path case. Table 2 shows the turbulence model simulation settings. Other settings are as in Table 1.

Parameters	Values
Reference altitude for turbulence scales	1000 [m]
Turbulence intensity	Moderate
Initial heading	0.0 [deg]
Averaged wind at 6.1 [m]	6.528 [m/s]
Averaged wind direction	East to West (Horizontal blow)

Table 2. Simulation Settings

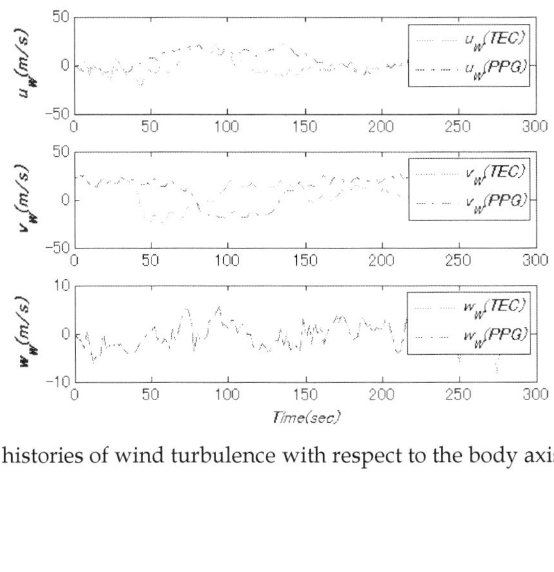

Figure 12. Time histories of wind turbulence with respect to the body axis components

Figure 12 presents an example of generated wind turbulence components related to the body axes, which are exerted to the UAV guided by the TEC method and the proposed PPG method. That figure shows that the wind turbulence has non-zero averaged velocity. The wind velocity components are frozen to the earth frame: the wind velocities in each simulation are equal when the UAVs fly through the same path, but each component is varied on every field, thereby producing the different velocity components and different result in each simulation.

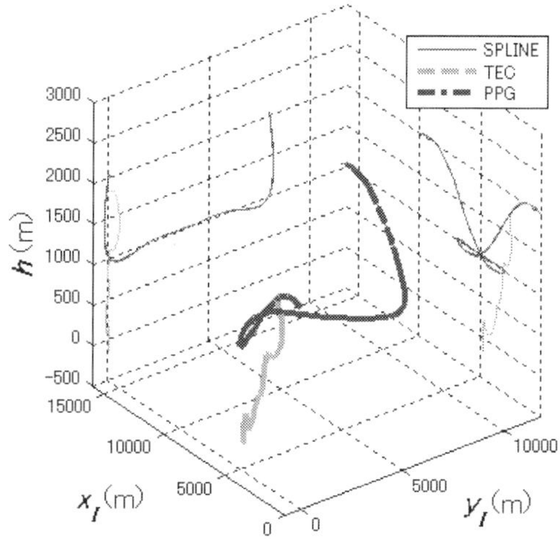

Figure 13. Tracking flight along a gentle curved path under wind turbulence

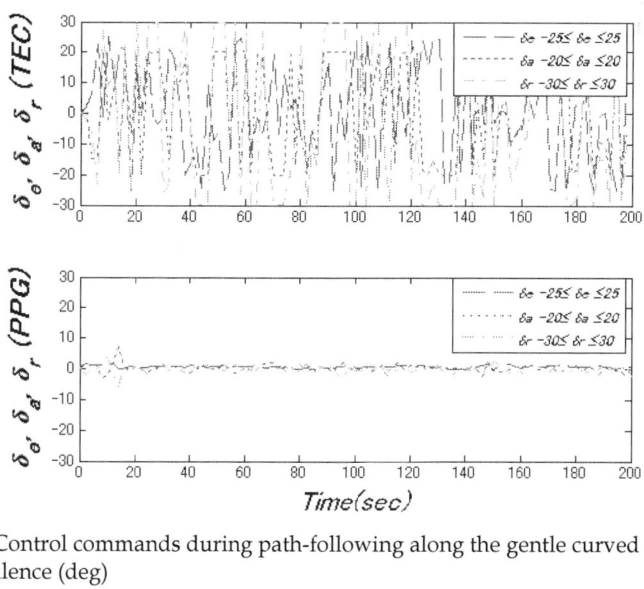

Figure 14. Control commands during path-following along the gentle curved path under wind turbulence (deg)

Figures 13-16 show simulation results under wind turbulence. The conventional TEC-based guidance method fails to follow the desired trajectory, even in gently curved trajectory in wind turbulence, whereas the proposed PPG method can recover from the deviation and finally follow the trajectory, even in the sharply curved path-following case. These figures show that the PPG method with the reseeding waypoint has robustness relative to external wind turbulence and to deviation from the desired trajectory, whereas the conventional TEC-based guidance method does not.

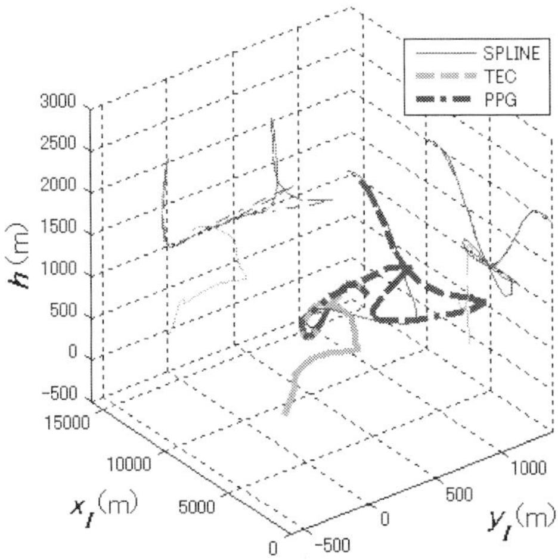

Figure 15. Tracking flight along a sharply curved path under wind turbulence

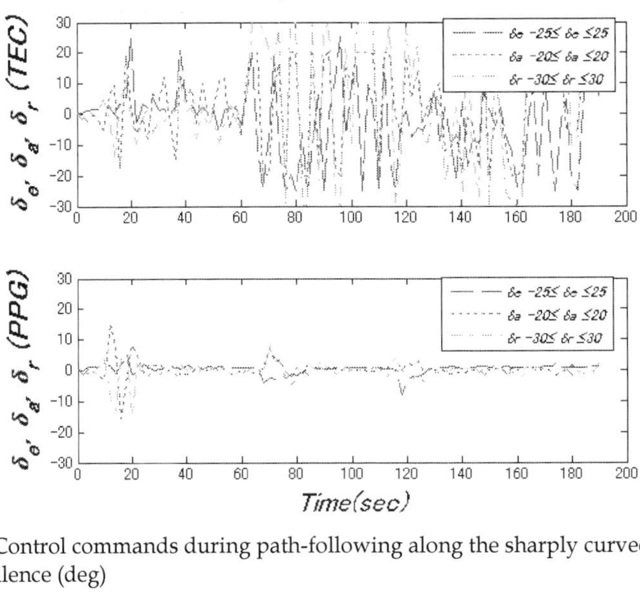

Figure 16. Control commands during path-following along the sharply curved path under wind turbulence (deg)

6. Conclusions

A robust path-following guidance and control system for a UAV is introduced. Pure pursuit guidance with a receding virtual waypoint is proposed to achieve robust path-following capability for the UAV. The system enables the UAV to recover from huge-tracking-error situations alleviating control saturation. The UAV can be equipped with robustness for any sharply curved path-following and wind turbulence circumstances. An optimal navigation constant according to a given path and a proper receding virtual waypoint position considering a UAV dynamics should be analyzed in the future.

7. References

Azam, M. and Singh, S. N. (1994). Invertibility and Trajectory Control for Nonlinear Maneuvers of Aircraft, *Journal of Guidance, Control, and Dynamics AIAA*, Vol. 17, No. 1, pp. 192–200

Baba, Y., Takano, H. and Sano, M. (1996). Desired Trajectory and Guidance Force Generators for an Aircraft, *Proc. Guidance, Navigation and Control Conference AIAA*, 96-3873

Baba, Y. and Takano, H. (1998). Robust Flight Trajectory Tracking Control Using Fuzzy Logic, *Proc. the 8th ISDG&A*, Maastricht, pp. 68–75

Baba, Y., Takano, H., Miyamoto, S. and Ono, K. (2002). Air Combat Guidance Law for an UCAV, *Proc. 1st UAV Conference*, AIAA-2002-3427, Virginia

Blajer, W. (1990). Aircraft program motion along a predetermined trajectory. Part II. Numerical simulation with application of spline functions to trajectory definitions, *Aeronautical Journal*, Vol. 94, pp. 53–58

Boyle, D. P. and Chamitoff, G. E. (1999). Autonomous Maneuver Tracking for Self-Piloted Vehicles, *Journal of Guidance, Control, and Dynamics*, Vol. 22, No. 1, AIAA, pp. 58–67

Brinker, J. S. and Wise, K. A. (1996). Stability and Flying Qualities Robustness of a Dynamic Inversion Aircraft Control Law, *Journal of Guidance, Control, and Dynamics*, Vol. 19, No. 6, pp. 1270–1277

Brockett, R. W. (1978). Feedback Invariants for Nonlinear Systems, *Proc. the 1978 IFAC World Congress*, pp. 1115–1120

Gilbert, W. P., Nguyen, L. T. and Van Gunst, R. W. (1976). Simulator Study of the Effectiveness of an Automatic Control System Designed to Improve the High-Angle-of-Attack Characteristics of a Fighter Airplane, TN D-8176, NASA

Jackson, J. W. and Crouch, P. E. (1991). Dynamic Interpolation and Application to Flight Control, *Journal of Guidance, Control and Dynamics*, Vol. 14, No. 4, pp. 814–822

Kaminer, I., Pascoal, A., Hallberg, E. and Silvestre, C. (1998). Trajectory Tracking for Autonomous Vehicles: An Integrated Approach to Guidance and Control, *Journal of Guidance, Control and Dynamics*, Vol. 21, No. 1, 29–38

Lane, S. H. and Stengel, R. F. (1988). Flight Control Design Using Non-linear Inverse Dynamic, *Automatica*, Vol. 24, pp. 471–483

Machol, R. E., Tanner, Jr., W. P. and S. N. Alexander (1965). *System Engineering Handbook*, Chap. 19 Guidance, R. E. Howe, McGraw-Hill Book Company

Military Specification, U. S. (1980). MIL-F-8785C, 5th November

Menon, P. K. A., Badgett, M. E., Walker, R. A. and Duke, E. L. (1987). Nonlinear Flight Test Trajectory Controller for Aircraft, *Journal of Guidance*, Vol. 10, No. 1, AIAA, pp. 67–72

Ochi, S., Takano, H. and Baba, Y. (2002). Flight Trajectory Tracking System applied to inverse control for aerobatic maneuvers, *Inverse Problems in Engineering Mechanics III*, Elsevier Science Ltd., pp. 337–344

Park, S., Deyst, J. and How, J. P. (2004). A New Nonlinear Guidance Logic for Trajectory Tracking, *Proc. Guidance, Navigation and Control Conference AIAA* 2004-4900

Park, S., Deyst, J. and How, J. P. (2007). Performance and Lyapunov Stability of a Nonlinear Path-Following Guidance Method, *Journal of Guidance, Control, and Dynamics AIAA*, Vol. 30, No. 6, pp. 1718–1728

Parris, B. L. (1975). Modeling Turbulence for Flight Simulations at NASA-AMES, CSCR No. 4, NASA

Rysdyk, R. (2006). Unmanned Aerial Vehicle Path Following for Target Observation in Wind, *Journal of Guidance, Control, and Dynamics*, Vol. 29, No. 5, pp. 1092–1100

Sato, Y., Yamasaki, T., Takano, H. and Baba, Y. (2006). Trajectory Guidance and Control for a Small UAV, *KSAS International Journal*, Vol. 7, No. 2, pp. 137–144

Snell, S. A. (1992). Preliminary Assessment of the Robustness of Dynamic Inversion Based Flight Control Laws, AIAA-92-4330-CP, pp. 206–216

Snell, S. A., Enns, D. F. and Garrard Jr., W. L. (1992). Nonlinear Inversion Flight Control for a Supermaneuverable Aircraft, *Journal of Guidance, Control, and Dynamics AIAA*, Vol. 15, No. 4, pp. 976–984

Tanaka, N., Suzuki, S., Masui, K. and Tomita, H. (2006). Restructurable Guidance and Control for Aircraft with Failures Considering Gust Effects, *Journal of Guidance, Control, and Dynamics*, Vol. 29, No. 3, AIAA, pp. 671–679

Yamasaki, T., Enomoto, K., Tanaka, D., Takano, H. and Baba, Y. (2006). Automatic Control for Chase Aircraft, *KSAS International Journal*, Vol. 7, No. 2, pp. 145–154

Yamasaki, T., Sakaida, H., Enomoto, K., Takano, H. and Baba, Y. (2007). Robust Trajectory-Tracking Method for UAV Guidance Using Proportional Navigation, *Proc. International Conference on Control, Automation and Systems*, Seoul, pp. 1404–1409

Zarchan, P. (2002). *Tactical and Strategic Missile* Guidance - Fourth Edition - Progress in Astronautics and Aeronautics *AIAA*, Vol. 176, ISBN-13: 978-1-56347-874-1, Virginia

33

Flapping Wings with Micro Sensors and Flexible Framework to Modify the Aerodynamic Forces of a Micro Aerial Vehicle (MAV)[1]

Lung-Jieh Yang
Tamkang University
Taiwan

1. Introduction

The flight of birds or insects has fascinated scholars and physicists for many centuries. Flapping motion, as shown by many nature flyers, is the most efficient way of flying objects whose size are smaller than or around 6 inches. In this chapter, the author introduced how to use modern technology to fabricate the flapping wings for micro aerial vehicles (MAVs) with flexibility and smartness. The terminology of MAV, defined by Defense Advanced Research Projects Agency (DARPA), denotes the size-limitation and the performance requirements of air vehicles (Ashley, 1998). The total wingspan of a MAV is expected to be less than 15 cm; the highest velocity is about 48 km/hr; the range of the flight mission is about 10 km; and the flight endurance is about 20-120 minutes.

The earliest flapping vehicle (or ornithopter) was made by Gustave Trouvé (Chanute, 1894). No ornithopter was developed using MEMS technology until the end of the last century. A lightest flapping MAV with a total mass of only 11.69 grams was made by Caltech micromachining lab in 1999-2002 (Pornsin-sirirak, 2001; 2002). They used titanium-alloy as the frame of flapping wings, and assigned parylene, a polymer material, as the covering skin of the airfoil. This integrated structure can withstand extreme vibration with frequency of more than 30 Hz and it weights only 0.3 gram. The Caltech MAV, "Micro-bat", can be remotely controlled as will and the flight endurance is more than 6 minutes. Besides the work of Caltech, several groups in other universities developed their flapping MAVs with different configurations and actuation principles (Website http://www.artificialmuscle.com/), (Website http://fourier.vuse.vanderbilt.edu/cim/projects/crawler.htm), (Website http://www.gtri.gatech.edu/atas/teams/proj_entomopter.html), (Sitti, 2001; Barrett, 2005; Jones, 2005). For instance, TU Delft's MAV "Delfly" composed of a pair of dragonfly-like flexible wings recently announced their successful hovering (Barrett, 2005). Otherwise, a

[1] The major content of this chapter is extracted from the author's following two papers:
L.J. Yang, C.K. Hsu, J.Y. Ho, and C.K. Feng, "Flapping Wings with PVDF Sensors to Modify the Aerodynamic Forces of a Micro Aerial Vehicle," *Sensors and Actuators A: Physical*, Vol. 139, pp. 95-103, 2007.
L.J. Yang, C.K. Hsu, F.Y. Hsiao, C.K. Feng, and Y.K. Shen, "A Micro-Aerial-Vehicle (MAV) with Figure-of-Eight Flapping Induced by Flexible Wing Frames," AIAA 2009-0875.

fixed-wing type MAV with a scissor-like clapping tail thruster made by Jones et al. of Naval Graduate School. They showed the long flight endurance over 20 min (Jones, 2005).

Although the previous interesting works of MAVs managed to fly via wireless remote control, none of the programs has been able to achieve a long and sustainable flight. Moreover, the detailed mechanism of the flapping flight is still under study and the unsteady aerodynamic characteristics of the flapping MAVs are unclear.

Predicting the lift force during flight is a very critical issue in the design of a MAV. Measuring the instantaneous aerodynamic force on a flapping model or live insect remains a great challenge in experimental aerodynamics. The early surveys of measuring averaged force were made using delicate balances (Hollick, 1940; Jensen, 1956). Other prior researches used piezoelectric probes (Cloupeau, 1979), strain gauges (Ho, 2003) and the laser interferometers (Dickinson, 1996). Restricted by the size of measuring tools, it is usually assumed that all the wings behave identically and have the equal force contribution to the flying body. Therefore, efforts should be focused on fabricating the very dedicate and identical wing structures in the conventional development of MAV.

In order to gather more information during the flapping maneuver, we measured the aerodynamic force of a Caltech-like MAV by the conventional load-cells in the wind-tunnel, and also proposed the integration of PVDF piezoelectric foils to the parylene flapping wings of the MAV to pick up the *in-situ* lift force. We used MEMS technology to fabricate a titanium-alloy wing frame and a set of gear-reduction transmission components, and deposited conformal parylene film on the wing frame as the airfoil skin (Pornsin-sirirak, 2001; 2002) (Website http://parylene.com). The actuation force or torque available for the flapping wings is drained from the gear-reduction transmission set coupled to a high-speed DC motor powered by commercial poly-lithium batteries. The parylene wing skin also serves as an electrical isolation layer between the PVDF piezoelectric sensing element and the titanium frame. The on-site lift information acquired from a PVDF sensing skin was done by the authors of this presented work. Herein, we employed four-linkage concept to design a transmission mechanism for the MAV. Although the two wings of the MAV have the same flapping angle, there exists unavoidably a mechanical phase lag between them by virtue of the transmission's mechanical principle of operation. The result in this chapter proposes a novel approach of integrating a parylene-PVDF hybrid wing in an *in-situ* way to monitor the lift force of a flapping MAV in the wind-tunnel test. We found a new design methodology to adjust the aerodynamic performance of MAV by changing the phase lag between the two flapping wings through fine tuning of the mechanism linkages.

Additionally, the figure-of-eight stroke of hovering hummingbirds is the ideal trajectory which many researchers of flapping MAVs hope to pursue. Several sophisticated mechanisms of flapping wings were claimed to fit this natural maneuver motion in the conceptual design, for instance, the Banala et al. in Delaware University employed a 5-bar mechanism for generating a prescribed wing motion taken from hawkmoth kinematic flight data (Banala, 2005). They designed a mechanism for biaxial rotation of a wing for a hovering MAV (McIntosh, 2006). Meanwhile, an insect-like flapping wing mechanism was proposed by Cranfield University (Żbikowski, 2005; 2005) through the novel idea of a double spherical Scotch yoke. All the above designs have still not availably been applied to the 20-cm size MAV and there exists scarcely successful examples of this kind artificial palm-size MAVs manufactured by conventional machining. Herein, a biomimetic figure-of-eight flapping induced by the flexible wing frames of the MAVs in this chapter were found accordingly.

The flight information obtained from a MAV with successful flight record can be as a reference compared to natural flapping animals, and give a guideline to the development of next generation MAV in an empirical manner.

2. DESIGN CONCEPT

The issue of reducing weights for an air vehicle, especially for a flapping-wing flight vehicle herein discussed in this chapter, is a very critical problem. The current trend is to employ small, palm-sized wing foils. However, lift and thrust forces produced by the movement of wing flapping also decreased with size reducing of the airfoil, and therefore the wing size should be sufficiently large. For this reason, adopting the titanium-alloy of high strength-to-mass ratio as the airframe material and parylene as the skin material of flapping-wing is an appropriate compromise. Note that a light-weighted and high-power battery is also needed. To obtain an accurate size of a titanium-alloy airframe with no residual stress is not easy with the conventional machining methods. Interior residual stress could cause a warping deformation in the structures of airframes, and easily distorts the geometry of the airfoil. Therefore, instead of the regular machining techniques, wet etching technique is employed to tailor the airframe structures from a titanium-alloy plate and no apparent residual stress is found. Parylene coating technique is applied to laying the wing skin attached on the titanium frame. The followings are the design details of the airframe and a gear transmission system associated with the airframe.

2.1 Design of the titanium-alloy airframe

The flapping-wing movement of a nature bird includes flapping, twisting, folding and gliding. These functions correlate to the high efficiency of flapping mechanism. In order to reach very light weight, the MAV in this work has flapping movement of only one degree-of-freedom on purpose. Therefore in order to achieve better flight performance, the flapping wing must be able to sustain higher flapping frequency and provide sufficient force for flight. Higher flapping frequency involves concerns with the structural integrity. Designing the flapping wings with high strength and low weight is a crucial issue.

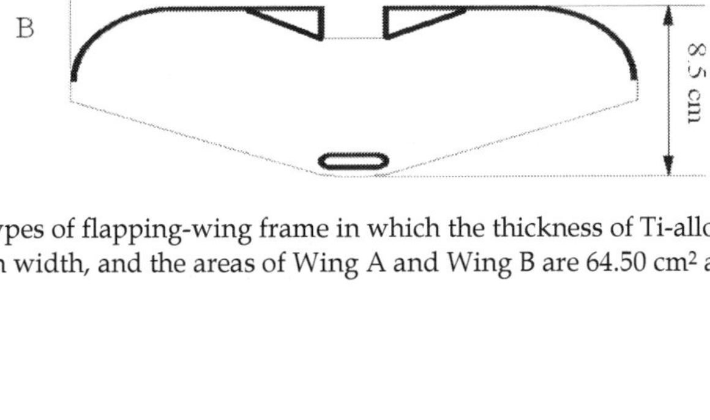

Figure 1. Two types of flapping-wing frame in which the thickness of Ti-alloy frame is 250 μm and 2 mm in width, and the areas of Wing A and Wing B are 64.50 cm^2 and 108.75 cm^2, respectively

We use existing MEMS technology for fabricating the wings to ensure the accurate size control and smartness of the flying system. The material of the airframe is titanium-alloy (the mechanical properties of titanium grade 4: density = 4.54 g/cm^3; Young's modulus = 104 GPa; tensile strength = 552 MPa). The detailed dimensions of two types of the flapping wings as well as their wing frames are shown in Fig. 1.

2.2 Design of the gear transmission system

The gear transmission system for the MAV is composed of a gear-reduction set and a four-bar linkage, as shown in Fig. 2. We use a 7mm-diameter DIDEL electric motor to drive the transmission system. The gear set with a gear ratio of 26.6 adopted in this work provides a sufficient torque for driving the flapping wings. The whole gear set and the motor are arranged on an aluminum base. OA is the driving linkage and the following linkage BC is connected to the wing structure. The driving linkage can perform a full revolution and the following linkage undergoes a rocking motion. The links OA, BC are made of aluminum by EDWC (electrical discharge wire cutting) (Weller, 1983), and the link BC is made of titanium-alloy by MEMS process. The nodal point is a 1mm-diameter stainless tube that connects the aluminum base rigidly.

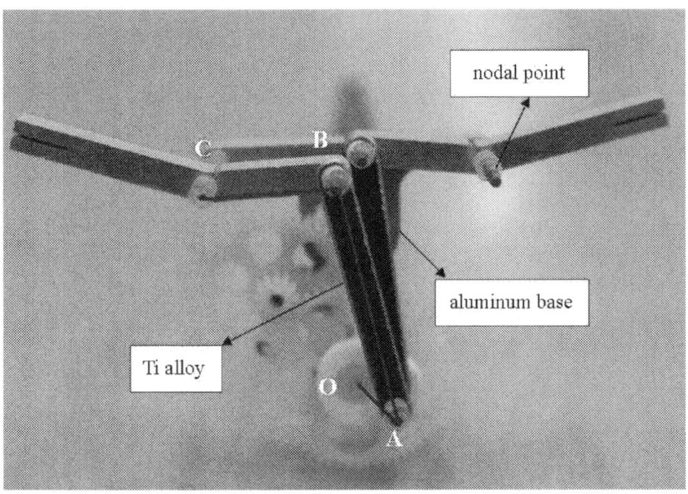

Figure 2. Gear transmission system. In wind-tunnel tests, the aluminum base is adopted and the light plastic base is selected for real flight

We can easily achieve different stroke angles and flapping symmetry (the lag phenomenon between the two wings in the flapping motion) by adjusting the dimensions of links OA, AB and BC. The selected lengths for each links are OA=4 mm and AB=21.5 mm. Link BC has a variable length, which varies from 8 mm to 12 mm, and with length variation of such a range, there are correspondingly large changes of stroke angles and phase angle lags (max) in flapping movement, as seen in table 1. Although the two wings have the same flapping angle of 39 to 61 degrees, there exists unavoidably a mechanical phase lag of 2.5 to 13 degrees between them due to the gear-transmission's principle of operation (see the simulated data in Fig. 3.)

BC (mm)	Stroke angle (degree)	Lag (max)
8	60.5	13.3
9	53.0	9.4
10	47.2	6.4
11	42.7	4.1
12	39.0	2.5

Table1. Designs of the transmission system

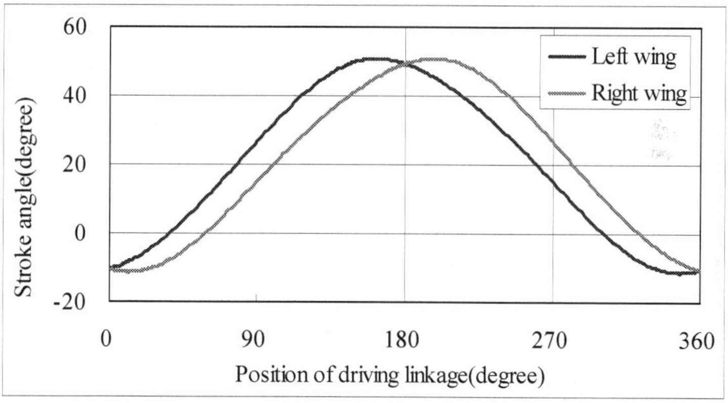

Figure 3. Stroke angles of two flapping wings vs. one period of time

2.3 Aerodynamic force measurement by PVDF

Piezoelectric materials have been wildly used in many key technologies, including microelectronics, signal processing, sensors and actuators. Among all piezoelectric materials, the PVDF material can be fabricated in any shape of thin film and integrated with parylene by MEMS technology. Thus, a flexible thin film of PVDF is suitable in studying the aerodynamic force of flapping wings.

The mathematical model for lift force measurement by PVDF is described as follows. We assume that there is a pressure difference distribution $\Delta p(x,y,z)$ between the upper and the lower sides of the flapping wing. Lift force is calculated by taking the integral of Δp over the wing surface S:

$$L = \iint_S \Delta p(x, y, z) dA \quad (1)$$

When the pressure difference Δp of the air flow field acts on the PVDF film, just like the case of giving a force on a shell or a plate, the PVDF film and wing skin become deformed and stretched. We assume the plane stress distribution on the flapping wing is $\sigma_x(x,y,z)$ and $\sigma_y(x,y,z)$ which are related to $\Delta p(x,y,z)$ by the following linear relationship (Hooke's low):

$$\Delta p(x, y, z) = k_x(x, y, z) \cdot \sigma_x(x, y, z) = k_y(x, y, z) \cdot \sigma_y(x, y, z) \quad (2)$$

where k_x and k_y are stiffness functions assumed by the authors. Regarding PVDF film as an air foil for sensing aerodynamic force, we get the charge density ρ_i $(i=x,y,z)$ by the following piezoelectric transformation relationship (Liu, 2006):

$$\begin{bmatrix} \rho_x \\ \rho_y \\ \rho_z \end{bmatrix} = \begin{bmatrix} 0 & 0 & 0 & 0 & 0 & 0 \\ 0 & 0 & 0 & 0 & 0 & 0 \\ d_{31} & d_{32} & d_{33} & 0 & 0 & 0 \end{bmatrix} \begin{bmatrix} \sigma_x & \sigma_y & \sigma_z & \tau_{xy} & \tau_{yz} & \tau_{zx} \end{bmatrix}^T \tag{3}$$

The none-zero coefficient d_{ij} of PVDF film are

$$\begin{cases} d_{31} = 20 \quad pC/N \\ d_{32} = 2 \quad pC/N \\ d_{33} = -30 \quad pC/N \end{cases} \tag{4}$$

We assume the flapping wing as a plane-stress problem, ($\sigma_z = 0$). The charge density along the Z-direction therefore can be derived as:

$$\rho_z(x,y,z) = d_{31} \cdot \sigma_x(x,y,z) + d_{32} \cdot \sigma_y(x,y,z) \tag{5}$$

Combining Eqs. (2 to 5), we get the total charge Q in terms of the pressure difference Δp and the stiffness function k as follows:

$$Q = \iint_S \rho_z(x,y,z)dA = \iint_S \Delta p(x,y,z) \cdot \left[\frac{d_{31}}{k_x(x,y,z)} + \frac{d_{32}}{k_y(x,y,z)} \right] dA \tag{6}$$

Because the stiff function k_x and k_y are not uniform over the wing skin area S, the lift force L in Eq. (1) can not be directly separated from Eq. (6). We temporarily and equivalently replace the bracket in Eq. (6) with d^*, and obtain the following expression:

$$Q = d^* \cdot L \tag{7}$$

In Eq. (7), the total charge Q can be also expressed by a voltage V across the piezoelectric (PVDF) capacitor C by the relationship

$$Q = C \cdot V \tag{8}$$

Combining Eqs. (7) and (8), the lift force can be determined linearly with the voltage V.

$$L = \frac{Q}{d^*} = (\frac{C}{d^*})V \tag{9}$$

The simplified model described by Eq (9) gives us a guideline in principle that the lift information can be obtained electrically from the voltage given by the PVDF piezoelectric film in an *in-situ* way. In addition, such an output signal from the PVDF film needs calibration or quantitative comparison with the load-call sensors installed in the wind-tunnel available for the test of MAVs in this work.

3. FABRICATION

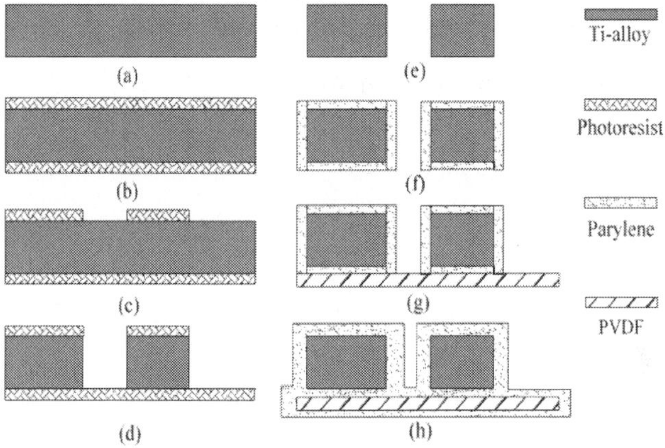

Figure 4. Process flow of a PVDF-parylene wing : (a)Clean the titanium-alloy substrate; (b)Coat the double sides of the titanium-alloy substrate with photoresist (PR); (c)Photo-lithographically pattern the PR on the top side as the etching mask; (d) Etch the titanium-alloy substrate to wing frames by HF acid; (e)Strip PR and clean; (f)Coat the first parylene film as the insulated layer; (g)Paste titanium-alloy frame with PVDF; (h)Coat the second parylene film all over the PVDF and wing-frame, and compete PVDF-parylene wings

The fabrication process of titanium-alloy MEMS wings with sensing PVDF film is shown in Fig. 4 and described as below.

- Step (a): A titanium-alloy substrate is cleaned with acetone and isopropyl alcohol (IPA). Then, it is flushed with de-ionized water (DI water.)
- Step (b): Both sides of the titanium-alloy substrate are coated with photoresist (PR), AZ4620, by a spin coater. Exact control of the rotation speed is exercised to get 10 μm thick PR, serving as a mask layer for subsequent operation.
- Step (c): The upside PR is patterned by an I-line UV contact aligner. This step defines a pattern as an etching masking for the titanium-alloy frames against chemical etchant. The developer for the AZ4620 PR is AZ400K.
- Step (d): The titanium-alloy substrate is dipped in hydrofluoric (HF) acid to etch uncovered titanium for 50 minutes. After the etching process, the geometry of the wing frames appears.
- Step (e): The substrate then put in acetone solution to strip PR from both sides of the titanium-alloy substrate. All the parts are cleaned and flushed with DI water.
- Step (f): The first parylene diaphragm is deposited on the titanium-alloy in the SCS PDS-2010 parylene coater. For this step, 15 g of parylene dimmer yields an approximately 11.5 μm thick film and this parylene film is used as an insulated layer.
- Step (g): A 25 μm thick piezoelectric film, PVDF, is pasted on the frames.
- Step (h): The PVDF film and the wing frame are coated with the second parylene layer.

Having done the process mentioned above, we can get the PVDF flapping wings shown in Fig. 5. The parylene flapping wings without PVDF sensing film are obtained by the similar process, except using another pasting tape as the supporting stuff in Step (g) and delaminating the tape after parylene coating. By careful fabrication and assembly, two types of MAVs in this work are shown in Fig. 6. The two-wing MAV named Wing A (Caltech-like MAV) has a mass of 7.91g, whereas the single-wing MAV Wing B has a mass of 7.52g. The total mass of the two MAVs are increased to 13.91 and 13.52 grams respectively after they are combined with empennage and poly-lithium batteries additionally.

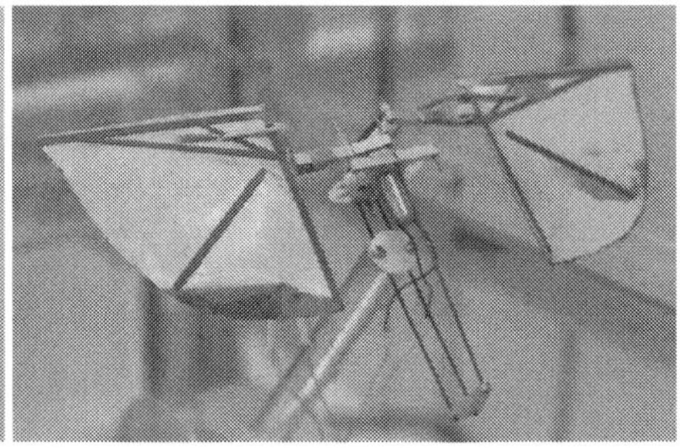

(a) (b)

Figure 5. MAV models in wind-tunnel: (a) The completed PVDF-parylene flapping wing. (b) Setup for testing the flapping wings with PVDF sensing skin in the wind-tunnel

(a) (b)

Figure 6. The appearance of the flapping MAVs (including gear transmission system, wings, DC motor, fuselage, and tail): (a) Wing A with mass of 7.91 g; (b) Wing B with mass of 7.52 g

4. TEST AND RESULTS

4.1 Wind-tunnel test

The aerodynamic testing of MAVs in this work was conducted in a small wind-tunnel. The dimension of the test section has a space of 30×30×100 cm^3 and the inlet contraction ratio is 6.25. The wind speed ranged from 0 to 7 m/s, measured with a hot-wire anemometer. The load-cell (Bertec, OH, USA) with specifications of 200 g and 100 g are responsible for the force measurement of lift and drag, respectively. It has the maximum error 0.2% of the full-scale signal due to the linearity or hysteresis. In the wind tunnel testing, the MAV is placed on the load-cell directly to obtain the data of lift and drag forces. Fig. 7 shows the experimental setup.

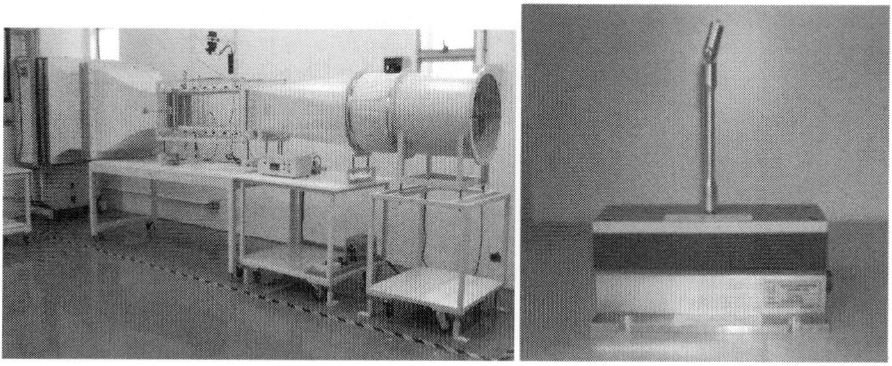

(a) (b)

Figure 7. The wind-tunnel test system: (a) Low speed wind-tunnel; (b) Load-cell (Bertec Corp.)

The scheduled process of collecting a data point by the setup in Fig.7 is defined as below:

1. 0 s: Turn on the load cell;
2. 2 s: Reset the load-cell;
3. 5 s: Turn on the wind-tunnel;
4. 12 s: Start the flapping of the MAV;
5. 18 s: Start the data collection of lift and thrust;
6. 30 s: Stop the data collection (the total data collection time is 12 s.)

The data-breeding rate of the load-cell is set as 1,000 points per second. Then we collected 12,000 points of data in every flapping condition, and integrate them into time-averaged values of lift ***L*** and thrust *T*. These 12,000 point data may not be divided into round number of cycles for different wingbeat frequencies (from 7.2 to 23.6 Hz). However, the error due to noninteger cycles of wingbeating is confined less than 0.47 %.

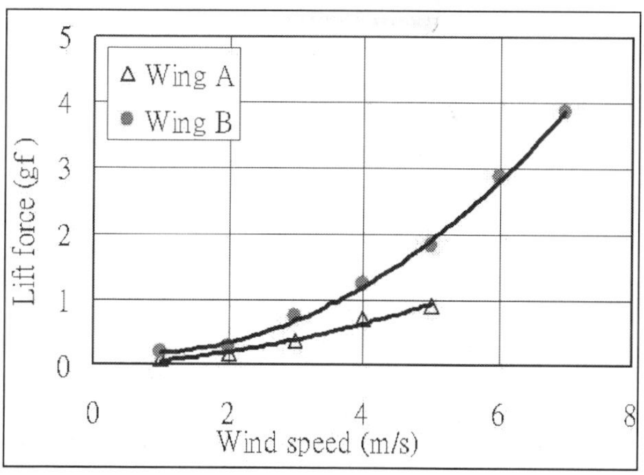

Figure 8. Relationship of lift-force vs. wind speed with no flapping

In order to study the relationship between the lift force and the wind speed. The MAVs with parylene wings are set at an attack angle to 20 degrees. Wing A and Wing B were measured successively with no flapping in the wind-tunnel first. With wind speed increased from 1 to 7 m/s, the lift force increase vigorously. The acquired data of lift force versus wind speed are shown in Fig. 8. This result reveals that Wing B has a greater lift force than Wing A (Caltech-like) does due to the larger wing area. Both of them have the right tendency that lift forces increase with the square of wind speed.

The aerodynamic performance of the two wings during flapping was also studied. In the flapping test, the flapping frequency *f* is controlled by a DC motor and the voltage applied to the motor was set from 3 V to 7 V to get the flapping frequency ranging from 7 Hz to 16 Hz. This dynamic experiment was conducted at the wind speed U of 1 to 5 m/s. Fig. 9 shows the experimental results. The lift and thrust coefficients can be expressed as follows:

$$C_L = \frac{2L}{\rho \cdot A \cdot U^2} \tag{10}$$

$$C_T = \frac{2T}{\rho \cdot A \cdot U^2} \tag{11}$$

where $L, T, \rho, U,$ and A are lift, trust, air density, flight speed, and wing area, respectively. The advance ratio *J* is also defined as the ratio of the flight speed to the speed of the wingtip:

$$J = \frac{U}{2 \cdot \phi \cdot f \cdot b} \tag{12}$$

where ϕ, f and b are stroke angle, flapping frequency, and semi-wing span, respectively. For the general case of unsteady-state flapping flight, the advance ratio *J* is less than 1. The advance ratio *J* approaches very large for the case of a fixed wing (no flapping). In this work, even with a constant angle of the attack is set constant (20 degrees), we can still adjust the flapping frequency and the wind speed to get various values of *J* in the flapping (dynamic) test. From the result illustrated in Fig. 9, in the regime where *J* less than one (unsteady state) the C_L is increased rapidly. With *J* increasing, the C_L approaches to the static value of 0.1 (J>4) which is almost equal to the value of C_L of no flapping mode in Fig. 8.

The result of Fig. 9 also shows that Wing A (Caltech-like MAV) is superior to Wing B both in generating lifts and trusts during the flapping flight. The data are rather scattering near *J*=1. The issue is due to the very long testing time (over 10 hours for extracting 50 data points of C_L and C_T) of our MAV in the wind tunnel. Consequently, the structure of our MAV encounters the aging problem. For example, many mechanical components of the MAV start to deviate from their original positions, and consequently degrading their aerodynamic performance. Fig. 9 also reveals that, as *J* is less than 1, Wing A (Caltech-like) generates somewhat larger lift and thrust coefficients than Wing B in the flapping mode of flight. It is for the excuse that Wing A has its wing spar, wing chord and wing rib stronger than Wing B, and such a high stiffness of flapping wing generates flight with high speed and flapping frequency.

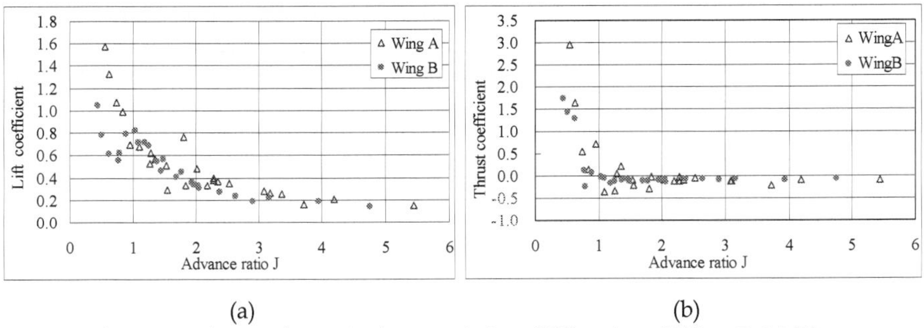

Figure 9. The unsteady aerodynamic characteristics of Wing A and Wing B. (a) lift coefficient ; (b) thrust coefficient

We also allowed the two types of flapping MAVs to perform their free flights, i.e., let them sustain the flight by poly-lithium batteries without remote flight control on the stability of the MAV in real time. Even though Wing A has a superior aerodynamic performance as mentioned above, it is hard to have a successful free flight. The reason may be due to that the difficulty in locating and adjusting the gravitation center as well as the aerodynamic center of this two-wing MAV that we cannot ensure the static stability of the Wing A (Caltech-like). On the contrary, Wing B has a successful free flight with a flight range of more than 10 meters.

The on-site lift output from PVDF sensing skin on the MAV is no doubt the primary viewpoint and contribution of this work. The Wing A with PVDF-parylene composite skin is adopted for this test. The aerodynamic signals are picked up successfully by PVDF and load-cell simultaneously. While the link BC is 11 mm, the collected lift data are shown in Fig. 10. It appears that there is a phase delay between the PVDF signal (Fig. 10(a)) and load-cell signal (Fig. 10(b)). However this is due to that the left wing is composed of a PVDF sensing skin and the right wing is pure parylene. The signal from the PVDF film denotes the lift information of the left wing, which does not include that of the right wing. Meanwhile, the signal from the load-cell sums up the global forcing effect from both wings. With variable lengths of link BC, the phase lag of the two wings resulted in the curve of PVDF, which is ahead or behind the load-cell.

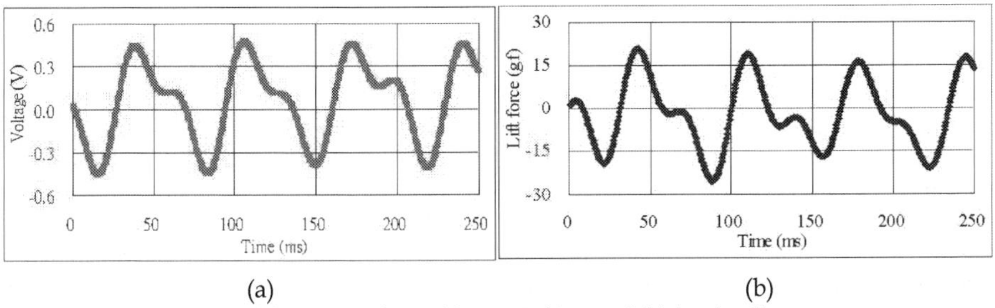

Figure 10. Lift signals of the MAV: from (a) PVDF film and (b) load-cell

By superimposing a similar curve of Fig. 10(a) with a proper phase lag (11 degree) into Fig. 10(a) itself, we can get the "pseudo" result in Fig. 11 which matched with Fig. 10(b) very well in a qualitative way. To arrive at the proper phase lag for superimposition, we accurately superimpose by mathematical method, from 0 to 27 degrees, and use the standard deviation to define the similarity between the pseudo PVDF and load-cell curve. We find the phase lag of 11 degrees is the most fitting case, as shown in Fig. 12.

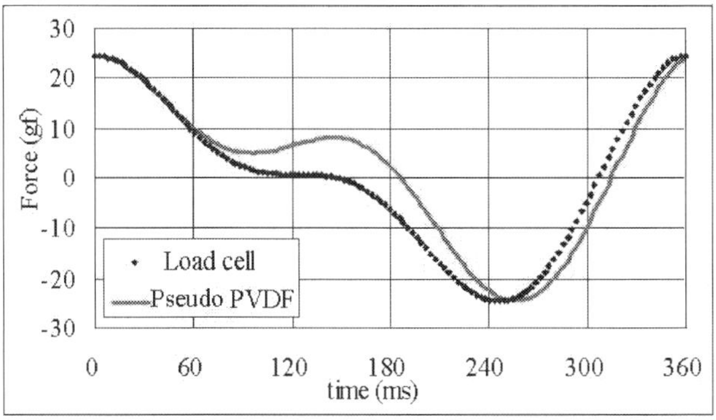

Figure 11. The "pseudo PVDF" signal comes from superimposing a similar curve of Fig. 10(a) with a proper phase lag (11 degree) into Fig. 10(a) itself to match well with load cell signal

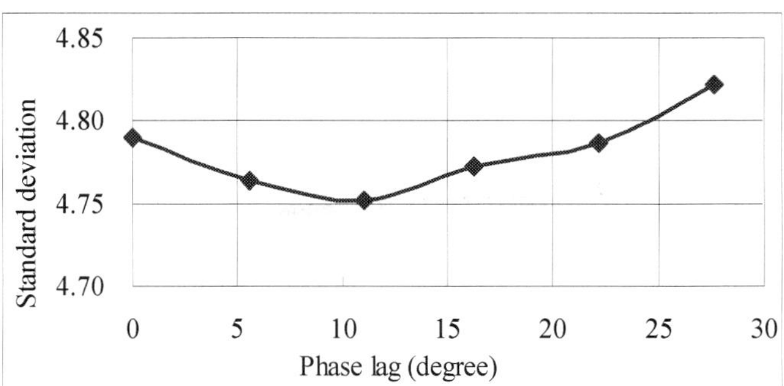

Figure 12. By using the standard deviation to judge the similarity

One of the significant findings from the new configuration of the flapping wings is that the phase lag of the gear transmission system is inevitable in the mechanical design. Therefore it is instead used to enhance the lift behavior of MAV. For example, we can make the lift curve smoother to sustain a better attitude of the MAV during the flapping cruise. After establishing the relationship between the phase lags from gear-transmission and the lift response in Fig. 13, we will adjust the aerodynamic performance of MAV reliably by changing the phase lag between the two flapping wings through fine tuning the mechanism linkages of the gear transmission system.

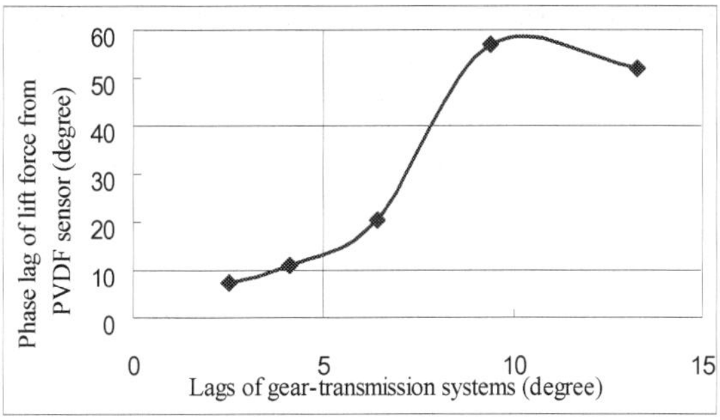

Figure 13. Deduced phase lag of lift force from the PVDF sensor vs. different lags of gear-transmission systems

4.2 Improved flight performance

Based on the experience of the preceding free flight tests of MAVs, we found two problems influencing the flight performance. First, the transmission system needs two poly-lithium batteries of 6.5 g weight to supply the power for sustaining the wing flapping with a frequency of 16Hz. The weight of the batteries is a critical issue since they contributed half of the MAV's total weight. The second issue involves the un-balanced angular momentum which yields the shaking phenomenon from the gear transmission system. It will alter the attitude of the MAV and result in a disastrous flight consequence, such as stall.

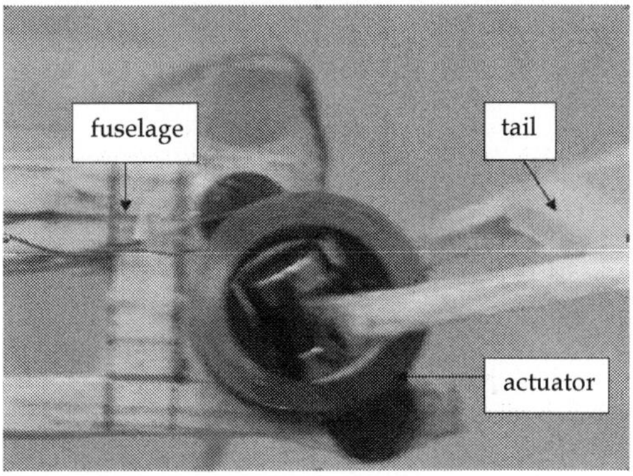

Figure 14. The actuator connects the fuselage and the tail

In order to get a desired flight via wireless control on purpose, light weight materials for the components of the MAV and re-design of the gear transmission system is required. We set the length of wingspan to be 30 cm and 8 cm in wing cord and there is no rib placed in the airfoil. By utilizing the previous fabricating experience of MAV, the new acrylic base as well as the gear transmission system that has a gear ratio of 26.6 is installed. In addition, light receiver and a magnetic actuator which are only 0.9 g and 1.1 g, respectively are employed in the MAV as well. The receiver has two channels, one to control the rotating speed of the DC motor and the other to adjust the tail actuator acting as a rudder for airplanes. In order to match the actuator on purpose, the construction design of the fuselage and the tail actuator is shown in Fig. 14.

Finally, we integrate all the components including the receiver for the remote control, tail, actuator, and one poly-lithium battery into the MAV. The total weight of MAV is 10.67g. The MAV named as "Eagle-II" is shown in Fig. 15. In the flight testing with wireless attitude control, the MAV has a successful flight with a range of 40 meters and an endurance of 10 seconds. The detailed flight trajectory of the MAV prototype is shown in Fig. 16. We additionally measured the averaged aerodynamic data of "Eagle-II" and shown in Fig. 17.

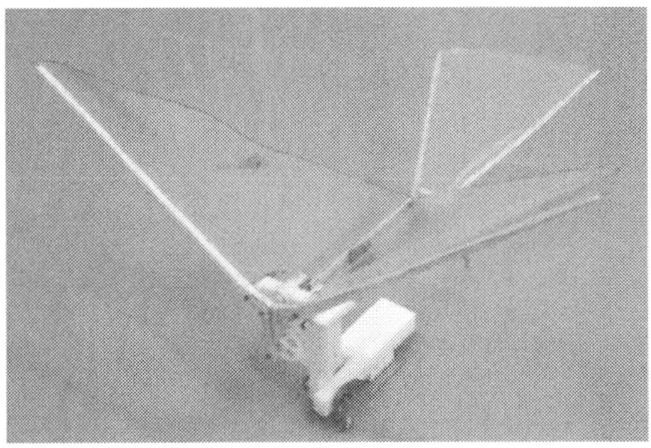

Figure 15. The appearance of modified MAV "Eagle-II" with the configuration of Wing B

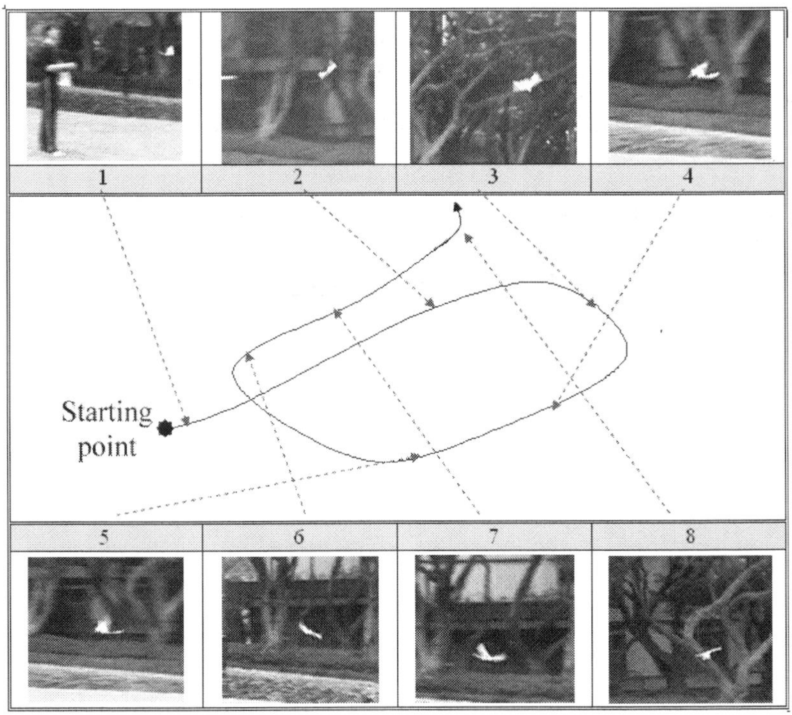

Figure 16. The flight trajectory of MAV

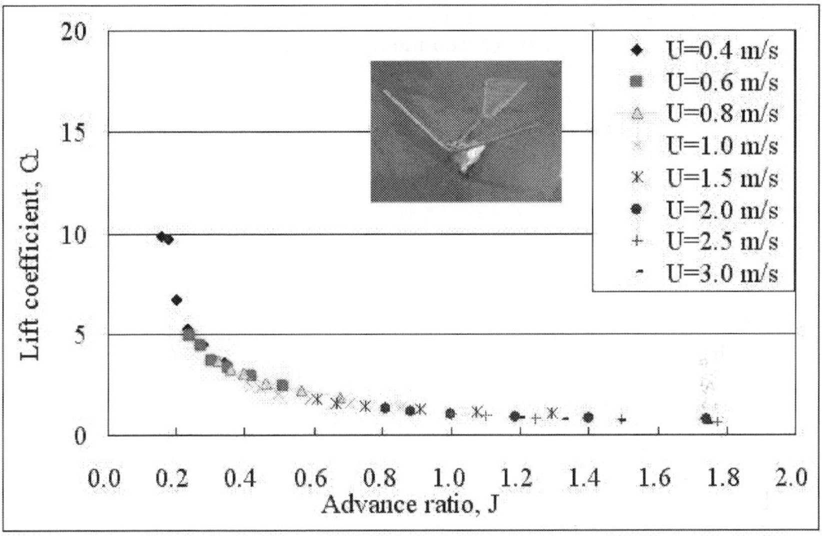

(a)

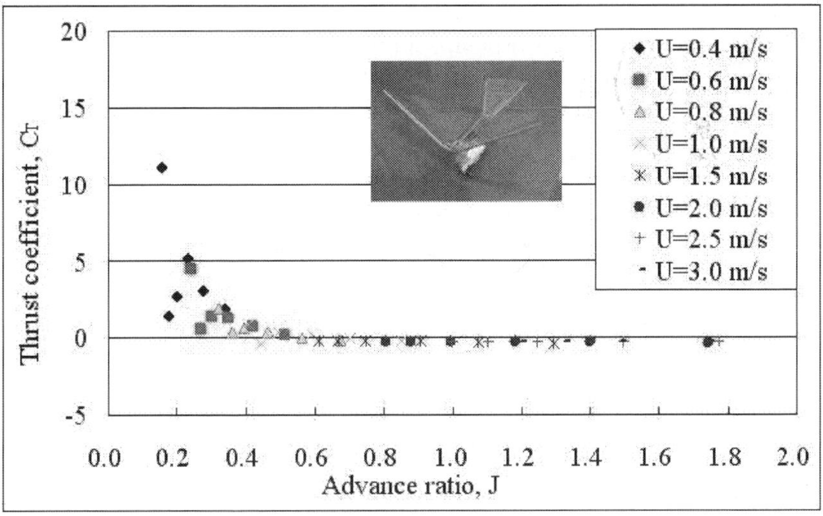

(b)

Figure 17. (a) Lift coefficient CL and (b) thrust coefficient CT of the MAV "Eagle-II" versus advance ratio J corresponding to different freestream velocities U. The wing skin of the assembled MAV "Eagle-II" has been changed as parylene film of 20-30 μm thick, and the leading edge beam is made of Balsa wood and it is quite stiffened compared to wing skin

4.3 Modification of MAVs

The flapping stroke angle ϕ is set as 35° for the case of "Eagle-II". It is really small compared to the stroke angle of 120° for hummingbirds (mentioned in the book of Norberg (Norberg, 1990)) or even the case of 60° or more for other medium-size birds. This angle limitation of our MAVs is due to the considerations of light weight and wear capability for the four-bar linkage module.

The lift and thrust coefficients C_L, C_T are extracted from a wind-tunnel test of the MAV "Eagle-II" with a semi-rigid leading edge subject to a fixed inclined angle of flapping plane of 20° (larger than the ordinary stall angle-of-attack 15° for the fixed-wing planes) and different advance ratios. Of course, the actual instantaneous angle of attack for the flapping flyers may be changing with time and larger than 20°. Even highest angle amplitude of attack, e.g., 50°, for the unsteady cases still does not cause catastrophic flow separation or stall, rather induce better unsteady lift force by paying the price of smaller thrust. This is the famous effect termed as "delayed stall" [13]. In general, these time-averaged lift coefficient C_L of the MAV larger than 3.0 in the regime of J below 0.4 resembles the enough counterbalance force to the weight of the MAV. The thrust coefficient C_T subject to J value below 0.4 is always positive also denotes the continuous forward pushing without speed decreasing during the real flight.

However, after a certain period of operation time the thrust data in the small advance-ratio or high flapping-frequency region reveal the trend of diversity or data fluctuation. Such a diversity of thrust data was traced back and suspected from the irregular operation of the worn gear transmission module. In other words, the wear of the power transmission mechanism impedes the smooth flapping and induces the corresponding unpredictable aerodynamic forces. Figure 18 shows the detailed four-bar (black dash line) gear transmission module for the MAV "Eagle-II". The seriously worn (reaming) portion is specifically marked with a red ellipse for the Balsa leading-edge frame (and the motor is removed in this figure.) This kind of reaming on the Balsa wood not only denotes an aging problem against good life-time of our MAVs but also reveals a certain uncommon "penalty" force resulting from the improper matching of the semi-rigid wing structures into our simple flapping mechanism.

The Knoller-Betz effect of thrust generated by means of a simply up-and-down flapping was observed in the last century (Knoller, 1909; Betz, 1912). For the conventional rigid symmetric wing with very large bending moment of inertia, the corresponding thrust force cannot react with apparent forward deformation on the whole wing structure. Additionally, the exact time location for the biggest thrust during a full flapping cycle has not been pointed out precisely by the Knoller-Betz effect. Under the constraint of the almost rigid wing frames, there seems no way but to design a complicated flapping mechanism to follow the 3-D flapping trajectory like hummingbirds (or even having the trajectory of "clap-and-fling" insects) for the artificial wingbeating vehicles. In that classical thinking of mechanism design, too much flexibility of the wing frames means the operation failure or malfunction of the flapping machines mimicking the biological gesture of natural flyers.

The design methodology in this study, on the contrary, is to use the lighter and more flexible materials as the key structures for our new version of MAVs. For instance, we tried to shorten the wingspan from 30 cm to 21.6 cm toward the miniaturization trend. The correlated body mass regulated from the scaling law of birds[13] is no more than 10 g. For the light weight consideration, we replaced the Balsa-wood leading edge frame with fine carbon-fiber rods. Meanwhile, the four-bar linkage transmission module is reconstructed by the electrical-discharge-wire-cutting (EDWC) of aluminum-alloy 7075. The new power transmission module made of aluminum alloy 7075 shown in Fig. 19 saves us greatly with 2 g in weight at least. This above belief or revelation from small technology is that the more miniaturized artifacts may render us with better structural stiffness and operation life time.

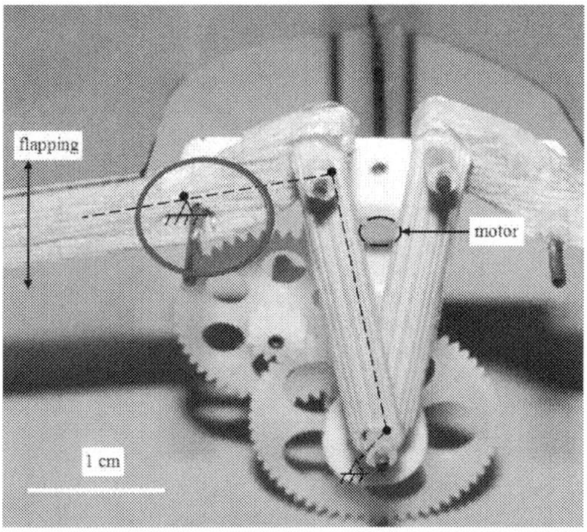

Figure 18. The four-bar (black dash line) gear transmission module for the MAV "Eagle-II". The seriously worn portion is marked with an ellipse for the Balsa leading-edge frame (and the motor is removed temporarily.)

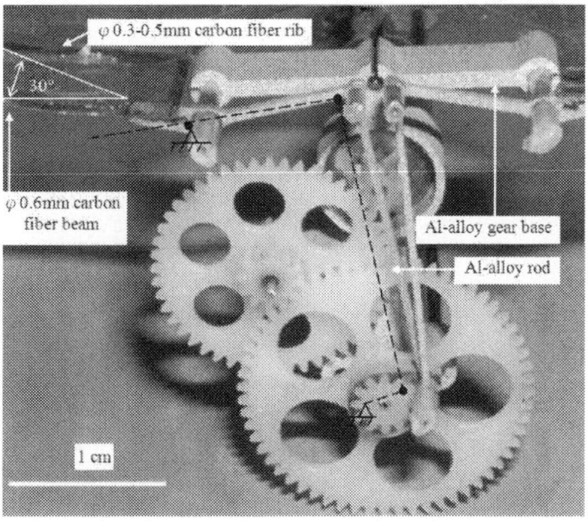

Figure 19. The weight-reduced four-bar (black dash line) gear transmission module for the MAV "Golden Snitch". The leading-edge beam is made of 0.6 mm-diameter carbon-fiber beams and aluminum-alloy bars. A rib with 30° back-swept to the leading edge beam is optionally installed. All the tiny aluminum parts were precisely manufactured by electrical-discharging-wire-cutting (EDWC)

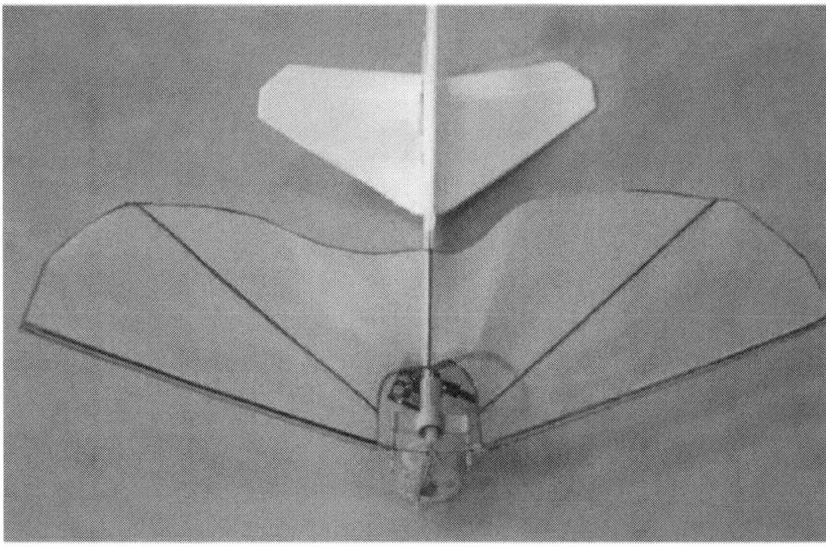

Figure 20. The completed MAV "Golden Snitch" with the carbon-fiber leading-edge beam/rib and parylene skin. A 7mm-diameter electrical motor mounted in the newly developed aluminum-alloy holder with the new transmission module is actuated by a poly-lithium battery of 30-50 mAh. The total mass including flexible wing frame, motor, gear assembly, wing, fuselage, tail, battery and receiver is only 5.9 g

We remodeled the MAV with the new flexible wing frames and the new gear transmission module mentioned above, kept reducing the total weight and wingspan, and finally obtained the modified MAV named as "Golden Snitch" (wingspan = 21.6 cm, body mass = 5.9 g) in Fig. 20. The aerodynamic coefficients versus advance ratio J for the MAV "Golden-Snitch" were measured (Fig. 21). The scheduled process of collecting data in the wind-tunnel testing is the same as the case of "Eagle-II". The new testing results of aerodynamic performance effectively improve the thrust data with 35 % increasing in magnitude and without the previous problem of unpredictable data fluctuation. The disappearance of the structure reaming on pivoting joints and the "penalty" force from the unsteady flapping prevent lots of friction loss and may benefit the real flight of our new MAVs. Furthermore, with this great decrease of friction loss in the mechanical transmission module, the wingbeat frequency of the MAV "Golden Snitch" can increase to 20 Hz or much faster than before.

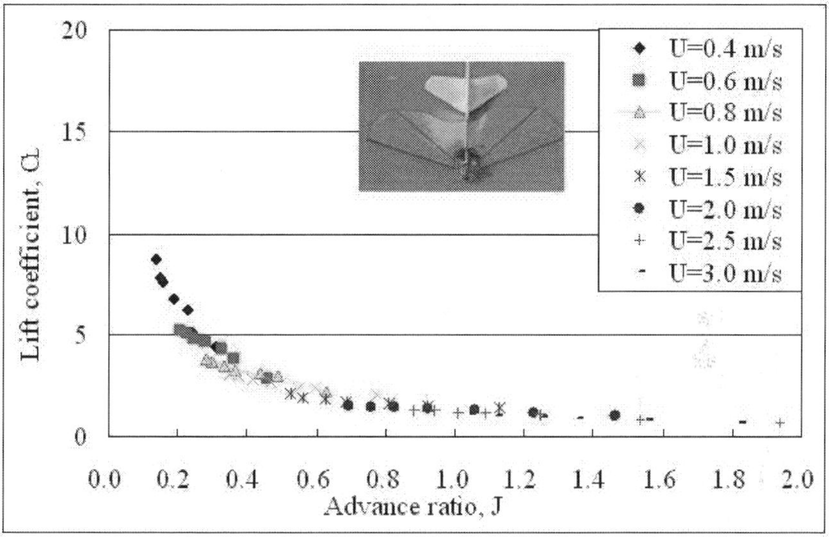

(a)

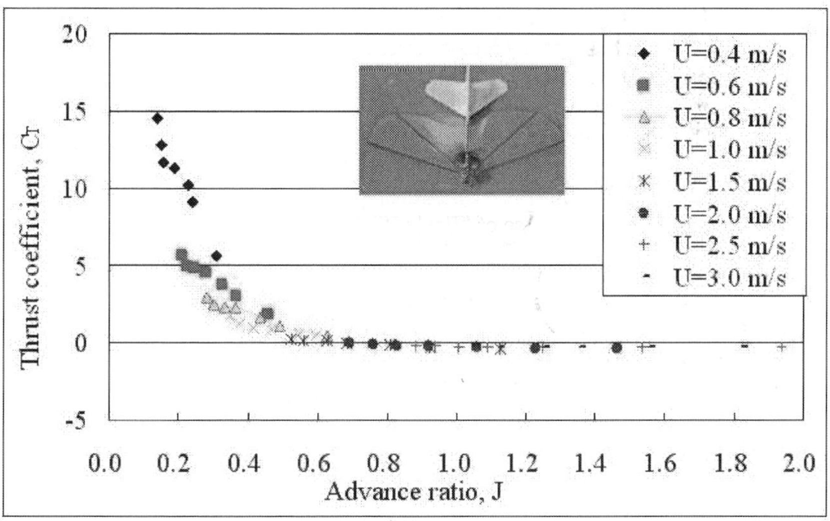

(b)

Figure 21. (a) Lift coefficient CL and (b) thrust coefficient CT of the MAV "Golden-Snitch" versus advance ratio J corresponding to different freestream velocities U (2b=21.6 cm; ϕ=38.9°; f =10.5-23.6 Hz.)

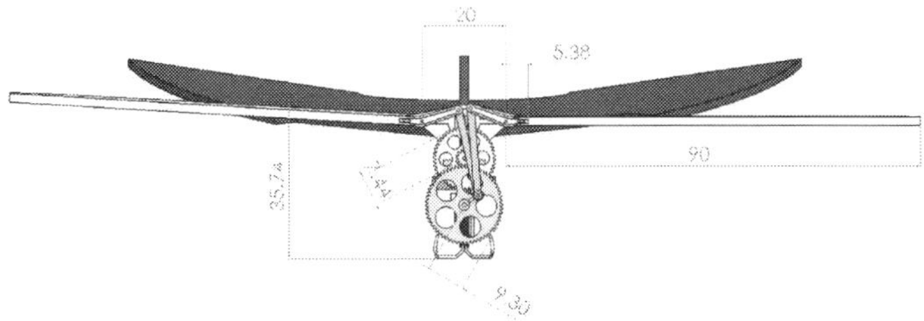

(a)

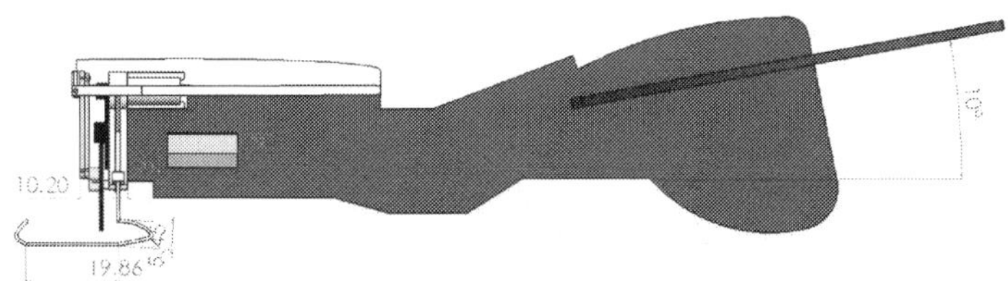

(b)

(c)

Figure 22. "Golden Snitch" with its flight endurance up to 122 s by enlarging the tail area and decreasing the corresponding tail angle: (a) the front view; (b) the side view; (c) the real plane

On the issue of evaluating the efficiency of mechanical power, the merit of removing the wing skin of our artificial MAV by the constructer can help a lot. Therefore, the flapping of the bare skeleton of wing frames can be regarded as the case for investigating the power dissipation due to the friction in the mechanical transmission mechanism. In this work, the friction dissipation of the MAV "Eagle-II" counts 70% of the total electrical power; whereas the modified mechanism of the MAV "Golden Snitch" only consumes 45% of the total power.

The new MAV "Golden Snitch" equipped with the above new configuration and characterized with light weight and high flexibility has a far more outstanding flight performance than the semi-rigid "Eagle-II." With scarcely real-time adjustment by remote control "Golden Snitch" actually made a successful flight record of 47 s at least, breaking the 11 s record of "Eagle-II" to a great extent. Without the negative effect of side gust wind, "Golden Snitch" would like to fly roughly along the virtual barrel surface of an imaginary cylindrical column with the diameter of several meters if the horizontal tail generating negative lift is only used to stabilize the MAV. Recently, we additionally found that "Golden Snitch" can prolong its flight endurance up to 122 s by increasing the stroke angle ϕ to be 52° and enlarging the tail area like Fig. 22. In this new tail design the center of gravity of the MAV moves backwards, and high lift occurs for the positive lift contributed by the flapping wing and the horizontal tail simultaneously. It additionally shows the high-lift mode as well as the high-AOA mode of "Golden Snitch" which flies up and down along the longitudinal direction vigorously.

4.4 MAVs with flexible wing frame

Thanks to the intrinsically flexible wing structure of "Golden Snitch" herein, this MAV can fly in free flight without much energy loss in resisting against the surrounding air. And we luckily found that a streamwise vibration of the carbon-fiber leading edge is characterized by wingbeat frequency from 15.6 to 21.7 Hz for the wings with 30°-ribs. That frequency is much smaller than the natural frequency of 85 Hz for the wing structure. (It means this special vibration has no matter with the resonance of the wing frame.) By combining this induced coherent streamwise vibration of the wing frame and the vertical reciprocating flapping motion in Fig. 23, a stereo figure-of-eight flapping motion of the MAV obviously shows up and can be seen even by human naked eyesight. This is justified preliminarily by a technique of LED illumination shown in Fig. 24. This streamwise vibration of the wing frame can be originated from the forementioned Knoller-Betz effect and interpreted as a coherent forward locomotion, but the exact moment of the biggest thrust or the vibration amplitude moreover occurs at the very instants of stroke reversals, i.e., the highest and lowest points of the flapping motion. We believe that this biomimetic figure-of-eight wing tip trajectory generated by a simple flapping mechanism in flexible MAVs is first time been found and reported herein.

The ordinary video camera with 30 frames of images per second only provides too slow speed for depicting detailed viewgraphs of a full-cycle flapping kinematics in this work. The continuous blinks of the figure-of-eight trajectory of the wing tip were then taken and certificated by a high speed CCD camera (Phantom v. 4.2) for the MAV "Golden Snitch" in quiet ambient. We installed this MAV in the wind-tunnel and captured the real-time images of the flapping wings subjected to cases with and without the 30°-rib in Fig. 25. The surface morphology of the wing skin shows the wavy profile from the leading edge to trailing edge and from the wing tip to inner wing root simultaneously, which accords the supposed flexible wing design with a sinusoidal change of the leading-edge angle-of-attack. More additionally, we verified the similar phenomena of figure-of-eight flapping for the MAV "Golden Snitch" subjected to different freestream velocity which is set from 0.4 to 3.0 m/s.

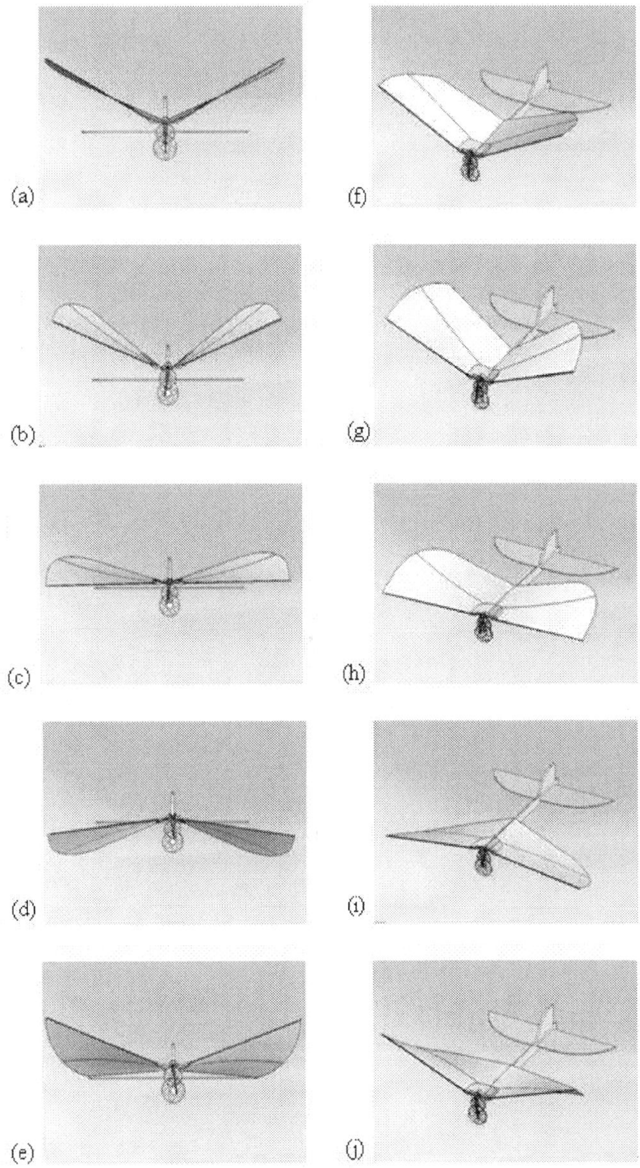

Figure 23. The supposed cartoon of the continuous full-cycle flapping motion of the MAVs with a rigid four-bar transmission module: (a) denotes the neutral position; (b) to (c) denote the downstroke; (d) to (e) denote the upstroke; the bird's-eye views from (f) to (j) are corresponding to the front side views from (a) to (e), respectively

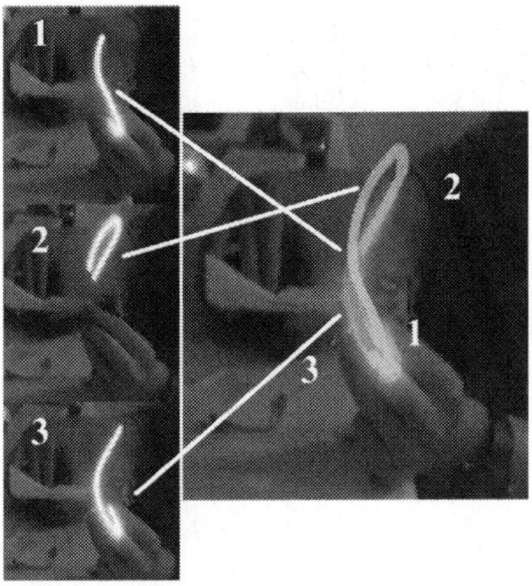

Figure 24. Wing tip trajectory by LED illumination. 1-3 are the instantaneous images of a figure-of-eight taken by a digital video. The right image combines the left 3 images

4.5 Analogous relationship in scaling laws for flapping flyer

Rather than depicting the anatomical analysis on the wing frames of flapping flyers, herein we prefer to categorize the MAV with figure-of-eight flapping into some animal groups and conventional MAVs according to their realistic wingspans and the wingbeat frequencies with respect to different body masses. Based on Norberg (Norberg, 1990), Greenewalt (Greenewalt, 1975), and Shyy's (Shyy, 1999) collecting information, the denoting spots of the MAVs "Eagle-II" (wingspan = 30 cm, body mass = 11 g, wingbeat frequency = 16Hz) and "Golden Snitch" (wingspan = 21.6 cm, body mass = 5.9 g, wingbeat frequency = 20Hz) in this work were inserted into the previous plots of wing-span versus body-mass as Fig. 26 and wingbeat-frequency versus body-mass as Fig. 27. "Eagle-II" is still allocated on the margin of conventional MAVs and small natural flyers, whereas "Golden Snitch" with figure-of-eight flapping is approaching to the characteristic region of hummingbirds. Apparently the inductive remark on this observation come to that "Golden Snitch" here has strong biomimetic relationship with hummingbirds due to the common property of figure-of-eight flapping, even though the inclined gesture of the flapping plane for these two flyers are quite different. With such an encouraging finding, the domain of MAVs marked as the shadow region in Figs. 26 and 27 extend from body mass of 11 g down to 5.9 g by this study. No matter how we followed the empirical design formulae deduced from hummingbirds, the figure-of-eight flapping is verified empirically and necessary to design a MAV which can fly successfully with wingspan less than 21.6 cm and body mass below 10 g. Subjected to this design requirement of the artificial novel figure-of-eight flapping, a flexible wing frame driven by a single DOF, four-bar linkage mechanism is provided herein as the neatest design so far as we know. In this work or the practice manner, the cross section of the leading edge carbon-fiber beam is suggested to be round, and the natural frequency of this leading edge beam should be so much larger than the wingbeat frequency without structural resonance as to sustain a successful coherent vibration out from the flap-sweeping plane due to the forward pushing by the periodic vigorous thrust.

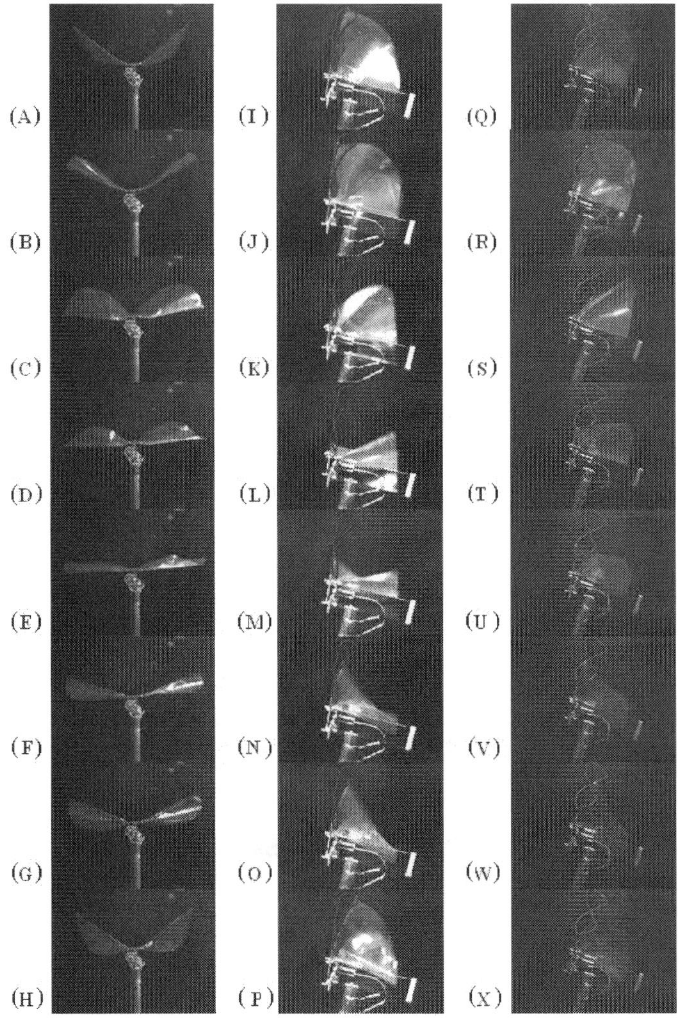

Figure 25. The continuous full-cycle blinks, taken by a high speed camera, of Golden Snitch's flapping wings. The front side views of Golden Snitch from (A) to (H) are corresponding to the side views of it without rib from (I) to (P), and to the side views of it with 30°-rib from (Q) to (X), respectively. The wingbeat frequency is 20.83 Hz. The angle of attack is 20°. The red contours added into the images show the figure-of-eight trajectories of wingtips. Even the parylene wing without the 30°-rib has the more flexible property than the stiffened 30°-rib wing, the figure-of-eight flapping still appears and the proper frequency interval for the better figure-of-eight is broader (about 9.4-25 Hz for the no-rib wing; 15.6-21.7 Hz for the stiffened 30°-rib wing.)

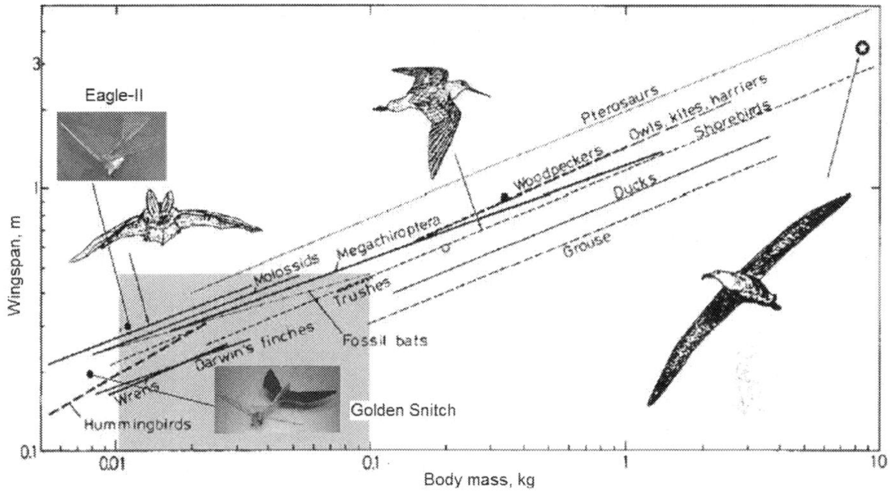

Figure 26. Wingspan plotted on logarithmic coordinates against body mass for some animal groups (Greenewalt, 1975; Norberg, 1990), conventional "shadow" region for MAVs (Shyy, 1999), and the MAVs (Eagle-II and Golden Snitch) in this chapter

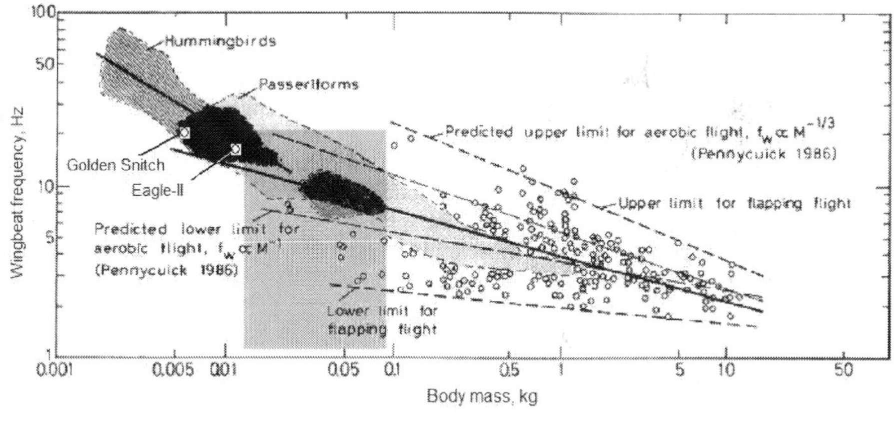

Figure 27. Wingbeat frequency (or flapping frequency) versus body mass for some animal groups (Greenewalt, 1975; Norberg, 1990), conventional "shadow" region for MAVs (Shyy, 1999), and the MAVs (Eagle-II and Golden Snitch) in this chapter

5. CONCLUSION

This study presents a smart wing with PVDF-parylene composite wing by MEMS fabricating process and a four-bar linkage transmission system with variable phase lags of the flapping wings. The signals from the PVDF film and the load-cell are acquired simultaneously and the curves are quite similar. This result demonstrates that the smart

PVDF skin has the promising capability to monitor aerodynamic information of flapping in the future. In the wind-tunnel tests, the collected data has a phase lag between the PVDF and load-cell curves due to the design of asymmetric flapping movement. By a superimposing method, the "pseudo" PVDF curve has higher similarity to the lift signals. The smart PVDF-parylene composite wing and the superimposing method can help to design a micro MAV with ideal aerodynamic characteristic by changing the phase lag between the two flapping wings through fine tuning of the mechanism linkages.

The performances of flapping wings are also studied in this chapter. The lift force increases with the square of the wind speed in the non-flapping mode. In the unsteady state or the flapping mode of our MAVs, the two-wing configuration has greater superiority in terms of aerodynamic performance than the single wing configuration. It is apparent that stiffness of flapping wing plays an important role in producing lift and trust forces. The first generation MAV which has wingspan of 30 cm, including one commercial lithium battery herein has a total weight less than 11 grams and it has a successful flight of more than 40 meters by wireless remote control.

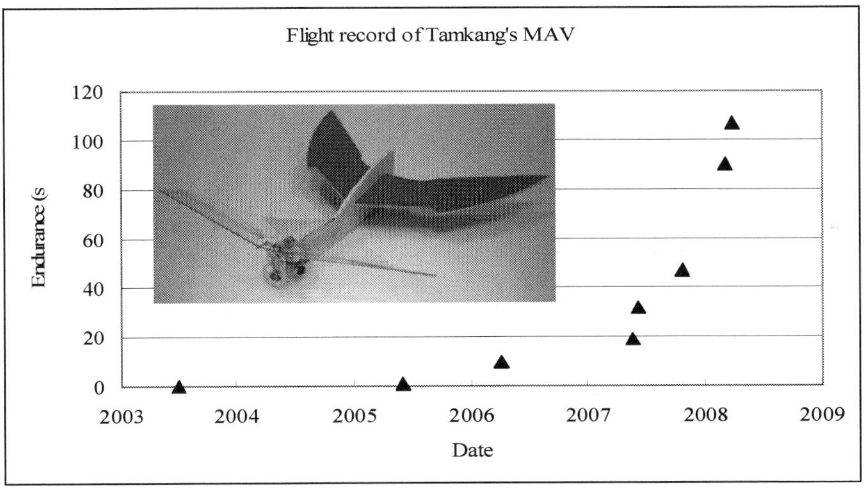

Figure 28. Flight record of MAVs developed in Tamkang University

Additionally, the passive design aspect of improving the performance of the wing frames for flapping MAVs "Eagle-II" and "Golden Snitch" in this chapter tends to provide a simpler flying platform with moderate gear-transmission module but to produce bountiful and interesting phenomena deserving furthermore biomimetic study. With the help of a high-speed camera, the figure-of-eight trajectories of the wing tips of "Golden Snitch" were verified by their proper function of the carbon-fiber wing frames and the parylene wing skin. The unsteady lift data and the highly enhanced thrusting locomotion acquired from the wind-tunnel testing did great contribution to the successful flight of our MAVs, even the implied unsteady flow mechanism still needs more theoretical explanations and more supporting evidences. This work also removed the technical barrier against the designers in making complicated multi-DOF actuating mechanisms for generating figure-of-eight flapping. Finally, with the advantage of light-weight and size miniaturization, we effectively extend the existing domain of MAVs and created new flight record depicted in Fig. 28. We believe this neatest design methodology for flapping MAVs will be promising in practice, and the correlated flow visualization of the figure-of-eight or the theoretical investigation of the solid-fluidic (aero-elastic) interaction deserve follow-up attentions.

6. ACKNOWLEDGEMENT

We would like to thank the financial support from National Science Council of Taiwan by the project numbers of NSC-93-2212-E-032-012, 94-2212-E-032-012, 96-2221-E-032-013, 96-2221-E-032-014, 97-2628-E032-015, 97-2221-E-032-016 and the experimental helps from Dr. C.-K. Hsu, Prof. S.-W. Kang, Prof. F.-Y. Hsiao, Mr. H.-M. Shih, Mr. M.-W. Gao, Dr. W.-C. Wei, Mr. C.-Y. Kao, and Mr. J.-W. Liao of Tamkang University. The discussions with Prof. Y.-C. Tai of Caltech, Prof. W.-P. Shih, Prof. A.-B. Wang of National Taiwan University, Prof. Y.-K. Shen of Taipei Medical University, and Prof. J.-M. Miao of National Defense University, are also highly acknowledged.

7. REFERENCES

Ashley, S. (1998). Palm-size spy plane, *The American Society of Mechanical Engineers*, February, 74.

Betz, A. (1912). Ein beitrag zur erklärung des segelfluges, *Zeitschrift für Flugtechnik und Motorluftschiffahrt*, 3, 269-272.

Barrett, R., McMurtry, R., Vos, R., Tiso, P., and De Breuker, R. (2005). Post-buckled precompressed (PBP) elements: a new class of flight control actuators enhancing high-speed autonomous VTOL MAVs, *Proceedings of SPIE - The International Society for Optical Engineering*, Vol. 5762, San Diego, USA, May, 7-9, pp. 111-122.

Banala, S.K. and Agrawal, S.K. (2005). Design and optimization of a mechanism for out-of-plane insect winglike motion with twist, *Journal of Mechanical Design/ Transactions of the ASME*, 127(4), 841-844.

Cloupeau, M. (1979). Direct measurements of instantaneous lift in desert locust: comparison with Jensen's experiments on detached wings, *J. Exp. Biol.*, 80, 1-15.

Chanute, O. (1894). *Progress in Flying Machines*, Dover Publications, Inc.

Dickinson, M.H. and Götz, K. (1996). The wake dynamics and flight forces of the fruit fly Drosophila melanogaster, *J. Exp. Biol.*, 199(99), 2085-2104.

Greenewalt, C.H. (1975). The flight of birds, *Trans. Am. Philos. Soc.*, 65, 1-67.

Ho, S., Nassef, H., Pornsinsirirak, T.N., and Tai, Y.C. (2003). Unsteady aerodynamics and flow control for flapping wing flyers, *Progress in Aerospace Science*, 39, 635-681.

Hollick, F.S.J. (1940). The flight of the dipterous fly Muscina stabulans fallen, *Phil. Trans.*, B 230, 357-90.

Jensen, M. (1956). Biology and physics of locust flight. III. The aerodynamics of locust flight, *Philosophical Transactions of the Royal Society of London, Series B, Biological Sciences*, 239(667), 511-552.

Jones, K.D., Bradshaw, C.J., Papadopoulos, J., and Platzer, M.F. (2005). Bio-inspired design of flapping-wing micro aerial vehicles, *Aeronautical Journal*, 109(1098), 385-393.

Knoller, R. (1909). Die Gesetze des luftwiderstands, *Flug-und Motortechnik (Wien)*, 3(21), 1-7.

Liu, C. (2006). Foundations of MEMS, Pearson/Prentice Hall, 1st ed, p.256.

McIntosh, S.H., Agrawal, S.K., and Khan, Z. (2006). Design of a mechanism for biaxial rotation of a wing for a hovering vehicle, *IEEE/ASME Transactions on Mechatronics,* 11(2), 145-153.

Norberg, U.M. (1990). *Vertebrate Flight: Mechanics, Physiology, Morphology, Ecology and Evolution,* Springer-Verlag, New York, Chapters 2, 8, 9, 10.

Pornsin-sirirak, T.N. et al. (2001). Titanium-alloy MEMS wing technology for a micro aerial vehicle application, *Sensors and Actuators A: Physical,* 89, 95-103.

Pornsin-sirirak, T.N. et al. (2002). Flexible parylene-valved skin for adaptive flow control, Proceedings of 15th IEEE MEMS conference, Las Vegas, USA, pp. 101-104.

Shyy, W., Berg, M., and Ljungqvist, D. (1999). Flapping and flexible wings for biological and micro air vehicles, *Progress in Aerospace Sciences,* 35, 455-505.

Sitti, M. (2001). PZT actuated four-bar mechanism with two flexible links for micromechanical flying insect thorax, Proceedings of the IEEE Robotics and Automation Conference, Korea, pp. 3893-3900.

Weller, E.J. (1983). *Nontraditional Machining Processes,* 2nd Edition, The Society of Manufacturing Engineers.

Yang, L.J., Hsu, C.K., Ho, J.Y. and Feng, C.K. (2007). Flapping wings with PVDF sensors to modify the aerodynamic forces of a micro aerial vehicle, *Sensors and Actuators A: Physical,* 139, 95-103.

Żbikowski, R., Galin´ski, C., and Pedersen, C.B. (2005). Four-bar linkage mechanism for insectlike flapping wings in hover: concept and an outline of its realization, *Journal of Mechanical Design/ Transactions of the ASME,* 127(4), 817-824.

Żbikowski, R. and Galin´ski, C. (2005). Insect-like flapping wing mechanism based on a double spherical Scotch yoke, *Journal of the Royal Society Interface,* 2, 223-235.

34

Autonomous Guidance of UAVs for Real-Time Target Tracking in Adversarial Environments

Ugur Zengin and Atilla Dogan
University of Texas at Arlington
USA

1. Introduction

While recent technological advances have enabled the development of unmanned vehicular systems and recent implementations have proven their benefits in both military and civilian applications, the full benefit of unmanned systems will be utilized when they can operate autonomously. The primary requirements of autonomy are the capabilities of detecting internal and external changes, and of reacting to them without human intervention in a safe and efficient manner. This can be achieved by developing and implementing autonomous guidance and control systems (AGCS) to "pilot" unmanned vehicles (Rathbun et al., 2002; Finke et al., 2003; Flint et al., 2002; Jun et al., 2002; Pongpunwattana & Rysdyk, 2004; Nikolas et al., 2003; Zhu et al., 2005; Waydo & Murray,2003). Tracking highly mobile targets is a type of mission that can significantly benefit from the use of UAVs (Unmanned Aerial Vehicles) with the capability of autonomy, especially when the pursuit is to take place in an environment where various sources of "threat", obstacles and restricted areas may exist. In a military scenario, multiple UAVs can be used to track enemy or escort a friendly convoy while avoiding no-fly zones and possible sites of SAMs (Surface-to-Air Missiles). In a border patrol application, UAVs can be employed to track intruders while staying within the border and avoiding high elevation. In a law-enforcement scenario, criminals might be pursued or a specific vehicle might be tracked for protection while avoiding high buildings or residential areas. As a wild-life protection effort, animals can be tracked while avoiding high elevation.

When threat exposure, obstacle and/or restricted region are not among the concerns of a tracking problem, UAV trajectories are commanded to fly directly over the moving target (Sengupta & Hedrick, 2003; Spry et al., 2005). When the target is evasive in an intelligent manner, the tracking problem is the subject of pursuit-evasion game theory (Jang & Tomlin, 2005; Antoniades et al., 2003; Vidal et al., 2002; Hespanha et al.,2000; Hespanha et al.,1999). Ground target tracking and required sensors are also particularly studied (Sengupta & Hedrick, 2003; Schumacher 2005; Sinha et al.,2004; Shea, 2000; Koch & Klem, 2001). For example, in (Sengupta & Hedrick, 2003), the tracking is performed by utilizing an offset vector. There are various new challenges when tracking needs to be done by mobile sensors in an area where there exist various threats, obstacles and/or restricted areas as well as other vehicles to avoid. There might be various and completely different types of "threats", obstacles and restricted areas, and the information regarding their presence and/or level

might contain some uncertainty (Jun & D`Andrea, 2003; Hespanha et al.,2001). "Threat" does not necessarily mean locations or objects that have the potential of inflicting damage on the pursuing vehicles. An area or object on which the proximity of the pursuing vehicles may pose some undesired risk is also referred to as threat.
The previous work by the authors (Zengin & Dogan, 2004; Dogan & Zengin, 2006) developed a rule-based (Negnevitsky, 2002) guidance algorithm for a UAV to follow a dynamic ground target while minimizing threat exposure level, eliminating the risk of flying into obstacles and avoiding no-fly zones. This was done by introducing and utilizing the probabilistic threat exposure map (PTEM), which quantifies threats, obstacles and restricted regions in single framework. This chapter presents a new and significantly improved approach that has resulted in a more systematic and analytic formulation of the guidance strategy, a computationally more efficient algorithm. Specifically, the algorithm described in this chapter uses a gradient search approach in minimizing threat exposure and avoiding restricted areas and systematically utilizes mathematical tools such as level curves, inertial and moving-rotating frames and rotation matrices, vector representations of directions and geometric relations between cones, circles and straight lines. Furthermore, strategy states of this algorithm are better organized, which makes the algorithm much easier for reading and programming. As compared to the previous algorithms, a computationally more feasible guidance algorithm is developed without compromising the performance of tracking, avoiding obstacles/restricted-areas and minimizing threat exposure.

2. Formulation of Adversarial Environment

Adversarial environment is an environment where threat exposure should be minimized, obstacles and restricted areas should be avoided. In this chapter, "threat" is used as a broad term to describe the risk or cost for a UAV to occupy a given location at a given time as well as obstacles and restricted regions in the area of operation. When a UAV is flying in an area with multiple threats, safety of the flight is characterized by the probability of the UAV becoming disabled at a certain location, specified by it's x -- and y -- coordinates relative to a frame of reference, (x, y) at a certain time t. To be able to construct the problem in a probabilistic framework, several events are defined and their probabilities are determined.

2.1 Formulation of Area of Operation

Let $\mathrm{E_i}(x, y, t)$ be the event that the UAV becomes disabled by $\mathrm{i^{th}}$ source of threat at the position of (x, y) at time t in the area of operation. $\mathrm{E}(x, y, t)$ is the event that the UAV becomes disabled by at least one of the threat sources at the position (x, y) at time t. Then, let $\mathrm{f}_{p,i}(x, y)$ and $\mathrm{f}_{t,i}(x, y)$ be probability density functions (*pdf*) such that the probability of the UAV becoming disabled by i^{th} threat source at the neighbourhood of (x, y) at time t is

$$\mathrm{p}_i(x, y) = \mathrm{f}_{p,i}(x, y)\, \mathrm{f}_{t,i}(t)\, \Delta x\, \Delta y\, \Delta t \tag{1}$$

where Δx and Δy are to define the area of a neighbourhood of (x, y) and Δt is to define a neighbourhood of t. Note that, $\mathrm{f}_{p,i}(x, y)$ models the dependency of becoming disabled on position and $\mathrm{f}_{t,i}(t)$ models the dependency of becoming disabled on time. As an example, $\mathrm{f}_{p,i}(x,y)$ can be characterized by a Gaussian *pdf*, which specifies the concentration point (location) of the threat by the mean value and the level of penalty of flying close to it by the variance. Regarding the time dependency of becoming disabled, various possible *pdfs* can be used for $\mathrm{f}_{t,i}(t)$. For example, a uniform *pdf* for $\mathrm{f}_{t,i}(t)$ means that the threat exposure level of the UAV at a given position does not depend on time itself but the amount of elapsed time in the neighbourhood of that position. If the level of exposure of the UAV to threats increases as it stays longer in the area of operation, then an increasing probability density function of time should be defined for $\mathrm{f}_{t,i}(t)$.

Now, let $\mathrm{S}(x,y,t)$ be a certain event that the UAV follows trajectory S to reach (x,y) at time t. Then, the conditional probability of the event that the UAV becomes disabled by i^{th} source of threat at the position of (x,y) at time t under the condition that the UAV follows trajectory S is defined as

$$\mathrm{p}_{\mathrm{S},i}(x,y,t) = \mathrm{P}[\mathrm{E}_i(x,y,t) \mid \mathrm{S}(x,y,t)] = \int_t \mathrm{f}_{p,i}(x,y)\, \mathrm{f}_{t,i}(\mathrm{t})\, l_1(x,y,t) l_2(x,y,t) dt \tag{2}$$

where l_1 and l_2 are used to define the neighbourhood at a point on trajectory S (e.g. radar signature area of the vehicle). Also note in (2) (x,y) are functions of time and thus $\mathrm{f}_{p,i}$ is also a function of time.

To better explain the dependency of the conditional probability on both position and time, (2) will be presented in a special case where the UAV enters the area of operation at time t_0, moves until time t_1, afterwards stops and hovers at the same position. In this case,

$$\begin{aligned} \mathrm{p}_{\mathrm{S},i}(x,y,t) = & \int_{t_0}^{t_1} \mathrm{f}_{p,i}(x,y)\mathrm{f}_{t,\mathrm{i}}(t) l_1(x,y,t) l_2(x,y,t)\mathrm{d}t \\ & + \mathrm{f}_{p,i}(x(t_1),y(t_1)) \int_{t_1}^{t} \mathrm{f}_{t,\mathrm{i}}(t) l_1(x(t_1),y(t_1),t) l_2(x(t_1),y(t_1),t)\mathrm{d}t \end{aligned} \tag{3}$$

where $\mathrm{f}_{p,i}$ is constant after time t_1 because the UAV does not change its position. If l_1 and l_2 do not explicitly depend on time, then

$$\begin{aligned} \mathrm{p}_{\mathrm{S},i}(x,y,t) = & \int_{t_0}^{t_1} \mathrm{f}_{p,i}(x,y)\mathrm{f}_{t,\mathrm{i}}(t) l_1(x,y,t) l_2(x,y,t)\mathrm{d}t \\ & + \mathrm{f}_{p,i}(x(t_1),y(t_1)) \mathrm{l}_1(x(t_1),y(t_1)) \mathrm{l}_2(x(t_1),y(t_1)) \int_{t_1}^{t} \mathrm{f}_{t,i}(t)\mathrm{d}t \end{aligned} \tag{4}$$

Note that in the above equation, the second term shows how the probability increases even when the position of the UAV does not change. Let us assume that $\mathrm{f}_{t,i}(t)$ is a uniform *pdf*, i.e. it is constant in the interval when it is not zero. Let us further assume that during the time the UAV stays in the area of operation, $\mathrm{f}_{t,i}(t)$ is nonzero. Then,

$$
\begin{aligned}
\mathrm{p}_{\mathrm{S},i}(x,y,t) &= \mathrm{f}_{t,\mathrm{i}}(t_0)\int_{t_0}^{t_1}\mathrm{f}_{p,i}(x,y)l_1(x,y,t)l_2(x,y,t)\mathrm{d}t \\
&+ \mathrm{f}_{p,i}(x(t_1),y(t_1))\mathrm{l}_1(x(t_1),y(t_1))\mathrm{l}_2(x(t_1),y(t_1))\mathrm{f}_{t,i}(t_0)(t-t_1)
\end{aligned} \tag{5}
$$

where note in the first term that $\mathrm{f}_{p,i}(x,y)$ is still implicitly a function of time since the UAV moves until t_1.

If there are N number of sources of threat in the area of operation, then the conditional probability of the UAV becoming disabled by at least any one of the sources of threat at the position of (x,y) at time t under the condition that it follows trajectory S is

$$\mathrm{p}_{\mathrm{S}}(x,y,t) = \mathrm{P}[\mathrm{E}(x,y,t)\,|\,\mathrm{S}(x,y,t)] \tag{6}$$

Since

$$\mathrm{E}(x,y,t) = \bigcup_{i=1}^{\mathrm{N}}\mathrm{E}_i(x,y,t) \tag{7}$$

and $\mathrm{E}_i(x,y,t)$ are not necessarily disjoint events,

$$\mathrm{p}_{\mathrm{S}}(x,y,t) \le \sum_{i=1}^{\mathrm{N}}\mathrm{p}_{\mathrm{S},i}(x,y,t) = \sum_{i=1}^{\mathrm{N}}\left[\int_{\mathrm{S}}\mathrm{f}_{p,i}(x,y)\mathrm{f}_{t,i}(t)\,l_1(x,y,t)\,l_2(x,y,t)\mathrm{d}t\right] \tag{8}$$

by Union Bound (Stark & Woods, 1994). Thus we can easily compute an upper bound on the probability of a UAV becoming disabled if it follows a certain trajectory in an area with multiple threat sources. If l_1 and l_2 are assumed to be constant for any position and time on the trajectory and $\mathrm{f}_{t,i}(t)$ is the same for all threat sources, then

$$\mathrm{p}_{\mathrm{S}}(x,y,t) \le l_1 l_2 \int_t \mathrm{f}_t(t)\left[\sum_{i=1}^{\mathrm{N}}\mathrm{f}_{p,i}(x,y)\right]\mathrm{d}t \tag{9}$$

since $\mathrm{f}_{p,i}(x,y)$ and $\mathrm{f}_t(t)$ are probability density functions and therefore integrable.

2.2 Probabilistic Threat Exposure Map (PTEM)

In the formulation of the probability of becoming disabled, introduced in the previous section, the dependency on position (the part of (9) in square brackets) is defined to be the Probabilistic Threat Exposure Map (PTEM), which quantifies the risk of exposure to sources of threat as a function of position. This concept is particularly useful in defining, in a single framework, various types of threats such as objects or locations that need to be avoided as far as possible, obstacles or restricted areas that should not be entered. This probabilistic map is not meant to provide the actual map of the area of operation, but to provide a way to plan the trajectory of a UAV to avoid obstacles and accomplish the given mission such as path planning and target tracking. All the threat sources (e.g. exposure to enemy radar, obstacles, and no-fly zones or restricted areas) are characterized in the same probabilistic framework using the sum of probability distributions of threats, obstacles and restricted areas. As mentioned earlier, if a threat is characterized by a Gaussian *pdf,* there are two parameters needed to fully specify the threat; the mean value specifies the concentration point (location) of the threat source and the variance specifies how the threat can have an area (or volume for 3-D case) of effectiveness. The mean values and variances of each Gaussian function are specified such that the obstacles, restricted areas, no-fly zones and threats in the area of operation are all represented. Gaussian distribution can be used to model enemy radars or missiles as well as obstacles and restricted areas. Note that there is not necessarily one-to-one correspondence between the actual obstacles/restricted-areas/threats and the Gaussian functions used in the construction. An actual obstacle, for example, may require multiple Gaussian functions while a Gaussian function may be enough to represent a threat and restricted area, together.

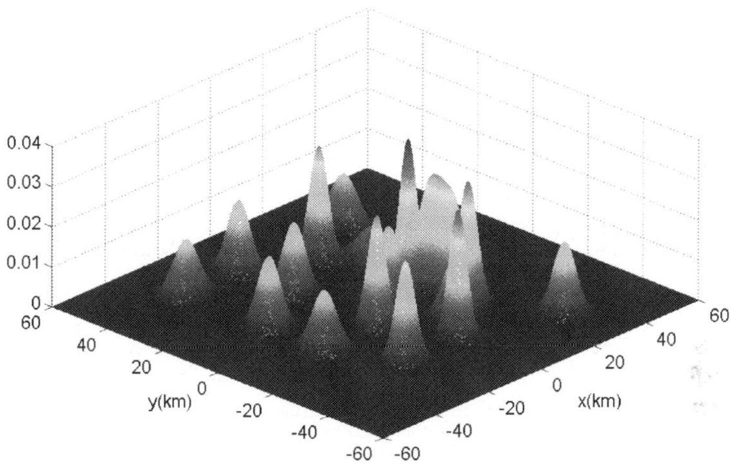

Figure 1. Probabilistic threat exposure map of an operation area

Once PTEM is constructed, there is no need to distinguish between the types of threats, obstacles, or no-fly zones and the use of the probabilistic map is sufficient for decision making. This is because the map already contains the information on the penalty of flying close to a source of threat or a restricted area. The PTEM quantifies the threat exposure level at a given position in the area of operation. A position in the area of operation is defined by its vector, $\underline{\mathrm{r}}$, relative to the origin of a reference coordinate system. Let r be the representation of vector $\underline{\mathrm{r}}$, i.e. $\underline{\mathrm{r}} = [x \quad y]^T$, where x and y are the components of vector $\underline{\mathrm{r}}$ along the x- and y- axes of the reference frame. The same notation will be used for all the other vectors and their representation throughout this chapter, i.e. $\underline{\mathrm{v}}$ is a vector with its representation v in the reference frame. By using this notation, PTEM equation of an area of operation, modeled by Gaussian distributions, can be written as:

$$f(\mathrm{r}) = \sum_{\mathrm{i}=1}^{\mathrm{N}} \frac{1}{2\pi\sqrt{\det(\mathrm{K}_\mathrm{i})}} \exp\left[-\frac{1}{2}(\mathrm{r} - \mu_\mathrm{i})^T \mathrm{K}_\mathrm{i}^{-1} (\mathrm{r} - \mu_\mathrm{i}) \right] \mathrm{dt} \tag{10}$$

where μ_i and $\mathrm{K_i}$ are the mean vector and the covariance matrix of the ith threat, respectively and defined as

$$\mu_i = \begin{bmatrix} \mu_{x,i} \\ \mu_{Y,i} \end{bmatrix}, \mathrm{K_i} = \begin{bmatrix} \sigma^2{}_{x,i} & 0 \\ 0 & \sigma^2{}_{y,i} \end{bmatrix} \tag{11}$$

Note that, threats modeled by Gaussian distributions are characterized by the well-known Multidimensional Gaussian Law. Fig.1 shows a sample PTEM constructed by a set of Gaussian *pdfs*. Note that if the parameters of *pdfs* are constants, then PTEM is time-invariant. Any other *pdf* can be used to construct the map as long as it is differentiable.

2.3 Restricted Regions Formulation

Based on the value of PTEM, restricted regions where the UAV should never enter can be defined in the area of operation. Such regions are quantified by a lower limit (f_r), where the value of PTEM is greater than or equal to f_r. Namely,

$$A_r(\mathrm{t}) = \{\mathrm{r} : f(\mathrm{r}) \geq f_r\} \tag{12}$$

Note that, these regions are not fixed over time in the area of operation if the probabilistic map itself is time-variant with respect to position and effectiveness area. Nevertheless, if the position and effectiveness area of the threats are fixed but the effects of the threats are still time-variant, e.g. the level of threat exposure of the UAV increases as it stays longer in the area of operation, these regions will be fixed over time.

2.4 Gradient Search on PTEM

If the PTEM is differentiable, i.e. the area of operation is all modeled by Gaussians or any other differentiable distribution functions, the gradient search approach, which is extensively used in robotics and optimization literature (Konolige, 1996; Choi & Lee, 1996; Mitchell & Sastry, 2003; Ogren et al.,2004) can be employed. This determines the direction of minimum increase or steepest descent of the PTEM. In other words, it can be easily determined in which direction the UAV should move to minimize the threat exposure level or maximize the likelihood of avoiding a restricted region or a collision. Since the PTEM is constructed as the sum of differentiable functions, the determination of gradient or the sharpest-descent direction can be carried out by utilizing the sum of "directional" derivatives of $f(\mathrm{r})$ given in (10) along the axes of the reference coordinate system of choice. Let u be a direction, at position $\underline{\mathrm{r}}$, whose angle from the positive x-axis is ψ. Thus, the vector defining direction u is

$$\underline{\mathrm{u}} = \cos\psi\,\hat{I} + \sin\psi\,\hat{J} \tag{13}$$

where $\hat{I}, \hat{J}$ are the unit vectors of x- and y-axes of the reference frame, respectively. Then, the directional derivative of $f(\mathrm{r})$ along direction $\underline{\mathrm{u}}$ at position $\underline{\mathrm{r}}$ is

$$D_{\mathrm{u}} f(\mathrm{r}) = f_x(\mathrm{r})\cos\psi + f_y(\mathrm{r})\sin\psi \tag{14}$$

where $f_x(\mathrm{r})$ and $f_y(\mathrm{r})$ are the partial derivatives of $f(\mathrm{r})$ with respect to x and y, respectively:

$$f_x(\mathrm{r}) = \sum_{i=1}^{N} -\frac{x-\mu_{x,i}}{2\pi\sigma_{x,i}^3\,\sigma_{y,i}} \exp\left\{-\frac{[(x-\mu_{x,i})^2\sigma_{y,i}^2 + (y-\mu_{y,i})^2\sigma_{x,i}^2]}{2\sigma_{x,i}^2\sigma_{y,i}^2}\right\} \tag{15}$$

$$f_y(\mathrm{r}) = \sum_{i=1}^{N} -\frac{y-\mu_{y,i}}{2\pi\sigma_{x,i}\,\sigma_{y,i}^3} \exp\left\{-\frac{[(x-\mu_{x,i})^2\sigma_{y,i}^2 + (y-\mu_{y,i})^2\sigma_{x,i}^2]}{2\sigma_{x,i}^2\sigma_{y,i}^2}\right\} \tag{16}$$

At a given position, to find the direction along which the threat exposure is reduced the most, i.e. the steepest descent, the directional derivative in (14) should be minimized over angle ψ. Namely, "the minimizing direction", $\underline{\mathrm{u}}_{\min}$, at position $\underline{\mathrm{r}}$ in terms of its angle from the positive x-axis is

$$\psi_{\min}(\mathrm{r}) = \arg\min_{\psi} D_{\mathrm{u}} f(\mathrm{r}) \tag{17}$$

which yields "the minimizing angle" at position $\underline{\mathrm{r}}$ as

$$\psi_{\min}(\mathrm{r}) = \tan^{-1}\left[\frac{-f_y(\mathrm{r})}{-f_x(\mathrm{r})}\right] \tag{18}$$

Then, the vector representing the minimizing direction is

$$\underline{\mathrm{u}}_{\min} = \cos\psi_{\min}\hat{I} + \sin\psi_{\min}\hat{J} \tag{19}$$

Note, further, that the gradient of $f(\mathrm{r})$ is (Larson et al., 2002)

$$\nabla f(\mathrm{r}) = f_x(\mathrm{r})\hat{I} + f_y(\mathrm{r})\hat{J} \tag{20}$$

In terms of the gradient, the minimum value of the directional derivative is given by

$$\min D_{\mathrm{u}} f(\mathrm{r}) = -\|\nabla f(\mathrm{r})\| \tag{21}$$

Note that this is, in fact, the value of the directional derivative along direction $\underline{\mathrm{u}}_{\min}$, i.e.

$$\min D_{\mathrm{u}} f(\mathrm{r}) = D_{\underline{\mathrm{u}}_{\min}} f(\mathrm{r}) = f_x(\mathrm{r})\cos\psi_{\min} + f_y(\mathrm{r})\sin\psi_{\min} \tag{22}$$

3. Gradient Search Guidance Algorithm

The main goal of gradient search guidance algorithm is to generate feasible speed and heading commands for UAVs to safely pursue a moving target in an area with multiple sources of threats and/or restricted zones. This goal requires a trade-off between three possibly conflicting objectives given in the order of their priorities:(i) to avoid restricted regions and obstacles (ii) to maintain the proximity of the target, (iii) to minimize the level of threat exposure. Thus, the strategy should guide the UAV autonomously to keep it within a pre-specified proximity of the target as well as trying to minimize the threat exposure at any time while avoiding restricted areas.

3.1 Target Following Strategy

The strategy guides a UAV by generating commanded heading, ψ_{cmd}, and speed, V_{cmd}. While generating these commands, the strategy takes the dynamic constraints of the UAV into account. Another requirement for the strategy, addressed in this algorithm, is that the strategy algorithm be executed on-line during the pursuit on an on-board computer/micro-processor. Thus, the strategy should be computationally feasible regarding the flight characteristics of the UAV and the configuration of the on-board processor/computer.

3.1.1 Preliminary Definitions

In this section, mathematical and geometric preliminaries are introduced to familiarize the reader with concepts, parameters and constraints used for the description and formulation of the strategy. Note that throughout the chapter C , D and V refer to circle, disk and cone, respectively. Let T_s be the "guidance update period" ', i.e., the commanded heading and speed are updated only when time $t = kT_s$, $k \in \{0,1,2,\ldots\}$. The positions of the UAV and the target are defined relative to the fixed reference frame where the PTEM is presented as a function of position.

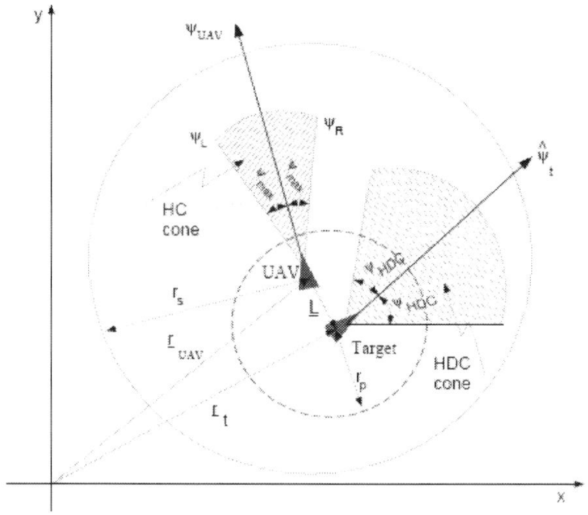

Figure 2. Proximity & sensor circles, HC & HDC cones, and LOS vector

As seen from Fig.2., the position vectors of the UAV and the target are defined to be $\underline{r}_{UAV}$ and $\underline{r}_t$ with (x_u, y_u) and (x_t, y_t) coordinates, respectively. The headings of the UAV and the target are angles measured from positive x-axis as ψ_{UAV} and ψ_t, respectively. Let

$\underline{L}$, ``line of sight'' (LOS) vector, be the position vector of target with respect to UAV and written as

$$\underline{L} = (x_t - x_u)\hat{I} + (y_t - y_u)\hat{J} \tag{23}$$

Thus, the LOS angle, ψ_{LOS}, can be computed as

$$\psi_{LOS}(k) = \tan^{-1}\left(\frac{y_t - y_u}{x_t - x_u}\right) \tag{24}$$

Then, the time rate of change of ψ_{LOS} is calculated as

$$\dot{\psi}_{LOS}(k) = \frac{(\dot{y}_t - \dot{y}_u)(x_t - x_u) - (\dot{x}_t - \dot{x}_u)(y_t - y_u)}{(x_t - x_u)^2 + (y_t - y_u)^2} \tag{25}$$

Proximity circle, $C_p(k)$, denotes a circle, at the kth (current) update instant, centered at the predicted target position at the $(k+1)^{th}$ (next) update instant and with a specified radius, r_p (see Fig.2.). Namely,

$$C_p(k) = \left\{\underline{z} \in \Re^{2:} : \left|\underline{z} - \hat{\underline{r}}_t(k+1)\right| = r_p\right\} \tag{26}$$

Similarly, *Proximity disk,* $D_p(k)$, is defined to be the region bounded by $C_p(k)$,i.e.

$$D_p(k) = \left\{\underline{z} \in \Re^{2:} : \left|\underline{z} - \hat{\underline{r}}_t(k+1)\right| \le r_p\right\} \tag{27}$$

$D_p(k)$ is introduced to define the proximity of the target and thus it is a design parameter that quantifies how close to the target the strategy should keep the UAV during the pursuit. Note that the proximity of the target is the objective of the strategy with the second highest priority. Once this objective is secured, the strategy should try to achieve the last objective, minimizing the threat exposure level. In this regard, $D_p(k)$ quantifies the trade-off between these two objectives. Similar to $C_p(k)$, *Reachability circle,* $C_r(k)$, is defined to be a circle, centered at the current UAV position (see Fig.3) and has a radius determined by the speed of the UAV and the guidance update period, T_s:

$$C_r(k) = \left\{\underline{z} \in \Re^{2:} : \left|\underline{z} - \underline{r}_{UAV}(k)\right| = T_s \times V_{UAV}(k)\right\} \tag{28}$$

where $\underline{r}_{UAV}(k)$ and $V_{UAV}(k)$ are the samples of the UAV position and speed at the kth update instant, respectively. Note that $C_p(k)$ is centered at the predicted target position at the next update instant while $C_r(k)$ is centered at the UAV position at the current update instant. Thus, these two circles are used to determine (i) whether it is possible stay within $D_p(k)$ until the next update instant and (ii) if not, the range of directions that would steer the UAV towards $D_p(k)$. Based on $\psi_{HDC}(k)$ from Fig.2, HDC (*Heading Difference Constraint*) Cone (V_{HDC}) is defined to be the range of headings that are considered to be close to the estimated heading of the target $\hat{\psi}_t$. Thus, as shown in Fig.2, V_{HDC} moves with the target, is centered at its estimated heading and expands in both clockwise and counter-clockwise directions by $\psi_{HDC}(k)$. Similarly, HC (*Heading Constraint*) Cone, V_{HC}, defines the range of headings that are admissible when only the dynamics of the UAV is considered. V_{HC}, as shown in Fig.2, moves with the UAV, is centered at its heading and expands in both clockwise and counter-clockwise directions by $\psi_{\max}$.

When the UAV is outside the proximity disk, $D_p(k)$, it is desirable to know (i) whether it is possible to move into $D_p(k)$ at the current speed and (ii) if it is, the range of feasible headings that would steer the UAV towards $D_p(k)$. To be able to answer the first question, the PR (Proximity Range) Cone, V_{PR}, is constructed, as shown in Fig.3, by the intersection points of $C_r(k)$ and $C_p(k)$.

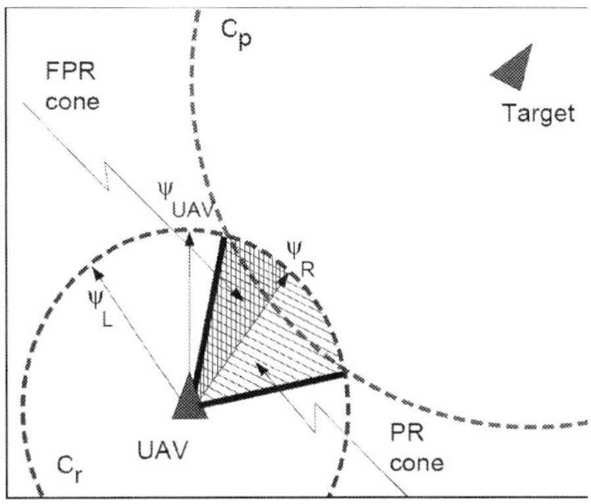

Figure 3. General proximity disk cone

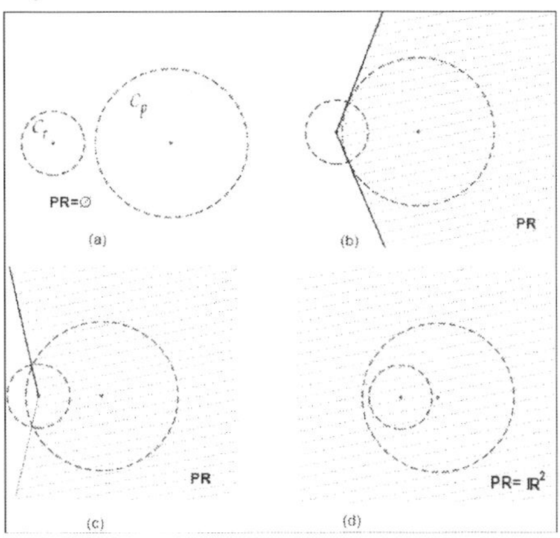

Figure 4. PR cone in various cases of position of C_r relative to C_p, (a) $C_r(k) \not\subset D_p(k)$ and $C_r(k) \cap C_p(k) = \varnothing$, (b),(c) $C_r(k) \not\subset D_p(k)$ and $C_r(k) \cap C_p(k) \neq \varnothing$, (d) $C_r(k) \subset D_p(k)$ and $C_r(k) \cap C_p(k) = \varnothing$

Obviously, if $C_r(k) \not\subset D_p(k)$ and $C_r(k) \cap C_p(k) = \varnothing$ (Note that $\varnothing$ denotes the empty set), then $V_{PR} = \varnothing$ and thus the answer to question (i) is negative (see (a) of Fig.4). When $C_r(k) \not\subset D_p(k)$ and $C_r(k) \cap C_p(k) \neq \varnothing$ (see (b) and (c) of Fig.4), however, the feasibility of V_{PR} should be investigated by determining its intersection with V_{HC}. If $V_{HC} \cap V_{PR} = \varnothing$, then the answer to question (i) is still negative. When $V_{HC} \cap V_{PR} \neq \varnothing$, the answer to question (i) is affirmative, and the answer to question (ii) is the intersection cone, which is defined to be FPR (Feasible Proximity Range) Cone, i.e. $V_{FPR} = V_{HC} \cap V_{PR}$ as shown in Fig.3. When $C_r(k) \subset D_p(k)$, as seen in (d) of Fig.4, $C_r(k) \cap C_p(k) = \varnothing$.

However, in this special case, $V_{PR} = \Re^2$ since any direction will lead to $D_p(k)$, and thus, $V_{FPR} = V_{HC}$. The objective of the strategy with the highest priority is to always avoid the restricted areas. To be able to do so, the range of the headings that would steer the UAV towards a restricted area, especially when the restricted area is close-by, should be determined. This can be accomplished by utilizing the directions that are tangent to the level curve of the PTEM that passes through a given UAV position. Note that the tangent directions are always normal to the gradient, $\psi_{\min}(k)$. The angles of the two tangent directions are referred to as $\psi_{\mathrm{tg_R}}(k)$ and $\psi_{\mathrm{tg_L}}(k)$ as shown in Fig.5. Then, SHR (Safe Heading Range) Cone, V_{SHR}, is defined to be the range of directions within the V_{HC} through which the directional derivative of the PTEM is zero or negative.

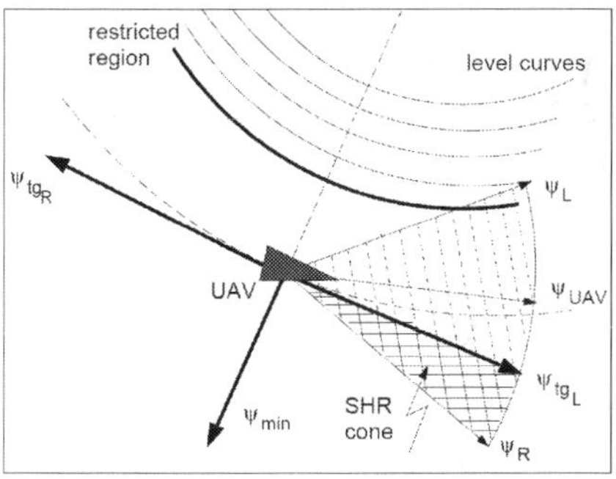

Figure 5. SHR cone

3.1.2 Decision Factors and Strategy States

The intelligent strategy first computes the "desired" heading and speed as well as the "admissible" ranges of heading and speed. The desired signals are computed in accordance with the three objectives of the strategy without considering the dynamic and strategy-imposed constraints. At the same time, the admissible ranges are determined based on the states of the UAV, the local PTEM and the constraints. Then, the strategy generates the commanded signals (heading and speed) by considering the desired signals and their respective admissible ranges. When a desired signal is within the respective admissible range, then obviously the desired signal is assigned as the commanded signal. Otherwise, the boundary of the admissible range that is closest to the desired signal is selected to be the commanded signal. This section of the chapter introduces the decision factors and decision states that are defined based on the strategy objectives and their assigned priorities and used by the strategy to infer the desired heading, desired speed and their admissible ranges. There are four decision factors that are used to determine the decision states:

Factor1: This factor determines whether the UAV is in a risk of getting into a restricted region. It is quantified by using V_{SHR} and evaluating the PTEM at three positions that would be the positions of the UAV in n update periods ahead if it flies with its current speed in the current, maximum right and maximum left heading directions ($\psi_i, i = \{1,2,3\}$). Thus, these three prospective positions are calculated as

$$x_i(k+n) = x_c + V_c(k)nT_s\cos\psi_i(k) \tag{29}$$

$$y_i(k+n) = y_c + V_c(k)nT_s\sin\psi_i(k) \tag{30}$$

where n is the number of update periods that would take for the UAV to make a 90 degree-turn by utilizing the maximum available turn rate. It is computed by $n = ceil[\pi/(2\psi_{\max})]$ where *ceil* is a function that rounds a real number to the nearest integer towards positive infinity.

Then, the PTEM is evaluated at these three positions. If all $f(x_i, y_i)$ are greater than or equal to f_r, then the UAV is in HIGH risk since all the headings steer the UAV to a restricted area. If only some of $f(x_i, y_i)$ is greater than f_r and $V_{SHR} = \varnothing$ then the UAV is still in HIGH risk since $V_{SHR} = \varnothing$ means that the entire V_{HC} is directed completely towards the restricted area. If $V_{SHR} \neq \varnothing$ and there are some $f(x_i, y_i)$ less than f_r, then the UAV is considered to be in LOW risk. If all $f(x_i, y_i)$ are less than f_r, then the risk is NONE since none of the directions steers the UAV to the restricted area.

Factor 2: This factor determines whether the UAV will be within the proximity of the target at the next update instant. This is quantified by employing V_{PR}, V_{HC} and V_{SHR} depending on the answer to Factor 1. When Factor 1 is NONE (i.e. no risk of getting into A_r), the intersection of V_{PR} and V_{HC} is taken. If

$$V_{HC} \cap V_{PR} \neq \varnothing \tag{31}$$

then the answer is YES, i.e. it is feasible for the UAV to be within $D_p(k+1)$. Otherwise, Factor 2 is NO. When Factor 1 is LOW, V_{HC} is replaced by V_{SHR} in (31). Note that in the case when $V_{PR} = \Re^2$, i.e. $C_r(k)$ is completely in $D_p(k)$, Factor 2 is always YES when Factor 1 is NONE or LOW. When Factor 1 is HIGH, the answer to Factor 2 is not defined.

Factor 3: This factor determines whether the headings of the UAV and the target are close. This is quantified by the intersection of V_{HC} and V_{HDC} when Factor 1 is NONE.

If

$$V_{HC} \cap V_{HDC} \neq \varnothing \tag{32}$$

then the answer is YES, i.e. the UAV and the target headings are close, otherwise the answer is NO. When Factor 1 is LOW, V_{HC} is replaced by V_{SHR} in (32).

Factor 4: This factor determines whether the UAV is heading towards $D_p(k)$ only when Factor 1 is NONE or LOW, Factor 2 is NO and Factor 3 is YES. This is quantified by comparing $\psi_{LOS}(k)$ with $(V_{HC} \cap V_{HDC})$ or $(V_{SHR} \cap V_{HDC})$. When Factor 1 is NONE and

$$\psi_{LOS}(k) \subset (V_{HC} \cap V_{HDC}) \tag{33}$$

then the answer is YES, namely the UAV is heading towards $D_p(k)$, otherwise the answer is NO. When Factor 1 is LOW, V_{HC} is replaced by V_{SHR} in (32) as done in Factors 2 and 3.

STATES	DECISION FACTORS			
	1	2	3	4
1	NONE	NO	YES	YES
2	NONE	NO	YES	NO
3	NONE	NO	NO	N/A
4	NONE	YES	N/A	N/A
5	LOW	NO	YES	YES
6	LOW	NO	YES	NO
7	LOW	NO	NO	N/A
8	LOW	YES	N/A	N/A
9	HIGH	N/A	N/A	N/A

Table 1. Decision factors and strategy states (N/A: Not Applicable)

Table 1 summarizes all the rules to determine the states of the strategy based on the decision factors. In *States-1* to -4, the UAV is not in any risk of flying into a restricted area. In *State-1,* the UAV is not in D_p, but the headings of the UAV and the target are close and the UAV is heading towards D_p. However, In *State-2,* the UAV is not heading towards D_p. Note that *Factor 4* is considered only when there is NONE or LOW risk of flying into a restricted region and the headings of the UAV and the target are close. In *State-3,* the UAV is not in the proximity disk and moreover the headings of the UAV and the target are not close. As the simulation results will reveal, this state occurs rarely due to the imposed heading difference constraint by the strategy. In *State-4,* the UAV is within $\tilde{D}_p$. Note also that this is the state that strategy tries to keep the UAV in as much as possible during the mission. In *States-5* to -8, the UAV is in low risk region. In both *States-5* and -6 , the UAV is outside the proximity disk and the headings of the UAV and the target are close. However, in *State-6,* the UAV is not heading towards D_p . In *State-7,* the UAV is not in the proximity disk and moreover the headings of the UAV and the target are not close. In *State-8,* the UAV is within D_p. Note that *Factors* 3 and 4 are not considered in *States-4* and -8 because the UAV is already in the proximity of the target. In *State-9,* the UAV is in high risk region. Thus, other decision factors are not considered.

3.1.3 Computation of Desired Heading and Admissible Range

As stated earlier, the commanded heading is computed based on the desired heading ψ_{des} and the AHR (Admissible Heading Range) cone, V_{AHR}, which is defined to be the cone that consists of the feasible heading directions. Depending on the strategy state at a given time, different criteria are used to determine ψ_{des} and V_{AHR}. Note that the rules in this section and the next section can be defined to be the rules that represent a strategy to select the commanded signals. Recall that the strategy has three objectives with different levels of priority. There are three headings, $\hat{\psi}_{\text{t}}(k), \psi_{\min}(k)$ and $\psi_{\text{LOS}}(k)$, one of which is chosen as the desired heading, $\psi_{\text{des}}(k)$, at the kth update instant, based on the objective of the strategy in a given strategy state.

In *States-1* and -2, the objective with the highest priority (avoidance of the restricted regions) is considered to be "achieved" since Factor 1 is NONE. However, the second objective needs to be targeted since Factor 2 is NO, i.e., D_p should be intercepted. Since the Factor 3 is YES, the headings of the UAV and the target are already close. To achieve the second objective, a pursuit guidance law is employed for the UAV to intercept D_p as soon as possible:

$$\psi_{\text{des}}(k) = \psi_{LOS}(k) + K_D \dot{\psi}_{LOS}(k) \tag{34}$$

where the first term represents the well-known velocity guidance (Pastrick et al., 1981) and the derivative term, with gain K_D, is added to improve the pursuit performance. The admissible heading range is selected such that the commanded heading does not violate the dynamic constraint of the UAV and the heading difference constraint imposed by the strategy itself, i.e.

$$V_{AHR} = V_{HC} \cap V_{HDC} \tag{35}$$

Note that the desired heading selection strategy is the same for *States-1* and -2. However, the gain, K_D, in (34) will be selected differently. Another difference originates from the selection of the desired speed as explained in the next section.

In *State-3*, similar to the first two states, the objective with the highest priority is achieved but the second objective is not. Furthermore, since the Factor 3 is NO, the headings of the UAV and the target are not close to each other. In a high speed pursuit, flying in a direction different from that of the target will lead the UAV to lose the proximity of the target very soon. In other words, it will lead to the failure of the second objective. To turn the UAV in the same direction with the target, $\psi_{\text{des}}(k)$ is selected to be $\hat{\psi}_{\text{t}}(k)$ and V_{AHR} is selected to be V_{HC} to allow the UAV to make the sharpest turn possible. Simulation experiments have shown that, during high speed pursuits, the strategy employed in *States-1* and -2 would not be as efficient particularly due to occurrence of restricted regions between $\tilde{\text{D}}_\text{p}$ and the UAV. This is partly the reason why a different strategy is employed in *State-3*. Further, note that as pursuit speed decreases, it will be less likely for this state to occur because, as to be explained in Section 3.1.5, the heading difference constraint will be relaxed.

In *State-4*, since the first two objectives of the strategy are achieved, the third one can be targeted. Thus, the strategy, in this state, should try to minimize the threat exposure level while ensuring that the other objectives are not compromised. Note that strategy should steer the UAV in a direction that would minimize threat exposure, which, in turn, implies that the UAV is guided away from any possible restricted area. In other words, objective with the highest priority has no conflict with the third objective. However, moving in a direction to minimize the threat exposure may be in conflict with the second objective, i.e. staying in D_p. Thus, in *State-4*, an efficient compromise between the second and the third objective should be formulated while considering that the second objective has a higher priority. As stated earlier, the third objective is quantified by the steepest descent direction, $\psi_{min}(k)$. Namely, if the third objective was the only concern, $\psi_{min}(k)$ would be the desired heading. On the other hand, if the UAV was commanded to fly tangent to C_p, the UAV would never leave D_p. Thus, a direction tangent to C_p is defined to quantify the second objective. Recall that the computation of $\psi_{min}(k)$ is already introduced in Section 2.4. Now, calculation of the tangent direction will be presented. Then, the method will be explained that computes $\psi_{des}(k)$ based on $\psi_{min}(k)$ and the tangent direction to quantify the trade-off between the two conflicting objectives. First, the admissible heading range is determined to ensure that the dynamic constraints of the UAV are not violated. Furthermore, any direction that would certainly move the UAV outside $D_p(k)$ should be eliminated. Thus,

$$V_{AHR} = V_{HC} \cap V_{PR} \tag{36}$$

V_{HDC} is not considered (i.e *Factor 3* is not considered) in this state because *Factor 2* is already YES. Note that the most likely heading of the UAV is $\psi_{min}(k)$ provided it is within V_{AHR}. Hence, a temporary heading, $\psi_{temp}(k)$ is defined to be $\psi_{min}(k)$ if $\psi_{min}(k) \in V_{AHR}$ and, otherwise, the boundary of V_{AHR} that is closest to $\psi_{min}(k)$. $\psi_{temp}(k)$ is used to define a local coordinate system, as shown in Fig.6, whose origin is at the current UAV position and y-axis, y_L, has angles $\psi_{temp}(k)$ from the positive x-axis of the inertial reference frame. The two points where y_L intersects C_p are defined to be y_{L_1} and y_{L_2}.

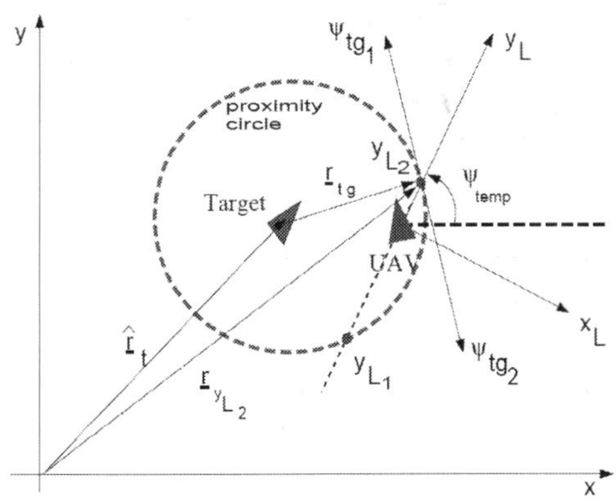

Figure 6. Tangent heading bounding the UAV turns

Transformation between the inertial reference frame and the local frame is used to facilitate the computation of $y_{L_1}(k)$ and $y_{L_2}(k)$. The rotation matrix from the inertial frame to the local frame is

$$\mathrm{R}_{LI}(k)=\begin{bmatrix}\sin\psi_{temp}(k) & -\cos\psi_{temp}(k)\\ \cos\psi_{temp}(k) & \sin\psi_{temp}(k)\end{bmatrix} \tag{37}$$

The predicted target position, then, can be written in the local frame by employing the rotation matrix as

$$\hat{\mathrm{r}}_{t,L}(k+1)=\mathrm{R}_{LI}(k)\left[\hat{\mathrm{r}}_t(k+1)-\mathrm{r}_u(k)\right] \tag{38}$$

where $\hat{\mathrm{r}}_t$ is the representation of $\underline{\hat{\mathrm{r}}}_t$ in the inertial frame, and $\hat{\mathrm{r}}_{t,L}$ is the representation in the local frame. Equation (38) gives the components of $\underline{\hat{\mathrm{r}}}_t$ in the local frame as

$$\hat{x}_{t,L}(k+1)=\left[\hat{x}_t(k+1)-x_u(k)\right]\sin\psi_{temp}(k)-\left[\hat{y}_t(k+1)-y_u(k)\right]\cos\psi_{temp}(k) \tag{39}$$

$$\hat{y}_{t,L}(k+1)=\left[\hat{x}_t(k+1)-x_u(k)\right]\cos\psi_{temp}(k)+\left[\hat{y}_t(k+1)-y_u(k)\right]\sin\psi_{temp}(k) \tag{40}$$

Now, $\mathrm{C_p}$ can be easily formulated in the (x_L, y_L) local frame as

$$\left[x_L-\hat{x}_{t,L}(k+1)\right]^2+\left[y_L-\hat{y}_{t,L}(k+1)\right]^2={r_p}^2 \tag{41}$$

As Fig.6 implies, $y_{L_1}(k)$ and $y_{L_2}(k)$ are solutions to (41) when $x_L=0$, since y_{L_1} and y_{L_2} are defined to lie on the y_L axis

$$y_{L_{1,2}}(k)=\hat{y}_{t,L}(k+1)\mp\sqrt{{r_p}^2-\hat{x}_{t,L}(k+1)} \tag{42}$$

Note, from Fig.6, that $y_{L_1}(k)$ is always negative and $y_{L_2}(k)$ is always positive in *State-4* since the UAV is in $\mathrm{D_p}$. Since the UAV is likely to head towards y_{L_2}, the two tangent directions of $\mathrm{C_p}$ at this point need to be computed. Recall that $\mathrm{C_p}$ is centered at the predicted target position, $\underline{\hat{\mathrm{r}}}_t(k+1)$, at the next update instant. If the radial direction, $\underline{\mathrm{r}}_{tg}(k)$, from the center of $\mathrm{C_p}$ to y_{L_2} is known, then the tangent directions are easily computed as they are normal to the radial directions, as shown in Fig.6. Note that,

$$\underline{\mathrm{r}}_{tg}(k)=\underline{\mathrm{r}}_{y_{L_2}}(k)-\underline{\hat{\mathrm{r}}}_t(k+1) \tag{43}$$

where $\underline{\mathrm{r}}_{y_{L_2}}(k)$ is the position vector of point y_{L_2} relative to the inertial reference frame. By employing the inverse of the transformation used in (38), the representation of $\underline{\mathrm{r}}_{y_{L_2}}(k)$ in the inertial frame is computed as

$$\mathrm{r}_{y_{L_2}}(k)=R_{LI}^T(k)\begin{bmatrix}0\\ y_{L_2}(k)\end{bmatrix}+\mathrm{r_u}(k) \tag{44}$$

After carrying out matrix multiplication and addition, (44) yields $\underline{r}_{y_{L_2}}(k)$, written as a vector,

$$\underline{r}_{y_{L_2}}(k) = \{x_u(k) + y_{L_2}(k)\cos[\psi_{temp}(k)]\}\hat{I} + \{y_u(k) + y_{L_2}(k)\sin[\psi_{temp}(k)]\}\hat{J} \tag{45}$$

Substituting (45) in (43) yields

$$\begin{aligned}\underline{r}_{tg}(k) &= \{x_u(k) - \hat{x}_t(k+1) + y_{L_2}(k)\cos[\psi_{temp}(k)]\}\hat{I} \\ &+ \{y_u(k) - \hat{y}_t(k+1) + y_{L_2}(k)\sin[\psi_{temp}(k)]\}\hat{J}\end{aligned} \tag{46}$$

Then the tangent angles, shown in Fig.6, can be calculated as:

$$\psi_{tg1,2}(k) = \tan^{-1}\left\{\frac{y_u(k) - \hat{y}_t(k+1) + y_{L_2}(k)\sin[\psi_{temp}(k)]}{x_u(k) - \hat{x}_t(k+1) + y_{L_2}(k)\cos[\psi_{temp}(k)]}\right\} \tag{47}$$

Among the two tangent directions, ψ_{tg1} and ψ_{tg2}, the one, referred to as $\psi_{tg}(k)$, that is closest to the current UAV heading, $\psi_{UAV}(k)$, is selected to quantify the second objective. For example, in Fig.6, $\psi_{tg}(k) = \psi_{tg1}(k)$ since it is closer to the UAV heading.

Once the two objectives are quantified by $\psi_{temp}(k)$ to minimize threat exposure and $\psi_{tg}(k)$ to stay with the proximity of the target, the next step is to consolidate these two possibly conflicting objectives in an efficient way. Let $\underline{u}_{temp}(k)$ and $\underline{u}_{tg}(k)$ be the unit vectors along $\psi_{temp}(k)$ and $\psi_{tg}(k)$ directions, respectively. Namely,

$$\underline{u}_{temp}(k) = \cos\psi_{temp}(k)\hat{I} + \sin\psi_{temp}(k)\hat{J} \tag{48}$$

$$\underline{u}_{tg}(k) = \cos\psi_{tg}(k)\hat{I} + \sin\psi_{tg}(k)\hat{J} \tag{49}$$

Then, the vector with the desired heading direction is constructed as the weighted vectorial sum of $\underline{u}_{temp}(k)$ and $\underline{u}_{tg}(k)$ as

$$\underline{r}_{des}(k) = [1 - \alpha(k)]\underline{u}_{temp}(k) + \alpha(k)\underline{u}_{tg}(k) \tag{50}$$

where $\alpha(k)$, a real number between 0 and 1, is used to quantify the level of trade-off between the two objectives. Note that when the UAV is close to the center of C_p, there is no immediate danger of leaving D_p and thus minimizing the threat exposure should be the primary objective. This implies that $\alpha(k)$ should be 0 so that $\underline{r}_{des}(k) = \underline{u}_{temp}(k)$. On the other hand, when the UAV is close to C_p, there is an immediate risk of leaving the proximity of target and thus $\underline{r}_{des}(k) = \underline{u}_{tg}(k)$ (i.e. $\alpha(k) = 1$) so that the UAV does not head towards outside of D_p. When the UAV is between these two extreme cases, $\alpha(k)$ should take a value between 0 and 1 so that the two objectives are consolidated. This logic is formulated by a scheduling scheme for $\alpha(k)$ as

$$\alpha(k)=\begin{cases}1, & \alpha_{sch}(k)\le\alpha_{tg}\\ \dfrac{\alpha_{sch}(k)(\alpha_{temp}-1)}{\alpha_{temp}-\alpha_{tg}}, & \alpha_{tg}<\alpha_{sch}(k)\le\alpha_{temp}\\ 0, & otherwise\end{cases} \tag{51}$$

In this scheduling, α_{tg} and α_{temp} are strategy design parameters and α_{sch} quantifies how close the UAV is to leaving $\mathrm{D_p}$ as

$$\alpha_{sch}(k)=\frac{y_{L_2}(k)}{2r_p} \tag{52}$$

Note that when $\alpha(k)$ is close to 0, the UAV is actually close to $\mathrm{C_p}(k)$ but heading towards inside of $\mathrm{D_p}(k)$ and thus this case is not considered as a danger of leaving $\mathrm{D_p}(k)$. By using the angle of the resultant vector in (50), the desired heading angle in *State-4* is computed as

$$\psi_{des}(k)=\tan^{-1}\left\{\frac{[1-\alpha(k)]\sin\psi_{temp}(k)+\alpha(k)\sin\psi_{tg}(k)}{[1-\alpha(k)]\cos\psi_{temp}(k)+\alpha(k)\cos\psi_{tg}(k)}\right\} \tag{53}$$

In *States-5* to *-8*, exactly the same strategies as in *States-1* to *-4*, respectively, for computing $\psi_{des}(k)$ have been implemented. The only difference in these states is that there is a LOW threat region instead of having NONE threat region. Thus, V_{HC} is replaced by V_{SHR} during the calculation of V_{AHR} in all these states.

In *State-9*, the objective with the highest priority is not considered to be "achieved" since Factor 1 is HIGH. Thus, the UAV should be commanded to make a turn to stop approaching the restricted area. Since $\psi_{\min}(k)$ is the direction of the sharpest descent, it is selected to be $\psi_{des}(k)$. Furthermore, the UAV should make the turn as fast as possible, thus, V_{AHR} is selected to be V_{HC} to allow the UAV to make the sharpest turn possible.

3.1.4 Computation of Desired Speed and Admissible Range

In the previous section, the computation of the commanded heading at each update instant is presented. Once the commanded heading, $\psi_{cmd}(k)$, is computed at the kth update instant, the strategy computes the desired speed, $V_{des}(k)$, based on $\psi_{cmd}(k)$ and the current decision state. Furthermore, an admissible speed range is determined based on the speed and acceleration constraints of the UAV. Note that during a pursuit mission, the target speed might be varying drastically. Further, the path of the UAV, determined by the commanded heading, might be significantly different from that of the target because the strategy steers the UAV to avoid restricted areas and minimize threat exposure level. Thus, the commanded speed is determined to address objective-2, i.e. to help maintain or obtain the proximity of the target in almost all decision states. The exception is *State-9* where the commanded speed is determined to help avoid restricted areas.

In *States-1* to *-8*, the desired speed is calculated by a proportional control algorithm based on two different error signals. The first one, speed error, is the difference between the speed of the UAV and the estimated target speed. This is used to ensure that the UAV speed will be adjusted as the target speed varies. The second one, the position error, is defined to quantify the distance between the UAV and $D_p(k)$. To compute the position error, a local reference frame is defined as shown in Fig.7, such that its origin is at the UAV position and its y-axis, y_L, has an angle of $\psi_{cmd}(k)$ from the positive x-axis of the inertial frame. Note that this local frame is similar to the one used in the previous section except the angle used to define the orientation relative to the inertial frame. Then, the position error, $e(k)$, is defined to be the arithmetic mean of the two intersection points of y_L-axis with C_p.

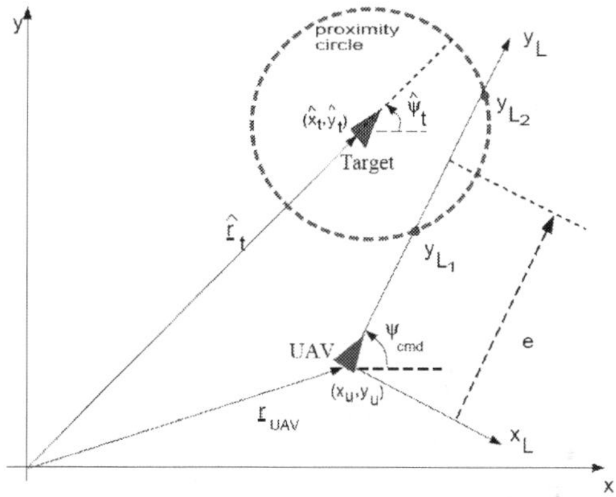

Figure 7. Inertial and UAV local coordinate systems

Note that y_{L_1} and y_{L_2} can be calculated by using the same approach detailed in previous section with $\psi_{cmd}(k)$ that replaces $\psi_{temp}(k)$. The proportional control algorithm is formulated as

$$V_{des}(k) = V_{UAV}(k) + K_s[\hat{V}_t(k) - V_{UAV}(k)] + K_e[e(k)/T_s] \tag{54}$$

where $\hat{V}_t(k)$ is the estimated target speed, K_s and K_e are the proportional gains for the speed and position errors, respectively. Note that different values can be assigned to the gains in different states. For example, for the simulations presented in Section 3.3, in *State-1* and *State-5*, the gains are twice as big as the values used in other states. This is because, in these states, the UAV is outside D_p but heading towards it (Recall that *Factor-4* is YES). Thus, a greater speed increase should be commanded so that it will take the UAV a shorter time to attain D_p. As stated earlier, in *State-9*, the objective with the highest priority needs to be addressed. Namely, there is a HIGH risk of incursion into a restricted area and the sharpest descent direction is commanded to turn the UAV away from the restricted area. The speed command is also utilized to improve the performance of the strategy. Note that the lower the speed of a UAV, the sharper turns it can make. Thus, the minimum speed possible, given the deceleration constraint of the UAV, is commanded, i.e.

$$V_{des}(k) = V_{UAV}(k) + a_{\min} T_s \tag{55}$$

Once the desired speed is determined, the admissible range for the speed should be determined to compute the commanded speed so that the speed and acceleration constraints of the UAV are not violated. The upper and the lower bounds of the admissible speed range are calculated as

$$V_{\max_{const}}(k) = \min\{V_{\max}, V_{UAV}(k) + a_{\max}T_s\} \tag{56}$$

$$V_{\min_{const}}(k) = \min\{V_{\min}, V_{UAV}(k) + a_{\min}T_s\} \tag{57}$$

3.1.5 Scheduling Scheme for Heading Difference Constraint

Recall that V_{HDC} is introduced to serve as a strategy imposed constraint on the heading of the UAV. This is necessary because the turning radius of a vehicle increases as its speed increases. If the heading difference is not bounded, the strategy may change the UAV heading drastically to minimize the threat exposure when the UAV is within $D_p(k)$. This, in turn, may increase the risk of the target getting outside $D_p(k)$ and eventually outside the sensor range. To restrict the motion of the UAV in and around the direction where $D_p(k)$ is heading, $\psi_{HDC}(k)$ is introduced. However, as stated earlier, this constraint, if it was fixed, would become a liability in the case of low speed and even more so when the target stops. Thus, a scheduling scheme is developed to impose a heading difference constraint in high speed pursuit and to relax it when the target moves slow or stops. The scheduling should be done in such a way that no discontinuity is introduced in the computation of the commanded heading. According to the scheduling scheme used (see Fig.8), $\psi_{HDC}(k)$ is set to a constant, $\psi^*_{HDC}(k)$, when the estimated target speed, $\hat{V}_t$, is high (i.e. greater than V_{thres_2}); when the target slows down, the constraint is gradually relaxed by linearly increasing $\psi_{HDC}(k)$. This increase is performed with such a slop that $\psi_{HDC}(k)$ becomes 180° when $\hat{V}_t$ is equal to another threshold, V_{thres_1}. When the target moves very slow or stops (i.e. $\hat{V}_t$ is less than V_{thres_1}), $\psi_{HDC}(k)$ is set to 180°, which effectively removes the constraint.

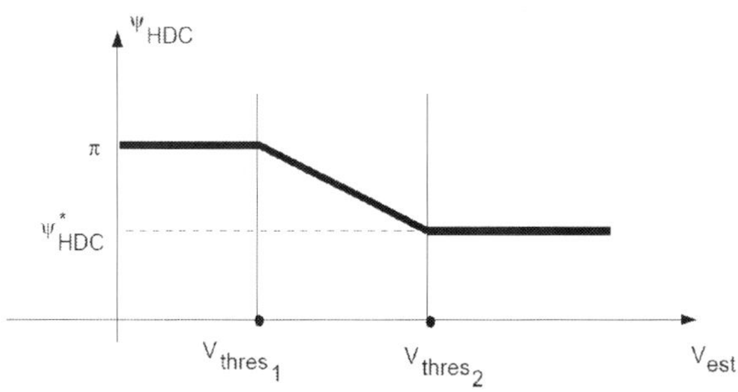

Figure 8. Heading difference constraint angle ψ_{HDC}

3.1.6 Detection of Local Minima

Note that $\underline{u}_{min}(k)$ is the sharpest descent direction at the current update instant, k. Thus, the directional derivative of PTEM in this direction

$$D_{u_{min}(k)} f(r(k)) \le 0 \tag{58}$$

Let $\underline{r}(k+1)$ be the position vector of the UAV at the next update instant, $(k+1)$ if it flies with the current speed in the sharpest descent direction. The directional derivative at $\underline{r}(k+1)$ in the direction of $\underline{u}_{min}(k)$ is referred to as $D_{u_{min}(k)} f(r(k+1))$. Thus, the local minima is considered to be present ahead if (see Fig.9)

$$D_{u_{min}(k)} f(r(k)) D_{u_{min}(k)} f(r(k+1)) < 0 \tag{59}$$

When this condition occurs there is a local minimum of the PTEM ahead if the UAV would fly in the sharpest descent direction. If the strategy keeps commanding ψ_{min} through such a local minima, the simulation experiments have shown that the commanded heading may show an unnecessary oscillation. To prevent this, during the occurrence of local minima, a vectorial sum of the two sharpest descent directions is computed as

$$\underline{u}_{min,avg}(k) = \underline{u}_{min}(k) + \underline{u}_{min}(k+1) \tag{60}$$

Thus, the angle of this new direction is

$$\psi_{min,avg}(k) = \tan^{-1}\left[\frac{\sin\psi_{min}(k) + \sin\psi_{min}(k+1)}{\cos\psi_{min}(k) + \cos\psi_{min}(k+1)}\right] \tag{61}$$

which replaces $\psi_{min}(k)$ in the current execution of the strategy.

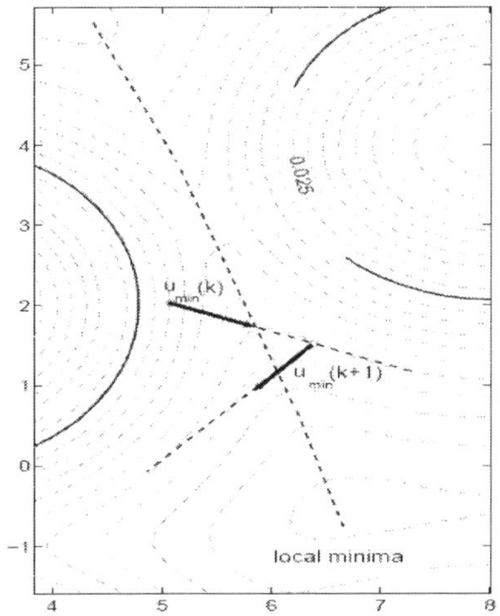

Figure 9. Detection of local minima

3.3 Simulation Results

To test the full capabilities of the strategy, 3 different tracking cases, which are representative of most likely scenarios, are simulated. In the $30\ km \times 30\ km$ area of operation, there are 17 threat sources along with the restricted regions. In all the simulations, the UAV is considered to have a minimum speed of $72\ km/h$, a maximum speed of $180\ km/h$, a maximum acceleration of $2\ m/s^2$, a maximum deceleration of $-2\ m/s^2$, and a turning rate of $10\ deg/s$ with a guidance and estimation update periods of 3 seconds. The first order transfer functions have time constants 0.5 and 0.01 for heading and speed responses of the UAV, respectively. Strategy design parameters f_r and $\psi^*_{HDC}(k)$ are selected to be 0.025 and 45 deg, respectively. Scheduling parameters for heading difference constraint V_{thres_1} and V_{thres_2} are $V_{\min}$ and $0.2V_{\min}$, respectively. The α-scheduling parameters α_{tg} and α_{temp} are 0.05 and 0.2, respectively. The proportional controller gains for the speed and position errors are 0.5 and 0.0004, respectively. The derivative gain, K_D, in (34) is 10 for States -1 and -5, and 60 in States -2 and -6.

The standard deviations of the noise added to the x and y positions of the target to obtain the measurements are selected to be $0.05km$ for Cases 1 and 2 and $0.1km$ for Case 3. The target position is measured during the time when the target is within a circle that is centered at the UAV and moves with it and whose radius r_s (see Fig.2) is equal to the sensor range. The position of the target is estimated from these measurements by employing a least-squares estimation technique with batch processing mode, based on a sliding window of measurements. Namely, only a specified number of measurements are stored and the measurement array is updated with new measurement by removing the oldest measurement and thus retaining the size of the array. Then, the kinematic equations are used to calculate the heading and speed of the target. The initial conditions, sensor and proximity ranges in each case are given in Table 2.

Case	Target			UAV				
	IP	IS	IH	IP	IS	IH	SR	PR
1	(7,-13)	80	97.4	(7,-13)	144	60	2.0	1.5
2	(2.85,-13)	144	23.0	(4,-12)	144	60	2.0	1.5
3	(4,-12)	80	57.3	(5,-11)	144	60	2.0	1.0

Table 2. Parameters: Initial Position (IP) [*km*], Speed (IS)[*km/h*], Heading (IH) [deg], Sensor Range (SR)[*km*], Proximity Range (PR) [*km*]

Fig.10 shows the first case where a target accelerates until it reaches its maximum speed and continues with this speed. As seen from Fig.10, target passes through three restricted regions. UAV, when guided by the algorithm, avoids these regions while continuing the pursuit of the target even if target gets outside the sensor range during the second and the third restricted regions.

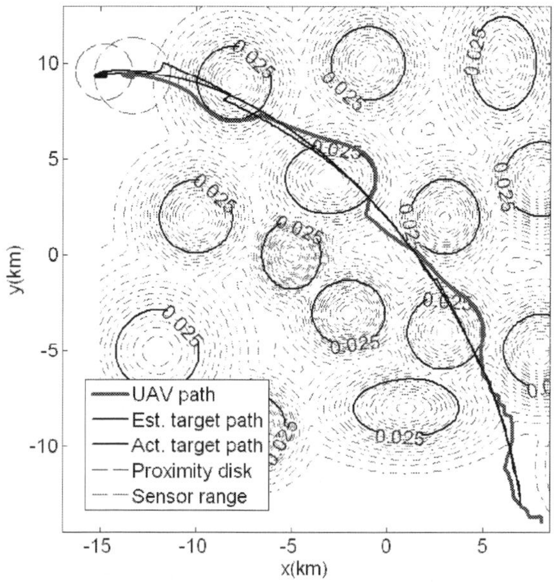

Figure 10. Case 1: UAV trajectory following an accelerating target

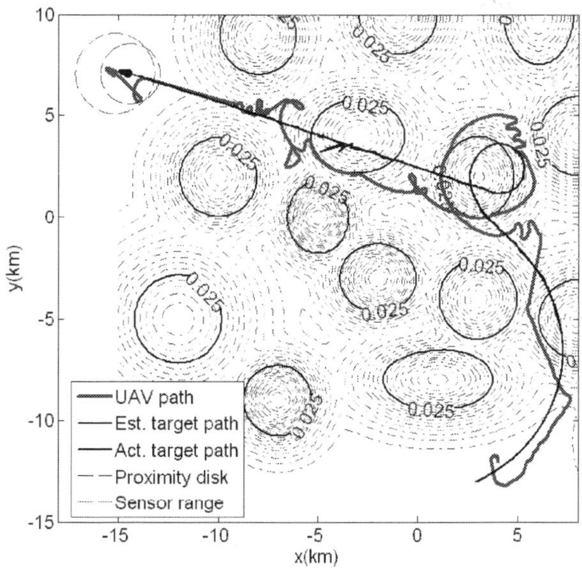

Figure 11. Case 2: UAV trajectory following a slowing target

Fig.11 shows the second case where the target, while moving on the path shown in Fig 11, reduces its speed to 36 km/h in 600 seconds and then maintains it in the rest of the pursuit. Note that the speed of the target is much less than the minimum speed of the UAV during the most of the pursuit. Thus, the UAV reduces its speed to the minimum speed and tries to loiter within the proximity disk when there is no restricted area while minimizing the threat exposure level as shown in Fig.11. Note that there is no pre-defined loitering mode in algorithm. In fact, there is no need for the strategy to have a separate loitering mode because the strategy puts the UAV in loitering autonomously based on the speed of the proximity circle and the local PTEM within the proximity disk.

The third case is shown in Figs.12 and 13. In this case target continuously reduces its speed and stops at 660 seconds in a region where there is no restricted area. After staying in this region for 140 seconds, the target, at 800 seconds, starts accelerating and reaches its maximum speed at 1000 seconds. At 1100 seconds, it starts decelerating again and stops at 1360 seconds in a restricted region during the rest of the pursuit. This case shows the full capability of the strategy for autonomously putting the UAV in loitering mode both with and without restricted areas close-by. As seen from Fig.13a, when the target stops in a region where there is no close-by restricted region, the algorithm puts the UAV in loitering around the local minimum of the PTEM inside the proximity circle. This shows the benefit of utilizing the proximity-circle tangent-direction. Also note from Fig.13b that when the target stops the second time within a restricted area, a small portion of the proximity circle is still outside the restricted area. The algorithm loiters the UAV around this portion of the proximity circle. This shows the benefit of utilizing the pursuit guidance, with the right choice of the gain, based on the LOS angle and its derivative in the algorithm.

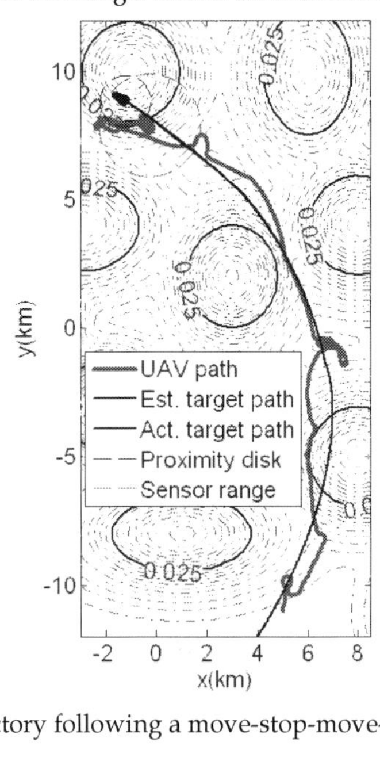

Figure 12. Case 3: UAV trajectory following a move-stop-move-stop target

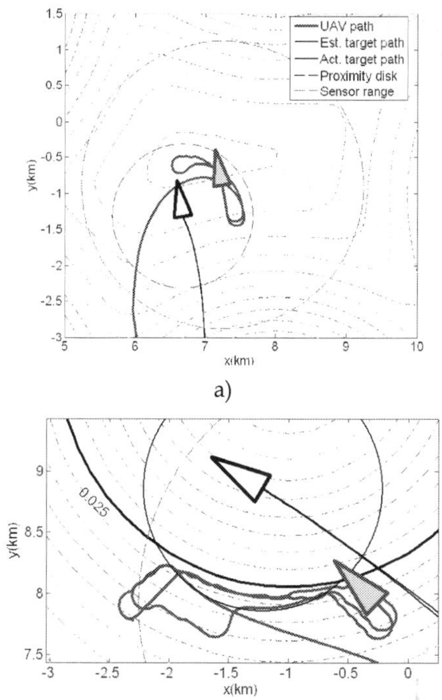

a)

b)

Figure 13. Case 3: When the target stops in a) non-restricted region and b) restricted region

4. Conclusion

A rule-based guidance strategy is developed for autonomous UAVs to track targets moving in an area with various types of threats, obstacles and restricted areas. The concept of PTEM (the Probabilistic Threat Exposure Map) is introduced as a mathematical formulation of the area of operation in terms of threats, obstacles and restricted areas. PTEM defines various types of threats, obstacles and restricted areas in a single framework that quantifies the threat exposure level as a function of position. A gradient search algorithm is applied on PTEM to determine the directions to avoid obstacles and restricted areas and to minimize threat exposure level. To keep the UAV within the proximity circle of the highly mobile target is an objective that is generally in conflict with the objectives of avoiding obstacles/restricted-areas and minimizing threat exposure. The rule-based guidance strategy is formulated to quantify the trade-off between these conflicting objectives and to generate the commanded heading and speed for the UAV. The rule-based intelligent decision approach has provided a very systematic method of developing the autonomous guidance strategy. First, the objectives of the guidance strategy and their priorities are determined. Then, based on the local threat information extracted from PTEM, position, heading and speed of the UAV relative to the target at a given time, the primary objective and/or the level of trade-off between the objectives are quantified. At the same time, admissible ranges for the heading and speed are determined based on the dynamic and strategy-imposed constraints. This approach has facilitated the formulation of the guidance strategy that takes into account all the objectives of the mission with defined priorities and the constraints of the host UAV. On the other hand, utilization of the pursuit-guidance techniques based on LOS angle, proportional control for speed command and the weighted vectorial summation of minimizing direction and proximity-circle tangent has enabled the algorithm to perform better. In all simulation cases, guided by the algorithm, the UAV safely avoided restricted-areas/obstacles while continuing the pursuit of the target.

5. References

Rathbun, D.; Kragelund, S.; Pongpunwattana, A.; & Capozzi, B. (2003). An evolution based path planning algorithm for autonomous motion of a uav through uncertain environment, *Proceedings of the 22nd Digital Avionics Systems Conference*, Indianapolis, Indiana, Oct. 2003.

Finke, J. ; Passino, K. M. ; & Sparks, A. (2003). Cooperative control via task load balancing for networked uninhabited autonomous vehicles, *Proceedings of the 42nd IEEE Conference on Decision and Control*, Maui, Hawaii, Dec. 2003.

Flint, M. ; Polycarpou, M. ; & Gaucherand, E. F. (2002). Cooperative control for multiple autonomous uav's searching for targets, *Proceedings of the 41st IEEE Conference on Decision and Control*, pp. 2823–2828, Las Vegas, Nevada, Dec. 2002.

Jun, M. ; Chaudry, A. I. & D'Andrea, R. (2002). The navigation of autonomous vehicles in uncertain dynamic environments: A case study, *Proceedings of the 41st IEEE Conference on Decision and Control*, pp. 3770-3775, Las Vegas, Nevada, Dec. 2002.

Pongpunwattana , A. & R. Rysdyk, (2004) Real-time planning for multiple autonomous vehicles in dynamic uncertain environments, *Journal of Aerospace Computing, Information, and Communication*, Vol. 1, Dec 2004, pp. 580–604.

Nikolas, I. K. ; Valavanis, K. P.; Tsourveloudis, N. C. & Kostaras, A. N. (2003). Evolutionary algorithm based offline/online path planner for uav navigation, *IEEE Transactions on Systems, Man, and Cybernetics-Part B: Cybernetics*, Vol. 33, No. 6, 2003, pp. 898–912.

Zhu, R.; Sun, D.; & Zhou, Z. (2005). Cooperation strategy of unmanned air vehicles for multitarget interception, *Journal of Guidance, Control and Dynamics*, Vol. 28, No. 5, 2005, pp. 1068–1072.

Waydo S. & Murray, R. M. (2003). Vehicle motion planning using stream functions, *Proceedings of the 2003 IEEE International Conference on Robotics&Automation*, pp. 2484–2491, Taipei, Taiwan, Sep. 2003.

Sengupta, R. ; Hedrick, J. K. & et al, (2003). Strategies of path planning for a uav to track a ground vehicle, *Proceedings of the Second Annual Symposium on Autonomous Intelligent Networks and Systems*, Menlo Park, CA, 2003.

Spry, S. C. ; Girard, A. R. & Hedrick, J. K. (2005). Convoy protection using multiple unmanned aerial vehicles:organization and coordination, *Proceedings of American Control Conference*, pp. 3524–3529, Portland, OR, June, 2005.

Jang, J. S. & Tomlin, C. J. (2005). Control strategies in multi-player pursuit and evasion game," *Proceedings of the AIAA Guidance, Navigation and Control Conference and Exhibit*, San Francisco, CA, Aug. 2005.

Antoniades, A. ; Kim, H. J. & Sastry, S. (2003). Pursuit-evasion strategies for teams of multiple agents with incomplete information, *Proceedings of the 42nd IEEE Conference on Decision and Control*, pp. 756–61, Maui, Hawaii, Dec. 2003.

Vidal, R. ; Shakernia, O. ; Kim, H. J. ; Shim, D. H. & Sastry, S. (2002). Probabilistic pursuit-evasion games: Theory, implementation, and experimental evaluation, *IEEE Transactions on Robotics and Automation*, Vol. 18, No.5, pp. 662–69, Oct. 2002.

Hespanha, J. P. ; Prandini, M. & Sastry, S. (2000). Probabilistic pursuit-evasion games:a one-step nash approach, *Proceedings of the 39th IEEE Conference on Decision and Control*, pp. 2272–2277, Sydney,Australia, Dec. 2000.

Hespanha, J. P. ; Kim, H. J. & Sastry, S. (1999). Multiple-agent probabilistic pursuit-evasion games, *Proceedings of the 38th IEEE Conference on Decision and Control*, pp. 2272–2277, Phoenix,AR, Dec. 1999.

Schumacher, C. (2005). Ground moving target engagement by cooperative uavs, *Proceedings of American Control Conference*, pp. 4502–4505, Portland, OR, June. 2005.

Sinha, A. ; Kirubarajan, T. & Bar-Shalom, Y. (2004). Optimal cooperative placement of gmti uavs for ground target tracking, *Proceedings of IEEE Aerospace Conference*, pp. 1859–1868, 2004.

Shea , P. J. & et al, (2000). Precision tracking of ground targets, *Proceedings of IEEE Aerospace Conference*, pp. 473–482, Vol. 3, 2000.

Koch , W. and Klemm, R. (2001). Ground target tracking with stap radar, *IEE Proc.-Radar, Sonar Navig.*, Vol. 148, pp. 173–185, June, 2001.

Jun , M. & D'Andrea, R. (2003). Probability map building of uncertain dynamic environments with indistinguishable obstacles, *Proceedings of the American Control Conference*, pp. 3417–3421, Denver, CO, June 2003.

Hespanha, J. P. ; Kizilocak, H. & Ateskan, Y. S. (2001). Probabilistic map building for aircraft-tracking radars, *Proceedings of the American Control Conference*, Arlington, VA, Jun. 2001.

Zengin , U. & Dogan, A. (2004). Dynamic target pursuit by uavs in probabilistic threat exposure map," *Proceedings of AIAA 3rd "Unmanned Unlimited" Technical Conference, Workshop and Exhibit*, Chicago, IL, Sep. 2004.

Dogan , A. & Zengin, U. (2006). Unmanned aerial vehicle dynamic-target pursuit by using a probabilistic threat exposure map, *AIAA Journal of Guidance, Dynamics and Control*, Vol. 29, No. 4, pp. 944–954, 2006.

Negnevitsky, M. (2002). *Artificial Intelligence: A Guide to Intelligent Systems*, Addison-Wesley, England.

Stark , H. & Woods, J. W. (1994). *Probability, Random Processes, and Estimation Theory for Engineers*, Uppper Saddle River, NJ : Prentice-Hall, Inc.

Konolige K. (1996). A gradient method for real-time robot control, *Proceedings of the International Conference on Intelligent Robots and Systems*, pp. 639-646, Takamatsu, Japan, Jan. 1996.

Choi, C. & Lee, J. (1996). Dynamical path planning algorithm of a mobile robot : Local minima problem and nonstationary environments, *Mechatronics*, Vol. 6, No. 1, pp. 81–100, June 1996.

Mitchell, I. M. & Sastry, S. (2003). Continuous path planning with multiple constraints, *Proceedings of the 42nd IEEE Conference on Decision and Control*, pp. 5502–5507,Maui,Hawai, Dec. 2003.

Pastrick, H. L. ; Seltzer, S. M. & Warren, M. E. (1981). Guidance laws for short-range tactical missiles, *Journal of Guidance, Control and Dynamics,* Vol. 4, No. 2, pp. 98-108, 1981.

Ogren , P.; Fiorelli, E. & Leonard, N.E. (2004). Cooperative control of mobile sensor networks: Adaptive gradient climbing in a distributed environment, IEEE *Transactions on Automatic Control,* Vol. 49, No. 8, pp. 1292-1302, 2004.

Larson, R.; Hostetler, R. P.; Edwards, B. H. & Heyd, D. E. (2002). *Calculus with Analytic Geometry,* Boston, NY: Houghton Mifflin Company.

Optic Flow Based Visual Guidance: From Flying Insects to Miniature Aerial Vehicles

Nicolas Franceschini, Franck Ruffier, Julien Serres and Stéphane Viollet
Biorobotics Lab, Institute of Movement Science, CNRS & Aix-Marseille University
France

1. Introduction

Insects and birds have been in the sensory-motor control business for more than 100 million years. The manned aircraft developed over the last 100 years rely on a similar lift to that generated by birds' wings. Aircraft designers have paid little attention, however, to the pilot's *visual sensor* that finely controls these wings, although it is definitely the most sophisticated avionic sensors ever known to exist. For this reason, the thinking that prevails in the field of aeronautics does not help us much grasp the visuo-motor contol laws that animals and humans bring into play to control their flight. To control an aircraft, it has been deemed essential to measure state variables such as barometric altitude, groundheight, groundspeed, descent speed, etc. Yet the sensors developed for this purpose - usually emissive sensors such as Doppler radars, radar-altimeters or forward-looking infrared sensors, in particular - are far too cumbersome for insects or even birds to carry and to power. Natural flyers must therefore have developed other systems for controlling their flight.

Flying insects are agile creatures that navigate swiftly through most unpredictable environments. Equipped with "only" about one million neurons and only 3000 pixels in each eye, the housefly, for example, achieves 3D navigation at an impressive 700 body-lengths per second. The lightness of the processing system at work onboard a fly makes us turn pale when we realize that this creature actually achieves just what is being sought for in the field of aerial robotics: dynamic stabilization, 3D autonomous navigation, ground avoidance, collision avoidance with stationary and nonstationary obstacles, tracking, docking, autonomous takeoff and landing, etc. Houseflies add insult to injury by being able to land gracefully on ceilings.

The last seven decades have provided evidence that flying insects guide themselves through their environments by processing the *optic flow* (OF) that is generated on their eyes as a consequence of their locomotion. In the animal's reference frame, the translational OF is the *angular speed* ω at which contrasting objects in the environment move past the animal (Kennedy, 1939; Gibson, 1950; Lee, 1980; Koenderink, 1986).

In the present chapter, it is proposed to summarize our attempts to model the visuomotor control system that provides *flying* insects with a means of autonomous guidance at close range. The aim of these studies was not (yet) to produce a detailed neural circuit but rather to obtain a more functional overall picture, that is, a picture that abstracts some basic control

principles (Marr, 1982). We attempted to determine the variables the insect really measures (see also: Taylor & Krapp, 2008), the actions it takes to control its flight and the causal and dynamic relationships between the sensory and motor variables involved.
Our progress on these lines was achieved by performing simulation experiments and testing our control schemes onboard miniature aircraft. Like our early terrestrial "robot-mouche" (robot Fly) (Pichon et al., 1989; Franceschini et al., 1992), these aerial robots are based on the use of an electronic *OF sensor* (Blanes, 1986) inspired by the housefly Elementary Motion Detectors (EMD), which we previously studied in our laboratory (Rev.: Franceschini, 1985; Franceschini et al., 1989).

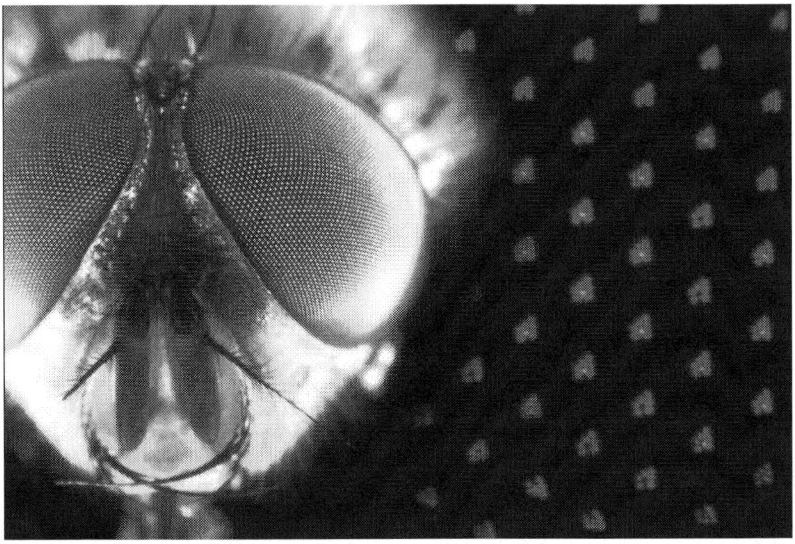

Figure 1. Head of the blowfly *Calliphora erythrocephala* with its panoramic compound eyes (left). The right part shows about 2% of the housefly retinal mosaic. A cluster of micrometer-sized photoreceptors is located in the focal plane of each facet lens : 6 outer receptors R1-6 surround a central cell R7 (prolonged by an R8, not seen here). These are the natural *autofluorescence* colors of the receptors observed *in vivo* under blue excitation (Franceschini et al., 1981a,b), after « optical neutralization of the cornea » (From Franceschini, 2007)

Section 2 recalls some aspects of the fly visual system. Section 3 focuses on the realisation of fly-inspired OF sensors. Section 4 examines the problem of ground avoidance. Section 5 introduces the *OF regulator*. In Section 6 the miniature helicopter we built is described, equipped with an electronic *OF sensor* and an *OF regulator*. The extent to which the *OF regulator* accounts for the actual behavioral patterns observed in insects is discussed in Section 7. Section 8 introduces the *dual OF regulator* that is able to control both speed and distance to the walls in a corridor. In Section 9, another visuomotor control system is presented which enables an aerial robot to stabilize in yaw by visually fixating a target with *hyperacuity*, on the basis of a principle inspired by the fly. Section 10 ends up by discussing the potential applications of these insect-derived principles to the navigation of aerial vehicles - in particular Micro Aerial Vehicles (MAVs) and Micro Space Vehicles (MSVs).

2. The fly visual system and its motion sensitive neurons

Each compound eye consists of an array of *ommatidia*. The frontend of each *ommatidium* is a facet lens that focusses light on a small group of photoreceptor cells (Fig. 1, right). The fly retina is among the most complex and best organized retinal mosaic in the animal kingdom. It has been described in great details, with its different spectral types of photoreceptor cells, polarization sensitive cells and sexually dimorphic cells. There exists a typical *division of labour* within the retina (Rev.: Franceschini, 1984; Hardie, 1985):

- The two central photoreceptor cells, R7-8, display various spectral sensitivities that are randomly scattered across the retinal mosaic, as attested by the various R7 autofluorescence colors (Fig. 1, right). R7 and R8 are thought to participate in *color vision*.
- The outer 6 photoreceptor cells (R1-R6) all have the same, wide-band spectral sensitivity. They participate, in particular, in *motion detection* (Rev.: Buchner, 1984, Heisenberg & Wolf, 1984; Riehle & Franceschini, 1984). In this visual pathway, signal-to-noise ratio is improved by the presence of an ultraviolet sensitizing pigment that enhances the quantum catch (Kirschfeld & al., 1977), and by an exquisite opto-neural projection called *"neural superposition"* (Braitenberg, 1967; Kirschfeld, 1967; Kirschfeld and Franceschini, 1968). The R1-R6 photoreceptors therefore make for a high sensitivity ("scotopic") system (Kirschfeld and Franceschini, 1968).

To estimate the OF, insects use motion sensitive neurons. In flies, part of the 3rd optic ganglion called the *Lobula Plate* (LP) appears as a genuine "visual motion processing center". It comprises approximately 60 *uniquely identifiable* neurons, the LP tangential cells (LPTC) that analyze the OF field resulting from the animal's walking or flying. Some of these neurons transmit their electrical signals via the neck to thoracic interneurons that will drive the wing-, leg-, and head-muscles. Other neurons (in particular H1, see Fig. 2b) send their signals to the contralateral eye. The LPTCs are actually large-field collator neurons that pool the electrical signals from many retinotopic input elements called 'Elementary Motion Detectors' (EMDs) (Rev.: Hausen & Egelhaaf, 1989; Egelhaaf & Borst, 1993; Hausen, 1993; Krapp et al., 1998; Borst & Haag, 2002, Taylor & Krapp, 2008). The cellular details underlying an EMD has not been fully identified in any insects - nor in any vertebrates.

Taking advantage of the micro-optical techniques we had developed earlier (Rev.: Franceschini, 1975), we were able, however, to activate a single EMD in the eye of the living housefly by stimulating single *identified* photoreceptor cells within a single ommatidium while recording the response of an *identified* motion sensitive neuron (H1) in the LP (Riehle & Franceschini, 1984, Franceschini, 1985, 1992; Franceschini et al., 1989). We applied pinpoint stimulation to two neighboring photoreceptors (diameter ≅ 1µm) within the selected ommatidium (Fig.2a) by means of an optical stimulation instrument (Fig. 2d) whose main objective was quite simply the facet lens itself (diameter ≅ 25 µm, focal length ≅ 50 µm). This exact instrument (a rotating triple beam incident light 'microscope-telescope') served to: (i) select a facet lens, (ii) select 2 out of the 7 receptors (R1 and R6), and (iii) stimulate them *successively* with 1µm-light spots. Sequential stimulation produced an 'apparent motion' that would *simulate* a real motion within the small visual field of the selected ommatidium. The H1-neuron responded by a conspicuous *increase* in spike rate, as long as the phase relationship between the two stimuli mimicked a movement occurring in the *preferred* direction (see Fig. 2c, top trace). The null response observed for the reverse sequence (Fig 2c, bottom trace) attests to the remarkable sequence discriminating ability of an EMD, which is sensitive to *directional motion* (Franceschini et al., 1989, Franceschini, 1992).

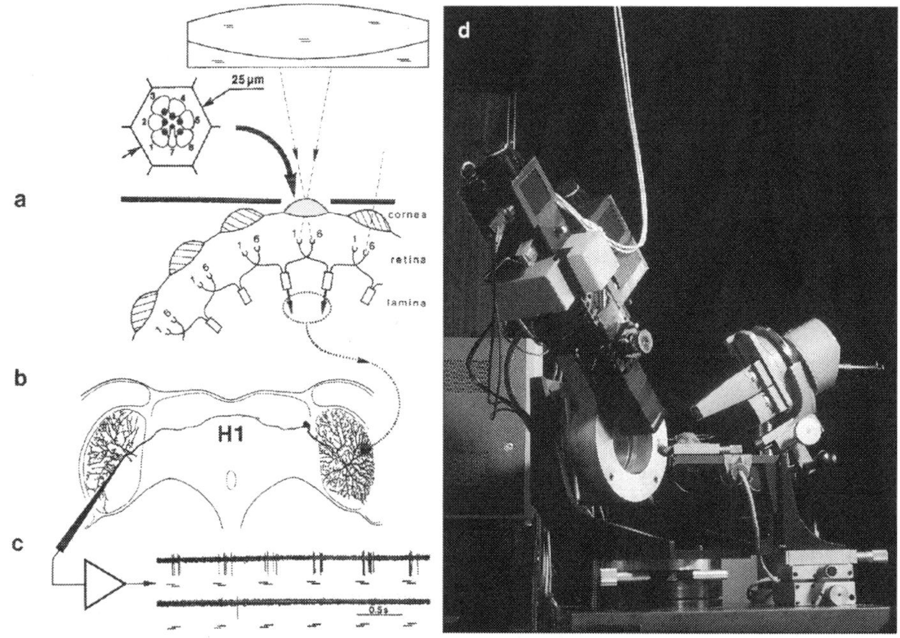

Figure 2. (left) : (a-c) Principle of the experiment aimed at deciphering the principle of motion vision in flies, using optical stimulation of single photoreceptors. (d) Triple-beam incident light "microscope-telescope" that delivers a 1µm light spot to two neighboring photoreceptor cells, R1 and R6, *successively* (see (*a*)). A microelectrode (c) records the electrical response (nerve impulses) of the motion sensitive neuron H1 to this "apparent motion" (From Franceschini et al., 1989)

From many experiments of this kind, in which various sequences of light steps and/or pulses were applied to selected receptor pairs, we established an EMD *block diagram* and characterized each block's dynamics and nonlinearity (Franceschini, 1985, 1992; Franceschini et al., 1989). While not unveiling the cellular details of the EMD circuit, our analysis allowed the EMD principle to be understood *functionally*, paving the way for its transcription into another, man-made technology.

3. From biological to electronic optic flow sensors

In the mid 1980's, we designed a neuromorphic optic flow sensor (Blanes, 1986, 1991, Franceschini et al., 1986), the signal processing scheme of which was inspired by what we had learned from the fly EMD. The OF is an angular speed $\boldsymbol{\omega}$ that corresponds to the inverse of the time Δt taken by a contrasting feature to travel between the visual axes of two adjacent photoreceptors, separated by an angle $\Delta\varphi$. Our OF sensor processes this delay Δt so as to generate a response ω_{meas} that grows monotonically with the inverse of Δt, and hence with the optic flow $\omega = \Delta\varphi / \Delta t$ (Fig. 3a). Short delays Δt give higher voltage outputs and vice versa.

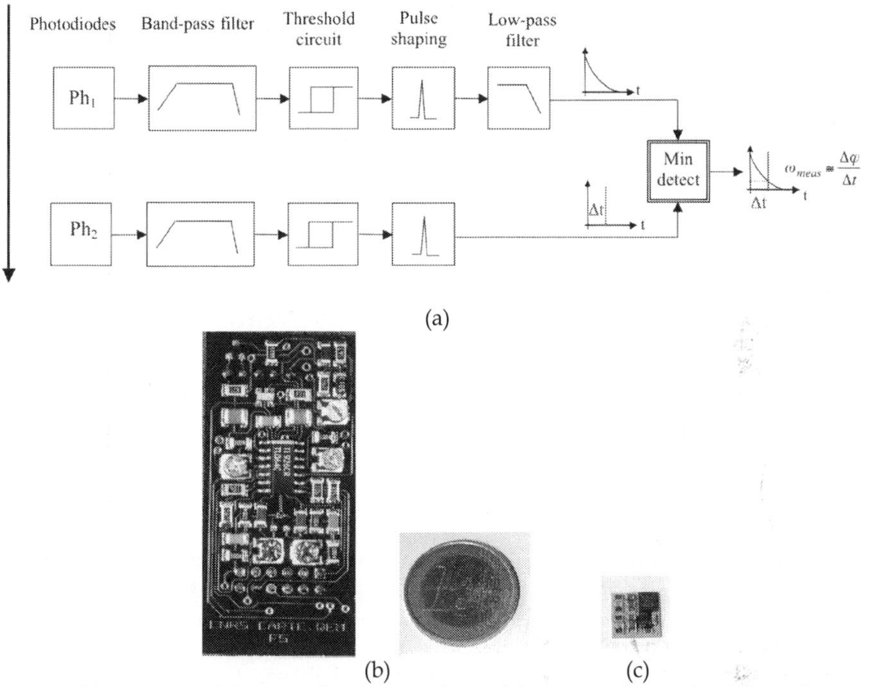

Figure 3. (a) Principle of the optic flow sensor derived from our electrophysiological analyses of the housefly's EMD (Blanes, 1986, Franceschini et al., 1986). (b) purely analogue version (weight 5 grams) built in 1989 for the Robot-Fly, whose compound eye housed a ring of 114 EMDs of this type (Pichon et al., 1989 ; Blanes, 1991 ; Franceschini & al., 1992) (c) hybrid (analogue + digital) version (size : 7mm x 7 mm, mass 0.2 grams) based on a microcontroller and built using Low Temperature Co-fired Ceramics technology (LTCC) (Pudas et al., 2007)

Our scheme is not a "correlator scheme" (cf Hassenstein & Reichardt, 1956, Reichardt, 1969) and corresponds to the class of "feature-matching schemes"(Ullman, 1981), where a given feature (here a change in intensity that may represent a moving edge) is extracted and tracked in time. The photodiode signal of each channel is first *bandpass filtered* (Fig. 3a) - resembling the analog signal measured in the large monopolar neurons of the fly lamina (Laughlin, 1984). The next step consists of hysteresis thresholding and generation of a unit pulse. In the EMD version built in 1989 for the Robot-Fly (Fig. 3b), the unit pulse from one channel was sampling a long-lived decaying exponential function generated by the other channel (Blanes, 1991), via a nonlinear circuit called a minimum detector, to give an output ω_{meas} that is virtually equal to the OF, $\omega = \Delta\varphi/\Delta t$ (Fig. 3a). The thresholding operation makes the voltage output virtually invariant to texture and contrast and the circuit responds as well to natural scenes (Portelli et al., 2008). A very similar EMD principle has been conceived, independently, a decade later by C. Koch's group at CALTECH, where it became known as the "facilitate and sample" velocity sensor

(Kramer et al., 1995). Yet another variant of our original "time of travel" principle was proposed another decade later (Moeckel &Liu, 2006).

Since our original analog design (Blanes, 1986, 1991; Franceschini et al, 1986), we have built various versions of OF sensors based on this very principle. In the EMD currently used onboard our aerial robots, the signals are processed using a mixed (analog + digital) approach (Ruffier et al., 2003). Such OF sensors can be small and lightweight (the smallest one weighs only 0.2 grams: Fig. 3c) to mount onto MAVs. Several OF sensors of this type were also recently integrated on a miniature FPGA (Aubépart et al. 2004, 2008; Aubépart & Franceschini, 2007). A different kind of OF sensor was designed recently (Barrows et al., 2001) and mounted on a model plane (Barrows et al, 2003; Green & Barrows, 2004).

4. The problem of ground avoidance

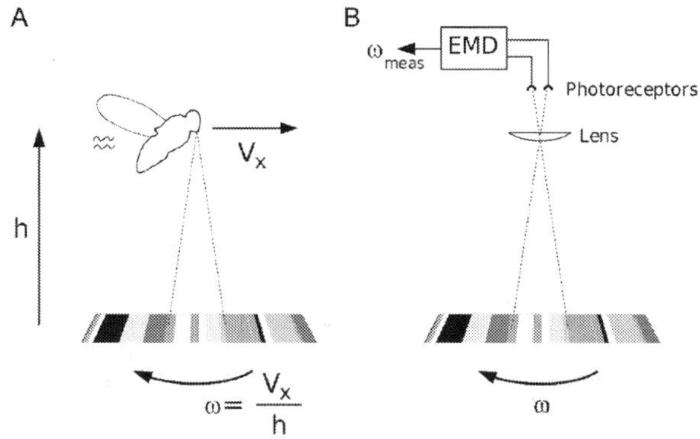

Figure 4. Definition (a) and measurement of the ventral optic flow ***ω*** experienced by an insect (or a robot) flying in translation in the vertical plane. (b) an EMD of the type shown in Fig. 3, like the EMDs driving some insects' neurons (Kien, 1975; Olberg, 1981; Ibbotson, 2001), is able to measure the ventral OF, i.e., the angular speed ***ω*** at which a contrasting feature moves under the flying agent (From Franceschini et al., 2007)

The ventral OF experienced in the vertical plane by flying creatures - including aircraft pilots - is the apparent *angular velocity* ***ω*** generated by a point directly below on the flight track (Gibson et al., 1955; Whiteside & Samuel, 1970). As shown in figure 4a, the ventral OF depends on both the groundspeed V_x and the groundheight h and is equal to the ratio between these two variables:

$$\omega = V_x / h \ [\text{rad.s}^{-1}] \tag{1}$$

We know that flies and bees are able to react to the *translational OF* independently of the spatial texture and contrast (David, 1982; Kirchner & Srinivasan, 1989, Srinivasan et al., 1991, 1993; Baird et al., 2005). We also know that some insects' visual neurons may be involved in this reaction because they respond monotonically to ***ω*** with little dependence on texture and contrast (Kien, 1975; Olberg, 1981; Ibbotson, 2001; Shoemaker et al., 2005). Neurons facing downwards can therefore act as ventral *OF sensors*, and thus assess the V_x/h ratio (figure 4).

Based on laboratory experiments on mosquitoes and field experiments on locusts, Kennedy put forward an "optomotor theory" of insect flight, according to which flying insects maintain a "preferred retinal velocity" with respect to the ground below (Kennedy, 1939, 1951). In response to wind, for example, insects may adjust their groundspeed or groundheight to restore the apparent velocity of the ground features. Kennedy's hypothesis has been repeatedly confirmed during the last 30 years: both flies and bees were found to maintain a constant OF with respect to the ground while cruising or landing (David, 1978; Preiss, 1992; Srinivasan et al., 1996, 2000, Baird et al., 2006).

The problem is *how* insects may achieve this feat, since maintaining a given OF is a kind of chicken-and-egg problem, as illustrated by Eq.1: an insect may hold its ventral OF, ω, constant by adjusting either its groundspeed (if it knows its groundheight) or its groundheight (if it knows its groundspeed). Besides, the insect could maintain an OF of 1rad/s (i.e., 57°/s), for instance, by flying at a speed of 1m/s at a height of 1 meter or by flying at a speed of 2m/s at a height of 2m: there is an infinitely large number of possible combinations of groundspeed and groundheight that will give rise to the same "preferred OF".

Drawing on the experience we had with OF-based visual navigation of a terrestrial robot (Pichon et al., 1989; Franceschini et al., 1992), we attempted early to develop an explicit flight control scheme for aerial navigation in the vertical plane. Our first tentative step on these lines was not particularly successful, because we were cornered by the general notion that prevailed in those days that insect navigation relies on *gauging range* (Kirchner & Srinivasan, 1989; Srinivasan et al., 1991; Franceschini et al., 1992; Srinivasan, 1993). In the experimental simulations we performed in 1994, for example (Mura & Franceschini, 1994), we assumed that the insect (or the robot) would know its groundspeed V_x (by whatever means), so that by measuring $\boldsymbol{\omega}$ it would be able to gauge the distance $\boldsymbol{h}$ from the ground (Eq.1) and react accordingly to avoid it. Although this procedure may be justified in robotics (see, e.g., Barber et al, 2005; Srinivasan et al., 2006; Garratt & Chahl, 2008), where groundspeed can be determined via GPS, this makes the way insects operate all the more elusive.

In 1999, we established (in simulation) *how* a rotorcraft (or an insect) might be able to follow a terrain (Fig. 5a) and land (Fig. 5b) on the sole basis of OF cues, *without measuring its groundspeed and groundheight* (Netter & Franceschini, 1999).

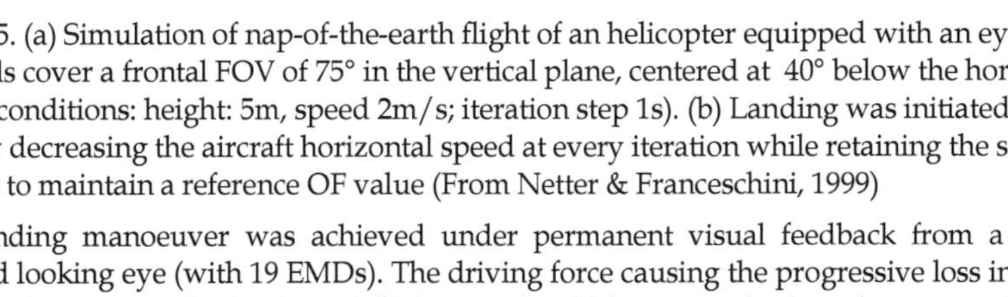

Figure 5. (a) Simulation of nap-of-the-earth flight of an helicopter equipped with an eye whose 20 pixels cover a frontal FOV of 75° in the vertical plane, centered at 40° below the horizon (initial conditions: height: 5m, speed 2m/s; iteration step 1s). (b) Landing was initiated by linearly decreasing the aircraft horizontal speed at every iteration while retaining the same request to maintain a reference OF value (From Netter & Franceschini, 1999)

The landing manoeuver was achieved under permanent visual feedback from a 20-pixel forward looking eye (with 19 EMDs). The driving force causing the progressive loss in altitude was the decrease in the horizontal flight speed, which occurred when the rotorcraft (or the insect) was about to land - either volontarily or because of a head wind. The landing trajectory obtained in this simulation (Fig. 5b) already resembles the final approach of bees landing on a flat surface (Srinivasan et al, 1996). The principle was first validated onboard FANIA, a miniature tethered helicopter having a single (variable pitch) rotor, an accelerometer and a forward looking eye with 20 pixels (and, therefore, 19 EMDs) arranged in the frontal meridian (Netter & Franceschini, 2002). This 0.8-kg rotorcraft had three degrees of freedom (surge, heave and pitch). Mounted at the tip of a flight mill, the robot lifted itself by increasing its rotor collective pitch. Upon remotely inclining the servo-vane located in the propeller wake, the operator made the helicopter pitch forward by a few degrees so that it gained speed and, therefore, climbed to maintain a reference OF with respect to the terrain below. FANIA jumped over contrasting obstacles by increasing its collective pitch as a function of the fused signals from its 19 EMDs (see Fig. 8 in Netter & Franceschini, 2002).

5. The "optic flow regulator"

In spite of this early success to explain how an insect could navigate on an OF basis, we considered that Kennedy's insightful "optomotor theory" was calling for a clear formalization that would bring to light:

- the flight variables really involved
- the sensors really required
- the dynamics of the various system components
- the causal and dynamic links existing between the sensory output(s) and the variable(s) to be controlled
- the points of application of the various disturbances that insects may experience
- the variables insects have to control to compensate for these disturbances.

We came up with an autopilot called OCTAVE (OCTAVE stands for Optical altitude Control sysTem for Autonomous VEhicles) that is little demanding in terms of neural (or electronic) implementation and could be just as appropriate for insects as it would be for aircraft (Ruffier & Franceschini, 2003; Franceschini et al., 2003). A ventral OF sensor was integrated into a feedback loop that would drive the robot's lift, and thus the groundheight, so as to compensate for any deviations of the OF sensor's output from a given set point (Ruffier & Franceschini, 2003, 2004a,b, 2005; Franceschini et al., 2007). This simple autopilot (Fig. 6a) enabled a miniature helicopter to perform challenging tasks such as take-off, terrain following, reacting suitably to wind, and landing (see Section 6).

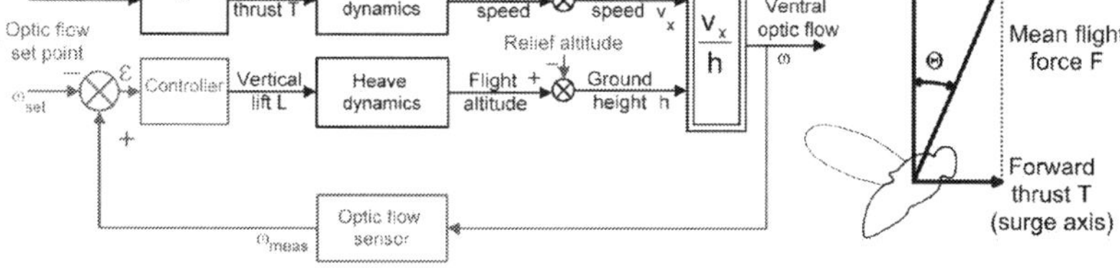

Figure 6. (a) The *OF regulator* OCTAVE (bottom red feedback loop) controls the lift ***L***, and consequently the groundheight, at all times so as to maintain the ventral optic flow $\boldsymbol{\omega}$ constant and equal to the set point $\boldsymbol{\omega}_{set}$. (b) Flies and bees, like helicopters, pitch forward to increase their forward thrust, and hence their airspeed. As long as they pitch forward by $\Theta < 10°$, the lift component L does not incur any major loss (From Franceschini et al., 2007)

The OCTAVE autopilot can be said to be an *OF regulator*. The word 'regulator' is used here in the strictest acceptation of the word in the context of control theory, where it denotes a *feedback control system* designed to maintain an output signal constantly equal to a given set point. The Watt flyball governor from the 18th century, for instance, was not only one of the first servomechanisms ever built: it was also the very first *angular speed regulator*. It served to maintain the rotational speed of a steam engine shaft at a given set point, whatever interferences occurred as the result of unpredictable load disturbances. The Watt regulator was based on a rotational speed sensor (meshed to the output shaft), whereas our *OF regulator* is based on a noncontact rotational speed sensor - the *OF sensor* - that measures the ventral OF - again in rad/s.

Specifically, the OF signal ω_{meas} delivered by the OF sensor (see Fig. 6a, bottom) is compared with the OF set point, ω_{set} . The comparator produces an error signal: $\varepsilon = \omega_{meas} - \omega_{set}$ which drives a controller adjusting the lift L, and thus the groundheight h, so as to minimize ε. All the operator does is to set the pitch angle Θ and therefore the airspeed (see figure 6a): the *OF regulator* does the rest, that is, it adjusts the groundheight h proportionally to the current groundspeed V_x, keeping the OF, i.e., the V_x/h ratio constant in the steady state:

$$V_x / h = \omega \cong \omega_{meas} \cong \omega_{set} = \text{constant} \tag{2}$$

Upon looking at figure 6b, one might be tempted to object that controlling F (via the rotor speed) will affect not only L but also T, and hence V_x . This *coupling* is negligible, however, because the ensuing change in T is *much smaller* than the change in L (by a factor of at least 5.67 = cotan 10°, because the maximum speed of 3m/s is already attained for Θ =10°).

6. A miniature helicopter equipped with an "optic flow regulator"

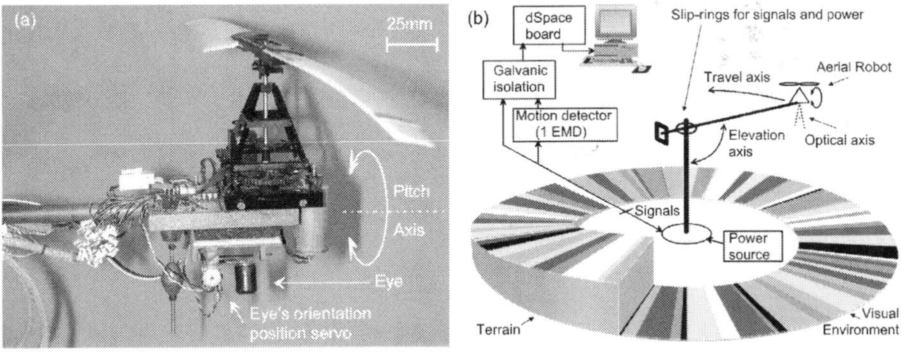

Figure 7. (a) Miniature helicopter (MH) equipped with an *OF sensor* (Fig. 3) and an *OF regulator* (Fig. 6). The operator sets the forward pitch, and thus the airspeed. The gaze remains automatically oriented downwards. (b) The 100-gram robot lifts itself and circles around CCW at speeds up to 3m/s and heights up to 3 m over the arena, giving rise to the flight patterns given in Fig. 8-10 (From Ruffier & Franceschini, 2003)

We tested the idea that insects may be equipped with a similar *OF regulator* by comparing the behavior of insects with that of a 'seeing helicopter' placed in similar situations. The robot we built (figure 7a) is a miniature helicopter (MH) equipped with a simple, 2-pixel ventral eye driving an EMD acting as an OF sensor (Fig. 3 and 4b). The 100-gram robot is tethered to an instrumented flight mill consisting of a light pantographic arm driven in terms of its elevation and azimuth by the MH's lift and forward thrust, respectively (Fig. 7b).

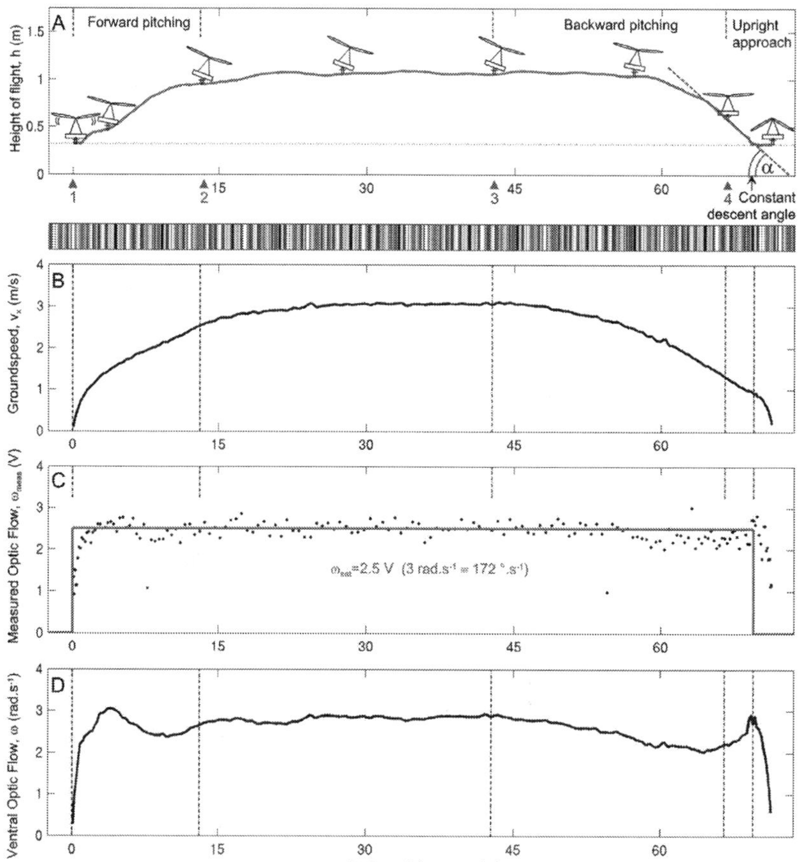

Figure 8. Flight parameters monitored during a 70-meter flight (consisting of about 6 laps on the test arena: figure 7b) performed by the miniature helicopter (MH) (Fig. 7a), equipped with an *OF regulator (*figure 6a). The complete journey over the randomly textured pattern (shown at the bottom of (a)) includes take-off, level flight and landing
(a) Vertical trajectory in the longitudinal plane. **(b)** Groundspeed V_x monitored throughout the journey. **(c)** Output ω_{meas} of the OF sensor showing the relatively small deviation from the OF set point ω_{set} (red curve). **(d)** Actual OF, ω (calculated as V_x/h) resulting from the behavioral reaction: ω is seen to have been held relatively - but not perfectly - constant throughout the journey, even during the takeoff and landing maneuvers where the groundspeed varies considerably as shown in (b) (From Franceschini et al., 2007)

Any increase in the rotor speed causes the MH to rise, and the slightest (operator mediated) forward ("nose-down") tilting by a few degrees produces a forward thrust component that causes the MH to gain forward speed. The flight mill is equipped with ground-truth azimuthal and elevation sensors that allow the position and speed of the MH to be monitored at high accuracy and in real time. Since the MH purpose was to demonstrate a basic principle, it was equipped with an elementary ventral eye composed of only two photoreceptors driving a single Elementary Motion Detector (EMD) built according to the "travel time" principle shown in Fig. 3 (Ruffier & Franceschini, 2003, Ruffier et al., 2003).

7. Comparison of robots' and insects' behavioral patterns

The OCTAVE control scheme (Fig. 6) produced the helicopter behavioral patterns shown in figures 8-10. The robot's behavior was found to account for a series of puzzling, seemingly unconnected flying abilities observed by many authors during the last 70 years in various species (fruitflies, honeybees, moths, mosquitoes, dung-beetles, migrating locusts and butterflies, and birds) (Franceschini et al., 2007). Most of the data published in these studies are qualitative, but recent quantitative findings on honeybees' landing performances can also be explained on the basis of this simple control scheme, including the constant descent angle observed in the bee's final approach (Srinivasan et al., 1996, 2000).
Let us now examine these various behavioral patterns, assuming the insect to be equipped with the *OF regulator* shown in Fig. 6 (Franceschini et al., 2007).

7.1 Take-off

An insect that increases its forward thrust by pitching forward (Fig. 6b) like a fly or a helicopter (David, 1978), is bound to rise into the air because the OF regulator (Fig. 6a) will constantly increase the groundheight $\boldsymbol{h}$ proportionally to the current groundspeed $\boldsymbol{V_x}$, to maintain the OF constant (Eq.2).
The performances of the MH illustrate this point quite clearly (figure 8a, left part). Starting at position 1 (with the rotor axis oriented vertically and the lift just balancing the weight), the MH is remotely commanded to pitch 'nose-down' (Θ from 0° to +10° rampwise, Fig. 6b). The ensuing increase in groundspeed (figure 8b) automatically causes the micro-flyer to rise (figure 8a) because the feedback loop slaves $\boldsymbol{h}$ to $\boldsymbol{V_x}$, effectively maintaining the actual ratio $\omega = V_x/h$ relatively constant *throughout take-off* (Fig. 8d). Upon reaching a steady groundspeed of 3 m.s^{-1} (Figure 8b), the MH can be seen to have cruised at a steady groundheight (of around one meter: figure 8a), which depends on the OF set point (here, ω_{Set} = 3 rad.s^{-1}: figure 8c).

7.2 Terrain following

An increase in relief absolute altitude is regarded as a "disturbance" that causes a reduction in the local groundheight h. This disturbance will be compensated for by an increase in flight altitude (see figure 6a). Figure 9a shows that the MH is able to clear a gentle slope. Figure 9b confirms the prediction made at the end of Section 5 that the groundspeed hardly changes in spite of the changes in the mean flight force F needed to change the altitude (see figure 6b).
Insects are able to follow a terrain or a canopy and the above scheme may explains *how* they achieve this feat. Migrating butterflies crossing narrow canyons fly down into the gully and

across the bottom. When traversing dense forests, they clear the canopy at a distance roughly equal to their previous height above the ground, and once they have crossed this large obstacle, they re-descend (Srygley & Oliveira, 2001).

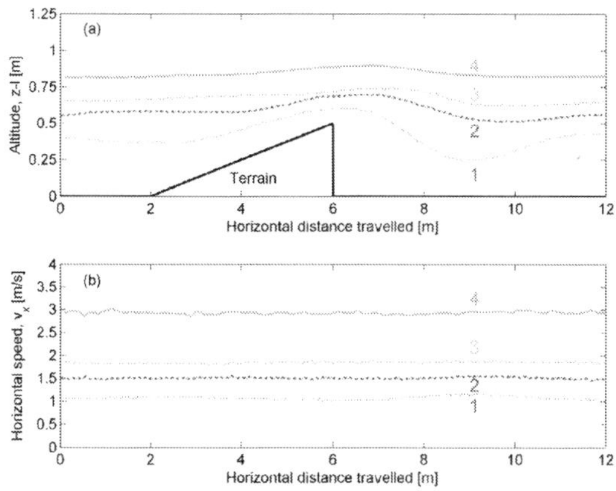

Figure 9. Terrain following by the miniature helicopter (Fig. 7a) recorded at 4 different groundspeeds (1m/s ; 1.5 m/s; 1.8 m/s; 3 m/s). The altitude **z-l** plotted in (a) is that of the landing gear which is at a distance l= 0.3 m under the eye. The higher the speed, the higher the clearance from the terrain (From Ruffier & Franceschini, 2003)

Figure 9 shows that when the MH is set to travel at a higher speed, it automatically maintains a greater height above the terrain, giving rise to a "safe" flight. This is precisely the flight pattern that was observed in dung-beetles: they fly higher at higher groundspeeds and lower at lower speeds (Steiner, 1953). Likewise, bees cruise at a height roughly proportional to their horizontal flight speed (see Fig. 7c in Srinivasan et al., 2000). In other words, there is a similarly tight link between groundspeed and groundheight in insects and our helicopter.

7.3 Flying against wind

Locusts and other migrating insects have long been known to descend under headwind and ascend during lull (Kennedy, 1951, Srygley & Olivera, 2001). Under headwind conditions, Braüninger (1964) observed that bees were flying at such a low speed and so close to the ground that he was able to accompany them on his bicycle all the way to the nectar source.
Windspeed disturbs the OF autopilot but the point of application of this disturbance differs from that of the relief disturbance (see Fig. 6a). A light headwind (produced by a fan) was observed to reduce the MH groundspeed (Fig. 10b, dotted curve). The reduction in speed forced the groundheight h to decrease in similar proportion (figure 10a, dotted curve). Since the speed increased again when the robot left the wind-swept region of the arena, the MH rose again. This MH flight pattern therefore accounts particularly well for the field data cited above.
A strong head wind reduced the groundspeed so much that it forced the MH to land (Fig. 10a, continuous curve). Landing was smooth, however, because h was decreased *under*

visually closed loop, in proportion to V_x. This finding is highly reminiscent of what occurs in many insects and birds, which descend by headwind and settle when the wind grows stronger (Kennedy, 1939, 1951, Williams, 1965; Alestam, 1990). Given the natural decrease in wind speed that occurs close to the earth in the boundary layer, these reactions can be said to be "ecological" because they allow the insects to travel farther for the same amount of energy (this point is discussed extensively in Ruffier & Franceschini, 2005). Whether for terrain following or for the reaction to head wind, the control scheme (Fig. 6) underlines the direction of causality: it is the change in groundspeed that induces the change in groundheight, and not the other way round (this point is discussed further in Section 8).

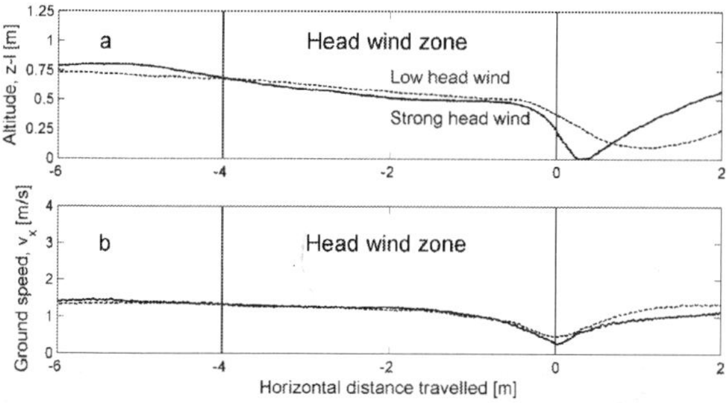

Figure 10. Reaction of the miniature helicopter (MH) to head wind (dashed curves: low head wind, V_w = 0.5m/s; continuous curves: strong head wind, V_w = 1.5m/s). The OCTAVE autopilot (Fig. 6a) makes the MH react to the reduction in speed (b) by reducing its altitude (a). The strong wind forced the MH to land at negligible forward speed (continuous curves). In these experiments, a 30 x 20cm planar airfoil was mounted perpendicular to the travel axis to catch the wind better, emphasizing the effect (From Ruffier & Franceschini, 2004)

7.4 Flying over mirror-smooth water

Bees crossing mirror-smooth water during foraging trips were reported to fly lower and lower until crashing head-first into the water. They were rescued from drowning only when the water surface was rippled or when a slatted bridge was placed on the water surface to provide sufficient contrast (Heran & Lindauer, 1963). Recent studies on bees trained to fly across a lake did not quite confirm these early results (Tautz et al., 2004), but the (large) lake may not have been as ripple-free as the (small) flooded quarry used in the original experiments.

A featureless expanse of water no longer provides the animal's eye with any contrasting features, and the OF sensor will no longer respond. If we make the feedback signal $\omega_{meas} = 0$ in figure 6a, the error signal ε becomes large and negative (equal to ω_{set}), which leads to a fatal decrease in groundheight h. The *OF regulator* therefore may account for the puzzling finding that bees plunge straight into calm water during their foraging trip. The closed feedback loop irrevocably pulls the insect down whenever the OF sensor fails to respond. A similarly disastrous tendency was observed in the MH when it was flying over a ground that was made deliberately blank over a distance of 1.5 meters. A simple way to rescue the MH flight from this dangerous situation consisted in holding the last OF measurement for some time (e.g., 0.5 seconds, see figure 9 in Ruffier & Franceschini, 2005).

7.5 Landing

Video recording of landing honeybees showed that their grazing landings on a flat surface occur with a constant slope (Srinivasan et al., 2000). Bees were assumed to meet two requirements to be able to land smoothly: "(i) adjusting the speed of forward flight to hold constant the angular velocity of the image of the surface as seen by the eye, (ii) making the speed of descent proportional to the forward speed"(Srinivasan et al., 2000). Our OF regulator scheme (Fig. 6a) controls neither the forward flight speed nor the descent speed, yet predicts that the bee should land with a constant slope, as actually observed in the MH flight pattern (Fig. 8a, right part).

Both MH's and bees' landing behavior can be explained in the same way because the surge dynamics can be described in both cases by similar (first order) linear differential equations (see Supplements in Franceschini & al., 2007).

Landing of the MH is initiated by commanding it to gradually pitch backwards to the vertical (at arrowhead 3 in figure 8a), which induces a gradual loss of speed (figure 8b). The groundheight h is then bound to decrease proportionally to V_x since the feedback loop strives to make V_x/h = constant (Eq. 2).

The *final approach* actually starts when the MH has regained its completely upright position (arrowhead 4 in Fig. 8a). From this moment onwards, V_x decreases exponentially with the "surge time constant" τ_{MH} =2.15s (see figure 4 in Franceschini & al., 2007) and the feedback loop forces h (and therefore its derivative dh/dt, i.e., the descent speed V_z) to decrease with the same time constant. The descent speed:groundspeed ratio $V_z(t):V_x(t)$, that is, the descent *slope*, will therefore remain constant throughout the final approach, as actually observed in both landing bees (Srinivasan et al., 2000) and the landing MH (figure 8a).

In (Franceschini & al., 2007), we established that the ensuing descent angle α [rad] can be calculated as follows :

$$\alpha = -\arctan\left(\frac{1}{\omega_{set}\tau}\right)$$

where ω_{set} = OF set point [rad.s^{-1}] , and τ = surge time constant [s] of the MH (τ_{MH} =2.15s).

This means that the slope of the final approach depends on only two parameters: the surge time constant and the OF set point.

If the constant-slope landing glide observed in bees reflects the presence of an *OF regulator* onboard these insects, then by substituting the landing data obtained on bees (Srinivasan et al., 2000): α_{BEE} = - 28° ; ω_{setBEE} = 500 deg.s^{-1}, into Eq. 3, we can determine the bee's *surge time constant* :

$$\tau_{BEE} = 0.22s$$

It is striking that this value predicted by our *OF regulator* model matches closely the time constant values (0.22s, 0.27s, 0.29s and 0.57s) that we derived from the data of Srinivasan and coworkers (2000) on the exponential decrease in bees' *groundheight* with time. This makes the OF regulator scheme (Fig. 6a) an appealing hypothesis for the control system that is implemented onboard the bee.

8. A dual optic flow regulator for joint speed control and lateral obstacle avoidance

The model presented above differs from another one where the OF would control the groundspeed V_x rather than the groundheight h (Preiss & Kramer, 1984; Srinivasan et al., 1996, 2000; Baird et al., 2005). If the ventral OF controlled V_x instead of h in figure 6a, this would lead to strikingly different flight patterns from those observed in insects by previous authors, as follows:

(i) instead of following a slanting terrain, as migrating butterflies (Williams, 1965; Srygley & Oliveira, 2001) and our MH do, the insect would gradually decelerate until touching the rising ground at a negligible speed, thus inopportunely interrupting its journey.

(ii) instead of descending in a headwind and rising in a tailwind, as occurs in honeybees (Braüninger, 1964), locusts (Kennedy, 1951), dung-beetles (Steiner, 1953), mosquitoes (Kennedy, 1939), and our MH (figure 10), the insect would compensate for the unfavourable headwind by increasing its airspeed without changing its groundheight.

These two models can be reconciled, however, if we add the hypothesis that another *OF regulator*, based on OF sensors from the *lateral* parts of the eyes, may be in charge of controlling the groundspeed V_x.

In line with the OCTAVE autopilot, we designed the LORA III autopilot (LORA III stands for Lateral Optic flow Regulator Autopilot, mark III), which is able to control both the forward speed V_x of an aerial vehicle and its lateral distances D_r and D_l to the two walls of a corridor (Fig. 11).

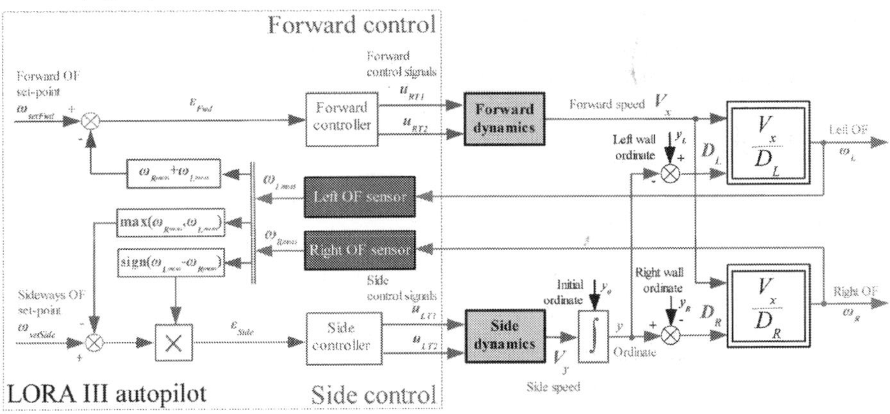

Figure 11. The LORA III autopilot is a *dual OF regulator* that enables a *fully actuated* hovercraft to navigate in a (straight or tapered) corridor by controlling its forward speed V_x and its distance D_r, D_l to the walls jointly, without requiring any speed or distance measurements (From Serres et al., 2008a)

We showed the feasibility of this scheme in simulation experiments where a miniature hovercraft navigates in a straight or tapered corridor (Serres et al., 2008a). Our hovercraft is equipped with two additional *lateral thrusters* that make it *fully actuated*. This means that it is capable of independent side-slip and forward slip, like hymenopteran insects (Zeil et al., 2008). The groundspeed is constrained by the environment, as observed on bees navigating in a straight or tapered corridor (Srinivasan et al., 1996). The clearance to the walls is constrained by the environment as well, but the robot does not show a systematic "centering behavior", in contrast with former observations on bees (Kirschner & Srinivasan, 1989). However, the robot's "wall-following behavior" observed is fully consistent with the results of our own behavioural experiments on bees navigating in a large corridor (Serres et al. 2008b).

LORA III is based on only two OF sensors (one looking to the right, one to the left). The forward speed V_x is controlled by the error signal ε_{Fwd} between the *sum* of the two OFs (right and left) and the forward OF set point ω_{setFwd} . The lateral clearance from the walls is controlled by the error signal ε_{Side} between the *larger* of the two OFs and the sideways OF set point $\omega_{setSide}$. The LORA III *dual OF regulator* accounts particularly well for both types of behaviours. The typical "wall-following behavior" that we observed on both the bees and the robot can be shown to degenerate into a "centering behavior" for particular *relative* values between the two OF set points ω_{setFwd} and $\omega_{setSide}$ (Serres et al., 2008a, Serres, 2008). The advantage of this control scheme is that it determines both the forward speed and the clearance from the walls on the sole basis of two parameters - two OF set points - without any needs for measuring forward speed, lateral distances and corridor width, that is, without using any velocimeters or range finders. The simulation experiments performed in straight or tapered corridors were shown to mimic the honeybee's behaviour remarkably well (Serres et al., 2008a).

9. OSCAR as a yaw control autopilot

Electrophysiological recordings on freely flying flies have revealed that the muscle attached to the base of the retina (Burtt & Patterson, 1970; Hengstenberg, 1971) causes *retinal microscanning* : the whole retinal mosaic (Fig.1a) *translates* repeatedly at about 5 Hz by a few micrometers underneath the facet mosaic, causing the visual axes to *rotate* by a few degrees (Franceschini & Chagneux, 1997). Such a purely *rotational OF* is exploited on-board another aerial robot we built: robot OSCAR (Fig. 12a).

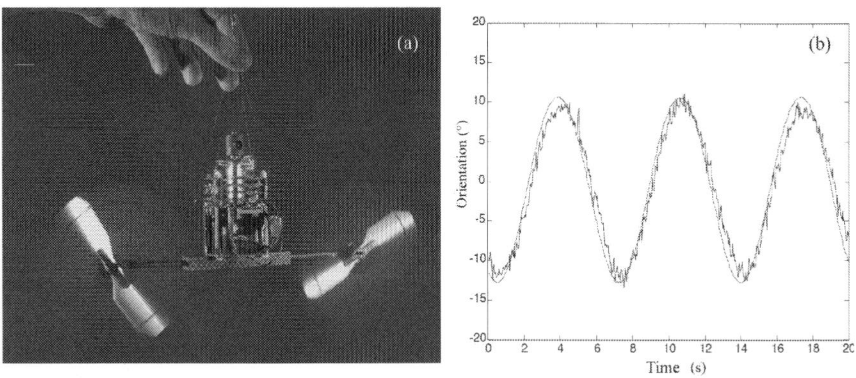

Figure 12. (a) OSCAR is a 100-gram twin-engine robotic demonstrator that sees. It is equipped with an elementary eye (visible on top), composed of a lens and only two photoreceptors driving an EMD of the type shown in Fig. 3 (Photo by H. Raguet) (b) Smooth pursuit (tracking) of a dark edge (contrast m = 0.4) displaced sinusoidally at 0.15 Hz in a frontal plane situated 1.3 meters from the eye (From Viollet & Franceschini, 1999b)

This project was based on the assumption that microscanning in flies operates in connection with motion detection. On the basis of this fly's retinal microscanning process, we developped an optronic sensor (Viollet & Franceschini, 1999a), and a novel aerial robot, equipped with this sensor (Viollet & Franceschini, 1999b, 2001). OSCAR is attached to a thin, 2-meter long nylon wire secured to the ceiling of the laboratory. It is free to adjust its yaw by driving its two propellers differentially, and it does so in spite of the many (aerodynamic or pendular) disturbances it encounters.

Thanks to its microscanning eye, this miniature (100-gram) robot is not only able to fixate a nearby "target" (a dark edge or a bar) steadily but also to track it at angular speeds of up to 30°/s - a value similar to the maximum tracking speed of the human eye. It cannot, however, estimate the distance from the target since it deals only with the *rotational OF*.

The OSCAR sensor is able to *detect* a bar subtending an angle much smaller than the interreceptor angle $\Delta\varphi = 4°$, making it applicable for nonemissive power line detection onboard fullscale helicopters (Viollet et al., 2006). The OSCAR sensor is also able to *locate* an edge with great positioning accuracy (0.1 degrees, which is 40 times smaller than $\Delta\varphi$): OSCAR has acquired *hyperacuity* (Viollet & Franceschini, 2001). This property makes the principle appealing for accurate stabilization of various platforms (Franceschini et al., 2005). Inspired by the fly's thoracic halteres that have long been known to act as gyroscopes, we equipped the OSCAR robot with an additional inertial feedback based on a MEMS *rate gyro*. The interplay of the two (visual and inertial) control loops enhances both the dynamic stability and performances of the tracking system (Viollet & Franceschini, 1999b, 2001). More recently, we introduced a mechanical *decoupling* between eye and body, and implemented a genuine *vestibulo-ocular reflex* that was key to maintaining the robot's gaze perfectly fixed on the target (Viollet & Franceschini, 2005, Kerhuel et al., 2007). Robot OSCAR II is a novel 100-gram aerial robot whose gaze, decoupled from the heading, remains robustly locked onto a target in spite of various perturbations that we deliberately applied to the body. It is the gaze orientation of OSCAR II that ultimately controls its heading (Viollet et al, this volume).

10. Summary and conclusion

In this chapter, we reported about our attempts to understand *how* flying insects could behave on the basis of OF cues. We presented several bio-inspired, OF based visual guidance principles and showed that they may be helpful in the context of autonomous guidance of MAVs over a terrain or in a corridor. The principles we came up with differ markedly from a current OF based navigation strategy. The latter requires that OF sensors be associated with a groundspeed sensor (based on GPS, for example) in order to determine the groundheight (see Eq 1) to achieve terrain following or landing (Barrows et al., 2003; Barber et al., 2005; Srinivasan et al., 2006; Garrat & Chahl, 2008).

OCTAVE and LORA III autopilots harness the power of the *translational OF* in a quite different manner. They do not need to estimate any speed or range, and they do not aim to achieve any "speed holding" or "range holding" abilities. They aim instead at producing safe behavior by a process that can be called "OF hold". These autopilots consist of feedback control loops, called « *optic flow regulators* ». Their block diagrams (Figures 6 & 11) pinpoint which variables are to be either *measured* or *controlled* or *regulated*, and give the *causal* and *dynamical* relationships between these variables. Three (ventral, forward and sideways) *OF regulators* strive to maintain a given OF $\boldsymbol{\omega}$ constant by acting upon the lift, the forward thrust or the side thrust, respectively. OF sensors are shown to be sufficient for these tasks and there is no need to measure (or estimate) groundspeed, groundheight and clearance to the walls in a corridor. Simulation experiments and robotic demonstrators showed that risky maneuvers such as automatic takeoff, ground avoidance, level flight, terrain following, landing, centering, wall following, and groundspeed control can all be successfully performed on the basis of these three *OF regulators*. Moreover, these control schemes were found to account for a series of puzzling, seemingly unconnected flying abilities observed by many authors over the last 70 years in numerous insect species, suggesting that *OF regulators* may well be implemented onboard insects (see Sections 7 & 8).

Three parameters - the three OF set points ω_{set} of Fig. 6 and 11 - would suffice to determine the flight behavior, at least in the simplified, random but richly contrasted environments considered here. Onboard the insect, these set points may depend on either innate, internal or external factors.

Electronic implementation of an *OF regulator* is not very demanding - nor is its neural implementation - since it relies upon a few linear operations (such as summation, difference and various filters) and nonlinear operations (such as minimum or maximum detection). The worthiest component of an OF regulator is the OF sensor itself. Like the honeybee's Velocity Tuned (VT) neurons (Ibbotson, 2001), the neuromorphic OF sensors we have been building since 1985 deliver an output that grows monotonically with the OF ω, with little dependence on spatial frequency and contrast (see Section 3). They respond over a 10-fold range in OF (from 40°s to 400°/s: Ruffier et al., 2003; Aubépart & Franceschini, 2007). This range is largely sufficient because each OF sensor needs only to detect a *deviation* ε from the OF set point. The miniature helicopter flight pattern illustrates this point: the ventral *OF regulator* produces a behavior such that the OF output ω_{meas} varies little throughout the journey, from takeoff to landing (Fig. 8c). A similar observation holds for the forward and sideways *OF regulators* (see Figs 7c,e & 12c,e in Serres et al., 2008a).

Each OF sensor at work on our proof of concept robots is characterized by a low resolution. The inter-receptor angle in the eyes of OCTAVE, LORA III and OSCAR robots is around 4°, a value close to the interommatidial angle (5°) of the fruitfly's eye (Franceschini, 1975). The field of view (FOV) of the eyes and the provocatively small number of pixels (2 pixels per ventral or lateral eye) will obviously need to be increased. Covering the frontal part of the FOV, in particular, will be essential to sense and avoid obstacles directly in the flight path, and here the retinal microscanning principle (Section 9) could be a great help (Mura & Franceschini, 1996; Viollet & Franceschini, 2001). A recent study showed that a more frontally oriented EMD may provide OCTAVE autopilot with a feedforward anticipatory signal that helps the robot climb steeper rises (Ruffier & Franceschini, 2008).

The control schemes we proposed are simple but thus far limited to the use of the maximal OF sensed either ventrally or laterally, perpendicular to the heading direction. Once developed beyond the state of the current minimalistic robotic demonstrators, OCTAVE and LORA III principles could potentially be harnessed to guide a MAV indoors or through complex terrains such as mountainous canyons or urban environments. Since the control systems are parsimonious and do not rely on GPS or any emissive (power-hungry) sensors such as RADARs or LADARs, they promise to match the draconian constraints imposed upon micro aerial vehicles, in terms of size, mass and consumption.

For the same reasons, these autopilots could potentially be developped for robotic missions on other planets or moons. A Martian lander could make use of the OCTAVE principle, since it ensures safe automatic landing (see Section 7.5). A rover flying in Martian canyons and gullies could make use of the LORA III principle, since both the craft's groundspeed and its clearance to the walls would automatically depend on the canyon's width (see Section 8, and Serres et al, 2008a). Whether on Earth or on another celestial body, the craft would need to receive from the Earth based (or orbital) station only three *low bandwidth* signals: the values of the OF set points.

11. Acknowledgments

We are grateful to F. Aubépart L. Kerhuel and G. Portelli for their fruitful comments and suggestions during this research. We are also grateful to Marc Boyron (electronics engineer), Yannick Luparini and Fabien Paganucci (mechanical engineers) for their expert technical assistance. Serge Dini (beekeeper) gave plenty of advices during the behavioral experiments. This research was supported by CNRS (Life Science ; Information and Engineering Science and Technology), by an EU contract (IST/FET-1999-29043) and a DGA contract (2005-0451037).

12. References

Alestam, T. (1990). *Bird Migration.* Cambridge University Press

Aubépart, F. & N. Franceschini., N. (2007). Bio-inspired optic flow sensors based on FPGA: application to micro-air vehicles. *Journal Microprocessors and Microsystems* 31, 408-419

Aubépart, F., El Farji, M. & Franceschini, N. (2004). FPGA implementation of elementary motion detectors for the visual guidance of micro-air vehicles. *Proceedings of IEEE International Symposium on Industrial Electronics (ISIE'2004)*, Ajaccio, France, pp. 71-76

Aubépart, F., Serres, J., Dilly, A., Ruffier, F. & Franceschini., N. (2008). Field programmable gate arrays (FPGA) for bio-inspired visuo-motor control systems applied to micro-air vehicles, In: *Aerial Vehicles*, published by In-Tech

Baird, E., Srinivasan, M.V., Zhang, S.W. & Cowling., A. (2005). Visual control of flight speed in honeybees. *Journal of Experimental Biology* 208, 3895-3905

Baird, E., Srinivasan, M.V., Zhang, S.W., Lamont R. & Cowling, A. (2006).Visual control of flight speed and height in the honeybee. In: *From Animals to Animats* 9, S. Nolfi et al. (Eds.), 4095, 40-51

Barber, D.B.; Griffiths, S..R.; McLain, T.W.;& Beard, R.W. (2005) Autonomous landing of miniature aerial vehicles. In: *Proceedings of American Institute of Aeronautics and Astronautics Conference*

Barrows, GL, Neely, C. & Miller, K.T. (2001). Optic flow sensors for MAV navigation. In : Fixed and flapping wing aerodynamics for micro-air vehicle applications. *Progress in Astronautics and Aeronautics,* Vol. 195, 557-574

Barrows, GL; Chahl, J.S.; Srinivasan MV. (2003) Biologically inspired visual sensing and flight control. *Aeronautic Journal* 107, 159-168

Blanès, C. (1986) Appareil visuel élémentaire pour la navigation à vue d'un robot mobile autonome. *MS thesis in Neuroscience,* University of Aix-Marseille II, Marseille, France

Blanès, C. (1991). Guidage visuel d'un robot mobile autonome d'inspitation bionique. *PhD thesis,* Polytechnical Institute, Grenoble, France (thesis work at the Neurocybernetics lab, CNRS, Marseille, France)

Borst, A. & Haag, J. (2002). Neural networks in the cockpit of the fly. *Journal of Comparative Physiology* A188, 419-437

Braitenberg, V. (1967). Patterns of projection in the visual system of the fly, I/Retina-Lamina projections. *Experimental Brain Research* 3, 271-298

Bräuninger, H.D. (1964). Uber den Einfluss meteorologischer Faktoren auf die Entfernungs-weisung im Tanz der Bienen. *Zeitschrift für Vergleichende Physiologie* 48, 1-130

Buchner E. (1984). Behavioral analysis of spatial vision in insects In: *Photoreception and Vision in Invertebrates,* Ali, M. (Ed.), New-York : Plenum, 561-621

Burtt, J. & Patterson, J. (1970). Internal muscle in the eye of an insect. *Nature* 228, 183-184

David, C. (1978). The relationship between body angle and flight speed in free-flying Drosophila. *Physiological Entomology* 3, 191-195

David, C. (1982). Compensation for height in the control of groundspeed by Drosophila in a new 'barber's pole' wind tunnel. *Journal of Comparative Physiology* A 147, 1432-1351

Egelhaaf, M. & Borst, A. (1993). Movement detection in Arthropods. In : *Visual motion an its role in the stabilization of gaze,* F. Miles & J. Wallman, Eds.,Amsterdam: Elsevier, 53-77

Franceschini N. (1975). Sampling of the visual environment by the compound eye of the fly: fundamentals and applications, in: *Photoreceptor Optics,* A. Snyder, R. Menzel (Eds), Berlin : Springer, Chapt. 17, 98-125

Franceschini, N. (1984). Chromatic organisation and sexual dimorphism of the fly retinal mosaic. In: *Photoreceptors.* Borsellino, A. and Cervetto, L. (Eds.), pp. 319-350

Franceschini, N. (1985). Early processing of colour and motion in a mosaic visual system. *Neuroscience Research,* Supplement 2, 517-549

Franceschini, N. (1992). Sequence-discriminating neural network in the eye of the fly. In: *Analysis and Modeling of Neural Systems,* F.H.K. Eeckman (Ed.), Norwell, USA, Kluwer Acad. Pub., 142-150.

Franceschini, N. (2007). Sa majesté des mouches. In: *Voir l'Invisible,* JP.Gex (Ed.), Paris: Omnisciences (2007), pp 28-29

Franceschini, N. (2008). Towards automatic visual guidance of aerospace vehicles: from insects to robots. *Acta Futura : Proceedings of the ESA International Symposium on Advanced Concepts : « A Bridge to Space »* Noordwick, (in press)

Franceschini, N. & Chagneux, R. (1997). Repetitive scanning in the fly compound eye. *Proceedings of the 25th Göttingen Neurobiology Conference* (N. Elsner, H. Wässle, Eds.) Stuttgart, : G. Thieme, 279

Franceschini, N., Blanes, C. & L. Oufar, L. (1986). Appareil de mesure, passif et sans contact, de la vitesse d'un objet quelconque. *Technical Report ANVAR/DVAR,* N°51549, Paris.

Franceschini, N., Pichon, J.M. & Blanès, C. (1992). From insect vision to robot vision. *Philosophical Transactions of the Royal Society London B* 337, 283-294

Franceschini, N., Kirschfeld, K., Minke, B. (1981a). Fluorescence of photoreceptor cells observed in vivo. *Science* 213, 1264-1267

Franceschini, N., Hardie, Ribi, W., Kirschfeld, K. (1981b). Sexual dimorphism in a photoreceptor. *Nature* 291, 241-244

Franceschini, N., Riehle, A. & Le Nestour A. (1989). Directionally selective motion detection by insect neurons. In: *Facets of Vision*, D.G. Stavenga and R.C. Hardie (Eds.), Berlin, Springer, 360-390.

Franceschini, N., Ruffier, F. & Serres, J. (2007). A bio-inspired flying robot sheds light on insect piloting abilities. *Current Biology* 17, 329-335

Franceschini, N.; Viollet, S. & Boyron, M. (2005). Method and device for hyperacute detection of an essentially rectilinear contrast edge, and system for fine following and fixing of said contrast edge. *International Patent* PCT/FR2005/11536

Franceschini, N., Ruffier, F., Viollet, S. & Boyron, M. (2003). Steering aid system for altitude and horizontal speed, perpendicular to the vertical, of an aircraft and aircraft equipped therewith. *International Patent* PCT/FR2003/002611

Garratt M.A. & Chahl, J.S. (2008). Vision-based terrain following for an unmanned rotorcraft. *Journal of Field Robotics* 25, 284-301

Gibson, J.J. (1950). *The perception of the visual world,* Boston: Houghton Mifflin

Gibson J.J., Olum, P. & Rosenblatt, F. (1955). Parallax and perspective during aircraft landings. *American Journal of Psychology* 68, 372-395

Green, WF, Oh, Y. & Barrows, G. (2004).Flying insect inspired vision for autonomous axial robot maneuvers in near earth environments. *Proceedings of the IEEE International Conference on Robotics and Automation* (ICRA04) 2343-2352

Hardie R.C. (1985). Functional organization of the fly retina. *Progress in Sensory Physiology* 5, Ottosson D. (Ed.), Springer, Berlin

Hassenstein, B. & Reichardt, W. (1956) Systemtheoretische Analyse der Zeitreihenfolgen undVorzeichenauswertung bei der Bewegungsperzeption des Rüsszelkäfers Chlorophanus. *Zeitschrift für Naturforschung* 11b, 513-524

Hausen , K. (1993). Decoding of retinal image flow in insects. In : *Visual motion an its role in the stabilization of gaze,* F.A. Miles and J. Wallman (Eds.) Amsterdam : Elsevier, 203-235

Hausen, K. & Egelhaaf, M. (1989) Neural mechanisms of course control in insects. In: *Facets of vision*; D.G. Stavenga & R.. Hardie, Berlin, Springer pp. 391-424

Hengstenberg, R. (1971). Das Augenmuskel System der Stubenfliege Musca domestica. *Kybernetik* 2 56-77

Heran, H. (1955). Versuche über die Windkompensation der Bienen. *Naturwissenschaften* 5, 132-133

Heran, P. & Lindauer, M. (1963). Windkompensation und Seitenwindkorrektur der Bienen Flug über Wasser. *Zeitschrift für vergleichende Physiologie* 47, 39-55

Heisenberg, M, Wolf ,R. (1984).*Vision in Drosophila.* Berlin, Springer.

Ibbotson, M.R. (2001). Evidence for velocity-tuned motion-sensitive descending neurons in the honeybee. *Proceedings of the Royal Society, London B* 268, 2195-2201

Kennedy, J.S. (1939). Visual responses of flying mosquitoes. *Proceedings of the Zoogical Society of London* 109, 221-242

Kennedy, J.S. (1951). The migration of the desert locust (Schistocerca gregaria Forsk.) I. The behaviour of swarms. *Phil. Transactions of the Royal Society, London, B* 235, 163-290

Kerhuel, L., Viollet, S. & Franceschini, N. (2007). A sighted aerial robot with fast gaze and heading stabilization. *Proceedings of the IEEE International Conference on Intelligent Robots and Systems* (IROS), San Diego, USA, pp. 2634-2641

Kien, J. (1975) Neuronal mechanism subserving directional selectivity in the locust optomotor system. *J. Comp. Physiol.* 113: 161-179

Kirchner, W.H. & Srinivasan, M.V. (1989). Freely moving honeybees use image motion to estimate distance. *Naturwissenchaften* 76, 281-282

Kirschfeld, K. (1967). Die Projektion der optischen Umwelt auf das Raster der Rhabdomere im Komplexauge von Musca. *Experimental Brain Research* 3, 248-270

Kirschfeld, K.; Franceschini, N. (1968). Optische Eigenschaften der Ommatidien im Komplexauge von Musca. *Kybernetik* 5 47-52

Kirschfeld, K.; Franceschini, N., Minke, B. (1977). Evidence for a sensitizing pigment in fly photoreceptors » *Nature* 269, 386-390

Koenderink, J.J. (1986). Optic flow. *Vision Research* 26, pp. 161-79

Kramer, J., Sarpeshkar, R. & Koch C. (1995). An analog VLSI velocity sensor. In *Proceedings of IEEE International ymposium on Circuits and Systems, Seattle,* USA, pp 413-416

Krapp, H, Hengstenberg, B, Hengstenberg, R.(1998). Dendritic structure and receptive-field organisation of optic flow processing interneurons in the fly, *Journal of Neurophysiology* 79, 1902-1917.

Laughlin, S. (1984). The role of parallel channels in early visual processing by the arthropod compound eye. In: *Photoreception and vision in invertebrates.* M.A. Ali Ed., Plenum, 457-481

Lee, D. N. (1980). The optic flow field: the foundation of vision. *Philosophical Transactions of the Royal Society London B* 290, 169-179

Marr, D. (1982). *Vision,* San Francisco: Freeman

Moeckel, R. & Liu S.C. (2007). Motion detection circuits for a time-to-travel algorithm. *Proceedings IEEE Int. Symposium Circuits and Systems* (ISCAS07), pp. 3079-3082

Mura, F. & Franceschini., N. (1994). Visual control of altitude and speed in a flying agent. In : *From Animals to Animats III,* D. Cliff et al. (Eds), Cambridge: MIT Press, 91-99

Mura, F.; Franceschini, N. (1996). Obstacle-avoidance in a terrestrial mobile robot provided with a scanning retina. *Proceedings IEEE International Symposium on Intelligent Vehicles.* N. Aoki and I. Masaki, eds., Seikei University, Tokyo, Japan, pp. 47-52.

Netter, T. & Franceschini, N. (1999). Neuromorphic optical flow sensing for nap-of-the-Earth flight. In : *Mobile Robots XIV,* D. Gage and H. Choset (Eds), SPIE, Vol. 3838, 208-216

Netter, T. & Franceschini, N. (2002). A robotic aircraft that follows terrain using a neuromorphic eye. In : *Proceedings of the IEEE International Conference on Intelligent Robots and Systems* (IROS) Lausanne, Switzerland, pp. 129-134

Olberg R.M. (1981) Object- and Self- Movement detectors in the ventral nerve cord of the dragonfly. *J. Comp. Physiol.* 141, 327-334

Pichon, J-M., Blanès, C. & Franceschini, N. (1989). Visual guidance of a mobile robot equipped with a network of self-motion sensors. In : *Mobile Robots IV,* W.J. Wolfe , W.H. Chun (Eds.) Bellingham, U.S.A : SPIE , Vol. 1195, 44-53

Portelli, G., Serres, J., Ruffier F. & Franceschini, N. (2008). An Insect-Inspired Visual Autopilot for Corridor-Following. *Proceedings of the 2nd Biennial IEEE Int. Conference on Biomedical Robotics and Biomechatronics,* BioRob 08, Scottsdale, USA, pp. 19-26

Preiss, R. (1992). Set point of retinal velocity of ground images in the control of swarming flight of desert locusts. *Journal Comparative Physiology A* 171, 251-256

Preiss, R. & Kramer, E. (1984). Control of flight speed by minimization of the apparent ground pattern movement. In : *Localization and Orientation in Biology and Engineering*, D. Varju and H. Schnitzler, eds. (Berlin, Springer), 140-142

Pudas, M. ; Viollet, S. ; Ruffier, F. ; Kruusing, A. ; Amic, S. ; Leppävuori, S. & Franceschini, N. (2007). A miniature bio-inspired optic flow sensor based on low temperature co-fired ceramics (LTCC) technology, *Sensors and Actuators A* 133, 88-95

Reichardt, W. (1969): Movement perception in insects. In: *Processing of Optical Data by Organisms and by Machines*, W. Reichardt (Ed.), New York: Academic Press , 465-493

Riehle, A. & Franceschini, N. (1984). Motion detection in flies: parametric control over ON-OFF pathways. *Experimental Brain Research* 54, 390-394

Ruffier, F. & Franceschini, N. (2003) OCTAVE, a bioinspired visuo-motor control system for the guidance of Micro-Air Vehicles, In: *Bioengineered and Bioinspired Systems*, A. Rodriguez-Vazquez et al., Eds., Bellingham, U.S.A , SPIE, Vol. 5119, 1-12

Ruffier, F. & Franceschini, N. (2004a). Visually guided micro-aerial vehicle: automatic take-off, terrain following, landing and wind reaction. *Proceedings IEEE International Conference on Robotics and Automation* (ICRA04), New Orleans, USA, pp. 2339-2346

Ruffier, F. & Franceschini, N. (2004b). Optic flow based AFCS for rotorcraft automatic maneuvring (terrain following, takeoff and landing). *Proceedings of the 30th European Rotorcraft Forum*, AAF/CEAS, Marseille, 71.1-71.9.

Ruffier, F. & Franceschini, N. (2005). Optic flow regulation: the key to aircraft automatic guidance. *Robotics and Automomous Systems* 50, 177-194

Ruffier, F. & Franceschini, N. (2008). Aerial robot piloted in steep relief by optic flow sensors. *Proceedings of the IEEE/RSJ International Conference on Intelligent Robots and Systems* (IROS08), Nice, France, pp. 1266-1271.

Ruffier, F.; Viollet, S.; Amic, S. & Franceschini, N. (2003). Bio-inspired optical flow circuits for the visual guidance of micro-air vehicles. *Proceedings the IEEE International Symposium on Circuits and Systems* (ISCAS), Bangkok, Thaïland, III, pp. 846-849

Serres, J. (2008). De l'abeille au robot : la « régulation du flux optique », *PhD Thesis*, University of Montpellier, France (Thesis work from the Biorobotics lab, CNRS, Marseille, France).

Serres, J.; Dray, D.; Ruffier, F. & Franceschini, N. (2008a). A vision-based autopilot for a miniature air vehicle: joint speed control and lateral obstacle avoidance. *Autonomous Robots* 25, 103-122

Serres, J.; Masson, G.; Ruffier, F. & Franceschini, N. (2008b). A bee in the corridor : centring and wall following. *Naturwissenschaften*, 95 (12) 1181-1187

Shoemaker, P.A. ; O'Caroll, D.C. & Straw, A.D. (2005). Velocity constancy and models for wide-field visual motion detection in insects. *Biological Cybernetics*, 93, 275-287

Srinivasan, M.V. (1993). How insects infer range from visual motion. In : *Visual motion and its role in the stabilization of gaze*. F.A. Miles and J. Wallman, Eds. Amsterdam : Elsevier, 139-156

Srinivasan, M.; Thurrowgood, S. & Soccol, D. (2006). An optical system for guidance of terrain following in UAVs. *Proceedings of the IEEE International Conference on video and signal based surveillance* AVSS06

Srinivasan, M.V.; Lehrer, M.; Kirchner, W.H. & Zhang, S.W. (1991). Range perception through apparent image speed in freely flying honeybees. *Visual Neuroscience* 6, 519-535

Srinivasan, M.V., Zhang, S.W. and Chandrashekara K. (1993). Evidence for two distinct movement-detecting mechanisms in insect vision. *Naturwissenchaften* 80, 38-41.

Srinivasan, M.V., Zhang, S.W., Lehrer, M. & Collett., T. (1996). Honeybee navigation en route to the goal: visual flight control and odometry. *Journal of Experimental Biology,* 199, 237-244

Srinivasan, M.V., Zhang, S.W., Chahl, J.S., Barth, E. & Venkatesh, S. (2000). How honeybees make grazing landings on flat surface. *Biological Cybernetics* 83, 171-183

Srygley, R.B. & Oliveira, E.G (2001). Orientation mechanisms and migration strategies within the flight boundary layer. *Insect movements: mechanisms and consequences.* T.P. Woiwod, D.R. Reynolds and C.D. Thomas (Eds.), CAB international, 183-206

Steiner, G. (1953). Zur Duftorientierung fliegender Insekten. *Naturwissenchaften* 19, 514-515

Tautz, J., Zhang, S., Spaethe J., Brockman A., Si, A. & Srinivasan M.V. (2004). Honeybee odometry: Performance in varying natural terrain. *PLOS Biology,* 2, 915-923

Taylor G.K. & Krapp, H.G. (2008). Sensory systems and flight stability : what do insects measure and why ? In : *Advances in Insect Physiol. 34 : Insects mechanisms and control,* J. Casas and S.J. Simpson (Eds.) Amsterdam : Elsevier, 231-316

Ullman, S. (1981). Analysis of visual motion by biological and computer systems, *Computer* 14, 57-69

Viollet, S. & Franceschini, N. (1999). Biologically-Inspired Visual Scanning Sensor for Stabilization and Tracking . *Proceedings of the IEEE International Conference on Intelligent Robots and Systems* (IROS 99) Kyon-gyu, Corea, pp. 204-209

Viollet, S. & Franceschini, N. (1999). Visual servo system based on a biologically-inspired scanning sensor. *Proceedings of the Conference on sensor fusion and decentralized control in robotics II,* Bellingham, U.S.A : SPIE Vol. 3839, 144-155

Viollet, S. & Franceschini, N. (2001). Super-Accurate Visual Control of an Aerial Minirobot In : *Autonomous minirobots for Research and Edutainment,* U. Rückert, J. Sitte, U. Witkowski (Eds.), Heinz Nixdorf Institut, Padderborn, Germany, 215-224

Viollet, S. & Michelis, J. & Franceschini, N. (2006). Toward a bio-inspired nonemissive powerline detection system. *Proceedings of the 32th European Rotorcraft Forum,* ERF06, Maastricht, Netherlands, Paper N° AD09

Viollet, S. & Franceschini, N. (2005). A high speed gaze control system based on the Vestibulo-Ocular Reflex. *Robotics and Autonomous Systems* 50 147-161

Viollet, S., Kerhuel, L. & Franceschini, N. (2008). A novel biomimùetic steering control system for sighted vehicles. In : *Aerial Vehicles,* published by In-Tech

Whiteside, T.C. & Samuel, G.D. (1970). Blur zone. *Nature* 225, 94–95

Williams, C.P. (1965). *Insect migration.* London: Collins, second edition

Zeil, J., Boeddeker, N. & Hemmi, J.M. (2008).Vision and the organization of behavior, *Current Biology* 18, 320-323